FUNDAMENTALS
OF
ECOLOGY

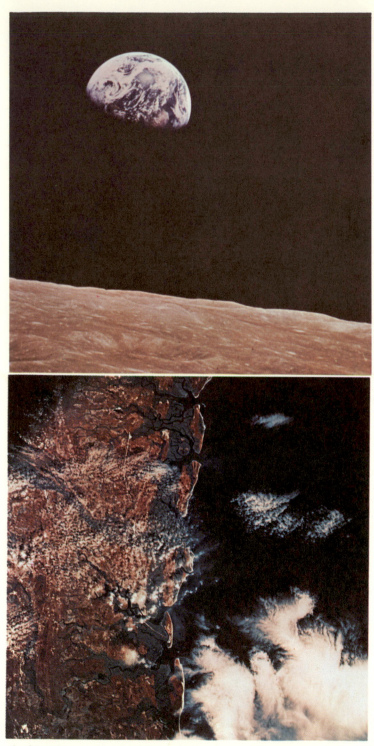

The biosphere and some of its ecosystems as seen from near and afar. This series of photographs illustrates how remote photography can aid in the study of interrelations between organism and environment. *Upper:* The biosphere level—a view of earth from a space capsule orbiting the moon (surface of moon in foreground). *Lower:* The ecosystem level—an Apollo 9 astronaut's view of the Georgia coast. Beaches fronting the Atlantic Ocean sparkle white in this infrared photograph, which turns the green of forest to shades of red. Between islands and mainland salt marshes provide a nursery for sea life. Many of the islands are uninhabited.

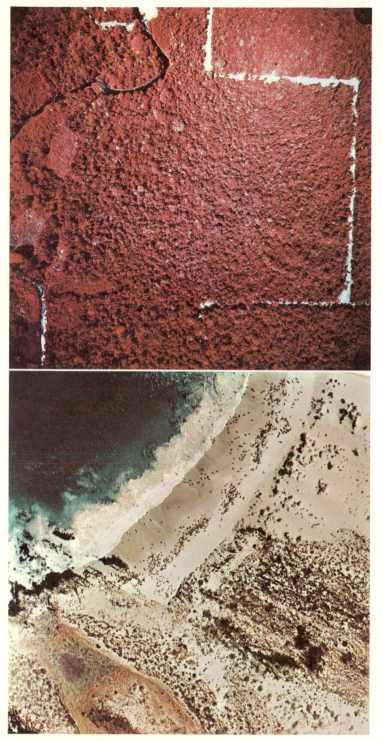

Upper: The community level—the successional stages of a seasonal tropical forest are demonstrated in this infrared photograph of the Development and Resources Transportation Co. test site in Tocache, Peru (note two small clearings in the forest and a roadway). Emergent trees, which lose their leaves in the dry season, show up as gray flecks. *Lower:* The population level—the desert meets the fertile sea, which provides food for the sea lions (black specks on beach), in this photograph of the California coastline. The contrast in fertility and infertility is clearly noticeable. (All four photographs courtesy of the National Aeronautics and Space Administration, Washington, D.C.)

EUGENE P. ODUM

Alumni Foundation Professor of Zoology
University of Georgia
Athens, Georgia

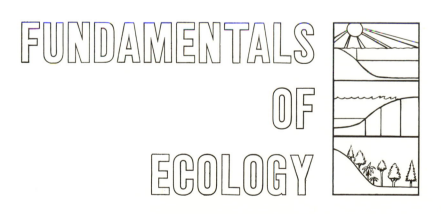

FUNDAMENTALS OF ECOLOGY

THIRD EDITION

W. B. SAUNDERS COMPANY • PHILADELPHIA • LONDON • TORONTO

W. B. Saunders Company: West Washington Square
Philadelphia, Pa. 19105

12 Dyott Street
London, WC1A 1DB

1835 Yonge Street
Toronto 7, Ontario

Listed here is the latest translated edition of this book together with the language of the translation and the publisher.

Polish (2nd Edition)   Rolnicze i Lesne,
                              Warsaw, Poland

Spanish (2nd Edition)   Nueva Editorial Interamericana, S.A., de C.V.,
                              Mexico

Fundamentals of Ecology   ISBN 0-7216-6941-7

Print No.:   9   8   7   6   5   4

PREFACE TO THE THIRD EDITION

Practice has caught up with theory in ecology. The holistic approach and ecosystem theory, as emphasized in the first two editions of this book, are now matters of world-wide concern. People in general have accepted the root meaning of the word "ecology," which refers to the whole environmental "house" in which we live. Thus, to many persons ecology now stands for the study of "the totality of man and environment." Although the same general format which students and teachers found useful in the previous edition has been retained, the third edition is greatly expanded and updated in light of the increasing importance of the subject in human affairs. All chapters from the second edition have been extensively revised; three completely new chapters have been added to Part 1, and Part 3 has been completely rewritten. Illustrative material and references have been more than doubled, and two-thirds of the figures and tables are new to the third edition.

In revising textbooks one worries about the "dinosaur syndrome." Sometimes textbooks become so enlarged in successive editions that the brevity and simplicity that made them successful in the early editions is lost. To avoid this contingency I have structured the third edition so that it is *three books in one*, each of which can serve a different purpose.

Book No. 1. This involves the macroscopic or "big picture" ecology as it relates to human affairs: Chapters 1 through 4, plus 9, 15, 16, and 21. These eight chapters provide a review of ecology for the concerned citizen, the student of the social sciences, the humanities, or the professions (law, medicine, engineering, and so forth), and for the specialist in science, government, or industry. Also this group of chapters provides a reference base for campus-wide courses in "man and environment" or "human ecology."

Book No. 2. For the undergraduate college course in ecology, Part 1 (Chapters 1 through 10) and Chapters 15, 16, and 21 (a total of 13 chapters), with Part 2 and other chapters in Part 3 as reference for specific field or laboratory work, is recommended.

Book No. 3. The whole book (21 chapters) is a comprehensive reference work on principles, environments, and ecological technology. It is also a textbook for graduate courses.

Numerous cross-references in all the chapters make it feasible to begin reading at any point in the book, or to select various combinations of chapters as needed.

As was true in the first two editions, this edition owes much to my brother, Howard T. Odum. His contributions to the second edition, especially to Chapter 3 (Principles and Concepts Pertaining to Energy in Ecological Systems), have been retained and expanded. His highly original approaches to systems ecology, now incorporated in a separate book (see H. T. Odum, 1971), are cited in a number of different chapters. I am also indebted to my son,

William E. Odum, for ideas and the use of his unpublished data. Without the understanding and encouragement of my wife, Martha Ann, who also assisted with the illustrations and index, I would never have been able to face the task of revising subject matter that is expanding so rapidly in scope. I am most appreciative of the very great personal encouragement provided through all three editions of this book by Tyler Buchenau, recently retired as College Editor at the W. B. Saunders Company. The staffs of the editorial and production departments of Saunders have been unfailing in their dedication to making this edition a reality despite my many false starts, delays, and changes in manuscript and proof. My thanks go also to Gail McCord, Ann Young, and Joseph Mahoney here at the University of Georgia for their dedicated work on manuscript and proof.

This book is very much a product of concepts and research of students and colleagues who have been associated with the Institute of Ecology at the University of Georgia during the past 25 years. In addition to special chapters and sections, which are credited, the work of Institute members and former students is cited in practically every chapter; their published research makes up an impressive part of the bibliography at the end of the book. It is with great pleasure that I dedicate the third edition to the following staff members (past and present) and former students:

Gary W. Barrett, Robert J. Beyers, Claude E. Boyd, U. Eugene Brady, Joel H. Braswell, Alicja Breymeyer, I. Lehr Brisbin, Paul R. Burkholder, Larry D. Caldwell, James L. Carmon, E. L. Cheatum, Edward Chin, David C. Coleman, Clyde E. Connell, G. Dennis Cooke, William B. Cosgrove, John W. Crenshaw, Jr., William H. Cross, D. A. Crossley, Jr., Rossiter H. Crozier, Armando A. de la Cruz, Michael D. Dahlberg, Howard E. Daugherty, Leslie B. Davenport, Jr., Robert Davis, Michael Dix, Richard Dugdale, Richard G. Eagon, Alfred C. Fox, Dirk Frankenberg, John B. Gentry, J. Whitfield Gibbons, Cameron E. Gifford, Frank B. Golley, C. Philip Goodyear, Robert W. Gorden, Robert E. Gordon, Albert G. Green, Jr., Carl W. Helms, David L. Hicks, Kinji Hogetsu, Milton N. Hopkins, Jr., James D. Howard, John H. Hoyt, Melvin T. Huish, Robert L. Humphries, Preston Hunter, Kermit Hutcheson, James H. Jenkins, Robert E. Johannes, A. Stephen Johnson, A. Sydney Johnson, Philip Johnson, David W. Johnston, Marvin P. Kahl, Herbert W. Kale II, Hiroya Kawanabe, Stephen H. King, Edward J. Kuenzler, George H. Lauff, Thomas L. Linton, Jack I. Lowe, Joseph J. Mahoney, Jr., R. Larry Marchinton, Frederick Marland, Timothy G. Marples, James A. Marsh, William H. Mason, Bernard S. Martof, J. Frank McCormick, Wayne McDiffett, John T. McGinnis, Terry A. McGowan, Edward F. Menhinick, Jiro Mishima, Carl D. Monk, Syuiti Mori, Daniel J. Nelson, Robert P. Nicholls, Robert A. Norris, Howard D. Orr, Bernard C. Patten, William J. Payne, George A. Petrides, Gayther L. Plummer, Lawrence R. Pomeroy, Steven E. Pomeroy, Marvin M. Provo, Ernest E. Provost, H. Ronald Pulliam, Robert A. Ragotzkie, Robert J. Reimold, Mervin Reines, David T. Rogers, Jr., Berton Roffman, Lech Ryszkowski, Herbert H. Ross, Masako Satomi, Claire L. Schelske, James E. Schindler, Jay H. Schnell, Donald C. Scott, Homer F. Sharp, L. Roy Shenton, John L. Shibley, Alfred E. Smalley, Michael H. Smith, Allen D. Stovall, Wallace A. Tarpley, John M. Teal, James P. Thomas, Robert L. Todd, Elliot J. Tramer, J. Bruce Wallace, Kenneth L. Webb, Harold E. Welch, William J. Wiebe, Richard G. Wiegert, William K. Willard, Richard B. Williams, John E. Wood, and J. David Yount.

EUGENE P. ODUM

Athens, Georgia

CONTENTS

Chapter 8

THE SPECIES AND THE INDIVIDUAL IN THE ECOSYSTEM 234

Chapter 9

DEVELOPMENT AND EVOLUTION OF THE ECOSYSTEM 251

Chapter 10

SYSTEMS ECOLOGY: THE SYSTEMS APPROACH AND MATHEMATICAL MODELS IN ECOLOGY 276

By Carl J. Walters

Part 2

THE HABITAT APPROACH

Part 3

APPLICATIONS AND TECHNOLOGY

Part 1

BASIC ECOLOGICAL PRINCIPLES AND CONCEPTS

Chapter 1

INTRODUCTION: THE SCOPE OF ECOLOGY

1. ECOLOGY—ITS RELATION TO OTHER SCIENCES AND ITS RELEVANCE TO HUMAN CIVILIZATION

Man has been interested in ecology in a practical sort of way since early in his history. In primitive society every individual, to survive, needed to have definite knowledge of his environment, i.e., of the forces of nature and of the plants and animals around him. Civilization, in fact, began when man learned to use fire and other tools to modify his environment. It is even more necessary than ever for mankind as a whole to have an intelligent knowledge of the environment if our complex civilization is to survive, since the basic "laws of nature" have not been repealed; only their complexion and quantitative relations have changed, as the world's human population has increased and as man's power to alter the environment has expanded.

Like all phases of learning, the science of ecology has had a gradual, if spasmodic, development during recorded history. The writings of Hippocrates, Aristotle, and other philosophers of the Greek period contain material which is clearly ecological in nature. However, the Greeks literally did not have a word for it. The word "ecology" is of recent coinage, having been first proposed by the German biologist Ernst Haeckel in 1869. Before this, many of the great men of the biological renaissance of the eighteenth and nineteenth centuries had contributed to the subject even though the label "ecology" was not in use. For example, Anton van Leeuwenhoek, best known as a pioneer microscopist of the early 1700s, also pioneered the study of "food chains" and "population regulation" (see Egerton, 1968), two important areas of modern ecology. As a recognized distinct field of biology, the science of ecology dates from about 1900, and only in the past decade has the word become part of the general vocabulary. Today, everyone is acutely aware of the environmental sciences as indispensable tools for creating and maintaining the quality of human civilization. Consequently, ecology is rapidly becoming the branch of science that is most relevant to the everyday life of every man, woman, and child.

The word ecology is derived from the Greek *oikos*, meaning "house" or "place to live." Literally, ecology is the study of organisms "at home." Usually ecology is defined as the study of the relation of organisms or groups of organisms to their environment, or the science of the interrelations between living organisms and their environment. Because ecology is concerned especially with the biology of *groups* of organisms and with *functional* processes on the lands, in the oceans, and in fresh waters, it is more in keeping with the modern emphasis to define ecology as the study of the structure and function of nature, it being understood that mankind is a part of nature. One of the definitions in Webster's Unabridged Dictionary seems especially appropriate for the closing decades of the 20th century, namely, *"the totality or pattern of relations between organisms and their environment."* In the long run the best definition for a broad subject field is probably the shortest and least technical one, as, for example, "environmental biology."

So much for definitions. To understand the scope and relevance of ecology, the subject must be considered in relation to other branches of biology and to *"ologies"* in general. In the present age of specialization in human endeavors, the inevitable connections between different fields are often obscured by the large masses of knowledge within the fields (and sometimes also, it must be admitted, by stereotyped college courses). At the other extreme, almost any field of learning may be so broadly defined as to take in an enormous range of subject material. Therefore, recognized "fields" need to have recognized bounds, even if these bounds are somewhat arbitrary and subject to shifting from time to time. A shift in scope has been especially note-worthy in the case of ecology as general public awareness of the subject has in-creased. To many, "ecology" now stands for "the totality of man and environment." But first let us examine the more tradi-tional academic position of ecology in the family of sciences.

For the moment, let us look at the divi-sions of biology, "the science of life." We traditionally cut the biology "layer cake," as it were, into small pieces in two distinct ways, as shown in Figure 1–1. We may di-vide it "horizontally" into what are usually called "basic" divisions because they are concerned with fundamentals common to all life, or at least are not restricted to partic-ular organisms. Morphology, physiology, genetics, ecology, evolution, molecular biology, and developmental biology are examples of such divisions. We may also divide the cake "vertically" into what may be called "taxonomic" divisions, which deal with the morphology, physiology, ecology, etc., of specific kinds of organisms. Zoology, botany, and bacteriology are large divisions of this type, and phycology, protozoology, mycology, entomology, ornithology, etc., are divisions dealing with more limited groups of organisms. Thus ecology is a basic division of biology and, as such, is also an integral part of any and all of the taxonomic divisions. Both approaches are profitable. It is often very productive to restrict work to certain taxonomic groups, because different kinds of organisms require different meth-ods of study (one cannot study eagles by the same methods used to study bacteria) and because some groups of organisms are

economically or otherwise much more im-portant or interesting to man than others. Ultimately, however, unifying principles must be delimited and tested if the subject field is to qualify as "basic." It is the pur-pose of Part 1 of this book to outline briefly this aspect of ecology.

Perhaps the best way to delimit modern ecology is to consider it in terms of the con-cept of *levels of organization* visualized as a sort of "biological spectrum" as shown in Figure 1–2. Community, population, organ-ism, organ, cell, and gene are widely used terms for several major biotic levels shown in hierarchial arrangement from large to small in Figure 1–2. Interaction with the physical environment (energy and matter) at each level produces characteristic func-tional systems. By a *system* we mean just what Webster's Collegiate Dictionary de-fines as "regularly interacting and interde-pendent components forming a unified whole." Systems containing living com-ponents (biological systems or biosystems) may be conceived at any level in the hier-archy illustrated in Figure 1–2, or at any intermediate position convenient or practi-cal for analysis. For example, we might con-sider not only gene systems, organ systems, and so on, but also host-parasite systems as intermediate levels between population and community.

Ecology is concerned largely with the right-hand portion of this spectrum, that is, the system levels beyond that of the organ-ism. In ecology the term *population*, origi-nally coined to denote a group of people, is broadened to include groups of individuals of any one kind of organism. Likewise, *com-*

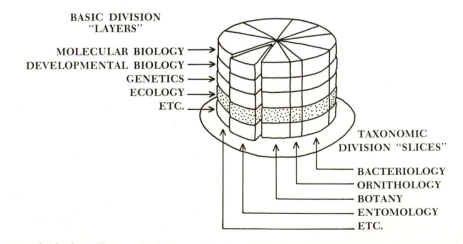

Figure 1–1.   The biology "layer cake," illustrating "basic" (horizontal) and "taxonomic" (vertical) divisions.

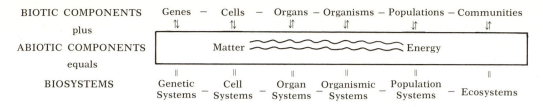

Figure 1–2. Levels of organization spectrum. Ecology focuses on the right-hand portion of the spectrum, that is, the levels of organization from organisms to ecosystems.

munity in the ecological sense (sometimes designated as "biotic community") includes all of the populations occupying a given area. The community and the nonliving environment function together as an ecological system or *ecosystem*. *Biocoenosis* and *biogeocoenosis*, terms frequently used in the European and Russian literature, are roughly equivalent to community and ecosystem respectively. The largest and most nearly self-sufficient biological system we know about is often designated as the *biosphere* or *ecosphere*, which includes all of the earth's living organisms interacting with the physical environment as a whole so as to maintain a steady-state system intermediate in the flow of energy between the high energy input of the sun and the thermal sink of space.

It is important to note that no sharp lines or breaks were indicated in the above "spectrum,"* not even between the organism and the population. Since in dealing with man and higher animals we are accustomed to think of the individual as the ultimate unit, the idea of a continuous spectrum may seem strange at first. However, from the standpoint of interdependence, interrelations and survival, there can be no sharp break anywhere along the line. The individual organism, for example, cannot survive for long without its population any more than the organ would be able to survive for long as a self-perpetuating unit without its organism. Similarly, the community cannot exist without the cycling of materials and the flow of energy in the ecosystem.

One reason for listing the levels of organization horizontally instead of vertically is to emphasize that, in the long run, no one level is any more or less important or any more or less deserving of scientific study than any other level. Some attributes, obviously, become more complex and variable as we proceed from the left to the right, but *it is an often overlooked fact that other attributes become less complex and less variable as we go from the small to the large unit.* Because homeostatic mechanisms, that is, checks and balances, forces and counter forces, operate all along the line, a certain amount of integration occurs as smaller units function within larger units. For example, the rate of photosynthesis of a forest community is less variable than that of individual leaves or trees within the community, because when one part slows down another may speed up in a compensatory manner. *When we consider the unique characteristics which develop at each level,* there is no reason to suppose that any level is any more difficult or any easier to study quantitatively. For example, growth and metabolism may be effectively studied at the cellular level or at the ecosystem level by using technology and units of measurement of a different order of magnitude. Furthermore, the findings at any one level *aid in the study of another level, but never completely explain the phenomena occurring at that level.* This is an important point because persons sometimes contend that it is useless to try to work on complex populations and communities when the smaller units are not yet fully understood. If this idea was pursued to its logical conclusion, all biologists would concentrate on one level, the cellular, for example, until they solved the problems of this level; then they would study tissues and organs. Actually, this philosophy was widely held until biologists discovered that each level had characteristics which knowledge of the next lower level explained only *in part.* In other words, not all attributes of a higher level are predictable if we know only the properties of the lower level. Just as

---

* Actually the "levels" spectrum, like a radiation spectrum or a logarithmic scale, theoretically can be extended infinitely in both directions.

the properties of water are not predictable if we know only the properties of hydrogen and oxygen, so the characteristics of ecosystems cannot be predicted from knowledge of isolated populations; one must study the forest (i.e., the whole) as well as the trees (i.e., the parts). Feibleman (1954) has called this important generalization the "theory of integrative levels."

In summary, the principle of *functional integration involving additional properties with increasing complexity of structure* is an especially important one for the ecologist to note. Advances in technology in the past 10 years have made it possible to deal quantitatively with large, complex systems such as the ecosystem. Tracer methodology, mass chemistry (spectrometry, colorimetry, chromatography, etc.), remote sensing, automatic monitoring, mathematical modeling, and computer technology are providing the tools. Technology is, of course, a two-edged sword; it can be the means of understanding the wholeness of man and nature or of destroying it.

2. THE SUBDIVISIONS OF ECOLOGY

In regard to subdivisions, ecology is sometimes divided into *autecology* and *synecology*. Autecology deals with the study of the individual organism or an individual species. Life histories and behavior as a means of adaptation to the environment are usually emphasized. Synecology deals with the study of groups of organisms which are associated together as a unit. Thus, if a study is made of the relation of a white oak tree (or of white oak trees in general) or a wood thrush (or of wood thrushes in general) to the environment, the work would be autecological in nature. If the study concerned the forest in which the white oak or the wood thrush lives, the approach would be synecological. In the former instance attention is sharply focused on a particular organism with the purpose of seeing how it fits into the general ecological picture, much as one might focus attention on a particular object in a painting. In the latter instance the picture as a whole is considered, much as one might study the composition of a painting.

For the purpose of this textbook the subject of ecology is subdivided in three ways. In Part 1 the chapters are arranged according to the levels of organization concept, as discussed in the previous section. We shall start with the ecosystem, since this is the level with which we must ultimately deal, and then consider in sequence communities, populations, species and individual organisms. Then we shall return to the ecosystem level in order to consider the development, evolution and modeling of nature.

In Part 2 the subject is subdivided according to the kind of environment or habitat, namely freshwater ecology, marine ecology, and terrestrial ecology. Although the basic principles are the same, the kinds of organisms, interrelationships with man, and the methods of study may be quite different for different environments. Subdivision by habitat is also convenient as preparation for field trips and for the presentation of descriptive data on the biota.

In Part 3 applications are considered under such subdivisions as "natural resources," "pollution," "space travel" and "applied human ecology" in order to relate basic principles to problems.

As was discussed for biology in general, ecology also may be subdivided along taxonomic lines, for example, plant ecology, insect ecology, microbial ecology and vertebrate ecology. Orientation within a restricted taxonomic group is profitable since attention is focused on the unique or special features in the ecology of that group, as well as on the development of detailed methods. In general, problems which pertain only to restricted groups are beyond the scope of this text since they are best considered after the general principles have been outlined.

Subdivisions in ecology, as in any other subject, are useful because they facilitate discussion and understanding as well as suggest profitable ways to specialize within the field of study. From the brief discussion in this section, we see that one might concentrate on processes, levels, environments, organisms, or problems and make valuable contributions to the overall understanding of environmental biology.

3. ABOUT MODELS

A model is a formulation that mimics a real-world phenomenon, and by means of which predictions can be made. In simplest form, models may be verbal or graphic (i.e., informal). Ultimately, however, models must be statistical and mathematical (i.e., formal) if quantitative predictions are to be reasonably good. For example, a mathematical formulation that mimics numerical changes taking place in a population of insects, and by means of which numbers in the population at some point in time could be predicted, would be considered a biologically useful model. If the population in question is a pest species, the model could also become economically important.

Computer operations of models make it possible to predict probable outcomes as parameters in the model are changed, new parameters added, or old ones removed. In other words, a mathematical formulation can often be "tuned" by computer operations to improve the "fit" to the real-world phenomenon. Above all, models are extraordinarily useful as summaries of what is understood about the situation being modeled, and thereby are useful in delimiting aspects needing new or better data or new principles. When a model does not work, i.e., is a poor mimic of the real world, computer operations often can provide clues to the refinements or changes needed. Once a model proves to be a useful mimic, then opportunities for experimentation are unlimited since one can introduce new factors or perturbations and see how they would affect the system.

Contrary to the feeling of many skeptics when it comes to modeling complex nature, information about only a relatively small number of variables is often a sufficient basis for effective models because "key factors" or "integrative factors" (as were discussed in Section 2 of this introductory chapter) often dominate or control a large percentage of the action. Watt (1963), for example, states that "We do not need a tremendous amount of information about a great many variables to build revealing mathematical models for population dynamics." When we move up to the level of whole nature, or the ecological system, this principle should still hold, provided the formulations used in the model are also brought up to that level. In summary, models are not intended to be exact copies of the real world but simplifications that reveal the key processes necessary for prediction.

In the chapters that follow in Part 1 of this book the paragraphs headed by the word **"Statement"** are, in effect, "word" models of the ecological principle in question. In many cases graphic or circuit models will also be presented and in some cases simplified mathematical formulations are included to clarify quantitative relationships. An introduction to the procedures used in mathematical modeling is presented as the final chapter in Part 1 under the title of "Systems Ecology." Most of all, what this text attempts to provide are the principles, simplifications, and abstractions that one must deduce from the real world of nature before one can even start to construct a mathematical model of it.

PRINCIPLES AND CONCEPTS PERTAINING TO THE ECOSYSTEM

Chapter 2

1. CONCEPT OF THE ECOSYSTEM

Statement

Living organisms and their nonliving (abiotic) environment are inseparably interrelated and interact upon each other. Any unit that includes all of the organisms (i.e., the "community") in a given area interacting with the physical environment so that a flow of energy leads to clearly defined trophic structure, biotic diversity, and material cycles (i.e., exchange of materials between living and nonliving parts) within the system is an ecological system or *ecosystem.* From the trophic (fr. *trophe* = nourishment) standpoint, an ecosystem has two components (which are usually partially separated in space and time), an *autotrophic component* (autotrophic = self-nourishing), in which fixation of light energy, use of simple inorganic substances, and buildup of complex substances predominate; and a *heterotrophic component* (heterotrophic = other-nourishing), in which utilization, rearrangement, and decomposition of complex materials predominate. For descriptive purposes it is convenient to recognize the following components as comprising the ecosystem: (1) *inorganic substances* (C, N, CO_2, H_2O, etc.) involved in material cycles; (2) *organic compounds* (proteins, carbohydrates, lipids, humic substances, etc.) that link biotic and abiotic; (3) *climate regime* (temperature and other physical factors); (4) *producers*, autotrophic organisms, largely green plants, which are able to manufacture food from simple inorganic substances; (5) *macroconsumers or phagotrophs* (*phago* = to eat), heterotrophic organisms, chiefly animals, which ingest other organisms or particulate organic matter; (6) *microconsumers*, *saprotrophs* (*sapro* = to decompose), or *osmotrophs* (*osmo* = to pass through a membrane), heterotrophic organisms, chiefly bacteria and fungi, which break

down the complex compounds of dead protoplasms, absorb some of the decomposition products, and release inorganic nutrients that are usable by the producers together with organic substances, which may provide energy sources or which may be inhibitory or stimulatory to other biotic components of the ecosystem. Items 1 through 3 comprise the abiotic components, and 4 through 6 constitute the *biomass* (= living weight).

Another useful two-category subdivision for heterotrophs suggested by Wiegert and Owens (1970) is as follows: *biophages*, organisms consuming other living organisms, and *saprophages*, organisms feeding on dead organic matter. As will be explained below, such a classification takes into consideration the time lag between consumption of living and dead matter.

From the functional standpoint an ecosystem may be conveniently analyzed in terms of the following: (1) energy circuits, (2) food chains, (3) diversity patterns in time and space, (4) nutrient (biogeochemical) cycles, (5) development and evolution, and (6) control (cybernetics).

The ecosystem is the basic functional unit in ecology, since it includes both organisms (biotic communities) and abiotic environment, each influencing the properties of the other and both necessary for maintenance of life as we have it on the earth.

Explanation

Since no organism can exist by itself or without an environment, our first principle may well deal with the "interrelation" and the principle of "wholeness" that are part of our basic definition of ecology given in Chapter 1, Section 1. The term ecosystem was first proposed by the British ecologist A. G. Tansley in 1935, but, of course, the concept is by no means so recent. Allusions to the idea of the unity of organisms and

8

environment (as well as the oneness of man and nature) can be found as far back in written history as one might care to look. However, it was not until the late 1800s that formal statements began to appear, interestingly enough, in a parallel manner in the American, European, and Russian ecological literature. Thus, Karl Mobius in 1877 wrote (in German) about the community of organisms in an oyster reef as a "biocoenosis," and in 1887 the American S. A. Forbes wrote his classic essay on the lake as a "microcosm." The Russian pioneering ecologist V. V. Dokuchaev (1846–1903) and his chief disciple G. F. Morozov (who specialized in forest ecology)* placed great emphasis on the concept of the "biocoenosis," a term later expanded by Russian ecologists to "geobiocoenosis" (see Sukachev, 1944). Thus, no matter what the environment (whether freshwater, marine, or terrestrial), biologists around the turn of the century began serious consideration of the idea of the unity of nature. Some other terms that have been used to express the holistic viewpoint are holocoen (Friederichs, 1930), biosystem (Thienemann, 1939), and bioenert body (Vernadsky, 1944). As already indicated in Chapter 1 (page 5), *ecosystem* is, as might be expected, the preferred term in English, while *biogeocoenosis* (or *geobiocoenosis*) is preferred by writers using the Germanic and Slavic languages. Some writers have attempted to make a distinction between the two words, but as far as this textbook is concerned the two are considered to be synonyms. "Ecosystem" has the very great advantage of being a short word which is easily assimilated into any language!

The concept of the ecosystem is and should be a broad one, its main function in ecological thought being to emphasize obligatory relationships, interdependence, and causal relationships, that is, the coupling of components to form functional units. A corollary to this is that since parts are operationally inseparable from the whole, the ecosystem is the level of biological organization most suitable for the appli-

cation of systems analysis techniques, a subject to be dealt with in Chapter 10. Ecosystems may be conceived of and studied in various sizes. A pond, a lake, a tract of forest or even a laboratory culture (microecosystem) provide convenient units of study. As long as the major components are present and operate together to achieve some sort of functional stability, even if for only a short time, the entity may be considered an ecosystem. A temporary pond, for example, is a definite ecosystem with characteristic organisms and processes even though its active existence is limited to a short period of time. The practical considerations involved in delimiting and classifying ecosystems will be considered later.

One of the universal features of all ecosystems, whether terrestrial, freshwater, or marine, or whether man-engineered (agricultural, etc.) or not, is the interaction of the autotrophic and heterotrophic components, as outlined in the Statement. Very frequently these functions and the organisms responsible for the processes are partially separated in space in that they are stratified one above the other with greatest autotrophic metabolism occurring in the upper "green belt" stratum in which light energy is available and the most intense heterotrophic metabolism taking place in the "brown belt" below in which organic matter accumulates in soils and sediments. Also, the basic functions are partially separated in time in that there may be a considerable delay in the heterotrophic utilization of the products of autotrophic organisms. For example, photosynthesis predominates in the canopy of a forest ecosystem. Only a part, often only a small part, of the photosynthate is immediately and directly used by the plant and by herbivores and parasites which feed on foliage and new wood; much of the synthesized material (in the form of leaves, wood, and stored food in seeds and roots) eventually reaches the litter and soil, which together constitute a well-defined heterotrophic system.

This space-time separation leads to a convenient classification of energy circuits into (1) a *grazing* circuit, in which the term grazing refers to the direct consumption of living plants or plant parts, and (2) an *organic detritus* circuit, which involves the accumulation and decomposition of dead materials. The term detritus (= a product of disintegration, from the Latin *deterere*, to

---

* Dokuchaev's chief work, reprinted in Moscow in 1948, was *Uchenie o zonax prirody* (*Teaching About the Zones of Nature*). Morozov's chief book is *Uchenie o lese* (*Teaching About Forests*). We are indebted to Dr. Roman Jakobson, Professor of Slavic Languages at Harvard University, for information on these two works which are but little known in the United States.

wear away) is borrowed from geology where it is traditionally used to designate the products of rock disintegration. As used in this text, "detritus," unless otherwise indicated, refers to all the particulate organic matter involved in the decomposition of dead organisms. Detritus seems the most suitable of many terms that have been suggested to designate this important link between the living and the inorganic world (see Odum and de la Cruz, 1963). There will be more about energy circuits in the next chapter, but it may be helpful at this point to take a preliminary look at Figure 3–8, page 66.

Further subdivision of the ecosystem into six "components" and six "processes" (as listed in the Statement) provides a convenient, if somewhat arbitrary, ecological classification, the former emphasizing structure and the latter emphasizing function. Although different methods are often required to delineate structure on the one hand and to measure rates of function on the other, the ultimate goal of study at any level of biological organization is the understanding of the relationships between structure and function. The next eight chapters are devoted to this task with regard to the ecological levels.

Abiotic components that limit and control organisms will be considered in detail in Chapter 5, and the role of organisms in controlling the abiotic environment will be considered later in this chapter. As a general principle we can point out that from the operational standpoint the living and nonliving parts of ecosystems are so interwoven into the fabric of nature that it is difficult to separate them (hence the operational classifications that do not make a sharp distinction between biotic and abiotic). Most of the vital elements (C, H, O, N, P, and so on) and organic compounds (carbohydrates, proteins, lipids, and so on) are not only found both inside and outside of living organisms, but they are in a constant state of flux between living and nonliving states. There are, however, some substances that appear to be unique to one or the other state. The high energy storage material, ATP (adenosine triphosphate), for example, is found only inside living cells (or at least its existence outside is very transitory), whereas *humic substances*, which are resistant end products of decomposition (see page 29), are never found inside cells yet

are a major and characteristic component of all ecosystems. Other key biotic complexes, such as the genetic material DNA (deoxyribonucleic acid) and the chlorophylls, occur both inside and outside organisms but become nonfunctional when outside the cell. As will be noted later, quantitative measurement of ATP, humus, and chlorophyll on an area or volume basis provides indices of biomass, decomposition, and production respectively.

The three living components (producers, phagotrophs, and saprotrophs) may be thought of as the three "functional kingdoms of nature" since they are based on the type of nutrition and the energy source used. These ecological categories should not be confused with taxonomic kingdoms, although there are certain parallels, as pointed out by Whittaker (1969) and as shown in Figure 2–1. In Whittaker's arrangement of the phyla into an evolutionary "family tree" all three types of nutrition are found in the Monera and Protista, while the three higher branches, namely "plants," Fungi, and "animals," specialize as "producers," "absorbers," and "ingesters" respectively. It should be emphasized that the ecological classification is one of function rather than of species as such. Some species of organisms occupy intermediate positions in the series and others are able to shift their mode of nutrition according to environmental circumstances. Separation of heterotrophs into large and small consumers is arbitrary but justified in practice because of the very different study methods required. The heterotrophic microorganisms (bacteria, fungi, etc.) are relatively immobile (usually imbedded in the medium being decomposed) and are very small with high rates of metabolism and turnover. Specialization is more evident biochemically than morphologically; consequently, one cannot usually determine their role in the ecosystem by such direct methods as looking at them or counting their numbers. Organisms which we have designated as macroconsumers obtain their energy by heterotrophic ingestion of particulate organic matter. These are largely the "animals" in the broad sense. They tend to be morphologically adapted for active food seeking or food gathering, with the development of complex sensory-neuromotor as well as digestive, respiratory, and circulatory systems in the higher forms. In earlier

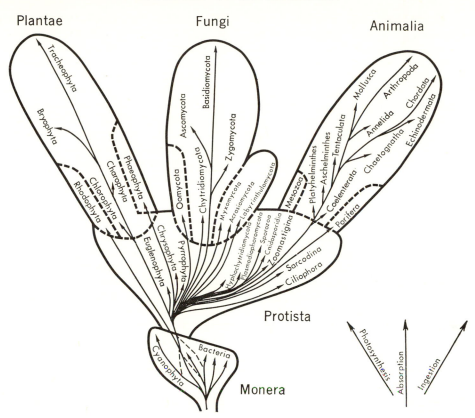

Figure 2–1. A five-kingdom system based on three levels of organization—the procaryotic (kingdom Monera), eucaryotic unicellular (kingdom Protista), and eucaryotic multicellular and multinucleate. On each level there is divergence in relation to three principal modes of nutrition—the photosynthetic, absorptive, and ingestive. Many biology and microbiology texts list four kingdoms by combining the "lower Protista" (i.e., Monera) with the "higher Protista" to form the "Protista." Evolutionary relations are much simplified, particularly in the Protista. Only major animal phyla are entered, and phyla of the bacteria are omitted. The Coelenterata comprise the Cnidaria and Ctenophora; the Tentaculata comprise the Bryozoa, Brachiopoda, and Phoronida, and in some treatments the Entoprocta. (From Whittaker, 1969.)

editions of this text the microconsumers, or saprotrophs, were designated as "decomposers," but recent work has shown that in some ecosystems animals are more important than bacteria or fungi in the decomposition of organic matter (see, for example, Johannes, 1968). Consequently, it seems preferable not to designate any particular organisms as "decomposers" but rather to consider "decomposition" as a process involving all of the biota and abiotic processes as well. For additional general discussions of the ecosystem concept see Forbes' (1887) classic essay, Tansley (1935), Evans (1956), and Cole (1958). Schultz (1967) and Van Dyne (1969) discuss the concept from the standpoint of resource management, and Stoddard (1965) from the viewpoint of a geographer. Every student of ecology and, indeed, every citizen should read Aldo Leopold's *Land Ethic* (1949), an eloquent, often quoted and reprinted essay on the special relevance of the ecosystem concept to man. We all also should reread "Man and Nature" by the Vermont prophet George Perkins Marsh (written in 1864, reprinted in 1965) who analyzed the causes of the decline of ancient civilizations and forecast a similar doom for modern ones unless man takes what we would call today an "ecosystematic" viewpoint of man and nature.

Examples

One of the best ways to begin the study of ecology is to go out and study a small pond, or a meadow or old-field. In fact, any area exposed to light, even a lawn, a window flower box, or a laboratory-cultured microcosm, can be the "guinea pig" for the beginning study of ecosystems, provided that the physical dimensions and biotic diversity are not so great as to make observations of the whole difficult. In other words, one does not begin the "practical" or "lab" study of ecology by tackling the great forest or an ocean! In order to illustrate as many aspects as possible, let us now consider five examples: a pond, a meadow, a watershed, a laboratory microecosystem, and a spacecraft.

The Pond

Let us consider the pond as a whole as an ecosystem, leaving the study of populations within the pond for the second section of this book. The inseparability of living organisms and the nonliving environment is at once apparent with the first sample collected. Not only is the pond a place where plants and animals live, but plants and animals make the pond what it is. Thus, a bottle full of the pond water or a scoop full of bottom mud is a mixture of living organisms, both plant and animal, and inorganic and organic compounds. Some of the larger animals and plants can be separated from the sample for study or counting, but it would be difficult to completely separate the myriad of small living things from the nonliving matrix without changing the character of the fluid. True, one could autoclave the sample of water or bottom mud so that only nonliving material remained, but this residue would then no longer be pond water or pond soil but would have entirely different appearances and characteristics.

Despite the complexities, the pond ecosystem may be reduced to the several basic units, as shown in Figure 2–2.

1. *Abiotic substances* (Fig. 2–2, I) are basic inorganic and organic compounds, such as water, carbon dioxide, oxygen, calcium, nitrogen and phosphorus salts, amino and humic acids, etc. A small portion of the vital nutrients is in solution and immediately available to organisms, but a much larger portion is held in reserve in particulate matter (especially in the bottom sediments), as well as in the organisms themselves. As Hayes (1951) has expressed it, a pond or lake "is not, as one might think, a body of water containing nutrients, but an equilibrated system of water and solids, and under ordinary conditions nearly all of the nutrients are in a solid stage." The rate of release of nutrients from the solids, the so-lar input, and the cycle of temperature, day length and other climatic regimes are the most important processes which regulate the rate of function of the entire ecosystem on a day-to-day basis.

2. *Producer organisms*. In a pond the producers may be of two main types: (1) rooted or large floating plants generally growing in shallow water only (Fig. 2–2, IIA) and (2) minute floating plants, usually algae, called *phytoplankton* (*phyto* = plant; *plankton* = floating) (Fig. 2–2, IIB), distributed throughout the pond as deep as light penetrates. In abundance, the phytoplankton gives the water a greenish color; otherwise, these producers are not visible to the casual observer, and their presence is not suspected by the layman. Yet, in large, deep ponds and lakes (as well as in the oceans) phytoplankton is much more important than is rooted vegetation in the production of basic food for the ecosystem.

3. *Macroconsumer organisms*, animals, such as insect larvae, crustacea, and fish. The primary macroconsumers (herbivores) (Fig. 2–2, III-1A, III-1B) feed directly on living plants or plant remains, and are of two types, namely *zooplankton* (animal plankton) and *benthos* (= bottom forms), paralleling the two types of producers. The secondary consumers (carnivores) such as predaceous insects and game fish (Fig. 2–2, III-2, III-3) feed on the primary consumers or on other secondary consumers (thus making them tertiary consumers). Another important type of consumer is the *detritivore* (III-1A), which subsists on the "rain" of organic detritus from autotrophic layers above.

4. *Saprotrophic organisms* (Fig. 2–2, IV). The aquatic bacteria, flagellates, and fungi are distributed throughout the pond, but they are especially abundant in the mud-water interface along the bottom where bodies of plants and animals accumulate. While a few of the bacteria and fungi are pathogenetic in that they will attack living organisms and cause disease, the great majority begin attack only after the organism dies. When temperature conditions are favorable, decomposition occurs rapidly in a body of water; dead organisms do not retain their identification for very long but are soon broken up into pieces, consumed by the combined action of detritus-feeding animals and microorganisms, and their nutrients released for reuse.

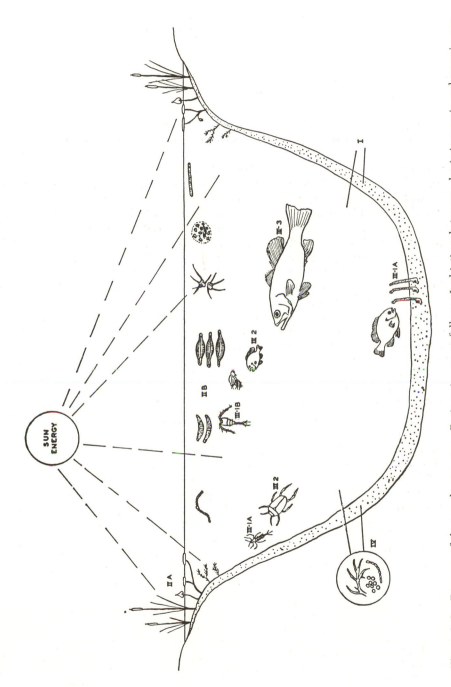

Figure 2–2. Diagram of the pond ecosystem. Basic units are as follows: I, abiotic substances – basic inorganic and organic compounds; IIA, producers – rooted vegetation; IIB, producers – phytoplankton; III-1A, primary consumers (herbivores) – bottom forms; III-1B, primary consumers (herbivores) – zooplankton; III-2, secondary consumers (carnivores); III-3, tertiary consumers (secondary carnivores); IV, saprotrophs – bacteria and fungi of decay. The metabolism of the system runs on sun energy, while the rate of metabolism and relative stability of the pond depend on the rate of inflow of materials from rain and from the drainage basin in which the pond is located.

Table 2–1. *Daily Community Metabolism in the Water Column of a Pond as Indicated by Mean Oxygen Changes at Successive Depths*

| DEPTH | O₂ CHANGE (GMS/M³) | | GROSS PRODUCTION (GMS O₂/M³) | COMMUNITY RESPIRATION (GMS O₂/M²) |
	Light Bottle	Dark Bottle		
Top m³	+3	−1	4	1
2nd m³	+2	−1	3	1
3rd m³	0	−1	1	1
Bottom m³	−3	−3	0	3
Total metabolism of water column, (gms O₂/m²/day)	—	—	8	6

The partial stratification of the pond into an upper "production" zone and a lower "decomposition - nutrient regeneration" zone can be illustrated by simple measurements of total diurnal metabolism of water samples. A "light-and-dark bottle" technique may be employed for this purpose, and also to provide a starting point for charting energy flow (one of the six processes listed in the ecosystem definition). As shown in Figure 2–3, samples of water from different depths are placed in paired bottles, one of which (the dark bottle) is covered with black tape or aluminum foil to exclude all light. Other water samples are "fixed" with reagents so that the original oxygen concentration at each depth can be determined.* Then the string of paired dark and light bottles is suspended in the pond so that the samples are at the same depth from which they were drawn. At the end of a 24-hour period the string of bottles is removed and the oxygen concentration in each sample is determined and compared with the concentration at the beginning. The decline of oxygen in the dark bottle indicates the amount of respiration by producers and consumers (i.e., the total community) in the

water, whereas oxygen change in the light bottle reflects the net result of oxygen consumed by respiration and oxygen produced by photosynthesis, if any. Adding respiration and net production together, or subtracting final oxygen concentration in the dark bottle from that in the light bottle (provided that both bottles had the same oxygen concentration to begin with) gives an estimate of the total or gross photosynthesis (food production) for the 24-hour period, since the oxygen released is proportional to dry matter produced.

The hypothetical data in Table 2–1 illustrates the kind of results one might expect to get with a light-and-dark bottle experiment in a shallow, fertile pond on a warm, sunny day. In this hypothetical case photosynthesis exceeds respiration in the top two meters* and just balances it in the third meter (zero change in light bottle); below three meters the light intensity is too low for photosynthesis so only respiration occurs. The point in a light gradient at which plants are just able to balance food production and utilization is called the *compensation level* and marks a convenient functional boundary between the autotrophic stratum (*euphotic zone*) and the heterotrophic stratum.

A daily production of 8 grams O₂ per m² and excess production over respiration would indicate a healthy condition for the ecosystem, since excess food is being produced in the water column that becomes available to bottom organisms as well as to all the organisms during periods when light

---

* The Winkler method is the standard procedure for oxygen measurement in water. It involves fixation with $MnSO_4$, H_2SO_4, and alkaline iodide, which releases elemental iodine in proportion to oxygen. The iodine is titrated with sodium thiosulphate (the "hypo" used to fix photographs) at a concentration calibrated to estimate milligrams of oxygen per liter, which, conveniently, is also grams per m³ and parts per million (ppm). Electronic methods employing oxygen electrodes are now being developed which will probably eventually replace the standard chemical methods, especially when continuous monitoring of oxygen changes is desirable. For details on methods see reference listed under "American Public Health Association" in the bibliography.

---

* Where water is clear, as in large lakes and the ocean, photosynthesis is actually suppressed by high light intensity near the surface so that the highest rate of photosynthesis usually occurs below the top meter (see Figure 3–3, page 50).

Figure 2–3. Measuring the metabolism of a pond with light-and-dark bottles. *A.* Filling a pair of light and dark (black) bottles with water collected at a specific depth with a water sampler (the cylindrical instrument with rubber stoppers at each end). *B.* Lowering a string of dark and light bottles to the depth at which the water was collected. The white plastic jug will serve as a float. See text for further explanation of the method. The energetics of the pond shown in these pictures is discussed and modeled in Chapter 3. (Photos by K. Kay for Institute of Ecology, University of Georgia.)

and temperature are not so favorable. If the hypothetical pond were being polluted with organic matter, O_2 consumption (respiration) would greatly exceed O_2 production, resulting in oxygen depletion and (should the imbalance continue) eventual anaerobic (= without oxygen) conditions which would eliminate fish and most other animals. In assaying the "health" of a body of water we need not only to measure the oxygen concentration as a condition for existence, but also to determine rates of change and the balance between production and use in the diurnal and annual cycle. Monitoring oxygen concentrations, then, is one convenient way of "feeling the pulse" of the aquatic ecosystem. Measurement of "biological oxygen demand" (B.O.D.) is also a standard method of pollution assay (see Chapter 16).

Enclosing pond water in bottles or other containers such as plastic spheres or cylinders has obvious limitations, and the bottle method used here as an illustration is not adequate for assaying the metabolism of the whole pond since oxygen exchanges of bottom sediments and the larger plants and animals are not measured. Other methods will be discussed in Chapter 3.

The Watershed Unit

Although the pond seems self-contained in terms of the biological components, its rate of metabolism and its relative stability over a period of years is very much determined by the input of sun energy and especially by the rate of inflow of water and materials from the watershed. A net inflow of materials often occurs particularly when bodies of water are small or outflow is restricted. When man increases soil erosion or introduces quantities of organic material (sewage, industrial wastes) at rates that cannot be assimilated, the rapid accumulation of such materials may be destructive to the system. The phrase *cultural eutrophication* (= cultural enrichment) is becoming widely used to denote organic pollution resulting from man's activities. Therefore, *it is the whole drainage basin, not just the body of water, that must be considered as the minimum ecosystem unit when it comes to man's interests.* The ecosystem unit for practical management must then include for every square meter or acre of water at least

20 times an area of terrestrial watershed.[*] The cause of and the solutions for water pollution are not to be found by looking only into the water; it is usually the bad management of the watershed that is destroying our water resources. The entire drainage or catchment basin must be considered as the management unit — but more about this later. The Everglades National Park in south Florida is a good example of this need to consider the whole drainage basin. Although it is large in area, the park does not now include the source of the freshwater that must drain southward into the park if it is to retain its unique ecology. The Everglades National Park, therefore, is completely vulnerable to reclamation, agricultural, and jetport developments north of the park boundary, which could divert or pollute the "life blood" of the "glades."

For a picture of a watershed manipulated and monitored for experimental study, see Figure 2–4.

The Meadow

Having stressed the pond's dependency on the land, we should now move to a brief consideration of the terrestrial ecosystem. Figure 2–5 shows some ecologists sampling the consumers associated with the vegetation (i.e., the autotrophic stratum) and also "feeling the pulse," as it were, of a grassland. Although the meadow looks quite different from the pond and different tools are needed to study it, the two types of ecosystems actually have the same basic structure and they function in the same fashion, when we consider the system as a whole. Of course, most of the species on land are different from those in water, but they can be grouped into comparable ecologic assemblages, as shown in Table 2–2.

The vegetation performs the same function as the phytoplankton, the insects and spiders in the vegetation are comparable to the zooplankton, the birds and mammals (and people who use the meadow for crops or cattle grazing!) are comparable to the fish, and so on. However, aquatic and terrestrial communities do differ in the relative size of some of the biological compartments, and, of course, in the relative impact that such

---

[*] The ratio of water surface to watershed area varies widely and depends on the rainfall, geological structure of underlying rocks, and topography.

Figure 2–4. Experimental watersheds at the Coweeta Hydrologic Laboratory in the mountains of western North Carolina. All the trees have been cleared from the watershed in the center of the picture in order to compare water input (rainfall) and output (stream run-off) with that of the undisturbed forested watersheds on either side. Insert shows the V-notch weir and recording equipment used to measure the amount of water flowing out of each watershed area. (Photo courtesy U. S. Forest Service.)

physical components as water have on the conditions of existence (and, therefore, on the adaptive physiology and behavior of the organisms).

The most striking contrast between land and water ecosystems is in the size of the green plants. The autotrophs of land tend to be fewer but very much bigger, both as individuals and as biomass per unit area (Table 2–2). The contrast is especially impressive when we compare the open ocean, where phytoplankton are even smaller than in the pond, and the forest with its huge trees. Shallow water communities (edges of ponds, lakes, oceans, as well as marshes), grasslands, and deserts are intermediate between these extremes. In fact, the whole biosphere can be visualized as a vast gradient of ecosystems with the deep oceans at one extreme and the large forests at the other.

Terrestrial autotrophs invest a large part of their productive energy in supporting tissue, which is necessary because of the much lower density (and hence lower supporting capacity) of air as compared to water. This supporting tissue has a high content of cellulose and lignin (wood) and requires little energy for maintenance because it is resistant to consumers. Accordingly, plants on land contribute more to the structural matrix of the ecosystem than do plants in water, and the rate of metabolism per unit volume or weight of land plants is correspondingly much lower. This is a good point to introduce the concept of *turnover,* which may be broadly defined as the *ratio of throughput to content.* Turnover can be conveniently expressed either as a rate fraction or as a "turnover time," which is the reciprocal of the rate fraction. Let us consider the productive energy flow as the "throughput" and the standing crop biomass (gms dry wt./m² in Table 2–2) as the "content." If we assume that the pond and the meadow have a comparable gross photosynthetic rate of 5 grams per m² per day, then the turnover rate for the pond would be ⁵/₅ or 1, and the turnover time would be one day. In contrast, the turnover rate for the meadow would be ⁵/₅₀₀ or 0.01, and the turnover time would be 100 days. Thus, the tiny plants in the pond may replace themselves in a day when the pond metabolism

Figure 2–5. Meadow and old-field ecosystems. *A.* Ecologists using a vacuum "sweeper" to sample arthropods in the above-ground stratum of a grassland community. Organisms trapped within the open-ended plastic cylinder are vacuumed into a net. *B.* Measuring the rate of photosynthesis of one of the plant species in an old-field ecosystem with an infrared gas analyzer, which determines the carbon dioxide uptake as air is drawn through the transparent chamber. By covering the chamber with a dark cloth (or making measurements at night) the respiration of the plant community can be measured; the total autotrophic metabolic rate can be estimated in the same manner as described for the "light-and-dark bottle" experiment (see text and Figure 2–3). (Photos courtesy of the Institute of Ecology, University of Georgia.)

Table 2-2. *Comparison of Density (Numbers/M²) and Biomass (as Grams Dry Weight/M²) of Organisms in Aquatic and Terrestrial Ecosystems of Comparable and Moderate Productivity*

ECOLOGIC COMPONENT	OPEN-WATER POND			MEADOW OR OLD-FIELD		
	Assemblage	No./M²	Gms Dry Wt./M²	Assemblage	No./M²	Gms Dry Wt./M²
Producers	Phytoplanktonic algae	10^8–10^{10}	5.0	Herbaceous angiosperms (grasses and forbs)	10^2–10^3	500.0
Consumers in the auto-trophic layer	Zooplanktonic crustaceans and rotifers	10^5–10^7	0.5	Insects and spiders	10^2–10^3	1.0
Consumers in the hetero-trophic layer	Benthic insects, mollusks, and crustaceans°	10^5–10^6	4.0	Soil arthropods, annelids, and nematodes†	10^5–10^6	4.0
Large roving consumers (permeants)	Fish	0.1–0.5	15.0	Birds and mammals	0.01–0.03	0.3‡ 15.0§
Microorganism consumers (saprophages)	Bacteria and fungi	10^{13}–10^{14}	1–10‖	Bacteria and fungi	10^{14}–10^{15}	10–100.0‖

° Including animals down to size of ostrocods.

† Including animals down to size of small nematodes and soil mites.

‡ Including only small birds (passerines) and small mammals (rodents, shrews, etc.).

§ Including 2 to 3 cows (or other large herbivorous mammals) per hectare.

‖ Biomass based on the approximation of 10^{13} bacteria = 1 gram dry weight.

is at a peak, whereas land plants are much longer-lived and "turnover" much more slowly. As we shall see in Chapter 4, the concept of turnover is especially useful in dealing with the exchange of nutrients between organisms and environment.

The large structural mass of land plants results in large amounts of resistant fibrous detritus (leaf litter, wood, etc.) reaching the heterotrophic layer. In contrast, the "rain of detritus" in the phytoplankton system constitutes small particles that are more easily decomposed and consumed by small animals. We would expect, therefore, to find larger populations of saprophagic microorganisms in soil than in sediments under open water (Table 2-2). However, as we have already emphasized, numbers and biomass of very small organisms do not necessarily reflect their activity; a gram of bacteria can vary manyfold in metabolic rate and turnover depending on the conditions. In contrast to the producers and microconsumers, numbers and weight of the macroconsumers tend to be more comparable in aquatic and terrestrial systems, if available energy in the system is the same. If we include large grazing animals on land, the numbers and biomass of large roving con-

sumers, or "permeants," tend to come out about the same in both systems (Table 2-2).

It should be emphasized that Table 2-2 is a tentative model. Strange as it may seem, no one has yet made a complete census of any one pond or meadow (or any other outdoor ecosystem for that matter)! We can only suggest approximations based on fragmentary information gleaned from many sites. Even in the simplest of natural ecosystems the number and variety of organisms and the complexity of the coupling of assemblages is bewildering. As would be expected, more is known about the large organisms (trees, birds, fish, and so on) than the small ones, which are not only more difficult to see but require technically difficult functional methods of assay. Likewise, we have lots of measurements of temperature, rainfall and other "macrofactors," but very little knowledge about micronutrients, vitamins, detritus, antibiotics, and other more difficult to assay "microfactors," which none-the-less have great importance to the maintenance of ecological balances. Better inventory techniques present a challenge to the next generation of ecologists because curiosity is no longer the only motivation for looking more deeply into nature; man's very existence is being threatened by his abysmal ignorance of what it takes to run a balanced ecosystem.

In both land and aquatic ecosystems a large part of solar energy is dissipated in the evaporation of water, and only a small part, generally less than 5 per cent, is fixed by photosynthesis. However, the role which this evaporation plays in the movement of nutrients and in maintaining temperature regimes is different. These contrasts will be considered in Chapters 3 and 4 (see also Smith, 1959, for interesting and speculative comparisons of aquatic and terrestrial ecosystems). We need only mention at this point that for every gram of CO_2 fixed in a grassland or forest ecosystem as much as 100 grams of water must be moved from the soil, through the plant tissues, and transpired (= term for evaporation from plant surfaces). No such massive water use is associated with production of phytoplankton or other submerged plants. This brings us back to man's minimum ecosystem unit, i.e., the drainage basin or watershed that includes terrestrial and aquatic systems together with man and his artifacts all functioning as a system. Watershed biohydrography (= the behavior of water in relation to

organisms) thus becomes a high priority subject in the analysis of man's ecosystem. In addition to an International Biological Program, which is planned to inventory man and the total environment, an International Hydrological Decade has also been proposed to encourage the systematic gathering of information on water and its movement in the biosphere.

The Microecosystem

Because outdoor ecosystems are complex, hard to delineate, and often difficult to study by traditional scientific means of "experiment and control," many ecologists are turning to laboratory and field microecosystems which can have discrete boundaries and can be manipulated and replicated at will. Figures 2–6 and 2–7 illustrate several types of systems that are being used to test ecological principles. These range from closed microcosms that require only light energy (miniature biospheres, as it were) to assemblages that are maintained in various kinds of chemostats and turbidostats with regulated inflow and outflow of nutrients and organisms. In terms of biological com-

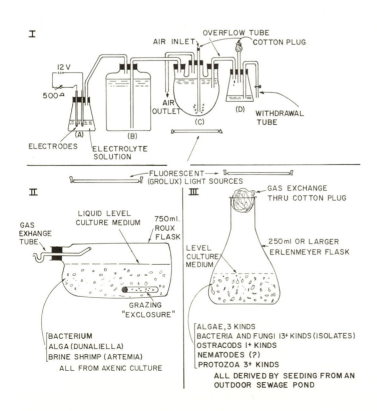

Figure 2–6. Three types of laboratory microecosystems. *I.* A simple, inexpensive *chemostat* in which a flow of culture media (B) through the culture chamber (C) and into an overflow vessel (D) is regulated by adjusting the electric current fed into an electrolysis pump (A). In the *turbidostat* a steady-state regulation is accomplished by a sensor, placed within the cultured community, that responds to the density (turbidity) of the organisms (internal regulation as contrasted to the external "constant input" regulation of the chemostat). (After Carpenter, 1969.) *II.* A gnotobiotic or "defined" microcosm containing three species from axenic (i.e., "pure") culture. The tube provides an area in which algae can multiply free from grazing by the shrimp (hopefully, preventing "overgrazing"). (After Nixon, 1969.) *III.* A microcosm "derived" from an outdoor system by multiple seeding (see Beyers, 1963). System *I* is "open," and systems *II* and *III* are "closed" except for the input of light energy and gas exchange with the atmosphere. Equilibrium in the closed systems, if achieved, results from nutrient cycle regulation by the community, rather than by mechanical control devices (as in the chemostat or turbidostat).

Figure 2–7. Laboratory micro-ecosystems. *A.* Aquatic microecosystems derived from nature (see Figure 2–6*III*). The large flasks are "climax" systems that maintain themselves indefinitely with only light input and gas exchange through cotton plug. When samples of equilibrium systems are inoculated into new culture medium (small flasks) the systems undergo a period of ecological succession or development that mimics ecosystem development in nature (see Chapter 9). Microecosystems shown in the photograph have been subjected to gamma irradiation to compare effect of stress on "youthful" and "mature" systems. (Courtesy of the Institute of Ecology, University of Georgia.) *B.* Terrestrial microecosystems maintained in plastic desiccators with transparent tops. Shown are forest floor herbaceous communities maintained under low light intensity as would be characteristic of forest floor vegetation and associated small organisms. (Courtesy of the Institute of Ecology, University of Georgia.)

ponents, two basic types may be distinguished: (1) microecosystems derived directly from nature by multiple seeding of culture media with environmental samples and (2) systems built up by adding species from axenic cultures (= free from other living organisms) until the desired combinations are obtained. The former systems represent nature "stripped down" or "simplified" to those organisms that are able to survive together within the limits of the container, the culture medium, and the light-temperature environment imposed by the experimenter. Such systems, therefore, are usually intended to simulate some specific outdoor situation. As in open nature the investigator can easily find out by observation what major biotic components (such as algae or invertebrates) are present, but it is difficult to determine the exact composition especially in regard to bacteria (see Gorden *et al.,* 1969). The ecological use of "derived" or "multiple-seeded" systems was pioneered by H. T. Odum and his students (Odum and Hoskins, 1957; Beyers, 1963). In the second approach, "defined" systems are "built up" by adding previously isolated and carefully studied components.

The resulting cultures are often called "gnotobiotic" (see Dougherty, 1959, for a discussion of terminology) because the exact composition, down to the presence or absence of bacteria, is known. Gnotobiotic cultures have been mostly used to study the nutrition, biochemistry, and other aspects of single species or strains, or for the study of two species interactions (see Chapter 7), but recently ecologists have begun to experiment with more complex "polyaxenic" cultures with the objective of building towards self-contained ecosystems (Nixon, 1969; Taub, 1969).

Actually these contrasting approaches to the laboratory microecosystem parallel the two long-standing ways ecologists have attempted to study lakes and other large systems of the real world. For example, in his 1964 essay "The Lacustrine Microcosm Reconsidered," the eminent American ecologist G. E. Hutchinson speaks of E. A. Birge's (1915) work on the heat budgets of lakes as pioneering the *holological* (fr. *holos* = whole) approach in which the lake is treated as a "black box"[*] with input and output. Hutchinson contrasts this with the *merological* (fr. *meros* = part) approach of Stephen Forbes (1887; see page 9) in which "we discourse on parts of the system and try to build up the whole from them."

At this point a word about a common misconception regarding the "balanced" fish aquarium. It is quite possible to achieve an approximate gaseous and food balance in an aquarium provided that the ratio of fish to water and plants remains small. Back in 1857, George Warington "established that wondrous and admirable balance between the animal and vegetable kingdoms" in a 12-gallon aquarium using a few goldfish, snails, and lots of eelgrass (*Vallisneria*), and, we might add, a diversity of associated microorganisms. He not only clearly recognized the reciprocal role of fish and plants but correctly noted the importance of the snail detritivore "in decomposing vegetation and confevoid mucus" thus "converting that which would otherwise act as a poisonous agent into a rich and fruitful pablum for vegetable growth." Most amateur attempts to balance aquaria fail because far too many fish are stocked for the available resources (diagnosis: an elemen-

tary case of gross overpopulation!). A glance at Table 2–2 shows that for complete self-sufficiency a medium-sized fish requires many cubic meters of water and attendant food organisms. Since "fish-watching" is the usual motivation for keeping aquaria in the home, office, or school, supplemental food, aeration, and periodic cleaning are necessary if large numbers of fish are to be crowded into small spaces. The home fish culturist, in other words, is advised to forget about ecological balance and leave the self-contained microcosm to the student of ecology. This is a good time to remind ourselves, however, that big "critters," such as fish and men, require more room than you might think!

Large outdoor tanks for aquatic systems (see H. T. Odum *et al.*, 1963) and various kinds of enclosures of terrestrial habitat (such as shown in Figure 2–5) represent increasingly used experimental setups that are intermediate between laboratory culture systems and the real world of nature.

In the next several chapters we shall have a number of occasions to illustrate how microecosystem research has helped to establish and clarify basic ecological principles. For a review of the "microcosm approach to ecosystem biology," the "balanced aquarium controversy," and for a useful bibliography, see Beyers (1964).

The Spacecraft as an Ecosystem

Perhaps the best way to visualize the ecosystem is to think about space travel, because when man leaves the biosphere he must take with him a sharply delimited, enclosed environment that must supply all vital needs using sunlight as the energy input from the surrounding and very hostile space environment. For journeys of a few weeks, such as to the moon and back, man does not need to take along a completely self-sustaining ecosystem since sufficient oxygen and food can be stored and CO_2 and other waste products can be fixed or detoxified for short periods of time. For long journeys, such as trips to the planets, man must engineer himself into a more closed or regenerative spacecraft. Such a self-contained vehicle must include not only all vital abiotic substances and the means to recycle them, but also the vital processes of production, consumption, and decomposi-

---

[*] A black box may be defined as any unit whose function may be evaluated without specifying the internal contents.

tion must be performed in a balanced manner by biotic components or their mechanical substitutes. In a very real sense the self-contained spacecraft is a microecosystem that contains man. It is interesting that the same two theoretical approaches as mentioned in the previous section, i.e., the "holological" and the "merological" approaches, are now being applied to the search for the "minimum ecosystem for man in space." The extent to which we can "short-cut" nature is unknown and highly controversial at present. Efforts to devise regenerative life-support systems for space travel, and their relevance to our continued survival on the large "earth spaceship," are discussed in detail in Chapter 20.

2. THE BIOLOGICAL CONTROL OF THE CHEMICAL ENVIRONMENT

Statement

Individual organisms not only adapt to the physical environment, but by their concerted action in ecosystems they also adapt the geochemical environment to their biological needs.

Explanation

Although everyone realizes that the abiotic environment ("physical factors") controls the activities of organisms, it is not always realized that organisms influence and control the abiotic environment in many ways. Changes in the physical and chemical nature of inert materials are constantly being effected by organisms which return new compounds and energy sources to the environment. The chemical content of the sea and of its bottom "oozes" is largely determined by the actions of marine organisms. Plants growing on a sand dune build up a soil radically different from the original substrate. A South Pacific coral island is a striking example of how organisms influence their abiotic environment. From simple raw materials of the sea, whole islands are built as the result of the activities of animals (corals, etc.) and plants. The very composition of our atmosphere is controlled by organisms, as will be detailed in the next section.

Man, of course, more than any other species attempts to modify the physical environment to meet his immediate needs, but in doing so he is increasingly disrupting, even destroying, the biotic components which are necessary for his physiological existence. Since man is a heterotroph and a phagotroph who thrives best near the end of complex food chains, his dependency on the natural environment remains no matter how sophisticated his technology becomes. The great cities are still only parasites in the biosphere when we consider what have been aptly called *the vital resources*, namely, air, water, and food. The bigger the cities the more they demand from the surrounding countryside and the greater the danger of damaging the natural environment "host." So far, man has been so busy "conquering" nature that he has yet given little thought or effort toward reconciling the conflicts in his dual role, that of manipulator of and inhabitant in ecosystems. The social, economic and legal steps that must now be taken to nurture man's ecological dependency on his environment are discussed in Chapter 21. The dictum "the greatest good for the greatest number" seemed like a good goal for society when we were not crowded, but not any more. For the quality of the individual, whether man in the city, cow in the pasture, or tree in the forest, the greatest good occurs when the numbers are not the greatest! The study of nature reveals many clues as to how quality controls can be established. Ecological principles, as we hope to show in this book, do provide a realistic basis for placing society's "pursuit of happiness" goal on a qualitative rather than on a quantitative basis.

Examples

One of the classic papers that should be a "must" on the reading list of every student of ecology is Alfred Redfield's summary-essay, published in 1958, entitled *The Biological Control of Chemical Factors in the Environment.* Redfield marshalls the evidence to show that the oxygen content of the air and the nitrate in the sea are produced and largely controlled by organic activity and, furthermore, that the quantities of these vital components in the sea are determined by the biocycling of phosphorus. This system is so intricate and beautifully organized as to render insignificant anything so far conceived by the engineering mind of man. Some idea of how it works will be presented in the next section of this chapter and in Chapter 4. If Redfield's estimates are correct, the oceans should be forever regarded as the great governor and protector of the biosphere, and not merely as a supply depot for unlimited exploitation. Thus, eutrophication and pollution of the oceans in a last-ditch effort to feed and supply the overcrowded land could very well lead to disastrous shifts in the content of the atmosphere and in the climates of the world.

The Copper Basin at Copperhill, Tennessee, and a similar area east of Butte, Montana, provide impressive demonstrations of the result of the *absence* of living organisms. In these regions fumes from copper smelters exterminated all the rooted plants over a wide area. All the soil eroded away, leaving a spectacular desert, as shown in Figure 2–8A. Although modern smelting methods no longer release fumes, vegetation has failed to return, and artificial reforestation attempts have largely failed. The area is too "raw" for plants to obtain an effective foothold and start the rebuilding process. No one can say how long it will take for natural processes to rebuild the soil and restore the forest, but it will not be within your lifetime or that of your children. Everyone should visit Copperhill as part of his general education. Or visit a badly eroded or strip-mined area (Fig. 2–8B) and ask: How much will it cost to rehabilitate such land that was needlessly abused?

3. PRODUCTION AND DECOMPOSITION IN NATURE

Statement

"Every year approximately 10^{17} grams (about 100 billion tons) of organic matter is produced on the earth by photosynthetic organisms. An approximately equivalent amount is oxidized back to CO_2 and H_2O during the same time interval as a result of the respiratory activity of living organisms. *But, the balance is not exact*" (Vallentyne, 1962). Over most of geological time (since the beginning of the Cambrian period 600 million years ago), a very small but significant fraction of the organic matter produced is buried and fossilized without being respired or decomposed, although there have been times in geological history when the balance shifted the other way. This excess

of organic production, which releases gaseous oxygen and removes CO_2 from air and water, over respiration, which accomplishes the reverse, has resulted in a buildup of oxygen in the atmosphere to the high levels of recent geological times, and this, in turn, has made possible the evolution and continued survival of the higher forms of life. About 300 million years ago especially large excess productions formed the fossil fuels that made man's industrial revolution possible. During the past 60 million years shifts in biotic balances coupled with variations in volcanic activity, rock weathering, sedimentation and solar input have resulted in an oscillating steady state in CO_2/O_2 atmospheric ratios. Oscillations in atmospheric CO_2 were associated with, and presumedly caused, alternate warming and cooling of climates. During the past half century human agroindustrial activities have had a significant effect in that the CO_2 concentration has been increased at least 13 per cent (Plass, 1959).

Explanation

Just how chlorophyll-bearing plants manufacture carbohydrates, proteins, fats, and other complex materials is not yet fully understood. However, the simplified photosynthesis formula is one of the first things learned in elementary biology. Written in word form, it goes something like this:

$$\begin{bmatrix} \text{Carbon} \\ \text{dioxide} \end{bmatrix} \text{plus [water] plus} \begin{bmatrix} \text{light energy in} \\ \text{presence of en-} \\ \text{zyme systems} \\ \text{associated with} \\ \text{chlorophyll} \end{bmatrix}$$

results in [glucose] plus [oxygen].

Chemically the photosynthetic process involves the storage of a part of the sunlight energy as potential or "bound" energy of food. Since this involves an oxidation-reduction reaction, and since there are two kinds of photosynthesis in nature, a general equation can be written as follows:

$$CO_2 + 2H_2A \xrightarrow{\text{light}} (CH_2O) + H_2O + 2A,$$

the oxidation being

$$2H_2A \rightarrow 4H + 2A \text{ and}$$

the reduction being

$$4H + CO_2 \rightarrow (CH_2O) + H_2O.$$

Figure 2–8. A. The Copper Basin at Copperhill, Tennessee is suggestive of what land without life would be like. A luxuriant forest once covered this area until fumes from smelters killed all of the vegetation. Although fumes are no longer released by modern methods of ore preparation, the vegetation has not become reestablished. (U. S. Forest Service Photo.) B. Farmland in Mississippi ruined by soil erosion. Such abuses leave abandoned houses and poor people. (U. S. Forest Service Photo.)

For green plants in general (algae, higher plants) the "A" is oxygen; it is water that is oxidized with release of gaseous oxygen, and the carbon dioxide is reduced to carbohydrate (CH_2O) with release of water. In bacterial photosynthesis, on the other hand, the H_2A (the "reductant") is not water but either an inorganic sulfur compound, such as hydrogen sulfide (H_2S) in the green and purple sulfur bacteria (Chlorobacteriaceae and Thiorhodaceae respectively), or an or-

ganic compound, as in the purple and brown nonsulfur bacteria (Athiorhodaceae). Consequently, oxygen is not released in bacterial photosynthesis.

The *photosynthetic bacteria* are largely aquatic (marine and freshwater) and in most situations play a minor role in the production of organic matter. However, they are able to function under conditions that are unfavorable for the general run of green plants and they play a role in the cycling of certain minerals in aquatic sediments. The green and purple sulfur bacteria, for example, are important in the sulfur cycle (see Figure 4–5). They are obligate anaerobes (able to function only in the absence of oxygen) and they occur in the boundary layer between oxidative and reduced zones in sediments or water where there is light of low intensity. Tidal mudflats are good places to observe these bacteria because they often form distinct pink or purple layers just under the upper green layers of mud algae (in other words, at the very upper edge of the anaerobic or reduced zone where light is available). In a recent study of Japanese lakes, Takahashi and Ichimura (1968) found that photosynthetic sulfur bacteria accounted for only 3 to 5 per cent of the total annual production in most lakes, but in stagnant lakes rich in H_2S these bacteria accounted for up to 25 per cent of the total photosynthesis. In contrast, the non-sulfur photosynthetic bacteria are generally facultative anaerobes (able to function with or without oxygen). They can also function as heterotrophs in the absence of light, as can many algae. Bacterial photosynthesis, then, can be helpful in polluted and eutrophicated waters, and hence is being increasingly studied; but it is no substitute for the "regular" oxygen-generating photosynthesis on which the world at large depends. For a detailed review of the photosynthetic bacteria see Pfennig (1967).

Another interesting group known as the *chemosynthetic bacteria* are often considered to be "producers" (i.e., chemoautotrophs), but in terms of their role in ecosystems they are intermediate between autotrophs and heterotrophs. These bacteria obtain their energy for carbon dioxide assimilation into cellular components not by photosynthesis but by the chemical oxidation of simple inorganic compounds, as for example, ammonia to nitrite, nitrite to nitrate, sulfide to sulfur, and ferrous to ferric iron. They can grow in the dark, but most require oxygen. The sulfur bacteria *Beggiatoa*, often abundant in sulfur springs, and the various nitrogen bacteria, which are important in the nitrogen cycle (see page 89), are examples of this type. One unique group of chemosynthetic bacteria, the hydrogen bacteria, is being seriously considered for life-support systems in spacecraft because on a weight basis they would be very efficient in removing CO_2 from the spacecraft atmosphere (see Chapter 20, page 505). Because of their ability to function in the dark recesses of sediments and soil, the chemosynthetic bacteria not only play a role in the "recovery" of mineral nutrients, but, as the Russian hydrobiologist Sorokin (1966) has pointed out, these bacteria rescue energy that would otherwise be lost for direct feeding by animals.

The photosynthetic process of food manufacture is often called the "business of green plants." It is now believed that the synthesis of amino acids, proteins, and other vital materials occurs simultaneously with the synthesis of carbohydrates (glucose), some of the basic steps involved being the same. A part of the synthesized food is used, of course, by the producers themselves. The excess as well as the producer protoplasm is then utilized by the consumers or, as already emphasized, part of it is frequently stored or transported into other systems. The dynamics of "production" will be considered in detail in Chapter 3.

Most species of the higher plants (Spermatophytes) and many species of the algae require only simple inorganic nutrients and are, therefore, completely autotrophic. Some species of algae, however, require a single complex organic "growth substance" which they themselves cannot synthesize; still other species require one, two, three, or many such growth substances and are, therefore, partly heterotrophic. See reviews by Provasoli (1958 and 1966), Hutner and Provasoli (1964), and Lewin (1963). In the land of "the midnight sun" in northern Sweden, Rodhe (1955) has presented evidence to indicate that during the summer, phytoplankton in lakes are producers; during the long winter "night" (which may last for several months), when they apparently are able to utilize the accumulated organic matter in the water, they are consumers.

Now let us turn our attention to respiration, the heterotrophic process that in the world at large approximately balances the autotrophic metabolism. If we consider respiration in the broad sense as "any energy-yielding biotic oxidation," then it is important to recognize the several types of respiration (which roughly parallel the types of photosynthesis):

1. Aerobic respiration—gaseous (molecular) oxygen is the hydrogen acceptor (oxidant).

2. Anaerobic respiration—gaseous oxygen not involved. An inorganic compound other than oxygen is the electron acceptor (oxidant).

3. Fermentation—also anaerobic but an organic compound is the electron acceptor (oxidant).

Type 1, aerobic respiration, is the reverse of the "regular" photosynthesis as shown earlier in the word formula. It is the means by which all of the higher plants and animals, and most of the Monerans and Protistans (see Figure 2–1) as well, obtain their energy for maintenance and for the formation of cell material. Complete respiration yields CO_2, H_2O and cell material, but the process may be incomplete, leaving organic compounds that still contain energy which may be used later by other organisms.

As a way of life respiration without O_2 is largely restricted to the saprophages (bacteria, yeasts, molds, protozoa), although it occurs as a dependent process within certain tissues of higher animals. The methane bacteria are good examples of obligate anaerobes which decompose organic compounds with the production of methane (CH_4) through reduction of either organic or carbonate carbon (thus employing both types 2 and 3 metabolisms). The methane gas, often known as "swamp gas," rises to the surface where it can be oxidized, or, if it catches fire, it may become an U.F.O. (unidentified flying object)! The methane bacteria are also involved in the breakdown of forage within the rumen of cattle and other ruminates. *Desulfovibrio* bacteria are ecologically important examples of type 2 anaerobic respiration because they reduce SO_4 in deep sediments to H_2S gas, which can rise to shallow sediments where it can be acted on by other organisms (the photosynthetic bacteria, for example). Yeasts, of course, are well-known examples of fermentors (type 3). They are not only commercially important to man but are abundant in soils where they play a key role in the decomposition of plant residues.

As already indicated, many kinds of bacteria are capable of both aerobic and anaerobic respiration (i.e., facultative anaerobes), but it is important to note that the end products of the two reactions will be different and the amount of energy released will be much less under anaerobic conditions. Figure 2–9 shows the results of an interesting study in which the same species of bacterium, *Aerobacter*, was grown under anaerobic and aerobic conditions with glucose as the carbon source. When oxygen was present almost all of the glucose was converted into bacterial protoplasm and CO_2, but in the absence of oxygen decomposition was incomplete, a much smaller portion of the glucose ended up as cell carbon, and a series of organic compounds was released into the environment. Additional bacterial specialists would be required to oxidize these. In general, complete aerobic respiration is many times more rapid than the partial process of anaerobic respiration in terms of energy

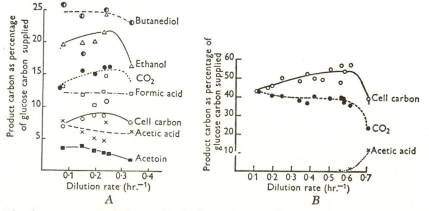

Figure 2–9. The decomposition of glucose by the bacterium *Aerobacter* under anaerobic (A) and aerobic (B) conditions. Note that under aerobic conditions decomposition is complete, and 40 to 50 per cent of the original carbon of glucose is converted into bacterial biomass. However, under anaerobic conditions only about 15 per cent glucose carbon is converted, and a number of incompletely decomposed organic substances remain in the environment. The decline in cell carbon and CO_2 and the beginning of acetic acid production in B indicate that oxygen is beginning to decrease, with the result that the situation shown in A is beginning to develop. (After Pirt, 1957.)

yield per unit of substrate acted upon. When the rate of input of organic detritus into soils and sediments is high, bacteria, fungi, protozoa, and other organisms create anaerobic conditions by using up the oxygen faster than it can diffuse into the media; decomposition does not stop then but continues at a slower rate, provided an adequate diversity of anaerobic microbial metabolic types is present.

In summary, although the anaerobic saprophages (both obligate and facultative) are minority components of the community, they are none-the-less important in the ecosystem because they alone can respire in the dark, oxygenless recesses of the system. By occupying this inhospitable habitat they "rescue" energy and materials for the majority of the aerobes. What would seem to be an "inefficient" method of respiration, then, turns out to contribute to the "efficient" exploitation of energy and materials in the ecosystem as a whole. For example, the efficiency of a sewage disposal system, which is a man-engineered heterotrophic ecosystem, depends on the partnership between anaerobic and aerobic saprophages. The general anaerobic-aerobic interaction in the sediment profile is diagramed in Figure 4–5 (page 91) and in Figure 12–13 (page 343).

Despite the fact that nature presents a broad spectrum of function, the simple autotroph-phagotroph-saprotroph classification is a good working arrangement for describing the ecological structure of a biotic community, while "production," "consumption," and "decomposition" are useful terms for describing overall functions. Since specialization in function tends to result in greater efficiency under the competitive conditions of nature, the evolutionally more advanced organisms tend to be restricted to a rather narrow range of function, leaving the job of filling in the gaps to less specialized organisms. The metabolic versatility of the one-celled bacteria, protozoa, and algae is not surprising when we recall that ATP is formed in the cell in the same way in both photosynthesis and respiration, namely by the transfer of elections from a reductant to an oxidant via cytochromes. In photosynthesis the reductant and oxidant are formed within the cell at the expense of light energy, whereas in respiration these are obtained ready-made from the environment. Although we regard microorganisms

as "primitive," man and other "higher" organisms cannot live without what LaMont Cole calls the "friendly microbes" (Cole, 1966); they synthesize necessary organics and provide the "fine tuning" in the ecosystem since they can adjust quickly to changing conditions.

As emphasized in the statement, it is the relationship between the total rate of production and the rate of decomposition, regardless of what organisms or abiotic processes are responsible, that is of overall importance in the biosphere as a whole. The interplay of these opposing functions controls our atmosphere and hydrosphere, and, as also emphasized, it is fortunate for man and his great oxygen-consuming machines that production has tended to exceed decomposition. But man now "taketh more than he giveth back" to an extent that threatens vital balances. *The delay in the complete heterotrophic utilization of the products of autotrophic metabolism* is, therefore, one of the most important features of the ecosystem, one that is gravely threatened by man's careless behavior. Accordingly, it will be well to review the highlights of the decomposition process at this point, even though many of the details will need to be developed later in the text.

Decomposition results from both abiotic and biotic processes. For example, prairie and forest fires are not only major limiting factors, as will be discussed in Chapter 5, but they are "decomposers" of detritus, releasing large quantities of CO_2 and other gases to the atmosphere and minerals to the soil. By and large, however, it is the heterotrophic microorganisms or saprophages that ultimately act on the dead bodies of plants and animals. This kind of decomposition, of course, is the result of the process by which bacteria and fungi obtain food for themselves. Decomposition, therefore, occurs through energy transformations within and between organisms. It is an absolutely vital function because, if it did not occur, all the nutrients would soon be tied up in dead bodies, and no new life could be produced. Within the bacterial cells and the fungi mycelia are sets of enzymes necessary to carry out specific chemical reactions. These enzymes are secreted into dead matter; some of the decomposition products are absorbed into the organism as food, and other products remain in the environment or are excreted from the cells. No single species of saprotroph can produce complete decomposition of a dead body. However, populations of decomposers prevalent in the biosphere consist of many species which, by their graduated action, can effect complete decomposition. Not all parts of the bodies of plants and animals are broken down at the same rate. Fats, sugars, and proteins are decomposed readily, but cellulose, lignin of wood, and the chitin, hair, and bones of animals are acted on very slowly. This is illustrated in Figure 2–10, which compares the rate of decomposition of dead marsh grass and fiddler crabs placed in nylon-mesh "litter bags" in a Georgia salt marsh. Note that about 25 per cent of the dry weight of the marsh grass was decomposed in a month and the remaining 75 per cent acted on more slowly. After 10 months 40 per cent of the grass still remained, but all of the crab remains had disappeared from the bag. As the detritus becomes finely divided and escapes from the bag, the intense activities of microorganisms often result in protein enrichment, as shown in Figure 2–10C, thus providing a more nutritious food for detritus-feeding animals (Odum and de la Cruz, 1967; Kaushik and Hynes, 1968). As will be emphasized in Chapter 6, this is part of nature's overall "strategy" of having its cake and eating it too—but more of this later!

The more resistant products of decomposition end up as *humus* (or *humic substances*), which, as already indicated, is a universal component of ecosystems. It is convenient to recognize three stages of decomposition: (1) the formation of particulate detritus by physical and biological action, (2) the relatively rapid production of humus and release of soluble organics by saprotrophs, and (3) the slower mineralization of humus. The slowness with which humus is decomposed is a factor in the decomposition lag and oxygen accumulation that we have stressed. In general appearance humus is a dark, often yellow-brown, amorphous or colloidal substance. According to Kononova (1961) no great difference in physical properties or chemical structure exists between humic substances in geographically scat-' tered or biologically different ecosystems. However, they are difficult to characterize chemically, a fact which is not surprising considering the great variety of organic matter from which they originate. In general terms, humic substances are condensa-

tions of aromatic compounds (phenols) combined with the decomposition products of proteins and polysaccharides. A model for the molecular structure of humus is shown on page 368. It is the phenolic type of benzene ring and the side chain bonding that makes these compounds recalcitrant to microbial decomposition. Fission of these structures apparently requires special deoxygenase enzymes (Gibson, 1968), which may not be present in the common soil and water saprotrophs. Ironically, many of the toxic materials that man is now adding .to the environment, such as herbicides, pesticides, and industrial effluents, are derivatives of benzene and are causing serious trouble because of their low degradability.

The pioneer microbial ecologist Winogradsky put forward, in 1925, the concept that organisms which decompose fresh organic matter are an ecologically separate "flora" from those that decompose humus; he called these groups *zymogenous* and *autochthonous* respectively (see Winogradsky, 1949, p. 473). To this day, however, it is not known for certain if humus is broken down by special organisms with special enzymes, or by abiotic chemical processes, or both. There has been little recent progress in the study of humus, perhaps because it does not lend itself to conventional study in the chemical laboratory. What is needed are more in situ studies in the environment such as those described by Tribe (1963), who observed humus formation in material placed between two glass slides which could be periodically removed from their positions in the soil for microscopic study and chemical analysis.

It is well known that detritus, humus, and other organic matter undergoing decomposition play an important role in soil fertility. In moderate quantity these materials provide texture that is favorable for plant growth. Furthermore, many organics form complexes with mineral nutrients that enhance uptake by plants. One way this occurs is by a process known as *chelation* (fr. *chele* = claw, referring to "grasping"), a complex formation with metal ions that keeps the element in solution and nontoxic, as compared with the inorganic salts of the metal. The following diagram shows how a copper ion can be held in "crab claws" by pairs of covalent ($\rightarrow$) and ionic ($-+$) linkages between two molecules of glycine, an amino acid:

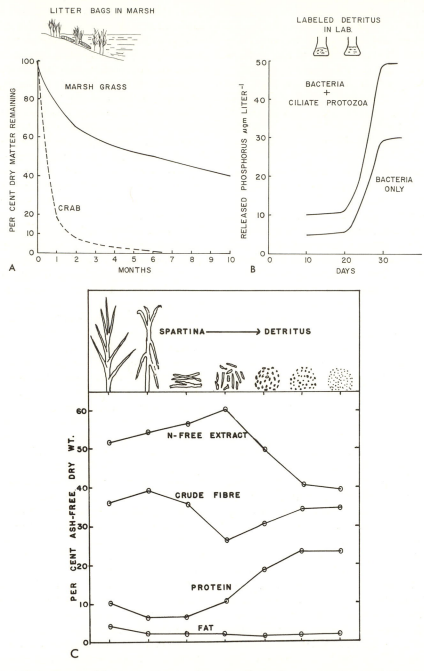

Figure 2–10. Three aspects of decomposition in a Georgia salt marsh, an ecosystem in which most of the productive energy follows the detritus pathway. A. Decomposition in terms of percentage of dead marsh grass (*Spartina alterniflora*) and fiddler crabs (*Uca pugnax*) remaining in nylon-mesh litter bags placed in the marsh where they would be subjected to daily tidal inundation. (After E. P. Odum and de la Cruz, 1967.) B. A laboratory experiment showing that the release of phosphorus (labeled with radioactive tracer) from decaying marsh grass is more rapid when the detritus is acted on by bacteria and protozoa, as compared with the action of bacteria alone. (After Johannes, 1965.) C. Protein enrichment resulting from microbial activity on marsh grass detritus in the final stages of particulate breakdown. (After E. P. Odum and de la Cruz, 1967.) Deposit-feeding animals (snails, worms, clams, shrimp, fish) often selectively feed on the more nutritious small particles (see Newall, 1965 and W. E. Odum, 1968).

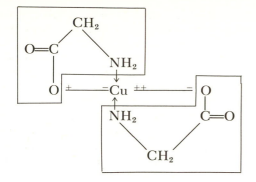

Chelators are nearly always added to culture media used in experiments in the laboratory systems shown in Figure 2–6.

Decomposition products can, of course, accumulate to the point of providing "too much of a good thing," a phrase we will have many occasions to use in regard to human excesses! The creation of anaerobic conditions has already been mentioned. In a recent study, Ghassemi and Christman (1968) found that the soluble yellow organic acids, decomposition products of plant origin, are beneficial in low concentration in lakes by virtue of their chelation of metals, especially iron. In high concentrations, however, they absorb light and reduce photosynthesis.

The bacteria, yeasts, and molds may work together or alternate in the breakdown process. Bacteria appear to be more important in the breakdown of animal flesh while fungi perhaps are more important in the breakdown of wood. Tribe (1957, 1961) describes an interesting succession of organisms which completely decomposed cellulose film which he buried in the soil. Fungi were the first to invade the film; only later did bacteria appear in numbers. As soon as the film had been broken up, nematodes and other soil invertebrates arrived on the scene and gobbled up the small pieces (including, of course, the microorganisms). Thus, microconsumers also provide food, indirectly or directly, for various macroconsumers, which in turn hasten the breakdown process. (For an example of a sequence of different bacterial and fungal groups involved in the decomposition of plant residues see page 367.) While large organisms are being slowly decomposed, a special habitat is provided for a variety of organisms. For example, a fallen log in a forest supports a well-developed subcommunity that changes with the state of decay.

Studies are now showing that the phagotrophs, especially the small animals (protozoa, soil mites, nematodes, ostracods, snails, and so on), play a more important role in decomposition than was previously suspected. When microarthropods (i.e., microscopic mites and collembolans) are removed from forest litter by an insecticide treatment that has little or no effect on bacteria and fungi, breakdown of the dead leaves and twigs is greatly slowed (see Figure 14–6, page 373). As shown in Figure 2–10B, the release of phosphorus tracer from decomposing marsh grass was reduced when protozoa were selectively filtered out of the sea water medium. Although most detritus-feeding animals (detritivores) cannot digest cellulose and obtain their food energy largely from microflora, they can speed up the decomposition of the cellulose-lignin matrix in a number of ways: (1) by breaking up detritus into small pieces thus increasing the surface area available for microbial action; (2) by adding proteins or growth substances that stimulate microbial growth; or (3), as has been suggested, by eating up some of the bacteria and thus stimulating the growth and metabolic activity of the population (i.e., maintaining a "log phase" growth rate, see page 183). Furthermore, many detritivores are *coprophagic* (fr. *kopros* = dung), that is, they regularly ingest fecal pellets after the pellets have been enriched by microbial activity in the environment (see Newell, 1965; Frankenberg and Smith, 1967). For example, the "betsy beetle" (*Popilius*), which lives in decaying logs, uses its tunnels as a sort of "external rumen" where fecal pellets and chewed wood fragments become enriched by fungi and then reingested (Mason and Odum, 1969). Coprophagy in this case is an insect-fungal partnership that enables the beetle to utilize the food energy in wood and also hastens the decay of the log.

When we compared land and water ecosystems in the preceding section of this chapter, we pointed out that since phytoplankton are more "edible" than land plants, the macroconsumers perhaps play a greater role in decomposition in the aquatic system, but more about this in Chapter 4. Finally, it is interesting to note that years ago it was suggested that invertebrate animals were beneficial in sewage beds (see the discussion by Hawkes, 1963), but there are few serious studies of the relationships between phago- and saprotrophs in sewage treatment since sanitary engineers generally assume that bacterial action is all that matters.

Although the "mineralization of organic matter" and the production of plant nutrients has been stressed as the primary function of decomposition, there is yet another function which is receiving increasing attention from ecologists. Apart from the use as food by other organisms, the organic substances released into the environment during decomposition may have profound effects on the growth of other organisms in the ecosystem. Julian Huxley, in 1935, suggested the term "external diffusion hormones" for those chemical substances which exert a correlative action on the system via the external medium; Lucas (1947) has proposed the term "ectocrine" (or "exocrine," as preferred by some writers). The term "environmental hormone" also clearly indicates what is meant. These substances may be inhibitory, as in the case of the "antibiotic" penicillin (which is produced by a fungus), or stimulatory, as in the case of various vitamins and other growth substances, for example, thiamin, vitamin B_{12}, biotin, histidine, uracil, and others, many of which have not been identified chemically. Organic compounds frequently combine with trace metals in the environment to form hormonelike substances.

Interest in environmental biochemicals goes back to the famous Pütter-Krogh argument (which still continues) about the role of dissolved organic substances which are abundant in water and soil. The former contended that these substances were extensively used by higher plants and animals as food; the latter presented experimental evidence that they were insignificant as food, at least for such organisms as zooplankton. Two recent contributions to this interesting question can be cited: (1) Radioactive tracer experiments have shown that soft-bodied marine animals can absorb and use dissolved sugars and amino acids present in sea water (see Stephens, 1967, 1968); (2) dissolved organics frequently form particulate aggregates that are easily ingested by animals. For example, what might be called "bubble detritus" is regularly formed from dissolved material by wave action in the sea (see Baylor and Sutcliffe, 1963; Riley, 1963). It is also now becoming clear that extracellular metabolites and many of the "waste products" of decomposition may be more important as chemical regulators

than as food as such. This is exciting to the ecologist because such regulators provide a mechanism for coordinating units of the ecosystem and help to explain both the equilibria and the succession of species which he so commonly observes in nature. Before we become too enthusiastic about this, there is much work to be done. Vitamin B_{12}, for example, has been much studied, but the investigators are not agreed on whether or not this substance is ecologically important. It is definitely required by many organisms, but it is also abundant and widely distributed; the question whether it ever becomes scarce enough to limit the growth of producers has not been satisfactorily determined (note, for example, the opposite views of Droop, 1957, and Daisley, 1957). The role of the inhibitory substances is perhaps more clearly defined. It should be mentioned at this point that although the saprotrophs seem to play the major role in the production of environmental hormones, the algae also excrete such substances (see the review by Fogg, 1962). The waste products of higher organisms, leaf and root excretions, for example, may also be important in this regard. C. H. Muller and his associates speak of such excretions as "allelopathic substances" (fr. *allelon* = of each other; *pathy* = suffering), and they have shown that the metabolites interact with fire in a complex manner in controlling desert and chaparral vegetation (see Figure 7–36, page 225). Also of undoubted regulatory function are volatile secretions called "pheromones" which control the behavior of insects and other organisms (see Wilson, 1965; Butler, 1967).

Some idea of the intense metabolic activity that accompanies microbial decomposition is obtained by observation of the increase in temperature in the substrate. Many people have observed the heat generated in an artificial compost pile. In the rapidly accumulating sediments of Lake Mead, which is formed by the huge Hoover Dam on the Colorado River, ZoBell, Sisler, and Oppenheimer (1953) found that the temperature of the bottom mud was as much as 6°C. above that of the adjacent water; they demonstrated that a part of this heat, at least, was due to the immense populations of microorganisms which were doing their best to keep up with the man-accelerated erosion in the watershed.

Because most biotic agents of decomposition are small and relatively undistinguished morphologically, they are difficult to study. One cannot analyze a microbial population simply by looking at it and counting the individuals as one might do with a stand of trees or a population of mammals, since microbes which look alike might have entirely different types of metabolism. Species not only have to be isolated and cultured but their activity must also be measured in situ and in the laboratory under conditions that mimic the natural environment (in contrast to the "enriched culture" techniques of conventional bacteriology). Because of the technical difficulties of study, microbial ecology is, unfortunately, often completely omitted from the general college course in ecology. This need not be the case. Certainly the cellulose film and marsh grass experiments described earlier could be adapted as a class exercise. Some idea of the activities of decomposers can be obtained by measuring the CO_2 evolution from soil, using fairly simple apparatus which can be designed for field use (see Witkamp, 1966). The prospects for breakthroughs in the study of microbial ecology are discussed in Chapter 20 (see also texts by Brock, 1966, and Wood, 1965).

In summary, we have seen how the long and complex process of degradation of organic matter controls a number of important functions in the ecosystem, for example: (1) the recycling of nutrients through mineralization of dead organic matter, chelation, and microbial recovery in the heterotrophic layer; (2) production of food for a sequence of organisms in the detritus food chain; (3) production of regulatory "ectocrine" substances; and (4) modification of the inert materials of the surface of the earth (to produce, for example, the unique earthly complex known as "soil"). Most of all, we have emphasized the importance of the overall balance between production and decomposition in regulating the conditions of existence for all life in the biosphere. We have pointed out that the lag in utilization of autotrophic production results not only in the buildup of a biological structure that mitigates the harsh physical environment, but that it also accounts for the oxygenic atmosphere to which man and the higher animals are especially adapted. Most of the organic matter that does escape decomposition becomes deposited in aquatic sediments; this is why oil is only found in areas that are or once were covered by water. Man is now unwittingly beginning to speed up decomposition (1) by burning the stored organic matter in fossil fuels and (2) by agricultural practices that increase the decomposition rate of humus. Although the amount of CO_2 injected into the atmosphere by man's agroindustrial activities is yet small compared to the total CO_2 in circulation (see Figure 4–4 for estimates of these amounts), it is beginning to have a perceptible effect because the atmospheric reservoir is small and the larger oceanic reservoir is not able to absorb the new CO_2 as fast as it is being produced by man. Revelle (1965) estimates a 25 per cent increase in atmospheric CO_2 by the year 2000 and a 170 per cent increase when all of the readily accessible fossil fuels have been consumed. Although there are aspects that are not understood, and are hence controversial, climatologists are generally agreed that relatively small changes in atmospheric CO_2 can and do have major effects on climate. Carbon dioxide is transparent to incoming visible sun energy, but, like glass, it absorbs infrared heat reradiated from the earth's surface. This so-called "greenhouse effect" means that an increase in CO_2 tends to result in an increase in the temperature of the biosphere as a whole. Melting of the polar ice caps and a subsequent world-wide tropical climate would be the most important effects of a rise in temperature. Revelle estimates that the sea level would rise 400 feet if all of the antarctic ice should melt, and this could occur within as short a time as 400 years. Such a change would, of course, be catastrophic for man since he would have to retreat from his coastal cities (or else build high dikes or live underwater!). While man is increasing the CO_2 content of the atmosphere, he is also increasing particulate pollution (i.e., "dust" in the air) which has the opposite effect of cooling the earth. We shall have more to say about the "uneasy balance" between gaseous and particulate pollution in Chapter 4.

4. HOMEOSTASIS OF THE ECOSYSTEM

Statement

Ecosystems are capable of self-maintenance and self-regulation as are their component populations and organisms. Thus, *cybernetics* (fr. *kybernetes* = pilot or governor), the science of controls, has important application in ecology especially since man increasingly tends to disrupt natural controls or attempts to substitute artificial mechanisms for natural ones. *Homeostasis* (*homeo* = same; *stasis* = standing) is the term generally applied to the tendency for biological systems to resist change and to remain in a state of equilibrium.

Explanation and Examples

The very elementary principles of cybernetics are modeled in Figure 2–11. In simplest form a control system consists of two black boxes (see page 22 for definition of black box) and a controlled quantity interconnected by output and input circuits or signals (Fig. 2–11A). In the familiar household heat control system the thermostat is the sensor (or "error detector" as it can also

be called), the furnace the effector (which receives its energy from the fuel), and the room temperature the controlled quantity. Control depends on *feedback*, which occurs when output (or part of it) feeds back as input. When this feedback input is positive (like compound interest, which is allowed to become part of the principal), the quantity grows. *Positive feedback* is "deviation-accelerating" and, of course, necessary for growth and survival of organisms. However, to achieve control—as for example, to prevent overheating a room or cancerous overgrowth of a population—there must also be *negative feedback*, or "deviation-counteracting" input, as shown in Figure 2–11A. Mechanical feedback mechanisms are often called servomechanisms by engineers; biologists use the phrase homeostatic mechanisms to refer to living systems. Cybernetics embraces both inanimate and animate controls (Wiener, 1948). The interaction of

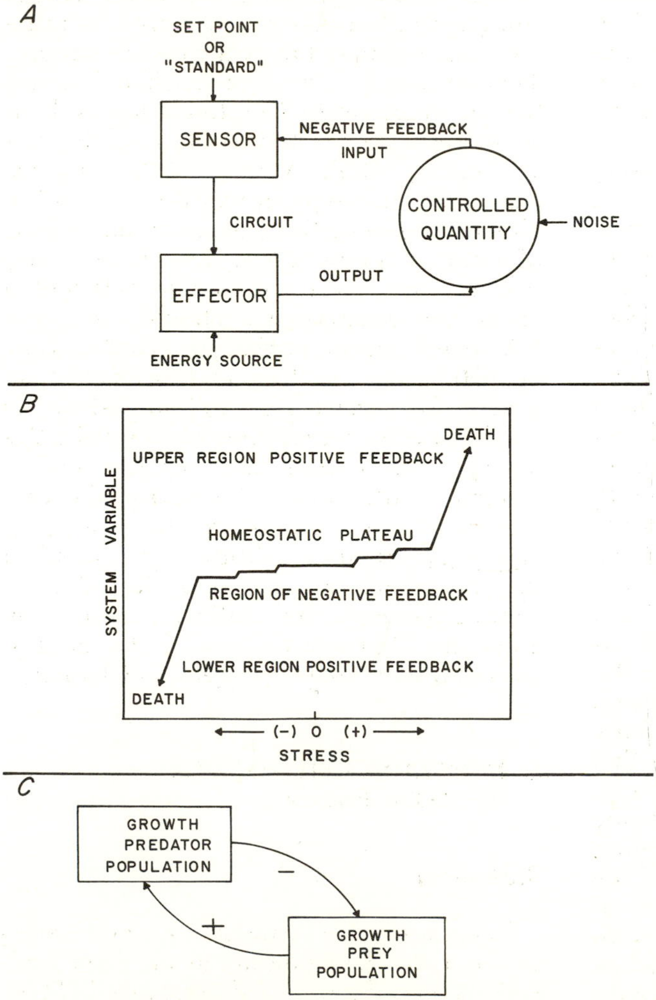

Figure 2–11. Elements of cybernetics. *A.* A simple control system, analogous to a household thermostat, in which some of the output is used as negative feedback to maintain some kind of equilibrium in a controlled quantity. *B.* The concept of the homeostatic plateau within which relative constancy is maintained by negative feedback despite tendency of stress to cause deviations. Beyond limits of homeostasis positive feedback results in rapid destruction of the system. (After Hardin, 1963.) *C.* The interaction of positive (+) and negative (−) feedback in a predator-prey "feedback loop" system. A period of evolutionary adjustments is usually required before such a system actually becomes stable. Newly associated predators and prey tend to oscillate violently (see Figure 7–32, page 219).

positive and negative feedback and the limits of homeostatic control are diagrammed in Figure 2–11B. As critics of human society are pointing out with increasing frequency (see Mumford, 1967, for example), the positive feedback involved in the expansion of knowledge, power, and productivity threatens the quality of human life and environment unless adequate negative feedback controls can be found. The science of controls, or cybernetics, thus becomes one of the most important subjects to be studied, understood, and practiced. A suggested reading list on this subject would include Ashby (1963), Langley (1965), Hardin (1963), and Maruyama (1963).

The existence of homeostatic mechanisms at different levels of biological organization was mentioned in the previous chapter. Homeostasis at the organism level is a well-known concept in physiology as outlined, for example, by Walter B. Cannon in his readable little book entitled *The Wisdom of the Body* (1932). We find that equilibrium between organisms and environment may also be maintained by factors which resist change in the system as a whole. Much has been written about this "balance of nature," but only with the recent development of good methods for measuring rates of function of whole systems has a beginning been made in the understanding of the mechanisms involved.

As in the turbidostat described in the legend of Figure 2–6, some populations are regulated by density, which "feeds back" by way of behavioral mechanisms to reduce or increase the reproductive rate (the "effector") and thus maintain the population size (the "controlled quantity") within set limits. Other populations do not seem to be capable of self-limitation but are controlled by outside factors (this may include man, but more about this later). As we have already described, control mechanisms operating at the ecosystem level include those which regulate the storage and release of nutrients and the production and decomposition of organic substances. *The interplay of material cycles and energy flows in large ecosystems generates a self-correcting homeostasis with no outside control or set-point required* (more about this in Chapters 3 and 4). The possible role of "ectocrine" substances in coordinating units of the ecosystem has been mentioned. In subsequent sections and chapters, we shall have frequent occasion to discuss these mechanisms and to present specific data demonstrating that the whole is often not as variable as the part.

It is important to note, as shown in Figure 2–11B, that homeostatic mechanisms have limits beyond which unrestricted positive feedback leads to death. Note also that we have shown the "homeostatic plateau" as a series of levels or steps. As stress increases, the system, although controlled, may not be able to return to the exact same level as before. We have already described how CO_2 introduced into the atmosphere by man's "industrial volcanos" is largely, but not quite, absorbed by the carbonate system of the sea; as the input increases, new equilibrium levels are slightly higher. In this case even a slight change may have far-reaching effects. We shall also have many occasions to note that *really good homeostatic control comes only after a period of evolutionary adjustment.* New ecosystems (such as a new type of agriculture) or new host-parasite assemblages tend to oscillate more violently and to be less able to resist outside perturbation as compared with mature systems in which the components have had a chance to make mutual adjustments to each other.

As a result of the evolution of the central nervous system, mankind has gradually become the most powerful organism, as far as the ability to modify the operation of ecosystems is concerned. So important is man's role becoming as "a mighty geological agent" that Vernadsky (1945) has suggested that we think of the "noosphere" (from Greek *noos*, mind), or the world dominated by the mind of man, as gradually replacing the biosphere, the naturally evolving world which has existed for billions of years. This is dangerous philosophy because it is based on the assumption that mankind is now not only wise enough to understand the results of all his actions but is also capable of surviving in a completely artificial environment. When the reader has finished with this book I am sure he will agree that we cannot safely take over the management of everything!

The idea of the ecosystem and the realization that mankind is a part of, not apart from, complex "biogeochemical" cycles with increasing power to modify the cycles are concepts basic to modern ecology and are also points of view of extreme importance in human affairs generally. Conservation of natural resources, a most important practical application of ecology, must be built around these viewpoints. Thus, if understanding of ecological systems and moral responsibility among mankind can keep pace with man's power to effect changes, the present-day concept of "unlimited exploitation of resources" will give way to "unlimited ingenuity in perpetuating a cyclic abundance of resources." Historian Lynn White, Jr. (1967) points out that the religious dogma that sharply separates man from nature unfortunately contributes to the present environmental crisis. Hutchinson (1948a), in another classic essay that should be widely read, has aptly expressed this viewpoint somewhat as follows: the ecologist should be able to show that it is just as much fun and just as important to repair the biosphere and keep it in good running or-der as to mend the radio or the family car.

We can summarize what has been presented in this chapter. The ecosystem is the central theme and most important concept of ecology. The two approaches to its study, the holological and the merological, must be integrated and translated into programs of action if man is to survive his self-generated environmental crisis. *It is man the geological agent, not so much as man the animal, that is too much under the influence of positive feedback, and, therefore, must be subjected to negative feedback.* Nature, with our intelligent help, can cope with man's physiological needs and wastes, but she has no homeostatic mechanisms to cope with bulldozers, concrete, and the kind of agroindustrial air, water, and soil pollution that will be hard to contain as long as the human population itself remains out of control.

Chapter 3

PRINCIPLES AND CONCEPTS PERTAINING TO ENERGY IN ECOLOGICAL SYSTEMS

1. REVIEW OF FUNDAMENTAL CONCEPTS RELATED TO ENERGY

Statement

Energy is defined as the ability to do work. The behavior of energy is described by the following laws. The *first law of thermodynamics* states that energy may be transformed from one type into another but is never created or destroyed. Light, for example, is a form of energy, for it can be transformed into work, heat, or potential energy of food, depending on the situation, but none of it is destroyed. The *second law of thermodynamics* may be stated in several ways, including the following: No process involving an energy transformation will spontaneously occur unless there is a degradation of the energy from a concentrated form into a dispersed form. For example, heat in a hot object will spontaneously tend to become dispersed into the cooler surroundings. The second law of thermodynamics may also be stated as follows: Because some energy is always dispersed into unavailable heat energy, no spontaneous transformation of energy (light, for example) into potential energy (protoplasm, for example) is 100 per cent efficient.

Organisms, ecosystems and the entire biosphere possess the essential thermodynamic characteristic of being able to create and maintain a high state of internal order, or a condition of low entropy (a measure of disorder or the amount of unavailable energy in a system). Low entropy is achieved by a continual dissipation of energy of high utility (light or food, for example) to energy of low utility (heat, for example). In the ecosystem, "order" in terms of a complex biomass structure is maintained by the total community respiration which continually "pumps out disorder."

Explanation

It is readily apparent how the fundamental concepts of physics outlined in the above paragraph are related to ecology. The variety of manifestations of life are all accompanied by energy changes, even though no energy is created or destroyed (first law of thermodynamics). The energy that enters the earth's surface as light is balanced by the energy that leaves the earth's surface as invisible heat radiation. The essence of life is the progression of such changes as growth, self-duplication, and synthesis of complex relationships of matter. Without energy transfers, which accompany all such changes, there could be no life and no ecological systems. We, as human beings, should not forget that civilization is just one of the remarkable natural proliferations that are dependent on the continuous inflow of the concentrated energy of light radiation. In ecology, we are fundamentally concerned with the manner in which light is related to ecological systems, and with the manner in which energy is transformed within the system. Thus, the relationships between producer plants and consumer animals, between predator and prey, not to mention the numbers and kinds of organisms in a given environment, are all limited and controlled by the same basic laws which govern nonliving systems, such as electric motors or automobiles.

Year in and year out light and other radiations associated with it leave the sun and pass into space. Some of this radiation falls on the earth, passes through the atmospheric film, and strikes forests, grasslands, lakes, oceans, cultivated fields, deserts, greenhouses, ice sheets, and many hundreds of other types of ecological systems which blanket the earth and compose the biosphere. When light is absorbed by some object which becomes warmer as a result, the light energy has been transformed into another kind of energy known as heat energy. Heat energy is composed of the vibrations and motions of the molecules that make up the object. The absorption of the sun's rays by land and water results in hot and cold areas, ultimately leading to the flow of air which may drive windmills and perform work such as the pumping of water against the force of gravity. Thus, in this case, light energy passes to heat energy of the land to *kinetic* energy of moving air which accomplishes work of raising water. The energy is not destroyed by lifting of the water, but becomes *potential energy,* because the latent energy inherent in having the water at an elevation can be turned back into some other type of energy by allowing the water to fall back down the well. As indicated in previous chapters, food resulting from photosynthetic activity of green plants contains potential energy which changes to other types when food is utilized by organisms. Since the amount of one type of energy is always equivalent to a particular quantity of another type into which it is transformed, we may calculate one from the other. For example, knowing the amount of light energy absorbed and knowing the conversion factor, we may determine the amount of heat energy which has been added.

The second law of thermodynamics deals with the transfer of energy toward an ever less available and more dispersed state. As far as the solar system is concerned, the dispersed state with respect to energy is one in which all energy is in the form of evenly distributed heat energy. That is, if left to itself, all energy where it undergoes a change of form will eventually tend to be transformed into the form of heat energy distributed at uniform temperature. This tendency has often been spoken of as "the running down of the solar system." Whether this tendency for energy to be leveled applies to the universe as a whole is not yet known.

At the present time the earth is far from being in a state of stability with respect to energy because there are vast potential energy and temperature differences which are maintained by the continual influx of light energy from the sun. However, it is the process of going *toward* the stable state that is responsible for the succession of energy changes that constitute natural phenomena on the earth as we know them. It is like a man on a treadmill; he never reaches the top of the hill, but his efforts to

do so result in well-defined processes. Thus, when the sun energy strikes the earth, it tends to be degraded into heat energy. Only a very small portion of the light energy absorbed by green plants is transformed into potential or food energy; most of it goes into heat, which then passes out of the plant, the ecosystem, and the biosphere. All the rest of the biological world obtains its potential chemical energy from organic substances produced by plant photosynthesis or microorganism chemosynthesis. An animal, for example, takes in chemical potential energy of food and converts a large part into heat to enable a small part of the energy to be reestablished as the chemical potential energy of new protoplasm. At each step in the transfer of energy from one organism to another a large part of the energy is degraded into heat.

The second law of thermodynamics, which deals with the dispersal of energy, is related to the *stability principle.* According to this concept any natural enclosed system with energy flowing through it, whether the earth itself or a smaller unit, such as a lake, tends to change until a stable adjustment, with self-regulating mechanisms, is developed. Self-regulating mechanisms are mechanisms which bring about a return to constancy if a system is caused to change from the stable state by a momentary outside influence, as was fully discussed in Chapter 2. When a stable adjustment is reached, energy transfers tend to progress in a one-way fashion and at characteristic steady rates, according to the stability principle.

H. T. Odum (1967), building on the concepts of A. J. Lotka (1925) and E. Schrödinger (1945), places the thermodynamic principles in the ecological context in the following manner.

Antithermal maintenance is the number one priority in any complex system of the real world. As Schrödinger has shown, the continual work of pumping out "disorder" is necessary if one wishes to maintain internal "order" in the presence of thermal vibrations in any system above absolute zero temperature. In the ecosystem the ratio of total community respiration to the total community biomass (R/B) can be considered as the maintenance to structure ratio, or as a thermodynamic order function. This "Schrödinger ratio" is an ecological turnover, a concept introduced in Chapter 2 (see page 17). If R and B are expressed in calories (energy units) and divided by absolute temperature, the R/B ratio becomes the ratio of entropy increase of maintenance (and related work) to the entropy of ordered structure. The larger the biomass, the greater the maintenance cost; but if the size of the biomass units (individual organisms, for example) is large (such as vegetation in a forest), the antithermal maintenance per unit of biomass structure is decreased. One of the theoretical questions now under debate is whether nature maximizes the ratio of structure to maintenance metabolism (see Margalef, 1968; Morowitz, 1968) or whether

it is energy flow itself that is maximized.

It is interesting to note that the words "economics" and "ecology" have the same root, *oikos*, which refers to "house." It can be said that economics deals with financial housekeeping and ecology deals with environmental housekeeping. While energy can be thought of as the "currency" of ecology, energy and money are not the same because they flow in opposite directions (i.e., exchange), and money circulates while energy does not (see H. T. Odum, 1971).

At this point, it would be well for the

Table 3–1. Units of Energy, and Some Useful Ecological Approximations

A. BASIC UNITS

Gram-calorie (gcal or cal) = amount of heat necessary to raise 1 gram (or milliliter) of water 1°C. at 15°C.

Kilogram-calorie (Kcal or Cal) = 1000 gcal (amount heat necessary to raise 1 kilogram (or liter) water 1°C. at 15°C.

British thermal unit (B.t.u.) = amount of heat necessary to raise 1 pound of water 1°F. = 252 gram-calories = 0.252 Kcal

Joule (J) = 0.24 gram calories = 10^7 ergs = 0.74 foot-pounds = 0.1 kilogram-meters

langley (ly) = 1 gram-calorie/cm²

watt (w) = 1 Joule/sec = 14.3 gcal/min = 3.7×10^{-7} horsepower hours

B. REFERENCE VALUES (Averages or Approximations)

Purified foods, Kcal/gm dry wt.: carbohydrate 4; protein 5; lipid 9.2

Biomass°	Kcal/Gm Dry Wt.	Kcal/Gm Ash-free Dry Wt.
Terrestrial plants (total)	4.5	4.6
seeds only	5.2	5.3
Algae	4.9	5.1
Invertebrates (excl. insects)	3.0	5.5
Insects	5.4	5.7
Vertebrates	5.6	6.3

Phytoplankton production: 1 gm carbon = 2.0 + gms dry matter = 10 Kcal

Daily food requirements (at nonstressing temperatures)
Man: 40 Kcal/kgm live body wt. = 0.04 Kcal/gm (about 3000 Kcal/day for 70 kgm adult)
Small bird or mammal: 1.0 Kcal/gm live body wt.
Insect: 0.5 Kcal/gm live body wt.

Gaseous exchange—caloric coefficients in respiration and photosynthesis

% Carbohydrate in Dry Matter Respired or Synthesized	Oxygen Kcal/Liter	Carbon Dioxide Kcal/Liter
100	5.0	5.0
66	4.9	5.5
33	4.8	6.0
0 (fat only)	4.7	6.7

° Since most living organisms are ⅔ or more water and minerals, 2 Kcal/gm live (wet) weight is a very rough approximation for biomass in general. Caloric values for dry weights listed in this table are based on Golley, 1961; Odum, Marshall and Marples, 1965; and Cummings, 1967.

reader to review and become thoroughly familiar with the units of energy. In Table 3–1 basic units are defined, and useful conversion factors and reference points are listed.

The behavior of energy in ecosystems can be conveniently termed the *energy flow* because, as we have seen, energy transformations are "one-way" in contrast to the cyclic behavior of materials. In subsequent sections of this chapter we shall consider first the total energy flow that constitutes the energy environment of an ecosystem and then turn to the study of that portion of the total energy flow that passes through the living components of the ecosystem.

2. THE ENERGY ENVIRONMENT

Statement

Organisms at or near the surface of the earth are immersed in a radiation environment consisting of solar radiation and longwave thermal radiation flux from nearby surfaces. Both contribute to the climatic environment (temperature, evaporation of water, movement of air and water, etc.), but only a small fraction of the solar radiation can be converted by photosynthesis to provide energy for the biotic components of the ecosystem. Extraterrestrial sunlight reaches the biosphere at a rate of 2 gcal per cm² per min,* but is attenuated exponentially as it passes through the atmosphere; at most, 67 per cent (1.34 gcal per cm² per min) may reach the earth's surface at noon on a clear summer day (Gates, 1965). Solar radiation is further attenuated, and the spectral distribution of its energy greatly altered as it passes through cloud cover, water, and vegetation. The daily input of sunlight to the autotrophic layer of an ecosystem varies mostly between 100 and 800, averaging about 300–400 gcal per cm² (= 3000 to 4000 kcal per m²) for an area in the temperate zone such as the United States (Reifsnyder and Lull, 1965). The 24-hour flux of heat energy within an ecosystem (or received by exposed organisms) can be several times as much as or considerably less than incoming solar radiation. The variation in total radiation flux within different strata of the ecosystem, as well as from one season or site to another on the

earth's surface, is enormous, and the distribution of individual organisms responds accordingly.

Explanation

In Figure 3–1 the spectral distribution of extraterrestrial solar radiation, coming in at a constant rate of 2 gcal per cm² per min (±3.5 per cent), is compared with (1) solar radiation actually reaching sea level on a clear day, (2) sunlight penetrating a complete overcast (cloudlight), and (3) light transmitted through vegetation. Each curve represents energy incident on a horizontal surface. For a picture of the effect of season (sun angle) and topographic slopes, see Figure 5–19. In hilly or mountainous country, south-facing slopes receive more and north-facing slopes receive much less solar radiation than do horizontal surfaces; this results in striking differences in local climates (microclimates) and vegetation.

Radiation penetrating the atmosphere is attenuated exponentially by atmospheric gases and dust but to varying degrees depending on the frequency or wavelength. Shortwave ultraviolet radiation below 0.3 μ is abruptly terminated by the ozone layer in the outer atmosphere (about 18 miles or 25 km altitude), which is fortunate since such radiation is lethal to exposed protoplasm. Visible light is broadly reduced and infrared radiation is irregularly reduced by adsorption in the atmosphere. Radiant energy reaching the surface of the earth on a clear day is about 10 per cent ultraviolet, 45 per cent visible, and 45 per cent infrared (Reifsnyder and Lull, 1965). Visible radiation is least attenuated as it passes through dense cloud cover and water, which means that photosynthesis (which is restricted to the visible range) can continue on cloudy days and at some depth in clear water. Vegetation strongly absorbs the blue and red visible wavelengths, the green less strongly, the near infrared very weakly, and the far infrared strongly (Gates, 1965). The cool, deep shade of the forest is thus due to the fact that most of the visible and far infrared are absorbed by the foliage overhead. The blue and red light (0.4–0.5 μ and 0.6–0.7 μ bands respectively) are especially absorbed by chlorophyll and the far infrared heat energy by the water in the leaves and the water vapor that surrounds them. Green plants thus efficiently absorb the blue and

---

* The "solar constant."

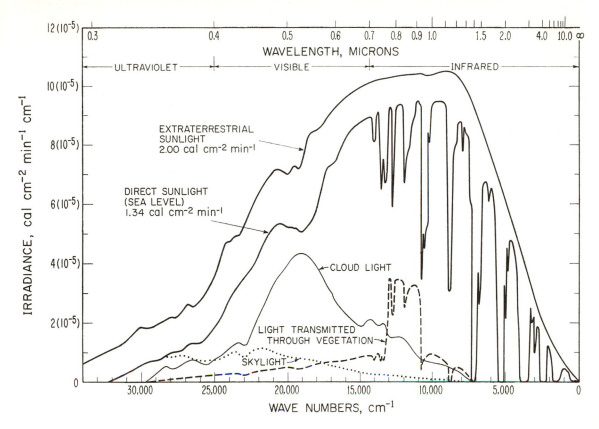

Figure 3–1. Spectral distribution of extraterrestrial solar radiation, of solar radiation at sea level for a clear day, of sunlight from a complete overcast, and of sunlight penetrating a stand of vegetation. Each curve represents the energy incident on a horizontal surface. (From Gates, 1965.)

red light most useful in photosynthesis. By rejecting, as it were, the near infrared band, in which the bulk of the sun's heat energy is located, leaves of terrestrial plants avoid lethal temperatures. In addition, aquatic plants are, of course, "water-cooled." Light as a limiting and controlling factor for organisms is discussed in Chapter 5. The use of multispectral imagery in the new technology of remote sensing is reviewed in Chapter 18, and the ecological impact of natural and man-made ionizing radiation is discussed in Chapter 17.

The other component of the energy environment, thermal radiation, comes from any surface or object that is at a temperature above absolute zero. This includes not only soil, water, and vegetation, but also clouds, which contribute a substantial amount of heat energy radiated downward into ecosystems. The "greenhouse effect" of reradiation has already been mentioned in connection with the CO_2 theory of climate. The longwave radiation fluxes, of course, are incident at all times and come from all directions, whereas the solar component is directional and is present only during the daytime. Thus, during a summer day an animal in the open or a leaf of a plant could be subjected to a total 24-hour upward and downward thermal radiation flux several times that of the direct downward solar input (1660 gcal per cm^2 as compared to 670 gcal per cm^2 in an example presented by Gates, 1963). Furthermore, thermal radiation is absorbed by biomass to a greater degree than is solar radiation. The daily variation is of great ecological significance. In places such as deserts or alpine tundras the daytime flux is manyfold greater than the nighttime flux, while in deep water or in the interior of a tropical forest (and, of course, in caves) the total radiation environment may be practically constant throughout the 24-hour period. Water and biomass, then, tend to mitigate fluctuations in the energy environment and thus make conditions less stressful for life.

Although the total radiation flux determines the "conditions of existence" to

Table 3–2. *Solar Radiation Received Regionally Over the United States on a Unit Horizontal Surface**

| | AVERAGE LANGLEYS (GCAL/CM²) PER DAY | | | | |
	Northeast	*Southeast*	*Midwest*	*Northwest*	*Southwest*
January	125	200	200	150	275
February	225	275	275	225	375
March	300	350	375	350	500
April	350	475	450	475	600
May	450	550	525	550	675
June	525	550	575	600	700
July	525	550	600	650	700
August	450	500	525	550	600
September	350	425	425	450	550
October	250	325	325	275	400
November	125	250	225	175	300
December	125	200	175	125	250
Mean-gcal/cm²/day	317	388	390	381	494
Mean-Kcal/m²/day (round figures)	3200	3900	3900	3800	4900
Estimated Kcal/m²/year (round figures)	1.17×10^6	1.42×10^6	1.42×10^6	1.39×10^6	1.79×10^6

* After Reifsnyder and Lull, 1965.

which organisms must adapt, it is the integrated direct solar radiation to the autotrophic stratum—that is, the sun energy received by green plants over days, months, and the year—which is of greatest interest in terms of productivity and nutrient cycling within the ecosystem. The average daily solar radiation received each month in five regions of the United States is shown in Table 3–2. In addition to latitude and season, cloud cover is a major factor, as shown in the comparison between the humid southeast and the arid southwest. A range between 100 and 800 gcal per cm² per day would probably cover most of the world most of the time, except in the polar regions or arid tropics in which conditions are so extreme anyway that there can be little biological output. Therefore, for most of the biosphere the radiant energy input is of the order of 3000–4000 kcal per m² per day and 1.1 to 1.5 million kilocalories per m² per year. In subsequent sections we shall see what happens to this input and how much of it can be converted to food to support the biotic community and man.

Of special interest is the so-called *net radiation* at the surface of the earth, which is "the difference between all downward streams of radiation minus all upward streams of radiation" (Gates, 1962). Between latitudes 40° north and south the annual net radiation is of the order of a million kcal per m² per year over the oceans and 0.6 million

kcal per m² per year over the continents (Budyko, 1955). This tremendous bundle of energy is dissipated in the evaporation of water and the generation of thermal winds and eventually passes into space as heat so that the earth as a whole may remain in an approximate energy balance. The differing impact of evaporative energy on terrestrial and aquatic ecosystems has already been mentioned (see page 20). Also, as already pointed out, any factor that delays the outward passage of this energy will cause the temperatures of the biosphere to rise.

At this point a word about measurements and units. Even though temperatures are recorded by a world-wide network of weather stations, direct measurements of energy fluxes have been made in only a few places. Consequently, the estimates we have given are based more on calculations than on direct measurements. Only in recent years have inexpensive, durable instruments been developed that promise to make possible large scale quantification of the energy environment. The solar component is usually measured by *pyrheliometers* or *solarimeters* which employ a thermopile, a junction of two metals that generates a current proportional to the light energy input. Instruments that measure the total flux of energy at all wavelengths are termed *radiometers*. The *net radiometer* has two surfaces, upward and downward, and records the difference in energy fluxes.

Finally, the units of irradiance energy we have been using, namely gcal per cm² (also called the *langley*) and Kcal per m², should not be confused with units of illumination, namely the foot candle (1 foot candle = 1 lumen per ft.²) or the lux (1 lux = 1 lumen per m² = approximately 0.1 foot candle), which apply only to the visible spectrum. Although one cannot convert irradiance rate to illuminescence accurately because of the variation in the brightness of different spectral regions, one gcal per cm² per min of sunlight multiplied by 6700 will give an approximate illuminance on a horizontal surface expressed as foot candles (Reifsnyder and Lull, 1965).

Of interest to the ecologist are David Gates's several excellent summaries of the energy environment in which we live (see Gates, 1962, 1963, 1965, and 1965*a*).

3. CONCEPT OF PRODUCTIVITY

Statement

Basic or *primary productivity* of an ecological system, community, or any part thereof, is defined as the rate at which radiant energy is stored by photosynthetic and chemosynthetic activity of producer organisms (chiefly green plants) in the form of organic substances which can be used as food materials. It is important to distinguish between the four successive steps in the production process as follows: *Gross primary productivity* is the total rate of photosynthesis, including the organic matter used up in respiration during the measurement period. This is also known as "total photosynthesis" or "total assimilation." *Net primary productivity* is the rate of storage of organic matter in plant tissues in excess of the respiratory utilization by the plants during the period of measurement. This is also called "apparent photosynthesis" or "net assimilation." In practice the amount of respiration is usually added to measurements of "apparent" photosynthesis as a correction in order to obtain estimates of gross production. *Net community productivity* is the rate of storage of organic matter not used by heterotrophs (that is, net primary production minus heterotrophic consumption) during the period under consideration, usually the growing season or a year. Finally, the rates of energy storage at consumer levels are referred to as *secondary productivities*. Since consumers only utilize food materials already produced, with appropriate respiratory losses, and convert to different tissues by one overall process, secondary productivity should not be divided into "gross" and "net" amounts. The total energy flow at heterotrophic levels which is analogous to gross production of autotrophs should be designated as "assimilation" and not "production." In all these definitions the term "productivity" and the phrase "rate of production" may be used interchangeably. Even when the term "production" is used to designate an amount of accumulated organic matter, a time element is always assumed or understood, as, for example, a year when we speak of agricultural crop production. To avoid confusion, the time interval should always be stated. In accordance with the second law of thermodynamics, as stated in Section 1, the energy flow is decreased at each step, as listed, by the heat loss that occurs with each transfer of energy from one form to another.

High rates of production, both in natural and cultured ecosystems, occur when physical factors are favorable and especially when there are *energy subsidies* from outside the system that reduce the cost of maintenance. Such energy subsidies may take the form of the work of wind and rain in a rain forest, tidal energy in an estuary, or the fossil fuel, animal, or human work energy used in the cultivation of a crop. In evaluating the productivity of an ecosystem it is important to take into consideration the nature and magnitude not only of the *energy drains* resulting from climatic, harvest, pollution, and other stress that divert energy away from the ecosystem, but also the *energy subsidies* that enhance productivity by reducing the respiratory heat loss (i.e., the "disorder pump-out") necessary to maintain the biological structure.

Explanation

The key word in the above definition is rate; the time element must be considered, that is, the amount of energy fixed in a given time. Biological productivity thus differs from "yield" in the chemical or industrial sense. In the latter case the reaction ends with the production of a given amount of material; in biological communities the process is continuous in time, so that it is necessary to designate a time unit; for example, the amount of food manufactured per day or per year. In more general terms, productivity of an ecosystem refers to its "richness." While a rich or productive community may have a larger quantity of organisms than a less productive community, this is by no means always the case. *Standing biomass* or *standing crop present at any given time should not be confused with productivity.* This point has been brought out in the earlier discussions, but it will not hurt to emphasize it again since students of ecology often confuse these two quantities. Usually, one cannot determine the primary productivity of a system, or the production of a population component, simply by counting and weighing (i.e., "censusing") the organisms which happen to be present at any one moment, although good estimates of net primary productivity may be obtained from standing crop data in situations in which organisms are large and living materials accumulate over a period of time

without being utilized (as in cultivated crops, for example). On the other hand, since small organisms "turnover" rapidly, and since organisms of all sizes are often "consumed" as they are being "produced," the size of the standing crop may bear little direct relation to productivity. For example, a fertile pasture which is being grazed by livestock would likely have a much smaller standing crop of grass than a less productive pasture which is not being grazed at the time of measurement. The "grazed pasture" situation is to be expected in a wide variety of natural communities in which heterotrophs are present and active, with the result that "consumption" occurs more or less simultaneously with "production." The contrast between the open-water pond and the meadow in this regard was stressed in Chapter 2. The interrelationships between standing crop, harvest procedure, and produc-

Table 3–3. *Relationships Between Solar Energy Input and Primary Productivity*

A. PERCENTAGE TRANSFERS

Steps	1 Total Solar Radiant Energy	2 Absorbed by Autotrophic Stratum	3 Gross Primary Production	4 Net Primary Production (Available to Heterotrophs)
Maximum	100	50	5	4
Average favorable condition	100	50	1	0.5
Average for biosphere	100	<50	0.2	0.1

B. PERCENTAGE EFFICIENCIES

Step	Maximum	Average Favorable Condition	Average for Whole Biosphere
1–2	50	50	<50
1–3	5	1	0.2
2–3	10	2	0.4
3–4	80	50	50
1–4	4	0.5	0.1

C. ON A KCAL/M²/YEAR BASIS (ROUND FIGURES)

Radiant Energy		Gross Primary Production	Net Primary Production
1,000,000	Maximum	50,000	40,000
	Average fertile regions*	10,000	5,000
	Open oceans and semi-arid regions†	1,000	500
	Mean for biosphere‡	2,000	1,000

* Moisture, nutrients, and temperature not strongly limiting; auxiliary energy input (see text for explanation).
† Moisture, nutrients, or temperature strongly limiting.
‡ Based on estimate of 10^{18} kcal gross productivity and 5×10^8/km² area of the whole biosphere (see Table 3–7).

*Table 3–4. Relationships Between Solar Radiation and Gross and Net Production on a Daily Basis in Crops Under Intensive Cultivation During Favorable Growing Season Conditions**

| | KCAL/M²/DAY | | | | | |
	Solar Radiation	Gross Produc- tion	Net Produc- tion	% GROSS/ SOLAR	% NET/ SOLAR	% NET/ GROSS
Sugar cane, Hawaii	4,000	306	190	7.6	4.8	62
Irrigated maize, Israel	6,000	405	190	6.8	3.2	47
Sugar beets, England	2,650	202	144	7.7	5.4	72

* From Montieth, 1965.

tivity become important considerations in evaluating the relative merits of algae culture and conventional land agriculture in food production for man, as we shall see later.

The different kinds of production, the important distinction between gross and net primary production, and their relationship to the solar energy input will become clear when Table 3–3 is carefully studied. Note that only about half of the total radiant energy (mostly in the visible portion) is absorbed, and at most about 5 per cent (10 per cent of the absorbed) can be converted as gross photosynthesis under the most favorable conditions. Then, the respiration of plants reduces to an appreciable extent (at least 20 per cent, usually about 50 per cent) the food (net production) available for the heterotrophs (including man).

It should be emphasized that Table 3–3 is a generalized model for energy transfers over the "long haul," that is, over the annual cycle or longer. During the peak of the growing season, especially during the long summer days of the north, more than 5 per cent of the total daily solar input may be converted into gross production, and more than 50 per cent of this may remain as net primary production during a 24-hour period. Such a situation is shown in Table 3–4. Even under the most favorable conditions, however, it is not possible to maintain these high daily rates over the annual cycle or to achieve such high yields over large areas of farmland, as is evident when we compare them with the annual yields (in terms of Kcal per day) shown in Table 3–9 (last column).

High productivity and high net/gross ratios in crops are maintained by virtue of large energy inputs involved in cultivation, irrigation, fertilization, genetic selection, and insect control. Fuel used to power farm machinery is just as much an energy input as is sunlight and it can be measured as Calories or horsepower diverted to heat in performance of the work of crop maintenance. Likewise, the energy of the tides may enhance the productivity of a natural coastal ecosystem by replacing part of the respiratory energy that would otherwise have to be devoted to mineral cycling and transportation of food and wastes. Any energy source that reduces the cost of internal self-maintenance of the ecosystem, and thereby increases the amount of other energy that can be converted to production, is called an *auxiliary energy flow* or an *energy subsidy* (H. T. Odum, 1967, 1967a).

The three crops listed in Table 3–4A make an interesting comparison among themselves. The high solar input of the irrigated desert resulted in a higher gross production but no greater net production than was achieved with less light in a more northern area. In general, high temperatures (and high water stress) require that the plant expend more of its gross production energy in respiration. Thus, it "costs" more to maintain the plant structure in hot climates. This may be one of the reasons (day length may be another) why yields of rice are consistently less in equatorial regions than in temperate regions (see Best, 1962).

Natural communities that benefit from natural energy subsidies are those with the highest gross productivity. The role of tides in coastal estuaries was mentioned earlier. The complex interaction of wind, rain, and evaporation in a tropical rain forest is another example of a natural energy subsidy that enables the leaves to make optimum use of the high solar input of the tropical day. As shown in Table 3–5, the gross primary productivity of a tropical rain forest

Table 3–5. *Annual Production and Respiration as Kcal/m²/year in Growth-Type and*
Steady-State Ecosystems

	ALFALFA FIELD (U.S.A.)[°]	YOUNG PINE PLANTA-TION (ENGLAND)[†]	MEDIUM-AGED OAK PINE FOREST (NEW YORK)[‡]	LARGE FLOWING SPRING (SILVER SPRINGS, FLORIDA)[§]	MATURE RAIN FOREST (PUERTO RICO)[‖]	COASTAL SOUND (LONG ISLAND, N.Y.)[¶]
Gross primary production (GPP)	24,400	12,200	11,500	20,800	45,000	5,700
Autotrophic respiration (R_A)	9,200	4,700	6,400	12,000	32,000	3,200
Net primary production (NPP)	15,200	7,500	5,000	8,800	13,000	2,500
Heterotrophic respiration (R_H)	800	4,600	3,000	6,800	13,000	2,500
Net community production (NCP)	14,400	2,900	2,000	2,000	Very little or none	Very little or none
Ratio NPP/GPP (per cent)	62.3	61.5	43.5	42.3	28.9	43.8
Ratio NCP/GPP (per cent)	59.0	23.8	17.4	9.6	0	0

[°] After Thomas and Hill, 1949. Heterotrophic respiration estimated as 5 per cent loss from insects and disease organisms.

[†] After Ovington, 1961. Mean annual production, 0–50 years, one-species plantation. GPP estimated from measurement of respiratory losses in young pines by Tranquillini, 1959. Part of NCP harvested (exported) as wood by man.

[‡] After Woodwell and Whittaker, 1968. 45 year old natural regeneration following fire; no timber harvest by man.

[§] After H. T. Odum, 1957.

[‖] After H. T. Odum and Pigeon, 1970

[¶] After Riley, 1956.

Note: Conversion factors from dry matter and carbon to Kcal as in Table 3–1. All figures rounded off to the nearest 100 Kcal.

can equal or exceed that of man's best agricultural efforts. It can be stated as a general principle that gross productivity of cultured ecosystems does not exceed that which can be found in nature. Man does, of course, increase productivity by supplying water and nutrients in areas where these are limiting (such as in deserts and grasslands). Most of all, however, man increases net primary and net community production through energy subsidies that reduce both autotrophic and heterotrophic consumption (and thereby increases the harvest for himself). H. T. Odum (1967) sums up this very important point as follows:

Man's success in adapting some natural systems to his use has essentially resulted from the process of applying auxiliary work circuits into plant and animal systems from such energy rich sources as fossil and atomic energy. Agriculture, forestry, animal husbandry, algal culture, etc. all involve huge flows of auxiliary energy that do much of the work that had to be self-served in former systems. Of course, when one provides the auxiliary support, the former species are no longer adapted since their internal programs would have them continue to duplicate the previous work effort and there would be no savings. Instead, species which do not have the machinery to self-serve have the edge and are selected either by man or by natural processes of survival.

Domestication in the extreme produces "organic matter machines," such as egg makers and milk-producing cows which can hardly stand up. All of the self-serving work of these organisms is supplied by new routes controlled and directed by man from auxiliary energy sources. In a real way the energy for potatoes, beef, and plant produce of intensive agriculture is coming in large part from the fossil fuels rather than from the sun. This lesson has probably been missed in the education of the general public. Many persons believe the great progress in agriculture, for example, is due to man's ingenuity in making new genetic varieties alone, whereas the use of such varieties is predicated on the enormous pumping of auxiliary energy. Those who attempt to improve foreign agriculture without supplying the auxiliary work from the industrial system do not understand the facts of life. Recommendations to underdeveloped countries based on experience of advanced countries cannot succeed if they are not accompanied by a tap-in to major auxiliary energy sources. . . .[°]

In other words, those who think that we can upgrade the agricultural production of the so-called "undeveloped countries" simply by sending seeds and a few "agricultural advisors" are tragically naive!

[°] From Odum, H. T., 1967. The Marine Systems of Texas. In *Pollution and Marine Ecology* (T. A. Olson and F. J. Burgess, eds.), p. 143. John Wiley & Sons.

Crops highly selected for industrialized agriculture must be accompanied by the fuel subsidies to which they are adapted! To put this in concrete energy terms, it has been estimated (see *The World Food Problem*, edited by I. L. Bennett and H. L. Robinson, Vol. II, pp. 397–398) that American agriculture employs an annual mechanical energy input of one horsepower per hectare of arable land as compared to an average of about 0.1 H.P. per hectare for Asia and Africa. The U.S. produces about three times as much food per hectare as Asia and Africa (see Table 3–9) but at a cost of 10 times as much very expensive auxiliary energy, which, incidentally, the "undeveloped countries" cannot possibly afford under current economic conditions! Estimated relationships between food yield and auxiliary energy work flows are shown in Figure 15–2 (page 412), while quantitative energy flow models for unsubsidized and subsidized agriculture are compared in Figure 10–7C–E (page 292).

There is one other important point to be made regarding the general concept of energy subsidy. A factor which under one set of environmental conditions acts as a subsidy in the sense that it enhances productivity may under another environmental condition act as an energy drain that reduces productivity. Thus, evapotranspiration may be an energy stress in dry climates but an energy subsidy in the humid climates (see H. T. Odum and Pigeon, 1970). Flowing water systems, such as the Florida spring listed in Table 3–5, tend to be more fertile than standing water systems, but not if the flow is too abrasive or irregular. The gentle ebb and flow of tides in a salt marsh, a mangrove estuary, or a coral reef contributes tremendously to the high productivity of these communities, but tides crashing against a northern rocky shore that is subjected to ice in winter and heat in summer can be a tremendous drain. Even in agriculture, man's attempts to assist nature often backfire. Thus, plowing the soil helps in the north but not in the south where the resulting rapid leaching of nutrients and loss of organic matter can severely stress subsequent crops. It is significant that agriculturists are now seriously considering "no-plow" crop procedures, a welcome trend toward a philosophy of "design with instead of against nature." Finally, certain types of pollution, such as treated sewage, can act as a subsidy or a stress, depending on the rate and periodicity of input (see Figure 16–2, page 435). Treated sewage released into an ecosystem at a steady but moderate rate can increase productivity, but massive dumping at irregular intervals can almost completely destroy the system as a biological entity. We will have more to say about the concept of "pulse stability" in Chapter 9.

In Table 3–5 selected ecosystems are listed in sequence from crop-type, rapid growth systems on the one hand to mature steady-state systems on the other hand. This arrangement brings out several important points regarding the relationships between gross, net primary, and net community production. Rapid growth or "bloom-type" systems (i.e., rapid production for short periods of time) such as the alfalfa field tend to have a high net primary production and, if protected from consumers, a high net community production. Reduced heterotrophic respiration, of course, can result either from evolved self-protection mechanisms (natural systemic insecticides or cellulose production, for example) or from outside assistance in terms of energy subsidies, as already pointed out. In steady-state communities the gross primary production tends to be totally dissipated by the combined autotrophic and heterotrophic respiration so that there is little or no net community production at the end of the annual cycle. Furthermore, communities with large biomasses or "standing crops," such as the rain forest, require so much autotrophic respiration for maintenance that there tends to be a low NPP/GPP ratio (Table 3–5). As a matter of fact, it is difficult, if not impossible, to distinguish by measurement between autotrophic and heterotrophic respiration in ecosystems such as forests. Thus, oxygen consumption or CO_2 production by a large tree trunk or tree root system is due as much to the respiration of associated microorganisms (many of which are beneficial to the tree) as to the living plant tissues. Consequently, the estimates of autotrophic respiration, and thereby the estimate of net primary production obtained by subtracting R_A from GPP, for the terrestrial communities listed in Table 3–5 are only rough approximations which are of more theoretical than practical value. This point needs to be emphasized because a number of summary papers on primary production published in the 1960s (see, for example, Westlake, 1963; Lieth, 1964) compare all kinds of communities ranging from low biomass aquatic communities and crops to high biomass forests on the basis of net primary production when, in actual fact, it was net community production (i.e., dry matter accumulation in the community) that was being compared.

The overall relationship between gross and net production can perhaps be explained by the graphic models in Figure 3–2. These diagrams are based on studies of crops, but the principles, we believe, are applicable to natural ecosystems as well. The leaf area index, as plotted along the x-axis of Figure 3–2A, can be considered as a measure of photosynthetic biomass. Maximum net production is obtained when the leaf area index is about 4 (that is, the leaf surface exposed to light is 4 times the ground surface), but maximum gross production is

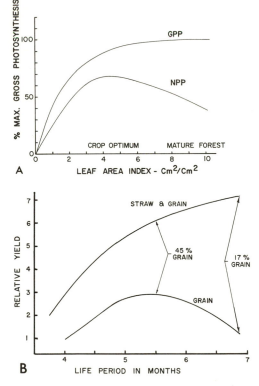

A. Relationships between gross primary production (GPP) and net primary production (NPP) with increasing leaf area index.

B. Effect of duration of the vegetative period on yield of grain.

reached when leaf area index reaches 8 to 10, which is the level found in mature forests (see Table 14–6, page 376). Net production is reduced at the higher levels because of the respiratory losses required to support the large volume of leaf and supporting tissue. In Figure 3–2B we see that maximum grain production comes at an earlier stage in the crop life period than does the maximum of total net production (dry matter accumulation). In recent years, considerable improvement in crop yields has resulted from attention to "crop architecture," which involves the selection of varieties that not only have high "grain-to-straw" ratios but also leaf out quickly to a leaf area index of 4 and remain at this level until harvested at the precise time of maximum food storage (see Loomis *et al.*, 1967; Army and Greer, 1967). Such artificial selection does not necessarily increase the total dry matter production of the whole plant; it mostly redistributes the production so that more goes into grain and less into leaves, stems, and roots (see Table 8–4, page 243).

In summary, it can be said that nature maximizes for gross production, whereas man maximizes for net production. The reasons why the strategies of man and nature differ in this manner and what this means in terms of ecosystem development theory and land use by man will be fully discussed in Chapter 9. Figure 3–2 may be one of the most important models (in terms of relevance to man) in this book; so consider it carefully!

We now come to the point where we can relate the principles of biogeochemical cycles, as outlined in Chapter 4, to principles pertaining to energy in ecosystems. The important point is: materials circulate, but energy does not. Nitrogen, phosphorus, carbon, water, and other materials of which living things are composed circulate through the system in a variable and complex manner. On the other hand, energy is used once by a given organism, is converted into heat, and is lost to the ecosystem. Thus, there is a nitrogen cycle, which means that nitrogen may circulate many times between living and nonliving entities; but there is no energy cycle. Life is kept going by the continuous inflow of sun energy from outside. Productivity should ultimately lend itself to precise measurement if energy flow can be measured. As we shall see, this is not easy to do in practice.

Figure 3–2. Productivity models for crops, illustrating that the maximum yield of edible parts does not coincide with maximum total production of the whole plant. *A.* Relationships between gross primary production (GPP) and net primary production (NPP) with increasing leaf area index (square centimeters of leaf surface exposed to light per square centimeter of ground surface). See text for explanation of comparison with mature forest. (From Monteith, 1965; after Black, 1963.) *B.* Effect of duration of the vegetative period on the yield of grain and total above-ground dry matter (grain and straw) in rice. (After Best, 1962.)

Examples

The stepwise flow of energy through an ecosystem is illustrated by the hypothetical model of a soybean crop ecosystem shown in Table 3–6. In the model a steady-state condition is assumed in which there is no net loss or gain in soil organic matter in the annual cycle. Cultures of crop plants in liquid nutrient media (reduced microbial populations) indicate that a minimum of 25 per cent of what the plant makes is required for its own use (i.e., plant respiration). At least an additional 5 per cent is estimated to be required by nitrogen-fixing bacteria in the root nodules and by mycorrhizal fungi associated with the uptake zone of the root system (see Chapter 7, page 232). It is important to emphasize that the energy required by beneficial microorganisms that fix nitrogen and assist in nutrient transport must come from the carbohydrates produced by the plant (see Allison, 1935). Therefore, such microbial respiration is as useful and necessary as that in the plant's own tissues. Even with intensive pest control another 5 to 10 per cent gross production will likely be lost to insects, nematodes, and disease microorganisms. This leaves about half (at most 65 per cent) of the total photosynthate available to macroconsumers such as man or his domestic animals. About half of this net community production (or about a third of the gross production) is in the form of stored energy of seeds (beans) that are readily edible and easily harvested. Man has the option, of course, of also harvesting the "hay" or leaving it in the field (as shown in Table 3–6) to maintain the soil structure and provide food energy for free-living microorganisms, many of which may fix nitrogen or perform other useful work. If man does remove all of the net production, it will "cost" him in terms of future work necessary to restore soil fertility. Also, if man attempts to remove all consumption by heterotrophic organisms (except for his own consumption) by very intensive use of broad-spectrum pesticides, he runs the risk of creating an "overkill" which will destroy beneficial as well as useful microorganisms (with a consequent loss in yield), and poisoning himself through contamination of his food and water.

Table 3–6. *Channeling of the Energy of Gross Production in a Soybean* (Glycine max) *Crop Ecosystem: Hypothetical Annual Budget**

ENERGY FLOW	PER CENT GROSS PRODUCTION UTILIZED	PER CENT GROSS PRODUCTION REMAINING
1. Plant respiration	25	—
Theoretical net primary production	—	75
2. Symbiotic microorganism (nitrogen-fixing bacteria and mycorrhizal fungi†)	5	—
Net primary production allowing for needs of beneficial symbionts	—	70
3. Root nematodes, phytophagous insects, and pathogens	5‡	—
Net community production allowing for minimum primary consumption by "pests"	—	65
4. Beans harvested by man (export)	32	—
Stems, leaves, and roots remaining in the field	—	33
5. Organic matter decomposed in soil and litter	33	—
Annual increment	—	0

° Adapted from Gorden, 1969.

† Mutualistic fungi that aid mineral uptake by roots (see page 232).

‡ Low per cent possible only with energy subsidy by man (fossil fuel, human and/or animal labor involved in cultivation, pesticide application, etc.).

In summary, the soybean ecosystem model illustrates two important points. First, since it is difficult to draw a hard and fast line between plant respiration and associated microorganism respiration, it is difficult to distinguish between net primary and net community production. As already stated in the previous section, this point is especially relevant when we consider forests or other large biomass communities in which a proportionally larger amount of energy goes into respiration. Second, a "prudent" consumer, such as man, should not count on harvesting more than one-third of the gross or one-half of the net production unless he is prepared to pay for substitutes for the "self-serving mechanisms" that nature has evolved to insure a long-term continuation of primary production in this biosphere.

The vertical distribution of primary production and its relation to biomass is il-

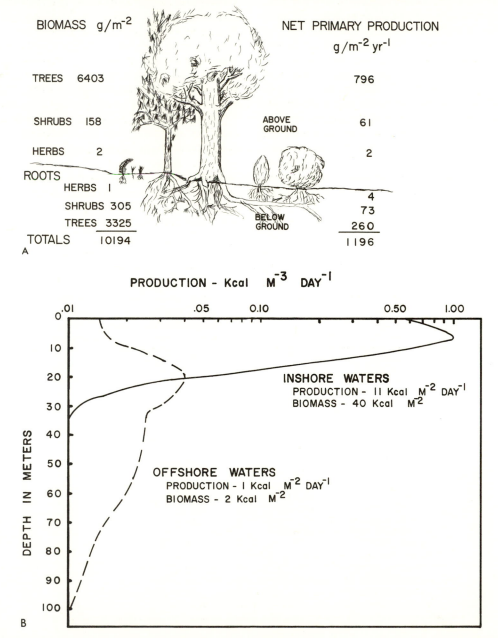

Figure 3–3. Comparison of the vertical distribution of primary production and biomass in the forest (*A*) and in the sea (*B*). These data also contrast the rapid turnover in the sea (*B/P* ratio is 2 to 4 days in this illustration) with slower turnover of the forest (*B/P* ratio is 9 years). (*A* based on data of Whittaker and Woodwell, 1969, for young oak-pine forest; *B* based on data of Currie, 1958, for the northeast Atlantic.)

lustrated in Figure 3–3. In this diagram the forest (Fig. 3–3A), in which turnover time (ratio of biomass to production) is measured in years, is compared with the sea (Fig. 3–3B), in which turnover time is measured in days. Even if we consider only the green leaves, which comprise 1 to 5 per cent of total forest plant biomass (see Figure 3–8; also page 66), as the "producers" comparable to the phytoplankton, replacement time would still be longer in the forest. In the more fertile inshore waters primary production is concentrated in the upper 30 meters or so, whereas in the clearer but poorer waters of the open sea the primary production zone may extend down to 100 meters or more. This is why coastal waters appear dark greenish and the ocean waters blue. In all water, the peak of photosynthesis tends to occur just under the surface because the circulating phytoplankton are "shade adapted" and tend to be inhibited by full sunlight. In the forest, where the photosynthetic units (i.e., the leaves) are permanently fixed in space, tree leaves are sun adapted and understory leaves are shade adapted (see Figure 3–5, page 62).

A number of attempts have been made to estimate the primary productivity of the biosphere as a whole (see Riley, 1944; Lieth, 1964). Conservative estimates of gross primary productivity of major ecosystem types, round figure estimates of the areas occupied by each type, and the total gross productivity for land and water are listed in Table 3–7. When we look at the estimated mean values for large areas, we see that *productivity varies by about two orders of magnitude (100-fold), from 200 to 20,000 kcal per m² per year, and that the total gross production of the world is on the order of 10^{18} Kcal per year.* The general pattern of distribution of world productivity is diagramed in Figure 3–4.

Table 3–7. *Estimated Gross Primary Production (Annual Basis) of the Biosphere and Its Distribution Among Major Ecosystems*

ECOSYSTEM	AREA (10^6 KM²)	GROSS PRIMARY PRODUCTIVITY (KCAL/M²/YEAR)	TOTAL GROSS PRODUCTION (10^{16} KCAL/YEAR)
Marine°			
Open ocean	326.0	1,000	32.6
Coastal zones	34.0	2,000	6.8
Upwelling zones	0.4	6,000	0.2
Estuaries and reefs	2.0	20,000	4.0
Subtotal	362.4	–	43.6
Terrestrial†			
Deserts and tundras	40.0	200	0.8
Grasslands and pastures	42.0	2,500	10.5
Dry forests	9.4	2,500	2.4
Boreal coniferous forests	10.0	3,000	3.0
Cultivated lands with little or no energy subsidy	10.0	3,000	3.0
Moist temperate forests	4.9	8,000	3.9
Fuel subsidized (mechanized) agriculture	4.0	12,000	4.8
Wet tropical and subtropical (broadleaved evergreen) forests	14.7	20,000	29.0
Subtotal	135.0	–	57.4
Total for biosphere (not including ice caps) (round figures)	500.0	2,000	100.0

° Marine productivity estimated by multiplying Ryther's (1969) net carbon production figures by 10 to get Kcal, then doubling these figures to estimate gross production and adding an estimate for estuaries (not included in his calculations).

† Terrestrial productivity based on Lieth's (1963) net production figure doubled for low biomass systems and tripled for high biomass systems (which have high respiration) as estimates of gross productivity. Tropical forests have been upgraded in light of recent studies, and the industrialized (fuel subsidized) agriculture of Europe, North America, and Japan have been separated from the subsistence agriculture characteristic of most of the world's cultivated lands.

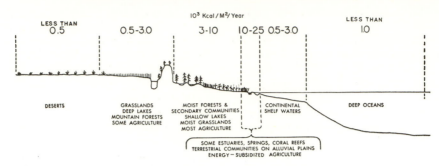

Figure 3–4. The world distribution of primary production in terms of annual gross production (in thousands of kilocalories per square meter) of major ecosystem types. Only a relatively small part of the biosphere is naturally fertile. (After E. P. Odum, 1963.)

A very large part of the earth is in the low production category because either water (in deserts, grasslands) or nutrients (in the open ocean) are strongly limiting. Although the land area comprises only about one-fourth of the earth's surface, it probably out-produces the oceans because so much of the ocean is essentially "desert." Even though mariculture (sea cultivation) is feasible in estuaries and along the edges of the sea, intensive "farming" of the open ocean is probably not practical (and would also be dangerous in terms of atmospheric balance; see page 24).

It should be noted that favorable sites in each broad ecosystem type may be capable of producing double (or more) the mean values shown (see Table 3–5). For all practical purposes a level of 50,000 Kcal per m² per year can be considered as the upper limit for gross photosynthesis. Until it can be conclusively shown that photosynthetic conversion of light energy can be substantially altered without endangering the balance of other, more important life cycle resources, man should plan to live within this limit. Most agriculture shows up low on an annual basis because annual crops are productive for less than half the year. Double cropping, that is, raising crops that produce throughout the year, can approach the gross productivity of the best of natural communities. Recall that net primary production will average about half of gross productivity and that the "yield to man" of crops will be one-third or less of the gross productivity. Also, as just pointed out, maximizing for yield may actually reduce gross production.

As of 1970, any estimate of total biospheric productivity can only be a very crude approximation, perhaps little better than a good guess. The number of measurements are not only inadequate, but the methods and the units of measurement employed are so variable that it is difficult to convert the data to the common "energy currency" of Calories. Again, it is hoped that research stimulated by the International Biological Program will show how close the 10^{18} Kcal estimate is to reality and to what extent man might raise total biospheric productivity without fouling his living space.

Additional data on the productivity of specific ecosystems are considered in Part II of this book. Reference may also be made to the following comprehensive summaries and symposia: Marine and aquatic productivity—Steemann-Nielsen, 1963; Goldman (ed.), 1965; Raymont, 1966; Strickland, 1965; Ryther, 1969. Terrestrial productivity—Lieth, 1962; Westlake, 1963; Newbould, 1963; Woodwell and Whittaker, 1968; Eckart (ed.), 1968. Crop productivity—Hart (ed.), 1962; Evans (ed.), 1963; San Pietro et al. (eds.), 1967. Some key references on production theory include Lindeman, 1942; Ivlev, 1945; Macfadyen, 1949; H. T. Odum, 1956, 1967a, and 1971.

A final example, an energy budget for a cattail marsh ecosystem as shown in Table 3–8, brings us back again to the consideration of the total solar energy environment as reviewed in section 1 of this chapter. The point needs to be made again that although the 1 per cent of solar radiation converted by photosynthesis is of paramount ecological interest because it directly supports all life, the other 99 per cent of sun energy is not "wasted"; it performs the work of driving hydrological and mineral cycles and maintaining environmental temperatures within the narrow range that can be tolerated by protoplasm. Such energy flows are as vital to survival as food. Thus, before we propose

Table 3–8. *Energy Budget for a Cattail* (Typha) *Marsh*[*]

| | YEAR | | GROWING SEASON | | | |
| | Total Radiation | | Total Radiation | | Visible Radiation | |
	Kcal/m²	Per Cent	Kcal/m²	Per Cent	Kcal/m²	Per Cent
Solar radiation	1,292,000	100	760,000	100	379,000	100
Photosynthesis (gross)	8,400	0.6	8,400	1.1	8,400	2.2
Reflection	439,000	34.0	167,000	22.0	11,400	3.0
Evapotranspiration	413,000	32.0	292,000	38.4 ⎫	359,000	94.8
Conduction-convection	431,000	33.4	293,000	38.5 ⎭		

[*] After Bray, 1962.

to strive for substantially greater "harvest of the sun," we must not forget about the first law of thermodynamics (conservation of energy). Diverting energy from one pathway means a reduction in the flow from some other pathway, which might turn out to be even more vital to survival of the ecosystem.

Man's Use of Primary Production

Primary production in terms of food for man is summarized in Tables 3–9 and 3–10. Yields and estimated net primary production of the major food crops in the "developed" and "undeveloped" countries are compared with world averages in Table 3–9. A developed country is defined as one with a per capita gross national product (GNP) of more than $600, usually more than $1000 (see Revelle, 1966). About 30 per cent of all human beings live in such countries, which also tend to have a low population growth rate (about 1 per cent per year). In contrast, 65 per cent of all human beings live in undeveloped countries that have a per capita GNP of less than $300, usually less than $100, and also a high growth rate (more than 2 per cent per year). As already pointed out, undeveloped countries have low per-hectare food production because they are too poor to afford the energy subsidies. The division between these two masses of humanity is a very sharp one (that is, the distribution of GNP is strongly bimodal) since only about 5 per cent of people live in what might be called "transitional countries" that have a per capita GNP between $300 and $600 (see Table 1 in Revelle, 1966). The sobering fact that must be faced is that the world average crop production is much closer to the lower than to the upper range, and in the undeveloped countries yields are not rising as fast as the population. Also, it is now generally recognized that protein, rather than total Calories, tends to limit the diet in the undeveloped world. Under equivalent conditions the yield of a high protein crop, such as soybean, must of necessity always be less (in terms of total Calories) than that of a carbohydrate crop, such as sugar cane (compare the means of these two crops in Table 3–9). In this regard it is interesting that sugar cane is often cited as the "champion" dry matter producer among cultivated plants. Annual yields of up to 30 tons of organic matter per acre per year (about 26,000 kcal per m² per year) have been recorded in Hawaii where crops can be grown on 8-year cycles with three cuttings before replanting (Burr *et al.*, 1957). Year-around growth from a perennial rhizome is one reason for such high yields; low nutritive quality of the growth is another. High protein annual crops could not possibly achieve such "bulk" productivity. As already pointed out, daily rates of net primary production tend to be less (and the protein content tends to be reduced) in hot climates, but longer growing seasons could more than compensate. It would seem to make "ecological common sense" to use perennial plants in tropical agriculture for two reasons; they can better utilize the long seasons and their culture avoids excessive leaching of nutrients that occurs as a result of the frequent plowing and replanting that is necessary to grow conventional "row-crop" annuals. Sustained yields of annual crops in the tropics require the expenditure of a great deal of work energy to maintain soil fertility, as man has learned by experience in the ancient art of paddy rice culture.

Table 3–10A is a more generalized model (see also Figure 15–2, page 412) of food production at three levels that exist today, and also at a theoretical level that might be ob-

Table 3–9. *Annual Yields of Edible Food and Estimated Net Primary Production of Major Food Crops at Three Levels: (1) Fuel Subsidized Agriculture (U.S., Canada, Europe, or Japan); (2) Little or No Fuel Subsidy (India, Brazil, Indonesia, or Cuba); and (3) World Average*

	EDIBLE PORTIONS		ESTIMATED NET PRIMARY PRODUCTION	
	Harvest Weight* *(Kg/ha)*	*Caloric Content†* *(Kcal/m²)*	*Dry Matter Production‡* *(Kcal/m²)*	*Rate, Growing Season§* *(Kcal/m²/day)*
Wheat—Netherlands	4,400	1,450	4,400	24.4
India	900	300	900	5.0
World average	1,300	430	1,300	7.2
Corn—U.S.	4,300	1,510	4,500	25.0
India	1,000	350	1,100	6.1
World average	2,300	810	2,400	13.3
Rice—Japan	5,100	1,840	5,500	30.6
Brazil	1,600	580	1,700	9.4
World average	2,100	760	2,300	12.8
White Potatoes—U.S.	22,700	2,040	4,100	22.8
India	7,700	700	1,400	7.8
World average	12,100	1,090	2,200	12.2
Sweet Potatoes and Yams—Japan	20,000	1,800	3,600	20.0
Indonesia	6,300	570	1,100	6.1
World average	8,300	750	1,500	8.3
Soybeans—Canada	2,000	800	2,400	13.3
Indonesia	640	260	780	4.3
World average	1,200	480	1,400	7.8
Sugar—Hawaii (from cane)	11,000	4,070	12,200	67.8
Netherlands (from beets)	6,600	2,440	7,300	40.6
Cuba (from cane)	3,300	1,220	3,700	20.6
World average (all sugar: beets and cane)	3,300	1,220	—	—

* Mean value 1962–1966 compiled from "Production Yearbook," Vol. 21 (1966), Food and Agricultural Organization, United Nations.

† Conversion, Kcal/gm harvested weight as follows: Wheat, 3.3; Corn, 3.5; Rice, 3.6; Soybeans, 4.0; Potatoes, 0.9; Crude sugar, 3.7 (see USDA Agriculture Handbook No. 8, 1963).

‡ Estimated on basis of 3X edible portion for grains, soybeans and sugar, 2X for potatoes (see text for explanation).

§ Estimated to be six months (180 days) except sugar cane where sugar yields are calculated on 12 months growing season (365 days).

tained with algae culture supported by massive subsidies of energy and money. The reason such high yields are theoretically possible with algae and perhaps not with larger plants is that the microscopic plants would require less of the gross production for their own respiration. However, the cost of machinery and fuel needed to operate such an algal system is so high that it is doubtful that such an agriculture will have any net value, except, perhaps, to a limited extent in crowded urban areas where there is no space for regular crops.

As of 1967, there were an estimated 3.5 × 10⁹ people in the world, each requiring about 10⁶ kcal per year, or a total of 3.5 × 10¹⁵ kcal of food energy needed to support the human "biomass." The origin of the 5.3 × 10¹⁵ kcal of food that was estimated as harvested for human consumption in 1967 is shown in Table 3–10B. This harvest comes to about 1 per cent of the net or 0.5 per cent of the gross primary production of the biosphere (as estimated in Table 3–7). Although it might seem that man is not yet making much of a dent in the photosynthetic capacity of the earth, there is much more to be considered than just man's own food intake. For example, what about the food requirements of the huge population of domestic animals (cows, pigs, horses, poultry, sheep, etc.), most of which are direct consumers of primary production, not only from agricultural lands but also from "wildlands" (grasslands, forests, etc.) as well? The standing crop of livestock in the world at large is five times that of human beings in terms of equivalent food requirements (see Borgstrom, 1965, for an explanation of the concept of "livestock population equivalent"). Thus, man and his domestic animals are already consuming at least 6 per cent of net production of the whole biosphere, or at least 12 per cent of that produced on land. Man also consumes huge quantities of primary production in the form of fibers (wood, paper, cotton, etc.), so that there is really very little of the earth's surface from which man does not harvest something, if only an occasional fish or stick of wood.

According to Borgstrom (1965), the ratio of "population equivalent" livestock to man varies from 43 to 1 in New Zealand to 0.6 to 1 in Japan, where fish largely replace terrestrial meat in the diet. One might say that the general ecology of the landscape, not to mention the culture and the economy, is determined by the sheep in New Zealand and by the fish in Japan! In large areas of the U.S. the landscape ecology is determined by beef cattle!

Looking at man's impact on the biosphere in another way, his density is now one person to about 4 hectares (10 acres) of land (i.e., 3.5×10^9 people on 14.0×10^9 hectares of land). When we add the domestic animals, the density is one population equivalent to about 0.7 hectare (i.e., 18.2×10^9 population equivalents on 14.0×10^9 hectares of land). This is less than 2 acres (0.7 ha) for every man and man-sized domestic animal consumer! If the population doubles in the next century and if we wish to continue to eat and use animals, then there will only be about one acre (0.4 ha) to supply all the needs (water, oxygen, minerals, fibers, living space, as well as food) for each 50-kilogram consumer, and this does not include pets and wildlife that contribute so much to the quality of human life!

The ecological basis for the present world food and population crisis will be considered again in Part 3, but the situation is rapidly becoming so critical that it merits a preliminary appraisal based on the principles and data reviewed in this chapter. The following points need immediate and serious consideration:

Table 3–10. *Food Yield (Production) for Man*

A. EDIBLE PORTION OF NET PRIMARY PRODUCTION ON A UNIT AREA BASIS

Level of Agriculture	Kgms Dry Matter/ha/year	Kcal/m²/year
Food gathering culture	0.4–20	0.2–10
Agriculture without energy (fuel) subsidy	50–2,000	25–1,000
Energy-subsidized grain agriculture*	2,000–20,000	1,000–10,000
Theoretical energy-subsidized algae culture	20,000–80,000	10,000–40,000

B. TOTAL FOR BIOSPHERE AS OF 1967 ($\times 10^{12}$ KCAL/YEAR)[†]

	Ocean	Land	Total
Plant	0.06	4,200	4,200.06
Animal	59.20	1,094	1,153.20
Totals	59.26	5,294[‡]	5,353.26

* Auxiliary fossil fuel (or other outside input) work energy flow equal at least to caloric yield of the crop (see H. T. Odum, 1967*a*, and Giles, 1967).

[†] Estimates based on Emery and Iselin (1967). Their "millions of tons wet weight" converted to 10^{12} Kcal by multiplying by 2 (1 gm wet wt. = approximately 2 Kcal; see Table 3–1).

[‡] Since approximately 10 per cent of the land area of the biosphere is cropland, 4.2×10^{15} Kcal comes from about 14×10^{12} m² of tilled land, or about 300 Kcal/m²/year. Since about 30 per cent of land is in crop and pasture 5.3×10^{15} total food comes from 40×10^{12} m² total agricultural land, or about 140 Kcal/m²/year, of which about one-fifth is secondary production (animal origin).

1. The public, and many professional specialists as well, have been misled by incomplete agricultural bookkeeping that failed to include the cost of energy subsidies and the cost to society of the environmental pollution that must necessarily accompany heavy use of machinery, fertilizers, pesticides, herbicides, and other potent chemicals (see Figure 15–2, page 412).

2. No more than 24 per cent of the land is truly arable in that it is really suitable for intensive farming (see report on *The World Food Problem* cited below). Irrigating huge areas of dry lands and farming the ocean would require large expenditures of money and have far reaching effects on global weather and atmospheric balances, with no assurance that some of these effects might not be disastrous.

3. The global impact of the domestic animal and man's need for animal protein have been underestimated.

4. As Ehrlich and Ehrlich (1970) point out, the undeveloped nations will become the "never-to-be-developed" nations unless the rate of population growth is greatly slowed. On the other hand, the quality of life in the developed countries is threatened by too much affluence that leads to pollution, crime, and a growing population of "undeveloped," poverty-stricken people within its borders. Thus, there must be a simultaneous global strategy aimed at leveling off the population growth in all the world, but especially in the undeveloped world, and, at the same time, leveling off per capita consumption and diverting more of the GNP to recycling resources and other strategies that maintain the quality of the environment in the developed world.

5. It is becoming increasingly evident that the optimum population density for man should be *keyed to the quality of the living space* (i.e., the "labenraum") *and not to food Calories.* The world can feed a lot more "warm bodies" than it can support quality human beings that have a reasonable chance for liberty and the pursuit of happiness. Kenneth Boulding (1966), an economist, has made what we believe to be an excellent ecological statement of the situation as follows: "The essential measure of success of the economy is not production and consumption at all, but the nature, extent, quality and complexity of the total capital stock, including in this the state of the human bodies and minds included in the system." Thus, should not man, like nature, maximize for the quality and diversity of the "biomass" rather than for the rate of production and consumption as such?

Three books and monographs are especially recommended for their thoroughgoing, ecologically oriented analysis of man's real impact on the productivity of the biosphere: *The Hungry Planet* by Borgstrom (1965); *The World Food Problem,* a 3-volume report of the Panel on World Food Supply, President's Scientific Advisory Committee, the White House (1967) (available from the Superintendent of Documents, Washington, D.C.); and *Population Resources and Environment; Issues in Human Ecology* by Ehrlich and Ehrlich (1970).

Measurement of Primary Productivity

Because of its great importance, brief attention should be given to the methods of measuring productivity in ecological systems, although detailed consideration of methods is outside the scope of this text. As already indicated, the ideal way to measure productivity would be to measure the energy flow through the system, but this has so far proved difficult to do. Most measurements have been based on some indirect quantity, such as the amount of substance produced, the amount of raw material used, or the amount of by-product released. One point to be emphasized is that no two of the various methods to be listed below measure exactly the same aspect of the complex process of autotrophic-heterotrophic metabolism! The simplified equation for photosynthesis given in Chapter 1 gives the overall reaction which takes place during the production of carbohydrates from raw materials as the result of light energy acting through chlorophyll. Since most kinds of production in nature result in new protoplasm, a more inclusive equation of productivity is as follows:

$$1{,}300{,}000 \text{ Cal. radiant energy} + 106CO_2 + 90H_2O$$
$$+ 16NO_3 + 1PO_4 + \text{mineral elements}$$
$$\text{equals}$$
$$13{,}000 \text{ Cal. potential energy in 3258 gms}$$
$$\text{protoplasm (106C, 180H, 46 O, 16N, 1P,}$$
$$815 \text{ gms mineral ash)} + 154O_2$$
$$+ 1{,}287{,}000 \text{ Cal. heat energy dispersed (99\%)}$$

This equation is based on the ratios of elements in, and the energy content of, protoplasm (Sverdrup *et al.,* 1942; Clarke, 1948).

It is evident that productivity can be measured, theoretically at least, by determining the amount of any one of the above items over the period of time that productivity is being measured. Equations such as this one can be used to convert (and to check against one another) productivity measurements between units of energy, carbon dioxide, nitrate or phosphate utilization, protoplasm weight (or the amount of carbon put in the form of food), and the amount of oxygen used. This is the theory; now let us see about the practice of measurement.

One of the greatest difficulties in determining the productivity of a particular ecological system is determining whether or not the system is in dynamic equilibrium or a *steady state*. In a "steady state" inflows balance outflows of material and energy. The rate of production is in equilibrium with the supply or the rate of inflow of the minimum limiting constituent (in other words, the law of the minimum is applying; see Chapter 5). For example, let us assume that carbon dioxide was the major limiting factor in a lake, and productivity was therefore in equilibrium with the rate of supply of carbon dioxide coming from the decay of organic matter. We shall assume that light, nitrogen, phosphorus, etc., were available in excess of use in this steady-state equilibrium (and hence not limiting factors at the moment). If a storm brought more carbon dioxide into the lake, the rate of production would change, and be dependent upon the other factors as well. While the rate is changing there is no steady state and no minimum constituent; instead, the reaction depends on the concentration of *all* constituents present, which in this transitional period differs from the rate at which the least plentiful is being added. The rate of production would change rapidly as various constituents were used until some constituent, perhaps carbon dioxide again, became limiting, and the lake system would once again be operating at the rate controlled by the law of the minimum. In most natural systems the production rate passes from one temporary steady-state equilibrium to another because of changes imposed on the system from the outside.

Some of the actual methods used to measure productivity may be briefly summarized as follows:

1. THE HARVEST METHOD. In situations in which herbivore animals are not important and in which a steady-state condition is never reached, the harvest method can be used. This is the usual situation in regard to cultivated crops that involve annual species since efforts are made to prevent insects and other animals from removing material, and the rate of production starts from zero at the time of planting seeds and reaches a maximum by the time of harvest. Weighing the growth produced by cultivated crops and determining the caloric value of the harvest is straightforward; the productivity of the crops, as given in Table 3–9, was determined in this manner. The harvest method can be used also in noncultivated terrestrial situations in which annual plants predominate, as in a ragweed field or other early stages of revegetation of abandoned fields, or where plants are little consumed until growth has been completed. In these cases it is best to take harvest samples at intervals during the season rather than to rely on a single terminal harvest since there usually will be a succession of annual species reaching maturity and dying during the growing season (see Penfound, 1956; E. P. Odum, 1960). Single harvests can be used to approximate net production in young forests or in croplike forest plantations (see Ovington, 1957, 1962). Harvest methods could not be used where the food produced is being removed as it is produced, as in many natural communities. If the consumers are large, long-lived animals one might determine productivity in these cases by harvesting the consumers, which are removing the food at a consistent rate, and thus estimate primary productivity from secondary productivity. A method of this sort, of course, is used frequently by animal husbandry men or range managers. The productivity of a western winter range, for example, may be expressed in terms of the number of cattle that can be supported by so many acres (or number of acres per one "animal unit"). The possible pitfalls in this procedure have already been suggested. Since food used by the plants themselves, and associated microorganisms and animals, is not included, the harvest method always measures *net community production*. If consumption by animals can be estimated, then a correction can be added to estimate *net primary production* (see Woodwell and Whittaker, 1968).

2. OXYGEN MEASUREMENT. Since there is a definite equivalence between oxygen and food produced, oxygen production can be a basis for determining productivity.

However, in most situations animals and bacteria (as well as plants themselves) are rapidly using up the oxygen; and often there is gas exchange with other environments. The "light-and-dark bottle" method of measuring oxygen production, and thereby estimating primary production, in aquatic situations has already been described in Chapter 2 (see page 14). The sum of the oxygen produced in the light bottle and oxygen used in the dark bottle is the total oxygen production, thus providing an estimate of primary production with appropriate conversion to Calories (see Table 3–1). If respiration by either plants or bacteria should differ in the dark and in the light, a source of error would be introduced since respiration in the dark bottle is assumed to be the same as respiration in the light bottle (where, of course, respiration and production are indistinguishable). By adding a heavy isotope of oxygen, which could be distinguished from ordinary oxygen, Brown (1953) found that the heavy isotope was used up at the same rate in the dark, at least for a number of hours, as in the light, thus indicating that respiration should be the same in both light and dark bottles during experiments of short duration. However, more recent research has shown that for many plants respiration in the dark is not the same as respiration in the light. Nevertheless, the "light-and-dark bottle method," pioneered by Gaarder and Gran in 1927, is widely used in both marine and freshwater environments. Dissolved oxygen is measured titrametrically by the Winkler method or electronically by one of several types of oxygen electrodes, and a given experiment is limited to one 24-hour cycle or less. The combination light-and-dark bottles measure *gross primary production,* and the light bottle measures *net community production* of whatever *part of the community* is in the bottle. The method obviously does not measure the metabolism of the part of the community on the bottom; also, the effects of enclosing the community in a bottle have not been clearly delimited. The use of large plastic spheres instead of small glass bottles reduces the inner surface-to-volume ratio and is presumed to reduce the effect of surface bacterial growth (see Antia, McAllister, Parsons, and Strickland, 1963).

Oxygen production may also be measured in certain aquatic ecosystems by the "diurnal curve method." In this instance, measurements of dissolved oxygen in the water at large are made at intervals throughout the day and night so that production of oxygen during the day and its use during the night may be estimated by determining the area under the diurnal curves. This method is particularly applicable to flowing water systems such as streams or estuaries (H. T. Odum, 1956) and is especially useful in dealing with polluted waters (Copeland and Dorris, 1962, 1964; H. T. Odum, 1960). If oxygen diffuses out of the body of water or into it from the atmosphere at an appreciable rate, a source of error is introduced; however, reasonable corrections can be made since diffusion is dependent on well-defined physical laws. The "diurnal curve method" measures *gross primary production* since oxygen used at night is added to that produced during the day (thus automatically including the respiration of the whole community). Instructions for analyzing diurnal oxygen curves may be found in H. T. Odum and Hoskins (1958) and H. T. Odum (1960).

In specialized situations, such as deep temperate lakes, productivity has been measured by a sort of reverse procedure, that is, by measuring the rate of oxygen disappearance in the deep waters (hypolimnion), which are producing no oxygen and are not in circulation with the upper waters during a major part of the production (summer) season (see Chapter 11). Thus, the greater the production in the upper, lighted waters (epilimnion), the more dead cells, bodies, feces, and other organic matter fall to the bottom where they decay under the action of bacteria and fungi with the use of oxygen. The rate of oxygen depletion is, therefore, proportional to productivity. Since decay of both plants and animals uses up oxygen, the "hypolimnetic method" measures the *net production of the whole community* (that is, both primary and secondary production) of the epilimnion. Edmondson and his associates (1968) have made good use of this method in following changes in productivity that have taken place over a period of years in Lake Washington, a large lake in the middle of the city of Seattle. This lake is considered in detail in Chapter 16 as a classic case of cultural eutrophication. Species of fish which require a cold water environment, as, for example, the cisco of the Great Lakes region, can live only in relatively unproductive lakes where cold bottom waters do not become depleted of oxygen during the summer.

3. CARBON DIOXIDE METHODS.   In terrestrial situations it is more practical to measure CO_2 rather than O_2 changes. Plant physiologists have long used CO_2 uptake to measure photosynthesis in leaves or single plants, whereas crop scientists and ecologists have made various attempts at measuring production in whole, intact communities by enclosing them in a transparent chamber, beginning with the pioneer experiments of Transeau (1926). A large bell jar or a plastic box or tent is placed over the community (see Figure 2–5B, page 18); air is drawn through the enclosure and the CO_2 concentration in incoming and outgoing air is measured with an infrared gas analyzer (or, in the older method, by absorption on a KOH column). For a description of a portable enclosure of this sort, see Musgrave and Moss (1961). As in the "light-and-dark bottle" aquatic method, gross minus respiration or *net community production* is measured during the day and community respiration during the night (or in a dark enclosure). *Gross primary production* can be estimated if both dark and light enclosures are used.

The difficulty with the enclosure method is that the terrestrial chamber, unlike the aquatic one, acts like a greenhouse, which quickly heats up unless a strong flow of air is maintained; this, in turn, may greatly change the rate of photosynthesis with respect to that occurring outside the enclosure. Refrigerating or air conditioning the chamber often becomes necessary if measurements are to extend over an appreciable period of time. Also, the sheer size and structural complexity of many terrestrial communities makes them very difficult to enclose. Ecologists have had some success in estimating the total metabolism of a forest by integrating simultaneous measurements made in separate chambers that enclose portions of branches, trunks, shrubs, soil, etc. (see Woodwell and Whittaker, 1968).

By far the most promising procedure for future measurements of terrestrial productivity is what may be termed the *aerodynamic method,* which is analogous to the aquatic "diurnal curve method" described above in that the community is not artificially enclosed. In principle, the flux of CO_2 above and within a community can be esti-

mated from periodic measurements of the vertical gradient of gas concentration and appropriate transfer coefficient with no disturbance to the community other than the erection of a mast on which is placed a series of CO_2 sensors arranged vertically from well above the vegetation canopy down to the ground level. In the profile thus obtained during daylight, the CO_2 concentration in the autotrophic layer as compared to that in the air above the community will be reduced proportionally to net photosynthesis, whereas the CO_2 concentration at ground level should be increased proportionally to soil and litter respiration. The nighttime gradient (like the "dark" enclosure) can be used to estimate total community respiration. As in the diurnal curve method, the accuracy of the aerodynamic method depends on the accuracy of the corrections that must be made for mass movements of the air and for gas evolution from soil that may contain CO_2, which is not a product of metabolism during the period of measurement. The aerodynamic method was pioneered by Huber (1952) in Germany and further developed by Monteith (1960, 1962) in England, Lemon (1960, 1967) in the United States, and Inoue (1958, 1965) in Japan. So far the method has been mostly applied to crops, grasslands, or other structurally simple communities, but Woodwell and Dykeman (1966) were able to estimate the respiration of a whole forest by measuring the accumulation of CO_2 in the vertical gradient during a period of temperature inversion that produced a temporary "enclosure," that is, horizontal and vertical air movements were at a minimum. H. T. Odum and Pigeon (1970) have used a compromise between the closed and open systems by encircling a small area of forest in a large tent open at the top and bottom and equipped with a fan at the bottom to bring about a one-way movement of air upward through the canopy. In such a setup net CO_2 flux can be estimated by measuring the concentration at top and bottom and the air flow rate.

Future success of aerodynamic methods will depend on improvements in remote sensing and continuous monitoring techniques (see Chapter 18) not only for CO_2 but for water vapor, air movement, heat transfer, and other factors that affect the complex production process.

4. THE pH METHOD.   In aquatic ecosystems the pH of the water is a function of the

dissolved carbon dioxide content, which, in turn, is decreased by photosynthesis and increased by respiration. However, to use pH as an index to productivity the investigator must first prepare a calibration curve for the water in the particular system to be studied because (1) pH and CO_2 content are not linearly related and (2) the degree of pH change per unit of CO_2 change depends on the buffering capacity of the water (thus, one unit of CO_2 removed by photosynthesis will bring about a greater pH increase in soft water from a mountain stream than in well-buffered sea water). Detailed instructions for preparing calibration graphs are given by Beyers *et al.*, 1963, and Beyers, 1964. The pH method has been especially useful in the study of laboratory microecosystems such as those pictured in Figure 2–7A, since with a pH electrode and recorder one can obtain a continuous record of net daytime photosynthesis and nighttime respiration (from which gross production can be estimated) without removing anything or otherwise disturbing the community (see McConnell, 1962; Beyers, 1963, 1965; Cooke, 1967 and Gorden *et al.*, 1969). In this regard, the use of the pH method in natural ecosystems has some of the same advantages and difficulties as does use of the aerodynamic method.

5. DISAPPEARANCE OF RAW MATERIALS. As indicated by the equation presented earlier, productivity can be measured not only by the rate of formation of materials (food, protoplasm, minerals) and by measuring gaseous exchange but also by the rate of the disappearance of raw material minerals. In a steady-state equilibrium, however, the amount used might be balanced by the amount being released or entering the system, and there would be no way of determining the actual rate of use by organisms. Where constituents such as nitrogen or phosphorus are not being supplied steadily, but perhaps once a year or at intervals, the rate at which their concentration decreases becomes a very good measure of productivity during the period in question. This method has been used in certain ocean situations in which phosphorus and nitrogen accumulate in the water during the winter and the rate of use can be measured during the period of spring growth of phytoplankton. The method must be used with caution since non-living forces may also cause the disappearance of materials. The disappear-

ance method measures the *net production of the whole community.*

6. PRODUCTIVITY DETERMINATIONS WITH RADIOACTIVE MATERIALS. As with many other fields of science, the use of radioactive tracers in ecology opens new possibilities in determining productivity. With a known amount of "marked material," which can be identified by its radiations, the rate of transfer can be followed even in the steady-state system as mentioned above, with the added advantage of less disturbance to the system.

One of the most sensitive and widely used methods for measuring aquatic plant production is done in bottles with radioactive carbon (^{14}C) added as carbonate. After a short period of time, the plankton or other plants are filtered from the water, dried, and placed in a counting device. With suitable calculations and a correction for "dark uptake" (^{14}C adsorption in a dark bottle) the amount of carbon dioxide fixed in photosynthesis can be determined from the radioactive counts made. When Steeman-Nielsen (1952), who first developed the method, made a series of measurements in the tropical oceans of the world, he found lower values than reported in many previous studies based on oxygen changes in bottled water (light-and-dark bottle method). Ryther (1954a) and others have since shown that the radioactive carbon method measures *net production* and not gross, as does the O_2 method, or at least it measures a quantity closer to net than to gross. The radioactive tracer uptake apparently measures that excess organic matter which is stored over and beyond the simultaneous needs for respiration. Tropical waters have a high respiration rate (recall our previous discussion of high respiration in tropical crop plants and tropical communities in general), which results in very little net production, thus explaining the low values obtained by Steeman-Nielsen. The estimates for total ocean productivity as given in Table 3–7 are based on the ^{14}C method. Detailed instructions for using this method are given by Strickland and Parsons (1968), and a critical evaluation of the method is provided by Thomas (1964).

When radioactive phosphorus, or ^{32}P, became generally available, it appeared to be a promising tool for applying the "disappearance" method to steady-state communities. While a great deal has been learned about the turnover rates in the phosphorus cycle, as will be described in Chapter 4, ^{32}P has not proved to be very satisfactory in the measurement of short-term productivity because phosphorus in any form is readily "adsorbed" by sediments and by organisms without being immediately incorporated into protoplasm. Thus, while ^{32}P added to a lake may be removed from the water at a rate perhaps proportional to primary productivity, it has been difficult to distinguish between biological assimilation and physical "uptake." Short-term uptake rate, however, may prove to be a good index to potential productivity since it seems to be proportional to the surface area in the ecosystem (see E. P. Odum *et al.*, 1958).

Radionuclide tracers other than ^{14}C and ^{32}P offer many possibilities which are yet to be investigated. As noted in Chapter 17 bioelimination of a variety of radionuclide tracers provides a means for measuring energy flow at the consumer population level (see also E. P. Odum and Golley, 1963).

7. THE CHLOROPHYLL METHOD. The possibilities of using the chlorophyll content of whole natural communities as a measure of productivity has been actively investigated. Offhand it might appear that chlorophyll would be a better measure of standing crop of plants than of productivity, but with proper calibration, the area-based chlorophyll content of a whole community can provide an index to its productivity. Gessner, in 1949, made the remarkable observation that the chlorophyll which actually develops on a "per square meter" basis tends to be similar in diverse communities, thus strongly suggesting that the content of the green pigment in whole communities is more uniform than in individual plants or plant parts. We apparently have here another striking example of "community homeostasis" in which the whole is not only different from but cannot be explained by the parts alone. In intact communities various plants, young and old, sunlit and shaded, are apparently integrated and adjust, as fully as local limiting factors allow, to the incoming sun energy, which, of course, impinges on the ecosystem on a "square meter" basis.

Figure 3–5 shows the amount of chlorophyll to be expected per square meter in four types of ecosystems that cover the range found in nature. The dots in the diagrams indicate the relative concentration of chlorophyll per cell (or per biomass). The relation of total chlorophyll to the photosynthetic rate is indicated by the *assimilation ratio*, or rate of production per gram of chlorophyll, shown as gms O_2 per hr per gm chlorophyll in the bottom row of numbers below the diagrams in Figure 3–5.

Shade-adapted plants or plant parts tend to have a higher concentration of chlorophyll than light-adapted plants or plant parts; this property enables them to trap and convert as many scarce light photons as possible. Consequently, efficiency of light utilization is high in shaded systems, but the photosynthetic yield and the assimilation ratio are low. Algae cultures grown in weak light in the laboratory often become shade-adapted. The high efficiency of such shaded systems has been sometimes mistakenly projected to full sunlight condition by those who are enthusiastic about the possibilities of feeding mankind from mass cultures of algae; when light input is increased in order to obtain a good yield, the efficiency goes down, as it does in any other kind of plant.

Total chlorophyll is highest in stratified communities, such as forests, and is generally higher on land than in water. In a given light-adapted system the chlorophyll in the autotrophic zone self-adjusts to nutrients and other limiting factors. Consequently, if the assimilation ratio and the available light are known, gross production can be estimated by the relatively simple procedure of extracting pigments and then measuring the chlorophyll concentration with a spectrophotometer. The chlorophyll method was first used in the study of the sea and other large bodies of water where extracting chlorophyll from water samples and measuring incident radiation are cheaper and less time consuming than ^{14}C or O_2 methods. For example, Ryther and Yentsch (1957) found that marine phytoplankton at light saturation has a reasonably constant assimilation ratio of 3.7 grams of carbon assimilated per hour per gram of chlorophyll. Calculated production rates based on this ratio and on chlorophyll-light measurements were very similar to those obtained by simultaneous use of the light-and-dark bottle oxygen method. Recently, Japanese ecologists have made extensive investigation of the relationship between area-based chlorophyll and dry matter production in terrestrial commu-

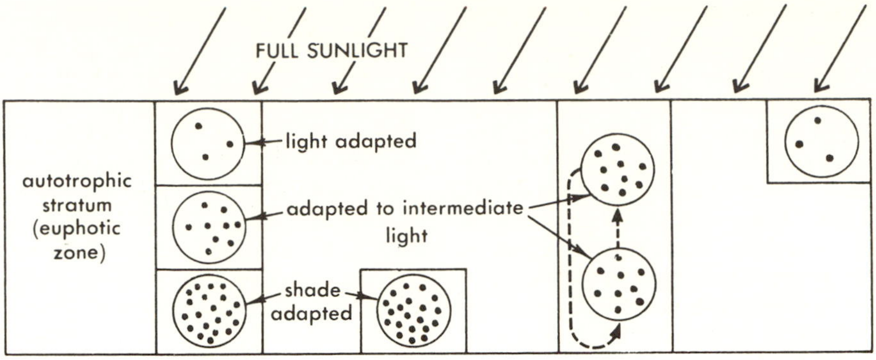

FULL SUNLIGHT

autotrophic stratum (euphotic zone)

light adapted

adapted to intermediate light

shade adapted

Community type	Stratified	Shaded	Mixing	Thin-bright
Examples	Forests; stratified grasslands and croplands	Winter, underwater or cave communities; lab cultures under low light intensity	Phytoplankton in lakes and oceans	Thin vegetation; algae mats on rocks; young crops; lab cultures under intense light (side-lighted)
Chlorophyll: g per square meter	0.4–3.0	0.001–0.5	0.02–1.0	0.01–0.60
Assimilation ratio: $\frac{\text{g } O_2 \text{ produced (per hour)}}{\text{g chlorophyll}}$	0.4–4.0	0.1–1.0	1–10	8–40

Figure 3–5. The amounts of chlorophyll to be expected in a square meter of four types of communities. The relation of area-based chlorophyll and photosynthetic rate is also indicated by the ratio between chlorophyll and oxygen production. (From E. P. Odum, 1963, after H. T. Odum, McConnell, and Abbott, 1958.)

nities (see Aruga and Monsi, 1963). Chlorophyll as an index of community function is fully discussed by H. T. Odum, McConnell, and Abbott (1958).

Of special interest is the possibility, now being investigated, that the ratio between the yellow carotenoids and the green chlorophylls can serve as a useful index to the ratio of heterotrophic to autotrophic metabolism in the community as a whole. When photosynthesis exceeds respiration in the community, the chlorophylls predominate, whereas carotenoids tend to increase as the community respiration increases. This is readily apparent when one looks down on a landscape from an airplane; rapidly growing young crops or forests appear bright green in contrast to the yellow-green color of older forests or mature crops. Margalef (1961, 1967) has found that the ratio of the optical density of acetone extracts of pigments at wavelength 430 mμ to that at 665

mμ provides a simple "yellow-green" index that is inversely correlated with the P/R ratio in cultures and in plankton communities. Thus, the yellow-green ratio tends to be low (values between 1 and 2, for example) in young cultures, or during the spring "bloom" in natural waters when respiration is low, and high (perhaps 3 to 5) in aging cultures or under late summer conditions in plankton communities when respiration is relatively high.

Improvements in multispectral photography and other remote sensing techniques from airplanes and satellites open up exciting possibilities for using the color of vegetation as an index to its metabolism. It should not be too many more years before the production/consumption ratio and its effect on the vital O_2/CO_2 balance might be monitored over large areas of the land surface. However, quantitation from above depends on accurate calibration, which in turn depends on improved pigment and productivity measurements on the ground below (i.e., "ground truth," see Chapter 18).

4. FOOD CHAINS, FOOD WEBS, AND TROPHIC LEVELS

Statement

The transfer of food energy from the source in plants through a series of organisms with repeated eating and being eaten is referred to as the *food chain*. At each transfer a large proportion, 80 to 90 per cent, of the potential energy is lost as heat. Therefore, the number of steps or "links" in a sequence is limited, usually to four or five. The shorter the food chain (or the nearer the organism to the beginning of the chain), the greater the available energy. Food chains are of two basic types: the *grazing food chain*, which, starting from a green plant base, goes to grazing herbivores (i.e., organisms eating living plants) and on to carnivores (i.e., animal eaters); and the *detritus food chain*, which goes from dead organic matter into microorganisms and then to detritus-feeding organisms (detritivores) and their predators. Food chains are not isolated sequences but are interconnected with one another. The interlocking pattern is often spoken of as the *food web*. In complex natural communities, organisms whose food is obtained from plants by the same number of steps are said to belong to the same *trophic level*. Thus, green plants (the producer level) occupy the first trophic level, plant-eaters the second level (the primary consumer level), carnivores, which eat the herbivores, the third level (the secondary consumer level), and secondary carnivores the fourth level (the tertiary consumer level). It should be emphasized that *this trophic classification is one of function and not of species as such;* a given species population may occupy one, or more than one, trophic level according to the source of energy actually assimilated. The *energy flow* through a trophic level equals the total assimilation (A) at that level, which, in turn, equals the production (P) of biomass plus respiration (R).

Explanation

Food chains are more or less familiar to everyone in a vague sort of way, at least,

because man himself occupies a position at or near the end of a chain of food items. For example, man may eat the big fish that eats the little fish that eats the zooplankton that eats the phytoplankton that fixes the sun energy; or he may eat the beef that eats the grass that fixes the light energy; or he may utilize a much shorter food chain by eating the grain that fixes the sun energy; or, as is usually the case, man may occupy an intermediate trophic position between primary and secondary consumers when his diet is composed of mixtures of plant and animal food. What is not usually recognized by the layman, however, is that potential energy is lost at each food transfer, and, as described in previous sections, only a very small percentage of the available sun energy was fixed by the plant in the first place. Consequently, the number of consumers, such as people, that can be supported by a given primary production output very much depends on the length of the food chain; each link in the chain decreases the available energy by about one order of magnitude (order of 10), which means that fewer people can be supported when large amounts of meat are part of the diet. Or, to put it more realistically, meat will disappear, or be very much reduced, in man's diet if he does not exercise his option of controlling his own population growth.

The principles of food chains and the workings of the two laws of thermodynamics can be clarified by means of flow diagrams as shown in Figures 3–6, 3–7, and 3–8. In these diagrams the "boxes" represent trophic levels and the "pipes" depict the energy flow in and out of each level. Energy inflows balance outflows as required by the first law of thermodynamics, and each energy transfer is accompanied by dispersion of energy into unavailable heat (i.e., respiration) as required by the second law.

Figure 3–6 is a very simplified energy flow model of three trophic levels. This diagram introduces standard notations for the different flows, which will be described in more detail later in this section, and illustrates how the energy flow is greatly reduced at each successive level regardless of whether we consider the total flow (I and A) or the components P and R. Also shown are the "double metabolism" of producers (i.e., gross and net production) and the approximately 50 per cent absorption-1 per cent conversion of light at the first trophic level,

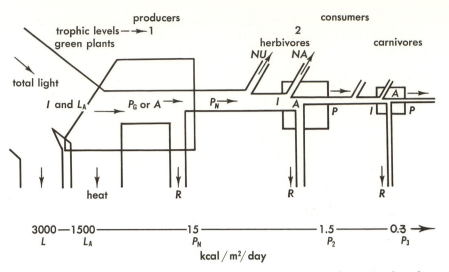

Figure 3-6. A simplified energy flow diagram depicting three trophic levels (boxes numbered 1, 2, 3) in a linear food chain. Standard notations for successive energy flows are as follows: I = total energy input; L_A = light absorbed by plant cover; P_G = gross primary production; A = total assimilation; P_N = net primary production; P = secondary (consumer) production; NU = energy not used (stored or exported); NA = energy not assimilated by consumers (egested); R = respiration. Bottom line in the diagram shows the order of magnitude of energy losses expected at major transfer points, starting with a solar input of 3000 kcal per square meter per day. (After E. P. Odum, 1963.)

as was discussed in the previous section (compare Table 3–3, page 44). Secondary productivity (P_2 and P_3 in the diagram) tends to be about 10 per cent at successive consumer trophic levels, although efficiency may be higher, say 20 per cent, at the carnivore levels as shown.

Figure 3–7 pictures one of the first published energy flow models as pioneered by H. T. Odum in 1956. In this model a community boundary is shown, and, in addition to light and heat flows, the import, export, and storage of organic matter are also included. Decomposer organisms are placed in a separate box as a means of partially separating the grazing and detritus food chains. As pointed out in Chapter 2, "decomposers" are actually a mixed group in terms of energy levels. For the purposes of modeling, natural features such as the shore of a lake or edge of a forest may be used to mark boundaries around ecosystems, or purely arbitrary limits, such as a road around a square mile of farmland, or a political unit such as a county may be established if convenient for the generation of necessary data. As long as the imports and exports are considered, the energetics of any area can be described in terms of an energy flow model. The smaller the area, of course, the greater the importance of exchanges with the surrounding areas.

In Figure 3–8 the grazing and detritus food chains are sharply separated in a Y-shaped or 2-channel energy flow diagram, which is a more practical working model than the single channel model because (1) it conforms to the basic stratified structure of ecosystems (see page 9), (2) direct consumption of living plants and utilization of dead organic matter is usually separated in both time and space, and (3) the macroconsumers (phagotrophic animals) and the microconsumers (saprotrophic bacteria and fungi) differ greatly in size-metabolism relations and in techniques required for study (see page 10). Figure 3–8 also contrasts the biomass-energy flow relationships in the sea and the forest as was first discussed in Chapter 2. In the marine community the energy flow via the grazing food chain is shown to be larger than via the detritus pathway, whereas the reverse is shown for the forest, in which 90 per cent or more of the net primary production is normally utilized in the detritus food chain. Such a difference is not necessarily inherent in aquatic and terrestrial systems. In a heavily grazed pasture or grassland 50 per cent or more of the net production may pass down the grazing path, whereas there are many aquatic systems, especially shallow water ones, that, like mature forests, operate largely as detritus systems. Since not all food eaten by grazers is actually assimilated, some (undigested material in feces, for example) is diverted to the detritus route; thus, the impact of the grazer on the community depends on the

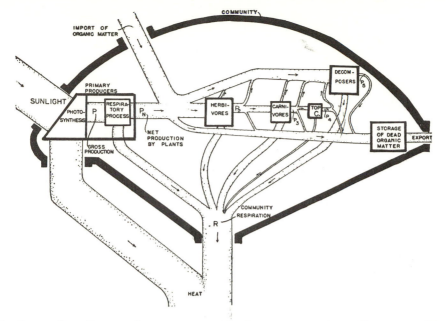

Figure 3-7. Energy-flow diagram of a community with a large import and a smaller export of organic matter, showing successive fixation and transfer by components and the large respiratory losses at each transfer. P = gross primary production, P_N = net primary production, and P_2, P_3, P_4, and P_5 = secondary production at the indicated levels. (Redrawn from H. T. Odum, 1956.)

rate of removal of living plant material as well as on the amount of energy in the food that is assimilated. Marine zooplankton commonly "graze" more phytoplankton than they can assimilate, the excess being egested to the detritus food chain (see Cushing, 1964). As already discussed in the section on primary productivity, direct removal of more than 30 to 50 per cent of the annual plant growth by terrestrial grazing animals or man can reduce the ability of the ecosystem to resist future stress.

One is impressed with the many mechanisms in nature that control or reduce grazing, just as one is unimpressed with man's past ability to control his own grazing animals, since it is increasingly evident that overgrazing has contributed to the decline of past civilizations. One has to be careful, of course, about the choice of words here. "Overgrazing," by definition, is detrimental, but what constitutes overgrazing in different kinds of ecosystems is only now being defined in terms of energetics as well as in terms of long-term economics. The "grazed pasture model" shown in Table 15-2 on page 419 is based on long-term studies on the Great Plains and is an example of the kind of model that needs to be put into practice. "Undergrazing" can also be detrimental. In the complete absence of direct consumption of living plants, detritus

could accumulate at a rate greater than microorganisms can decompose it, thereby delaying mineral recycling and perhaps making the system vulnerable to destructive fires.

At this point it may be well to examine the basic component of an energy flow model in detail. Figure 3-9 presents what might be called a "universal" model, one that is applicable to any living component whether it be plant, animal, microorganism, or individual, population, or trophic group. Linked together, such graphic models can depict food chains, as already shown, or the bioenergetics of an entire ecosystem. In Figure 3-9, the shaded box labeled "B" represents the living structure or "biomass" of the component. Although biomass is usually measured as some kind of weight (living [wet] weight, dry weight, or ash-free weight), it is desirable to express biomass in terms of calories so that relationships between the rates of energy flow and the instantaneous or average standing-state biomass can be established. The total energy input or intake is indicated by "I" in Figure 3-9. For strict autotrophs this is light, and for strict heterotrophs it is organic food. As discussed in Chapter 2, some species of algae and bacteria can utilize both energy sources, and many may require both in certain proportions. A similar situation holds

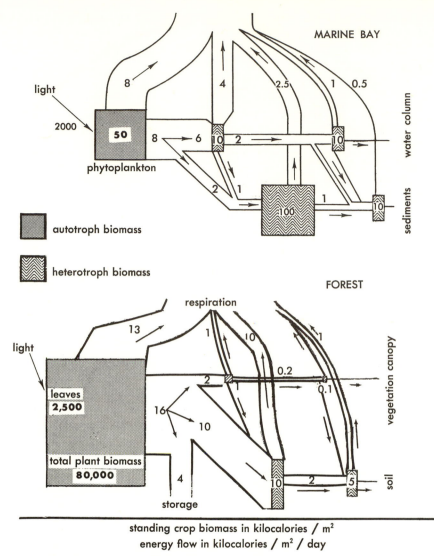

Figure 3–8. A Y-shaped or 2-channel energy flow diagram that separates a grazing food chain (water column or vegetation canopy) from a detritus food chain (sediments and in soil). Estimates for standing crops (shaded boxes) and energy flows compare a hypothetical coastal marine ecosystem (upper diagram) with a hypothetical forest (lower diagram). (Modified from E. P. Odum, 1963.)

for invertebrate animals and lichens, which contain mutualistic algae. In such cases the input flow in the energy flow diagram can be subdivided accordingly to show the different energy sources, or the biomass can be subdivided into separate boxes if one wishes to keep everything in the same box at the same energy level (i.e., the same trophic level).

Such flexibility in usage can be confusing to the beginner. It is important to emphasize again that the *concept of trophic level is not primarily intended for categorizing species.* Energy flows through the community in stepwise fashion according to the second law of thermodynamics, but a given population of a species may be (and very

often is) involved in more than one step or trophic level. Therefore, the universal model of energy flow illustrated in Figure 3–9 can be used in two ways. The model can represent a species population, in which case the appropriate energy inputs and links with other species would be shown as a conventional species-oriented food-web diagram (see Figure 3–10); or the model can represent a discrete energy level, in which case the biomass and energy channels represent all or parts of many populations supported by the same energy source. Foxes, for example, usually obtain part of their food by eating plants (fruit, etc.) and part by eating herbivorous animals (rabbits, field mice, etc.). A single box diagram could be

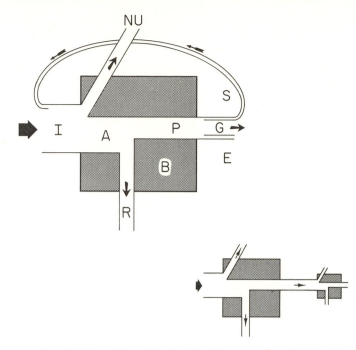

Figure 3–9. Components for a "universal" model of ecological energy flow. *I* = input or ingested energy; *NU* = not used; *A* = assimilated energy; *P* = production; *R* = respiration; *B* = biomass; *G* = growth; *S* = stored energy; *E* = excreted energy. See text for explanation. (After E. P. Odum, 1968.)

used to represent the whole population of foxes if our objective is to stress intrapopulation energetics. On the other hand, two or more boxes (such as shown in the lower right of Figure 3–9) would be employed should we wish to apportion the metabolism of the fox population into two trophic levels according to the proportion of plant and animal food consumed. In this way we can place the fox population into the overall pattern of energy flow in the community. When an entire community is modeled one cannot mix these two usages unless all species happen to be restricted to single trophic levels (e.g., a highly simplified grass-cow-man ecosystem).

So much for the problem of the source of the energy input. Not all of the input into the biomass is transformed; some of it may simply pass through the biological structure, as occurs when food is egested from the digestive tract without being metabolized or when light passes through vegetation without being fixed. This energy component is indicated by "NU" ("not utilized"). That portion which is utilized or assimilated is indicated by "A" in the diagram. The ratio between these two components, i.e., the efficiency of assimilation, varies widely. It may be very low, as in light fixation by plants or food assimilation in detritus-feeding animals, or very high as

in the case of animals or bacteria feeding on high energy food such as sugars and amino acids. There will be more on efficiencies later.

In autotrophs the assimilated energy ("A") is, of course, "gross production" or "gross photosynthesis." Historically, the term "gross production" has been used by some authors for the analogous component in heterotrophs, but as already pointed out (see page 43), the "A" component in heterotrophs represents food already "produced" somewhere else. Therefore, the term "gross production" should be restricted to primary or autotrophic production. In higher animals the term "metabolized energy" is often used for the "A" component (see Kleiber and Dougherty, 1934; Kendeigh, 1949).

A key feature of the model is the separation of assimilated energy into the "P" and "R" components. That part of the fixed energy ("A") which is burned and lost as heat is designated as respiration ("R"), and that portion which is transformed to new or different organic matter is designated as production ("P"). This is "net production" in plants or "secondary production" in animals. It is important to point out that the "P" component is energy available to the next trophic level, as opposed to the "NU" component that is still available at the same trophic level.

The ratio between "P" and "R" and between "B" and "R" varies widely and is of great ecological significance from the thermodynamic standpoint as explained in Section 1 of this chapter and also in Chapter 2 (pages 17 and 24). In general, the proportion of energy going into respiration, that is, maintenance, is large in populations of large organisms, such as men and trees, and in mature (i.e., "climax") communities. As already pointed out, "R" goes up when a system is stressed. Conversely, the "P" component is relatively large in active populations of small organisms, such as bacteria or algae, in the young or "bloom" stages of ecological succession, and in systems benefiting from energy subsidies. The relevance of P/R ratios to food production for man was mentioned in Section 3 of this chapter and will be noted again in Chapter 9.

Production may take a number of forms. Three subdivisions are shown in Figure 3–9. "G" refers to growth or additions to the biomass. "E" refers to assimilated organic matter that is excreted or secreted (e.g., simple sugars, amino acids, urea, mucus, etc.). This "leakage" of organic matter, often in dissolved or gaseous form, may be appreciable but is too often ignored because it is hard to measure. Finally, "S" refers to "storage," as in the accumulation of fat which may be reassimilated at some later time. The reverse "S" flow shown in Figure 3–9 may also be considered a "work loop" since it depicts that portion of production that is necessary to insure a future input of the new energy (e.g., reserve energy used by a predator in the search for prey). As already pointed out, artificial selection for domestication of plants and animals too often results in breeding out the "self protective" work loop, with the result that man himself must provide outside energy input for maintenance.

Figure 3–9 shows only a few of the ecologically useful subdivisions of the basic pattern of energy flow. In practice, we are often hampered by the difficulties of measurement, especially in field situations. A primary purpose of a model, of course, is to define components that we want to measure in order to stimulate research into methodology. Even if we are not yet able to chart all the flows, measurement of gross inputs and outputs alone may be revealing. Because energy is the ultimate limiting factor, the amounts available and actually utilized must be known if we are to evaluate the importance of other potentially limiting or regulating factors. Many of the controversies about food limitation, weather limitation, competition, and biological control could be resolved if we had accurate data on energy utilization by the populations in question.

The "boxes-and-pipes" graphic models can be readily transformed into *compartment models,* in which energy flow between compartments is expressed as transfer coefficients, or into *circuit models,* in which each structure and function is indicated by a symbolic module interconnected by a network of energy circuits. Such models are especially suitable for digital and analog computer manipulation. Examples of both compartment and circuit models will be presented later in this chapter and again in Chapter 10.

Examples

Many people think of the arctic as a barren region of no interest or value to man. Regardless of the outcome of present efforts by man to utilize the arctic, this vast region is of great interest if for no other reason than that its ecology is simplified. Because temperature exerts such a powerful limiting effect, only relatively few kinds of organisms have become successfully adapted to the far northern conditions. Thus, the entire living part of the ecosystem is built around relatively few species. Studies in the arctic aid in the understanding of more complex conditions elsewhere, since such basic relationships as food chains, food webs, and trophic levels are simplified and readily comprehended. Charles Elton early realized this and spent much time in the 1920's and 1930's studying the ecology of arctic lands, with the result that he was one of the first to clarify the above-mentioned principles and concepts. Consequently, we might well look to the arctic for our first examples of food chains.

The region between the limit of trees and the perpetual ice is generally known as the *tundra.* One of the important groups of plants on the tundra is the reindeer lichens (or "moss"), *Cladonia,* which represent a partnership between algae and fungi, the former being the producers, of course (see Figure 7–39). These plants, together with the grasses, sedges, and dwarf willows, form the diet of the caribou of the North American tundra and of its ecological counterpart, the reindeer of the Old World tundra. These animals, in turn, are preyed upon by wolves and man. Tundra plants are also eaten by lemmings — shaggy haired voles with short tails and a bearlike appearance — and the ptarmigan or arctic grouse. Throughout the long winter, as well as during the brief summer, the arctic white fox and the snowy owl may depend largely on the lemming and related rodents. In each of these cases the food chain is relatively short and any radical change in numbers at any of the three trophic levels has violent repercussions on the other levels, because there is often little in the way of alternate choice of food. As will be discussed later, this may be one reason, at least, why some groups of arctic organisms undergo violent fluctuations in numbers, running the gamut from superabundance to near extinction. It is rather interesting to note, in passing, that the same sort of thing often happened to human civilizations that depended on a single or on relatively few local food items (recall the Irish potato famine). In Alaska, man has inadvertently caused severe oscillation by introducing the domestic reindeer from Lapland. Unlike the native caribou, the reindeer do not migrate. In Lapland, reindeer are herded from one place to another to prevent overgrazing, but herding is not a part of the culture of Alaskan Indians and Eskimos (since caribou did their own herding). As a result, reindeer have severely overgrazed many areas, reducing the carrying capacity for caribou as well. This is a good example of where only a part of the adapted "system" was introduced. We shall have many occasions to note that introduced animals often become severe pests when their natural or man-made control mechanisms are not also introduced into the new locality.

During the brief arctic summer, insects emerge and migratory birds may be locally abundant. Food chains become longer and definite food webs develop, as in more southern regions. Summerhayes and Elton (1923) describe the interesting situation on the island of Spitsbergen where there are no lemmings. Here foxes are able to feed on birds, insects, or plants in summer but are forced to spend the winter out on the ice feeding on the remains of seals killed by polar bears and on the dung of the bears. The foxes thus become a part of the food web of the sea, which in arctic regions may be more productive than the land.

A food web which has been worked out for small organisms of a stream community is shown in Figure 3–10. This diagram not

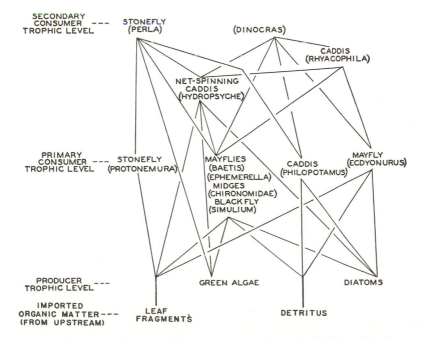

Figure 3–10. A portion of a food web in a small stream community in South Wales. The diagram illustrates: (1) the interlocking of food chains to form the food web, (2) three trophic levels, (3) the fact that organisms, such as *Hydropsyche*, may occupy an intermediate position between major trophic levels, and (4) an "open" system in which part of the basic food is "imported" from outside the stream. (Redrawn from Jones, 1949.)

only illustrates the interlocking nature of food chains and three trophic levels, but also brings out the fact that some organisms occupy an intermediate position between the three major trophic levels. Thus, the net-spinning caddis feeds on both plant and animal material and is therefore intermediate between primary and secondary consumer levels.

A farm pond managed for sport fishing, thousands of which have been built all over the country, provides an excellent example of food chains under fairly simplified conditions. Since the object of a fish pond is to provide the maximum number of fish of a particular species and a particular size, management procedures are designed to channel as much of the available energy into the final product as possible. This is done by reducing the number of food chains by restricting the producers to one group, the floating algae or phytoplankton (other green plants such as rooted aquatics and filamentous algae are discouraged). Figure 3–11 is a compartment model of a sport-fishing pond in which transfers at each link in the food chain are quantitated in terms of kcal per m² per year. In this model only

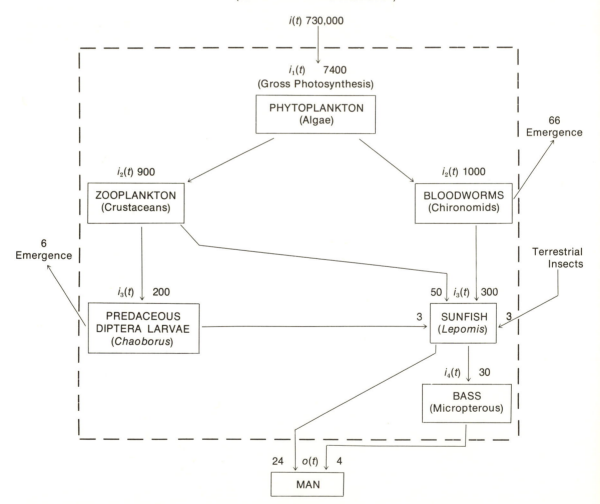

Figure 3–11. Compartment model of the principal food chains in a Georgia pond managed for sports fishing. Estimated energy inputs (i) with respect to time (t) are kilocalories per square meter per year. $i_1(t)$, $i_2(t)$, $i_3(t)$, and $i_4(t)$ represent ingested food energy at successive trophic levels; losses during assimilation and respiration are not shown. $o(t)$ is output from the pond in terms of caloric value of fish caught by man. The model suggests an interesting possibility that fish production might be increased if the "side food chain" through *Chaoborus* were eliminated, but the possibility must also be considered that this side chain enhances the stability of the system. (Data from Welch, 1967, with his estimate of assimilated energy at the i_2 level changed to estimated ingested energy on the basis of a 60 per cent assimilation efficiency for zooplankton and 40 per cent for bloodworms.)

the successive inputs of ingested energy with respect to time, i(t), are shown; losses during respiration and assimilation are not shown. As shown, the phytoplankton is fed upon in turn by the zooplankton crustacea in the water column, and the plankton detritus is taken by certain benthic invertebrates, notably blood worms, or chironomids, which are the preferred food of sunfishes; these in turn are fed upon by bass. The balance between the last two food chain groups (i.e., sunfish-bass) is very important as far as harvest by man is concerned. Thus, a pond with sunfish as the only fish could actually produce a greater total weight or biomass of fish than one with bass and sunfish, but most of the sunfish would remain small because of high reproduction rate and competition for available food. Fishing by hook and line would soon

be poor. Since man wants his fish in large sizes, and not as "sardines," the final predator is necessary for a good sport-fishing pond (see page 416).

Fish ponds are good places to demonstrate how secondary productivity is related to (1) the length of the food chain, (2) the primary productivity, and (3) the nature and extent of energy imports from outside the pond system. As shown in Table 3–11, large lakes and the sea yield less fish on an acre, or square meter, basis than do small, fertilized and intensively managed ponds not only because primary productivity is less and food chains longer, but because man only harvests a part of the consumer population in large bodies of water. Likewise, yields are several times greater when herbivores, such as carp, are stocked as when carnivores, such as bass, are harvested; the

Table 3–11. Secondary Productivity as Measured in Fish Production

	MAN'S HARVEST	
ECOSYSTEM AND TROPHIC LEVEL	*Lbs/acre/year*	*Kcal/m²/year*
I. Unfertilized Natural Waters		
Mixed Carnivores (Natural Populations)		
World marine fishery (average)*	1.5	0.3
North Sea†	27.0	5.0
Great Lakes‡	1–7	0.2–1.6
African lakes§	2–225	0.4–50
U. S. small lakes ‖	2–160	0.4–36
Stocked Carnivores		
U. S. fish ponds (sports fishery)§	40–150	9–34
Stocked Herbivores		
German fish ponds (carp)¶	100–350	22–80
II. Peru Current Upwelling Area (Anchovies)		
Heavy Natural Fertilization#	1,500	335
III. Artificially Fertilized Waters		
Stocked Carnivores		
U. S. fish ponds (sports fishery)**	200–500	45–112
Stocked Herbivores		
Philippine marine ponds (milkfish)§	500–1,000	112–202
German fish ponds (carp)¶	1,000–1,500	202–336
IV. Fertilized Waters—*Outside Food Added*		
Carnivores		
One-acre pond, U.S. ¶	2,000	450
Herbivores		
Hong Kong§	2,000–4,000	450–900
South China§	1,000–13,500	202–3,024
Malaya§	3,500	785

* 60×10^6 metric tons harvested (FAO 1967 Prod. Handb.) from 360×10^6 km² total ocean.
† FAO Statistics.
‡ Rawson, 1952.
§ Hickling, 1948.
‖ Rounsefell, 1946.
¶ Viosca, 1936.
# World's most productive natural fishery, 10^7 metric tons from 6×10^{10} m² (Ryther, 1969).
** Swingle and Smith, 1947.

latter, of course, require a longer food chain. The high yields in Section IV of Table 3–11 are obtained by adding food from outside the ecosystem, that is, plant or animal products which represent energy fixed somewhere else. Actually, such yields should not be expressed on an area basis unless one adjusts the area to include the land from which the supplemental food was obtained. Many people have misinterpreted the high yields obtained in the orient, thinking that they could be compared directly with fish pond yields in the United States where outside food is not usually provided. As might be expected fish culture depends on the human population density. Where man is crowded and hungry, ponds are managed for their yields of herbivores or detritus consumers; yields of 1000 to 1500 pounds per acre are easily obtainable without supplemental feeding. Where man is not crowded nor hungry, sport fish are desired; since these fish are usually carnivores produced at the end of a long food chain, yields are much less — 100 to 500 pounds per acre. Finally, the 300 kcal per m² per year fish yield from the most fertile natural waters or ponds managed for short food chains approaches the 10 per cent conversion of net primary production to primary consumer production (compare Tables 3–3C and 3–7 with Table 3–11) as suggested by the generalized model of Figure 3–6.

In addition to the operation of the second law of thermodynamics, size of food is one of the main reasons underlying the existence of food chains, as Elton (1927) pointed out. This is because there are usually rather definite upper and lower limits to the size of food that can efficiently support a given animal type. The matter of size is involved, also, in the difference between a predator chain and a parasite chain; in the latter, organisms at successive levels are smaller and smaller instead of being generally larger and larger. Thus, roots of vegetable crops are parasitized by nematodes which may be attacked by bacteria or other smaller organisms. Mammals and birds are commonly parasitized by fleas, which in turn have protozoan parasites of the genus *Leptomonas*, to cite another example. However, from the energy standpoint there is no fundamental difference between predator and parasite chains, since both parasites and predators are "consumers." For this reason no distinction has been made in the energy flow diagrams; a parasite of a

green plant would have the same position in this diagram as a herbivore, whereas animal parasites would fall in the various carnivore categories. Theoretically, parasite chains should be shorter on the average than predator chains since the metabolism per gram increases sharply with the diminishing size of the organism, resulting in a rapid decline in the biomass that can be supported, as will be discussed in the next section of this chapter.

A good example of a detritus food chain is one based on mangrove leaves as worked out in considerable detail by Heald (1969) and W. E. Odum (1970). In the brackish zone of southern Florida, leaves of the red mangrove (*Rhizophore mangle*) fall into the warm, shallow waters at an annual rate of 9 metric tons per hectare (about 2.5 gms or 11 kcal per m² per day) in those areas occupied by stands of mangrove trees. Since only 5 per cent of the leaf material was found to have been removed by grazing insects before leaf abscission, most of the annual net production becomes widely dispersed by tidal and seasonal currents over many square miles of bays and estuaries. As shown in Figure 3–12A, a key group of small animals, comprising only a few species but very large numbers of individuals, ingest large quantities of the vascular plant detritus along with the associated microorganisms — and also smaller quantities of algae. The particles ingested by the detritus consumers (detritivores or saprotrophs) range from sizable leaf fragments to tiny clay particles on which organic matter has been sorbed. These particles pass through the guts of many individuals and species in succession (i.e., the process of coprophagy; see page 31), resulting in repeated removal and regrowth of microbial populations (or repeated extraction and reabsorption of organic matter) until the substrate has been exhausted. Protein enrichment, as previously described (see Figure 2–10, page 30), was found to occur as the dead mangrove leaves were converted to detritus. Stomach contents of more than 100 species of fish were examined in detail during the study. Almost without exception small fishes were found to be feeding on the detritus consumers and the larger game fishes, in turn, to be feeding on the smaller ones (Fig. 3–12A). The study demonstrated that mangroves, which have been generally considered to have little economic value, actually make a substantial contribution to the food chain that

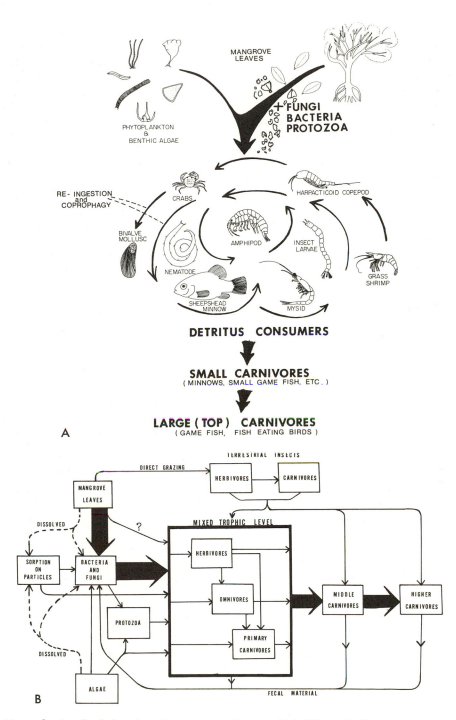

Figure 3–12.   A detritus food chain based on mangrove leaves which fall into shallow estuarine waters of south Florida. Leaf fragments acted on by saprotrophs and colonized by algae are eaten and re-eaten (coprophagy) by a key group of small detritus consumers which, in turn, provide the main food for game fish, herons, storks and ibis. A "picture model" of the food chain is shown in *A* (upper diagram) and a "compartment model" in *B* (lower diagram). (Redrawn from W. E. Odum, 1970.)

supports the fisheries that are important to the economy of southern Florida (Heald and Odum, 1970). In a similar manner, detritus from seagrasses, saltmarsh grasses and seaweeds (large algae) support fisheries in other estuarine areas (Darnell, 1958, 1967; E. P. Odum and de la Cruz, 1967; Wood *et al.*, 1970). Teal (1962) has constructed a detailed energy flow chart for the Georgia salt marsh estuary based on coordinated studies by a number of investigators. The key link between primary production and fish production in this ecosystem is a group of algal-detritus feeders comprising small crabs, nematodes, polychaete worms, grass shrimp, and snails. As is so often the case, the true nature of the ecology of a complex system is not evident from superficial examination!

Thus, we see that the detritus chain ends up in a manner similar to the grazing pathway (big fish eat little fish, so to speak), but the way in which the two chains begin is quite different. As shown in the more detailed model of the mangrove system, Figure 3–12*B*, detritus consumers, in contrast to grazing herbivores, are a mixed group in terms of trophic levels. As a group, the detritus feeders obtain some of their energy directly from plant material, most of it secondarily from microorganisms, and some tertiarily through carnivores (for example, by eating protozoa or other small invertebrates that have fed on bacteria that have digested plant material). Since detritus feeders may assimilate only a small portion of the bulky and chemically complex materials they in-

gest, apportioning the energy sources utilized at the individual and species level presents a difficult technical problem which has not been solved. Such information is not necessary to model the system, however, since we can consider the whole group as a convenient "black box." So far as is now known, the model of Figure 3–12*B* can serve equally well for a forest or grassland and for an estuary: the flow patterns would be expected to be the same; only the species would be different.

The distribution of energy, of course, is not the only quantity influenced by food chain phenomena, as will be made evident in subsequent chapters. At this point we need only emphasize that some substances become concentrated instead of dispersed with each link in the chain. What has come to be known as *food chain concentration*, or in the popular press as *biological magnification*, is dramatically illustrated by the behavior of certain persistent radionuclides and pesticides. Examples of the accumulation of radioactive wastes along food chains are given in Chapter 17, and an example of similar buildup of DDT is shown in Table 3–12. To control mosquitoes on Long Island, DDT was for many years sprayed on the marshes. Insect control specialists were careful to use spray concentrations that were not directly lethal to fish and other wildlife, but they failed to reckon with ecological processes and with the fact that DDT residues remain toxic for long periods of time. Instead of being washed out to sea, as some predicted, the poisonous residues

Table 3–12. *An Example of Food Chain Concentration of a Persistent Pesticide, DDT**

	PPM† DDT RESIDUES
Water	0.00005
Plankton	0.04
Silverside Minnow	0.23
Sheephead Minnow	0.94
Pickerel (predatory fish)	1.33
Needlefish (predatory fish)	2.07
Heron (feeds on small animals)	3.57
Tern (feeds on small animals)	3.91
Herring Gull (scavenger)	6.00
Fish Hawk (Osprey) egg	13.8
Merganser (fish-eating duck)	22.8
Cormorant (feeds on larger fish)	26.4

* Data from Woodwell, Wurster, and Isaacson, 1967.

† Parts per million (ppm) of total residues, DDT + DDD + DDE (all of which are toxic), on a wet weight, whole organism basis.

adsorbed on detritus, became concentrated in the tissues of detritus feeders and small fishes, and again concentrated in the top predators such as fish-eating birds. The concentration factor (ratio of ppm in organism to ppm in water) is about half a million times for fish-eaters in the case shown in Table 3–12. In hindsight, a study of the detritus food chain model, such as shown in Figure 3–12, would indicate that anything which sorbs readily on detritus and soil particles and is dissolved in guts would become concentrated by the ingestion-reingestion process that goes on at the beginning of the detritus food chain. Such a buildup of DDT on detritus has been documented by W. E. Odum, Woodwell, and Wurster (1969). The magnification is compounded in fish and birds by the extensive deposition of body fat in which DDT residue accumulates. The end result of the widespread use of DDT is that whole populations of predatory birds such as the fish hawk (osprey) and of detritus feeders such as fiddler crabs are being wiped out. It has been shown that birds are especially vulnerable to DDT poisoning because DDT (and other chlorinated hydrocarbon insecticides as well) interferes with egg shell formation by causing a breakdown in steroid hormones (see Peakall, 1967; Hickey and Anderson, 1968), which results in fragile eggs that break before the young can hatch. Thus, very small amounts that are not lethal to the individual turn out to be lethal to the population! It was scientific documentation of this sort of frightening buildup (frightening because man is also a "top carnivore"!) and unanticipated physiological effects that finally marshalled public opinion for restricting the use of DDT and similar pesticides. As we shall see in Chapter 16, there are other and better alternatives for insect control that do not require that we poison a whole food chain in order to control one pest species!

The principle of biological magnification (see review by Woodwell, 1967) is a good one to consider in any waste management strategy. It should be pointed out, however, that many nonbiological factors may either reduce or augment the concentration factor. Thus, man ingests less DDT than fish hawks in part because food processing and cooking removes some of the materials. On the other hand, a fish is in double jeopardy because it may become contaminated by direct absorption from the environment through the gills or skin as well as by way of its food.

Although the buildup of radionuclide wastes or fallout in food chains is a matter for concern, the experimental use of radionuclide tracers has proved to be a great help in charting network details in complex food webs of intact ecosystems. By labeling a given energy source or a species population, that portion of the food web supported by the source can be "traced out" or "isolated," as it were, by following the transfer of radioactive material in the intact, whole system without very much disturbance of the system's function (E. P. Odum and Kuenzler, 1963; Wiegert, Odum, and Schnell, 1968; Wiegert and Odum, 1969). Examples of such studies are described in Chapter 17 (see especially Figure 17–6, page 462).

Ecological Efficiencies

Ratios between energy flow at different points along the food chain are of considerable ecological interest. Such ratios, when expressed as percentages, are often called "ecological efficiencies." In Table 3–13 some of these ratios are listed and defined in terms of the energy flow diagram. For the most part, these ratios are meaningful in reference to component populations as well as to whole trophic levels. Since the several types of efficiencies are often confused, it is important to define exactly what relationship is meant; the energy flow diagram (Figs. 3–6 and 3–9) is a great help in this clarification.

Most importantly, efficiency ratios are meaningful in comparison only when they are dimensionless, that is, when the numerator and denominator of each ratio are expressed in the same unit. Otherwise, statements about efficiency can be very misleading. For example, poultrymen may speak of a 40 per cent efficiency in the conversion of chicken feed to chickens (the P_t/I_t ratio in Table 3–13), but this turns out to be a ratio of "wet" chicken (worth about 2 Kcal per gm) to dry feed (worth 4 + Kcal per gm). The true growth efficiency in terms of Kcal/Kcal in this case is more on the order of 20 per cent. Wherever possible, ecological efficiencies should be expressed in the "energy currency" (i.e., Calories to Calories).

The general nature of transfer efficiencies between trophic levels has already been discussed, that is, the 1 to 5 per cent P_g/L,

Table 3–13. *Various Types of Ecological Efficiencies*

Symbols are as follows (see Figure 3–6): L—light (total); L_A—absorbed light; P_G—total photosynthesis (gross production); P—production of biomass; I—energy intake; R—respiration; A—assimilation; NA—ingested but not assimilated; NU—not used by trophic level shown; t—trophic level; t-1—preceding trophic level.

RATIO	DESIGNATION AND EXPLANATION
A. Ratios Between Trophic Levels	
$\dfrac{I_t}{I_{t-1}}$	Trophic level energy intake (or Lindeman's) efficiency. For the primary level this is $\dfrac{P_G}{L}$ or $\dfrac{P_G}{L_A}$
$\dfrac{A_t}{A_{t-1}}$	Trophic level assimilation efficiency ⎫
$\dfrac{P_t}{P_{t-1}}$	Trophic level production efficiency ⎬ For the primary level P and A may be in terms of either L or L_A as above; $A_t/A_{t-1} = I_t/I_{t-1}$ for the primary level, but not for secondary levels.
$\dfrac{I_t}{P_{t-1}}$ or $\dfrac{A_t}{P_{t-1}}$	Utilization efficiencies
B. Ratios Within Trophic Levels	
$\dfrac{P_t}{A_t}$	Tissue growth efficiency
$\dfrac{P_t}{I_t}$	Ecological growth efficiency
$\dfrac{A_t}{I_t}$	Assimilation efficiency

the 2 to 10 per cent P_g/L_A and the 10 to 20 per cent production efficiencies between secondary trophic levels. Assimilation and growth efficiencies (ratios within trophic levels) are often on the order of 10 to 50 per cent but they can be much higher. Organisms feeding on very highly nutri-

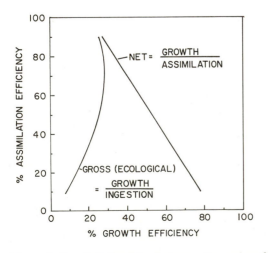

Figure 3–13. Relationships between tissue growth efficiency, ecological growth efficiency, and assimilation efficiency in animal populations. (After Welch, 1968.)

tious food can assimilate up to 100 per cent of what they ingest. In other cases a large percentage of assimilated energy can go into growth if "antithermal" maintenance requirements imposed by the environment are greatly reduced and if the organisms are small. In nature, there is a tendency for an inverse relationship to exist between tissue growth efficiency and assimilation efficiency in animals, as shown in Figure 3–13 (see E. P. Odum and A. L. Smalley, 1957; Welch, 1968). In contrast, ecological growth efficiency tends to remain more constant, at around 20 per cent (Slobodkin, 1961, 1962). More study is needed to determine the significance of these trends, especially since the literature reveals quite a few apparent exceptions to the model.

The very low primary efficiencies which are characteristic of intact natural systems have puzzled many persons in view of the relatively high efficiencies obtained in electric motors and other mechanical systems. This has, quite naturally, led many to consider seriously ways of increasing nature's efficiency. Actually, long-time, large scale ecosystems are not directly comparable to short-time mechanical systems in this regard. For one thing, a considerable portion of fuel goes for repair and maintenance in living systems, and depreciation and repair are not included in calculating fuel efficiencies of engines. In other words, a lot of energy (human or otherwise) other than fuel is required to keep machines running, repaired, and replaced; it is not fair to compare engines and biological systems unless this is considered, because the latter are self-repairing and self-perpetuating. Secondly, it is likely that more rapid growth per unit time has greater survival value than maximum efficiency in the use of fuel. Thus, to use a simple analogy, it might be more important to reach a destination quickly at fifty miles per hour than to achieve maximum efficiency in fuel consumption by driving slowly! It is important for engineers to understand that any increase in the efficiency of a biological system will be obtained at the expense of maintenance. There always comes a point where a gain from increasing the efficiency will be lost in increased cost, not to mention the danger of increased disorder that may result from oscillations.

For further discussion of the idea that low efficiency is necessary for maximum power output, see Odum and Pinkerton (1955) and H. T. Odum (1971).

5. METABOLISM AND SIZE OF INDIVIDUALS

Statement

The standing crop biomass (expressed as the total dry weight or total caloric content of organisms present at any one time) which can be supported by a steady flow of energy in a food chain depends to a considerable extent on the size of the individual organisms. The smaller the organism, the greater its metabolism per gram (or per Calorie) of biomass. Consequently, the smaller the organism, the smaller the biomass which can be supported at a particular trophic level in the ecosystem. Conversely, the larger the organism, the larger the standing crop biomass. Thus, the amount of bacteria present at any one time would be very much smaller than the "crop" of fish or mammals even though the energy utilization was the same for both groups.

Explanation and Examples

The metabolism per gram of biomass of the small plants and animals such as algae, bacteria, and protozoa is immensely greater than the metabolic rate of large organisms such as trees and vertebrates. This applies to both photosynthesis and respiration. In many cases the important parts of the community metabolically are not the few great conspicuous organisms but the numerous tiny organisms which are often invisible to the naked eye. Thus, the tiny algae (phytoplankton), comprising only a few pounds per acre at any one moment in a lake, can have as great a metabolism as a much larger volume of trees in a forest or hay in a meadow. Likewise, a few pounds of small crustacea (zooplankton) "grazing" on the algae can have a total respiration equal to many pounds of cows in a pasture.

The rate of metabolism of organisms or association of organisms is often estimated by measuring the rate at which oxygen is consumed (or produced, in the case of photosynthesis). There is a broad general tendency for the metabolic rate per organism in animals to increase as the two-thirds power of the volume (or weight) increases, or the metabolic rate per gram biomass to decrease inversely as the length (Zeuthen, 1953; Bertalanffy, 1957; Kleiber, 1961). A similar relationship appears to exist in plants, although structural differences in plants and animals (see below) make direct comparisons in terms of volume and length difficult. These relations are shown in Figure 3–14 by the smoothed lines which indicate in an approximate manner the relation between size and metabolism. Various theories proposed to account for these trends have centered around diffusion processes; larger organisms have less surface area per gram through which diffusion processes might occur. However, the real explanation for the relationship between size and metabolism has not been agreed upon. Comparisons, of course, should be made at similar temperatures because metabolic rates are usually greater at higher temperatures than at lower temperatures (except with temperature adaptation; see page 109).

It should be pointed out that when organisms of the same general order of magnitude in size are compared the linear relationship shown in Figure 3–14 does not always hold. This is to be expected since there are many factors, secondary to size,

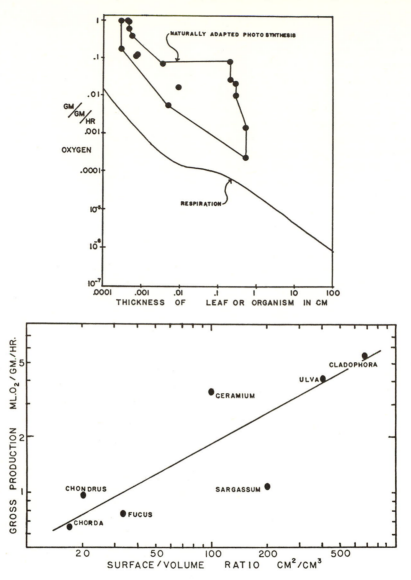

Figure 3–14. Metabolism per gram biomass as a function of organismal size. From top to bottom the curves represent: photosynthetic rate per gram biomass in a variety of algae and leafy plants in relation to length or thickness of leaf, respiration per gram biomass in heterotrophic organisms (animals and bacteria) in relation to the length of the organism, and gross production of several species of large marine seaweeds in relation to surface-to-volume ratio. In each case small organisms or thin organisms with a high surface-to-volume ratio have higher rates of metabolism per gram than large or thick organisms. Upper graph after H. T. Odum (1956a) from data of Verduin and Zeuthen. Lower graph after E. P. Odum *et al.* (1958).

which affect the rate of metabolism. For example, it is well known that warmblooded vertebrates have a greater respiration rate than coldblooded vertebrates of the same size. However, the difference is actually relatively small as compared with the difference between a vertebrate and a bacterium. Thus, given the same amount of available food energy, the standing crop of coldblooded herbivorous fish in a pond may be of the same order of magnitude as that of warmblooded herbivorous mammals on land. However, as already pointed out in Chapter 2, oxygen is less available in water than in air and is therefore more likely to be limiting in water. In general, aquatic animals seem to have a lower weight-specific respiratory rate than terrestrial animals of the same size. Such an adaptation may well affect the trophic structure (see Misra *et al.*, 1968).

In the study of size-metabolism in plants,

it is often difficult to decide what constitutes an "individual." Thus, we may commonly regard a large tree as one individual, but actually the leaves may act as "functional individuals" as far as size-surface area relationships are concerned (recall the leaf area index concept, Figure 3–2). In a study of various species of seaweeds (large multicellular algae), we found (see Figure 3–14) that species with thin or narrow "branches" (and consequently a high surface-to-volume ratio) had a higher rate per gram biomass of food manufacture, respiration, and uptake of radioactive phosphorus from the water than did species with thick branches (E. P. Odum, Kuenzler, and Blunt, 1958). Thus, in this case the "branches" or even the individual cells were "functional individuals" and not the whole "plant," which might include numerous "branches" attached to the substrate by a single holdfast.

The inverse relationship between size and metabolism may also be observed in the ontogeny of a single species. Eggs, for example, usually show a higher rate per gram than the larger adults. In data reported by Hunter and Vernberg (1955), the metabolism per gram of trematode parasites was found to be ten times less than that of the small larval cercariae.

To avoid confusion, it must be reiterated that it is the weight-specific metabolic rate that decreases with increasing size, not the total metabolism of the individual. Thus, an adult man requires more total food than a small child, but less food per pound of body weight.

6. TROPHIC STRUCTURE AND ECOLOGICAL PYRAMIDS

Statement

The interaction of the food chain phenomena (energy loss at each transfer) and the size-metabolism relationship results in communities having a definite *trophic structure*, which is often characteristic of a particular type of ecosystem (lake, forest, coral reef, pasture, etc.). Trophic structure may be measured and described either in terms of the standing crop per unit area or in terms of the energy fixed per unit area per unit time at successive trophic levels. Trophic structure and also trophic function

may be shown graphically by means of *ecological pyramids* in which the first or producer level forms the base and successive levels the tiers which make up the apex. Ecological pyramids may be of three general types: (1) the *pyramid of numbers* in which the number of individual organisms is depicted, (2) the *pyramid of biomass* based on the total dry weight, caloric value, or other measure of the total amount of living material, and (3) the *pyramid of energy* in which the rate of energy flow and/or "productivity" at successive trophic levels is shown. The numbers and biomass pyramids may be inverted (or partly so), that is, the base may be smaller than one or more of the upper tiers if producer organisms average smaller than consumers in individual size. On the other hand, the energy pyramid must always take a true upright pyramid shape, provided all sources of food energy in the system are considered.

Explanation and Examples

Ecological pyramids are illustrated in Figure 3–15, and the three kinds are compared in a hypothetical model in Figure 3–16.

The pyramid of numbers is actually the result of three phenomena which usually operate simultaneously. One of these phenomena is the familiar geometrical fact that a great many small units are required to equal the mass of one big unit, regardless of whether the units are organisms or building blocks. Thus, even if the weight of large organisms were equal to the weight of the smaller ones, the number of smaller organisms would be vastly greater than that of the larger ones. Because of the geometry, therefore, the existence of a valid pyramid of numbers in a natural group of organisms does not necessarily mean that there are fewer of the larger organisms on a weight basis.

The second phenomenon contributing to the pattern of many small organisms and few large ones is the food chain. As pointed out in Section 2, useful energy is always lost (into the form of heat) in the transfer through each step in the food chain. Consequently, except where there are imports of organic matter (as in Table 3–11, IV) there is much less energy available to the higher trophic levels. The third factor involved in the pyra-

A. PYRAMID OF NUMBERS. Individuals (exclusive of microorganisms and soil animals) per 0.1 hectare

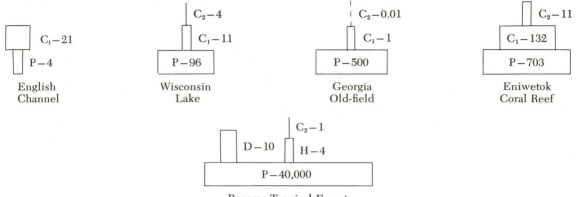

B. PYRAMID OF BIOMASS. Grams dry weight per square meter

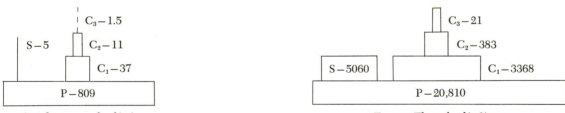

C. COMPARISON OF STANDING CROP AND ENERGY-FLOW PYRAMIDS FOR SILVER SPRINGS, FLORIDA

D. SEASONAL CHANGE IN BIOMASS PYRAMID IN THE WATER COLUMN (NET PLANKTON ONLY) OF AN ITALIAN LAKE. Milligrams dry weight per cubic meter

E. "SUBPYRAMIDS" OF NUMBERS, BIOMASS, AND RESPIRATORY ENERGY FOR SOIL ARTHROPODS IN A MICHIGAN OLD-FIELD

Figure 3–15. See opposite page for legend.

mid of numbers is the inverse size-metabolic rate pattern discussed in the previous section.

Actually, the pyramid of numbers is not very fundamental or instructive as an illustrative device since the relative effects of the "geometric," "food chain," and "size" factors are not indicated. The form of the numbers pyramid will vary widely with different communities, depending on whether producing individuals are small (phytoplankton, grass) or large (oak trees). Likewise, numbers vary so widely that it is difficult to show the whole community on the same numerical scale. This does not mean that the number of individuals present is of no interest, but rather that such data are probably best presented in tabular form.

The pyramid of biomass is of more fundamental interest since the "geometric" factor is eliminated, and the quantitative relations of the "standing crop" are well shown. In general, the biomass pyramid gives a rough picture of the overall effect of the food chain relationships for the ecological group as a whole. When the total weight of individuals at successive trophic levels is plotted, a gradually sloping pyramid may be expected as long as the size of the organisms does not differ greatly. However, if organisms of lower levels average much smaller than those of higher levels, the biomass pyramid may be inverted. For example, where size of the producers is very small and that of the consumers large, the total weight of the latter may be greater at any one moment. In such cases, even though more energy is being passed through the producer trophic level than through consumer levels (which must always be the case), the rapid metabolism and turnover of the small producer organisms accomplish a larger output with a smaller standing crop biomass. Examples of inverted biomass pyramids are shown in Figure 3–15B, D. In lakes and in the sea the plants (phytoplankton) usually outweigh their grazers (zooplankton) during periods of high primary productivity, as during the spring "bloom," but at other times, as in winter, the reverse may be true, as is shown in the example taken from an Italian lake (see also Pennak, 1955; Fleming and Laevastu, 1956).

If, for the moment, we may assume that the examples in Figure 3–15 are representative of the range of situations to be expected, we may make the following generalizations: (1) In terrestrial and shallow water ecosystems, where producers are large and relatively long-lived, a broad-based, relatively stable pyramid is to be expected. Pioneer or newly established communities will tend to have fewer consumers in proportion to producers (i.e., the apex of the biomass pyramid will be small) as illustrated by the "old-field" pyramid in comparison with that of the coral reef in Figure 3–15B. Generally speaking, consumer animals in terrestrial and shallow water communities have more complicated life histories and habitat requirements (specialized shelter, etc.) than do green plants; hence animal populations may require a longer period of time for maximum development. (2) In open water or deep water situations where producers are small and short-lived the standing crop situation at any one moment may be widely variable and the biomass pyramid may be inverted. Also, the overall size of the total standing crop will likely be smaller (as indicated graphically by the area of the biomass pyramid) than that of land or shallow water communities, even if the total energy fixed annually is the same. Finally, (3) lakes and ponds, where both large rooted plants and tiny algae are important, may be expected to have an intermediate arrangement of standing crop units (as illustrated by the Wisconsin Lake, Figure 3–15B).

Of the three types of ecological pyramids, the energy pyramid gives by far the best overall picture of the functional nature of communities since the number and weight of organisms that can be supported at any level in any situation depends not on the amount of fixed energy present at any one

Figure 3–15. Ecological pyramids of numbers, biomass, and energy in diverse ecosystems ranging from open-water types to large forests. P = producers; C_1 = primary consumers; C_2 = secondary consumers; C_3 = tertiary consumers (top carnivores); S = saprotrophs (bacteria and fungi). D = "decomposers" (bacteria, fungi + detritivores). Pyramids are somewhat generalized, but each is based on specific studies as follows: A. Grassland plant data from Evans and Cain, 1952; animal data from Wolcott, 1937; temperate forest is based on Wytham woods, near Oxford, England, as summarized by Elton, 1966, and Varley, 1970. B. English Channel, Harvey, 1950; Wisconsin lake (Weber Lake), Juday, 1942; Georgia old-field, E. P. Odum, 1957; coral reef, Odum and Odum, 1955; Panama forest, F. B. Golley and G. Child (unpublished). C. Silver Springs, H. T. Odum, 1957. D. Italian lake (Lago Maggiore), Ravera, 1969. E. Soil arthropods, Engelmann, 1968.

time in the level just below but rather on the *rate* at which food is being produced. In contrast with the numbers and biomass pyramids, which are pictures of the standing states, i.e., organisms present at any one moment, the energy pyramid is a picture of the rates of passage of food mass through the food chain. Its shape is not affected by variations in the size and metabolic rate of individuals, and, if all sources of energy are considered, it must always be "right side up" because of the second law of thermodynamics.

Comparison of biomass and energy pyramids (see Figure 3–15) for the rich and beautiful Silver Springs, Florida, which is visited by thousands of tourists annually, is especially interesting since an estimate of all of the community, including the decomposers, is shown. Beds of freshwater eelgrass (*Sagittaria*) and attached algae make up the bulk of the standing crop of producers in this spring in which numerous aquatic insects, snails, herbivorous fish and turtles comprise the primary consumers. Other fish and invertebrates form the smaller "crop" of secondary consumers, and bass and gar are the chief "top carnivores." Animal parasites were included in the latter level. Since the decomposers, or saprovores, are primarily concerned with the breaking down of the large bulk of plants but also decompose all other levels as well, it is logical to show this component as a tall bar resting on the primary trophic level but extending to the top of the pyramid as well. Actually the biomass of bacteria and fungi is very small in rela-

tion to their importance in the energy flow of the community. Thus, the pyramid of numbers greatly overrates the microscopic saprovores and the pyramid of biomass greatly underrates them. Neither numbers nor weights, in themselves, have much meaning in determining the role of microorganism decomposers in community dynamics; only measurements of actual energy utilization, as shown on the energy pyramid, will place the microconsumers in true relationship with the macroscopic components.

The concept of energy flow not only provides a means of comparing ecosystems with one another, but also a means of evaluating the relative importance of populations. Table 3–14 lists estimates of density, biomass and energy flow rates for six populations that differ widely in size of individual and in habitat. In this series, numbers vary 17 orders of magnitude (10^{17}) and biomass about 5 (10^5), whereas energy flow varies only about fivefold. Similarity of energy flow indicates that all six populations are functioning at approximately the same trophic level (primary consumers), even though neither numbers nor biomass indicate this. We can state a sort of "ecological rule" something like this: *Numbers overemphasize the importance of small organisms, and biomass overemphasizes the importance of large organisms;* hence, we cannot use either as a reliable criterion for comparing the functional role of populations that differ widely in size-metabolism relationships, although of the two bio-

Table 3–14. *Density, Biomass, and Energy Flow of Five Primary Consumer Populations Differing in the Size of Individuals Comprising the Population**

	APPROXIMATE DENSITY (M^2)	BIOMASS (G/M^2)	ENERGY FLOW ($KCAL/M^2/DAY$)
Soil bacteria	10^{12}	0.001	1.0
Marine copepods (*Acartia*)	10^5	2.0	2.5
Intertidal snails (*Littorina*)	200	10.0	1.0
Salt marsh grasshoppers (*Orchelimum*)	10	1.0	0.4
Meadow mice (*Microtus*)	10^{-2}	0.6	0.7
Deer (*Odocoileus*)	10^{-5}	1.1	0.5

* After E. P. Odum, 1968.

Table 3–15. Comparison of Total Metabolism and Population Density of Soil Microorganisms Under Conditions of Low and High Organic Matter[*]

	NO MANURE ADDED TO THE SOIL	MANURE ADDED TO THE SOIL
Energy Dissipated: Kilocalories × 10⁶/acre/year	1	15
Average Population Density: Number/Gram of Soil		
Bacteria, ×10⁸	1.6	2.9
Fungi mycelia, ×10⁶	0.85	1.01
Protozoa, ×10³	17	72

[*] Data from Russell and Russell, 1950.

mass is generally more reliable than numbers. *However, energy flow (i.e., P + R) provides a more suitable index for comparing any and all components of an ecosystem.*

The data in Table 3–15 provide a further illustration of how the activities of decomposers and other small organisms may bear very little relation to the total numbers or biomass that are present at any one moment. Note that a fifteenfold increase in energy dissipated, which resulted from the addition of organic matter, was accompanied by less than a twofold increase in the number of bacteria and fungi. In other words, these small organisms merely "turnover" faster when they become more active, and do not increase their standing crop biomass proportionally as do large organisms. The protozoa, being somewhat larger than the bacteria, exhibited a somewhat greater increase in numbers.

In Figure 3–16 data for a hypothetical alfalfa-calf-boy food chain situation are arranged in the form of all three types of pyramids. In the energy pyramid respiration is not included, only production of new biomass. While cows do not usually subsist entirely on alfalfa, nor boys entirely on meat, these diagrams are realistic *models* of the kind of ecosystem in which man is not merely an interested observer (as in Silver Springs) but is himself a vital part. Thus, most of man's domestic food animals are herbivores, and man could satisfy his nutritional requirements on a diet of meat (especially if he consumed various parts of the animal, not just muscle tissue). He could also be healthy as a vegetarian and assume the place of the calf, in which case more people could be supported with the same basic primary fixation of energy. Actually, of course, man usually occupies a variable position intermediate between a herbivore and a carnivore. The diagrams in Figure 3–16 also indicate the general situation to be expected in those terrestrial communities in which producers and consumers are relatively large in individual size.

As graphic devices, ecological pyramids can also be used to illustrate quantitative relationships in specific parts of ecosys-

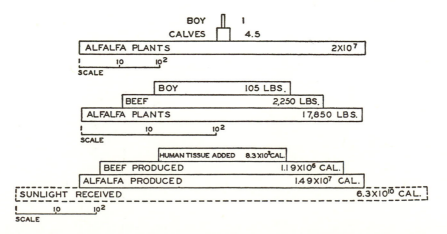

Figure 3–16. The three types of ecological pyramids illustrated for a hypothetical alfalfa-calf-boy food chain computed on the basis of 10 acres and one year and plotted on a log scale. Compiled from data obtained as follows: Sunlight: Haurwitz and Austin (1944), "Climatology." Alfalfa: "USDA Statistics, 1951"; "USDA Yearbook 1948"; Morrison (1947), "Feeds and Feeding." Beef calf: Brody (1945), "Bioenergetics and Growth." Growing boy: Fulton (1950), "Physiology"; Dearborn and Rothney (1941), "Predicting the Child's Development."

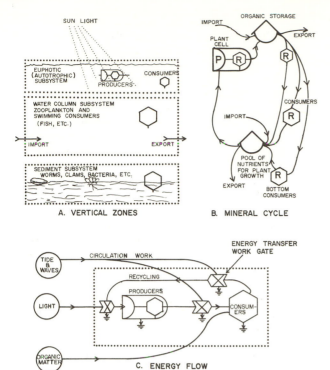

Figure 3–17. Three aspects of an estuarine ecosystem. *A.* Vertical zonation with photosynthetic production (*P*) above and most of the respiratory consumption (*R*) below. *B.* Mineral cycle with circulation of plant nutrients upward and organic matter food downward. *C.* Energy flow circuit diagram showing three main sources of potential energy input into the system. See text for explanation of symbols. (After H. T. Odum, Copeland and McMahon, 1969.)

tems in which one might have a special interest, for example, predator-prey or host-parasite groups. As already indicated a "parasite pyramid of numbers" would generally be reversed in contrast to biomass and energy pyramids. Unfortunately, almost no measurements have been made on entire populations of parasites and hyperparasites (parasites living on or in other parasites). One thing seems certain, however: one cannot take literally the well-known jingle by Jonathan Swift, or the whimsical diagram* of Hegner:

Big fleas have little fleas
Upon their backs to bite 'em
And little fleas have lesser fleas
And so, ad infinitum.

The number of levels or steps in the parasite chain or pyramid is not "ad infinitum" but is definitely limited, both by size relations and by our friend, the second law of thermodynamics!

---

* Reproduced from "Big Fleas Have Little Fleas, or Who's Who among the Protozoa," by Robert Hegner, Williams & Wilkins Co., 1938.

7. SUMMARIZATION: ECOSYSTEM ENERGETICS

Figure 3–17 can serve as a quick visual summary of the major ecosystem principles that have been reviewed in some detail in Chapters 2 and 3. The diagrams relate the vertical zonation (*A*), the cycling of materials (*B*), and the one-way flow of energy (*C*) in an estuary, which, as we have already pointed out, is an ecosystem-type intermediate between the great extremes of nature, the open sea and the forest. The diagrams introduce an energy circuit language devised by H. T. Odum (see Figure 10–7*A*, page 292) in which special symbols indicate biological structures and specific functions. The bullet-shaped modules represent the producers with their double metabolism, that is, P (production) and R (respiration). The hexagons are populations of consumers which possess storage, self-maintenance, and self-reproduction. The storage bins represent nutrient pools in and out of which move nitrogen, phosphorus, and other vital substances. In diagrams *B* and *C*, the lines represent "the invisible wires of nature" that link the components into a functional network. In diagram *C*, the

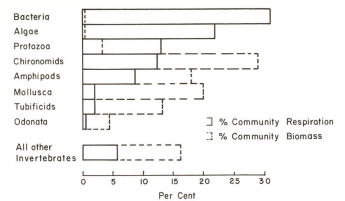

Figure 3–18. Partitioning of community respiration and community biomass (dry weight) between various members of the benthic community in a small lake in western Canada. "In situ" total community respiration at 15°C. was estimated to be 16 ml O_2 per m² per hr and the total benthic biomass to be 4.2 gm per m². (Drawn from data of Efford, 1969.)

"ground" symbol, or the "arrow into the heat sink," indicates where energy is dispersed and no longer available to the food chain. The circles show three types of energy sources that function as inputs into the ecosystem. The "work gate" symbols indicate where a flow of work energy along one pathway assists a second flow to pass over energy barriers. Note that some of the lines of flow of potential energy loop back from "downstream" energy sources to "upstream" inflows, there serving various roles including control functions (saprotrophs controlling photosynthesis by controlling the rate of mineral regeneration, for example). The diagram shows how the energy subsidy of the tide assists in (1) recycling nutrients from consumer to producer and (2) speeding up movement of plant food to the consumer. Reducing tidal flow by diking the estuary (see Figure 13–6, page 361) may reduce productivity just as surely as would cutting out some of the light. Stress resulting from pollution and harvest could be shown in the diagram by adding circles enclosing a negative sign linked with appropriate heat sinks to show where energy is diverted away from the ecosystem. As already indicated, both subsidies (+) and stresses (−) can be quantitated in terms of Calories added or diverted per unit of time and space. The application of energy circuit models to mathematical modeling will be explained in detail in Chapter 10.

The importance of considering size-metabolism relationships when it comes to evaluating the biological components of an ecosystem is emphasized by Figure 3–18. Without exception, any sizable, self-sustaining ecosystem will contain an array of organisms ranging in size from tiny microbes to large plants or animals, or both. As shown in the diagram, which is based on data from a pond community, the small organisms, that is, the bacteria, algae, and protozoa, account for most of the community respiration, whereas the larger invertebrate animals comprise most of the biomass. As we have emphasized many times, *any holistic or total appraisal of ecosystems must be based on coordinated measurements of standing crop structure and rates of function*, with measurements of the latter becoming increasingly necessary as the size of organism decreases. Some of the problems, and the prospects for improved technology, in dealing with the "microbial" component of the ecosystem are reviewed in Chapter 19.

PRINCIPLES AND CONCEPTS PERTAINING TO BIOGEOCHEMICAL CYCLES

Chapter 4

1. PATTERNS AND BASIC TYPES OF BIOGEOCHEMICAL CYCLES

Statement

The chemical elements, including all the essential elements of protoplasm, tend to circulate in the biosphere in characteristic paths from environment to organisms and back to the environment. These more or less circular paths are known as *biogeochemical cycles.* The movement of those elements and inorganic compounds that are essential to life can be conveniently designated as *nutrient cycling.* For each cycle it is also convenient to designate two compartments or pools: (1) the *reservoir pool,* the large, slow-moving, generally non-biological component, and (2) the *exchange or cycling pool,* a smaller but more active portion that is exchanging (i.e., moving back and forth) rapidly between organisms and their immediate environment. From the standpoint of the biosphere as a whole biogeochemical cycles fall into two basic groups: (1) *gaseous types,* in which the reservoir is in the atmosphere or hydrosphere (ocean), and (2) *sedimentary types,* in which the reservoir is in the earth's crust.

Explanation

As was emphasized in Section 2, Chapter 2, it is profitable in ecology to study not only organisms and their environmental relations but also the basic nonliving environment in relation to organisms. Of the 90-odd elements known to occur in nature, between 30 and 40 are known to be required by living organisms. Some elements such as carbon, hydrogen, oxygen, and nitrogen are needed in large quantities; others are needed in small, or even minute, quantities. Whatever the need may be, essential elements (as well as nonessential elements) exhibit definite biogeochemical cycles.

"Bio" refers to living organisms and "geo" to the rocks, air, and water of the earth. Geochemistry is an important physical science concerned with the chemical composition of the earth and with the exchange of elements between different parts of the earth's crust and its oceans, rivers, and other bodies of water (see review by Vallentyne, 1960). Biogeochemistry, a term made prominent by G. E. Hutchinson's early monographs (1944, 1950), thus becomes the study of the exchange or flux (that is, the back-and-forth movement) of materials between living and nonliving components of the biosphere.

In Figure 4–1 a biogeochemical cycle is superimposed on a simplified energy-flow diagram to show how the one-way flow of energy drives the cycle of matter. Elements in nature are never, or almost never, homogeneously distributed nor are they present in the same chemical form throughout the ecosystem. In Figure 4–1 the reservoir pool, that portion that is chemically or physically remote from organisms, is indicated by the box labeled "nutrient pool," whereas the cycling portion is designated by the stippled circle going from autotrophs to heterotrophs and back again. Sometimes the reservoir portion is called the "unavailable" pool and the active cycling pool the "available" or "exchangeable" pool; such a designation is permissible provided it is clearly understood that the terms are relative. An atom in the reservoir is not necessarily permanently unavailable to organisms, because there are slow fluxes between available and unavailable components.

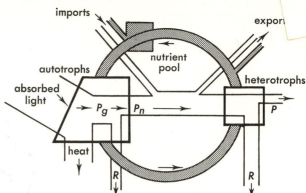

Figure 4-1. A biogeochemical cycle (stippled circle) superimposed upon a simplified energy-flow diagram, contrasting the cycling of material with the one-way flow of energy. P_g = gross production; P_n = net primary production, which may be consumed within the system by heterotrophs or exported from the system; P = secondary production; R = respiration. (After E. P. Odum, 1963.)

Some cycles, such as those involving carbon, nitrogen, or oxygen, self-adjust rather quickly to perturbations because of the large atmospheric reservoir. Local increases in CO_2 production by oxidation or combustion, for example, tend to be quickly dissipated by air movement and the increased output compensated for by increased plant uptake and carbonate formation in the sea. Gaseous-type cycles, then, can be considered to be relatively "perfect" in the global sense because of natural negative feedback control (recall page 34 for discussion of feedback). However, as already indicated, man is finding that local disruptions can be troublesome and that there are definite limits to the self-adjustment capacity of the atmosphere as a whole! Sedimentary cycles, which involve elements such as phosphorus or iron, tend to be much less perfect and more easily disrupted by local perturbations because the great bulk of material is in a relatively inactive and immobile reservoir in the earth's crust. Consequently, some portion of the exchangeable material tends to get "lost" for long periods of time when "downhill" movement is more rapid than "uphill" return. As we shall see, return or "recycle" mechanisms in many cases are chiefly biotic.

Hutchinson (1948a) points out that man is unique in that not only does he require the 40 essential elements but also in his complex culture he uses nearly all the other elements and the newer synthetic ones as well. He has so speeded up the movement of many materials that the cycles tend to become imperfect, or the process becomes "acyclic," with the result that man increasingly suffers from the paradoxical situation of too little here and too much there. For example, we mine and process phosphate rock with such careless abandon that severe local pollution results near mines and phosphate mills. Then, with equally acute myopia we increase the input of phosphate fertilizers in agricultural systems without controlling in any way the inevitable increase in run-off output that severely stresses our waterways and reduces water quality through eutrophication (see page 16 for definition of this term). The concept of man as "the mighty geological agent" was introduced in Chapter 2 (see page 35).

The aim of conservation of natural resources in the broadest sense is to make acyclic processes more cyclic. The concept of "recycle" must become a major goal for society. Recycling water is a good place to start, because if we can maintain and repair the hydrological cycle, we will have a better chance of controlling nutrients that move along with the water. More about this in Chapters 16 and 21.

Examples

Three examples will suffice to illustrate the principle of cycling. The nitrogen cycle (Fig. 4-2) is an example of a very complex gaseous-type cycle; the phosphorus cycle (Fig. 4-3) is an example of a simpler, possibly less perfect sedimentary type. As will be outlined in Chapter 5, both these elements are often very important factors, limiting or controlling the abundance of organisms, and hence they have received much attention and study. The sulfur cycle (Fig. 4-5) is a good one to illustrate linkage between air, water, and the earth's crust. Both the nitrogen and the sulfur cycles illustrate the key role played by microorganisms, and also the complications caused by industrial air pollution.

As shown in Figure 4-2A the nitrogen

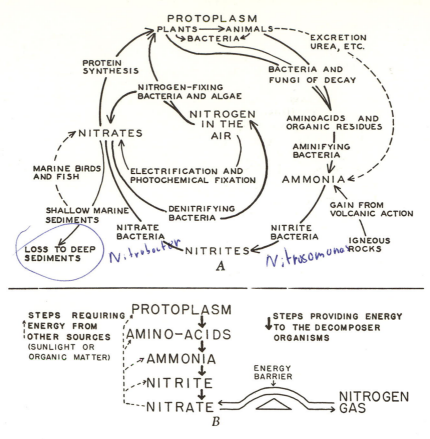

Figure 4–2. Two ways of picturing the nitrogen biogeochemical cycle, an example of a relatively perfect, self-regulating cycle with a large gaseous reservoir. In *A* the circulation of nitrogen between organisms and environment is depicted along with microorganisms which are responsible for key steps. In *B* the same basic steps are arranged in an ascending-descending series, with the high energy forms on top to distinguish steps which require energy from those which release energy.

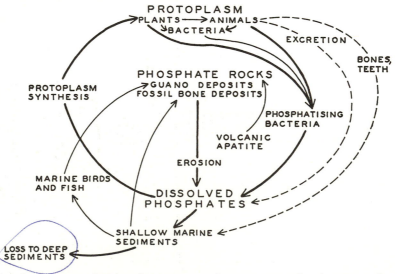

Figure 4–3. The phosphorus cycle. Phosphorus is a rare element compared with nitrogen. Its ratio to nitrogen in natural waters is about 1 to 23 (Hutchinson, 1944*a*). Chemical erosion in the United States has been estimated at 34 metric tons per square kilometer per year. Fifty-year cultivation of virgin soils of the Middle West reduced the P$_2$O$_5$ content by 36 per cent (Clarke, 1924). As shown in the diagram, the evidence indicates that return of phosphorus to the land has not been keeping up with the loss to the ocean.

in protoplasm is broken down from organic to inorganic form by a series of decomposer bacteria, each specilized for a particular part of the job. Some of this nitrogen ends up as nitrate, the form most readily used by green plants (although some organisms can use nitrogen in other forms as illustrated), thus completing the cycle. The air, which contains 80 per cent nitrogen, is the greatest reservoir and safety valve of the system. Nitrogen is continually entering the air by the action of denitrifying bacteria and continually returning to the cycle through the action of nitrogen-fixing bacteria or algae and through the action of lightning (i.e., electrification). In Figure 4–2*B* the components of the nitrogen cycle are shown in terms of the energy necessary for the operation of the cycle. The steps from proteins down to nitrates provide energy for organisms which accomplish the breakdown, whereas the return steps require energy from other sources, such as organic matter or sunlight. For example, the chemosynthetic bacteria (see page 26) *Nitrosomonas* (which converts ammonia to nitrite) and *Nitrobacter* (which converts nitrite to nitrate) obtain energy from the breakdown, while denitrifying and nitrogen-fixing bacteria require energy from other sources to accomplish their respective transformations. The carbohydrate requirements for nitrogen fixation were estimated in the model of the soybean ecosystem (Table 3–6, page 49). The trace element molybdenum is also required as part of the N-fixing enzyme system and may sometimes be a limiting factor, as will be discussed in the next chapter.

Until about 1950, the capacity to fix atmospheric nitrogen was thought to be limited to these few, but abundant, kinds of microorganisms:

Free-living bacteria—*Azotobacter* (aerobic) and *Clostridium* (anaerobic)
Symbiotic nodule bacteria on legume plants—*Rhizobium* (see Figure 4–4)
Blue-green algae—*Anabaena, Nostoc,* and other members of the order Nostocales (see review by Fogg and Stewart, 1966)

It was then discovered that the purple bacterium *Rhodospirillum* and other representatives of the photosynthetic bacteria are nitrogen fixers (see Kamen and Gest, 1949; Kamen, 1953), and that a variety of pseudomonas-like soil bacteria also have this ca-

Figure 4–4. Root nodules on a legume, the location of nitrogen-fixing bacteria of the symbiotic or mutualistic type (see also Figure 4–2). The legume shown is blue lupine, a cultivated variety used in southeastern U.S. (U.S. Soil Conservation Service Photo.)

pacity (see Anderson, 1955). The use of the isotopic tracer ^{15}N and the acetylene reduction method (the nitrogen-fixing enzyme, nitrogenase, reduces acetylene to ethylene, providing a sensitive indicator for nitrogen fixation; see Stewart *et al.*, 1967) has resulted in a "measurement breakthrough" which is revealing that the ability to fix nitrogen is widespread among photosynthetic, chemosynthetic, and heterotrophic microorganisms. There is even evidence that algae and bacteria growing on leaves and epiphytes in humid tropical forests fix appreciable quantities of atmospheric nitrogen, some of which may be used by the trees themselves. However, no higher plant has been shown to be able to fix nitrogen without some help. Legumes and a few genera of other families of vascular plants (for example, *Alnus, Casuarina, Coriaria, Ceonothus, Myrica, Araucaria, Ginkgo, Elaegnus*) do so only with the aid of symbiotic bacteria. Likewise, some lichens are able to fix nitrogen because of symbiotic blue-green algae. In short, it appears that biological nitrogen fixation by free-living and symbiotic microorganisms goes on in both the autotrophic and heterotrophic strata of ecosystems and in both aerobic and anaerobic zones of soils and aquatic sediments.

In 1944, Hutchinson estimated that the amount of nitrogen fixed from the air lies between 140 and 700 mg per m² per year (about 1 to 6 lbs per acre) for the biosphere as a whole. Most of this fixation is thought to be biological with only a small portion (not more than 35 mg per m² per year in temperate regions) resulting from electrification and photochemical fixation. Recent estimates (Delwiche, 1965, 1970) place biological N-fixation on the earth's terrestrial surface as at least 1 gm per m² per year (about 10 lbs per acre). Biological fixation in fertile areas may be much greater than these biospheric averages, up to 20 gm per m² per year (about 200 lbs per acre) according to Fogg (1955). Estimates of annual fixation rates for large lakes and the oceans are not yet available. Dugdale (1966) has reported N-fixation in the photic zones of small lakes as ranging from about 1 to 50 μg per liter per day; the high values were recorded from somewhat polluted lakes with large populations of blue-green algae. Although the rate of N-fixation on a square-meter basis is undoubtedly less in the ocean than on

land (because of the generally lower productivity), the total amount of nitrogen fixed in the oceans must be large and very important to the global cycle. Throughout the biosphere, rainfall is important in the rapid recycling of available nitrogen. In moist regions, there is enough nitrogen (and other nutrients) in rainwater to support epiphytic plants that have no other source of mineral nutrients, although most epiphytes get leachate dripping down from tree leaves located above them. The annual rainfall input of nitrogen into ecosystems has been found to be as much as 0.75 gm per m² in moist temperate areas and as much as 3.0 gm per m² in the wet tropics (see Goldschmidt, 1954). Much of this is ammonia or other volatile components released by biological communities, while an unknown fraction is nonbiological fixation in the atmosphere.

The importance of the nitrogen-fixing bacteria associated with legumes (Fig. 4–5) is well known, of course, and in modern agriculture continuous fertility of a field is maintained as much by crop rotation involving legumes as by the application of nitrogen fertilizers. Secretion from the legume root stimulates growth of the nodule bacteria, and bacterial secretions cause root hair deformation, the first step in nodule formation (Nutman, 1956). Strains of bacteria have evolved which will grow only on certain species of legumes. Making use of biological N-fixation has advantages over heavy nitrate fertilization in terms of maintaining soil structure and avoiding runoff pollution. Unfortunately, such "natural" recycling requires fallowing and crop rotation which does not produce the short-term yield of continuous grain culture. However, as discussed in the previous chapter, there is strong evidence that land needs to be "rested" at intervals if its quality and long-term productivity are to be maintained. In the Orient it has been found that the blue-green algae which occur naturally in the rice paddies are very important in maintaining fertility under intensive cropping. Seeding the rice fields with extra algae often results in increased yields (Tamiya, 1957). An important point to emphasize, in summary, is that the chief mechanism for moving nitrogen from the air reservoir into the productivity cycle is biological nitrogen fixation by bacteria and algae. If man will but share food with these "friendly mi-

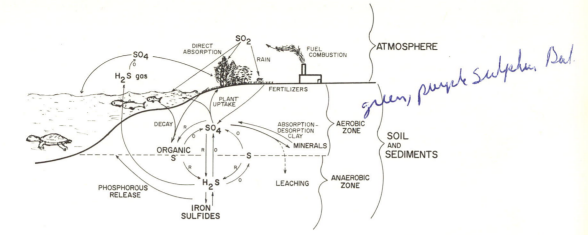

green, purple Sulphur Bal.

Figure 4–5. The sulfur cycle linking air, water, and soil. The center "wheel-like" diagram shows oxidation (O) and reductions (R) that bring about key exchanges between the available sulfate (SO_4) pool and the reservoir iron sulfide pool deep in soils and sediments. Specialized microorganisms are largely responsible for the following transformations: $H_2S \rightarrow S \rightarrow SO_4$, colorless, green, and purple sulfur bacteria; $SO_4 \rightarrow H_2S$ (anaerobic sulfate reduction), desulfovibrio bacteria; $H_2S \rightarrow SO_4$ (aerobic sulfide oxidizers), thiobacilli bacteria; organic $S \rightarrow SO_4$ and H_2S, aerobic and anaerobic heterotrophic microorganisms, respectively. The metabolism of these various sulfur bacteria is described in Chapter 2. Primary production, of course, accounts for the incorporation of sulfate into organic matter, while animal excretion is a source of recycled sulfate (see Figure 4–11). Sulfur oxides (SO_2) released into the atmosphere on burning of fossil fuels, especially coal, are becoming increasingly bothersome components of industrial air pollution.

crobes" and not poison their soil and water environment, they will carry out this vital work free of charge! (See Burris, 1969.)

The self-regulating, feedback mechanisms, shown in a very simplified way by the arrows in the diagram (Fig. 4–2), make the nitrogen cycle a relatively perfect one, when large areas or the biosphere as a whole is considered. Some nitrogen from heavily populated regions of land, fresh water, and shallow seas is lost to the deep ocean sediments and thus gets out of circulation, at least for a while (a few million years perhaps). This loss is compensated for by nitrogen entering the air from volcanic gases. Thus, volcanic action is not to be entirely deplored but has some use after all! If nothing else, ecology teaches us not to make snap judgments as to whether a thing or an organism is "useful" or "harmful." One must consider the "totality" of a problem before arriving at a judgment. Capping the world's volcanos, even if technically possible, could very well starve more people to death than it could save from damage by eruptions. There will be many other examples of this in subsequent chapters.

The phosphorus cycle, in comparison to the nitrogen cycle, appears to be somewhat simpler. As shown in Figure 4–3, phospho-rus, an important and necessary constituent of protoplasm, tends to "circulate," the organic compounds being broken down eventually to phosphates which are again available to plants. The great reservoir of phosphorus is not the air, however, but the rocks or other deposits which have been formed in past geological ages. These are gradually eroding, releasing phosphates to ecosystems, but much phosphate escapes into the sea, where part of it is deposited in the shallow sediments and part of it is lost to the deep sediments. The means of returning phosphorus to the cycle may presently be inadequate to compensate for the loss. In some parts of the world there is no extensive uplifting of sediments at present, and the action of marine birds and fish (being brought to land by animals and man) is not adequate. Sea birds have apparently played an important role in returning phosphorus to the cycle (witness the fabulous guano deposits on the coast of Peru). This transfer of phosphorus and other materials by birds from the sea to land is continuing, but apparently not at the rate at which it occurred in some of the past ages. Man, unfortunately, appears to hasten the rate of loss of phosphorus and thus to make the phosphorus cycle less perfect. Although man harvests a lot of marine fish, Hutchinson estimates that only about 60,000 tons of elementary phosphorus per year is returned in this manner, compared with one to two million tons of phosphate rock which is mined and most of which is washed away and lost. Agronomists tell us there is no immediate cause for concern, since the known reserves of phosphate rock are large. Just now, man is more concerned with the "traffic jam" of dissolved phosphate in the waterways resulting from increased "erosion" that cannot be compensated for by "protoplasm synthesis" and "sedimentation" (see diagram, Figure 4–3). However, man may ultimately have to go about completing the phosphorus cycle on a large scale if he is to avoid famine. Of course a few geological upheavals raising the "lost sediments" might accomplish it for us, who knows? One procedure for recycling phosphorus "uphill" that is being experimented with involves spraying waste water on upland vegetation instead of piping it directly into waterways. At any rate, take a good look at the diagram of the phosphorus cycle; its importance may loom large in the future.

A comprehensive diagram of the sulfur cycle is shown in Figure 4–5. Many of the main features of biogeochemical cycling are illustrated by this diagram, as, for example, the large reservoir pool in soil and sediments and a smaller reservoir in the atmosphere; the key role in the rapidly fluxing pool (the center "wheel" in Figure 4–5) played by specialized microorganisms that function like a relay team, each carrying out a particular chemical oxidation or reduction (see legend, Figure 4–5); "microbial recovery" from deep sediments resulting from an upward movement of a gaseous phase (H_2S), as was discussed on page 27; the interaction of geochemical and meteorological processes (erosion, sedimentation, leaching, rain, adsorption-desorption, etc.) and biological processes (production and decomposition); and the interdependence of air, water, and soil in regulation of the cycle at the global level. Sulfate (SO_4), like nitrate and phosphate, is the principal available form that is reduced by autotrophs and incorporated into proteins, sulfur being an essential constituent of certain amino acids. Not as much sulfur is required by the ecosystem as nitrogen and phosphorus, nor is it as often limiting to the growth of plants and animals; nevertheless, the sulfur cycle is a key one in the general pattern of production and decomposition, as was alluded to in Chapter 2 (Section 3). For example, when iron sulfides are formed in the sediments, phosphorus is converted from insoluble to soluble form (note "phosphorus release" arrow in Figure 4–5) and thus becomes available to living organisms. Here is an excellent illustration of how one cycle regulates another. The interesting metabolism of the several kinds of sulfur bacteria has already been discussed (pages 25–26), and the part played by the sulfur cycle in the zonation of marine sediments is shown in Figure 12–13, Chapter 12.

Both the nitrogen and the sulfur cycles are increasingly being affected by industrial air pollution. The oxides of nitrogen (NO and NO_2) and sulfur (SO_2) are normally but transitory steps in their respective cycles and present in most environments in very low concentrations. The combustion of fossil fuels, however, has greatly increased the concentrations of the volatile oxides in the air, especially in urban areas, to a point at which they become poisons to biotic components of ecosystems. In 1966, these oxides constituted about a third of the estimated 125 million tons of industrial air pollutants discharged into the air over the United States. Coal-burning electrical generating plants constitute a major source of SO_2, and the automobile is a major source of NO_2. Sulfur dioxide is damaging to the photosynthetic process (the destruction of vegetation around copper smelters is caused by this pollutant; see Figure 2–8A), and the oxides of nitrogen can be hard on the respiratory processes of higher animals and man. Furthermore, chemical reactions with other pollutants produces a synergism (= total effect of the interaction exceeds the sum of the effects of each substance) that increases the danger. For example, in the presence of ultraviolet radiation in sunlight NO_2 reacts with unburned hydrocarbons (both produced in large quantities by automobiles) to produce the eye-watering "photochemical smog" (see page 445 for the chemical composition of this type of air pollution). Redesign of the internal combustion engine, removal of sulfur from fuel, and a switch to atomic energy for electric power generation will all hopefully relieve these very serious perturbations in the nitrogen and sulfur cycles. (However, such shifts in man's strategy in the use of power will produce other problems that must be anticipated; see Chapter 16.)

2. QUANTITATIVE STUDY OF BIOGEOCHEMICAL CYCLES

Statement

The rates of exchange or transfers from one place to another are more important in determining the structure and function of an ecosystem than the amounts present at any one time in any one place. To understand and thereby better control man's role in the cycles of materials, cycling *rates* as well as *standing states* must be quantitated. During the past 25 years improvements in tracers, mass chemistry, monitoring, and remote sensing techniques (four of the six measurement "breakthroughs" mentioned in Chapter 1; see page 6) have made it possible to measure cycling rates in sizable units such as lakes and forests and to begin the all-important task of quantitating biogeochemical cycles at the global level.

Examples

Diagrams such as those in Figures 4–2, 4–3, and 4–5 show only the broad outlines of biogeochemical cycles. Quantitative relations, that is, how much material passes along the routes shown by the arrows, and how fast it moves, are still poorly known. Radioactive isotopes, which have become generally available since 1946, are providing a tremendous stimulus for such studies since these isotopes can be used as "tracers" or "tags" to follow the movement of materials. It should be emphasized that tracer studies in ecosystems, as in organisms, are designed so that the amount of radioactive element introduced is extremely small in comparison with the amount of nonradioactive element already in the system. Therefore, neither the radioactivity nor the extra ions disturb the system; what happens to the tracer (which can be detected in extremely small amounts by the telltale radiations which it emits) simply reflects what is normally happening to the particular material in the system.

Ponds and lakes are especially good sites for study since their nutrient cycles are relatively self-contained over short periods of time. Following the pioneer experiments of Coffin, Hayes, Jodrey, and Whiteway (1949), and Hutchinson and Bowen (1948, 1950), numerous papers have appeared reporting the results of the use of radio-phosphorus (^{32}P) and other sophisticated techniques in studies of phosphorus circulation in lakes. Hutchinson (1957) and Pomeroy (1970) have summarized studies on the rate of cycling of phosphorus and other vital elements.

It has been generally found that phosphorus does not move evenly and smoothly from organism to environment and back to organism as one might think from looking at the diagram in Figure 4–3, even though, as we have already indicated, a long-term equilibrium tends to be established. At any one time, most of the phosphorus is tied up either in organisms or in solids (i.e., organic detritus and inorganic particles which make up the sediments). In lakes, 10 per cent is the maximum likely to be in a soluble form at any one time. Rapid back-and-forth movement or exchange occurs all of the time, but extensive movement between solid and dissolved states is often irregular or "jerky," with periods of net release from the sediments followed by periods of net uptake by organisms or sediments, depending on seasonal temperature conditions and activities of organisms. Generally, uptake rate is more rapid than release rate. Plants readily take up phosphorus in the dark or under other conditions when they cannot use it. During periods of rapid growth of producers, which often occur in the spring, all of the available phosphorus may become tied up in producers and consumers. The system may then have to "slow down" until the bodies, feces, etc., can be decomposed and nutrients released. However, the concentration of phosphorus at any one time may bear little relation to productivity of the ecosystem. A low level of dissolved phosphate could mean that the system is impoverished or that it is very active metabolically; only by measuring the flux rate can the true situation be determined. Pomeroy (1960) summarizes this important point as follows: "Measurement of the concentration of dissolved phosphate in natural waters gives a very limited indication of phosphate availability. Much or virtually all of the phosphate in the system may be inside living organisms at any given time, yet it may be overturning every hour with the result that there will be a constant supply of phosphate for organisms able to concentrate it from a very dilute solution. Such systems may remain stable biologically for considerable periods in the apparent absence of available phosphate. The observations presented here suggest that a rapid flux of phosphate is typical of highly productive systems, and that the flux rate is more important than the concentration in maintaining high rates of organic production."

The concept of turnover, as first introduced in Chapter 2 (page 17), is a useful one for comparing exchange rates between different components of an ecosystem. In terms of exchanges after equilibrium has been established the *turnover rate* is the fraction of the total amount of a substance in a component which is released (or which enters) in a given length of time, whereas *turnover time* is the reciprocal of this, that is, the time required to replace a quantity of substance equal to the amount in the component (see Robertson, 1957, for a discussion of these concepts). For example, if 1000 units are present in the component and 10 go out or enter each hour, the turnover rate is 10/1000 or 0.01 or 1 per cent per hour. Turnover time would then be 1000/10 or 100 hours. *Residence time*, a term widely used in the geochemical literature, is a concept similar to turnover time in that it refers to the time a given amount of substance remains in a designated compartment of a system. Data on turnover time for two large components, the water and the sediments, in three lakes are given in Table 4–1. The smaller lakes have a shorter turnover time presumedly because the ratio of bottom "mud" surface to the volume of water is greater. In general, the turnover time for the water of small or shallow lakes is a matter of days or weeks; for large lakes it may be a matter of months.

Studies with ^{32}P-tagged fertilizers in land ecosystems have revealed similar patterns; much of the phosphorus is "locked up" and unavailable to plants at any given time (see

Comar, 1957, for a summary of some of these experiments). One very practical result of intensive studies of nutrient cycles has been the repeated demonstration that overfertilization can be just as "bad" from the standpoint of man's interest as underfertilization. When more materials are added than can be used by the organisms active at the time, the excess is often quickly tied up in soil or sediments or even lost completely (as by leaching), and is unavailable at the time when increased growth is most desired. The "blind dumping" of fertilizers in ecosystems such as fish ponds is not only wasteful insofar as the desired results are concerned but is likely to create unanticipated changes in the system as well as "downstream" pollution. Since different organisms are adapted to specific levels of materials, continued excess fertilization may result in a change in the kinds of organisms, perhaps discouraging the one man wants and encouraging the kinds he does not want. Among the algae, for example, *Botryococcus braunii* exhibits optimum growth at phosphorus concentration of 89 mg per m^3, while *Nitzschia palea* grows best at 18 mg per m^3. Increasing the amount of P from 18 to 89 would likely result in *Botryococcus* replacing *Nitzschia* (assuming other conditions favorable for both species), and this could have considerable effect on the kinds of animals that could be supported. The complete destruction of an oyster industry caused by a change in phytoplankton populations resulting from increased fertilization by phosphorus and nitrogen materials is described in Chapter 5 (see page 112).

As was emphasized in Chapter 2, bodies of water are not closed systems but need to be considered as parts of larger drainage basins or "watershed" systems. As was also pointed out, the watershed systems provide

Table 4–1. *Estimates of the Turnover Time of Phosphorus in Water and Sediments of Three Lakes as Determined with the Use of ^{32}P** [*]

LAKE	AREA KM^2	DEPTH M	TURNOVER TIME IN DAYS		RATIO MOBILE P TO TOTAL P IN WATER
			WATER	SEDIMENTS	
Bluff	0.4	7	5.4	39	6.4
Punchbowl	0.3	6	7.6	37	4.7
Crecy	2.04	3.8	17.0	176	8.7

[*] After Hutchinson, 1957.

a sort of minimum ecosystem unit insofar as practical management by man is concerned. Figure 4–6 pictures a quantitative model of the calcium cycle for small mountainous and forested watersheds in New Hampshire. The data are based on studies of six watersheds ranging in size from 12 to 48 hectares (Bormann and Likens, 1967). Precipitation, which averaged 123 cm or 58 inches per year, was measured by a network of gaging stations, and the amount of water leaving the watershed in the drainage stream of each watershed unit was measured by a V-notched weir similar to that shown in Figure 2–4, page 17. (By determining the concentration of minerals in the input and output water, the input-output mineral budget could be calculated.) The calcium content of the biotic and soil pools was estimated from data of Ovington (1962).

Retention by and recycling within the undisturbed forest proved to be so effective that the estimated loss from the ecosystem was only 8 kg per ha per yr of calcium (and equally small amounts of other nutrients). Since 3 kg of this was replaced in rain, only an input of 5 kg per ha would be needed to achieve a balance, and this is thought to be easily supplied by the normal rate of weathering from the underlying rock that constitutes the "reservoir" pool. A recent study by Thomas (1969), who used the radionuclide ^{45}Ca to measure turnover, demonstrates how understory trees, such as dogwood, act as calcium "pumps," which

counter the downward movement in soil and thus keep calcium in circulation between organisms and the active upper layers of litter and soil.

On one of the experimental watersheds all the vegetation was felled and regrowth the next season suppressed by aerial application of herbicides. Even though the soil was little disturbed and no organic matter was removed by this procedure, the loss of mineral nutrients in stream outflow was increased 3 to 15 times over losses from the undisturbed control watersheds (Likens, Bormann, and Johnson, 1969). Increased stream flow out of the cut-over ecosystem resulted primarily from the elimination of plant transpiration, and it was the additional stream flow that carried out additional minerals. As pointed out in the previous chapter (see page 47), transpiration, traditionally considered to be an unwanted stress in agricultural systems, seems to be a beneficial nutrient conservation mechanism as well as an energy subsidy in a forest. Again, quantitative studies of this type reveal the true costs to society as a whole of increasing water flow downstream to meet what in the long run may be unreasonable and unnecessary demands by a wasteful industrial society. Removing forests from the hills will increase water "yield" to the valleys (see Hibbert, 1967) but at the expense of water quality as well as the productive and air regenerating capacity of the watershed. More about this in the next section.

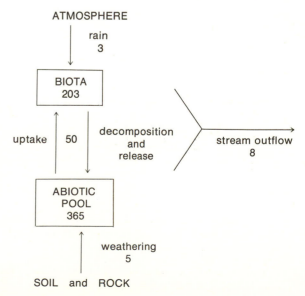

Figure 4–6. Calcium cycle in a northern forested watershed based on data of Bormann and Likens (1967) and Ovington (1962). All figures are kilograms per hectare per year. Figures in boxes are estimates of the content of major pools or "standing states"; all others represent estimates of annual rates of movement. Single arrows show inputs and outputs to the watershed system as a whole; double arrows show the exchange between biotic and abiotic pools within the system.

Figure 4–7 is a compartment model of the phosphorus exchange within a Georgia salt marsh estuarine watershed, which is a much more open system than the mountain stream watershed. In this model seasonal pulses are built into the circuits, since, as already indicated, movement of materials is neither continuous nor a linear function with time. In the marsh there is a major pulse in the growth of the marsh grass during the warmer months when phosphorus is "pumped" up through roots of grass that penetrate deeply into the anaerobic sediments. This "recovery" process was experimentally verified by injecting ^{32}P into deep sediments. There are two major pulses of decomposition that release phosphorus into the water, one in the hottest part of the summer and another in the fall when large quantities of dead grass are washed out of the marsh by high seasonal tides. When systems of differential equations were set up to describe the manner in which the content of phosphorus in one compartment affects that of adjoining parts of the system, and the whole model then tested with an analog computer, it was found that some adjustments had to be made in several of the "throughputs" in order to keep the "contents" of the smaller compartments stable. These procedures illustrate the two points emphasized in Section 3, Chapter 1, namely: (1) A realistic, although greatly simplified, model can be constructed from a relatively small amount of field data and (2) such models can be "tuned" by com-

puter manipulations to determine what transfer coefficients will make the model a better mimic of the real world (see also Chapter 10).

The importance of filter-feeder and detritus-feeder animals in recycling phosphorus in this estuarine system was verified by the model. Earlier work by Kuenzler (1961) had shown that the population of mussels (*Modiolus demissus*) alone "recycles" from the water a quantity of particulate phosphorus every two and a half days equivalent to the amount present in the water (i.e., a turnover time of only 2.5 days for particulate phosphorus in the water). Kuenzler (1961a) also measured the energy flow of the population and concluded that the mussel population is more important to the ecosystem as a biogeochemical agent than as a transformer of energy (i.e., as a potential source of food for other animals or man). This is an excellent illustration of the fact that a species does not have to be a link in the food chain of man to be valuable to him; many species are valuable in indirect ways that are not apparent on superficial examination. In subsequent chapters we will come back again to the often asked question: "What good are all those species in nature that man can't eat or sell?"

The Global Cycling of CO_2 and H_2O

Moving now to the global level, Figure 4–8 illustrates tentative attempts to quantitate what are probably the two most impor-

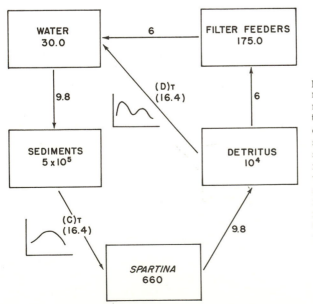

Figure 4–7. Compartmental flux diagram of phosphorus in a Georgia salt marsh ecosystem (see Pomeroy, Johannes, Odum, and Roffman, 1967). Two large reservoirs of phosphorus (sediments and detritus) and the three most active compartments (water, *Spartina* or marsh grass, and detritus-feeding animals) are shown. Quantities within compartments are standing stocks in mg P per m²; transfer fluxes are expressed in mg P per m³ per day. This is a linear model with two variable coefficients, (D)t and (C)t, that mimic seasonal pulses in release of phosphorus from decomposing detritus and uptake of phosphorus by marsh grass, respectively. A graph of each seasonal pulse is shown. Quantities in parentheses are integrated means of the variable transfers.

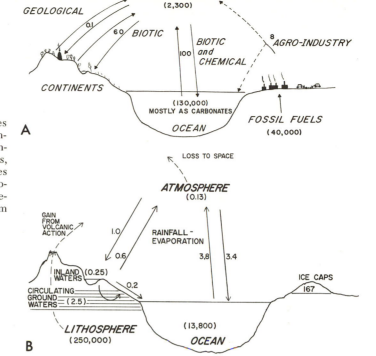

Figure 4–8. A. The CO_2 cycle. Figures are 10^9 tons CO_2 in major biosphere compartments and circulating between compartments (arrows). (Data from Plass, 1959.) *B.* The hydrological cycle. Figures are geograms (10^{20} grams) H_2O in major biosphere compartments and circulating between compartments (arrows). (Data from Hutchinson, 1957.)

tant cycles as far as man is concerned, namely the CO_2 cycle and the hydrological cycle. Both are characterized by small, but very active, atmospheric pools that are very vulnerable to man-made perturbations, which, in turn, can advertently or inadvertently change weather and climates. Regardless of whether one's thinking is oriented towards avoiding disaster or towards purposeful weather modification for the presumed benefit of somebody somewhere, it is extremely important that a network of worldwide measurements be set up to detect significant changes in the CO_2 and H_2O cycles that can quite literally affect man's future on earth.

Looking first to the CO_2 cycle (Fig. 4–8A), the CO_2 theory of climate and the recent gradual rise in the CO_2 content of the atmosphere was discussed in Chapter 2 (see page 33). Of the estimated 8 billion tons of CO_2 injected annually into the air by man's agro-industrial activity in 1970, 6 billion tons come from the burning of fossil fuels and 2 billion tons from cultivation of land for agriculture. Most, but not quite all, of this is quickly passed into the sea and "stored" in the form of carbonates. A net loss of CO_2 in agriculture may seem surprising, but it results from the fact that the CO_2 fixed by the crops (many of which are active for only a part of the year) does not compensate for

the CO_2 released from the soil, especially that resulting from frequent plowing. The rapid oxidation of humus and release of gaseous CO_2 normally held in the soil has other more subtle effects that are just now being recognized, including effects on the cycling of other nutrients. For example, Nelson (1967), in an elegant study, used clam shells to demonstrate that deforestation and agriculture have resulted in a decline in the amount of certain trace elements in soil water run-off. He found that clam shells from Indian middens 1000 to 2000 years old contained 50 to 100 per cent more manganese and barium than contemporary shells (a statistically significant difference). By a process of elimination Nelson concluded that reduced flow of CO_2-charged acidic water percolating deeply in the soil has decreased the rate of dissolution of these elements from the underlying rocks. In other words, water now tends to run off rapidly over the surface instead of filtering down through humus layers in the soil. In ecological terms we would say that the flux between the reservoir and the exchangeable pool has been altered in a rather fundamental way by man's present management of the landscape. As long as we recognize what has happened and learn to compensate, such changes do not have to be detrimental. It is perhaps no coincidence that agriculturalists are discovering that they must now add trace minerals to fertilizers in order to maintain yields in many areas.

At this point it would be well to recall how the earth's atmosphere came to have its present very low CO_2 content and its very high O_2 content. The evolution of the atmosphere is described in some detail in Chapter 9. When life began on earth two billion years ago, the atmosphere, like that of the planet Jupiter today, was composed of volcanic gases (atmospheric formation by "crustal outgassing," as a geologist would phrase it!). It contained quantities of CO_2 but little if any oxygen, and the first life was anaerobic. As described in Chapter 2, the buildup of oxygen and decline in CO_2 over geological time have resulted from the fact that P (production) has, on the average, slightly exceeded R (respiration). Also, according to Hutchinson (1944), the formation of reduced nitrogen compounds and the production of hydrogen from water and its escape from the atmosphere into space have also contributed to oxygen accumulation. It is perhaps significant that both the low concentration of CO_2 and the high concentration of O_2 are now limiting to photosynthesis. This is to say that most plants increase their rate of photosynthesis if either the CO_2 concentration is increased or O_2 concentration decreased experimentally. This makes green plants very responsive regulators! (See page 125.)

Even though the earth's photosynthetic green belt and the carbonate system of the sea have been very efficient in removing CO_2 from the atmosphere, the spiraling increase in consumption of fossil fuels (consider the tremendous amount of CO_2 that would be released if all of the fossil fuel pool was burned; see Figure 4–8A) coupled with the decrease in the "removal capacity" of the "green belt" is beginning to have an effect on the atmospheric compartment. It is the content of the small, active compartments that are most affected by changes in fluxes or "throughputs." What we apparently can expect in the next few decades is a new, but uncertain, balance between increasing CO_2 (which holds in reradiated heat) and increasing dust or "particulate" pollution (which reflects incoming radiant energy). Any significant *net* change in the heat budget, *either way*, will affect climates. For more about the global CO_2 cycle, see Revelle and Suess, 1956; Plass, 1955; Möller, 1963; Revelle, 1965.

As shown in the diagram of the hydrological cycle in Figure 4–8B, the H_2O atmospheric compartment is small, and it has a more rapid turnover rate and shorter residence time in the atmosphere than CO_2. The water cycle, like the CO_2 cycle, is beginning to be affected by man on the global scale. While we are now doing a fair job of keeping up with rainfall and stream-flows on a worldwide basis, there is an urgent need for more complete monitoring of all of the major fluxes. Hence, an "International Hydrological Decade" comparable to the "International Biological Program" is in the planning stages.

Two other aspects of the H_2O cycle need special emphasis. (1) Note that more water evaporates from the sea than returns via rainfall, and vice versa for the land. In other words, a part of the rainfall that supports land ecosystems, including most of man's food production, comes from water evaporated over the sea. In many areas, such as the Mississippi valley, for example, as much as 90 per cent of the rainfall is estimated to come from the sea (Benton, Blackburn, and Snead, 1950). (2) Since an estimated 0.25 geograms (1 geogram = 10^{20} grams) of water are in freshwater lakes and rivers, and 0.2 geograms runs off each year, the turnover time is about a year. The difference between the annual rainfall (1.0 geogram) and the run-off (0.2 geogram), or 0.8, is an estimate of the annual recharge rate of ground waters. As already indicated, man's tendency to increase the rate of run-off may soon reduce the very important ground water compartment. We should be putting more water back into the "aquifers" instead of trying to store it all in lakes (where evaporation is high).

As is detailed later in this chapter (and Chapter 17), the development of atomic energy has created hazardous contamination of biogeochemical cycles on a global scale. On the other hand, some unique opportunities for tracer research have also resulted, a good example being tritium in the hydrological cycle. Tritium (^3H), the radioactive isotope of hydrogen, is formed in small quantities at a constant rate by the action of cosmic rays in the upper atmosphere and is naturally present in the form of water throughout the hydrological cycle. It is removed from the cycle by radioactive decay according to the 12.3 years half life of the isotope (see page 453 for explanation of the concept of half life). It is fortunate that the rate of formation of natural tritium and the equilibrium levels in different water compartments were measured before the testing of hydrogen bombs, which during the late 1950s and the early 1960s introduced an amount of artificial tritium into the atmosphere about 10 times that produced naturally. Since the 1964 moratorium on nuclear testing, artificial tritium has not been added in noticeable amounts. In the meantime, the added charge of tritium has moved from the atmosphere and surface waters into ground water, ice masses, and the deeper water masses of the oceans, thus providing an opportunity to determine sources, recharge rates, and residence times of water in these reservoir pools. Suess (1969) reviews the possibilities along these lines and calls for coordinated international study. Partly because the radioactive emissions from tritium are weak or "soft" (see page 452 for explanation), the artificial injection has not had widespread effect on the biota or man; at least, no effect has been yet detected. However, as the uses of atomic energy expand, especially the use of fusion power (see page 461 for explanation), tritium remains one of the by-products of concern. It is to be hoped that tritium in the water cycle will never be more than a useful tracer!

For an excellent summary of the hydrological cycle, see Chapter 4 in Hutchinson's *Treatise on Limnology* (1957).

3. THE SEDIMENTARY CYCLE

Statement

Most elements and compounds are more earthbound than nitrogen, oxygen, carbon dioxide, and water, and their cycles follow a basic sedimentary cycle pattern in that erosion, sedimentation, mountain building, and volcanic activity, as well as biological transport, are the primary agents effecting circulation.

Explanation

A generalized picture of the sedimentary cycle of earthbound elements is shown in Figure 4–9. Some estimates of the amounts of material that pass through the cycle are marked on the arrows. Of course, very little is known about the flow of materials in the

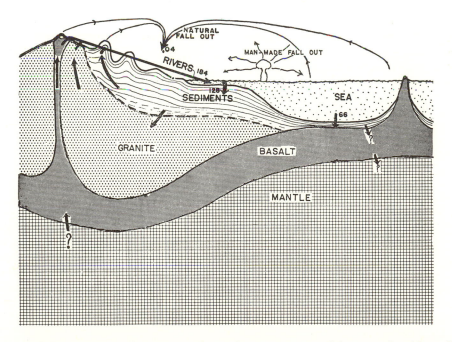

Figure 4–9. A diagram of the sedimentary cycle involving movement of the more "earthbound" elements. Where estimates are possible, the amounts of material are estimated in geograms per million year (one geogram = 10^{20} grams). The continents are sediment covered blocks of granite floating like corks on a layer of basalt which underlies the oceans. Below the black basalt is the mantle layer which extends 2900 km down to the core of the earth. Granite is the light colored very resistant rock often used for tombstones; basalt is the black rock found in volcanoes. (Diagram prepared by H. T. Odum.)

deep earth. The movement of solid matter through the air as dust is indicated as "fall-out." To the natural fallout, man is adding additional materials, relatively small in amount but significant biologically because of their poisonous effects or their action in blocking the incoming solar radiation. The chemical elements that are available to the communities of the biosphere are those which by their geochemical nature tend to be enclosed within the types of rocks that come to the surface. Elements that are abundant in the mantle are scarce at the surface. As already indicated, phosphorus is one of the elements whose scarcity on the earth's surface often limits plant growth.

The general "downhill" tendency of the sedimentary cycle is well shown in Figure 4–9. Estimated annual flow of sediments from each major continent to the oceans is shown in Table 4–2. It is significant that the continent with the oldest civilizations and most intensive human pressure, namely Asia, loses the most soil. Although the rate varies, the lowlands and the oceans tend to gain soluble or usable mineral nutrients at the expense of the uplands during periods of minimum geological activity. Under such conditions, local biological recycling mechanisms are extremely important in keeping the downhill loss from exceeding the re-generation of new materials from underlying rocks, as was emphasized in the discussion of the watershed calcium cycle. In other words, the longer vital elements can be kept within an area and used over and over again by successive generations of organisms, the less new material will be needed from the outside. Unfortunately, as already mentioned in the discussion of phosphorus, man tends to disrupt this ho-

meostasis, often unwittingly, through a lack of understanding of the symbiosis between life and matter, which may have taken thousands of years to evolve. For example, it is now suspected, although not yet proved, that the stopping of salmon runs by dams is resulting in a decline not only of salmon but also of nonmigratory fish, game, and even timber production in certain high altitude regions of western United States. When salmon spawn and die in the uplands, they deposit a load of valuable nutrients recovered from the sea. The removal of large volumes of timber without the return of the contained minerals to soil (as would normally occur during the decay of logs) undoubtedly also contributes to the impoverishment of uplands in situations where the pool of nutrients was already limited. One can readily visualize how the destruction of such biological recycling mechanisms could result in the impoverishment of the whole ecosystem for many years to come, since it would require time for the re-establishment of a circulating pool of minerals. In such a case, devising means of returning limiting materials (and keeping them in situ) would be far more effective than the stocking of fish or planting of tree seedlings. It should also be emphasized that the sudden rush of materials into the lowlands, resulting from man-accelerated erosion, does not necessarily benefit the lowland ecosystems, because these systems may not get a chance to assimilate the nutrients before they pass into the sea beyond the range of light and, thus, completely out of biological circulation (at least for a time).

4. CYCLING OF NONESSENTIAL ELEMENTS

Statement

The nonessential elements pass back and forth between organisms and environment, in the same general manner as do the essential elements, and many of them are involved in the general sedimentary cycle even though they have no known value to the organism. Many of these elements become concentrated in certain tissues, sometimes because of chemical similarity to specific vital elements. Chiefly because of the activities of man the ecologist must now be concerned with the cycling of many of the nonessential elements.

Table 4–2. Estimated Annual Sediment Flow from Continents to Oceans*

CONTINENT	DRAINAGE AREA 10^6 MI^2	SEDIMENT DISCHARGE TONS/MI^2†	TOTAL, 10^9 TONS
North America	8.0	245	1.96
South America	7.5	160	1.20
Africa	7.7	70	0.54
Australia	2.0	115	0.23
Europe	3.6	90	0.32
Asia	10.4	1,530	15.91
Total	39.2	–	20.16

* After Holeman, 1968.
† Conversion to volume: 1330 tons = 1 acre-foot.

Explanation

Most nonessential elements have little effect in concentrations normally found in most natural ecosystems probably because organisms have become adapted to their presence. Therefore, the ecologist would have little interest in most of the nonessential elements were it not for the fact that mining and industrial wastes contain high concentrations of mercury, lead, and other potentially very toxic materials, whereas atomic bombs and nuclear power operations produce radioactive isotopes of some of these elements, which then find their way into the environment. Consequently, the new generation of ecologists must be concerned with the cycling of just about everything! Even a very rare element, if in the form of a highly toxic metallic compound or a radioactive isotope, can become of biological concern, because a small amount of material (from the biogeochemical standpoint) can have marked biological effect.

Example

Strontium is a good example of a previously almost unknown element that must now receive special attention, because radioactive strontium appears to be particularly dangerous to man and other vertebrates. Strontium behaves like calcium, with the result that radioactive strontium gets into close contact with blood-making tissue, which is very susceptible to radiation damage. Since various aspects of radioactivity in the environment are considered in some detail in Chapter 17, we need only consider strontium in relation to the calcium cycles in this chapter.

One of the most abundant elements that make up the sedimentary cycle is calcium; it washes down the rivers, deposits as limestone, is raised in mountain ranges, and again washes down to the sea. About 7 per cent of the total sedimentary material flowing down the rivers is calcium. In comparison, the amount of phosphorus in the cycle is only about 1 per cent of the calcium. For every 1000 atoms of calcium, 2.4 atoms of strontium move to the sea like black sheep in a flock of white ones. As a result of nuclear weapons tests and production of waste materials from experimental and industrial uses of atomic energy,

radioactive strontium is becoming widespread in the biosphere. This strontium is new material added to the biosphere since it is the result of the fission of uranium. Thus, a few new black sheep are being added, but these sheep are wolves in sheep's clothing, so to speak! Tiny amounts of radiostrontium have now followed calcium from soil and water into vegetation, animals, human food, and human bones. In 1970, 1 to 5 picocuries (see page 453 for a definition of this unit of radioactivity) of radioactive strontium were present per gram of calcium in the bones of people. Studies on the effect of radiostrontium in producing cancer have led some scientists to suggest that these levels produce harmful effects.

5. CYCLING OF ORGANIC NUTRIENTS

Statement

Heterotrophs and many autotrophs as well (many species of algae, for example) require vitamins or other organic nutrients which they must obtain from their environment. Such organic nutrients "cycle" between organisms and environment in the same general manner as do inorganic nutrients, except that for the most part they are of biotic rather than abiotic origin.

Explanation and Example

While the chemical nature of vitamins and other growth-promoting organic compounds and the amounts required by man and his domestic animals have long been known, the study of such substances at the ecosystem level has hardly begun. The concentration of organic nutrients in water or soil is so small that they might be appropriately called the "micro-micronutrients" in contrast to "macronutrients" such as nitrogen or "micronutrient" such as trace metals (see next chapter); often the only way to measure them is by *biological assay* (or bioassay), using special strains of microorganisms whose growth is proportional to concentration. As emphasized in the previous section of this chapter, concentration often gives little hint as to importance or rate of flux. It is becoming evident that organic nutrients do play an important role in community metabolism and that they can be limiting. This is a challenging field of study that will undoubtedly receive increasing attention in the near future. The following description of the vitamin B_{12} (cobalamins) cycle in the sea, restated from Provasoli (1963), illustrates how little is known about the cycling of organic nutrients.

The main producers of B_{12} are the microorganisms (mostly bacteria), though it is not excluded that autotrophic algae may be important either as direct producers of vitamins or, after their death, as food for vitamin-producing microorganisms. Bacteria and algae are the main direct consumers, although recent experiments indicate that animals, especially filter-feeders with extensive and highly permeable gills, may absorb vitamins directly as solutes. The non-living particles (clay, organic and inorganic micelles, detritus, etc.) absorb large quantities of vitamin B_{12} and on ingestion supply phagotrophs with vitamins. How much the removal of vitamins from particles affects vitamin cycles is unknown; we do not know whether these particles fix vitamins in a stable way or only transiently. Does partial elution maintain a certain level of vitamins as solutes during high consumption of vitamins by phytoplankton? Elution in deep muds might fertilize upwelling waters (see Chapter 12, page 325).

Vitamin B_{12}-like growth factors thus behave like other ecologically significant nutritional variables, but are they limiting, and where? Probably not in coastal waters where the concentration is usually well over 1 $m\mu g$/liter, but perhaps in the open ocean where the concentration is often less than 0.1 $m\mu g$/liter. In the Sargasso Sea, where the concentration of B_{12} is often so low as to be unmeasurable, the dominant organism is a flagellate that does not require this vitamin and has a very low requirement for other nutrients (here citing work of Menzel and Spaeth, 1962).

From what has so far been learned it is clear that the combining of chemical and bioassay analysis with enrichment experiments (see page 114 for an account of such experiments) and the determination *in vitro* of the nutritional characteristics of the dominant species permits good insight into the ecological events.

The role played by organic growth substances in regulating the distribution and the succession of organisms is discussed in Chapters 5 and 9.

6. NUTRIENT CYCLING IN THE TROPICS

Statement

The pattern of nutrient cycling in the tropics is, in several important ways, different from that in the temperate zone. In cold regions a large portion of the organic matter and available nutrients is at all times in the soil or sediment; in the tropics a much larger percentage is in the biomass and is recycled within the organic structure of the system. For this reason agricultural strategy of the temperate zone, involving the monoculture of short-lived annual plants, may be quite inappropriate for tropical regions. An ecological reevaluation of tropical agriculture in particular, and environmental management in general, is urgent if man is to correct past mistakes and avoid future disasters as he quite literally "bulldozes" his way into the jungles seeking more food and living space.

Explanation

Figure 4–10 compares the distribution of organic matter in a northern and a tropical forest. Interestingly enough, in this comparison both ecosystems contain about the same amount of organic carbon, but over half is in litter and soil in the northern forest while more than three-fourths is in the vegetation in the tropical forest. Another comparison of northern and tropical forests is provided by Table 14–5B (Chapter 14, see page 375). About 58 per cent of total nitrogen is in the biomass of a tropical forest—44 per cent of it above ground—as compared with 6 per cent and 3 per cent respectively, in a British pine forest.

When a temperate forest is removed, the soil retains nutrients and structure and can be farmed for many years in the "conventional" manner, which involves plowing one or more times a year, planting annual species, and applying inorganic fertilizers. During the winter freezing temperatures help hold in nutrients and control pests and parasites. In the humid tropics, on the other hand, forest removal takes away the land's ability to hold and recycle nutrients (as well as to combat pests) in the face of high year-around temperatures and periods of leaching rainfall. Too

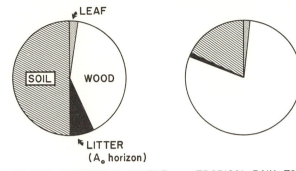

Figure 4–10. Distribution of organic carbon accumulated in abiotic (soil, litter) and biotic (wood, leaves) compartments of a northern and a tropical forest ecosystem. A much larger percentage of total organic matter is in the biomass in the tropical forest. (Redrawn from Kira and Shidei, 1967.)

often crop productivity declines rapidly and the land is abandoned, creating the pattern of "shifting agriculture" about which so much has been written. Nutrient cycling in particular and community control in general, then, tend to be more "physical" in the north and more "biological" in the south. This brief account, of course, oversimplifies complex situations, but the contrast underlies what now appears to be the basic ecological reason why sites in the subtropics or tropics that support luxurious and highly productive forests yield so poorly under northern-style crop management.

Studies of natural ecosystems that obviously have evolved mechanisms to solve the nutrient cycling "problem" should hopefully provide clues for designing agricultural and forestry systems that are more appropriate for warm climates. Two such ecosystem-types that are highly productive are coral reefs and tropical rain forests. Studies conducted during the past 20 years have revealed that intricate symbiosis between autotrophs and heterotrophs involving special microorganism intermediaries may be the key to success in both types of ecosystem. On the coral reef it may be the unique coral-algae symbiosis, as described in some detail in Chapter 12, page 345. An intensive study of a Puerto Rican rain forest by a team of scientists, as reported in an important new monograph (H. T. Odum and Pigeon, 1970), reveals a number of nutrient conservation mechanisms (see Chapter 14). Among them are the fungal mycorrhiza associated with the root system (see Chapter 7, page 232) that act as living "nutrient traps." On the basis of observations in the Amazon basin, Went and Stark (1968) propose what they call a "direct mineral cycling theory" as follows:

The theory is based on the fact that the bulk of minerals available in the tropical rain forest ecosystems is tied up in dead and living organic systems. Little available mineral ever occurs free in the soil at one time. Mycorrhiza which is extremely abundant in the surface litter and thin humus of the forest floor is believed to be capable of digesting dead organic litter and passing minerals and food substances through their hyphae to living root cells. In this manner little soluble mineral leaks into the soil where it can be leached away.

While such a sweeping theory may, again, be an oversimplification, the evidence certainly suggests that developing and testing crop plants with well-developed mycorrhizal root systems should be undertaken. The use of large perennial plants and the nutrient cycling efficiency of paddy rice culture in the tropics have already been mentioned in Chapter 3. It is perhaps significant that rice paddies have been cultivated on the same site for over a 1000 years in the Philippines (Sears, 1957), a record of success that few agricultural systems in use today can claim! According to Sears, these rice terraces are interspersed with patches of forests that are preserved by "religious taboos." In order to avoid what Hutchinson (1967a) has called "the technological quick-fix with an ecological backlash," we should first find out if the intermixture of forest and paddies has something to do with this admirable balance before we rush in and recommend that the forests be bulldozed in order to plant more rice.

One thing is certain—the following message must be communicated to well-meaning foreign aid bureaus of northern countries: *Temperate-zone industrialized agro-technology cannot be transferred unmodified to tropical regions.*

7. RECYCLE PATHWAYS

Statement

Paralleling the 2-channel energy flow diagram (Fig. 3–8) two major pathways for the recycling or regeneration of nutrients within the food chain may be distinguished as follows: (I) return by way of primary animal excretion and (II) return by way of microbial decomposition of detritus. While both function in any ecosystem, recycle pathway (I) may be expected to predominate in plankton and other communities in which the major energy flow is by way of the grazing food chain. Conversely, recycle pathway (II) will predominate in grasslands, temperate-zone forests and other communities in which the principal flow of energy is by way of the detritus food chain. A third pathway (III) has also been proposed that involves a direct cycling from plant to plant through symbiotic microorganisms.

Explanation

Having outlined the mass movements of nutrients on the local and global scale, it is appropriate that we close this chapter by focusing on nutrient cycling within the biologically active portion of the ecosystem. Recall that the same approach was used in Chapter 2 in regard to energy; the total energy environment was considered first, and then attention was focused on the fate of that small portion which is involved in the food chain. Also, a final emphasis on biological regeneration is relevant because, as already emphasized, "recycle" must increasingly become a major goal for human society.

Basic recycle pathways are diagramed in Figure 4–11. This compartment model is displayed in the same general manner as comparable models for food chains, only here we are concerned with the circular behavior of nutrients rather than the one-way flow of energy. The classic assumption has been that bacteria and fungi are the major

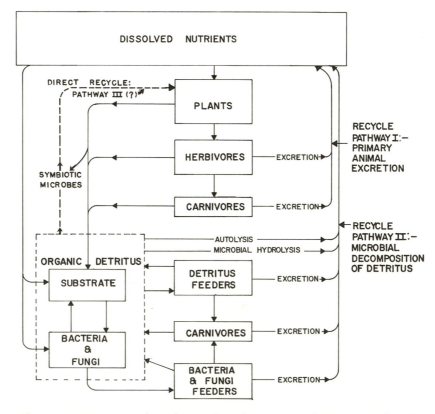

Figure 4–11. The major nutrient recycle pathways. Recycle routes I and II correspond to the grazing and detritus food chains (compare Figure 3–8). A third, more direct recycle pathway (III) has been proposed, as discussed in Sections 6 and 7. The autolysis pathway shown in the diagram could be considered a fourth pathway. (Model adapted and enlarged from Johannes, 1968—Figure 1, page 209.)

agents in nutrient regeneration, because the recycle pathway (II) in Figure 4–11 is certainly a dominant one in soils of the temperate zone where the regeneration process has been most studied. With regard to the sea, however, Rittenberg (1963) and Johannes (1968) have challenged this assumption. Rittenberg pointed out that the importance of bacteria in nutrient regeneration in the sea had never been successfully demonstrated, while Johannes postulates that in the water column the regeneration of nitrogen and phosphorus, at least, results primarily from animal excretion, as shown by recycle pathway (I) in Figure 4–11. Excretions from very small animals or "microzooplankton," which are too small to be collected in plankton nets (and thus were overlooked in the early studies of marine communities), appear to be especially important (Johannes, 1964). Recent measurements of turnover rate indicate that zooplankton in general (the herbivores and carnivores in Figure 4–11) release during their life span many times the amount of soluble nutrients as would be present for microbial decomposition after their death (Harris, 1959; Rigler, 1961; Pomeroy *et al.*, 1963, and many others). These excretions include dissolved inorganic and organic compounds of phosphorus and nitrogen (and, of course, CO_2), which are directly usable by producers without the necessity of any further chemical breakdown by bacteria. Presumedly, other vital nutrients are regenerated in this manner, although quantitative studies have not yet been made. In summary, where producers are restricted to phytoplankton in the water column, as in deep-water ecosystems, recycle pathway (I) can be expected to predominate, which may explain why microbiologists have generally found bacterial populations in sea water to be very low. It must be admitted, however, that methods of "censusing" bacteria and assaying their activities in nature are still primitive and not quantitatively accurate. In other ecosystems the importance of this pathway should be proportional to the importance of the grazing food chain. Since animals (i.e., phagotrophs as contrasted with saprotrophs) may thus be important decomposers and nutrient regenerators, the term "decomposer," formerly synonymous with saprophagic bacteria and fungi, needs to be broadened, as was alluded to in Chapter 2 (see page 11).

Although extensively studied, microbial decomposition involves such a variety of organisms and so complicated a network of exchanges and feedbacks that the totality of recycle pathway (II) is actually not yet well comprehended. Some of the more clearly understood processes, and the importance of the time lag, have been reviewed in previous chapters. For the time being, at least, one must consider the whole detritus complex as a "black box" (as enclosed by the dashed lines in Figure 4–11) having internal workings that are but vaguely known. Some suggestions of how to deal with this problem in field studies of soil and sediment subsystems are reviewed by Wiegert, Coleman, and E. P. Odum (1970), while some additional aspects are considered in Chapters 14 and 19.

As outlined in the preceding section, studies of tropical ecosystems, especially the tropical rain forest, have suggested the existence of a third pathway—from plant back to plant perhaps by way of symbiotic fungi or other microorganism links. Such a route is indicated by pathway (III) in Figure 4–11, but more study is needed to determine if this possibility is operationally significant, and if so in what ecosystems.

Finally, it should be pointed out that nutrients may be released from the dead bodies of plants and animals, and from fecal pellets, without being attacked by microorganisms, as can be demonstrated by placing such materials under sterile conditions; this recycle possibility is indicated by the "autolysis" pathway in Figure 4–11. In aquatic or moist environments, especially where the bodies or dead particles are small (thus exposing a large surface-to-volume ratio), 25 to 75 per cent of the nutrients may be released by autolysis before microbial attack begins, according to Johannes's (1968) review of the literature. We might well consider autolysis as a *fourth major recycle pathway,* one that does not involve metabolic energy. As already emphasized in Chapter 3 (Figure 3–17C), recycle work accomplished by mechanical or physical means can provide an energy subsidy for the system as a whole. In designing waste disposal systems man frequently finds it profitable to provide an input of mechanical energy to pulverize organic matter and thus hasten its decomposition. Physical breakdown by the activities of large animals is also of undoubted importance in the release of nutrients from large, resistant pieces of detritus such as leaves or logs.

PRINCIPLES PERTAINING TO LIMITING FACTORS

1. LIEBIG'S "LAW" OF THE MINIMUM

Statement

To occur and thrive in a given situation, an organism must have essential materials which are necessary for growth and reproduction. These basic requirements vary with the species and with the situation. Under "steady-state" conditions the essential material available in amounts most closely approaching the critical minimum needed will tend to be the limiting one. This "law" of the minimum is less applicable under "transient-state" conditions when the amounts, and hence the effects, of many constituents are rapidly changing.

Explanation

The idea that an organism is no stronger than the weakest link in its ecological chain of requirements was first clearly expressed by Justus Liebig in 1840. Liebig was a pioneer in the study of the effect of various factors on the growth of plants. He found, as do agriculturists today, that the yield of crops was often limited not by nutrients needed in large quantities, such as carbon dioxide and water, since these were often abundant in the environment, but by some raw material, as boron, for example, needed in minute quantities but very scarce in the soil. His statement that "growth of a plant is dependent on the amount of foodstuff which is presented to it in minimum quantity" has come to be known as Liebig's "law" of the minimum. Many authors (see, for example, Taylor, 1934) have expanded the statement to include factors other than nutrients (e.g., temperature) and to include the time element. To avoid confusion it seems best to restrict the concept of the minimum to chemical materials (oxygen, phosphorus, etc.) necessary for physiological growth and reproduction, as was originally intended, and to include other factors

and the limiting effect of the maximum in the "law" of tolerance. Both concepts can then be united in a broad principle of limiting factors, as outlined below. Thus, the "law" of the minimum is but one aspect of the concept of limiting factors which in turn is but one aspect of the environmental control of organisms.

Extensive work since the time of Liebig has shown that two subsidiary principles must be added to the concept if it is to be useful in practice. The first is a constraint that Liebig's law is *strictly applicable only under steady-state conditions,* that is, when inflows balance outflows of energy and materials. To illustrate, let us recall the following from Chapter 3 (page 57):

For example, let us assume that carbon dioxide was the major limiting factor in a lake, and productivity was therefore in equilibrium with the rate of supply of carbon dioxide coming from the decay of organic matter. We shall assume that light, nitrogen, phosphorus, etc. were available in excess of use in this steady-state equilibrium (and hence not limiting factors for the moment). If a storm brought more carbon dioxide into the lake, the rate of production would change, and be dependent upon other factors as well. While the rate is changing there is no steady state and no minimum constituent; instead, the reaction depends on the concentration of *all* constituents present, which in this transitional period differs from the rate at which the least plentiful is being added. The rate of production would change rapidly as various constituents were used up until some constituent, perhaps carbon dioxide again, became limiting, and the lake system would once again be operating at the rate controlled by the law of the minimum.

The example of carbon dioxide is especially interesting in view of current controversies in the water pollution literature as to whether carbon dioxide or phosphorus is the major limiting factor in fresh water and, therefore, the key nutrient in the process of cultural eutrophication (see Kuentzel, 1969). Since cultural eutrophication usually produces a highly "unsteady" state, involving severe oscillations (i.e., heavy blooms of algae followed by die-offs, which in turn trigger another bloom on release of nutrients), then the "either/or" argument may be highly irrelevant because phosphorus, nitrogen, carbon dioxide and many other constituents may rapidly replace one another as limiting factors during the course of the transitory oscillations. Accordingly, there is no theoretical basis for any "one factor" hypothesis under such transient-state conditions. The strategy of pollution control designed to prevent eutrophication must involve reducing the input of *both* organic matter (which generates CO_2 and probably growth promoting organics) and mineral nutrients that are also required for cancerous rates of production.

The second important consideration is *factor interaction.* Thus, high concentration or availability of some substance, or the action of some factor other than the minimum one, may modify the rate of utilization of the latter. Sometimes organisms are able to substitute, in part at least, a chemically closely related substance for one that is deficient in the environment. Thus, where strontium is abundant, mollusks are able to substitute strontium for calcium to a partial extent in their shells. Some plants have been shown to require less zinc when growing in the shade than when growing in full sunlight; therefore, a given amount of zinc in the soil would be less limiting to plants in the shade than under the same conditions in sunlight.

2. SHELFORD'S "LAW" OF TOLERANCE

Statement

The presence and success of an organism depend upon the completeness of a complex of conditions. Absence or failure of an organism can be controlled by the qualitative or quantitative deficiency or excess with respect to any one of several factors which may approach the limits of tolerance for that organism.

Explanation

Not only may too little of something be a limiting factor, as proposed by Liebig, but also too much, as in the case of such factors as heat, light, and water. Thus, organisms have an ecological minimum and maximum, with a range in between which represents the *limits of tolerance.* The concept of the limiting effect of maximum as well as minimum was incorporated into the "law" of tolerance by V. E. Shelford in 1913. From about 1910, much work has been done in "toleration ecology," so that the limits within which various plants and animals can exist are known. Especially useful are what can be termed "stress tests," carried out in the laboratory or field in which organisms are subjected to an experimental range of conditions (see Hart, 1952). Such a physiological approach has helped us to understand the distribution of organisms in nature; however, we should hasten to say, it is only part of the story. All physical requirements may be well within the limits of tolerance for an organism and the organism may still fail as a result of biological interrelations. As will be indicated by a number of examples that follow, studies in the intact ecosystem must accompany experimental laboratory studies, which, of necessity, isolate individuals from their populations and communities.

Some subsidiary principles to the "law" of tolerance may be stated as follows:

1. Organisms may have a wide range of tolerance for one factor and a narrow range for another.

2. Organisms with wide ranges of tolerance for all factors are likely to be most widely distributed.

3. When conditions are not optimum for a species with respect to one ecological factor, the limits of tolerance may be reduced with respect to other ecological factors. For example, Penman (1956) reports that when soil nitrogen is limiting, the resistance of grass to drought is reduced. In other words, he found that more water was required to prevent wilting at low nitrogen levels than at high levels.

4. Very frequently it is discovered that organisms in nature are not actually living at the optimum range (as determined experimentally) with regard to a particular physical factor. In such cases some other factor or factors are found to have greater importance. Certain tropical orchids, for example, actually grow better in full sunlight than in shade, provided they are kept cool (see Went, 1957); in nature they grow only in the shade because they cannot tolerate the heating effect of direct sunlight. In many cases population interactions (such as competition, predators, parasites, and so on), as will be discussed in detail in Chapter 7, pre-

vent organisms from taking advantage of optimum physical conditions.

5. The period of reproduction is usually a critical period when environmental factors are most likely to be limiting. The limits of tolerance for reproductive individuals, seeds, eggs, embryos, seedlings, and larvae are usually narrower than for nonreproducing adult plants or animals. Thus, an adult cypress tree will grow on dry upland or continually submerged in water, but it cannot reproduce unless there is moist unflooded ground for seedling development. Adult blue crabs and many other marine animals can tolerate brackish water, or fresh water that has a high chloride content; thus, individuals are often found for some distance up rivers. The larvae, however, cannot live in such waters; therefore, the species cannot reproduce in the river environment and never become established permanently. The geographical range of game birds is often determined by the impact of climate on eggs or young rather than on the adults.

To express the relative degree of tolerance, a series of terms have come into general use in ecology that utilize the prefixes "steno-," meaning narrow and "eury-," meaning wide. Thus,

stenothermal—eurythermal	refers to temperature
stenohydric–euryhydric	refers to water
stenohaline—euryhaline	refers to salinity
stenophagic—euryphagic	refers to food
stenoecious—euryecious	refers to habitat selection

As an example, let us compare the conditions under which brook trout (*Salvelinus*)

eggs and leopard frog (*Rana pipiens*) eggs will develop and hatch. Trout eggs develop between 0° and 12°C. with optimum at about 4°C. Frog eggs will develop between 0° and 30°C. with optimum at about 22°. Thus, trout eggs are stenothermal, low-temperature tolerant, compared with frog eggs, which are eurythermal, high-temperature tolerant. Trout in general, both eggs and adults, are relatively stenothermal, but some species are more eurythermal than the brook trout. Likewise, of course, species of frogs differ. These concepts, and the use of terms in regard to temperature, are illustrated in Figure 5–1. In a way, the evolution of narrow limits of tolerance might be considered a form of specialization, as discussed in the ecosystem chapter, which results in greater efficiency at the expense of adaptability, and contributes to increased diversity in the community as a whole (see Sections 4 and 5, Chapter 6).

The Antarctic fish *Trematomus bernacchi* and the desert pupfish *Cyprinodon macularius* provide an extreme contrast in limits of tolerances related to the very different environments in which they live. The Antarctic fish has a limit of temperature tolerance of less than 4°C. in the range of −2° to +2°, and is thus extremely stenothermally cold-adapted. As temperature rises to 0° the rate of metabolism increases but then declines as the temperature of the water rises to +1.9°, at which point the fish become immobile with heat prostration (see Wohlschlag, 1960). In contrast, the desert fish is eurythermal and also euryhaline, tolerating temperatures between 10° and 40°C. and salinities

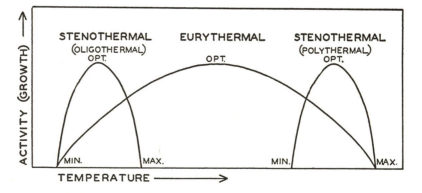

Figure 5–1. Comparison of the relative limits of tolerance of stenothermal and eurythermal organisms. Minimum, optimum, and maximum lie close together for a stenothermal species, so that a small difference in temperature, which might have little effect on a eurythermal species, is often critical. Note that stenothermal organisms may be either low-temperature tolerant (oligothermal), high-temperature tolerant (polythermal), or in between. (After Ruttner, 1953.)

ranging from fresh water to that greater than sea water. The ecological performance, of course, is not equal throughout such a range; food conversion, for example, is greatest at 20° and 15‰ salinity (Lowe and Heath, 1969).

Factor Compensation and Ecotypes

As we have emphasized many times in this book, organisms are not just "slaves" to the physical environment; they adapt themselves and modify the physical environment so as to reduce the limiting effects of temperature, light, water, and other physical conditions of existence. Such *factor compensation* is particularly effective at the community level of organization, but also occurs within the species. Species with wide geographical ranges almost always develop locally adapted populations called *ecotypes* that have optima and limits of tolerances adjusted to local conditions. Compensation along gradients of temperature, light, or other factors may involve genetic races (with or without morphological manifestations) or merely physiological acclimation. Reciprocal transplants provide a convenient method of determining to what extent genetic fixation is involved in ecotypes. McMillan (1956), for example, found that prairie grasses of the same species (and to all appearances identical) transplanted into experimental gardens from different parts of their range responded quite differently to light. In each case the timing of growth and reproduction was adapted to the area from which the grasses were transplanted, and the adapted growth behavior persisted when the grasses were transplanted. The possibility of genetic fixation in local strains has often been overlooked in applied ecology; restocking or transplanting of plants and animals may fail because individuals from remote regions were used instead of locally adapted stock. Factor compensation in local or seasonal gradients can also involve genetic races but often is accomplished by physiological adjustments in organ functions or by shifts in enzyme-substrate relationships at the cellular level. Somero (1969), for example, points out that immediate temperature compensation is promoted by an inverse relationship between temperature and enzyme-substrate affinity, while longer-term evolutionary adaptation is more likely to involve changes in enzyme-substrate affinity itself. Animals, especially larger ones with well-developed powers of movement, compensate through adapted behavior that avoids the extremes in local environmental gradients. Examples of such behavior regulation (which can be just as effective as internal physiological regulation) are cited in Chapter 8, Section 8.

At the community level, factor compensation is most frequently accomplished by species replacement in the environmental gradient. Since many examples of this will be described in Part 2 of this text, we need only cite one example here. In coastal waters copepods of the genus *Acartia* are often dominant forms in the zooplankton. Commonly, the species present in winter will be replaced in summer by other species that are more precisely adapted to warmer temperatures (see Hedgepeth, 1966).

Figure 5–2 illustrates two instances of temperature compensation, one at the species level and one at the community level. As shown in Figure 5–2A, northern jellyfish can swim actively at low temperatures that would completely inhibit individuals from the southern populations. Both populations are adapted to swim at about the same rate, and both function, to a remarkable extent, independently of the temperature variations in their particular environment. In Figure 5–2B we see that the rate of respiration of the whole community in a balanced microcosm is less affected by temperature than is the respiratory rate of one species (*Daphnia*). In the community many species with different optima and temperature responses develop reciprocal adjustments and acclimations enabling the whole to compensate for ups and downs in temperature. In the example shown (Fig. 5–2B) temperatures 8 to 10°C. above or below the temperature to which the microcosm was acclimated did result in a slight decrease in respiration; however, the effect was negligible compared to the more than two-fold effect of this range of temperature on *Daphnia. In general, then, metabolic rate-temperature curves will be flatter for ecosystems than for species,* another example, of course, of community homeostasis.

For reviews of the physiological basis for factor compensation, see Bullock (1955), Fry (1958), and Prosser (1967).

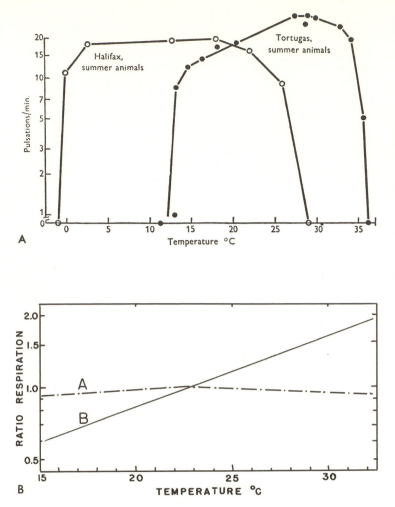

Figure 5–2. Temperature compensation at the species and community levels. *A.* The relation of temperature to swimming movement in northern (Halifax) and southern (Tortugas) individuals of the same species of jellyfish, *Aurelia aurita*. The habitat temperatures were 14° and 29°C., respectively. Note that each population is acclimated to swim at a maximum rate at the temperature of its local environment. The cold-adapted form shows an especially high degree of temperature independence. (From Bullock, 1955, after Mayer.) *B.* The effect of temperature on respiration of **A**, a balanced laboratory microcosm community, and **B**, a single species component, *Daphnia* (a small crustacean; see Figure 11–7, page 305). The relative change in the rate of CO_2 production is plotted as a ratio of the rate at 23°C., the temperature to which the microcosm was adapted. (Redrawn from Beyers, 1962.)

3. COMBINED CONCEPT OF LIMITING FACTORS

Statement

The presence and success of an organism or a group of organisms depends upon a complex of conditions. Any condition which approaches or exceeds the limits of tolerance is said to be a limiting condition or a limiting factor.

Explanation

By combining the idea of the minimum and the concept of limits of tolerance we arrive at a more general and useful concept of limiting factors. Thus, organisms are controlled in nature by (1) the quantity and variability of materials for which there

is a minimum requirement and physical factors which are critical and (2) the limits of tolerance of the organisms themselves to these and other components of the environment.

The chief value of the concept of limiting factors lies in the fact that it gives the ecologist an "entering wedge" into the study of complex situations. Environmental relations of organisms are apt to be complex, so that it is fortunate that not all possible factors are of equal importance in a given situation or for a given organism. Some strands of the rope guiding the organism are weaker than others. In a study of a particular situation the ecologist can usually discover the probable weak links and focus his attention, initially at least, on those environmental conditions most likely to be critical or "limiting." If an organism has a wide limit of tolerance for a factor which is relatively constant and in moderate quantity in the environment, that factor is not likely to be limiting. Conversely, if an organism is known to have definite limits of tolerance for a factor which also is variable in the environment, then that factor merits careful study, since it might be limiting. For example, oxygen is so abundant, constant, and readily available in the terrestrial environment that it is rarely limiting to land organisms, except to parasites or those living in soil or at high altitudes. On the other hand, oxygen is relatively scarce and often extremely variable in water and is thus often an important limiting factor to aquatic organisms, especially animals. Therefore, the aquatic ecologist has his oxygen determination apparatus in readiness and takes measurements as one of his procedures in the study of an unknown situation. The terrestrial ecologist, on the other hand, would less often need to measure oxygen even though, of course, it is just as vital a physiological requirement on land as in water.

To sum up, the first and primary attention should be given to factors that are "operationally significant" to the organism at some time during its life cycle. It is particularly important for the beginning ecologist to realize that the aims of environmental analysis are not to make long uncritical lists of possible "factors," but rather to achieve these more significant objectives: (1) to discover, by means of observation, analysis and experiment, which factors are "operationally significant" and (2) to determine how these factors bring about their effects on the individual, population or community, as the case may be. For an example of a predictive model based on singling out a few operationally significant factors, see Figure 12–5, page 333.

As shown in Figure 5–3, Fry (1947) has presented a graphic model that summarizes the general principle of limiting factors. This diagram brings out the important point that the actual range of tolerance in nature (as shown by the solid lines in the figure) is almost always narrower than the potential range of activity (dotted lines in the figure) such as might be indicated, for example, by noting short-term behavioral response in the laboratory. Usually accessory factors (the factor interaction previously mentioned) and the metabolic cost of physiological regulation at extreme conditions reduces the limits of tolerance at both upper and lower limits. As shown in Figure 5–3,

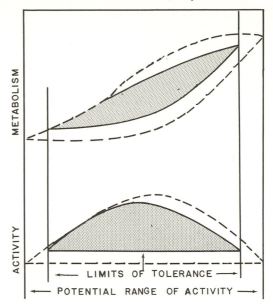

Figure 5–3. A model summarizing the general principles of limiting factors. The potential upper and lower (basal) rates of metabolism and the potential range of activity are indicated by the dotted lines. The actual ranges of metabolism or activity are indicated by the solid lines that enclose the stippled areas, depicting the reduced limits of tolerance; resulting from (1) the cost of physiological regulation that raises the minimum (or basal) metabolism rate, and (2) accessory factors in the environment that lower the upper metabolic capability, especially at the upper limits of tolerance. (Redrawn from F. E. J. Fry, 1947.)

both the scope (horizontal dimension) and the range (vertical dimension) of metabolic activity may be reduced by these interactions. Also, the optimum may be shifted, in this case to the left. Thus, the limits of tolerance of fish to thermal pollution cannot be determined simply by noting survival in a tank. If a fish has to devote all its metabolic energy to physiological adaptation it will have insufficient energy for food getting and the reproduction activities required for survival in nature. Adaptation becomes increasingly costly, energy-wise, as extreme conditions are approached. Anything that reduces this cost frees energy that can be used for growth or reproduction or for increased activity of other kinds (see concept of energy as a subsidy or stress, Chapters 3 and 16).

Examples

Several examples will serve to illustrate both the importance of the concept of limiting factors and the limitations of the concept itself.

1. As one drives along the broad highways of America from the Mississippi River to the Colorado Rockies, rainfall decreases gradually as one goes west. Water becomes the all-important limiting factor to plants, animals, and man. Trees give way to grassland as the amount of available water drops below the limits of tolerance for forests. Likewise, with increasing aridity, tall grass gives way to short grass species (see page 389). Thus, an annual rainfall of 16 inches (40 cm) is below the limit necessary for the little blue-stem grass, *Andropogon scoparius*, but adequate for grama grass, *Bouteloua gracilis*. However, under certain soil conditions which increase the availability of water to the plant, blue-stem grass is able to survive and compete locally in regions of 16 inches of rainfall (Rübel, 1935).

2. Ecosystems developing on unusual geological formations often provide instructive sites for limiting factor analysis since one or more of the important chemical elements may be unusually scarce or unusually abundant. Such a situation is provided by serpentine soils (derived from magnesium-iron-silicate rocks) which are low in major nutrients (Ca, P, N) and high in magnesium, chromium and nickel, with concentrations of the latter two approaching toxic levels for organisms. Vegetation growing on such soils has a characteristic stunted appearance, which contrasts sharply with adjacent vegetation on nonserpentine soils, and comprises an unusual flora with many endemic species and ecotypes (see the symposium edited by Whittaker, 1954). In attempting to single out the significant limiting factors Tadros (1957) experimented with two species of shrubs of the genus *Emmeranthe*, one restricted to serpentine soils in western United States and the other never found on such soils. He found that the nonserpentine species would not grow on serpentine soils, but that the serpentine species would grow quite well on normal garden soil provided that the soil was first sterilized, indicating that it is restricted to the peculiar soils by its inability to tolerate biotic competition of some kind. For a comparison of ecosystem function on serpentine and nonserpentine soils, see McNaughton, 1968 (see also page 267). The role which soil limitations play in community development is discussed in Chapter 9, Section 2 (see especially Figures 9–6 and 9–7).

3. Great South Bay in Long Island Sound, New York, provides a dramatic example of how too much of a good thing can completely change an ecosystem, to the detriment of man's interest in this case (recall from Chapter 2 that "too much of a good thing" is getting to be a general problem for mankind; see also Section 3, Chapter 9). This story, which might be titled "The Ducks vs. the Oysters," has been well documented and the cause and effect relations verified by experiment (Ryther, 1954). The establishment of large duck farms along the tributaries leading into the bay resulted in extensive fertilization of the waters by duck manure and a consequent great increase in phytoplankton density. The low circulating rate in the bay allowed the nutrients to accumulate rather than to be flushed out to sea. The increase in primary productivity might have been beneficial were it not for the fact that the organic form of the added nutrients and the low nitrogen-phosphorus ratio produced a complete change in the type of producers; the normal mixed phytoplankton of the area consisting of diatoms, green flagellates and dinoflagellates was almost completely replaced by very small, little-known green flagellates of genera *Nannochloris* and *Stichococcus*. (The most common species was so little known to marine botanists that it had to be described as a new species.) The famous "blue-point" oysters, which had been thriving for years on a diet of the normal phytoplankton and supporting a profitable industry, were unable to utilize the newcomers as food and gradually disappeared; oysters were found starving to death with a gut full of undigested green flagellates. Other shellfish were also eliminated and all attempts to reintroduce them failed. Culture experiments demonstrated that the green flagellates grow well when nitrogen is in the form of urea, uric acid, and ammonia, while the diatom *Nitzschia*, a "normal" phytoplankter, requires inorganic nitrogen (nitrate). It was clear that the flagellates could "short-circuit" the nitrogen cycle, that is, they did not have to wait for organic material to be reduced to nitrate (see Chapter 4, Section 7, Figure 4–11). This case is perhaps a good example of how a "specialist" which is normally rare in the usual fluctuating environment "takes over" when unusual conditions are stabilized.

This example also points up the common experience among laboratory biologists (reemphasized in Chapter 19) who find that the common species of unpolluted nature are often hard to culture in the laboratory under conditions of constant temperature and enriched media because they are adapted to the opposite, i.e., low nutrients and variable conditions. On the other hand, the "weed" species, normally rare or transitory in nature, are easy to culture because they are stenotrophic and thrive on enriched (i.e., "polluted") conditions. A good example of such a weed species is *Chlorella*, the alga highly touted for space travel and for solving man's world food problem (see page 500).

4. *Cordylophora caspa* is apparently an example of a euryhaline organism which does not actually live in waters of a salinity which is optimum for its growth. Kinne (1956) has made a detailed study of this species of marine hydroid (coelenterate) under laboratory conditions of controlled salinity and temperature. He found that a salinity of 16 parts per thousand resulted in the best growth, yet the organism was never found at this salinity in nature, but always at a much lower salinity; the reason for this has not yet been discovered.

At this point it would be well to comment upon the importance of combining field observation and analysis with laboratory experimentation, since the value of this approach is evident from the last three examples listed above. In the case of the serpentine soils, for example, detailed field analysis uncovered some of the probable limiting factors, but experimental work revealed a possibility which would not be uncovered by field observation alone. In the duck-oyster example, laboratory experiments verified the findings of the field analysis; these findings, of course, could not have been proved by field study alone. In case of the hydroid, the experimental approach revealed a degree of tolerance which would not have been suspected from field observation; in this case it is clear that the experimentation must be followed by field analysis if the natural situation is to be understood. In fact, it seems probable that no situation in nature can be really understood from either observation or experimentation alone, since each approach has obvious limitations. In the training of biologists during the past 40 years, there has been an unfortunate cleavage between the laboratory and the field, with the result that one group tended to be trained entirely in laboratory

philosophy (which developed little appreciation or tolerance for field work), while another group tended to be trained quite as narrowly in field techniques. Modern ecology, of course, has become especially relevant to our times because it breaks through this artificial barrier and provides a meeting ground for the biochemist and physicist on the one hand and the range, forest, or crop manager on the other!

Incidentally, the study of a series of situations in which environmental factors vary along a gradient is a good way to determine which factors are actually the limiting ones (see pages 145 and 154). One may be easily fooled or may jump to a premature conclusion as the result of a limited observation of a single situation. Hunters, fishermen, amateur naturalists, and laymen who are much interested in nature's complexes, and who are often keen observers, nevertheless are too often guilty of "jumping to conclusions" in regard to limiting factors. Thus, a sportsman may see an osprey catch a fish or a hawk catch a quail, and conclude that predators are the principal limiting factors in fish and quail populations. Actually, when the situation is well studied, more basic but less spectacular factors are generally found to be more important than large predators. Unfortunately, much time and money is wasted on predator control without the real limiting factors ever being discovered or the situation improved from the standpoint of increased yield.

5. Often a good way to determine which factors are limiting to organisms is to study their distribution and behavior at the edges of their ranges. If we accept Andrewartha and Birch's (1954) contention that distribution and abundance are controlled by the same factors, then study at range margins should be doubly instructive. However, many ecologists think that quite different factors may limit abundance in the center of ranges and distribution at the margins, especially since geneticists have reported that individuals in marginal populations may have different gene arrangements from central populations (see Carson, 1958). In any event the biogeographical approach (see page 363) becomes especially important when one or more environmental factors undergoes sudden or drastic change, thus setting up a natural experiment that is often superior to a laboratory experiment, because factors other than the one under consideration continue to vary in a normal man-

ner instead of being "controlled" in an abnormal, constant manner. Certain birds which have extended their breeding ranges within the last 50 to 100 years provide other examples of fortuitous field tests which aid in the determination of limiting factors. For example, when such song birds as the American robin (*Turdus migratorius*), the song sparrow (*Melospiza melodia*), and the house wren (*Troglodytes aedon*) extended their range southward, analyses indicated that man's alteration of the vegetation was the cause, and, therefore, temperature (or other climatic factors) was not the limiting factor in establishing the southern border in the original range (Odum and Burleigh, 1946; Odum and Johnston, 1951). In most cases, an appreciable lag occurs between the time of widespread change and the actual occupation of new territory because it takes time for population increase to occur. Once begun, however, invasion sometimes is very rapid, almost explosive in nature (see Elton, 1958). Studies in extreme environments such as the Antarctic or hot springs provide clues not only about the limits of physiological adaptation but also about the role that community organization may play in reducing physical limits. For example, certain flies in hot springs that are not particularly well adapted to high temperatures live in cooler "microclimates" created by the mat of algae that are tolerant (see page 302).

The Quantitative Expression of Limiting Factors

If a principle is to become firmly established and to prove useful in practice, it must eventually be subject to quantitative as well as qualitative analysis. Klages (1942) has developed a simple method of determining the optimum regions for agricultural crops. He considers not only the average yield over a period of years but also the coefficients of variation of the yields. The region with the highest average yield and the lowest coefficient of variation (hence the fewest crop failures) is the optimum region. As shown in Figure 5–4, Wisconsin and Ohio proved to be the optimum states for barley, as determined by this method. While yields are high further west, the variation was much greater due to greater year to year uncertainty in the rainfall.

Ever since the time of Liebig the most widely used approach to the determination of limiting factors is what might be called the "artificial enrichment experiment." This broad category includes the very unquantitative "trial and error" fertilization experiments that characterize the early development of agriculture, the unplanned cultural eutrophication about which we have already spoken, as well as more carefully designed experiments. As emphasized in Section 1 of this chapter, the problem with any enrichment experiment is that it creates a temporary transient, or unsteady, state which may make the interpretation of results difficult. Nevertheless, if the background knowledge about the ecosystem is adequate and if accessory factors are considered, then the enrichment approach can be useful and quantitative. Experiments of Menzel and Ryther (1961) and Menzel, Hulbert, and Ryther (1963) can be cited as an example. These investigators were interested in finding out what nutrients limit phytoplankton productivity in the Sargasso Sea, which is a kind of "marine desert." Their experiments brought out the importance of the time factor. Experiments lasting 1 hour, 24 hours, and several days often gave different results because the species composition sometimes changed during the longer experiments in response to the enrichment. It was concluded that experiments should not be much longer than the generation or turnover time of the organisms if the objective is to determine what limits the original populations. On the other hand, if the experiment is too short, the conclusions may also be misleading. For example, iron enrichment produced increased carbon uptake by the phytoplankton during the first 24 hours, but in order to sustain the increased production rate for several days it was necessary also to increase nitrogen and phosphorus. Menzel and coworkers concluded that the chief value of enrichment experiments was to determine what populations were capable of evolving into more productive ones in the presence of more nutrients.

As we have emphasized already in this chapter, it is highly desirable to carry out experiments in the field (the experiments described above were carried out on board an oceanographic ship). Goldman (1962) has described a method of studying nutrient limiting factors in situ in water columns isolated by polyethylene film. The im-

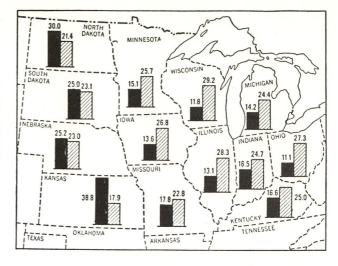

Figure 5–4. A method of determining the optimum regions for agricultural crops by comparing average yields and variability in yields from year to year. Cross-hatched columns indicate average yields in bushels per acre, and solid columns the coefficient of variation in the yields of barley in the states of the upper Mississippi valley. (After Klages, 1942.)

portance of this sort of field enclosure was mentioned in Chapter 2.

4. CONDITIONS OF EXISTENCE AS REGULATORY FACTORS

Statement

Light, temperature, and water (rainfall) are ecologically important environmental factors on land; light, temperature, and salinity are the "big three" in the sea. In fresh water other factors such as oxygen may be of major importance. In all environments the chemical nature and cycling rates of basic mineral nutrients are major considerations. All these physical conditions of existence may not only be limiting factors in the detrimental sense but also regulatory factors in the beneficial sense — that adapted organisms respond to these factors in such a way that the community of organisms achieves the maximum homeostasis possible under the conditions.

Explanation and Examples

Organisms not only adapt to the physical environment in the sense of tolerating it, but they "use" the natural periodicities in the physical environment to time their activities and "program" their life histories so as to benefit from favorable conditions. When we add interactions between organisms and reciprocal natural selection between species (coevolution, see page 271),

the whole community becomes programmed to respond to seasonal and other rhythms. The biological literature is laden with examples of adaptive responses. Usually these are described with reference to a particular group of organisms (for example, *Environmental Control of Plant Growth*, edited by Evans, 1963) or with regard to a particular habitat (for example, *Adaptations of Intertidal Organisms*, edited by Lent, 1969). Detailed consideration of regulatory adaptations is beyond the scope of this text, but perhaps two examples will suffice to bring out points of special ecological interest.

One of the most dependable cues by which organisms time their activities in temperate zones is the day-length period, or *photoperiod*. In contrast to most other seasonal factors, day length is always the same for a given season and locality. The amplitude in the annual cycle increases with increasing latitude, thus providing latitudinal as well as seasonal cues. At Winnipeg, Canada, the maximum photoperiod is 16.5 hours (in June) and the minimum is 8 hours (in late December). In Miami, Florida, the range is only 13.5 to 10.5 hours. Photoperiod has been shown to be the timer or trigger that sets off physiological sequences which bring about growth and flowering of many plants, molting, fat deposition, migration and breeding in birds and mammals, and the onset of diapause (resting stage) in insects. Photoperiodicity is coupled with what is now widely known as the organism's *biological clock* to create a timing mechanism of great versatility. The two contrasting theories concerning the mechanism of this coupling are considered in Section 6, Chapter 8. Day length acts through a sensory receptor, such as an eye in animals or a special pigment in the leaves of a plant, which, in turn, activates one or more back-to-back hormone and enzyme systems that bring about the physiological or behavioral response. It is not known just where in this sequence time is actually measured. Although the higher plants and animals are widely divergent in morphology, the linkage with environmental photoperiodicity is quite similar.

Among the higher plants some species bloom on increasing day length and are called long-day plants, while others that bloom on short days (less than 12 hours) are known as short-day plants. Animals likewise may respond to either long or short days. In many, but by no means all, photoperiod-sensitive organisms the timing can be altered by experimental or artificial manipulation of the photoperiod. As shown in Figure 5–5, an artificially speeded up light regime can bring brook trout into breeding condition four months early. Florists often are able to get flowers to bloom out of season by altering the photoperiod. In migratory birds there is a period of several months after the fall migration when the birds are refractory to photoperiod stimulation. The short days of fall are apparently necessary to "reset" the biological clock, as it were, and prepare the endocrine system for a response to long days. Anytime after late December an artificial increase in day length will bring on the sequence of molting, fat deposition, migratory restlessness, and gonad enlargement that normally occurs in the spring. The physiology of this response in birds is now fairly well known (see reviews by Farner, 1964, 1964a), but it is uncertain whether the fall migration is brought on by the direct stimulus of short days or is timed by the biological clock that was set by the long photoperiods of the spring.

Photoperiodism in certain insects is noteworthy because it provides a sort of "birth control." Long days of late spring and early summer stimulate the "brain" (actually a nerve cord ganglion) to produce a neurohormone that brings on the production of a diapause or resting egg that will not hatch until next spring no matter how favorable are temperatures, food, and other conditions (see Beck, 1960). Thus, population growth is halted before, rather than after, the food supply becomes critical.

It has even been shown that the number of underground nitrogen-fixing root nodules on legumes (see page 89, Figure 4–4) is controlled by photoperiod acting through the leaves of the plant. Since nitrogen-fixing bacteria in the nodules require food energy manufactured by the plant leaves to do their work, the more light and chlorophyll, the more food that becomes available to the bacteria; maximum coordination between the plant and its microbial partners is thus enhanced by the photoperiod regulator.

In striking contrast to day length, rainfall in a desert is highly unpredictable; yet desert annuals, which constitute the largest number of species in many desert floras (see page 393), use this factor as a regulator. The seeds of many such species contain a germination inhibitor that must be washed out by a certain minimum rainfall shower (for example, a half inch or more) that will provide all the water necessary to complete the life cycle back to seeds again. If such seeds are placed in moist soil in the greenhouse, they fail to germinate, but do so quickly when treated with a simulated shower of the necessary magnitude (see Went, 1955).

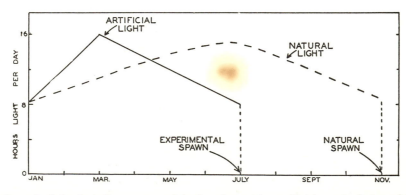

Figure 5–5. Control of the breeding season of the brook trout by artificial manipulation of the photoperiod. Trout, which normally breed in the autumn, spawn in summer when day length is increased artificially in the spring and then decreased in the summer to simulate autumn conditions. (Redrawn from Hazard and Eddy, 1950.)

Seeds may remain viable in the soil for many years, "waiting," as it were, for the adequate shower; this explains why deserts "bloom," that is, become quickly covered by flowers, a short time after heavy rainfall.

For reviews of photoperiodism see Evans, 1963; Salisbury, 1963; and Searle, 1965 (on plants); and Withrow, 1959 (on animals).

5. BRIEF REVIEW OF PHYSICAL FACTORS OF IMPORTANCE AS LIMITING FACTORS

As emphasized several times in the preceding discussions, the broad concept of limiting factors is not restricted to physical factors, since biological interrelations ("coactions" or "biological factors") are just as important in controlling the actual distribution and abundance of organisms in nature. However, the latter are best considered in subsequent chapters dealing with populations and communities, leaving the physical and chemical aspects of the environment to be reviewed in this section. To present all that is known in this field would require a book in itself, and would be beyond the scope of the present outline of ecological principles. Also, if we should become involved with the details, we would become sidetracked from our objective of obtaining an overall picture of the subject matter of ecology. Therefore, we need only to make a brief roll call of the items that ecologists have found important and worth studying.

1. TEMPERATURE. Compared with the range of thousands of degrees known to occur in our universe, life, as we know it, can exist only within a tiny range of about 300 degrees centigrade—from about −200° to 100°C. Actually, most species and most activity are restricted to an even narrower band of temperatures. Some organisms, especially in a resting stage, can exist at very low temperatures at least for brief periods, whereas a few microorganisms, chiefly bacteria and algae, are able to live and reproduce in hot springs where the temperature is close to the boiling point (see page 321). In general the upper limits are more quickly critical than the lower limits, despite the fact that many organisms appear to function more efficiently toward the upper limits of their tolerance ranges. The range of temperature variation tends to be less in water than on land, and aquatic organisms generally have a narrower limit of tolerance to temperature than equivalent land animals. Temperature, therefore, is universally important and is very often a limiting factor. Temperature rhythms, along with rhythms of light, moisture, and tides, largely control the seasonal and daily activities of plants and animals. Temperature is often responsible for the zonation and stratification which occur in both water and land environments (as will be described in Part 2). It is also one of the easiest of environmental factors to measure. The mercury thermometer, one of the first and most widely used precision scientific instruments, has more recently been supplemented by electrical "sensing" devices, such as platinum resistance thermometers, thermocouples (bimetallic junctions), and thermistors (metallic oxide thermally sensitive resistors), which not only permit measurement in "hard-to-get-at" places but also permit continuous and automatic recording of measurements. Furthermore, advances in the technology of telemetry now make it feasible to radiotransmit temperature information from the body of a lizard deep in its burrow or from a migratory bird flying high in the atmosphere (see Chapter 18).

Temperature variability is extremely important ecologically. A temperature fluctuating between 10 and 20°C. and averaging 15° does not necessarily have the same effect on organisms as a constant temperature of 15°C. It has been found that *organisms which are normally subjected to variable temperatures in nature* (as in most temperate regions) *tend to be depressed, inhibited or slowed down by constant temperature.* Thus, to give the results of one pioneer study, Shelford (1929) found that eggs and larval or pupal stages of the codling moth developed 7 or 8 per cent faster under conditions of variable temperature than under a constant temperature having the same mean. In another experiment (Parker, 1930), grasshopper eggs kept at a variable temperature showed an average acceleration of 38.6 per cent and nymphs an acceleration of 12 per cent, over development at comparable constant temperature.

It is not certain whether variation in itself is responsible for the accelerating effect, or whether the higher temperature causes more growth than is balanced by the low temperature. In any event, the stimulating effect of variable temperature, in the temperate zone at least, may be accepted as a well-defined ecological principle, and one that might be emphasized, since the tendency has been to conduct experimental work in the laboratory under constant temperature conditions.

Because organisms are sensitive to temperature changes and because temperature is so easy to measure, it is sometimes overrated as a limiting factor. One must guard against assuming that temperature is limiting when other, unmeasured factors may be more important. The widespread ability of plants, animals, and especially communities to compensate or acclimate to temperature has already been mentioned. Good advice for the beginning ecologist might be something as follows: In the study of a particular organism or problem, by all means consider temperature, but do not stop there.

2. RADIATION: LIGHT. As aptly expressed by Pearse (1939), organisms are on the horns of a dilemma in regard to light. Direct exposure of protoplasm to light results in death, yet light is the ultimate source of energy, without which life could not exist. Consequently, much of the structural and behavioral charac-

teristics of organisms are concerned with the solution of this problem. In fact, as described in some detail in Chapter 9, the evolution of the biosphere as a whole has chiefly involved the "taming" of incoming solar radiation so that its useful components can be exploited and its dangerous components mitigated or shielded out. Light, therefore, is not only a vital factor but also a limiting one, both at the maximum and minimum levels. There is no other factor of greater interest to ecologists!

The total radiation environment and something of its spectral distribution was considered in Chapter 3, as was the primary role of solar radiation in ecosystem energetics. Consequently, we need only consider light as a limiting and controlling factor in this chapter. Radiation consists of electromagnetic waves of a wide range in length. As shown in Figure 5–6, two bands of wave lengths readily penetrate the earth's atmosphere, namely, the visible band, together with some parts of adjacent bands, and the low-frequency radio band, having wave lengths greater than 1 centi-

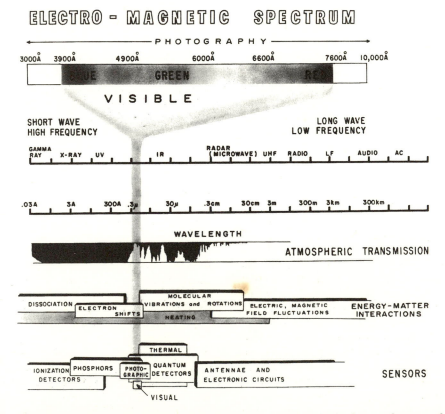

Figure 5–6. The electromagnetic spectrum in relation to the visible light, with an indication of atmospheric transmission, energetics, and methods of detection of different wave-frequency bands. Å = angstrom = 0.1 millimicron (mμ) = 0.0001 micron (μ). (Modified from Colwell *et al.*, 1963.)

meter. It is not known whether the long radio waves have ecological significance, despite assertions regarding positive effects on migrating birds or other organisms. As shown in Figure 3–1, the solar radiations which penetrate the upper atmosphere and reach the earth's surface consist of electromagnetic waves ranging in length from about 0.3 to 10 microns (μ); this equals from 300 to 10,000 mμ, or 3000 to 100,000 Å.[*] To the human eye, visible light lies in the range of 3900 to 7600 Å (390 to 760 mμ), as shown in Figure 5–6. This figure also shows the energy-matter interaction of different bands and the kind of sensors used to detect and measure them (see Chapter 18 for additional details). The role of ultraviolet (below 3900 Å) and infrared (above 7600 Å) has been considered in Chapter 3. The role which the high-energy, very short wave gamma radiation, as well as other types of ionizing radiation, may play as ecological limiting factors in the atomic age involves many special and complex considerations, which are best treated in Chapter 17.

Ecologically, the quality of light (wave length or color), the intensity (actual energy measured in gram-calories or foot candles), and the duration (length of day) are known to be important. Both animals and plants are known to respond to different wave lengths of light. Color vision in animals has an interesting "spotty" occurrence in different taxonomic groups, apparently being well developed in certain species of arthropods, fish, birds, and mammals but not in other species of the same groups (among mammals, for example, color vision is well developed only in primates). The rate of photosynthesis varies somewhat with different wave lengths. In terrestrial ecosystems the quality of sunlight does not vary enough to have an important differential effect on the rate of photosynthesis, but as light penetrates water, the reds and blues are filtered out and the resultant greenish light is poorly absorbed by chlorophyll. The red algae, however, have supplementary pigments (phycoerythrins) enabling them to utilize this energy and to live at greater depths than would be possible for the green algae (see page 333).

As was discussed in Chapter 3 the intensity of light (i.e., the energy input)

impinging on the autotrophic layer controls the entire ecosystem through its influence on primary production. The relationship of intensity to photosynthesis in both terrestrial and aquatic plants follows the same general pattern of a linear increase up to an optimum or *light saturation* level, followed in many instances by a decrease at very high intensities (Rabinowitch, 1951; Thomas, 1955). Light intensity-photosynthesis curves for several populations and communities are shown in Figure 5–7. As would be expected, factor compensation enters the picture since individual plants as well as communities adapt to different light intensities by becoming "shade-adapted" (i.e., reach saturation at low intensities) or "sun-adapted" (see Chapter 3, especially Figure 3–5). The extremes of light saturation are shown in Figure 5–7. Diatoms that live in beach sand or on intertidal mudflats are remarkable in that they reach a maximum rate of photosynthesis when light intensity is less than 5 per cent full sunlight, and they can maintain a net production at less than 1 per cent (Taylor, 1964). Yet these diatoms are only slightly inhibited by high intensities. Marine phytoplankton are shade-adapted but are very much inhibited by high intensities, which accounts for the fact that peak production in the sea usually occurs below, rather than at, the surface (see Figure 3–3). At the other extreme the "sun-loving" corn plant does not become light saturated until full sunlight intensities are reached (Fig. 5–7).

It is not so generally recognized that normal sunlight can be limiting when at full intensity, as well as when at low intensities. At high intensities, photo-oxidation of enzymes apparently reduces synthesis, and rapid respiration uses up the photosynthate. Protein synthesis is especially reduced so that high percentages of carbohydrates are produced at high intensities, which is one reason it is difficult to get good yields from high-protein crops in the tropics.

The role of light duration, or photoperiodicity, has been considered in Section 4.

3. WATER. A physiological necessity for all protoplasm, water, from the ecological viewpoint, is chiefly a limiting factor in land environments or in water environments in which the amount is subject to great fluctuation, or in which high salinity fosters water loss from organisms by osmosis (see Figure 11–3). Rainfall, humidity, the evaporating power of the air, and the available surface

---

[*] A micron (μ) is one-thousandth of a millimeter (10^{-3} mm); a millimicron (mμ) is one-millionth of a millimeter (10^{-6} mm); an angstrom (Å) is one-tenth of a millimicron (10^{-7} mm).

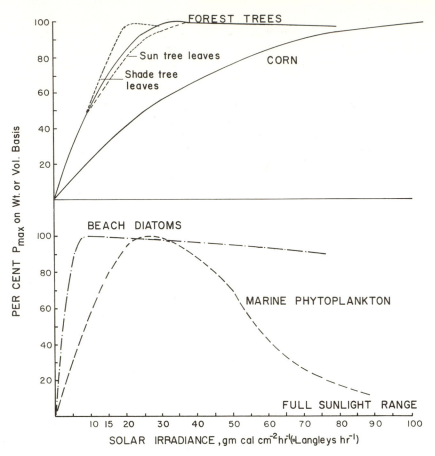

Figure 5–7. Light-photosynthesis curves for different plant populations, showing a wide range of response in terms of light saturation (light energy of maximum photosynthesis). (Upper diagram drawn from data tabulated by Hesketh and Baker, 1967; beach diatoms, data of Taylor, 1964; marine phytoplankton, data of Ryther, 1956.)

water supply are the principal factors measured. A brief résumé of each of these aspects follows.

Rainfall is largely determined by geography and the pattern of large air movements or "weather systems." A relatively simple example is shown in Figure 5–8. Moisture-laden winds blowing off the ocean deposit most of their moisture on the ocean-facing slopes, with a resulting "rain shadow" producing a desert on the other side; the higher the mountains the greater the effect, in general. As the air continues beyond the mountains, some moisture is picked up and rainfall may again increase somewhat. Thus deserts are usually found "behind" high mountain ranges or along the coast where winds blow from large, interior dry land areas rather than off the ocean as is the Pacific coast example shown in Figure 5–8. Rainfall distribution over the year is an extremely important limiting factor for organisms. The situation provided by a 35-

inch rainfall evenly distributed is entirely different from that provided by a 35-inch rainfall which falls largely during a restricted part of the year. In the latter case, plants and animals must be able to survive long periods of drought. In general, rainfall tends to be unevenly distributed over the seasons in the tropics and subtropics, often with well-defined wet and dry seasons resulting. In the tropics, this seasonal rhythm in moisture regulates the seasonal activities (especially reproduction) of organisms in much the same manner as the seasonal rhythm of temperature and light regulates temperate zone organisms. In temperate climates, rainfall tends to be more evenly distributed throughout the year, with many exceptions. The following tabulation gives a rough approximation of the climax biotic communities (see Chapter 14, Section 7) that may be expected with different annual amounts of rainfall evenly distributed in temperate latitudes:

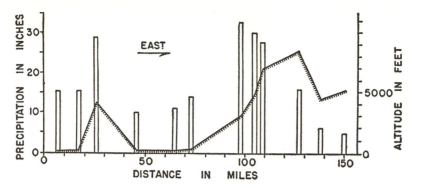

Figure 5–8. Mean annual rainfall (vertical columns) in relation to altitude (ornamented line) at a series of stations extending from Palo Alto in the Pacific coast eastward across the Coast Range and the Sierra Nevada to Oasis Ranch in the Nevada desert. The diagram shows (1) the approach effect on the west edge of the Sierra, (2) the zone of maximum rainfall on the middle western slope of the Sierra, and (3) rain shadows to the landward of the two mountain ranges. (After Daubenmire, *Plants and Environment*, John Wiley & Sons, Inc., 1947.)

0–10 inches per year..........Desert
10–30 inches per year..........Grassland, savanna,* or open woodland
30–50 inches per year..........Dry forest
over 50 inches per year........Wet forest

Actually, the biotic situation is not determined by rainfall alone but by the balance between rainfall and potential evapo-transpiration, the latter being loss of water by evaporation from the ecosystem, as will be discussed below.

Humidity represents the amount of water vapor in the air. Absolute humidity is the actual amount of water in the air expressed as weight of water per unit of air (grams per kilograms of air, for example). Since the amount of water vapor air can hold (at saturation) varies with the temperature and pressure, *relative humidity* represents the percentage of vapor actually present compared with saturation under existing temperature-pressure conditions. Relative humidity is usually measured by noting the difference between a wet and a dry bulb thermometer mounted on an instrument called a psychrometer. If both thermometers read the same, the relative humidity is 100 per cent; if the wet bulb thermometer reads less than the dry bulb one, as is usually the case, the relative humidity is less than 100 per cent, the exact value being determined by consulting prepared tables. Relative hu-

midity may also be conveniently measured by means of a hygrograph which provides a continuous record. Human hair, especially long blond hair, expands and contracts in proportion to relative humidity, and strands of it can thus be made to operate a lever writing on a moving drum. As is the case with temperature measurement, various electrical sensing devices are coming into widespread use. One such device utilizes the ability of a film of lithium chloride to change its electrical resistance in proportion to changes in relative humidity. Other hygroscopic materials are under experimentation.

In general, relative humidity has been the measurement most used in ecological work, although the converse of relative humidity, or *vapor pressure deficit* (the difference between partial pressure of water vapor at saturation and the actual vapor pressure), is often preferred as a measure of moisture relations, because evaporation tends to be proportional to vapor pressure deficit rather than to relative humidity.

Since there is generally a daily rhythm in humidity in nature (high at night, low during the day, for example), as well as vertical and horizontal differences, humidity along with temperature and light has an important role in regulating the activities of organisms and in limiting their distribution. Humidity has an especially important role in modifying the effects of temperature, as will be noted in the next section.

The evaporative power of the air is an important ecological factor, especially for land plants, and is usually measured by evaporimeters, which measure evaporation from pans, or atmometers, which measure

---

* A savanna is a grassland with scattered trees or scattered clumps of trees, a type of community intermediate between grassland and forest (see Figure 14–14).

evaporation from the surface of a porous bulb filled with water. Animals may often regulate their activities so as to avoid dehydration by moving to protected places or becoming active at night (see page 249); plants, however, have to "stand and take it." From 97 to 99 per cent of water which enters plants from the soil is lost by evaporation from the leaves, this evaporation being called *transpiration* which, as mentioned in Chapter 2, is a unique feature of the energetics of terrestrial ecosystems. When water and nutrients are nonlimiting, growth of land plants is closely proportional to the total energy supply at the ground surface, as we have already indicated. Since most of the energy is heat and since the fraction providing latent heat for transpiration is nearly constant, then growth is also proportional to transpiration (Penman, 1956). The relationships between evapotranspiration and primary productivity is expressed in the form of an equation model, shown on page 377. The ratio of growth (net production) and water transpired is called the *transpiration efficiency* and is usually expressed as grams of dry matter produced per 1000 grams of water transpired. Most agricultural crop species, as well as a wide range of noncultivated species, have a transpiration efficiency of 2 or less, that is, 500 grams or more of water are lost for every gram of dry matter produced (Briggs and Shantz, 1914; Norman, 1957). Drought resistant crops, such as sorghum and millet, may have efficiencies of 4. Strangely enough, desert plants can do little, if any, better; their unique adaptation involves not the ability to grow without transpiration but the ability to become dormant when water is not available (instead of wilting and dying as would be the case in non-desert plants). Desert plants that lose their leaves and expose only green buds or stems during dry periods do show a high transpiration efficiency (Lange *et al.*, 1969) but, of course, they do not make much food under these conditions (recall that high efficiency in light utilization also accompanies low yield; see page 61). It should be emphasized that desert plants, like all organisms, have upper as well as lower limits of tolerance, and each species may have a somewhat different range. When moisture is greatly increased in arid areas, as by irrigation, the primary productivity of the whole ecosystem is increased, but most, if not all, of the desert species of plants

die and are replaced by other species better adapted to high moisture. (Water can be "too much of a good thing" for the desert plant!)

The available surface water supply is, of course, related to the rainfall of the area, but there are often great discrepancies. Thus, owing to underground sources or supplies coming from nearby regions, animals and plants may have access to more water than that which falls as rain. Likewise, rain water may quickly become unavailable to organisms. Wells (1928) has spoken of the North Carolina sandhills as "deserts in the rain," because abundant rain of the region sinks so quickly through the porous soil that plants, especially herbaceous ones, find very little available in the surface layer. The plant and the small animal life of such areas resemble those of much dryer regions. Other soils in the western plains retain water so tenaciously that crops can be raised without a single drop of rain falling during the growing season, the plants being able to utilize the water stored from winter rains.

The general nature of the hydrological cycle has been considered in detail in Chapter 4, in which attention was drawn to the important, and still poorly understood, relationships between surface water and ground water and between rainfall and the atmospheric and ocean pools. Ecologists are in general agreement that we need to know more about water resources and to do a better job of managing them before we seriously consider manipulation of rainfall, as this becomes technically possible. We could really get ourselves into a mess by bungling both rainfall and surface water control! As it is, too severe removal of vegetative cover and poor land-use practices, with resultant destruction of soil texture and increased erosion, have often increased run-off to such an extent that local deserts are produced in regions of adequate rainfall. On the more positive side, irrigation and artificial impoundment of streams have aided in increasing local water supplies. However, these mechanical engineering devices, useful though they usually are, should never be regarded as substitutes for sound agricultural and forestry land-use practices, which trap the water at or near its sources for maximum usefulness to plants, animals, and man. The ecological viewpoint, considering *water as a cyclic commodity within the whole ecosystem,* is very important. People who think all our floods and erosion and water-use problems can be solved by building big dams or any other mechanical device alone may have a good appreciation of engineering, but they need to brush up on their ecology (see Chapters 9, 11, and 15). The idea that the watershed unit is a sort of minimum eco-system for management was first mentioned in Chapter 2, and examples of important research based on total watershed study are frequently cited in this text. Studies on experimental watersheds (see Figure 2–4 for a picture of such an experimental setup) provide very important life- and profit-sav-ing information if only political and business decision-makers will take notice! For ex-ample, logging access roads, rather than logging itself, often results in the chief dam-age to a forested watershed. If logging roads are planned and built properly (Fig. 15–6), damage from erosion is reduced and re-generation speeded up, and the operation is more profitable as well. Other experi-ments have shown that the important thing is not to destroy the vigorous interaction of plants, animals, and microorganisms that maintain the surface of the ground as a "living sponge," able to hold water and re-lease it gradually without excessive loss of valuable materials. Thus, the natural vegeta-tion can be modified in various ways to yield desirable products or to serve useful pur-poses without destroying the essential health of the biotic community, which is necessary for future productivity.

Dew may make an appreciable, and, in areas of low rainfall, a very vital contribu-tion to precipitation. Dew and "ground fog" may be important not only in coastal forests, where more total water may be precipitated in these forms than in rainfall (see page 385) but also in deserts. Recent development of better dew gauges and automatic-weighing lysimeters, which give the actual water up-take of vegetation, have resulted in better measurements. In many cases, the annual amount and the ability of plants to utilize the water have been greater than was thought formerly.

4. TEMPERATURE AND MOISTURE ACTING TOGETHER.   By considering the ecosystem concept first among ecological concepts we have avoided creating the impression that environmental factors operate independ-ently of one another. In this chapter we are attempting to show that consideration of individual factors is a means of getting at complex ecological problems but not the ultimate objective of ecological study, which is to evaluate the relative importance of various factors as they operate together in actual ecosystems. Temperature and mois-ture are so generally important in terrestrial environments and so closely interacting that they are usually conceded to be the most important part of climate. Thus, it may be well to consider them together before pro-ceeding to other factors.

The interaction of temperature and mois-ture, as in the case of the interaction of most factors, depends on the *relative* as well as the *absolute* values of each factor. Thus, temperature exerts a more severe limiting effect on organisms when moisture condi-tions are extreme, that is, either very high or very low, than when such conditions are moderate. Likewise, moisture plays a more critical role in the extremes of temperature. In a sense, this is another aspect of the prin-ciple of factor interaction, which was dis-cussed earlier in the chapter. For example, the boll weevil can tolerate higher tempera-tures when the humidity is low or moderate than when it is very high. Hot, dry weather in the cotton belt is a signal for the cotton farmers to have their spray guns ready and to be on the lookout for an increase in the weevil population. Hot, humid weather is less favorable for the weevil, but, unfortu-nately, not so good for the cotton plant! Large bodies of water greatly moderate land climates because of the high latent heat of evaporation and melting characteristic of water (see page 295). In fact, we may speak of two basic types of climate, namely, the continental climates, characterized by ex-tremes of temperature and moisture, and the marine climates, characterized by less extreme fluctuation due to the moderating effect of large bodies of water (large lakes thus produce local "marine climates").

Modern classifications of climate, such as those of Köppen or Thornthwaite (1931; 1948), are based largely on quantitative measures of temperature and moisture, tak-ing into consideration the effectiveness of precipitation and temperature (as deter-mined by growing seasons) and seasonal distribution, as well as the mean values. The comparison of precipitation and poten-tial evapo-transpiration (which is dependent on temperature) provides a particularly good picture of climates, as shown in Figure 5–9, which contrasts climates of three distinctly different biological regions or biomes (see

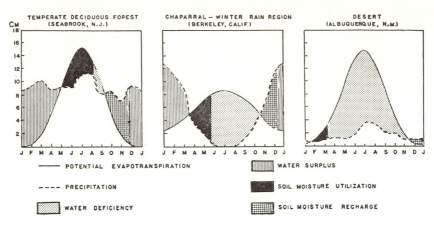

Figure 5–9. Relationship between rainfall and potential evapotranspiration (evaporation from soil plus transpiration from vegetation). in three localities representing three distinctly different ecological regions. The dotted area in the charts ("water deficiency") indicates the season during which water may be expected to be a limiting factor, whereas the vertical extent of this area indicates the relative severity of this limitation. (After Thornthwaite, 1955.)

Figures 14–7 and 14–8 for maps of the geographical extent of these biomes). The period of "soil moisture utilization" represents the principal period of primary production for the community as a whole and thus determines the supply of food available to the consumers and decomposers for the entire annual cycle. Note that in the deciduous forest region, water is likely to be severely limiting only in late summer, more so in the southern than in the northern portion of the region. Native vegetation is adapted to withstand the periodic summer droughts but some agricultural crops grown in the region are not. After bitter experiences with many late summer crop failures, farmers in the southern United States, for example, are finally beginning to provide for late summer irrigation. In the winter rain region, the main season of production is late winter and spring, while in the desert the effective growing season is much reduced.

In general, the classification of climatic types based on temperature-moisture indices correlates well with crop zones (Klages, 1942) and general vegetative zones. However, the climatic types set up by climatologists are often too broad to be useful to the ecologist in local situations, so it is often necessary to set up subdivisions based on the biotic community (see Daubenmire, 1959).

Climographs, or charts in which one major climatic factor is plotted against another, represent another useful method of graphic representation of temperature and moisture in combination. In temperature-rainfall or temperature-humidity charts mean monthly values are plotted with the temperature scale on the vertical axis and either humidity or rainfall on the horizontal axis, as shown in Figure 5–10. Months are indicated by numbers, starting with January. The resulting 12-sided polygon gives a "picture" of the temperature-moisture conditions and makes possible graphic comparison of one set of conditions with another. Climographs have been useful in the comparison of one area with another (Fig. 5–10*C*) and as an aid in testing the importance of temperature-moisture combinations as limiting factors (as in Figure 5–10*A* and *B*). Plots of other pairs of factors may also be instructive, for example, plots of temperature and salinity in marine environments (see Figure 12–3, page 328).

Climate chambers provide another useful approach to the study of combinations of physical factors. These vary from simple temperature-humidity cabinets in use in most laboratories to large controlled greenhouses, such as the "phytotron" in which any desired combination of temperature, moisture, and light can be maintained. These chambers are often designed to control environmental conditions in order that the investigator can study the genetics and physiology of cultivated or domesticated species. However, they can be useful for ecological studies, especially when natural rhythms in temperature and humidity can be simulated. As already stressed, experiments of this sort help single out factors which may be "operationally significant," but they can reveal only part of the story since many significant aspects of the eco-

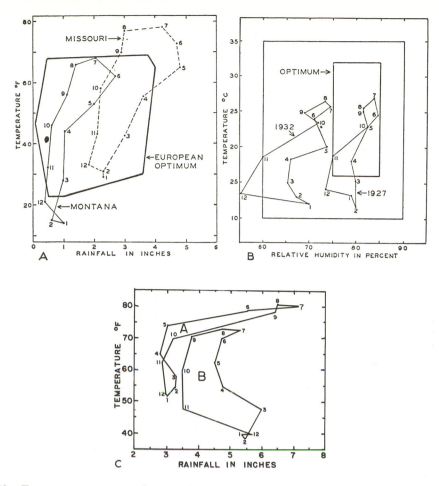

Figure 5–10. Temperature-moisture climographs. *A.* Temperature-rainfall monthly averages for Havre, Montana, where the Hungarian partridge has been successfully introduced, and Columbia, Missouri, where it has failed, compared with average conditions in the European breeding range. (Redrawn from Twomey, 1936.) *B.* Temperature-humidity conditions at Tel Aviv, Israel, for two different years, compared with optimum (inner rectangle) and favorable (outer rectangle) conditions for Mediterranean fruit fly. Damage to oranges was much greater in 1927. (After Bodenheimer, 1938.) *C.* Temperature-rainfall climographs for **A**, coastal Georgia (sea level), and **B**, northern Georgia (altitude 2000 to 3000 feet). In **A** a pronounced wet and dry season is the conspicuous seasonal variant as is characteristic of a subtropical climate, while in **B** seasonal differences in temperature are more pronounced than seasonal changes in rainfall. The climatic climax vegetation (see page 264 for definition of this term) in the coastal locality is a broad-leaved evergreen forest (see Figure 14–19), while in the more northern area it is a temperate deciduous forest (see Figure 14–12A).

system cannot be duplicated indoors, but must be studied or experimented with outdoors.

5. ATMOSPHERIC GASES. Except for the large variations in water vapor already discussed under the heading of water, the atmosphere of the major part of the biosphere is remarkably homeostatic. Interestingly enough, the present concentration of carbon dioxide (0.03 per cent by volume) and oxygen (21 per cent by volume) is somewhat limiting to many higher plants. It is well known that photosynthesis in many plants can be increased by moderate increases in CO_2 concentration, but it is not so

well known that decreasing the oxygen concentration experimentally can also increase photosynthesis. Björkman and collaborators[*] have reported that beans and many other plants increased their rate of photosynthesis by as much as 50 per cent when the oxygen concentration around their leaves was lowered to 5 per cent. It is speculated that oxygen inhibition is caused by a back-reaction between a highly reduced photosynthetic intermediate and molecular

---

[*] See Carnegie Institution of Washington Annual Report of the President, 1966–67, page 66; also Björkman, 1966.

oxygen that increases with O_2 concentration. Certain tropical grasses, including corn and sugar cane, did not show oxygen inhibition perhaps because they use a different "pathway" for carbon dioxide fixation. One can speculate that the reason for this situation is that broad-leaved plants evolved at a time when the CO_2 concentration was higher and the O_2 concentration lower than it is now, as described in Section 4, Chapter 9.

Oxygen becomes limiting to aerobes and CO_2 increases as one goes deeper in soils and sediments (also in bodies of large animals, the rumen of cattle being an anaerobic system); this results in a slow-down in rate of decomposition, the importance of which was discussed fully in Chapter 2 (see also Figures 9–9 and 12–13). Man's role in the CO_2 cycles was fully discussed in Chapter 4.

The situation in aquatic environments is quite different from that in the atmospheric environment because amounts of oxygen, carbon dioxide, and other atmospheric gases dissolved in water and thus available to organisms are quite variable from time to time and place to place. Oxygen is an A-1 limiting factor, especially in lakes and in waters with a heavy load of organic material. Despite the fact that oxygen is more soluble in water than is nitrogen, the actual quantity of oxygen that water can hold under the most favorable conditions is much less than that constantly present in the atmosphere. Thus, if 21 per cent by volume of a liter of air is oxygen, there will be 210 cc. of oxygen per liter. By contrast, the amount of oxygen per liter of water does not exceed 10 cc. Temperature and dissolved salts greatly affect the ability of water to hold oxygen, the solubility of oxygen being increased by low temperatures and decreased by high salinities. The oxygen supply in water comes chiefly from two sources, by diffusion from the air and from photosynthesis by aquatic plants. Oxygen diffuses into water very slowly unless helped along by wind and water movements, while light penetration is an all-important factor in the photosynthetic production of oxygen. Therefore, important daily seasonal and spatial variations may be expected in the oxygen concentration of aquatic environments. The details of oxygen distribution and its relation to aquatic communities will be covered in Part 2.

Carbon dioxide, like oxygen, may be present in water in highly variable amounts, but its behavior in water is rather different and its ecology is not as well known. It is therefore difficult to make general statements about its role as a limiting factor. Although present in low concentrations in the air, carbon dioxide is extremely soluble in water, which also obtains large supplies from respiration, decay, and soil or underground sources. Thus, the "minimum" is less likely to be of importance than is the case with oxygen. Furthermore, unlike oxygen, carbon dioxide enters into chemical combination with water to form H_2CO_3, which in turn reacts with available limestones to form carbonates ($-CO_3$) and bicarbonates ($-HCO_3$). As is shown in Figure 4–8A, a major reservoir pool of biospheric CO_2 in the carbonate system of the oceans. These compounds not only provide a source of nutrients but also act as buffers, helping to keep the hydrogen ion concentration of aquatic environments near the neutral point. Moderate increases in CO_2 in water seem to speed up photosynthesis and the developmental processes of many organisms. That CO_2 enrichment may be a key to cultural eutrophication has been mentioned (see page 106; also Lange, 1967; Kuentzel, 1969). High CO_2 concentrations may be definitely limiting to animals, especially since such high carbon dioxide concentrations are associated with low oxygen concentrations. Fishes respond vigorously to high concentrations and may be killed if the water is too heavily charged with unbound CO_2.

Hydrogen ion concentration, or pH, is closely related to the carbon dioxide complex (see page 60), and being relatively easy to measure, it has been much studied in natural aquatic environments. During the revival of interest in organized ecological study at the beginning of the twentieth century, pH was considered to be an important limiting factor and thus to be a promising overall indicator for determining the general ecological condition of aquatic environments, but this has not proved to be the case. The picture of the ecologist faring forth armed only with a pH kit was quite common. The physiologist, looking out of his laboratory window, probably smiled tolerantly, because the early interest in pH in ecological work was due to the discovery by the physiologist that pH was very important in regulating respiration and enzyme systems within the body, very small differences being critical. Unless values are extreme, communities compensate for differences by mechanisms already described in this chapter and show a wide tolerance for the naturally occurring range. However, when the total alkalinity is constant, pH change is proportional to CO_2 change, and therefore is a useful indicator of the rate or rates of total community metabolism (photosynthesis and respiration), as described in Section 3, Chapter 3. Soils and waters of low pH (i.e., "acid") are quite frequently deficient in nutrients and low in productivity.

6. BIOGENIC SALTS: MACRONUTRIENTS AND MICRONUTRIENTS. Dissolved salts vital to life may be conveniently termed *biogenic salts*. We have already presented several examples of their importance as limiting factors in soil and water. In fact, Liebig's original "law" of the minimum was based largely on the limiting action of vital raw materials which are scarce and variable in the environment. As already indicated, nitrogen and phosphorus salts are of major importance, and the ecologist may do well to consider these first as a matter of routine. Hutchinson (1957) states the case for phosphorus as the A-1 limiting factor as follows: "Of all the elements present in living organisms, phosphorus is likely to be the most important ecologically, because the ratio of phosphorus to other elements in organisms tends to be considerably greater than the ratio in the primary sources of the biological elements. A deficiency of phosphorus is therefore more likely to limit the productivity of any region of the earth's surface than is a deficiency of any other material except water." As discussed in Chapter 4 man creates ever-widening conditions when too much rather than too little becomes limiting.

Following closely on the heels of nitrogen and phosphorus, potassium, calcium, sulfur, and magnesium merit high consideration. Calcium is needed in especially large quantities by the mollusks and the vertebrates, and magnesium is a necessary constituent of chlorophyll, without which no ecosystem could operate. Elements and their compounds needed in relatively large amounts are often known as *macronutrients*.

In recent years great interest has developed in the study of elements and their compounds which are necessary for the operation of living systems, but which are required only in extremely minute quantities often as components of vital enzymes. These elements are generally called trace elements or *micronutrients*. Since minute requirements seem to be associated with an

equal or even greater minuteness in environmental occurrence, the micronutrients have importance as limiting factors. The development of modern methods of microchemistry, spectrography, x-ray diffraction, and biological assay has greatly increased our ability to measure even the very smallest amounts. Also, the availability of radioactive isotopes of many of the trace elements has greatly stimulated experimental studies. Deficiency diseases due to the absence of trace elements have been known, in a general way at least, for a long time. Pathological symptoms have been observed in laboratory, domestic, and wild plants and animals. Under natural conditions deficiency symptoms of this sort are sometimes associated with peculiar geological history and sometimes with a deteriorated environment of some sort, often a direct result of man's poor management. An example of peculiar geologic history is found in southern Florida. The potentially productive organic soils of this region did not come up to expectation (for crops and cattle) until it was discovered that this sedimentary region lacked copper and cobalt, which are usually present in most regions. A possible case of micronutrient deficiency resulting from changes in land management was discussed on page 97.

Eyster (1964) lists ten micronutrients that are definitely known to be essential to plants. These are iron, manganese, copper, zinc, boron, silicon, molybdenum, chlorine, vanadium and cobalt. In terms of function these can be arranged in three groups as follows: (1) those required for photosynthesis: Mn, Fe, Cl, Zn, and V; (2) those required for nitrogen metabolism: Mo, B, Co, Fe; (3) those required for other metabolic functions: Mn, B, Co, Cu, and Si. Most of these are also essential for animals, and a few others, such as iodine, are essential for certain animals, such as vertebrates. The dividing line between macro- and micronutrients, of course, is not sharp nor the same for all groups of organisms; sodium and chlorine, for example, would be needed in larger amounts by vertebrates than by plants. Sodium, in fact, is often added to the above list as a micronutrient for plants. Many of the micronutrients resemble vitamins in that they act as catalysts. The trace metals often combine with organic compounds to form "metallo-activators"; cobalt, for example, is a vital constituent of vitamin B_{12}. Goldman (1965) documents a case where molybdenum is limiting to a whole ecosystem when he found that addition of 100 parts-per-billion to the water of a mountain lake increased the rate of photosynthesis. He also found that in this particular lake cobalt concentration was high enough to be inhibitory to the phytoplankton. As with macronutrients, too much can also be limiting.

7. CURRENTS AND PRESSURES. The atmospheric and hydrospheric media in which organisms live are not often completely still for any period of time. Currents in water not only greatly influence the concentration of gases and nutrients, but act directly as limiting factors. Thus, the differences between a stream and a small pond community (see Chapter 11, Section 7) may be due in large part to the big difference in the current factor. Many stream plants and animals are specifically adapted morphologically and physiologically to maintaining their position in the current and are known to have very definite limits of tolerance to this specific factor. On land, wind exerts a limiting effect on the activities and even the distribution of organisms in the same manner. Birds, for example, remain quiet in protected places on windy days, which are, therefore, poor times for the ecologist to attempt to census a bird population. Plants may be modified structurally by the wind, especially when other factors are also limiting, as in alpine regions. Whitehead (1956) has demonstrated experimentally that wind limits the growth of plants in exposed mountain locations. When he erected a wall to protect the vegetation from wind, the height of plants increased. Storms are of major importance, even though they may be only local in extent. Hurricanes (as well as ordinary winds) transport animals and plants for great distances and, when they strike land, the winds may change the composition of the forest communities for many years to come. It has been observed that insects spread faster in the direction of the prevailing winds than in other directions to areas which seem to offer equal opportunity for the establishment of the species. In dry regions, wind is an especially important limiting factor for plants, since it increases the rate of water loss by transpiration. Good critical studies of the effect of wind are needed. In their preoccupation with temperature and moisture, ecologists have neglected this important factor.

Barometric pressure has not been shown to be an important direct limiting factor for organisms, although some animals appear able to detect differences, and, of course, barometric pressure has much to do with weather and climate, which are directly limiting to organisms. In the ocean, however, hydrostatic pressure is of importance because of the tremendous gradient from the surface to the depths. In water the pressure increases one atmosphere for every 10 meters. In the deepest part of the ocean the pressure reaches 1000 atmospheres. Many animals can tolerate wide changes in pressure, especially if the body does not contain free air or gas. When it does, gas embolism may develop. In general, great pressures such as are found in the depth of the ocean exert a depressing effect, so that the pace of life is slower.

8. SOIL. It is sometimes convenient to think of the biosphere as being made up of the atmosphere, the hydrosphere, and the pedosphere, the latter being the soil. Each of these divisions owes many of its characteristic features to the ecological reactions and coactions of organisms and to the interplay of ecosystems and basic cycles between them. Each is composed of a living and a nonliving component more easily separated on theoretical than on practical grounds. Biotic and abiotic components are especially intimate in soil, which by definition consists of the weathered layer of the earth's crust with living organisms and products of their decay intermingled. Without life, the earth would have a crust and might have air and water, but the air and water, and especially the "soil," would be entirely different from these components as we know them. Thus, soil not only is a "factor" of the environment of organisms but is produced by them as well (soil organisms and soil metabolism are considered in Chapter 14, Section 4). In general we may think of soil as the net result of the action of climate and organisms, especially vegetation, on the parent material of the earth's surface. Soil thus is composed of a parent material, the underlying geologic or mineral substrate, and an organic increment in which organisms and their products are intermingled with the finely divided particles of the modified parent material. Spaces between the particles are filled with gases and water. The texture and porosity of the soil are highly important characteristics and largely determine the availability of nutrients to plants and soil animals.

If we examine the cut edge of a bank or a trench (Fig. 5–11), it will be found that soil is composed of distinct layers, which often differ in color. These layers are called soil horizons, and the sequence of horizons from the surface down is called a soil profile. The upper horizon, or *A horizon* ("top soil"), is composed of the bodies of plants and animals which are being reduced to finely divided organic material by the process known as *humification*, a concept introduced in Chapter 2 (page 29) and described in more detail in Chapter 14 (page 367). In a mature soil this horizon is usually subdivided into distinct layers representing progressive stages of humification. These layers (Figures 5–11 and 5–12) are designated (from the surface downward) as litter (A-0), humus (A-1), and leached (light-colored) zone (A-2). The A-0 layer is sometimes subdivided as A-1 (litter proper), A-2 (duff), and A-3 (leaf-mold). The litter (A-0) horizon represents the detritus component and is considered as a sort of ecological subsystem in Chapter 14 (page 368). The next major horizon, or *B horizon,* is composed of mineral soil in which the organic compounds have been converted by decomposers into inorganic compounds by the process of *mineralization* and thoroughly mixed with finely divided parent material. The soluble materials of B horizon are often formed in the A horizon and deposited, or leached by downward flow of water, in B horizon. The dark band in Figure 5–11 represents the upper part of B horizon where materials have accumulated. The third horizon, or *C horizon*, represents the more or less unmodified parent material. This parent material may represent the original mineral formation which is disintegrating in place or it may have been transported to the site by gravity (colluvial deposit), water (alluvial deposit), glaciers (glacial deposit), or wind (eolian deposit, or loess). Transported soils are often extremely fertile (witness the deep loess soils of Iowa and the rich soils of the deltas of large rivers).

The soil profile and the relative thickness of the horizons are generally characteristic for different climatic regions and for topographic situations (Figs. 5–12 and 5–13). Thus, grassland soils differ from forest soils in that humification is rapid, but mineralization is slow. Since the entire grass plant, including roots, is short-lived, with each year are added large amounts of organic material which decays rapidly, leaving little litter or duff, but much humus. In the forest litter and roots decay slowly and, since mineralization is rapid, the humus layer remains narrow (Fig. 5–12). The average humus content of grassland soil, for example, is 600 tons per acre, compared with 50 tons per acre for forest soils (Daubenmire, 1947). In the forest-grassland buffer zone (see Figure 5–13) in Illinois, one can easily tell by the color of the soil which cornfield was once prairie and which was forest: the prairie soil is much blacker, due to its high

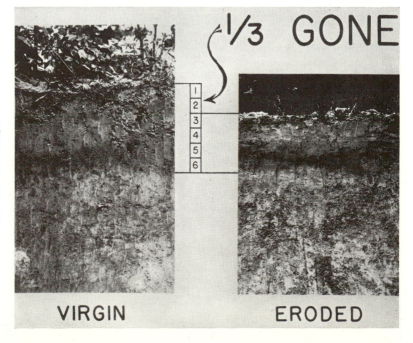

Figure 5–11. Soil profile of a virgin area compared with that of an eroded area in the deciduous forest region. In the left picture, 1–2 represents the A_1 horizon, 3–4 the A_2 horizon, and 5–6 the B_1 layer (accumulation of leached material). Compare with Figure 5–12. (U. S. Soil Conservation Service Photo.)

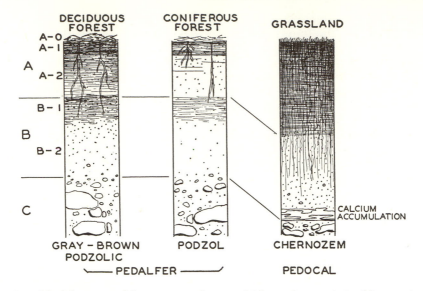

Figure 5–12. Simplified diagrams of three major soil types which are characteristic of three major biotic regions. See legend, Figure 5–14.

humus content. Given adequate rainfall, it is no accident that the "granaries of the world" are located in grassland regions.

Topographic conditions greatly influence the soil profile within a given climatic region. Hilly or well-drained land, especially if misused by man, will tend to have thin A and B horizons owing to erosion (Fig. 5–12). In flat land water may leach materials rapidly into the deeper layers, sometimes forming a "hardpan" through which plant roots, animals, and water cannot penetrate. Figure 9–7 illustrates an extreme case of hardpan condition and the resultant effect on the vegetation. Poorly drained situations such as bogs favor the accumulation of humus, since poor aeration slows down decay. The lack of oxygen and the accumulation of carbon dioxide and other toxic products become severe limiting factors. Sometimes soils developing on poorly drained sites are extremely productive if they are properly drained—witness the muck soils of Florida previously mentioned, which rate as some of the most productive soils in the world, when properly handled. The catch is that "proper handling" involves difficult ecological, as well as engineering, problems, because cultivation of such soils greatly increases the oxidation of accumulated organic matter and, in time, destroys some of the structural characteristics that made the soils so favorable at the outset. How to avoid "mining" such soils has not received adequate attention because agriculturists are often concerned only with the imme-

diate yields that can be obtained. The special problems of and the need for new approaches to use of tropical soils has been discussed in Chapter 4, Section 6 (see also Chapter 14).

Classification of soil types has become a highly empirical subject. The soil scientist may recognize dozens of soil types as occurring within a county or state. Local soil maps are widely available from country and state soil conservation agencies and from state universities. Such maps and the soil descriptions that accompany them provide useful background for studies of terrestrial ecosystems. The name for a local soil type is compounded from the name of the locality from which it was first described and a phrase indicating the textural class. For example, Norfolk sandy-loam refers to a soil first described near Norfolk that has a coarse texture (high ratio sand to silt-clay, i.e., "loam"). The ecologist, of course, should do more than merely name the soil on his study area. At the very minimum, measurements should be made of three important attributes in at least the A and B horizons as follows: (1) texture, that is, the per cent of sand, silt and clay (or more detailed determination of particle size), (2) per cent organic matter, and (3) exchange capacity, that is, an estimate of the amount of exchangeable nutrients; as emphasized in Chapter 4, the "available" minerals rather than the total amount determine potential fertility, other conditions being favorable.

Since soil is the product of climate and

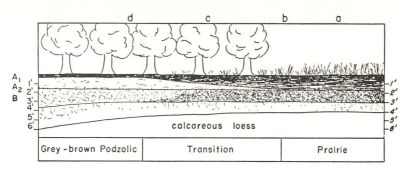

Figure 5–13. Soil-vegetation relationships in a prairie-forest transition zone. Distinctly different soils develop from the same parent material (calcareous loess, or wind-transported C horizon in this case) under influence of different vegetation and climate. The decrease in organic matter, the development of a podzolic A horizon (with narrow humus layer, see Figure 5–12), and increased structural development of the B horizon are the main features differentiating the forest soils from the prairie soils. (After Crocker, 1952.)

vegetation, a map of major (zonal) soil types of the world (Fig. 5–14) becomes a composite map of climate and vegetation (Fig. 5–15). Given a favorable parent material the action of organisms and climate will tend to build up a soil characteristic of the region (compare Figure 5–14 with map of biotic communities, Figure 14–7). From a broad ecological viewpoint, the soils of a given region may be lumped into two groups, those which are largely controlled by climate and vegetation of the region and those which are largely controlled by local or edaphic conditions of topography, water level, or type of parent material. As will be discussed in more detail in Chapter 9, local or edaphic conditions are more important (1) in regions which are geologically young and in which climate and vegetation have not yet been able to "build" a uniform soil cover and (2) in regions where the climate is extreme, as in deserts, for example, where small differences in soil may make big differences in the resulting biotic community. Consequently, the degree of soil maturity, (i.e., the extent to which equilibrium between soil, climate, and vegetation has been reached) varies greatly with the region. Wolfanger (1930), for example, estimated that 83 per cent of the soils in Marshall County, Iowa, are mature as compared with only 15 per cent of the soils in Bertie County, North Carolina, which is located on the "geologically young" Coastal Plain. *It is important to emphasize that the role that the "soil type" plays in ecosystem function depends on the stage in geological and ecological development* (more about this in Chapter 9).

For good general discussions of soil, reference may be made to Russell (1957), Eyre (1963), and Black (1968), while books dealing more specifically with soil organisms are listed on page 373.

9. FIRE AS AN ECOLOGICAL FACTOR. Research during the last 40 years has made necessary a rather drastic reorientation of our ideas about fire as an ecological factor. It is now evident that fire is not a minor or abnormal factor, but a major factor which is, and has been for centuries, almost a part of the normal "climate" in most of the terrestrial environments of the world. Consequently biotic communities adapt and compensate for this factor just as they do for temperature or water. As with most environmental factors, man has greatly modified its effect, increasing its influence in many cases and decreasing it in others. *Failure to recognize that ecosystems may be "fire adapted" has resulted in a great deal of "mismanagement" of man's natural resources.* Properly used, fire can be an ecological tool of great value. Fire is thus an extremely important limiting factor if for no other reason than that mankind is able to control it to a far greater extent than he can many other limiting factors.

Fire is most important in forest and grassland regions of temperate zones and in tropical areas with dry seasons. Archaeologists commonly consider evidence of a purposeful use of fire by man as the first "sign" of human development. In many parts of western or southeastern United States, it would be difficult to find a sizable area which does not give evidence of fire having occurred on it during the last fifty years at least. In many sections fires are started naturally by lightning. Primitive man, the North American Indian, for example, regularly burned woods and prairies for practical

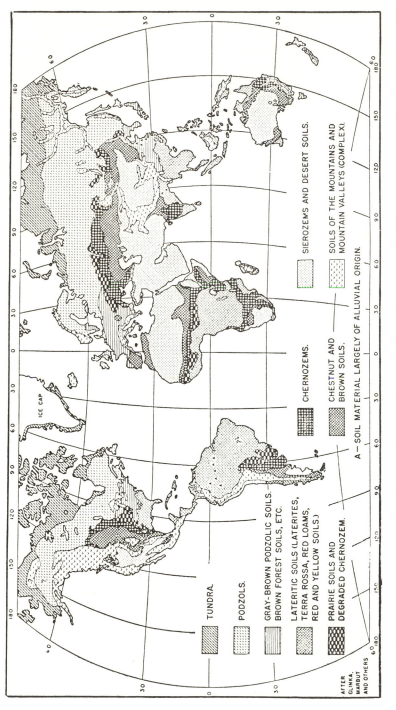

Figure 5–14. Schematic map of the primary soil types of the world. Not only north-south differences are evident, but also east-west differences which are related to rainfall. Podzolic and lateritic soils of humid regions are often known as "pedalfers" because of the accumulation of iron and aluminum in the B horizons, while the chernozems and other soils of more arid regions are "pedocals," because of the accumulation of calcium (see Figure 5–12). In recent years a more uniform and descriptive set of names has been proposed for zonal soil types, each name having the same root—"sol." Thus, northern podsol becomes spodosol (= ashy), the temperate gray-brown podzolic type becomes alfizol (Al-Fe, referring to B₂ mineral accumulation), the prairie soil becomes aridisol, and tropical lateritic soil becomes oxisol (= oxidized). (Map from U. S. Department of Agriculture Yearbook, 1938.)

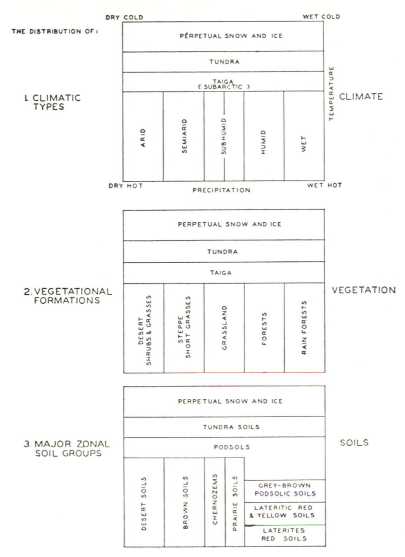

Figure 5–15. Schematic representation showing the interrelation of climate, vegetational formations, and zonal soil groups. (From Oosting, 2nd edition, 1952, after Blumenstock and Thornthwaite.)

reasons. Fire thus was a limiting factor long before modern man began to drastically modify his environment in an attempt to improve his status. Inadvertently, through careless behavior modern man has often so increased the effect of fire that the very thing he is seeking—a productive environment—is destroyed or injured. On the other hand, complete protection from fire has not always resulted in what was expected, namely, a more productive environment for man's purposes. Thus, it has become clear that fire should be considered an ecological factor along with such other factors as temperature, rainfall and soil, and should be studied with an open mind.

Whether fire is a friend or a foe of civilization, now as in the past, will depend on intelligent knowledge and control.

In speaking of fire as an ecological factor, the first thing to emphasize is that there are several types of fire in nature which are different in effect. Two extreme types are shown in Figure 5–16. For example, *crown fires* often destroy all the vegetation, whereas *surface fires* have entirely different effects. The former is limiting to most organisms; the biotic community must start to develop all over again, more or less from scratch, and it may be many years before the area is productive from man's viewpoint. Surface fires, on the other hand, exert a

Figure 5–16.   See legend on opposite page.

Figure 5–16.   The two extremes of fire. *Upper left:* Result of a severe crown fire in Idaho with subsequent severe erosion of the watershed. *Lower left:* A controlled burning operation of a long-leaf pine forest in southwest Georgia that removes hardwood competition, stimulates growth of legumes, and improves reproduction of valuable pine timber. Burning is done under damp conditions late in the afternoon (fire is stopped at night by dewfall). Note that the smoke is white (indicating little loss of nutrients) and that the thin line of fire can be stepped over at many points. Ants, soil insects, and small mammals are not harmed by such light surface fires. *Above:* A mature long-leaf pine forest that results from controlled burning. Shown in the picture is E. V. Komarek, pioneer fire ecologist lecturing to students. *Insert:* Long-leaf pine seedling showing terminal bud well protected from fire by long needles. (*Upper left,* U. S. Forest Service Photo; *lower left,* photo by Leon Neel, Tall Timbers Research Station; *above,* photos by E. P. Odum.)

selective effect; they are more limiting to some organisms than to others and thus favor the development of organisms with high tolerance to the fire factor. Also, light surface fires aid bacteria in breaking down the bodies of plants and in making mineral nutrients more quickly available to new plant growth (see page 28). Nitrogen-fixing legumes often thrive after a light burn. In regions especially subject to fire, regular light surface fires greatly reduce the danger of severe crown fires by keeping the combustible litter to a minimum. As was mentioned on page 32, fire often combines with plant-produced antibiotics to bring about rhythmic changes in vegetation (the "cyclic climax," see page 266) that results in alternate stabilization and rejuvenation of primary production and species diversity. In examining an area in regions where fire is a factor, the ecologist usually finds some evidence of the past influence of fire. Whether fire should be excluded in the future (assuming that it is practical) or should be used as a management tool will depend entirely on the type of community that is

desired or seems best from the standpoint of regional land use.

A single example taken from a well-studied situation will illustrate how fire acts as a limiting factor and how fire is not necessarily "bad" from the human viewpoint. On the coastal plain of southeastern United States the long-leaf pine is more resistant to fire than any other tree species. The terminal bud of seedling long-leaf pines is well protected by a bunch of long fire-resistant needles (Figure 5–16, insert). Thus, ground fires selectively favor this species. In the complete absence of fire, scrub hardwoods grow rapidly and choke out the long-leaf pines. Grasses and legumes are also eliminated, and the bob-white and other animals dependent on legumes do not thrive in the complete absence of fire in forested lands. Ecologists are generally agreed that the magnificent virgin, open stands of pine of the coastal plain and the abundant game associated with them (Figures 14–12 and 5–16, above) are part of a fire-controlled, or a "fire climax," ecosystem.

A good place to observe the long-term

Figure 5–17.   British heather moor burned in strips and patches (the light-colored areas of about 1 hectare each) to increase game production. This photo illustrates a desirable combination of young and mature vegetation (as discussed in detail in Section 3, Chapter 9), and also the principle of the "edge effect" (see Chapter 6). (After Picozzi, 1968; reproduced with the author's permission.)

effects of intelligent use of fire is the Tall Timbers Research Station in Northern Florida and the adjacent plantations of southwestern Georgia where for many years Herbert Stoddard and E. V. and Roy Komarek have been studying the relation of fire to the entire ecological complex. As a result of these studies, Stoddard (1936) was one of the first to advocate the use of controlled or "prescribed" burning for increasing both timber and game production, at a time when the "official" or "professional" viewpoint of foresters was that all fire was bad. For years high densities of both quail and wild turkeys have been maintained on land devoted to highly profitable timber crops through the use of a system of "spot" burning aided by a diversification in the land use. Since 1963, an annual "fire ecology conference" has been held at the Tall Timbers Station. The Proceedings* from these conferences not only review local experience but also fire-soil-vegetation-climate interrelationships all over the world.

The nature of other fire-adapted vegetation types is reviewed in Chapter 14. Fire is

especially important in grassland. Under moist conditions fire favors grass over trees, and under dry conditions fire is often necessary to maintain grassland against the invasion of desert shrubs. Fire has its use in grassland agriculture, as would be expected. Morris (1968), for example, found that late winter or early spring burning of Bermuda grass (*Cynodon dactylon*) increased forage yields when high levels of fertilization were employed.

An example of the use of fire in game management on the British heather moors is depicted in Figure 5–17. Extensive experimentation over the years has shown that burning in patches or strips of about 1 hectare each, with about 6 such patches per square kilometer, results in highest grouse populations and game yields. The grouse, which are herbivores that feed on buds, require mature (unburned) heather for nesting and protection against enemies but find more nutritious food in the regrowth on burned patches. This is a good example of a compromise between maturity and youth in an ecosystem that is very relevant to man himself (see Section 3, Chapter 9).

The question of when "to burn or not to burn" can certainly be confusing to the citizen. Because man, by his carelessness,

* Copies of these publications may be obtained from the Tall Timbers Research Station, Tallahassee, Florida.

tends to increase "wildfire," it is necessary to have a strong campaign for fire protection in forests and recreation areas. The citizen should recognize that he, as an individual, should never start or cause fires anywhere in nature; however, he should also recognize that the use of fire as a tool by trained persons is part of good land management. (See Garren, 1943; Sweeney, 1956; Ahlgren and Ahlgren, 1960; Cooper, 1961; Komarek, 1964, 1967).

Microenvironment

Regional differences in temperature, moisture, and other factors are important, but so are local horizontal and vertical differences. Organisms occupying the same general habitat may actually be living under very different conditions. The concept of microenvironment, the environment of small areas in contrast to large ones, has been developed only within the last decade. Other terms, commonly applied to this concept but much more restrictive in scope, are microclimate and bioclimate. Since the term microenvironment is a relative one, it may signify the immediate environmental area occupied by a pine stand, or equally well of that occupied

by a lichen within the stand. Critical studies at the microenvironmental level have significantly sharpened our approach to the study of individual organisms as well as to communities of organisms; the data are especially valuable in calculating energy flow of various populations within the community.

The study of microenvironments requires that considerable attention be given to instrumentation, since a series of measurements in vertical and horizontal gradients have far greater significance than single point measurements. For example, the conventional weather station measures temperature and other factors in an artificial shelter "chest high" above the ground in the open —a microenvironment especially relevant to man but not to a soil organism or a canopy tree leaf. The ecologist, therefore, has to set up his own environmental surveillance system. A number of books and manuals which describe the type and use of sensors are now available, as, for example, Platt and Griffiths (1964) and Wadsworth *et al.* (1967) for the terrestrial environment; Barnes (1959) and the "Standard Methods" manuals of the American Public Health Association (1965) for the aquatic environment.

Figure 5–18 rather spectacularly illustrates the microenvironmental variation on

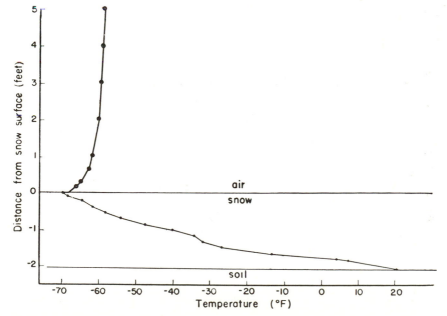

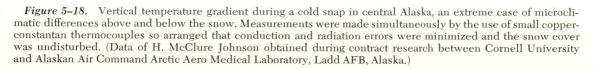

Figure 5–18. Vertical temperature gradient during a cold snap in central Alaska, an extreme case of microclimatic differences above and below the snow. Measurements were made simultaneously by the use of small copper-constantan thermocouples so arranged that conduction and radiation errors were minimized and the snow cover was undisturbed. (Data of H. McClure Johnson obtained during contract research between Cornell University and Alaskan Air Command Arctic Aero Medical Laboratory, Ladd AFB, Alaska.)

the tundra, showing differences in temperature recorded simultaneously at different points above and below the snow during a winter cold snap in central Alaska. Note that although temperatures ranged between 60 and 70 degrees below zero above the snow, the temperature at the surface of the soil below the snow was some 80 degrees higher! Animals living above the snow, such as the caribou or the arctic fox, would be subjected to the low temperatures, whereas animals living under the snow would virtually be living in a more southerly climatic zone! At the station where the measurements graphed in Figure 5–18 were made, voles of the genus *Microtus* were quite active in the two-inch air space between the snow and soil. These voles have rather thin fur and are not especially well adapted for extreme cold (compared with the arctic hare, for example); yet they are able to survive in a region of very cold winter climate because of the favorable subnivean microenvironment.

Very important microclimatic differences are created by topographic features of the landscape. As shown in Figure 5–19, a south-facing watershed slope receives much more solar radiation (more than 2.5 times as much in winter) than a north-facing slope at the Coweeta Hydrological Research Area in the mountains of North Carolina (for a photograph of this area, see Figure 2–4).

The differences in temperature, moisture, evapotranspiration, and other factors often result in entirely different communities on opposite slopes of the same valley, with more northern-type communities and organisms on the north-facing side and more southern ones on the south-facing side. A comparative study of north and south-facing slopes makes a good field exercise for an ecology class.

As already indicated, animals often use environmental gradients to "regulate" their own microenvironment in order to maintain it at a constant level. Other examples are mentioned in Chapter 8 (see also the "thermal niche" of a small bird, Figure 8–4).

6. ECOLOGICAL INDICATORS

Since, as we have seen, specific factors often determine rather precisely what kinds of organisms will be present, we can turn the situation around and judge the kind of physical environment from the organisms present. Often it is useful to employ a biological assay of this sort, particularly if we have something specific in mind and if the factor or factors in question are difficult or inconvenient to measure directly. In fact, the ecologist constantly employs organisms as indicators in exploring new situations or evaluating large areas. Ter-

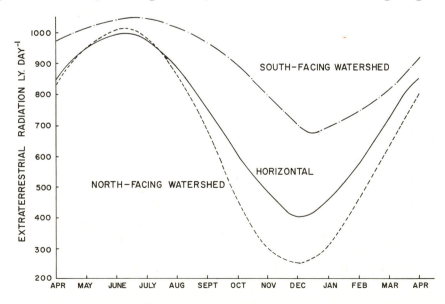

Figure 5–19. Solar radiation received by horizontal, north-facing and south-facing slopes at the Coweeta Watershed Experimental Area in the mountains of western North Carolina. The curves are theoretical radiation rates received before atmospheric and cloud attenuation. The actual energy received at the autotrophic layer would be about half that shown (see Section 1, Chapter 3). (Redrawn from Swift and van Bavel, 1961.)

restrial plants are especially useful in this regard. In western United States, for example, plants have been much used as indicators of water and soil conditions (especially as they affect grazing and agricultural potentials) since the early work of Shantz (1911) and Clements (1916). The use of vertebrate animals, as well as plants, as indicators of temperature zones — developed by Merriam, 1894, 1899 — has also been much studied. More recently, the use of functional assay methods, described on page 251, has received attention.

Some of the important considerations which should be borne in mind when dealing with ecological indicators follow:

1. In general, "steno" species make much better indicators than "eury" species, for reasons which should be obvious. Such species are often not the most abundant ones in the community. Additional discussion along this line and the concept of "fidelity" are included in Chapter 14, Section 5.

2. Large species usually make better indicators than small species because, as demonstrated in Chapter 3, a larger and more stable biomass or standing crop can be supported with a given energy flow. The turnover rate of small organisms may be so great (here today, gone tomorrow) that the particular species present at any one moment may not be very instructive as an ecological indicator. Rawson (1956), for example, found no species of algae which could serve as an indicator of lake types.

3. Before relying on single species or groups of species as indicators, there should be abundant field evidence, and, if possible, experimental evidence that the factor in question is limiting. Also, ability to compensate or adapt should be known; if marked ecotypes exist, the occurrence of the same series of taxa in different localities does not necessarily mean that the same conditions exist, as was emphasized earlier in this chapter.

4. Numerical relationships between species, populations, and whole communities often provide more reliable indicators than single species, since a better integration of conditions is reflected by the whole than by the part. This has been particularly well brought out in the search for biological indicators of various sorts of pollution, as will be considered in some detail in the next chapter. In Europe Ellenberg (1950) has demonstrated that the floristic makeup of weed communities provides excellent quantitative indicators of potential agricultural productivity of land.

An interesting atomic age slant on the subject of indicators is provided by the discovery that certain plants are useful in prospecting for uranium (Cannon, 1952, 1953, 1954). When deep rooted plants such as pines and junipers grow over uranium deposits, the above-ground plant parts contain a higher concentration of uranium than normal. Foliage may be easily collected, ashed, and examined fluorimetrically; over two ppm uranium in the ash is considered indicative of commercially usable deposits underground. Since selenium is often associated with uranium ore, selenium-indicating plants — for example, species of *Astragalus* in the Rocky Mountain Region — may also be useful in locating deposits. Likewise, where sulfur and uranium are associated, the sulfur-accumulating members of the mustard and lily family provide useful indicators.

PRINCIPLES AND CONCEPTS PERTAINING TO ORGANIZATION AT THE COMMUNITY LEVEL

Chapter 6

1. THE BIOTIC COMMUNITY CONCEPT

Statement

A biotic community is any assemblage of populations living in a prescribed area or physical habitat; it is an organized unit to the extent that it has characteristics additional to its individual and population components (see principle of integrative levels, page 6) and functions as a unit through coupled metabolic transformations. It is the living part of the ecosystem, as indicated in Statement 1, Chapter 2. "Biotic community" is, and should remain, a broad term which may be used to designate natural assemblages of various sizes, from the biota of a log to that of a vast forest or ocean. *Major communities* are those which are of sufficient size and completeness of organization that they are relatively independent; that is, they need only to receive sun energy from the outside and are relatively independent of inputs and outputs from adjacent communities. *Minor communities* are those which are more or less dependent on neighboring aggregations. Communities not only have a definite functional unity with characteristic trophic structures and patterns of energy flow but they also have compositional unity in that there is a certain probability that certain species will occur together. However, species are to a large extent replaceable in time and space so that functionally similar communities may have different species compositions.

Explanation

The community concept is one of the most important principles in ecological thought and in ecological practice. It is important in ecological theory because (it emphasizes the fact that diverse organisms usually live together in an orderly manner, not just haphazardly strewn over the earth as independent beings.) Like an ameba, the biotic community is constantly changing its appearance (visualize the forest in autumn and in spring), but it has structures and functions which can be studied and described, and which are unique attributes of the group. Victor E. Shelford, a pioneer in the field of biotic community ecology, has defined the community as an "assemblage with unity of taxonomic composition and a relatively uniform appearance." To this we might add: "and with a definite trophic organization and metabolic pattern." Communities may be sharply defined and separated from each other; this would be the case, of course, when the community habitat exhibits abrupt changes, but relatively sharp boundaries may also be the result of community interaction itself. Very frequently, however, communities blend gradually into one another so that there are no sharply defined boundaries. As is brought out in Section 3, continuity or discontinuity is largely a function of the steepness of the environmental gradient.

The community concept is important in the practice of ecology because "as the community goes, so goes the organism." Thus, often the best way to "control" a particular

140

organism, whether we wish to encourage or discourage it, is to modify the community, rather than to make a direct "attack" on the organism. For example, it has been demonstrated time and again that we have a better quail population by maintaining the particular biotic community in which the quail is most successful than by raising and releasing birds or manipulating any one set of limiting factors (such as predators, for example). Mosquitoes can often be controlled more efficiently and cheaply by modifying the entire aquatic community (as by fluctuating the water levels, for example) than by attempting to poison the organisms directly. Since "weeds" thrive under conditions of continual disturbance of the soil, the best way to control weeds along a roadside, for example, is to stop scraping and plowing up road shoulders and waysides, and encourage the development of a stable vegetation in which the weeds cannot compete (see Figure 15–6E). Man's welfare, like that of the quail, mosquito, or weed, depends ultimately on the nature of the communities and ecosystems upon which he superimposes his culture.

Examples

Perhaps the best way to illustrate the community concept is to give two examples of specific community studies, one made primarily from the descriptive or "standing crop" standpoint and the other made from the functional or "community metabolism" standpoint. For the descriptive study I have selected an early study of the biotic communities of the northern desert shrub region in western Utah made by R. W. Fautin (1946). In this research a number of "study areas" were selected, that is, areas large enough to be characteristic of the region but small enough to facilitate quantitative study. Over a period of three years the density and frequency of plants, the numbers and kinds of vertebrate animals (birds, mammals, and reptiles) and the invertebrate populations were carefully measured by the best sample methods then known, and data on the biota correlated with information on the communities' physical habitat, i.e., climate and soil-water relationships. A major weakness of a descriptive study of this sort is that the microorganism com-

ponent is not amenable to "sample collecting" procedures. The study revealed that there were two major communities (see definition in the Statement above). These were the sagebrush community and the shad scale community, each named after its most conspicuous organism, a plant. Temperatures were the same for both communities, the chief community limiting factor being water. The sagebrush community (for a picture of sagebrush desert see Figure 14–15C, page 395) occupied areas where "the precipitation is greater (about twice as great) and/or where the soil is deep, more permeable, and relatively saline free." The shad scale community, on the other hand, occurred where it was drier and where the soil was often impregnated with mineral salts. Within the shad scale major community there were a number of well-developed minor communities determined by differences in the availability of soil moisture. All communities, major and minor, had rather sharp boundaries. Large animals, especially predators, were found to range throughout major communities and from one major community to another. Smaller animals and many plants, on the other hand, were restricted to or had their greatest abundance in particular major or minor communities. Birds, rodents, lizards, ants, spiders, and tenebrionid beetles showed particular adaptations for living under dry conditions and consequently comprised the bulk of the animal population. Rodents were particularly important to the community as a whole because of their "grazing" and burrowing activities. Thus, the study pinpointed "operationally significant" factors and organisms, providing a basis for functional and systems analysis (as will be discussed in Chapter 10).

One of the best known examples of the community metabolism approach is the study of Silver Springs, Florida, by H. T. Odum (1957). This investigation of a very popular tourist attraction was referred to several times in Chapter 3 and is cited again in another context in Chapter 10. Standing crop analyses were combined with in situ measurements of the flow of energy and the exchange of materials. The study is summarized in the two diagrams of Figure 6–1. Of the 40-odd conclusions that were drawn from the study, the following can be cited as being of general interest to the study of community ecology.

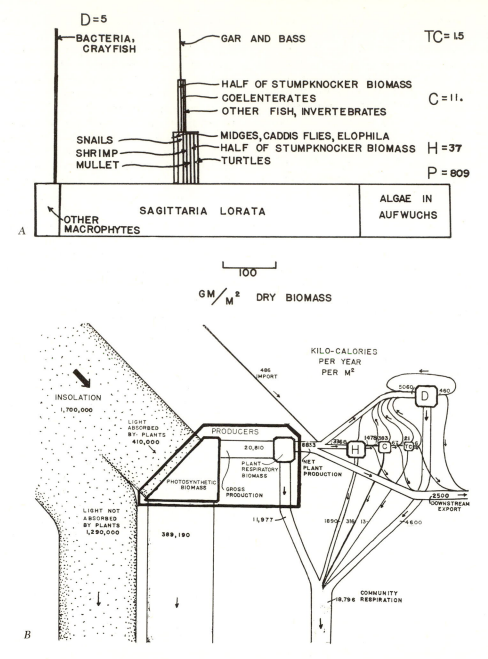

Figure 6–1. Standing crop (A) and energy flow (B) of the Silver Springs community. Mean biomass and annual energy flow are partitioned into five trophic groups as follows: producers (P), herbivores (H), carnivores (C), top carnivores (TC), and decomposers (D). Note: "Stumpknockers" are centrarchid fish, *Lepomis punctatus*, the most common small fish in the spring. (After H. T. Odum, 1957.)

1. Silver Springs is a thermostatic, chemostatic, and biostatic ecological community in seasonally pulsing steady-state climax.
2. The rate of primary production of the whole community is linearly proportional to the light intensity under natural conditions.
3. The efficiency of primary production relative to incoming light of usable wave lengths reaching plant level is about 5.3 per cent.
4. The community experiences an annual turnover of 8 times.
5. Most of the production goes into respiration, but 12 per cent is exported

downstream in the form of particulate matter.

6. A large aquatic insect emergence occurs in the evening at all times of the year, but only a small proportion of larval production ever emerges.

7. Some of the main invertebrate and fish components exhibit a strong photoperiodism in breeding cycles in this constant temperature spring.

8. The community metabolic quotient (O_2/CO_2) averaged 1.38 in summer and 0.95 in winter, indicating 39 per cent protein in primary production where nitrogen fixation is involved.

9. Bacteria constitute a relatively small part of the standing crop biomass, but next to green plants are the main consumers in terms of energy utilization.

10. Succession of small organisms (see Figure 13–5 on page 359) on individual eel grass (*Sagittaria*) blades occurs continually as the leaves grow out from the bottom, but the whole area is in steady state with a kind of stable succession distribution possibly analogous to a stable age distribution (see page 176).

2. INTRACOMMUNITY CLASSIFICATION, AND CONCEPT OF ECOLOGICAL DOMINANCE

Statement

Not all organisms in the community are equally important in determining the nature and function of the whole community. Out of the hundreds or thousands of kinds of organisms that might be present in a community, a relatively few species or species groups generally exert the major controlling influence by virtue of their numbers, size, production, or other activities. Relative importance in the community is not indicated by taxonomic relations, since major controlling or "ruling" organisms often belong to widely different taxonomic groups that have synergistic rather than competitive relationships (as will be discussed in the next chapter). Intracommunity classification therefore goes beyond taxonomic (floral and faunal) listing and attempts to evaluate the actual importance of organisms in the community. The most logical primary classification from this viewpoint is based on trophic or other functional levels as discussed in Chapters 2 and 3. Communities, at least major ones, have producers, macroconsumers, and microconsumers. Within these groups species or species groups which largely control the energy flow and strongly affect the environment of all other species are known as *ecological dominants*. The degree to which dominance is concentrated in one, several, or many species can be expressed by an appropriate *index of dominance* that sums each species' importance in relation to the community as a whole.

Explanation and Examples

The problem of classification within the biotic community may be clarified by taking a simplified example. Suppose we took a walk over a pasture and made a note of the important organisms which we observed. After such a "census" we might list:

bluegrass	beef cattle	turkeys
white clover	dairy cattle	sheep
oak trees	chickens	horses

Such a "taxonomic" listing alone would not give a very good picture of the pasture. Adding a quantitative estimate would help:

bluegrass	48 acres
white clover	2 acres
oak trees	2 individuals
beef cattle	2 individuals
dairy cattle	48 individuals
chickens	6 individuals
turkeys	2 individuals
sheep	1 individual
horses	1 individual

From this it would be clear that bluegrass is the "dominant" among the "producers," and dairy cattle among the "consumers." The community is essentially a dairy cattle pasture. A more complete picture, of course, would be obtained if we learned from the farmer the seasonal variation in use, the annual hay and milk production, etc., and if we knew something about the activities of microorganisms in the soil.

Actually, of course, there are many other kinds of organisms in a pasture, but the bluegrass and the cattle and the soil microorganisms are the most important from the viewpoint of controlling influence (aside from man, the ultimate dominant in this case). Natural communities may have an even larger number of species. Even so, a relatively few species often control the community and are said to be dominant. This does not mean that the more numerous rare species are not important; they are because they primarily determine diversity, an equally important aspect of community structure that will be considered in Section 4. Removal of the dominant would result in important changes not only in the biotic community but also in the physical environment (microclimate, for example), whereas removal of a nondominant species would produce much less change. Generally, dominants are those species in their trophic groups which have the largest productivity. In the Silver Springs community pictured in Figure 6–1 the eel grass, *Sagittaria*, can be seen to be the dominant autotroph. For large organisms, but not necessarily for small organisms, biomass may be an indicator of dominance, as we have had occasion to emphasize previously. Ecologists have used a wide variety of measurements to evaluate relative importance of species, as for example, "basal area" in forest communities (the cross section area of tree trunks), or "cover" in a grassland (area ground surface occupied). As with numbers and biomass, such specialized indices are applicable only if the populations being compared have approximately the same size-metabolism relationships (see Chapter 3, Section 5).

In land communities, spermatophytes usually are major dominants not only among the autotrophs but in the community as a whole because they provide shelter for the great bulk of the organisms in the community, and they modify physical factors in various ways. In fact, the term dominant has been largely used by plant ecologists to mean the "overstory" or tallest plants in the community. Clements and Shelford (1939) have pointed out that animals (consumers) may also control communities. Where plants are small in size, animals may produce relatively greater changes on the physical habitat. The concept of dominance has not been applied to the saprotrophic level, but there is every reason to assume that among the bacteria, etc., some kinds are more important than others. (See page 367 for an example of how the microbial population is dominated by different groups in a decomposition gradient in the soil.)

Northern communities almost always have fewer species which may be classed as dominants than have southern communities. Thus, a northern forest may have one or two species of trees which comprise 90 per cent or more of the stand. In a tropical forest, on the other hand, a dozen or more species may be dominant by the same cri-

Table 6–1. *Some Useful Indices of Species Structure in Communities*

A. INDEX OF DOMINANCE (c)[*]

$$c = \Sigma(ni/N)^2$$

where ni = importance value for each species (number of individual, biomass, production, and so forth)
N = total of importance values

B. INDEX OF SIMILARITY (S) BETWEEN TWO SAMPLES[†]

$$S = \frac{2C}{A + B}$$

where A = number of species in sample A
B = number of species in sample B
C = number of species common to both samples

Note: Index of dissimilarity = $1 - S$.

C. INDICES OF SPECIES DIVERSITY

(1) *Three species richness or variety indices (d)* [‡]

$$d_1 = \frac{S - 1}{\log N} \qquad d_2 = \frac{S}{\sqrt{N}} \qquad d_3 = S \text{ per 1000 ind.}$$

where S = number of species
N = number of individuals, etc.

(2) *Evenness index (e)*[§]

$$e = \frac{\bar{H}}{\log S}$$

where $\bar{H}$ = Shannon index (see below)
S = number of species

(3) *Shannon index of general diversity* $(\bar{H})$[‖]

$$\bar{H} = -\Sigma \left(\frac{ni}{N}\right) \log \left(\frac{ni}{N}\right)$$
or
$$-\Sigma P_i \log P_i$$

where ni = importance value for each species
N = total of importance values
P_i = importance probability for each species = ni/N

[*] See Simpson (1949).
[†] See Sørenson (1948); for a related index of "% difference," see E. P. Odum (1950).
[‡] d_1 see Margalef (1958a). d_2 see Menhinick (1964). d_3 see H. T. Odum, Cantlon and Kornicker (1960).
[§] See Pielou (1966); for another type of "equitability" index, see Lloyd and Ghelardi (1964).
[‖] See Shannon and Weaver (1949); Margalef (1968).
Note: In d_1, e and $\bar{H}$ natural logarithms ($\log_e$) are usually employed, but $\log_2$ is often used to calculate $\bar{H}$ so as to obtain "bits per individual."

terion (see page 401). Also, dominants are fewer where physical factors are extreme, as indicated in the previous section. Thus, dominance in all ecological groups is much more clear-cut on deserts, tundras, and other extreme environments. Or, to put it another way, the controlling influence in communities in extreme environments is divided among fewer species.

An equation for a simple index (c) for "concentration of dominance" is shown in Table 6–1. To illustrate, let us suppose that we have a community composed of five species, each equally important, and that we assign an importance value of 2 to each species (perhaps based on a density of 2 per m²). In another community of five species let us assume that one species has an importance value of 6 and the others only 1 each. If we calculate c from the quation in Table 6–1 we get a value of 0.2 for the first community and 0.4 for the second, in which dominance was more "concentrated" (in one species in this example). Whittaker (1965) has calculated dominance indices for a number of forests based on the contribution of each tree species to total net primary production; values ranged from 0.99 for a redwood forest (almost total dominance by one species) to 0.12 and 0.18 for two cove forests in the Smoky Mountains where dominance is shared by a large number of species, as in a tropical forest.

3. COMMUNITY ANALYSIS

Statement

Communities may be conveniently named and classified according to (1) major structural features such as dominant species, life forms or indicators, (2) the physical habitat of the community, or (3) functional attributes such as the type of community metabolism. No precise rules for naming communities have been formulated, as has been done for naming or classifying organisms, if indeed such is desirable or possible. Classifications based on structural features are rather specific for certain environments; attempts to set up a universal classification on this basis have largely been unsatisfactory. Functional attributes offer a better basis for the comparison of all communities in widely different habitats, for instance, terrestrial, marine, or freshwater.

Community analysis within a given geographical region or area of landscape has featured two contrasting approaches: (1) the zonal approach, in which discrete communities are recognized, classified, and listed in a sort of check-list of community types, and (2) the gradient analysis approach, which involves the arrangement of populations along a uni- or multi-dimensional environmental gradient or axis with community recognition based on frequency distributions, similarity coefficients, or other statistical comparisons. The term *ordination* is frequently used to designate the ordering of species and communities along gradients, and the term *continuum* to designate the gradient containing the ordered species or communities. In general, the steeper the environmental gradient, the more distinct or discontinuous are communities, not only because of greater probability of abrupt changes in the physical environment, but because boundaries are sharpened by competition and coevolutionary processes between interacting and interdependent species.

Explanation and Examples

First, let us consider naming and classification on a structural basis. Since the community is composed of organisms, many ecologists feel that communities should always be named for important organisms, generally the dominants. This works well where there are but one or two dominant species, or species groups (perhaps genera), as, for example, in the sagebrush and shadscale desert communities, described in Section 1, which remain conspicuous at all times. In many cases dominance is not so conveniently concentrated, as indicated in Section 2, or species may change continually with the seasons, as in many plankton-type communities. As was pointed out in the very first principle (Chapter 2), the ecosystem, rather than the community, is the real basic unit. Therefore, there is no logical reason why a community cannot be named after some nonliving community habitat feature, if this procedure results in conveying a clear picture of the community to someone else. Community names, like names for anything else, should be meaningful but kept as short as possible; otherwise, the names will not be used. The best way to name a community, therefore, is to pick some conspicuous, stable feature, whether living or not, and use that in the name. On land, major plants usually provide a conspicuous and stable point of reference. In aquatic communities, however, the physical habitat may also serve the purpose, as, for example, a stream–rapids community, mud-flat community, pelagic (open water) ocean community, or sand-beach community. If animals are conspicuous and highly characteristic, such as sessile bottom animals in the marine intertidal habitat, the community may conveniently be named after them, as, for example, an oyster-bed or barnacle community. In general, the inclusion of highly motile animals into the community name is not satisfactory because the animal component is generally too variable from time to time; rarely is there clear-cut long-term dominance by one or two species. Shelford has advocated including animals in the community name in order to emphasize that animals as well as plants are integral parts of the biotic community. However, this emphasis is accomplished by the description of the community which accompanies the name. Communities, as well as organisms, need to have descriptions as well as names.

Deciding where to draw boundaries is one of the interesting problems in community classification, just as it is in any other kind of classification. A parallel, perhaps, exists between communities and species in that both are more sharply delimited from one another when environmental and evolutionary gradients have discontinuities.

During the past 20 years (1950 to 1970) plant ecologists have engaged in a lively controversy as to whether land plant communities are to be thought of as discrete units with definite boundaries, as suggested by Clements (1905, 1916), Braun-Blanquet (1932, 1951), and Daubenmire (1966), or whether populations respond independently to environmental gradients to such an extent that communities overlap in a continuum so that recognition of discrete units is arbitrary, as viewed by Gleason (1926), Curtis and McIntosh (1951), Whittaker (1951), Goodall (1963), and others. Whittaker (1967) illustrates these contrasting viewpoints with the following example. If at the peak of autumn coloration in the Great Smoky Mountain National Park we were to select a vantage point along the highway so as to obtain a view of the altitudinal gradient from valley floor to ridge top, we would observe five zones of color as follows: (1) a multi-hued cove forest, (2) a dark green hemlock forest, (3) a dark red oak forest, (4) a reddish-brown oak-heath vegetation, and (5) a light green pine forest on the ridges. We could consider each of these five zones as discrete community types and analyze them accordingly, or we could consider all five as part of a *single continuum* to be subjected to some form of gradient analysis that would emphasize the distribution and response of individual species populations to changing environmental conditions in the gradient. This situation is illustrated in Figure 6–2, which shows the frequency distribution (as hypothetical bell-shaped

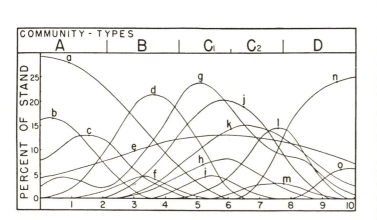

Figure 6–2. Distribution of populations of dominant trees along a hypothetical gradient, 0 to 10, illustrating the arrangement of component populations within a "continuum" type of community. Each species shows a "bell-shaped" distribution with a peak of relative abundance (per cent of stand) at a different point along the gradient; some species show a wider range of tolerance (and usually a lesser degree of dominance) than other species. Within the large community, subcommunities may be delimited (as indicated by A to D above the graph) on the basis of combinations of two or more dominants, indicators or other features. Such divisions will be somewhat arbitrary but useful for description and comparison. The curves have been patterned after data of several studies of tree distribution along an altitude gradient. (After Whittaker, 1954*a*.)

curves) of 15 species of dominant trees (*a* through *o*) that overlap along the gradient, and the somewhat arbitrary designation of five community types (*A, B, C₁, C₂* and *D*) based on the peaks of one or more dominants. There is much to be said for considering the whole slope as one major community, since all of the forests are linked together by exchanges of nutrients, energy, and animals as a watershed ecosystem which, as emphasized in Chapter 2, is the minimum-sized ecosystem unit amenable to functional studies and overall management by man. On the other hand, recognition of the zones as separate communities is useful to the forester or land manager, for example, since each community type differs in timber growth rate, timber quality, recreational value, vulnerability to fire and disease, and other aspects.

As is so often the case, concepts and approaches are a function of geography; thus, ecologists working in areas of gentle gradients and uniform, mature substrates (see page 131 for a discussion of the concept of mature versus geologically young soils) favor the continuum concept and various *ordination* techniques (i.e., statistical means of ordering species populations and communities along gradients), while ecologists working in areas of steep gradients or topographic discontinuities prefer the zonal con-

cept. Beals (1969) has made a direct comparison of vegetational changes along a steep and a gentle altitudinal gradient in Ethiopia, and two aspects of this study are shown in Figure 6–3. There was more discontinuity in the steep gradient as shown by the several sharp peaks in the plot of dissimilarity indices calculated for pairs of adjacent samples (Fig. 6–3, *upper*). (The formulas for coefficients of similarity and dissimilarity are shown in Table 6–1.) Furthermore, there was a distinct trend toward more sudden appearance and disappearance of species along the steep gradient than along the gentle one. As shown in Figure 6–3, *lower,* the frequency distribution curves for dominant species tend to be normal and bell-shaped (resembling the hypothetical ones in Figure 6–2) along the gentle gradient, but they were steep-sided and truncated in the steep gradient, indicating sharper delimitation of populations. Beals concluded that "along a steep gradient the vegetation itself can impose disjunctions on an extrinsically continuous environmental gradient, whereas along a gentle gradient it may not do so" (see also page 158). Three important processes, to be discussed in later chapters, can contribute to setting off one community from another, namely, (1) competitive exclusion, (2) symbiosis between groups of species

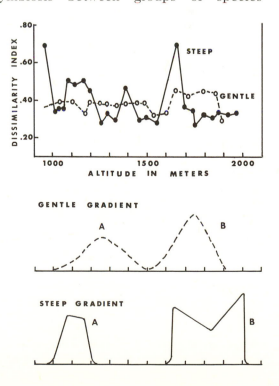

Figure 6–3. Gradient analysis of vegetative changes along a steep slope (solid lines) as compared with a gentle slope (broken lines), both having the same altitudinal range, in Ethiopia. *Upper:* A plot of dissimilarity indices (see Table 6–1) for adjacent segments along the gradients. *Lower:* Frequency distribution of two species, **A**—*Acacia senegal* and **B**—*Carissa edulis.* Species and community groups are more sharply delimited along the steep gradient. (Redrawn from Beals, 1969.)

that depend on one another (both these aspects are treated in Chapter 7), and (3) coevolution of groups of species (see Chapter 9, Section 5). Also, such factors as fire and antibiotic production can create sharp boundaries (see Figure 5–17). Buell (1956) describes a situation at Itasca Park, Minnesota; here, within the general matrix of maple-basswood forest, islands of spruce-fir forest maintain rather sharp boundaries that are not associated with changes in topography. Marine benthic communities show rather sharp zonation in steep gradients as described in Chapter 12 (see Figures 12–9 and 12–14*D*).

In any analysis of communities it is extremely important to consider the time gradient, both at the developmental level and on the longer evolutionary scale. These aspects are fully discussed in Chapter 9. Suffice it to say here that communities of different ages are often very sharply delimited from one another.

To summarize, gradient analyses (see Whittaker, 1967, and McIntosh, 1967, for comprehensive reviews) together with more complex and as yet little used matrical, pattern, or multi-axial approaches can detect discontinuities in an objective manner and bring out relationships between component populations. Just how communities are to be delimited and classified should depend on the study objectives and projected applications. One is reminded again of the two basic approaches to ecology as mentioned in Chapter 2, namely, the merological approach, which starts with the parts (species populations) and builds up a system from them, and the holistic approach, which starts with the whole. Thus, a whole continuum or any subunit, however delimited, can be a starting point.

4. SPECIES DIVERSITY IN COMMUNITIES

Statement

Of the total number of species in a trophic component, or in a community as a whole, a relatively small per cent are usually abundant (represented by large numbers of individuals, a large biomass, productivity, or other indication of "importance") and a large per cent are rare (have small "impor-

tance" values). While the few common species, or dominants (see Section 2, this chapter), largely account for the energy flow in each trophic group, it is the large number of rare species that largely determine the *species diversity* of trophic groups and whole communities. Ratios between the number of species and "importance values" (numbers, biomass, productivity, and so on) of individuals are called *species diversity indices*. Species diversity tends to be low in physically controlled ecosystems (i.e., subjected to strong physiochemical limiting factors) and high in biologically controlled ecosystems. In general, diversity increases with a decrease in the ratio of antithermal maintenance to biomass (the R/B or "Schrödinger ratio," or ecological turnover—see page 38). It is directly correlated with stability, but it is not certain to what extent this relationship is a cause-and-effect one.

Explanation

The general relationship between species and numbers (or other indications of "importance") can be pictured as a concave or "hollow" curve, as shown in Figure 6–4. While the pattern of a few common species with large numbers of individuals associated with many rare species with few individuals is characteristic of community structure everywhere, the quantitative species abundance relationships vary widely. As shown in Figure 6–4, stress will tend to flatten the curve as the number of rare species is reduced and the importance of, or concentration of dominance in, a few common species (that are tolerant of, or especially adapted to, the stress) increases. Two broad approaches are used to analyze species diversity in different situations, namely, (1) comparisons based on the shapes, patterns, or equations of species abundance curves and (2) comparisons based on *diversity indices*, which are ratios or other mathematical expressions, of species-importance relationships. Examples of several types of useful diversity indices are formalized in Table 6–1 and their use is illustrated in Figure 6–5; examples of graphic treatments are shown in Figure 6–6.

It is important to recognize that species diversity has a number of components which may respond differently to geographical, developmental, or physical factors. One

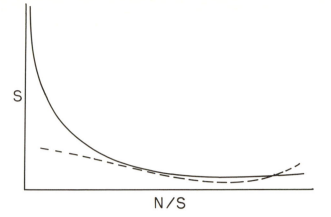

Figure 6–4. General relationship between the number of species (S) and the number of individuals per species (N/S). Most natural communities contain a few species with large numbers of individuals (the common or dominant species) and many species, each represented by a few individuals (the rare species). Rigorous physical environment, pollution, or other stresses will tend to flatten the curve, as shown by the dotted line.

major component might be called the *species richness or variety component,* as is expressed by simple ratios between total species, *S,* and total numbers (or importance values), *N,* such as the three "*d*" indices shown in Table 6–1. These indices can be used to compare one community or group of populations with another, provided it is first determined that the *S* is a linear function of the log or square root of *N.* A species/area index (that is, number of species per unit area) is also used to express variety, but in making comparisons one must make certain the sample sizes are comparable. Often, but by no means always, both species-numbers and species-area relationships are logarithmic or exponential (that is, new species are added in proportion to the logarithm or exponential function of the number of individuals or area sampled).

A second major component of diversity is what has been called *evenness* or *equitability* in the apportionment of individuals among the species. For example, two systems each containing 10 species and 100 individuals have the same *S/N* index, but these could have widely different evenness indices depending on the apportionment of the 100 individuals among the 10 species, for example, 91–1–1–1–1–1–1–1–1–1 at one extreme (minimum evenness) and 10 individuals per species (perfect evenness) at the other extreme. The "*e*" index in Table 6–1 is an example of a convenient expression for this component. The widely used Shannon function or *H̄* index, which is a mimic of the so-called information theory formula that contains hard-to-calculate factorials, combines the variety and evenness components as one *overall index of diversity.* This index is one of the best for mak-

ing comparisons where one is not interested in separating out diversity components (as shown in Figure 6–5C) because it is reasonably independent of sample size (which means that in practice fewer samples are required to obtain a reliable index for the purposes of comparison). It is also normally distributed (Bowman *et al.,* 1970; Hutcheson, 1970), so that routine statistical methods can be used to test for significance of differences between means. It can be noted that both *e* and *H̄* behave inversely to the index of dominance mentioned in Sections 2 (Table 6–1A) since high values indicate a low concentration of dominance.

There are a number of important ecological principles associated with diversity concepts, as outlined in the summary statement above. As Margalef (1968) expresses it, "the ecologist sees in any measure of diversity an expression of the possibilities of constructing feedback systems." Higher diversity, then, means longer food chains and more cases of symbiosis (mutualism, parasitism, commensalism, and so forth; see Chapter 7), and greater possibilities for negative feedback control, which reduces oscillations and hence increases stability. Where the antithermal maintenance costs imposed by the physical environment are reduced (that is, when the R/B ratio is low; see Chapter 3, page 38), more of the community energy can go into diversity. Consequently, communities in stable environments such as the tropical rain forest, have higher species diversities than communities subjected to seasonal or periodic perturbations by man or nature. What has not yet been measured is the extent to which an increase in community diversity in the same habitat can, in itself, increase the stability

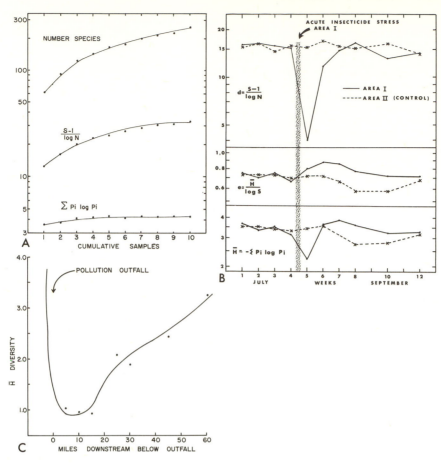

Figure 6–5. Some examples of the use of species diversity indices. *A.* The effect of increasing sample size on two diversity indices plotted on semilog graph paper together with the number of species cumulated in successive 0.1 m² samples of the arthropods in the vegetation of a millet (*Panicum*) field in Georgia. (See Figure 2–5.) *B.* The effect of a single application of the insecticide Sevin (an organophosphate insecticide that remains toxic for only about 10 days) on the arthropod population on a one-acre plot of the millet field. Two components of diversity (d and e) and a general index of total diversity (H̄) are based on 10 0.1 m² samples taken from the treated area and a control area at weekly or biweekly intervals from early July through September. The semilog plots facilitate a direct comparison of relative deviations resulting from the acute insecticide stress. *C.* Changes in the Shannon index of diversity (H̄) of the benthos (organisms living on the bottom) downstream from a pollution outfall (mixed domestic and industrial sewage from a small city), illustrating the marked effect of chronic pollution of a stream by inadequately treated wastes. (*A* and *B* plotted by E. P. Odum from data of Barrett, 1969; *C* redrawn from Wilhm, 1967.)

of the ecosystem in the face of external oscillations in the physical habitat. The significance of diversity trends in the development and the evolution of ecosystems is considered more fully in Chapter 9. Suffice it to say here that diversity tends to be high in older communities and low in newly established ones. While productivity or total energy flow certainly affects species diversity, the two quantities are not related in any simple linear manner. Very productive communities can have either very high (example, a coral reef) or very low species diversities (example, a temperate estuary).

As already indicated, stability seems more directly correlated with diversity than does productivity.

When we come to look at trophic levels, well-studied taxonomic components (such as birds, insects, or benthic aquatic populations), or other *parts* of communities, we find that species diversity is very much influenced by the functional relationships between the trophic levels. For example, the amount of grazing or predation greatly affects the diversity of the grazed or prey populations. Moderate "predation" often reduces the density of dominants, thus

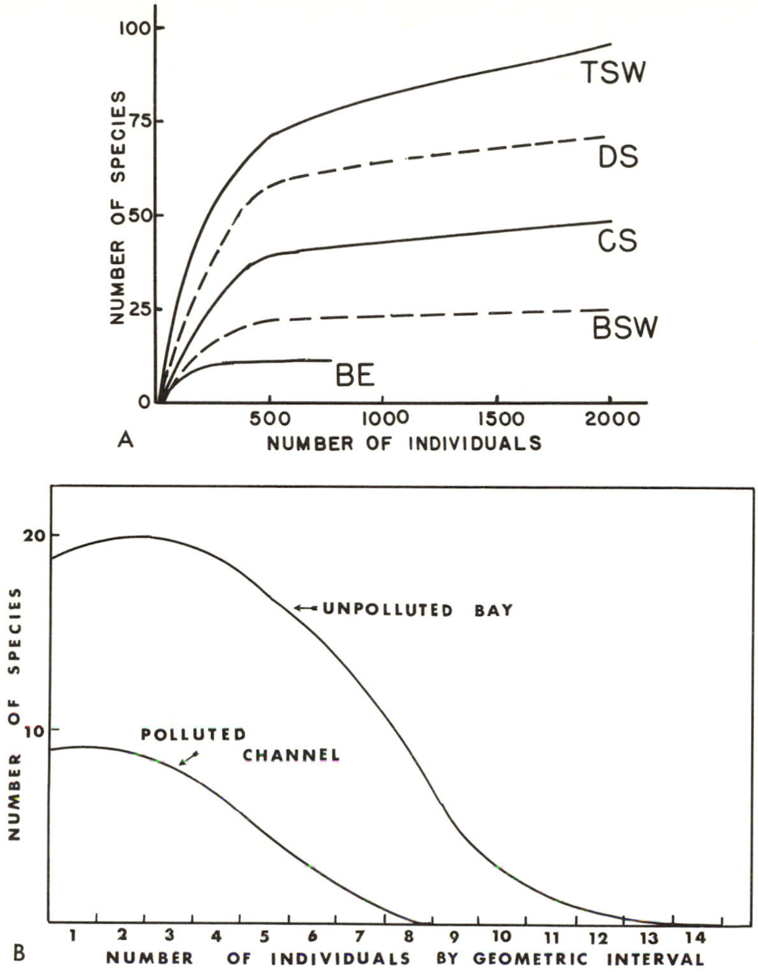

Figure 6–6. The use of species abundance curves to compare diversity in different habitats. *A.* The "rarefaction" method, involving the construction of curves based on using the percentage composition of species in single samples of marine sediments large enough to contain 500 to 3000 individuals of polychaetes and bivalves to determine the hypothetical number of species in successively smaller samples. The habitats in sequence from high to low species diversity are: tropical shallow water (TSW), deep sea (DS), continental shelf (CS), boreal shallow water (BSW), and boreal estuary (BE). *B.* The species structure of the diatom component of two estuarine communities in Texas as shown by truncated normal curves obtained by plotting the number of species in successive geometric intervals of abundance; that is, 1 to 2 individuals comprise the first interval, 2 to 4 the second, 4 to 8 the third, 8 to 16 the fourth, and so on. In the polluted channel (the Houston ship channel) the number of species in all abundance classes was sharply reduced. (*A* redrawn from Sanders, 1968; *B* redrawn from Patrick, 1967.)

providing less competitive species with a better chance to use space and resources. Harper (1969) reports that the diversity of herbaceous plants on the English chalk downs declined when grazing rabbits were fenced out. Severe grazing, on the other hand, acts as a stress and reduces the number of species to a few that are unpalatable. Paine (1966) found that species diversity of sessile organisms in the rocky intertidal habitat (where space is generally more limiting than food) was higher in both temperate and tropical regions in locations where first and second order predators were active. Experimental removal of predators in such situations reduced species diversity of all sessile organisms whether directly preyed upon or not. Paine concluded that "local species diversity is directly related to the efficiency with which predators prevent monopolization of major environmental requisites by one species." This conclusion does not necessarily hold for habitats in which competition for space is less severe. Man, the predator, produces the opposite effect in that he tends to reduce diversity and encourage monocultures.

During the past ten years (1960 to 1970) interest in species diversity has greatly increased among ecologists, perhaps because man is so rapidly, and often so heedlessly, reducing natural diversity to the point of raising a serious doubt whether this trend is in his own best interest. Accordingly, there has been a "literature explosion" on the subject of diversity that is hard to summarize or evaluate at the close of the decade. Hutchinson's essay on "Why are there so many species..." (1959) should certainly be recommended reading as a take-off point for subsequent literature study. The symposium edited by Woodwell and Smith (1969) provides a good sample of viewpoints and approaches and is especially useful to the beginner because the authors of the 19 papers in the symposium concern themselves with relationships between diversity and stability, and with the relevance of concepts to human affairs. Briefer reviews include Connell and Orias (1964), Whittaker (1965), MacArthur (1965), Pianka (1967), and McIntosh (1967a).

Examples

Examples of the use of diversity indices to evaluate man-made stresses are illustrated in Figure 6–5, while two types of graphic analyses are shown in Figure 6–6. Before diversity indices are used to compare one situation with another, the effect of sample size should be determined. One way to do this is shown in Figure 6–5A, in which indices are calculated for cumulative 0.1 m² samples of arthropods in a uniform grain field with use of a vacuum sweeper, as shown in Figure 2–5A. As may be seen in Figure 6–5A, the Shannon index leveled off at 4 to 5 samples (even though only a small portion of the species present had been sampled), but the variety index (S − 1/log N) did not begin to level off until 9 to 10 samples had been pooled. In Figure 6–5B the effect of an acute insecticide stress is shown. Note that although the variety index was greatly reduced by the treatment, evenness increased and remained elevated for most of the growing season. In this example the diversity of the pretreatment arthropod population was low, with strong dominance by a few species in each trophic level. When the insecticide killed large numbers of the dominant species, a greater evenness in the abundance of the surviving populations resulted. Had the diversity been high rather than low in the original population, both variety and evenness components would have been reduced. The Shannon index expresses the interaction of these two components of diversity and, therefore, exhibits an intermediate response (Fig. 6–5B). Although the insecticide used in this experiment remained toxic for only 10 days and the acute depression lasted only about two weeks, overshoots and oscillations in the diversity ratios were evident for many weeks. This example illustrates the desirability of separating the species richness and relative abundance components. Tramer (1969), who analyzed these components

in bird populations, found that differences in overall diversity in various community types from arctic to tropics were primarily the result of differences in the species richness component, the relative abundance component remaining stable at a high value (probably because of the territorial behavior of many species; see page 209). Tramer further suggests that communities from rigorous environments will vary in diversity according to their relative abundance component, while diversity in nonrigorous (biologically controlled) environments will be a function of number of species. The effect of municipal sewage (containing a mixture of domestic and industrial wastes) on the diversity of benthic (bottom) organisms in an Oklahoma stream is graphed in Figure 6–5C. Benthic diversity was depressed for more than 60 miles downstream. As discussed in Chapter 16, such wastes should be introduced first into a series of holding ponds, or otherwise treated, so that the final effluent discharged into the general environment does not produce such a drastic effect. Diversity indices provide one of the best ways to detect and evaluate pollution (see Wilhm, 1967; Wilhm and Dorris, 1968). For this purpose one needs only to be able to recognize species, not identify them by name. Errors resulting from failure to distinguish between closely similar species or counting life history stages as separate species are not critical because (1) closely related species are not apt to be found in the same sample (see page 213) and (2) different life history stages are, in themselves, part of diversity. Also, species are not the only useful units. A ratio between chlorophyll and microbial biomass on slides, for example, can be used as a pollution index (low in a "polluted" stream, high in a "clean" one).

Two graphic treatments of species diversity are shown in Figure 6–6. In the upper graph the diversity of major invertebrate components ("infauna," see page 335) of sediments in different marine habitats are compared by what has been called the "ratification" method, which involves plotting the cumulative number of species against cumulative number of individuals counted and evaluating differences in diversity from the shape of the curve (Sanders, 1968). Surprisingly, the deep sea sediments proved to house a high diversity of animals even though the density (number per m²) was quite low. The sequence of decreasing diversity from tropical shallow water (TSW) to boreal estuary (BE) supports the theory that habitat stability is a major factor in regulating species diversity. Another type of graphic analysis is shown in Figure 6–6B. When the number of species is plotted against number of individuals by geometric interval (that is, 1–2, 2–4, 4–8, 8–16, and so on), a truncated normal curve is obtained, which is different from the concave curves shown in Figure 6–4. Again, pollution or other stress tends to lower and flatten the curve (i.e., reduce the height of the mode in this case).

Graphic analyses have two advantages over indices: (1) sampling bias is reduced and (2) no specific mathematical relationship is assumed. However, fitting equations to such curves may help to reveal if there are basic mathematical "laws" governing relationships between "S" and "N." This is a subject for advanced texts, but we can open the door by noting that there are four main hypotheses, namely, that the relationship is (1) geometric (Motomura, 1932), (2) lognormal (Preston, 1948), (3) logarithmic (Fisher, Corbet, and Williams, 1943), or (4) random niche controlled (MacArthur, 1957; see also Chapter 8, page 236).

Table 6–2 presents a comparison of density and diversity of arthropod populations in a grain field and a natural herbaceous community that replaced it one year later.

The values shown are means of ten samples taken over the growing season. After only one year under "nature's management," so to speak, the following changes had occurred (compare 1966 and 1967): (1) The number of herbivorous (phytophagous) insects was greatly reduced, as was total arthropod density. (2) The variety component of diversity and the index of total diversity were significantly increased in each component, as well as for the total arthropods community. (3) The evenness component increased in the total sample and in some of the trophic components. (4) The number, diversity, and per cent composition of predators and parasites were greatly increased; predators and parasites made up only 17 per cent of the population density in the grain field as compared to 47 per cent in the natural field (where they actually outnumbered the herbivores). This comparison gives some clues as to why artificial communities often require chemical or other control of insects, while natural areas should not require such control if only we would give nature a chance to develop her own self-protection!

It should be emphasized that we have been concerned with diversity *within communities* or functional portions thereof (producers, for example) and not with the floral and faunal diversity of geographical areas containing a variety of habitats and mixed communities, although samples from

Table 6–2.   *Mean Density and Diversity of Arthropod Populations in the 1966 Unharvested Grain Crop* as Compared with the Natural Successional Community Which Replaced It in 1967* †

		INSECT HERBIVORES	INSECT PREDATORS	SPIDERS	TOTAL PREDATORS	INSECT PARASITES	MISC. ARTHROPODS	TOTAL ARTHROPODS
Density no./m²	1966	482 ± 36.0§	64 ± 8.9	18 ± 1.9	82 ± 8.8	24 ± 4.3	36 ± 2.5	624 ± 36.6§
	1967	156 ± 13.6	79 ± 12.2	38 ± 8.0‖	117 ± 19.0	51 ± 6.4§	31 ± 3.1	355 ± 35.8
Variety Index (d)‡	1966	7.19 ± 0.14	2.75 ± 0.13	1.56 ± 0.22	3.92 ± 0.20	6.32 ± 0.29	3.29 ± 0.29	15.57 ± 0.27
	1967	10.56 ± 0.46§	5.96 ± 0.52§	7.21 ± 0.58§	11.42 ± 0.93§	12.40 ± 1.21§	4.35 ± 0.30§	30.88 ± 1.92§
Evenness Index (e)‡	1966	0.65 ± 0.021	0.76 ± 0.014	0.56 ± 0.083	0.77 ± 0.018	0.89 ± 0.019	0.72 ± 0.014	0.68 ± 0.019
	1967	0.79 ± 0.017§	0.76 ± 0.024	0.86 ± 0.030‖	0.80 ± 0.020	0.90 ± 0.013	0.85 ± 0.017§	0.84 ± 0.006§
Total Diversity Index $(\bar{H})$‡	1966	2.58 ± 0.082	2.04 ± 0.059	1.12 ± 0.216	2.37 ± 0.071	2.91 ± 0.072	1.96 ± 0.066	3.26 ± 0.092
	1967	3.28 ± 0.095§	2.60 ± 0.067§	2.96 ± 0.101§	3.32 ± 0.092§	3.69 ± 0.134§	2.54 ± 0.071§	4.49 ± 0.075§

* A stand of Millet, *Panicum*, in a one-acre enclosure (see Figure 2–5A, page 18) stocked with three species of small mammals. Fertilizer applied at time of planting in prescribed agricultural manner, but no insecticide or other chemical treatment used (the control quadrat for the experiment graphed in Figure 6–5B).

† Unpublished data, E. P. Odum, G. Barrett, R. Pulliam, Inst. of Ecology, University of Georgia.   Figures are means ± one standard error.

‡ Formulas for indices shown in Table 6–1; these calculations based on natural logs.

§ Larger mean significantly different from smaller one at 99 per cent level.

‖ Difference significant at 95 per cent level.

mixed habitats (for example, insects attracted to a light trap) may show similar trends. Regional samples, however, will tend to reflect the variety of habitats present rather than the variety within any one habitat. Geographical diversity is a function of isolation and size of area; little islands have fewer species than big islands, which in turn have lower biotic diversity than continents.

5. PATTERN IN COMMUNITIES

Statement

The structure that results from the distribution of organisms in, and their interaction with, their environment can be called *pattern* (Hutchinson, 1953). Many different kinds of arrangements in the standing crop of organisms contribute to *pattern diversity** in the community, as, for example: (1) stratification patterns (vertical layering), (2) zonation patterns (horizontal segregation), (3) activity patterns (periodicity), (4) food-web patterns (network organization in food chains), (5) reproductive patterns (parent-offspring associations, plant clones, etc.), (6) social patterns (flocks and herds), (7) coactive patterns (resulting from competition, antibiosis, mutualism, etc.), and (8) stochastic patterns (resulting from random forces).

Explanation and Examples

Variety of species and their relative abundance (as discussed in the previous section) are by no means the only things involved in community diversity. Arrangement patterns and programmed activities also contribute to community function and stability. Since details of community patterns in land and water habitats are presented in Part 2, it will only be necessary here to illustrate the principle with several examples.

In a forest, the two basic layers—the autotrophic and heterotrophic strata—that are characteristic of all communities (see Chap-

ter 2) are frequently distinctly stratified into additional layers. Thus, the vegetation may exhibit herb, shrub, understory tree and overstory tree layers, while the soil is strongly layered, as shown in Figures 5–11 and 5–12. Such stratification is not limited to sessile plants or small organisms but is also characteristic of larger and more mobile animals. Thus, Dowdy (1947) sampled the arthropod populations of five major strata of an oak–hickory forest in Missouri throughout the year. He found that of 240 species of insects, spiders, and myriapods, 181 species (or about 75 per cent) were collected from one stratum only, 32 from two strata, 19 from three, and only 3 to 5 species were found in as many as four or all five of the strata. This indicates a rather remarkable adherence to strata by a highly motile group of organisms.

Even birds, which can easily fly from the ground to the tops of the highest trees in a few seconds if they so desire, often demonstrate close adherence to certain layers, especially during the breeding season. Not only the nests but also the entire feeding areas are often restricted to a surprisingly narrow vertical range. Stratification in highly motile groups such as birds is most marked where there are a number of similar species (often closely related) in potential competition. For example, in the evergreen forests of New England, the magnolia warbler occupies the low levels, the black-throated green warbler the middle levels, and the blackburnian warbler the high levels in the forest. All these species are members of the family Parulidae and have similar feeding habits. Other warbler species occupy different horizontal positions in the forest. MacArthur and MacArthur (1961) have shown that in the temperate zone at least bird species diversity is correlated with the height of the vegetation and its degree of stratification. As already indicated, additional factors, other than pattern diversity, are involved in the high species diversity observed in the tropics.

For a contrast in forest stratification, compare the multilayered tropical rain forest in Figure 14–20 with the essentially two-layered fire-controlled pine forest in Figure 14–12C.

An interesting example of population stratification in water is illustrated by the depth distribution of three species of game fish in TVA impoundments in midsummer (Fig. 6–7). As in many deep lakes in temper-

---

* Pielou (1966a) has used the term "pattern diversity" in a much more restricted sense, namely, to refer to the degree of segregation of individuals of one population from those of another.

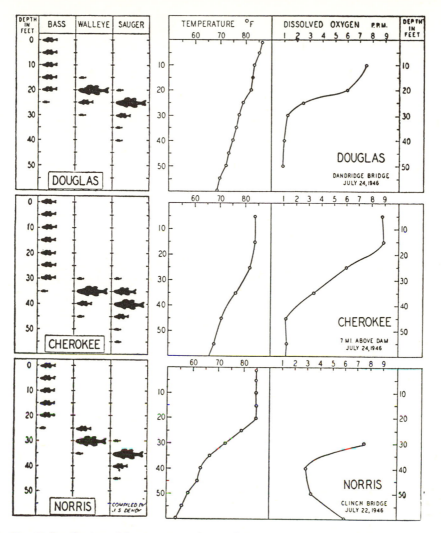

Figure 6-7. Depth distribution of three species of game fish in three TVA impoundments in midsummer. Oxygen and temperature conditions which determine the level at which different species aggregate are shown on the right. (After Dendy, 1945.)

ate regions, a distinct physical stratification develops during the summer with a layer of warm, oxygen-rich, circulating water lying over a deeper, colder, noncirculating layer which often becomes depleted of oxygen. (See Chapter 11, Section 5, for additional discussion of lake stratification.) As indicated in the figure, the large-mouth bass is the most tolerant of high temperatures (as is also indicated by the fact that it occurs in nature farther south than the other two species) and is found near the surface. The other two species aggregate in deeper waters, the sauger selecting the deepest (and, therefore, the coldest) water which still contains an adequate supply of oxygen. By determining the depth distribution of

oxygen and temperature it is thus possible to predict where the fish will be found in greatest numbers. Diagrams such as that in Figure 6-7 have, in fact, been published in local papers in order to aid the fisherman in deciding how deep he should fish to catch the desired species. As every fisherman knows, however, simply knowing where the fish are does not guarantee success; but it might help!

In the ocean, schools of fish are often remarkably stratified, so much so that they create sharply delineated "sound scattering layers," as shown in Figure 12-16.

One of the shortcomings of species diversity studies as described in the preceding section is that such analyses do not reveal

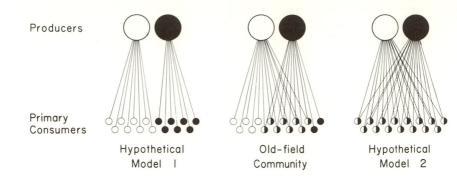

Figure 6–8. The center diagram models the actual food web as mapped by ³²P tracer transfers for a 1-year old-field community having two major plant dominants and 15 species of insects feeding on one or both plant species. The actual food chain network is intermediate in food chain diversity between the two hypothetical extremes shown. (After Wiegert and E. P. Odum, 1969.)

how species populations are linked together functionally. A high ratio of species to individuals is only *assumed* to indicate a complex food web with consequent interactions and feedback stability. A more advanced approach would be to determine directly the diversity of links or pathways in the network pattern. Precipitin tests (see Dempster, 1960) or, as described in detail in Chapter 17, radionuclide tracers make it possible to chart the actual material and energy flows from one species population to another (Wiegert, E. P. Odum, and Schnell, 1967). An example of a portion of a food web mapped by phosphorus-32 tracer is shown in Figure 6–8 (center diagram). In this example of a one-year-old field community there were two major dominant plant species and more than 120 species of arthropods associated with the above-ground vegetation stratum. Of the latter only 15 species actually fed on the two plants, as indicated by appreciable tracer transfer during a 6-week period of study in midsummer. Thus, most of the animal species present were not actually involved in the first link in the grazing food chain; many, of course, were predators or parasites, but many species were using the vegetation mostly as shelter, or were feeding on the nonlabeled, less common plants or on detritus, microorganisms, and so forth. As shown in Figure 6–8, one species of dominant plant (which is known to produce antibiotics) was "grazed" by fewer species than the other. The network diversity (as can be quantitated by counting the number of pathways and interactions) proved to be intermediate between a low diversity and a high diversity hypo-

thetical model shown on either side of the model for the actual community.

Since the daily progression of light, temperature, and other physical factors makes itself felt in all but deep water, soil, or cave communities, it is to be expected that the majority of populations in most communities will exhibit periodicities that are related either directly or indirectly to changes occurring during the 24-hour day–night period. The term *diel* (= day) *periodicity* refers to events which recur at intervals of 24 hours or less, while the term *circadian* (= about a day) *rhythm* refers to the persistent periodicity regulated by the biological clock that couples environmental and physiological rhythms (see page 245). Community periodicity results when whole groups of organisms exibit synchronous activity patterns in the day–night cycle. Some, for example, are active only during the period of darkness (*nocturnal*), others during the day (*diurnal*), and still others only during twilight periods (*crepuscular*).

A striking example of diel periodicity in aquatic habitats is to be found in the vertical "migration" of zooplankton organisms which regularly occurs in both lakes and oceans. Copepods, cladocerans, larval forms, etc., which make up the vast floating life in the open waters generally move upward to or toward the surface at night and downward during the daylight hours (Fig. 6–9). While light is clearly the controlling factor here, these diel movements are complex, and the physiological mechanisms have not yet been fully elucidated. Each species, and sometimes different stages of the same species, responds in a different manner so that

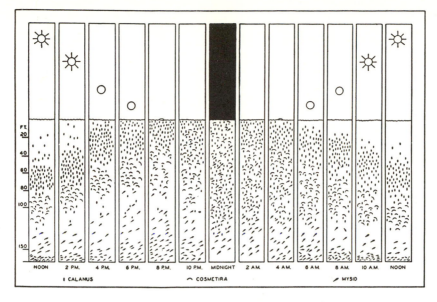

Figure 6–9.   Diagram of vertical diel migration of three species of marine plankters, namely, a copepod (*Calanus finmarchicus*), a small jellyfish (*Cosmetira pilosesella*), and a mysid crustacean (*Leptomysis gracilis*) in the North Atlantic. (From Allee *et al.*, 1949, after Russell and Yonge.)

all organisms do not attempt to crowd into the same region, but stratification is much more marked at midday than at midnight.

Seasonal periodicities are likewise nearly universal in communities and often result in almost complete change in the community structure during the annual cycle. The roles of temperature, photoperiod, wet and dry seasons, fire, and other seasonal periodicities in regulating community structure and function have been reviewed in the previous chapter. In the temperate zone temperature interacts with length of day. Thus, Leopold and Jones (1947) found that variation from year to year in the time of flowering of plants and the arrival of migratory birds is greater in early spring when temperatures are critical than in late spring, even though experiments (by other workers) show that the length of day is the controlling factor in many of the cases.

While we conventionally think in terms of four seasons (spring, summer, autumn, winter), ecologists studying terrestrial and freshwater communities in temperate regions have found that (as suggested in the above paragraph) early and late spring and early and late summer are as different from each other as autumn and winter. Hence, six seasons seem more representative of community periodicities: *hibernal* (winter or hiemal), *prevernal* (early spring), *vernal* (late spring), *aestival* (early summer), *serotinal* (late summer), and *autumnal*.

6. ECOTONES AND THE CONCEPT OF EDGE EFFECT

Statement

An ecotone is a transition between two or more diverse communities as, for example, between forest and grassland or between a soft bottom and hard bottom marine community. It is a junction zone or tension belt which may have considerable linear extent but is narrower than the adjoining community areas themselves. The ecotonal community commonly contains many of the organisms of each of the overlapping communities and, in addition, organisms which are characteristic of and often restricted to the ecotone. Often, both the number of species and the population density of some of the species are greater in the ecotone than in the communities flanking it. The tendency for increased variety and density at community junctions is known as the *edge effect*.

Explanation

As already outlined in Section 3 of this chapter, communities frequently change very gradually, as along a gradient, or they may change rather abruptly. In the latter case a tension zone would be expected between two competing communities. What may not be so evident from casual observation is the fact that the transition zone often supports a community with characteristics additional to those of the communities which adjoin the ecotone. Thus, unless the ecotone is very narrow, some habitats and, therefore, some organisms are likely to be found in the region of the overlap which are not present in either community alone. Since well-developed ecotonal communities may contain organisms characteristic of each of the overlapping communities plus species living only in the ecotone region, we would not be surprised to find the variety and density of life greater in the ecotone (edge effect). Furthermore, some species actually require as part of the habitat, or as part of their life history, two or more adjacent communities that differ greatly in structure. (See the grouse habitat example illustrated in Figure 5–17.) Organisms which occur primarily or most abundantly or spend the greatest amount of time in junctions between communities are often called "edge" species.

Examples

In terrestrial communities the concept of edge effect has been shown to be especially applicable to bird populations. For example, Beecher (1942) made a thorough attempt to locate all bird nests on a tract of land which contained a number of marsh and upland communities. He found that there were fewer nests in a large block of cattail marsh, for example, as compared with an equivalent acreage composed of numerous small blocks of the same plant community. For the study as a whole it was demonstrated that population density increased with the increase in the number of feet of edge per unit area of community. On a broader scale it is common knowledge that the density of song birds is greater on estates, campuses, and similar settings, which have mixed habitat and consequently much "edge," as compared with tracts of uniform forest or grassland.

As emphasized in the statement above, ecotones may have characteristic species not found in the communities forming the ecotones. For example, in a study of bird populations along a community developmental gradient (see Table 9–3) study areas were selected so as to minimize the influence of junctions with other communities. Thirty species of birds were found to have a density of at least 5 pairs per 100 acres in some one of these stages. However, about 20 additional species were known to be common breeding birds of upland communities of the region as a whole; 7 of these were found in small numbers, whereas 13 species were not even recorded on the uniform study areas. Among the latter were included such common species as robin, bluebird, mockingbird, indigo bunting, chipping sparrow, and orchard oriole. Many of these species require trees for nest sites or observation posts, yet feed largely on the ground in grass or other open areas; therefore, their habitat requirements are met in ecotones between forest and grass or shrub communities, but not in areas of either alone. Thus, in this case about 40 per cent (20 out of 50) of the common species known to breed in the region may be considered primarily or entirely ecotonal.

One of the most important general types of ecotones as far as man is concerned is the *forest edge*. A forest edge may be defined as an ecotone between forest and grass or shrub communities. Wherever man settles he tends to maintain forest edge communities in the vicinity of his habitations. Thus, if he settles in the forest he reduces the forest to scattered small areas interspersed with grasslands, croplands, and other more open habitats. If he settles on the plains, he plants trees, creating a similar pattern. Some of the original organisms of the forest and plains are able to survive in the man-made forest edge, whereas those organisms especially adapted to the forest edge, notably many species of "weeds," birds, insects, and mammals, often increase in number and expand their ranges as a result of creation by man of vast new forest edge habitats (see pages 114 and 242).

Before leaving the subject it should be emphasized that an increase in density at ecotones is by no means a universal phenomenon. Many organisms, in fact, may show the reverse. Thus, the density of trees is obviously less in a forest-edge ecotone than in the forest. Although many economic species of animals, such as game animals, are "edge" species, or utilize ecotones to a large extent, Barick (1950) has shown that for deer and grouse in the Adirondack region the edge-effect concept may be overrated. In fact, it seems likely that ecotones assume greater importance where man has greatly modified natural communities. In Europe, for example, where most of the forest has been reduced to forest edge, thrushes and other forest birds live in cities and suburbs to a greater extent than do related species in North America.

7. PALEOECOLOGY: COMMUNITY STRUCTURE IN PAST AGES

Statement

Since we know from fossil and other evidence that organisms were different in past ages and have evolved to their present status, it naturally follows that the structure of communities and the nature of environments must have been different also. Knowledge of past communities and climates contributes greatly to our understanding of present communities. This is the subject of paleoecology, a borderline field between ecology and paleontology that has been defined by Stanley Cain (1944) as "the study of past biota on a basis of ecological concepts and methods insofar as they can be applied," or, more broadly, as the study "of the interactions of earth, atmosphere, and biosphere in the past." The basic assumptions of paleoecology are: (1) that the operation of ecological principles has been essentially the same throughout various geological periods and (2) that ecology of fossils may be inferred from what is known about equivalent or related species now living.

Explanation

Since Charles Darwin brought the theory of evolution to the forefront of man's thinking, reconstruction of life in the past through the study of the fossil record has been an absorbing scientific pursuit. The evolutionary history of many species, genera, and higher taxonomic groups has now been pieced together. For example, the story of the skeletal evolution of the horse from a four-toed animal the size of a fox to its present status is pictured in most elementary biology textbooks. But what about the associates of the horse in its developmental stages? What did it eat, and what was its habitat and density? What were its predators and competitors? What was the climate like at the time? How did these ecological factors contribute to the natural selection which must have had a part to play in shaping the structural evolution? Some of these questions, of course, may never be answered. However, given quantitative information on fossils associated together at the same time and place, it should be possible for scientists to determine something of the nature of communities, and of their dominants, in the past. Likewise, such evidence, together with that of a purely geological nature, may aid in determining climatic and other physical conditions existing at the time. The development of "radioactive dating" and other new geological tools has greatly increased our ability to establish the precise time when a given group of fossils lived.

Until recently little attention was paid to the questions listed in the above paragraph. Paleontologists were busy describing their finds and interpreting them in the light of evolution at the taxonomic level. As such information accumulated and became more quantitative, however, it was only natural that interest in the evolution of the group should develop, and thus a new branch of science, paleoecology, was born. In summary, then, the paleoecologist attempts to determine from the fossil record how organisms were associated in the past, how they interacted with existing physical conditions, and how communities have changed in time. The basic assumptions of paleoecology are much the same as for paleontology, that is, that "natural laws" were the same in the past as they are today, and that organisms with structures similar to those organisms living today had similar behavior patterns and ecological characteristics. Thus, if the fossil evidence indicated that a spruce forest once occurred 10,000 years ago where an oak–hickory forest now is present, we have every reason to think that the climate was colder 10,000 years ago, since species of spruce as we know them today are adapted to colder climates than are oaks and hickories.

Illustrations

Since the general pattern of community evolution through the whole span of geological time is reviewed in Chapter 9, we can select illustrations from studies of communities in recent geological periods.

Fossil pollen provides excellent material for the reconstruction of terrestrial communities which have existed since the Pleistocene period. Figure 6–10 is a diagram showing how the nature of postglacial communities and climates can be reconstructed by determining the dominant trees. As the glacier retreats it often leaves scooped-out places which become lakes. Pollen from plants growing around the lake sinks to the

bottom and becomes fossilized in the bottom mud. Such a lake may fill up and become a bog. If a vertical core is taken from the bog or lake bottom, a chronological record is obtained from which the percentage of various kinds of pollen can be determined. Thus, in Figure 6–10, the "oldest" pollen sample is composed chiefly of spruce, fir, larch, birch, and pine, indicating a cold climate. A change to oak, hemlock, and beech indicates a warm moist period several thousand years later, whereas oak and hickory suggest a warm dry period still later with a return to slightly cooler and wetter conditions in the most recent part of the profile. Finally, the pollen "calendar" clearly reflects recent effects of man in opening up the forest as shown by an increase in herbaceous pollen. According to Davis (1969) pollen profiles in Europe even show the effects of the black plague when agriculture declined resulting in a decrease

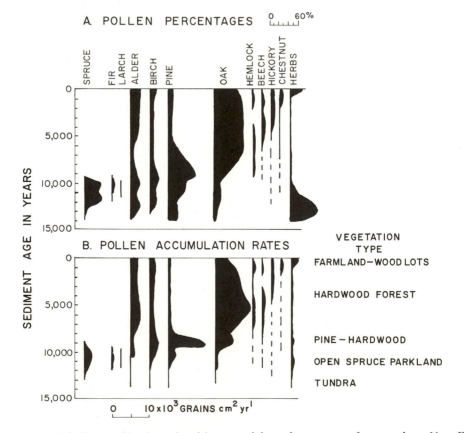

Figure 6–10. Fossil pollen profiles from dated layers in lake sediment cores from southern New England. In A the number of pollen grains of each species group is plotted as a percentage of the total number in the sample, while in B the estimated rate of pollen deposition for each plant group is plotted. The "rate" profile gives a better indication of the quantitative nature of the post-pleistocene vegetation than does the percentage profile. (Redrawn from Davis, 1969.)

Table 6–3. *Comparison Species Diversity of Benthic Shelled Mollusca on the Marine Continental Shelf in an Equator-to-North Pole Gradient During Warm (I) and Cool (II) Geological Periods**

| LATITUDINAL ZONES | I POLES WARM | | II POLES COOL | |
	No. Species	Cumulative Species (from equator)	No. Species	Cumulative Species (from equator)
6 (pole)	850	1,000	250	2,050
5	875	1,000	400	2,000
4	900	1,000	550	1,825
3	925	1,000	700	1,625
2	950	1,000	850	1,350
1 (equator)	1,000	1,000	1,000	1,000

* After Valentine, 1968.

in herbaceous pollen in sediment layers dated at the same time as the widespread die-off of human beings.

Figure 6–10 also illustrates how improved quantitation can change interpretations of the fossil record. When pollen abundance is plotted as a percentage of the total amount in the sample (the conventional approach, at least until recently), one gets the impression that New England was covered with a dense spruce forest 12,000 to 10,000 years ago. However, when carbon dating made it possible to determine the *rate of pollen deposition in dated layers* and rates plotted as in Figure 6–10B, it became evident that trees of all kinds were scarce 10,000 years ago and that the vegetation existing then was actually an open spruce parkland probably not unlike that present today along the southern edge of the tundra. This is a good example of what statisticians often admon-

ish: beware of percentage analyses; they may be misleading!

In the ocean, the shells and bones of animals often provide the best record. Shell deposits are especially good for diversity analyses, just as are present day populations as graphed in Figure 6–6A. Valentine (1968) presents an interesting model (Table 6–3) which brings out the importance of distinguishing between "in-community" and "geographical" or "in-gradient" diversity. Thus, in past ages when there was no ice at the poles (Table 6–3, column I) there were many more species in northern sea bottoms than there are now (Table 6–3, column II). However, in the whole pole-to-equator gradient there are twice as many species of benthic mollusks now — when the poles are ice-covered — because the sharper gradient increases speciation.

PRINCIPLES AND CONCEPTS PERTAINING TO ORGANIZATION AT THE POPULATION LEVEL

Chapter 7

1. POPULATION GROUP PROPERTIES

Statement

The population, which has been defined as a collective group of organisms of the same species (or other groups within which individuals may exchange genetic information) occupying a particular space, has various characteristics which, although best expressed as statistical functions, are the unique possession of the group and are not characteristic of the individuals in the group. Some of these properties are density, natality (birth rate), mortality (death rate), age distribution, biotic potential, dispersion, and growth form. Populations also possess genetic characteristics directly related to their ecology, namely, adaptiveness, reproductive (Darwinian) fitness, and persistence (i.e., probability of leaving descendents over long periods of time). See Dobzhansky (1968).

Explanation

As has been well expressed by Thomas Park (in Allee *et al.*, 1949), a population has characteristics or "biological attributes" which it shares with its component organisms, and it has characteristics or "group attributes" unique to the group. Among the former, the population has a life history in that it grows, differentiates, and maintains itself as does the organism. It has a definite organization and structure that can be de-

scribed. On the other hand, group attributes, such as birth rate, death rate, age ratio and genetic fitness, apply only to the population. Thus, an individual is born and dies and has age, but it does not have a birth rate, death rate, or an age ratio. These latter attributes are meaningful only at the group level. As indicated in the Statement, population attributes can be considered in two categories: (1) those dealing with numerical relationships and structure and (2) the three general genetic properties. In the following sections the important group attributes will be considered and examples given.

In simple laboratory populations observed under controlled conditions many of the above-mentioned group attributes can be measured, and the effect of various factors on them studied. In natural populations, however, it is often difficult or impossible to measure all the attributes. Some of this difficulty is being overcome as methods of population study improve. The development of better methods for measuring population properties of various important organisms is thus a very fruitful line of ecological research today. Even with great improvement in methods, it is doubtful that all population attributes can be measured with equal accuracy in nature. Fortunately, it is often not necessary actually to measure all these in order to obtain a useful picture of the population. Frequently one characteristic of a population can be calculated from data on another characteristic. Thus, accurate measurement of one or two properties may be more valuable than poor measurement of several.

162

2. POPULATION DENSITY AND INDICES OF RELATIVE ABUNDANCE

Statement

Population density is population size in relation to some unit of space. It is generally assayed and expressed as the number of individuals, or the population biomass, per unit area or volume—for example, 200 trees per acre, 5 million diatoms per cubic meter of water, or 200 pounds of fish per acre of water surface. As suggested in Chapter 2, a wide variety of attributes can serve as biomass units, ranging from dry weight to DNA or RNA content. Sometimes it is important to distinguish between *crude density*, the number (or biomass) per unit total space, and *specific* or *ecological density*,* the number (or biomass) per unit of habitat space (available area or volume that can actually be colonized by the population). Often it is more important to know whether a population is changing (increasing or decreasing) than to know its size at any one moment. In such cases, indices of *relative abundance* are useful; these may be "time-relative," as, for example, the number of birds seen per hour, or they may be percentages of various kinds, such as the percentage of sample plots occupied by a species of plant.

Explanation

In undertaking a study of a population, density would often be the first population attribute to receive attention. It might be said that natural history becomes ecology when "how many" as well as "what kinds" are considered. The effect that a population exerts on the community and the ecosystem depends not only on what kind of organism is involved but also on how many—in other words, on population density. Thus, one crow in a 100-acre cornfield would have little effect on the ultimate yield and cause the farmer no concern, but 1000 crows per 100 acres would be something else!

As is true of some of the other population attributes, population density is quite variable. However, it is by no means infinitely

variable; *there are definite upper and lower limits to species population sizes that are observed in nature or that theoretically could exist for any length of time.* Thus, a large area of forest might show an average of 10 birds per hectare and 20,000 soil arthropods per square meter, but there would never be as many as 20,000 birds per square meter or as few as 10 arthropods per hectare! As has been brought out in Chapter 3, the upper limit of density is determined by the energy flow (productivity) in the ecosystem, the trophic level to which the organism belongs, and the size and rate of metabolism of the organism. The lower limit may not be so well defined, but in stable ecosystems, at least, homeostatic mechanisms operate to keep density of the common or dominant organisms within rather definite limits.

Examples

In Figure 7–1 the range of density reported for common mammals is shown. Density (expressed as biomass per hectare) is that of the species within its normal geographical range, in its preferred habitat (i.e., ecological density) and under conditions where man or other "outside" forces are not unduly restrictive. Species are arranged in the chart according to trophic level, and within the four levels according to individual size. We see that while density of mammals as a class may range over nearly five orders of magnitude, the range for any given species or trophic group is much less. The influence of trophic level is, of course, striking, and the effect of size is also indicated since larger mammals in each level tend to maintain a larger biomass than the small mammals. The effect of trophic level on density of fish in ponds is illustrated by the data in Table 7–1A: the lower the trophic level, the higher the density. The point to emphasize is that the first order of population control is the energy flow–physical factor complex; the second order of control involves the subject matter of this chapter, namely, the interaction of populations with one another.

When the size and metabolic rate of individuals in the population is relatively uniform, density expressed in terms of number of individuals is quite satisfactory as a measure, but so often the situation exists

---
* Also called *economic density* by Elton (1933).

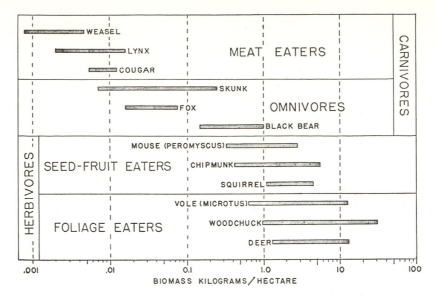

Figure 7–1.   The range of population density (as biomass per hectare) of various species of mammals as reported from preferred habitat of the species in localities where man is not unduly restrictive. Species are arranged according to trophic levels and according to individual size within the four levels to illustrate the limits imposed by trophic position and size of organism on the expected standing crop. (Graph prepared from data collected by Mohr, 1940, plus results of later studies.)

as shown in Table 7–1*B*. The relative merits of numbers, biomass, and energy flow parameters as indices were discussed in Chapter 3 (see page 82, and especially Table 3–14). It will be well to recall the following statement from that chapter: "Numbers overemphasize the importance of small organisms, and biomass overemphasizes the importance of large organisms," but components of energy flow "provide a more suitable index for comparing any and all populations in an ecosystem."

There are numerous special measures and terms in wide use that apply only to specific populations or groups of populations. Foresters, for example, determine "board feet per acre" or other measures of the commercially usable part of the tree. This, and many others, is a density measure as we have broadly defined the concept, since it expresses in some manner the size of the "standing crop" per unit area.

One of the greatest difficulties in measuring and expressing density arises from the fact that individuals in populations are often unevenly distributed in space, i.e., show a "clumped" distribution. Therefore, care must be exercised in choice of size and number of samples used to determine density. This problem will be discussed later in this chapter.

As was indicated in the Statement at the beginning of this section, relative abundance is often a useful measure when it is important to know how the population is changing, or when conditions are such that the absolute density cannot be determined. The terms "abundant," "common," "rare," and so forth, are most useful when tied to something that is measured or estimated in a manner that makes comparison meaningful. Some ways of assaying the importance or dominance (and the concept of "importance value") of species populations in the community were reviewed in Sections 4 and 5 of the previous chapter. As might be imagined, relative abundance "indices" are widely used with populations of larger animals and terrestrial plants, where it is imperative that a measure applicable to large areas be obtained without excessive expenditure of time and money. For example, federal administrators charged with setting up annual hunting regulations for migratory waterfowl must know if the populations are smaller, larger, or the same as in the previous year if they are to adjust the hunting regulations to the best interest of both the birds and the hunters. To do this they must rely on relative abundance indices obtained from field checks, hunter surveys, questionnaires, and nesting censuses. Percentage indices are widely used in the study of vegetation, and specially defined terms

Table 7–1. Density in Fish Populations Illustrating the Effect of Trophic Level on Biomass Density (A) and the Relationships Between Numbers and Biomass in a Population with a Rapidly Changing Age and Size Structure

A. MIXED POPULATIONS, BIOMASS PER UNIT AREA:

Fish in artificial ponds in Illinois (data from Thompson and Bennett, 1939). Fish groups arranged in approximate order of food chain relations with "rough fish" occupying the lowest trophic level and "game fish" the highest.

	FISH IN POUNDS PER ACRE		
	Pond No. 1	Pond No. 2	Pond No. 3
Game and pan fish (bass, bluegills, etc.)	232	46	9
Catfish (bullheads and channel cats)	0	40	62
Forage fish (shiners, gizzard shad, etc.)	0	236	3
Rough fish (suckers, carp, etc.)	0	87	1,143
Totals	232	409	1,217

B. COMPARISON OF INDIVIDUAL AND BIOMASS DENSITY WHERE SIZE OF ORGANISM UNDERGOES PRONOUNCED CHANGE WITH AGE:

Fingerling sockeye salmon in a British Columbia lake. The salmon hatch in streams and in April enter the lake, where they remain until mature. Note that between May and October the fish grew rapidly in size, with the result that biomass increased three times, even though the number of fish was greatly reduced. From October to the next April very little growth occurred, and continued death of fish reduced the total biomass. (Data from Ricker and Forester, 1948.)

	May	October	April
Individuals, thousands in the lake	4,000	500	250
Biomass, metric tons in the lake	1.0	3.3	2.0

have come into general use; for example, frequency = per cent of sample plots in which the species occurs, abundance = per cent of individuals in a sample, cover = per cent of ground surface covered as determined by projection of areal parts. Other such indices are discussed in Chapter 14. One should be careful not to confuse indices of relative abundance with true density, which is always in terms of a definite amount of space.

The contrast between *crude density* and *ecological density* can be illustrated by Kahl's study (1964) of the wood stork in the Florida Everglades. As shown in Figure 7–2, the density of small fish in the area as a whole goes down as the water levels drop during the winter dry season, but the ecological density in the contracting pools of water increases as fish are crowded into smaller and smaller water areas. The stork times its egg laying so that young will be hatched when ecological density of fish is at a peak and it is easy for the adults to catch the fish that provide the chief food for the young.

Many different techniques for measuring

population density have been tried, and methodology comprises an important field of research in itself. There would be little point in going into detail on methods here since methods are best detailed in field and methods manuals. Also, the man in the field will generally find that he will first have to review the original literature applying to his situation and then develop modifications and improvements of existing methods to fit his specific case. There is no substitute for experience when it comes to field censusing! It can be pointed out, however, that methods fall into several broad categories: (1) total counts, sometimes possible with large or conspicuous organisms or with those which aggregate into colonies. For example, Fisher and Vevers (1944) were able to estimate the size of the world population of gannets (*Sula bassana*), a large white sea bird that nests in a few densely populated colonies on north Atlantic shores of North America and Europe, as 165,600 ± 9500 individuals. (2) Quadrat sampling methods involving counting and weighing of organisms in plots or transects of appropriate size and number to get an estimate

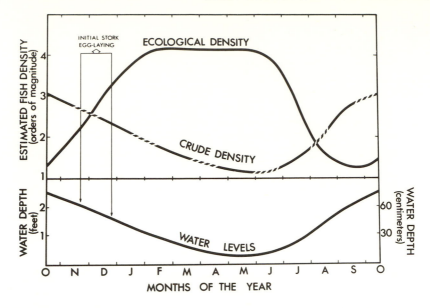

Figure 7–2. Ecological and crude density of fish prey in relation to the breeding of the stork predator. As water levels fall during the dry season in southern Florida, the crude density of small fish declines (i.e., the number of fish per square mile of total area declines because size and number of ponds is reduced), but the ecological density (i.e., number per square meter of remaining water surface) increases as fish are crowded into smaller water areas. Nesting of the stork is timed so that maximum food availability coincides with greatest food demand by the growing nestlings. (After Kahl, 1964.)

density in the area sampled. (3) Marking-recapture methods (for mobile animals), in which a sample of population is captured, marked, and released, and proportion of marked individuals in a later sample used to determine total populations. For example, if 100 individuals were marked and released and 10 out of a second sample of 100 were found to be marked, then population would be figured as follows: 100/P = 10/100, or P = 1000. A variant called the "calendar" method involves recording an individual present between first and last trapping whether actually recorded in the intervening period or not. Marking-recapture methods are most reliable when turnover is low; the method does not work well when density is undergoing rapid change. (4) Removal sampling, in which the number of organisms removed from an area in successive samples is plotted on the *y*-axis of a graph and the number previously removed plotted on the *x*-axis. If the probability of capture remains reasonably constant, the points will fall on a straight line that can be extended to the zero point (*x*-axis), which would indicate theoretical 100 per cent removal from the area (see Zippin, 1958; Menhinick, 1963). (5) Plotless methods (applicable to sessile organisms such as trees). For example, the quarter method—from a series of

random points the distance to the nearest individual is measured in each of four quarters. The density per unit area can be estimated from the mean distance (see Phillips, 1959).

3. BASIC CONCEPTS REGARDING RATES

Statement

Since a population is a changing entity, we are interested not only in its size and composition at any one moment, but also in how it is changing. A number of important population characteristics are concerned with rates. A rate may be obtained by dividing the change by the period of time elapsed during the change; such a rate term would indicate the rapidity with which something changes with time. Thus, the number of miles traveled by a car per hour is the speed rate, and the number of births per year is the birth rate. The "per" means "divided by." For example, the growth rate of a population is the number of organisms added to the population per time and is obtained by dividing the population increase by the time elapsed. In dealing with average rates of population change the standard notation is $\Delta N/\Delta t$, where N = population size (or other measure of importance) and t = time. For instantaneous rates the notation is dN/dt.

Explanation

If time is plotted on the horizontal axis (x-axis, or abscissa) and the number of organisms on the vertical axis (y-axis, or ordinate) of a graph, a population growth curve is obtained. In Figure 7–3, growth curves for colonies of two kinds of honeybees raised in the same apiary are shown. Also, the approximate growth rate at weekly intervals is plotted against time. Note that growth rate increases and decreases as the slope of the growth curve increases and decreases. Population B's growth rate is considerably less than A's during the first eight weeks or so, but eventually population B grows as rapidly as A. Not only do population growth curves provide a means of summarizing time phenomena, but the type of curve may give hints as to the underlying processes controlling population changes. Certain types of processes give characteristic types of population curves. As we shall see in Section 8, S-shaped growth curves and "humped-backed" growth rate curves are often characteristic of populations in the pioneer stage.

For convenience, it is customary to abbreviate "the change in" something by writing the symbol Δ (delta) in front of the letter representing the thing changing. Thus, if N represents the number of organisms and t the time, then:

ΔN = the change in the number of organisms.

$\dfrac{\Delta N}{\Delta t}$ or $\Delta N/\Delta t$ = the average rate of change in the number of organisms per (divided by, or with respect to) time. This is the growth rate.

$\dfrac{\Delta N}{N\Delta t}$ or $\Delta N/(N\Delta t)$ = the average rate of change in the number of organisms per time per organism (the growth rate divided by the number of organisms initially present or, alternately, by the average number of organisms during the period of time). This is often called the specific growth rate and is useful when populations of different sizes are to be compared. If multiplied by 100 (i.e., $\Delta N/(N\Delta t) \times 100$), it becomes the per cent growth rate.

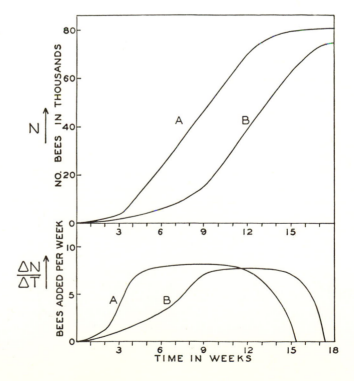

Figure 7–3. A population growth curve (upper) and growth rate curve (lower) for two bee colonies in the same apiary. **A**—Italian bees; **B**—Cyprian bees. (Redrawn from Bodenheimer, 1937.)

Often we are interested not only in the average rate over a period of time but in the theoretical instantaneous rate at particular times; in other words, the rate of change when Δt approaches zero. In the language of calculus, which is the branch of mathematics dealing (in part) with the study of rates, the letter d (for derivative) replaces the Δ when instantaneous rates are being considered. In this case the above notations become:

$dN/dt =$ the rate of change in the number of organisms per time at a particular instant.

$dN/(Ndt) =$ the rate of change in the number of organisms per time per individual at a particular instant.

In terms of the growth curve the slope (straight line tangent) at any point is the growth rate. Thus, in the case of population A in Figure 7–3, the growth rate was at a maximum between 4 and 11 weeks and zero after 16 weeks. The $\Delta N/\Delta t$ notation serves to illustrate the model for the usual purposes of measurement, but the dN/dt notation must be substituted in many types of actual mathematical manipulations of the models.

Example

Suppose a population of 50 protozoa in a pool is increasing by division. Suppose the population of 50 has increased to 150 after an hour has passed. Then

$$
\begin{aligned}
N \text{ (the initial number)} &= 50 \\
\Delta N \text{ (change in number)} &= 100 \\
\Delta N/\Delta t \text{ (average rate of change} \\
\text{per time)} &= 100 \text{ per hour} \\
\Delta N/(N\Delta t) \text{ (average rate of change} \\
\text{per time per individ-} \\
\text{ual)} &= 2 \text{ per hour} \\
&\quad \text{per individ-} \\
&\quad \text{ual (a 200\%} \\
&\quad \text{increase)} \\
&\quad \text{per hour}
\end{aligned}
$$

The instantaneous rate, dN/dt, cannot be measured directly, nor can $dN/(Ndt)$ be calculated directly from population counts. We would need to know the type of population growth curve exhibited by the population and then calculate the instantaneous rate from equations, as will be explained in Section 7. We cannot attach a "speedometer" to a population and determine its instan-

taneous speed rate as we can with a car. An approximation could be obtained, of course, by making a census of the population at very short intervals. We would likely discover that the rate of growth varies from time to time, something that the value for the average gives no hint. In the above example, the specific growth rate is in terms of the original population present at the beginning of the measurement (i.e., 50 protozoa). In other words the 50 protozoa gave rise in some manner or other to 2 additional protozoa for each of the 50 original ones. There are various ways in which this could come about; some individuals could have divided twice, some not at all, and some could have disappeared from the population. Our before-and-after census does not tell us just how it happened. The specific growth rate could also be expressed in terms of average population during the period of time. In estimating the annual per cent growth rate for human populations it is customary to use the estimated density at midyear as a base. Thus, a 1 per cent per annum rate means that one new person was added to the population for every 100 persons present at midyear. Expressing the rate in terms of the number present makes it possible to compare growth rates in populations of very different sizes, as, for example, the populations of a small and a large country.

4. NATALITY

Statement

Natality is the inherent ability of a population to increase. Natality rate is equivalent to the "birth rate" in the terminology of human population study (demography); in fact, it is simply a broader term covering the production of new individuals of any organism whether such new individuals are "born," "hatched," "germinated," "arise by division," or what not. *Maximum* (sometimes called absolute or physiological) *natality* is the theoretical maximum production of new individuals under ideal conditions (i.e., no ecological limiting factors, reproduction being limited only by physiological factors); it is a constant for a given population. *Ecological* or *realized natality* (or just plain "natality," without qualifying adjective) refers to population increase under an actual or specific environmental condition. It is not a constant for a population but may vary with the size and composition of the population and the physical environmental conditions. Natality is generally expressed as a rate determined by dividing the number of new individuals produced by time ($\Delta N_n / \Delta t$, the absolute natality rate) or as the number of new individuals per unit of time per unit of population [$\Delta N_n / (N\Delta t)$, the specific natality rate].

Explanation

Natality can be measured and expressed in a number of ways. Following the notation in the preceding section we have:

ΔN_n = production of new individuals in the population.

$\dfrac{\Delta N_n}{\Delta t}$ = B or natality rate.

$\dfrac{\Delta N_n}{N\Delta t}$ = b or natality rate per unit of population.

N may represent the total population or only the reproductive part of the population. With higher organisms, for example, it is customary to express natality rate per female. The specific natality rate, b, can be defined as *age-specific natality rates* for different age groups in the population. For example, Lord (1961), in his study of a wild rabbit population, found that 1- and 2-year-old females averaged a little over four young per year per female, while this age-specific rate was only 1.5 for females less than 1 year old. Considering all the different "kinds" of birth rates it is clear that confusion can easily result unless the concept used is clearly defined, preferably by using standard mathematical notations as above. Which concept is used will depend on the available data and the type of comparisons or predictions which one wishes to make.

Although the same notations may be used in referring to natality rate and population growth rate, the two are not the same because ΔN represents somewhat different quantities in the two cases. With respect to natality, ΔN_n represents *new individuals* added to the population. *Natality rate is zero or positive, never negative.* With respect to growth rate ΔN represents *the net increase or decrease* in the population, which is the result not only of natality but of mortality, emigration, immigration, etc.

Growth rate may be either negative, zero, or positive, since the population may be either decreasing, standing still, or increasing. Population growth rate will be considered again in Sections 7 and 8.

Maximum natality, as indicated in the above Statement, is the theoretical upper limit which the population, or the reproductive portion of the population, would be capable of producing under ideal conditions. As might be imagined, it is difficult to determine but is of interest for two reasons. (1) It provides a yardstick for comparison with the realized natality. Thus, a statement that natality of a population of mice was 6 young per female per year would mean more if it was known to what extent this might be higher if conditions were less limiting. (2) Being a constant, maximum natality is useful in setting up equations to determine or to predict the rate of increase in a population, as we shall see in subsequent sections. For practical purposes maximum natality can be approximated by experimental methods. For example, the highest average seed production achieved in a series of experiments with alfalfa, in which the most favorable known conditions of moisture, temperature, and fertilizer were combined, could be taken as maximum natality for that particular population. Another method of establishing a base is to observe the reproductive rate of a population when it is placed in a favorable environment, or when major limiting factors are temporarily nonoperative. If a small group of paramecia, for example, is placed in a new batch of favorable medium, the maximum reproductive rate achieved would be a good measure of maximum natality rate. This method can often be used in the field, if the ecologist is alert, since in nature there are often fortuitous circumstances when limiting factors are temporarily relaxed. As we shall point out later, many natural populations regularly exhibit maximum natality for brief seasonal or other periods. Reproductive performance during such favorable periods would be a practical approximation of maximum natality. *Since the value of maximum natality concept lies in its use as a constant against which variable observed natalities may be compared, any reasonable estimate could be used so long as conditions under which it was made are defined.*

It should be repeated that natality, and the other concepts discussed in this section, refer to the population and not to isolated individuals. The average reproductive capacity should be taken as the measure of natality, and not the capacity of the most productive or least productive individual. It is well known that occasional individuals in a population will exhibit unusual reproductive rates, but the performance of such individuals would not be a fair measure of the maximum possibilities of the population as a whole. Furthermore, in some populations the highest reproductive rate may occur when the population density is low, but in others — some of the higher vertebrates with complicated reproductive patterns, for example — the highest rate may occur when the population is medium sized or even relatively large (Allee effect, see Section 14 and Figure 7–26). Thus, the best estimate of the maximum natality should be made not only when physical factors are not limiting but also when population size is optimum. We shall say more about this business of a population acting as a limiting factor on itself in the next few sections.

Examples

In Table 7–2 two examples are worked out, one from field and one from laboratory data, to illustrate concepts of maximum-realized natality and absolute-specific natality rate. Maximum natality is based on somewhat arbitrary conditions, as must always be the case, but these conditions are clearly defined in the table. The ecological natalities are actual measurements. From these examples we see that the insect considered here has an enormously greater natality than does the vertebrate, but the latter realizes a greater percentage of its potential under the conditions listed. This explains, of course, why the grain beetle can so easily become a pest, while the bluebird is unlikely to overpopulate; when temperature and other conditions in stored grain are favorable, the insect "realizes" a large portion of its potential natality, thus generating a population explosion. For comparisons, it is more satisfactory to use the specific rate (i.e., so many eggs or young per female per time unit). In the case of the bluebird, the exact number of females in the population was not known so that rate calculations shown in Part III of Table 7–2 are approximations. It should also be pointed out that bluebird females

may actually lay more than 15 eggs per season, if one or more of her sets are destroyed. Under ideal conditions, however, no eggs would be lost and three broods of 5 each would theoretically be all that a female could physiologically manage during a season. The bluebird data also demonstrate a striking seasonal variation in natality; fewer eggs are laid in late broods and more of the eggs laid are infertile, fail to hatch, or are lost in other ways. Seasonal variation in natality is almost a universal phenomenon, as is also variation due to differences in the age and sex distribution in the population.

One problem in comparing the natality of different species populations rests in the difficulty of measuring natality at comparable stages in the life history; this is especially true of organisms such as insects and birds which have complicated life histories. Thus, in one case the number of eggs might be known whereas in another case only the number of larvae or independent young could be determined. In comparing one species and population with another, it is, therefore, important to be sure there is a comparable basis.

5. MORTALITY*

Statement

Mortality refers to death of individuals in the population. It is more or less the antithesis of natality with some parallel subconcepts. Mortality rate is equivalent to "death rate" in human demography. Like natality, mortality may be expressed as the number of individuals dying in a given period (deaths per time), or as a specific rate in terms of units of the total population or any part thereof. *Ecological or realized mortality* — the loss of individuals under a given environmental condition — is, like ecological natality, not a constant but varies with population and environmental conditions. There is a theoretical *minimum mortality,* a constant for a population, which represents the loss under ideal or nonlimiting condi-

---

* The author is indebted to Dr. W. T. Edmondson of the University of Washington for suggesting materials to be included in this and subsequent sections based on his long experience in teaching basic ecology.

Table 7–2. *Comparison of Maximum and Realized Natality of Two Species. Expressed as Natality Rate per Time (B) and Natality Rate per Time per Female (b)*

I. Field population of bluebirds (*Sialia sialis*) in city park, Nashville, Tenn., 1938 (from Laskey, 1939):

	MAX. NATALITY	ECOLOGICAL OR REALIZED NATALITY RATE			
	Total Eggs Laid	*Eggs Produced*		*Young Fledged*	
		No.	*% Max.*	*No.*	*% Max.*
1st brood	170°	170	100	123	72
2nd brood	175°	163	93	90	51
3rd brood	165°	122	74	52	32
Total for year	510°	455	89	265	52

II. Laboratory population of flour beetles (*Tribolium confusum*), 18 pairs for 60 days (approx. one generation) (from Park, 1934):

	MAX. NATALITY	ECOLOGICAL OR REALIZED NATALITY RATE			
	Total Eggs Laid	*Eggs Produced*		*Larvae Produced*	
		No.	*% Max.*	*No.*	*% Max.*
Fresh flour	11,988†	2,617	22	773	6
Old or "conditioned" flour‡	11,988†	839	7	205	2

III. Natality of bluebird and flour beetle populations expressed as specific rates ($\Delta N_n / N \Delta t$):

BLUEBIRD:

MAXIMUM SPECIFIC NATALITY RATE	REALIZED SPECIFIC NATALITY PER FEMALE	
Eggs per Female per Year	*Eggs*	*Fledged Young*
15	13.4	7.8

FLOUR BEETLE: MAXIMUM NATALITY RATE	REALIZED NATALITY PER FEMALE	
Eggs per Female per Day	*Eggs*	*Larvae*
Fresh flour 11.1	2.40	0.61
Condition flour 11.1	0.73	0.19

° Calculated by multiplying 5 times number nests attempted; five eggs per set is average number which the population is able to produce in the most favorable part of the season.

† Determined by average rate of 11.1 eggs per female per day which is average of two 60 day cultures held under optimum conditions (see Table 4, in Park, Ginsburg, and Horwitz, 1945).

‡ Flour in which a previous culture has been living; contains metabolic or "waste products." This situation might be similar to one in nature where the organism does not continually have the benefit of "unused" environment.

tions. That is, even under the best conditions individuals would die of "old age" determined by their *physiological longevity,* which, of course, is often far greater than the average *ecological longevity.* Often it is the *survival rate* that is of greater interest than the death rate. If the latter is expressed as a fraction, M, then survival rate is $1 - M$.

Explanation and Examples

Natality and mortality are complex population characteristics that may be expressed in a number of ways. To prevent confusion, therefore, the general term "mortality" needs to be qualified and, wherever pos-

sible, expressed by definite mathematical symbols, as indicated in previous sections. Generally, *specific mortality* is expressed as a percentage of the initial population dying within a given time.* Since we are often more interested in organisms that survive than in those that do not, it is often more meaningful to express mortality in terms of the reciprocal survival rate. As with natality, both the minimum mortality rate (theoretical constant) and the actual or ecological

* As with other rates we have been discussing, mortality rate can be expressed as a per cent of average population instead of the initial population; this would be of interest in situations in which density changed greatly during the period of measurement.

mortality rate (variable) are of interest, the former to serve as a base or "measuring stick" for comparisons. Since even under ideal conditions individuals of any population die of "old age," there is a minimum mortality that would occur under the best possible conditions that could be devised, and this would be determined by the average physiological longevity of the individuals. In most populations in nature the average longevity is far less than the physiologically inherent life span, and, therefore, actual mortality rates are far greater than the minimum. However, in some populations, or for brief periods in others, mortality may, for all practical purposes, reach a minimum and thus provide an opportunity for practical measurement under natural conditions.

Since mortality varies greatly with age, as does natality, especially in the higher organisms, specific mortalities at as many different ages or life history stages as possible are of great interest inasmuch as they enable us to determine the forces underlying the crude, overall population mortality. A complete picture of mortality in a population is given in a systematic way by the *life table*, a statistical device developed by students of human populations. Raymond Pearl first introduced the life table into general biology by applying it to data obtained from laboratory studies of the fruit fly, *Drosophila* (Pearl and Parker, 1921). Deevey (1947 and 1950) has assembled data for the construction of life tables for a number of natural populations, ranging from rotifers to mountain sheep. Since Deevey's reviews, numerous life tables have been published for a variety of natural and experimental populations. In a laboratory population one can start with a cohort of, say, 1000 young individuals and keep track of the number surviving at intervals of time until the end of that generation. In the field it is also often possible to determine at regular intervals the individuals surviving out of an initial marked population. Approximate life tables may also be constructed for natural populations if the age at death is known or if the age structure (that is, proportion of different ages) can be determined at intervals.

As an example, let us take the Dall mountain sheep (Table 7–3). The age of these sheep can be determined from the horns. When a sheep is killed by a wolf or dies for any other reason, its horns remain pre-

served for a long period. Adolph Murie spent several years in intensive field study of the relation between wolves and mountain sheep in Mt. McKinley National Park, Alaska. During this period he picked up a large series of horns, thus providing admirable data on the age at which sheep die in an environment subject to all the natural hazards, including wolf predation (but not including predation by man, as sheep were not hunted in the McKinley National Park). As shown in Table 7–3, the life table consists of several columns, headed by standard notations, giving (l_x), the number of individuals out of a given population (1000 or any other convenient number) that survive after regular time intervals (day, month, year, etc. — see column x); (d_x), the number dying during successive time intervals; (q_x), the death or mortality rate during successive intervals (in terms of initial population at beginning of period); and (e_x), the life expectancy at the end of each interval. As may be seen from Table 7–3, the average age is better than seven years, and if a sheep can survive the first year or so, its chances of survival are good until relative old age, despite the abundance of wolves and the other vicissitudes of the environment.

Curves plotted from life-table data may be very instructive. When data from column l_x are plotted with the time interval on the horizontal coordinate and the number of survivors on the vertical coordinate, the resulting curve is called a *survivorship curve*. If a semilogarithmic plot is used, with the time interval on the horizontal coordinate expressed as a percentage of the mean length of life (see column x', Table 7–3) or as a percentage of the total life span, species of widely different life spans may be compared. Furthermore, a straight line on a semilogarithmic plot indicates a constant specific rate of survival.

Survivorship curves are of three general types, as shown in Figure 7–4. A highly convex curve (A in Figure 7–4) is characteristic of species such as the Dall sheep, in which the population mortality rate is low until near the end of the life span. (We suggest you plot column l_x, Table 7–3, on semilog graph paper and compare the shape of the curve with the models in Figure 7–4.) Many species of large animals and, of course, man, exhibit this type of survivorship. At the other extreme a highly concave curve (C in Figure 7–4) results when mortality is high

*Table 7–3. Life Table for the Dall Mountain Sheep (Ovis d. dalli).**
Based on Known Age at Death of 608 Sheep Dying Before
1937 (both sexes combined).† Mean Length of Life 7.09 Years

AGE (YEARS)	AGE AS PER CENT DEVIATION FROM MEAN LENGTH OF LIFE	NUMBER DYING IN AGE INTERVAL OUT OF 1000 BORN	NUMBER SURVIVING AT BEGINNING OF AGE INTERVAL OUT OF 1000 BORN	MORTALITY RATE PER THOUSAND ALIVE AT BEGINNING OF AGE INTERVAL	EXPECTATION OF LIFE, OR MEAN LIFETIME REMAINING TO THOSE ATTAINING AGE INTERVAL (YEARS)
x	x'	d_x	l_x	$1000\ q_x$	e_x
0–0.5	−100	54	1000	54.0	7.06
0.5–1	−93.0	145	946	153.0	—
1–2	−85.9	12	801	15.0	7.7
2–3	−71.8	13	789	16.5	6.8
3–4	−57.7	12	776	15.5	5.9
4–5	−43.5	30	764	39.3	5.0
5–6	−29.5	46	734	62.6	4.2
6–7	−15.4	48	688	69.9	3.4
7–8	−1.1	69	640	108.0	2.6
8–9	+13.0	132	571	231.0	1.9
9–10	+27.0	187	439	426.0	1.3
10–11	+41.0	156	252	619.0	0.9
11–12	+55.0	90	96	937.0	0.6
12–13	+69.0	3	6	500.0	1.2
13–14	+84.0	3	3	1000	0.7

* From Deevey (1947); data from Murie (1944).
† A small number of skulls without horns, but judged by their osteology to belong to sheep nine years old or older, have been apportioned *pro rata* among the older age classes.

during the young stages. Oysters or other shellfish and oak trees would be good examples; mortality is extremely high during the free-swimming larval or the acorn-seedling stages, but once an individual is well established on a favorable substrate, life expectancy improves considerably! Intermediate are patterns in which the age-specific survival is more nearly constant so that curves approach a diagonal straight line on a semilog plot (B in Figure 7–4). A "stair-step" type of survivorship curve may be expected if survival differs greatly in successive life history stages as is often the case in holometabolous insects (i.e., insects with complete metamorphosis such as butterflies). In the model shown (B-1 in Figure 7–4), the steep segments represent egg, pupation, and short-lived adult stages, while the flatter segments represent larval and pupal stages that suffer less mortality (see Itô, 1959). Probably no population in the real world has a constant age-specific survival rate throughout the whole life span (B-2, Figure 7–4), but a slightly concave or

sigmoid curve (B-3, Figure 7–4) is characteristic of many birds, mice, and rabbits. In these cases, the mortality rate is high in the young but lower and more nearly constant in the adult (1 year or older).

Itô (1959) has pointed out that the shape of the survivorship curve is related to the degree of parental care or other protection given to the young. Thus, survivorship curves for honeybees and robins (which protect their young) are much less concave than are those for grasshoppers or sardine fish (which do not protect their young). The latter species, of course, compensate by laying many more eggs.

The shape of the survivorship curve may vary with the density of the population, as shown in Figure 7–5. Data on age structure were used by Taber and Dasmann (1957) to construct the curves for two stable deer populations living in the chaparral region of California (see Chapter 14, page 396, for a description of this type of habitat). As may be seen in the figure, the survivorship curve of the denser population is quite concave. In other words, deer living in the managed area, where the food supply was increased by controlled burning, have a shorter life expectancy than deer in the unmanaged area, presumedly because of increased hunting pressure, intraspecific competition, and so on. From the viewpoint of the hunter, the managed area is most favorable, but from the viewpoint of the individual deer, the less crowded area offers a better chance for a long life. History records a number of parallels to this situation in human populations where high density was not always favorable to the individual. Many ecologists believe that the present human population explosion is not so much a threat to man's survival as it is a threat to the quality of life for the individual, even though civilized man has greatly increased his own "ecological" longevity as a result of modern medical knowledge, better nutrition, and so forth, so that man's curve approaches the sharp-angled minimum mortality curve. However,

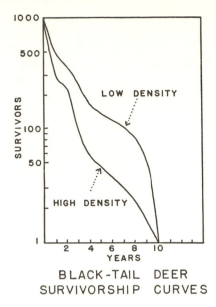

BLACK-TAIL DEER
SURVIVORSHIP CURVES

Figure 7-5. Survivorship curves for two stable deer populations living in the chaparral region of California. The high density population (about 64 deer per square mile) is in a managed area where an open shrub and herbaceous cover is maintained by controlled burning, thus providing a greater quantity of browse in the form of new growth. The low density population (about 27 per sq. mi.) is in an unmanaged area of old bushes unburned for 10 years. Recently burned areas may support up to 86 deer per sq. mi., but the population is unstable and hence survivorship curves cannot be constructed from age distribution data. (After Taber and Dasmann, 1957.)

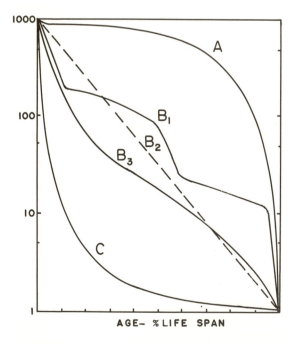

Figure 7-4. Several types of survivorship curves plotted on the basis of survivors per thousand log scale (vertical coordinate) and age as a per cent of the life span (horizontal coordinate). A. Convex type, in which most of the mortality occurs towards the end of the life span. B_1. Stairstep type, in which survival rate undergoes sharp changes in transition from one life history stage to another. B_2. Theoretical curve (straight line), in which age-specific survival remains constant. B_3. Slightly sigmoid type that approaches B_2. C. Concave type, in which mortality is very high during the young stages.

man apparently has not increased his maximum or "physiological" longevity, since no more people live now to be 100 years old than did in past centuries. More about these paradoxes in Chapter 21.

It is self-evident that neither population size nor trend (whether increase or decrease) to be expected can be determined from knowledge of mortality alone. Natality and other population characteristics must also be considered.

To prepare the way for mathematical models of population growth to be considered in subsequent sections it is instructive to add age-specific natality to the life table so that it is not just a "death" table (as is Table 7–3). Two abbreviated life tables that include data on natality as well as on survival are included in Table 7–4. Table 7–4A presents a simplified, hypothetical set of data to illustrate the principles and the calculations; Table 7–4B is based on data from an actual study of a laboratory population of grain weevils. The notations

Table 7–4. Life Tables That Include Both Age-specific Survival and Natality

A. LIFE TABLE FOR A HYPOTHETICAL POPULATION WITH A SIMPLE LIFE HISTORY

Age	Age-specific Survival Rate (as Fractions)	Age-specific Death Rate	Age-specific Natality Rate (Offspring per Female Aged x)	
x	l_x	d_x	m_x	$l_x m_x$°
0	1.00			
		0.20	0	0.00
1	0.80			
		0.20	0	0.00
2	0.60			
		0.20	1	0.60
3	0.40			
		0.20	2	0.80
4	0.20			
		0.10	2	0.40
5	0.10			
		0.05	1	0.10
6	0.05			
		0.05	0	0.00
7	0.00			

$\Sigma l_x m_x$ = net reproductive rate (R_o) = 1.90

B. LIFE TABLE FOR LABORATORY POPULATION OF RICE WEEVILS (*Calandra oryzae*) UNDER OPTIMUM CONDITIONS (29°C. and 14% moisture content of rice), SEX RATIO EQUAL†

Pivotal Age in Weeks	Age-specific Survival (as Fractions)	Age-specific Natality (No. Female Offspring per Female Aged x)	
x	l_x	m_x	$l_x m_x$
4.5	0.87	20.0	17.400
5.5	0.83	23.0	19.090
6.5	0.81	15.0	12.150
7.5	0.80	12.5	10.000
8.5	0.79	12.5	9.875
9.5	0.77	14.0	10.780
10.5	0.74	12.5	9.250
11.5	0.66	14.5	9.570
12.5	0.59	11.0	6.490
13.5	0.52	9.5	4.940
14.5	0.45	2.5	1.125
15.5	0.36	2.5	0.900
16.5	0.29	2.5	0.800
17.5	0.25	4.0	1.000
18.5	0.19	1.0	0.190
			R_o = 113.560

° Based on l_x at the start of the age period.
† After Birch, 1948.

x, d_x and l_x are as previously described; age-specific natality rates (as offspring per female produced in a unit of time) are indicated by the columns headed m_x. If we multiply l_x and m_x and obtain the sum of the values for the different age classes, we get what may be called a net reproductive rate (R_o). Thus:

$$R_o \text{ (net reproductive rate)} = \Sigma l_x m_x$$

(In this case l_x refers to females only.) In Table 7–4A the net reproductive rate of 1.9 indicates that for every one female we start with, 1.9 offspring are produced. If the sex ratio was equal, this would mean that the population just about replaces itself during the generation. In the grain weevil laboratory population under the most optimum conditions the population multiplies 113.6 times in each generation (i.e., R_o = 113.6)! Under stable conditions in nature, R_o in terms of the total population should be around one. Paris and Pitelka (1962), using the life table approach and data on l_x and m_x of year classes, calculated R_o for a sowbug population in a grassland as 1.02, indicating an approximate balance between births and deaths.

It is important to emphasize that the reproductive schedule greatly influences population growth and other population attributes. Natural selection can affect various kinds of change in the life history that will result in adaptive schedules. Thus, selection pressure may change the time at which reproduction begins without affecting the total number of offspring produced, or it may affect production or "clutch size" without changing the timing of the reproduction. These and many other aspects can be revealed by life table analyses.

6. POPULATION AGE DISTRIBUTION

Statement

Age distribution is an important population characteristic which influences both natality and mortality, as has been shown by the examples discussed in the preceding section. Consequently, the ratio of the various age groups in a population determines the current reproductive status of the population and indicates what may be expected in the future. Usually a rapidly expanding population will contain a large proportion of young individuals, a stationary population a more even distribution of age classes, and a declining population a large proportion of old individuals. However, a population may pass through changes in age structure without changing in size. There is evidence that populations have a "normal" or stable age distribution toward which actual age distributions are tending. Once a stable age distribution is achieved, unusual increases in natality or mortality result in temporary changes, with spontaneous return to the stable situation.

Explanation

In so far as the population is concerned, there are three ecological ages, which have been listed by Bodenheimer (1938) as *prereproductive, reproductive,* and *postreproductive*. The relative duration of these ages in proportion to the life span varies greatly with different organisms. In modern man, the three "ages" are relatively equal in length, about a third of his life falling in each class. Primitive man, by comparison, had a much shorter postreproductive period. Many plants and animals have a very long prereproductive period. Some animals, notably insects, have extremely long prereproductive periods, a very short reproductive period, and no postreproductive period. The mayfly (*Ephemeridae*) and the seventeen-year locust are classic examples. The former requires from one to several years to develop in the larval stage in the water and lives but a few days as an adult; the latter has an extremely long developmental history (not necessarily seventeen years, however), with adult life lasting less than a single season. It is obvious, therefore, that the duration of the ecological ages needs to be considered in interpreting data on age distribution.

Lotka (1925) has shown on theoretical grounds that a population tends to develop a stable age distribution, that is, a more or less constant proportion of individuals of different ages, and that if this stable situation is disrupted by temporary changes in the environment or by temporary influx from or egress to another population, the age distribution will tend to return to the previous

situation upon restoration of normal conditions. More permanent changes, of course, would result in development of a new stable distribution. A direct quotation from Lotka is perhaps the best way to clarify the important concept of stable age distribution:

> . . . the force of mortality varies very decidedly with age, and it might therefore be supposed that any discussion of the rate of increase of a population of organisms must fully take into account the age distribution. This supposition, however, involves an assumption, namely, the assumption that age distribution itself is variable. Now in point of fact, age distribution is indeed variable, but only within certain restricted limits. Certain age distributions will practically never occur, and if by arbitrary interference or by a catastrophe of nature, some altogether unusual form were impressed upon the age distribution of an isolated population, the irregularities would tend shortly to become smoothed over. There is, in fact, a certain stable type of age distribution about which the actual age distribution varies and toward which it returns if through any agency disturbed therefrom. The form of this distribution in an isolated population (i.e., with immigration and emigration negligible) is easily deduced. . . .

The mathematical proof of this parameter and the method of calculating the theoretical stable age distribution is beyond the scope of this presentation (see Lotka, 1925, Chapter IX, pages 110–115). Suffice it to say that life-table data and knowledge of the specific growth rate (see Sections 2 and 7) are needed to determine the stable age distribution.

The idea of a stable age distribution is an important one. Again, as in the case of the maximum natality constant, it furnishes a base for evaluating actual age distributions as they may occur. It is one more constant that may help us untangle the seemingly bewildering array of variables that occur in nature. The whole theory of a population, of course, is that it is a real biological unit with definite biological constants and definite limits to variations that may occur around or away from these constants.

Examples

A convenient way to picture age distribution in a population is to arrange the data in the form of a polygon or age pyramid (not to be confused with the ecological pyramids discussed in Chapter 3), the number of individuals or the percentage in the different age classes being shown by the relative widths of successive horizontal bars. The upper pyramids in Figure 7–6 illustrate three hypothetical cases: (left) a pyramid with broad base, indicating a high percentage of young individuals; (middle) a bell-shaped polygon, indicating a moderate proportion of young to old; and (right) an urn-shaped figure, indicating a low percentage of young individuals. The latter would generally be characteristic of a senile or declining population. The vole pyramids in Figure 7–6 show stable age distributions under conditions of maximum rate of population increase (left) and with no growth (right), i.e., natality equaling mortality. The rapidly growing population has the much greater proportion of young individuals.

In game birds and fur-bearing mammals, the ratio of first year animals to older animals, determined during the season of harvest (fall or winter) by examination of samples from the population taken by hunters or trappers, provides an index to the population trends. In general, a high ratio of juveniles to adults, as shown in the bottom diagrams in Figure 7–6, indicates a highly successful breeding season and likelihood of a larger population the next year, provided juvenile mortality is not excessive. In the muskrat example (lower right, Figure 7–6) the highest percentage of juveniles (85 per cent) occurred in a population which had been heavily trapped for the previous few years; reduction in total population in this manner had apparently resulted in increased natality for those individuals surviving. Thus, as Lotka would say, the population was "spontaneously" returning to a more stable age distribution which would presumably lie somewhere between the extremes shown.

As aging techniques have improved, so has knowledge of age distribution in wild populations. In rabbits it has been found that the weight of the lens of the eye is a good indicator of age. Using this indicator, Meslow and Keith (1968) investigated age

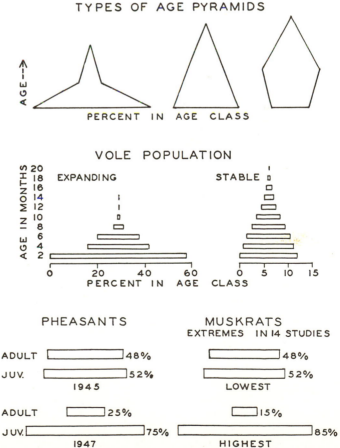

Figure 7–6. Age pyramids. *Upper:* Three types of age pyramids representing a large, moderate, and small percentage of young individuals in the population. *Middle:* Age pyramids for laboratory populations of the vole, *Microtus agrestis* (left), when expanding at an exponential rate in an unlimited environment, and (right) when birth rates and death rates are equal (data from Leslie and Ransom, 1940). *Lower:* Extremes in juvenile-adult ratios in pheasants in North Dakota (data from Kimball, 1948) and in muskrats in eastern United States (data from Petrides, 1950).

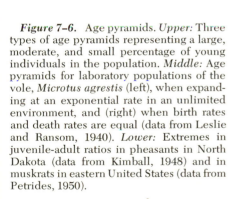

distribution in snowshoe hare populations, which, as described in Section 9, are famous for pronounced cycles of abundance. When the population was expanding rapidly (a year or two before the peak abundance) as high as 76 per cent of the summer population were one year olds with only 4 per cent three years or older. In contrast, 35 per cent were three years or older, and only about the same percentage was in the one-year class during the declining years just before the low point in the abundance cycle.

A phenomenon known as the "dominant age class" has been repeatedly observed in fish populations that have a very high potential natality rate. When a large year class occurs as a result of unusual survival of eggs and larval fish, reproduction is suppressed for the next several years. Herring in the North Sea provide a classic case, as shown in Figure 7–7. Fish of the 1904 year class dominated the catch from 1910 (when this age class was 6 years old and large enough to be caught effectively in commercial fish nets) until 1918 when, at 14 years of age, they still outnumbered fish of younger age groups! We observe here something of a compensatory mechanism in which high survival is followed by a high probability of low survival in subsequent years. Fishery ecologists are currently trying to find out what environmental conditions result in the unusual survival which occurs every now and then.

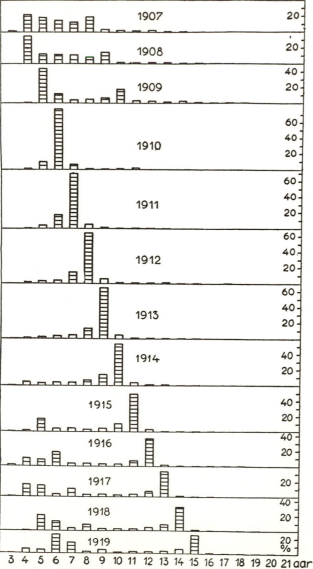

Figure 7–7. Age distribution in the commercial catch of herring in the North Sea between 1907 and 1919, illustrating the dominant age class phenomenon. The 1904 year class was very large and dominated the population for many years. Since fish younger than 5 years are not caught in the nets, the 1904 class did not show up until 1909. Age of the fish was determined by growth rings on scales, which are laid down annually in the same manner as growth rings on trees. (After Hjort, 1926.)

As final examples, Figure 7–8 shows a series of interesting age pyramids for human populations in two localities in Scotland. Both populations in 1861 were young and vigorous, and by 1901 they had assumed an age distribution of a stationary population. By 1931 one population had assumed a top-heavy age structure (relatively few children, large proportion of old people) as a result of a deteriorated habitat. These figures are interesting, also, in that they show how sex distribution can be pictured along with age distribution.

7. THE INTRINSIC RATE OF NATURAL INCREASE

Statement

When the environment is unlimited (space, food, other organisms not exerting a limiting effect), the specific growth rate (i.e., the population growth rate per individual) becomes constant and maximum for the existing microclimatic conditions. The value of the growth rate under these favorable

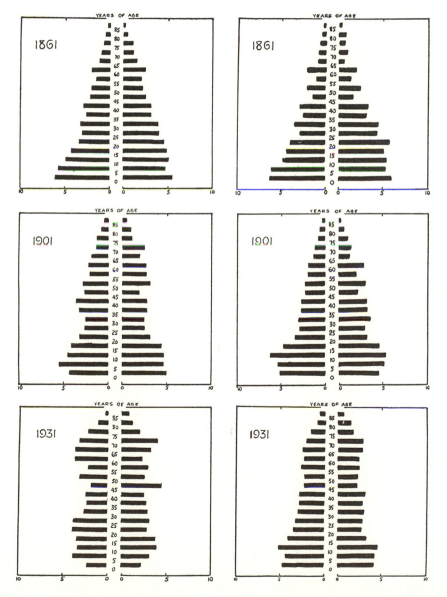

Figure 7–8. Age pyramids for man in two localities in Scotland. Series at the left is for a parish with top-heavy age structure by 1931 in a deteriorated habitat. Series at right is for a population in a somewhat healthier environment. Age classes have been reduced to percentage of total population, with males represented at left and females at right of each pyramid. (After Darling, 1951.)

population conditions is maximal, is characteristic of a particular population age structure, and is a single index of the inherent power of a population to grow. It may be designated by the symbol r, which is the exponent in the differential equation for population growth in an *unlimited environment* under specified physical conditions:

$$dN/dt = rN; \quad r = dN/(Ndt) \qquad (1)$$

Note that this is the same form as used in Section 3. The parameter, r, can be thought of as *an instantaneous coefficient of population growth*. The exponential integrated form follows automatically by calculus manipulation:

$$N_t = N_o e^{rt} \qquad (2)$$

where N_o represents the number at time zero, N_t the number at time t and e the base of natural logarithms. By taking the natural logarithm (ln or $\log_e$) of both sides, the equation is converted into a form used in making actual calculations. Thus:

$$\ln N_t = \ln N_o + rt; \quad r = \frac{\ln N_t - \ln N_o}{t} \qquad (3)$$

In this manner the index r can be calculated from two measurements of population size (N_o and N_t or at any two times during the unlimited growth phase in which case N_{t_1} and N_{t_2} may be substituted for N_o and N_t and $(t_2 - t_1)$ substituted for t in the above equations).

The index r is actually the difference between the instantaneous specific natality rate (i.e., rate per time per individual) and the instantaneous specific death rate, and may thus be simply expressed:

$$r = b - d \qquad (4)$$

The overall population growth rate under unlimited environmental conditions (r) depends on the age composition and the specific growth rates due to reproduction of component age groups. Thus, there may be several values of r for a species depending upon population structure. When a stationary and stable age distribution exists, the specific growth rate is called the *intrinsic rate of natural increase* or r_{max}. The maximum value of r is often called by the less specific but widely used expression *biotic potential*, or reproductive potential. The difference between the maximum r or biotic potential and the rate of increase

which occurs in an actual laboratory or field condition is often taken as a measure of the *environmental resistance*, which is the sum total of environmental limiting factors which prevent the biotic potential from being realized.

Explanation

We have now come to the point where we wish to put together natality, mortality, and age distribution—each important but each admittedly incapable of telling us very much by itself—and come out with what we really want to know, namely, how is the population growing as a whole and what would it do if conditions were different, and what is its best possible performance against which we may judge its everyday performances? Chapman (1928) proposed the term *biotic potential* to designate maximum reproductive power. He defined it as "the inherent property of an organism to reproduce, to survive, i.e., to increase in numbers. It is sort of the algebraic sum of the number of young produced at each reproduction, the number of reproductions in a given period of time, the sex ratio and their general ability to survive under given physical conditions." The concept of biotic potential, or *reproductive potential*, suggested by some as more descriptive (see Graham, 1952), has gained wide usage. However, as might be imagined from the very generalized definition given above, biotic potential came to mean different things to different people. To some it came to mean a sort of nebulous reproductive power lurking in the population, terrible to behold, but fortunately never allowed to come forth because of the forthright action of the environment (i.e., "if unchecked the descendants of a pair of flies would weigh more than the earth in a few years"). To others it came to mean simply and more concretely the maximum number of eggs, seeds, spores, and so forth, the most fecund individual was known to produce, despite the fact that this would have little meaning in the population sense in most cases, since most populations do not contain individuals all of which are continually capable of peak production.

It remained for Lotka (1925), Dublin and Lotka (1925), Leslie and Ranson (1940), Birch (1948), and others to translate the rather broad idea of biotic potential into mathematical terms that can be understood in any language (with, sometimes, the help of a good mathematician!). Birch (1948) expressed it well when he said: "If the 'biotic potential' of Chapman is to be given quantitative expression in a single index, the parameter r would seem to be the best measure to adopt since it gives the intrinsic capacity of the animal to increase in an unlimited environment." The index r is also frequently used as a quantitative expression of "reproductive fitness" in the genetic sense, as will be noted later.

In terms of the growth curves discussed in Section 3, r is the specific growth rate ($\Delta N/N\Delta t$) when population growth is exponential. It will be recognized that equation (3) in the Statement above is an equation for a straight line. Therefore, the value for r can be obtained graphically. If growth is plotted as logarithms or on semilogarithmic paper, the log of population number plotted against time will give a straight line if growth is exponential; r is the slope of this line. The steeper the slope, the higher the intrinsic rate of increase. In Figure 7–9 the same growth curve is plotted in two ways, with numbers (N) on an arithmetic scale (left-hand graph) and with N on a logarithmic scale (right-hand graph on semilog paper). In this example a hypothetical population of microorganisms is experiencing 6 days of exponential growth in which the population increases by a factor of 10 every two days. The slope of the line on the semilog plot is 1.15, which is the value for r. One can check this by plugging into equation (3) any two population values, for example, density at day 2 and day 4 as follows:

$$r = \frac{\ln N_{t_2} - \ln N_{t_1}}{(t_2 - t_1)}$$

$$r = \frac{\ln 100 - \ln 10}{2}$$

From the table of natural logarithms,

$$r = \frac{4.6 - 2.3}{2} = 1.15$$

The coefficient of population growth, r, should not be confused with the net reproductive rate, R_o, as discussed in Section 5 (see Table 7–4); the latter is strictly related to generation time and is not suitable for comparing different populations unless their generation times are similar. However, mean generation time (T) is related to R_o and r in the following manner:

$$R_o = e^{rT}; \text{ therefore, } T = \frac{\log_e R_o}{r}$$

The relationships between T, R_o, and r for a variety of animal populations is graphed on page 52 of Slobodkin's (1962) book. Populations in nature often exhibit exponential growth for short periods when there is ample food and no crowding effects, enemies, and so forth. Under such conditions the population as a whole is expanding at a terrific rate even though each organism is reproducing at the same rate as before, i.e., the specific growth rate is constant. Plankton "blooms," mentioned in previous chapters, pest eruptions, or growth of bacteria in new culture medium are examples of situations in which growth may be logarithmic. Many other phenomena such as absorption of light, monomolecular chemical reactions, and compound interest behave in the same manner. It is obvious that this exponential increase cannot continue for very long; often it is never realized. Interactions within the population as well as external environmental resistances soon slow down the rate of growth and play a part in shaping population growth form in various ways, as will be considered in the next sections.

Examples

Table 7–5 gives the intrinsic rate of natural increase for several insects, rodents, and man together with other related data. The values for the rodents and for the insects at 29°C. are presumed to be "maximum r" (biotic potential), since populations on which calculations were based were living under as nearly optimum conditions as could be devised experimentally. The extremely wide differences in potential ability of different populations to increase are especially emphasized in the last two columns, which show the number of times the population would multiply itself if the exponential rate continued, and the time required to double the population. These two parameters are mathematically derived from the intrinsic or "population interest" rate, r, as follows:

Finite rate of increase, $\lambda = e^r$; $\log_e \lambda = r$;
$\lambda = \text{antilog}_e r$

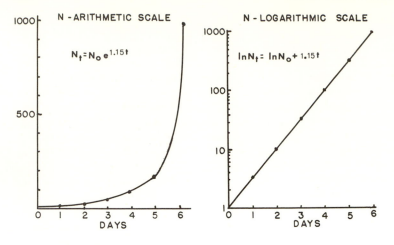

Figure 7–9. The same population growth curve shown in two ways: *Left:* Numbers (N) on an arithmetic scale. *Right:* Numbers on a logarithmic scale. In this hypothetical example, a population of microorganisms is experiencing six days of exponential growth during which the population increases tenfold every two days. See text for explanation of equations.

Doubling time, $t = \log_e 2/r = 0.6931/r$ (derived from the exponential equation 2, page 180, by setting $N_t/N_o = 2$).

In the case of the rice weevil it will be noted that varying the temperature above and below the optimum has marked effects on the ability of the population to increase. The difference between maximum r and that found at non-optimum temperature (but still within limits of tolerance) is an example of environmental control in the absence of crowding. Even under these less favorable temperatures, it will be noted, the population is still able to increase several hundredfold in the course of a year, *if* biotic conditions remain non-limiting. Since insects are often considered man's most serious competitors, it is indeed fortunate that we do not have to depend solely on physical factors

Table 7–5. *Comparison of Data Pertaining to Rate of Increase of Certain Insects, Rodents, and Man*

ORGANISM AND CONDITION	INTRINSIC RATE (r)		MEAN LENGTH OF GENERATION (T) Weeks	NET REPRODUCTION RATE (R_o)	FINITE RATE° (e^r)		DOUBLING TIME† Weeks (w) or Years (yrs)
	Week	Year			Week	Year	
Calandra (rice weevil)‡ at optimum temp. 29°C.	0.76	39.6	6.2	113.56	2.14	1.58×10^{16}	0.91 (w)
Calandra‡ at 23°C.	0.43	22.4	10.6	96.58	1.54	5.34×10^8	1.61 (w)
Calandra‡ at 33.5°C.	0.12	6.2	9.2	3.38	1.13	493	5.78 (w)
Tribolium castaneum§ (flour beetle) at optimum temp. 28.5°C.	0.71	36.8	7.9	275.0	2.03	1.06×10^5	0.96 (w)
Pediculus humanus‖ (human louse)	0.78	40.6	4.4	30.93	2.18	4.27×10^{16}	0.88 (w)
Microtus agrestis¶ (vole) optimum lab.	0.088	4.5	20.2	5.9	1.09	90	7.90 (w)
Rattus norvegicus# (brown rat) optimum lab.	0.104	5.4	31.1	25.66	1.11	221	6.76 (w)
Man, white, USA, 1920°°	–	0.0055	–	–	–	1.0055	126.0 (yrs)
Man, USA, 1968††	–	0.0077	–	–	–	1.0077	90.0 (yrs)
Man, world, 1968††	–	0.02	–	–	–	1.0202	34.7 (yrs)

° Number of times population would multiply in time indicated.
† Time required to double population = 0.6931/r.
‡ From Birch (1948).
¶ From Leslie and Ranson (1940). § From Leslie and Park (1949). ‖ From Evans and Smith (1952).
†† From Ehrlich and Ehrlich (1970). # From Leslie (1945). °° From Dublin and Lotka (1925).

(or spray guns!) to keep insects in check. Parasites, disease, predators, inter- and intraspecific competition, and other biological factors offer powerful checks also if we will but encourage them to operate. The more we can learn about these "living environmental resistances," the better equipped we should be in the "war" with insects.

For organisms smaller than insects, the intrinsic rate may be even higher than shown in Table 7–5. Smith (1963) found that r for *Daphnia* (a small crustacean) in an unlimited food environment was 0.44 per day or 3.08 per week. In another paper, Smith (1954) suggests that the range of r in the biotic kingdom may extend through 6 log cycles.

8. POPULATION GROWTH FORM AND CONCEPT OF CARRYING CAPACITY

Statement

Populations have characteristic patterns of increase which are called population growth forms. For the purposes of comparison we may designate two basic patterns based on the shapes of arithmetic plots of growth curves, namely, the *J-shaped growth form* and the *S-shaped* or *sigmoid growth form*. These contrasting types may be combined or modified, or both, in various ways according to the peculiarities of different organisms and environments. In the J-shaped form density increases rapidly in exponential or compound interest fashion (as shown in Figure 7–9) and then stops abruptly as environmental resistance or another limit becomes effective more or less suddenly. This form may be represented by the simple model based on the exponential equation considered in the preceding section:

$$\frac{dN}{dt} = rN \quad \textit{with a definite limit on N}$$

In the sigmoid form, the population increases slowly at first (establishment or positive acceleration phase), then more rapidly (perhaps approaching a logarithmic phase); but it soon slows down gradually as the environmental resistance increases percentagewise (the negative acceleration phase), until a more or less equilibrium level is reached and maintained. This form may be represented by the simple logistic model:

$$\frac{dN}{dt} = rN \frac{(K - N)}{K}$$

The upper level, beyond which no major increase can occur, as represented by the constant K, is the *upper asymptote* of the sigmoid curve and has been aptly called the *carrying capacity*. In the J-form there may be no equilibrium level, but the limit on N represents the upper limit imposed by the environment. The two growth forms and certain variants are shown schematically in Figure 7–10.

Explanation

When a few individuals are introduced into or enter an unoccupied area (for example, at the beginning of a season), characteristic patterns of population increase have often been observed. When plotted on an arithmetic scale, the part of the growth curve which represents population increase often takes the form of an S or a J, as shown in Figures 7–10*A* and *B*. It is interesting to note that these two basic growth forms are similar to the two metabolic or growth types that have been described in the case of individual organisms (Bertalanffy, 1957). However, it is not known if there is a causal relationship between growth of individuals and growth of populations; it is not safe at this point to do more than call attention to the fact that there are some similarities in patterns.

It will be noted that the equation given above as a simple model for the J-shaped form is the same as the exponential equation discussed in the previous section, except that a limit is imposed on N; that is, the relatively unrestricted growth is suddenly halted when the population runs out of some resource (such as food or space), or when frost or any other seasonal factor intervenes, or when the reproductive season suddenly terminates (perhaps, for example, because of development of diapause as described in Chapter 5, page 116). When the upper limit of N is reached, the density may remain at this level for a time, or, as is often the case, an immediate decline occurs, producing a "relaxation oscillation" pattern in density, as shown in Figure 7–10, A-1 and A-2. This type of pattern, which Nicholson (1954) has called "density triggered," seems to be characteristic of many populations in nature such as algal blooms,

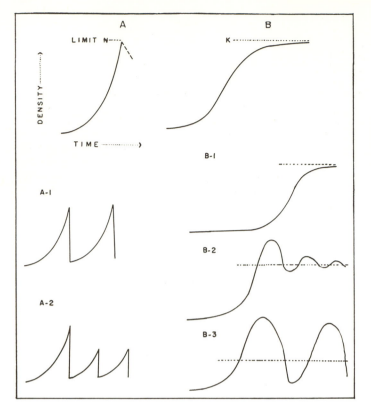

Figure 7–10. Some aspects of population growth form, when plotted on arithmetic scales, showing the J-shaped (exponential) (A) and the S-shaped (sigmoid) (B) forms and some variants. A-1 and A-2 show oscillations that would be inherent in the J-shaped form. B-1, B-2, and B-3 show some possibilities (but by no means all) where there is a delay in density effect, which occurs when time elapses between production of young individuals and full influence of the individuals (the case in higher plants and animals). When nutrients or other requisites accumulate prior to population growth, an "overshoot" may occur as shown in A-2 and B-2. (This explains why new ponds or lakes often provide better fishing than old ones!) (Curves adapted from Nicholson, 1954.)

annual plants, some insects, and perhaps lemmings on the tundra.

A type of growth form which is also observed frequently follows an S-shaped or sigmoid pattern when density and time are plotted on arithmetic scales. The sigmoid curve is the result of greater and greater action of detrimental factors (environmental resistance) as the density of the population increases, in contrast to the previous model in which environmental resistance was delayed until near the end of the increase. For this reason Nicholson (1954) has spoken of the sigmoid type as "density conditioned." The simplest case that can be conceived is the one in which detrimental factors are linearly proportional to the density. Such simple or "ideal" growth form is said to be logistic and conforms to the logistic equation[*] that we have used as a basis for our model of the sigmoid pattern. The equation may be written in a number of ways as follows (three forms plus the integrated form are shown in next column):

---

[*] The logistic equation was first proposed by P. F. Verhulst in 1838; it was extensively used by Lotka and "rediscovered" by Pearl and Reed (1920). For an account of mathematical derivations and curve-fitting procedures, see Pearl (1930).

$$\frac{dN}{dt} = rN\,\frac{(K-N)}{K} \text{ or } = rN - \frac{r}{K}\,N^2 \text{ or}$$

$$= rN\left(1 - \frac{N}{K}\right)$$

$$N = \frac{K}{1 + e^{a-rt}}$$

where dN/dt is the rate of population growth (change in number in time), r the specific growth rate or intrinsic rate of increase as discussed in Section 6, N the population size (number), K the maximum population size possible, or "upper asymptote," e the base of natural logarithm and a the constant of integration defining the position of the curve relative to the origin; it is the value of $\log_e (K-N)/N$ when $t = 0$.

As will be noted, this is the same equation as the exponential one written in the previous section with the addition of the expression $(K-N)/K$, $(r/K)N^2$, or $(1-N/K)$. The latter expressions are three ways of indicating the environmental resistance created by the growing population itself, which brings about an increasing reduction in the potential reproduction rate as population size approaches the carrying capacity. In word form, these equations simply mean:

$$\begin{bmatrix} \text{Rate of} \\ \text{population} \\ \text{increase} \end{bmatrix} \text{ equals } \begin{bmatrix} \text{maximum possible} \\ \text{rate of increase} \\ \text{(unlimited specific} \\ \text{growth rate times} \\ \text{numbers in the} \\ \text{population)} \end{bmatrix} \begin{array}{c} \text{times} \\ \\ \text{or} \\ \text{minus} \end{array} \begin{bmatrix} \text{degree of real-} \\ \text{ization of max-} \\ \text{imum rate} \end{bmatrix} \\ \begin{bmatrix} \text{unrealized} \\ \text{increase} \end{bmatrix}$$

In summary, then, this simple model is a product of three components: (1) a rate constant (r), a measure of population size (N), and a measure of the portion of available limiting factor not utilized by the population $(1 - N/K)$.

The logistic equation may also be written in terms of the rate of increase per generation, R, as follows:

$$\frac{dN}{dt} = N \log_e R \left(\frac{K - N}{K} \right)$$

It should now be emphasized that although the growth of a great variety of populations—representing microorganisms, plants, and animals—including both laboratory and natural populations, have been shown to follow the sigmoid pattern, it does not follow necessarily that such populations increase according to the logistic equation. There are many mathematical equations which will produce a sigmoid curve. Almost any equation in which the negative factors increase in some manner with density will produce sigmoid curves. Mere curve-fitting is to be avoided. One needs to have evidence that the factors in the equation are actually operating to control the population before an attempt is made to compare actual data with a theoretical curve. The simple situation in which environmental resistance increases linearly with density seems to hold for populations of organisms that have very simple life histories, as for example, yeasts growing in a limited space (as in a culture; see Figure 7–11). In populations of higher plants and animals, which have complicated life histories and long periods of individual development, there are likely to be delayed responses which greatly modify the growth form, producing what Nicholson (1954) has called "tardy density conditioned" patterns. In such cases a more concave growth curve may result (longer period required for natality to become effective), and almost always the population "overshoots" the upper asymptote and undergoes oscillations before settling down at the carrying capacity level (see Figure 7–10, curve B-2). Wangersky and Cunningham (1956 and

1957) have suggested a modification of the logistic equation to include two kinds of time lags: (1) the time needed for an organism to start increasing when conditions are favorable and (2) the time required for organisms to react to unfavorable crowding by altering birth and death rates. Letting these time lags be $t - t_1$ and $t - t_2$ respectively we get:

$$\frac{dN(t)}{dt} = rN_{(t-t_1)} \frac{K - N_{(t-t_2)}}{K}$$

When this kind of equation is studied by the use of an analog computer, the density "overshoots" and oscillates with decreasing amplitude with time, very much as shown in the graphic model in Figure 7–10, B-2. Deevey (1958) suggests that this growth form is perhaps the most common and the type most likely to be exhibited by the human population if controlled only by its own "self-crowding" effects (that is, if no "external" control is applied, such as systematic family planning). The reason for this is simple, and we actually observe the "overshoot" frequently on a local scale. Thus, when economic conditions are favorable (space and resources inexpensive) people may respond by having a lot of babies; then 10 to 20 years later (the time lag) schools and housing become overcrowded since people rarely "anticipate" needs before they actually occur. Fast-growing cities (where immigration is often the increase factor) almost always experience this kind of overshoot and subsequent oscillation. As long as the oscillation is mild, and if we would plan ahead, no great harm would be done. In fact, based on engineering experience it may be that a "damped oscillation" is easier to deal with than a completely asymptotic situation. What is frightening is the increasing probability that all cities might overshoot their resources simultaneously, and relief could no longer be obtained by the surplus population by simply moving elsewhere; or by input of relief capital (taxes, etc.) from elsewhere (since the "elsewhere" is equally stressed).

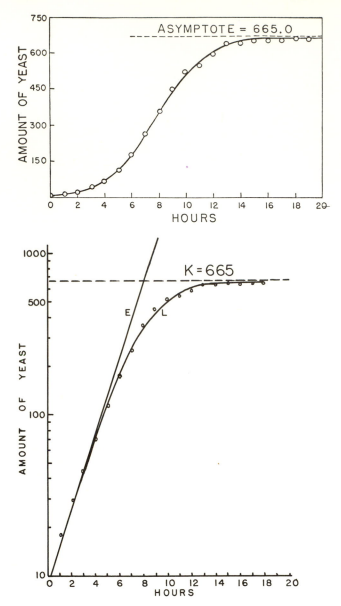

Figure 7–11. Growth of yeast in a culture. A simple case of the sigmoid growth form in which environmental resistance (in this case, detrimental factors produced by organisms themselves) is linearly proportional to the density. Open circles are observed growth values; solid lines are curves drawn from equations. In the upper graph growth of yeast is plotted on an arithmetic scale and a logistic curve fitted to the data. In the lower graph these same data are plotted (L) with the amount of yeast on a logarithmic scale; an exponential curve (E) is included to illustrate what growth would be like without self-limiting constraints. See Table 7–6 for data and equations. (Drawn from data of Pearl, 1927; upper graph from Allee *et al.*, 1949.)

Many modifications of the basic logistic model have been suggested in attempts to take into account complex life-history interactions that characterize higher organisms. Smith (1963), for example, suggested adding a maintenance constant, c (= rate of replacement at saturation) so that an r/c (growth to maintenance) ratio modifies the self-limiting component of the equation; such a form provided a more realistic model for the growth of *Daphnia* populations in a food-limited environment. Slobodkin (1962, Chapter 9) discusses in detail the feasibility of adding first-, second-, and third-order coefficients that might represent competition,

metabolic alteration of the environment, and social interactions, respectively.

Examples

Figure 7–11 illustrates the sigmoid growth form in its simplest form; Figure 7–12 illustrates the J-shaped pattern. In the latter the thrips (small insects) increase rapidly during favorable years until the end of the season calls a halt, after which an equally rapid decline in density occurs. In less favorable years the growth form is more sigmoid. Broadly speaking, the J-shaped form may be

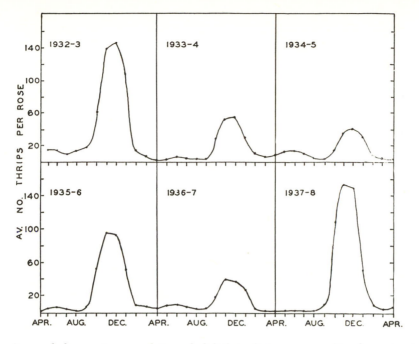

Figure 7–12. Seasonal changes in a population of adult thrips living on roses. (Graph constructed from data by Davidson and Andrewartha, 1948.)

considered an incomplete sigmoid curve since a sudden limiting effect is brought to bear before the self-limiting effects within the population become important.

In Figure 7–11 the growth of the yeast is plotted both on arithmetic (upper graph) and log (lower graph) scales, and a logistic curve fitted to the data. Note that in the semilog plot the latter takes the shape of an inverted "J" rather than an "S." In the lower graph an exponential curve (straight line-E) is also shown to illustrate what growth would be like if not limited by the size of the culture vessel and the density of the population. The data and equations are shown in Table 7–6; note that the equations are the integrated forms for the logistic and the exponential (J-shaped growth form) equations as already described, with actual values plugged in for the constants K, a, and r. The actual (observed) growth of the yeast was essentially exponential for the first three hours but thereafter slowed down appreciably in conformance to the logistic growth form. The area between the two curves (lower diagram, Figure 7–11) could be taken as a quantitative measure of the environmental resistance. The advantage of the semilog plot is that any deviation from a straight line indicates a change in the popu-

Table 7–6. *Growth of Yeast in a Culture**

TIME IN HOURS	OBSERVED YEAST BIOMASS	BIOMASS CALCULATED FROM LOGISTIC EQUATION[†]	BIOMASS CALCULATED FROM EXPONENTIAL EQUATION[‡] (ROUND FIGURES)
0	9.6	9.9	9.6
1	18.3	16.8	17
2	29.0	28.2	28
3	47.2	46.7	48
4	71.1	76.0	82
5	119.1	120.1	139
6	174.6	181.9	238
7	257.3	260.3	408
8	350.7	348.2	694
9	441.0	433.9	1,238
10	513.3	506.9	2,042
11	559.7	562.3	3,504
12	594.8	600.8	5,904
13	629.4	625.8	
14	640.8	641.5	
15	651.1	651.0	
16	655.9	656.7	
17	659.6	660.7	
18	661.8	662.1	14,765

* After Pearl (1927) from data of Carlson.

[†] $N = \dfrac{665}{1 + e^{4.1896-0.5355t}}$.

[‡] $N_t = N_o e^{0.5355t}$.
 ($N_o = 9.6$).

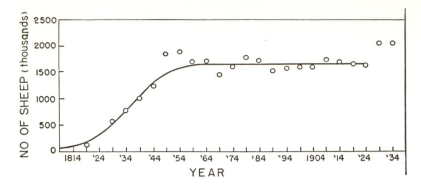

Figure 7–13. Growth of a sheep population introduced into a new environment on the island of Tasmania. Circles are averages for 5-year periods. (From Davidson, 1938.)

lation growth rate (*dN/dt*); the more the curve bends, the greater the change.

Even though simple logistic growth is probably restricted to small organisms or those with simple life histories, the sigmoid growth pattern may be observed in larger organisms when they are introduced into a new and temporarily unlimited environment. Overshoots and oscillations are likely as indicated in the Statement. A good example of growth of such a population is shown in Figure 7–13. Sheep were introduced on the island of Tasmania near Australia sometime around 1800 and reasonably good records were kept of the number present. The number of sheep followed the sigmoid curve with a mild overshoot and oscillation around a general asymptote of about 1,700,000 sheep. Some of the fluctuations after 1860, of course, were probably due to variations in climatic factors.

The general validity of the sigmoid growth form is indicated by the fact that growth rate is usually a decreasing function of density within the limits of "normal" population sizes found in stable ecosystems. Tanner (1966) found a statistically significant negative correlation between growth rate and density in two-thirds of 63 laboratory and field populations of insects and vertebrates for which data were available in the literature. Interestingly enough, *the human population was the only one showing a significant positive correlation!* However, many animal populations do "escape" from density control from time to time (as man is now doing) in that growth rate becomes independent, or an increasing function, of density. The situation appears to be similar in plant populations. Harper *et al.* (1965) report that establishment of seedlings is density limited when "micro-sites" in the soil were

scarce but not when they were abundant. In the next section we shall examine cases of oscillations, eruptions, and overshoots in the hopes that such examples will reveal mechanisms operating in the more "normal" situation in which population size deviates less from the asymptote.

9. POPULATION FLUCTUATIONS AND SO-CALLED "CYCLIC" OSCILLATIONS

Statement

When populations complete their growth, and $\Delta N / \Delta t$ averages zero, population density tends to fluctuate above and below the upper asymptotic or carrying capacity level even in populations that are subject to self-limited growth form or other forms of feedback control. Such fluctuations may result from changes in the physical environment which, in effect, raise and lower the asymptotic level, or interactions within the population, or both, or between closely interacting populations. Thus, fluctuations may occur even in a constant environment such as might be maintained in the laboratory. In nature, it is important to distinguish between (1) seasonal changes in population size, largely controlled by life-history adaptations coupled with seasonal changes in environmental factors, and (2) annual fluctuations. For the purposes of analysis the latter may be considered under two headings: (*a*) fluctuations controlled primarily by annual differences in the physical environment of the population, or extrinsic factors (i.e., outside sphere of population interactions), and (*b*) oscillations primarily controlled by population dynamics, or intrinsic factors (i.e., factors within populations). In general, the former tend to be irregular and clearly correlated with the variation in one or more major physical limiting factors (such as temperature or rainfall), while the latter often exhibits such regularity that the term "cycles" seems appropriate (and species exhibiting such regular variation in population size are often known as "cyclic" species). As we have seen in the previous section, violent oscillations would be inherent in populations exhibiting the J-shaped growth form, and damped oscillations would be characteristic of the S-shaped growth form in which there are life-history related time lags. It is, of course, understood that both extrinsic and intrinsic factors influence all fluctuations; the fundamental problem is to evaluate the importance of each, or at least to determine which is the major cause of variation in specific cases. It has been repeatedly observed that species population density fluctuates from year to year most violently in relatively simple ecosystems in which communities are composed of a relatively few species populations (i.e., where species diversity and pattern diversity are low; see Chapter 6), as, for example, in arctic communities and artificial pine forest communities.

Explanation and Examples

1. SAMPLES OF VARIOUS TYPES OF POPULATION DENSITY VARIATIONS. The mechanism behind the ebb and flow of population size is a fascinating and only partially solved problem in ecology. As indicated in the preceding paragraph, fluctuations could logically result from variations in the external physical environment, as, for example, climatic variations. We would classify such fluctuations as being extrinsic in nature, because they were caused by factors outside the population or community —that is, by nonbiological factors. On the other hand, it is also logical that pronounced fluctuations could be the result of intrinsic factors, that is, of events partly within the community, such as predation, disease and inherent type of growth form. Often it is difficult to distinguish quantitatively between these possibilities. Sometimes we can only state that certain definite types of fluctuations or oscillations are characteristic of certain populations in certain regions without

being able to state the causes. Since, as has been stressed in earlier chapters, populations modify and compensate for physical factor perturbations, we can reaffirm the following basic principle: The more highly organized and mature the community and the more stable the environment, the lower will be the amplitude of fluctuations in population density with time. We shall go into the development of community homeostasis in Chapter 9.

We are all familiar with seasonal variations in population size. We come to expect that at certain times of the year mosquitoes or gnats will be abundant, or the woods will be full of birds, or the fields full of ragweed; at other seasons populations of these organisms may dwindle to the vanishing point. Although it would be difficult to find in nature populations of animals, microorganisms, and herbaceous plants that do not exhibit some seasonal change in size, the most pronounced fluctuations occur with organisms which have limited breeding seasons, especially those with short life cycles and those with pronounced seasonal dispersal patterns (e.g., migration). Figure 7–12, already referred to, illustrates not only J-shaped growth form (no equilibrium level at higher densities) but also seasonal and annual fluctuations, as revealed by systematic study over a period of years. A pattern of this sort is probably typical of most insects and of most "annuals" among plants and animals. In aquatic situations, both freshwater and oceanic, seasonal cycles in plankton populations have been very intensively studied and modeled, as described in Chapter 11 (Figure 11–8, page 308).

Pronounced seasonal variations in population density are as characteristic of the tropics as of temperate and arctic regions. For example, Bates (1945) found that the relative abundance of only one out of seven tropical species of mosquitoes failed to show seasonal variations (Fig. 7–14). Fluctuations in the tropics are often related to rainfall, but they may be related to inherent periodicities within the community. Insect and bird densities, for example, may fluctuate with the periodic flowering and fruiting of plants that occur even in a constant rain forest environment.

A typical example of a rather irregular variation in population size that appears to be correlated with weather is shown in Figure 7–15. During most years the heron population in the two areas of Great Britain remains

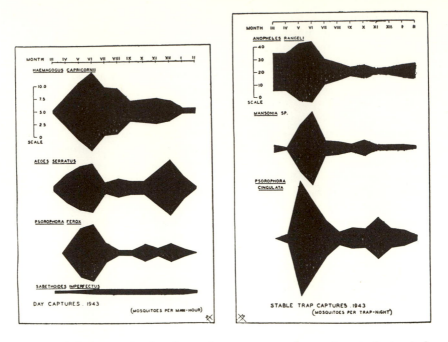

Figure 7–14. Seasonal changes in the abundance of seven species of mosquitoes in the tropical environment of eastern Colombia. (After Bates, 1945.)

relatively constant; apparently the local environments provided a rather stable carrying capacity for herons. However, a sharp decrease in density with subsequent recovery occurred following each of three series of severe winters (as indicated along the top of Figure 7–15). The fact that changes in density in the two areas were synchronous lends credibility to the correlation with winter mortality. Having more than one study area is a good rule for ecological field study! Incidentally, bird populations have been among the most intensively studied, and results of such studies have contributed a great deal to population theory. For a readable summary of bird population work, see Lack (1966).

Interesting and only partly understood

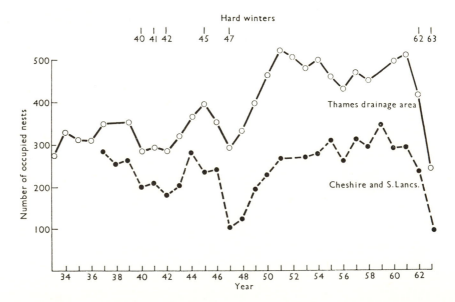

Figure 7–15. Changes in the abundance of the heron *Ardea cinerea* in two areas of Great Britain between 1933 and 1963. A relation between cold winters and a decline in abundance is indicated. (From Lack, 1966.)

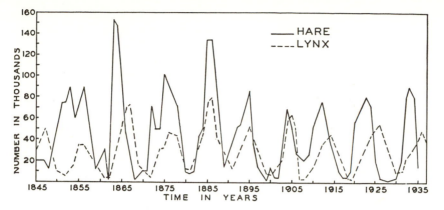

Figure 7–16. Changes in the abundance of the lynx and the snowshoe hare, as indicated by the number of pelts received by the Hudson Bay Company. This is a classic case of cyclic oscillation in population density. (Redrawn from MacLulich, 1937.)

density variations are those which are not related to seasonal or obvious annual changes, but which involve regular oscillations or cycles of abundance with peaks and depressions every few years, often occurring with such regularity that population size may be predicted in advance. The best known cases concern mammals, birds, insects, fish, and seed production in plants in northern environments. Among the mammals the best-studied examples exhibit either a 9- to 10-year or a 3- to 4-year periodicity. A classic example of a 9- to 10-year oscillation is that of the snowshoe hare and the lynx (Fig. 7–16). Since about 1800 the Hudson Bay Company of Canada has kept records of pelts of

Figure 7–17. The snowshoe or varying hare, famous in ecological annals for its spectacular cyclic abundance (see Figure 7–16). The individual shown is in its white winter pelage. The change from the brown summer pelage to the white one of winter has been shown to be controlled by photoperiodicity. (U. S. Soil Conservation Service Photo.)

fur-bearers trapped each year. When plot-ted, these records show that the lynx, for ex-ample, has reached a population peak every 9 to 10 years, averaging 9.6 years, through-out this long stretch of time. Peaks of abun-dance often are followed by "crashes," or rapid declines, the lynx becoming exceed-ingly scarce for several years. The snowshoe hare follows the same cycle, with a peak abundance generally preceding that of the lynx by a year or more. Since the lynx is largely dependent on the hare for food, it is obvious that the cycle of the predator is re-lated to that of the prey, but the two cycles are not strictly a cause-and-effect predator-prey interaction since the hare "cycles" in areas in which there are no lynxes. The shorter, or 3- to 4-year cycle, is character-istic of many northern murids (lemmings, mice, voles) and their predators (especially the snowy owl and foxes). The cycle of the lemming of the tundra and the arctic fox and the snowy owl are well-documented classic instances (Elton, 1942). Every three or four years, over enormous areas of the northern tundra of two continents, the lemmings (two species in Eurasia and one in North America of the genus *Lemmus*, and one species of *Dicrostonyx* in North America) become ex-tremely abundant only to "crash," often within a single season. Foxes and owls, which increase in numbers as their food in-creases, decrease very soon afterward. The owls may migrate south into the United States (sometimes as far south as Georgia) in search of food. This eruptive migration of surplus birds is apparently a one-way move-ment; few if any owls ever return. The owl population thus "crashes" as a result of the dispersal movement. So regular is this oscil-lation that bird students in the United States can count on a snowy owl invasion every 3 or 4 years. Since the birds are conspicuous and appear everywhere about cities, they attract a lot of attention and get their pictures in the paper or their skins mounted in local taxi-dermy shops. In years between invasions, few or no snowy owls are seen in the United States or southern Canada. Gross (1947) and Shelford (1943) have analyzed the invasion records and have shown that they are corre-lated with the periodic decrease in abun-dance of the lemming, their chief food item. In Europe, but apparently not in North America, the lemmings themselves become so abundant at the crest of the cycle that they may emigrate from their overcrowded

haunts. Elton (1942) vividly describes the famous lemming "migrations" in Norway. The animals pass through villages in such numbers that dogs and cats get tired of kill-ing them and just ignore the horde. On reaching the sea, many drown. Thus, simi-lar to the owl eruption, the lemming move-ment is one-way. These spectacular emigra-tions do not occur at every 4-year peak in density, but only during exceptionally high peaks. Often the population subsides with-out the animals leaving the tundra or mountains.

Two examples from long-term records of violent oscillations in foliage insects in European forests are shown in Figure 7–18. Such pronounced cycles have been reported mostly from northern forests, especially pure stands of conifers. As shown in Figure 7–18, density may vary over five orders of magni-tude (log cycles). For example, the density of *Bupalus* pupae (the caterpillars drop to the forest floor litter to pupate, thus provid-ing a convenient point in the life cycle for a census) varied from less than one per 1000 square meters to over 10,000 per 1000 square meters (0.001 to 10 per square meter). One can well imagine that with 10,000 potential moths emerging for every 1000 square me-ters, and several generations in a season, enough caterpillars could be produced to de-foliate and even kill the trees, as frequently happens. The cycles of defoliating cater-pillars are not as regular as the snowshoe hare oscillations, and their periodicity seems to be somewhere between 4 and 10 years (peaks of abundance average 7.8 and 8.8 years in the two examples shown). Nor are the cycles of different species synchronous. Periodic outbreaks of the spruce budworm and tent caterpillars are well-known exam-ples of similar patterns in the northern part of North America. As will be noted in Chap-ter 9, the "coevolution" of the insect–tree interaction results in a cyclic behavior of the whole ecosystem.

Probably the most famous of all insect population oscillations are those involving locusts or grasshoppers. In Eurasia records of outbreaks of the migratory locust (*Lo-custa migratoria*) go back to antiquity (Car-penter, 1940a). The locusts live in desert or semiarid country and in most years are non-migratory, eat no crops, and attract no atten-tion. At intervals, however, population den-sity increases to an enormous extent; the locusts actually become morphologically

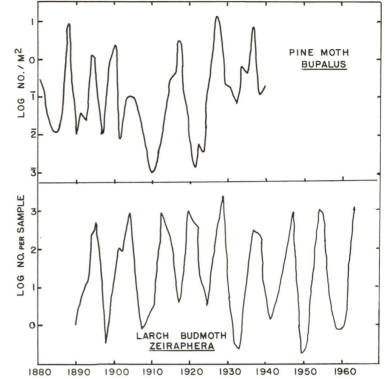

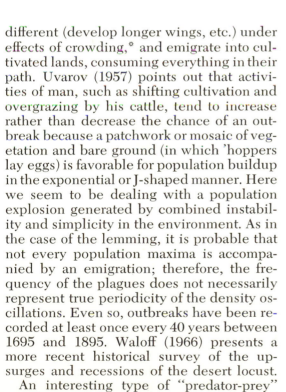

Figure 7–18. Fluctuations in numbers of two species of moths whose larvae feed upon foliage of conifers. *Upper:* The pine moth *Bupalus* in German pine forests at Litzlingen, which are managed as pure stands. *Lower:* The larch budmoth *Zeiraphera griseana* in larch forests in the Swiss valley of Engadin. (Upper graph redrawn from Varley, 1949; lower graph redrawn from Baltensweiler, 1964.)

different (develop longer wings, etc.) under effects of crowding,* and emigrate into cultivated lands, consuming everything in their path. Uvarov (1957) points out that activities of man, such as shifting cultivation and overgrazing by his cattle, tend to increase rather than decrease the chance of an outbreak because a patchwork or mosaic of vegetation and bare ground (in which 'hoppers lay eggs) is favorable for population buildup in the exponential or J-shaped manner. Here we seem to be dealing with a population explosion generated by combined instability and simplicity in the environment. As in the case of the lemming, it is probable that not every population maxima is accompanied by an emigration; therefore, the frequency of the plagues does not necessarily represent true periodicity of the density oscillations. Even so, outbreaks have been recorded at least once every 40 years between 1695 and 1895. Waloff (1966) presents a more recent historical survey of the upsurges and recessions of the desert locust.

An interesting type of "predator-prey" oscillation involves plants and animals. Seed production in conifers is often "cyclic," and seed-eating birds and other animals show corresponding oscillations.

2. THEORIES AS TO MECHANISM OF THE VIOLENT OSCILLATORY TYPE OF POPULATION FLUCTUATIONS. As previously indicated, fluctuations above and below an equilibrium level appear to be characteristic of most populations, and this would indeed be expected under nature's varied influence, but the seemingly regular cycles of abundance are almost paradoxical in the face of nature's notorious irregularity! Two striking features of these oscillations are: (1) they are most pronounced in the less complex ecosystems of northern regions, as is evident from the examples described above; and (2) although peaks of abundance may occur simultaneously over wide areas, peaks in the same species in different regions do not always coincide by any means. Theories that have been advanced to explain cycles in density can be grouped under several headings as follows: (1) meteorological theories, (2) random fluctuations theory, (3) population interaction theories, and (4) trophic level interaction theories. A very brief summary of each of these will be presented in the next several paragraphs.

Attempts to relate these regular oscilla-

---

* Solitary and migratory forms occur in a number of species of locusts and often were described as separate species, before their true relations were known.

tions to climatic factors have so far been unsuccessful despite the fact that their synchrony and prominence in northern latitudes would seem to suggest some cyclic event outside the local ecosystem. At one time the cycle of sunspots, which cause major weather changes, was considered by many as an adequate explanation for the cycle of the lynx and other 10-year cycles. However, MacLulich (1937) and others have demonstrated that, actually, there is no correlation. So far, no widespread climatic periodicities which would fit the 3- to 4-year interval have been demonstrated. Even if there were good correlations, the problem would still not be settled until it could be demonstrated just how sunspots or ultraviolet or other climatic factors affect natality, mortality, dispersal, or other population characteristics. As indicated before in our discussion, correlations (or "cycle-fitting") without evidence of mechanism are dangerous indeed.

Palmgren (1949) and Cole (1951 and 1954) have suggested that what appear to be regular oscillations could result from random variations in the complex biotic and abiotic population environment (recall that in Section 5, Chapter 6, we listed "stochastic patterns resulting from random forces" as one of several contributors to community patterns); if this is so, no one factor can be singled out as more important than numerous other factors. Keith (1963) has made a detailed statistical analysis of the northern bird and mammal cycles and concludes that the 10-year cycle is "real" (i.e., nonrandom), even though it would be difficult to prove that the shorter cycles might not be due to random fluctuations. In addition to the snowshoe hare and lynx, the ruffed grouse was shown to exhibit a 7- to 10-year cycle that is synchronous over large areas of Canada and the lake states and also generally synchronous with the cycles of the hare. Prairie grouse, ptarmigan, muskrats, and foxes are other species that experience cycles of similar length, but often on a more local or irregular scale. Interestingly enough, the introduced Hungarian partridge seems to be "adopting" a 10-year cycle in its new environment in northern North America, although the species has not been established long enough for sufficient data to be obtained. Keith concludes that the "10-year" cycle is restricted to the northern coniferous biome and its ecotones in North America (see Figure 14–8 for a map of this region) and is not evident in Europe. In contrast, 7- to 10-year cycles in forest insect pests are well documented in Europe, as we have seen.

If climatic factors, random or otherwise, should prove not to be the major cause of the violent oscillations, then we would naturally look for causes within the populations themselves (i.e., for "intrinsic factors"). Here we have some evidence of possible mechanisms which might operate in conjunction with weather or other changes in physical factors. As already indicated in the discussion of population growth form, populations which undergo more or less unrestricted exponential growth for a time may be expected to oscillate in density, since such populations may exceed the bounds of some limiting factor rather than achieve a steady state. The simpler the ecosystem (that is, the fewer the species and major limiting factors), the greater the likelihood of temporary imbalances. Simple overpopulation (followed by an inevitable decline), then, could generate oscillations such as are shown in Figure 7–10 B-3 with a period that might be proportional to the intrinsic rate of increase; or they could be generated, or augmented, by predator-prey interaction. As will be discussed later on, a system containing a single prey and a single predator specialized to feed upon it can oscillate even in a constant physical environment of the laboratory (see Utida, 1957), but such simple predator-prey systems usually either become extinct or the oscillations become dampened with time (see Figure 7–33). As indicated in the previous section, populations often "escape" from carrying capacity control (whether abiotic or biotic); what can be called the "overpopulation theory of cycles" is based on the idea that certain populations in northern ecosystems "escape" at regular rather than occasional or irregular intervals.

There is mounting evidence that physiological and genetic changes within the individuals in the populations accompany the violent oscillations in population density. Whether these changes cause, or are merely adaptations to, the oscillations is a matter of controversy at the present time, but most investigators are agreed that such changes can at least greatly modify the periodicity or amplitude of fluctuations. Building on Hans Selye's epoch-making medical theory of stress (i.e., the general adaptation syndrome), Christain and coworkers (see Christain, 1950, 1961, and 1963; Christain and Davis, 1964) have amassed considerable evidence both from the field and the laboratory that crowding in higher vertebrates results in enlarged adrenal glands, which are symptomatic of shifts in the neural-endocrine balance that, in turn, bring about changes in behavior, reproductive potential, and resistance to disease or other stress. Such changes often combine to cause a precipitous "crash" in population density. For example, snowshoe hares at the peak of density often die suddenly from "shock disease" that has been shown to be associated with enlarged adrenals and other evidence of endocrine imbalance. In the cyclic insects Wellington (1957, 1960) found that on the upswing of the cycle tent caterpillars (*Malacosoma*) build elongated tents that are shifted about, and the individuals are active in moving out into the foliage to feed. At peak density the caterpillars become inactive, build compact tents, feed less, and are more subject to disease. Adults reared from sluggish larvae were also sluggish and did not move far to lay their eggs. In the larch budmoth (whose abundance cycles are shown in Figure 7–17B) "strong" and "weak" physiological races that are "almost certainly genetic" (see detailed summary of the ecology of this species in Clark, Grier, Hughes, and Morris, 1967, pages 124–136) alternate in the low and high density phases of the cycle. Chitty (1960, 1967) suggests that similar genetic shifts account for differences in aggressive behavior and survival that are observed at different phases of vole cycles. Such "adaptation syndromes" would certainly seem to be mechanisms for "dampening" oscillation so as to prevent too great a fluctuation that might damage the ecosystem and endanger the survival of the species.

A fourth group of theories revolve around the idea that abundance cycles are intrinsic at the ecosystem level rather than at the population level. Certainly density changes that range over several orders of magnitude must involve not only secondary trophic levels such as predators and prey, but the primary plant-herbivore interactions as well. An example of an ecosystem-level theory is Schultz's (1964, 1969) "nutrient-recovery hypothesis," proposed to explain microtine cycles in the tundra. According to this hypothesis, which is supported by data from mineral cycling studies, heavy grazing by lemmings during the peak year ties up and reduces the availability of mineral nutrients (especially phosphorus) the following year, so that the food of the lemmings is low in nutritional quality; growth and survival of the young is accordingly greatly reduced. During the third or fourth year the recycle of nutrients is restored, plants recover, and the ecosystem can again support a high density of lemmings.

We have devoted a rather large amount of space to a discussion of large amplitude abundance cycles not because they are particularly common in the world in general but because their study reveals functions and interaction that probably have general application, but which might not be so evident in populations whose density is under better control. The problem of cyclic oscillation in any specific case may well boil down to determining whether one to several factors are primarily responsible or whether causes are so numerous as to be difficult to untangle, even though the total interaction may be understood to be what Cole (1957) calls "secondary simplicity" in that regularity may be "no greater than that encountered in a sequence of random numbers." As we have indicated, the former is certainly possible in simple ecosystems, whether experimental or natural; the latter may be more likely in complex ecosystems.

Having considered a number of very interesting specifics, we are now in a good position to consider the more general problem of population regulation.

10. POPULATION REGULATION AND THE CONCEPTS OF DENSITY-INDEPENDENT AND DENSITY-DEPENDENT ACTION IN POPULATION CONTROL

Statement

In low-diversity, physically stressed ecosystems, or in those subject to irregular or unpredictable extrinsic perturbations, populations tend to be regulated by physical components such as weather, water currents, chemical limiting factors, pollution, and so forth. In high-diversity ecosystems, or in those not physically stressed, populations tend to be biologically controlled. In all ecosystems there is a strong tendency for all populations to evolve through natural selection towards self-regulation (since overpopulation is not in the best interests of any population!) even though this is difficult to achieve under extrinsic stress. From the standpoint of the population itself, the following concepts are corollaries to this general theory.

Any factor, whether limiting or favorable (negative or positive) to a population, is (1) *density-independent* (also called density-legislative) if the effect or action is independent of the size of the population, or (2) *density-dependent* (also called density-governing) if the effect on the population is a function of density. Density-dependent action is usually direct in that it intensifies as the upper limit is approached, but it may also be inverse (decrease in intensity as density increases): Direct density-dependent factors act like governors on an engine (hence the term "density-governing") and for this reason are considered as one of the chief agents in preventing overpopulation and thereby responsible for the achievement of a steady state. Climatic factors often, but by no means always, act in a density-independent manner, whereas biotic factors (competition, parasites, pathogens, and so forth) often, but not always, act in a density-dependent manner.

Explanation

A general theory for population regulation, as stated in the first paragraph above, comes logically from our discussion of biotic potential, growth form, and variation around the carrying-capacity level. As emphasized in the previous chapter (see page 147), theories are often correlated with the environment of the theorist! Thus, ecologists working in stressful environments (such as arid regions) or with small organisms (such as insects or plankton which have short life cycles, high biotic potentials, and high rates of metabolism per gram and, hence, relatively small standing crop per unit space at any one time) have been impressed with the following: (1) the importance of the period of time when the rate of increase (r) is positive, (2) the importance of density-independent factors such as weather in determining the length of favorable periods, (3) the secondary rather than primary importance of self-limiting forces within the population, and (4) the general lack of stability in the density of any one species even when the ecosystem seems stable. These points have been particularly well brought out in the well-known book by Andrewartha and Birch (1954). Ecologists working in benign environments (such as English gardens, coral reefs, or tropical forests) or with larger organisms (such as birds, mammals or forest trees, whose life cycles are longer and numbers and biomass more clearly reflect energy flow) have been impressed with the following: (1) the importance of density-dependent factors, especially self-limiting intraspecific competition (as in the sigmoid growth equation) and interspecific checks of various sorts (interspecific competition, parasites, and so forth), (2) the stability or at least consistency in the density patterns, and (3) the general importance of biological control mechanisms. Interestingly enough, persons working with confined populations of small organisms, such as bacteria or flour beetles in culture, are also impressed with these latter aspects, which is perhaps not surprising since standing crop per unit space in cultures is greater than would be usual in nature. Likewise, it comes as no surprise that persons working with monoculture crops and forests are unimpressed with the efficiency of biological control! For an analysis of density-dependent control theories, see the three important papers by Nicholson (1954, 1957, and 1958), Lack (1954), and the appendix in Lack (1966).

Above all, when we become interested in intensive study of specific populations, we must not forget the integration that occurs at the community and ecosystem level, because (as emphasized in Chapter 2) the part (population) can never be completely explained without considering the whole (ecosystem), or vice versa. Where there are a large number of species occupying a trophic level (herbivorous insects in a forest, for example), the population ecology of any one species may make little sense unless we know what the "coworkers" in the community are doing at the same time.

In the final analysis, then, population regulation must be a function of the kind of ecosystem of which the population in question is a part. While contrasting the "physically controlled" and the "biologically controlled" ecosystem, as we have done, is arbitrary and produces an oversimplified "model," it is a relevant approach, especially since man seems to be in the business of creating ever more stressed systems that are incapable of either self-maintenance or self-regulation. Natural ecosystems in the temperate zone, which, of course, are the ones best known from the ecological standpoint, are intermediate in terms of the importance of physical and biotic regulators. Milne (1957 and 1962) emphasizes the interaction of density-independent and density-dependent control, and points out that the latter is rarely "perfect," as in the sigmoid or logistic growth model, but often "imperfect" in that limitations may be more effective at one density than at another. It is frequently observed in the field that a control factor, such as a predator, may be quite effective at low prey densities but quite ineffective at high densities when there is a time lag in the response of predator density to prey density. In fact Holling (1961) concludes that in insects predation is usually an increasing function of prey density at low density of prey, but a decreasing function (inverse density-dependent) at high densities.

Strictly speaking, an effect would be density-independent only if a constant influence were exerted on the population regardless of its size. For example, if the same number of hunters are allowed two deer per season and they hunt until they obtain their deer, then the same number of deer would be removed each year, regardless of the total number of deer present. It has been customary, however, also to speak of a factor as "density-independent" if the effect *percentagewise* is the same regardless of density, as would be the case, for example, if 10 per cent of the deer were removed each year, even though the number affected (removed) would depend on density! If we wish to make a distinction, we may speak of the constant percentage effect as density-proportional.

One way to clear up these matters is to think in terms of the growth equations previously given (another example of usefulness of mathematical models). A constant added to the equation, of course, represents absolute density-independent action. A constant multiplied by N (the number in the population) represents a constant percentage density-independent effect. Any factor [such as $(K-N)/K$] that increases or decreases percentagewise as the number increases would be density-dependent. It follows from this that direct density-dependence must be effective if a population is to be well regulated, if we define *population regulation as the tendency to return to equilibrium size, and population stability as the tendency to remain at a constant size.*

Perhaps the most important part of our general theory is the statement that "populations evolve towards self-regulation." In the preceding section we described how physiological and genetic shifts, or the alternation of ecotypes in time, as it were, can dampen oscillations and hasten return of density to lower levels. During the 1960s evidence has been piling up to indicate that populations do not just avoid the suicidal extremes of going too far above or below saturation levels; rather they evolve towards *regulating their density at a level well below the upper asymptote or carrying capacity that could be attained if all energy and space resources were fully utilized.* In this manner natural selection operates to maximize the *quality* of the individual's environment and to reduce the probability of extinction of the population. The case for *self-regulation* of populations is forcefully brought out by Wynne-Edwards (1962 and 1965), while regulation through interspecific interactions is viewed by Pimentel (1961 and 1968) as occurring by a process of *genetic feedback.* Wynne-Edwards documents two mechanisms that can stabilize density at a lower-than-saturation level: (1) *territoriality,* an exaggerated form of intraspecific competition that limits growth through "land-use" control, as will be discussed in Section 15; and (2) *group behavior,* such as "peck-orders," "sexual dominance," and other aspects, to be discussed in Chapter 8. The concept of genetic feedback refers to a mechanism whereby closely interacting species populations "power down" their demands on each other. In the words of Pimentel (1961), "In a herbivore-plant system, animal density influences selective pressure on plants; this selection influences the genetic makeup of plants, and in turn the genetic make-up of the plant influences animal density. The actions and reactions in interacting populations in the food chain cycling in the genetic feedback mechanism result in the evolution and regulation of populations." This sort of "coevolution" between ecologically linked populations will be considered again, and in more detail, in Chapter 9, Section 5.

To summarize, density-independent aspects of the environment tend to bring about variations, sometimes drastic, in population density and to cause a shifting of upper asymptotic or carrying-capacity levels, while density-dependent natality and mortality tend to maintain a population in a "steady state" or to hasten the return to such a level. The former play a greater role in control in physically stressed ecosystems, the latter becoming more important as extrinsic stress is reduced. As in a smoothly functioning cybernetic system (see page 34, Figure 2–11), additional negative feedback control is provided by interactions (both phenotypic and genetic) between populations of different species that are linked together in food chains, or by other important ecological relationships.

Examples

Severe storms, sudden drops in temperature, and other drastic changes in physical factors generally provide the most clear-cut examples of density-independent action. In a three-year study of a snail (*Acmaea*) that lives on rocks in the intertidal zone, Frank (1965) found that most changes in the population were density-dependent except for mortality following severe winter frosts when portions of the rock surface crumble away, removing snails regardless of the number present. We have already presented a good case of "perfect" density-dependence in the growth of yeast (Fig. 7–11). As indicated in the discussion, interspecific interactions are less likely to show linear density-dependence. Varley's (1947) intensive study of the knapweed gallfly shows that the action of a major parasitic insect, *Eurytoma curta,* can be density-dependent, since it killed a much greater percentage as well as a larger total number of the host when the population of the host was high:

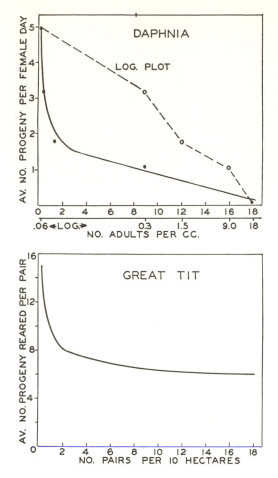

Figure 7–19. Density-dependent natality in a laboratory population of cladocera (upper graph) and a wild population of great tits (lower graph). The curve showing the decline in production of young with increasing density is strikingly similar in these two dissimilar populations. (Upper graph redrawn from Frank, 1952; lower graph redrawn from Kluijver, 1951.)

YEAR AND POPULATION LEVEL	LARVAL POPULATION AT BEGINNING OF SEASON (NO./M²)	LARVAE KILLED BY PARASITE (NO./M²)	PER CENT POPULATION KILLED BY PARASITE
1934 (low population)	43	6	14
1935 (high population)	148	66	45

However, in another study Varley and Edwards (1957) report that when the "area of discovery" is low, as was the case with the parasitic wasp *Mormoniella,* the action of the parasite on its dipteran hosts was not necessarily density-dependent. Thus, differences in behavior can be important. Holling (1965, 1966) has incorporated be-

havioral characteristics into a series of elegant mathematical models that predict how effective a given insect parasite will be in the control of the insect host at different densities. More about this in Chapter 10.

That natality as well as mortality may vary with density is shown by two examples, one taken from the laboratory and one from the field, illustrated in Figure 7–19. From these graphs we see that the production of eggs and young per female decreased with increasing density in laboratory cultures of cladocera ("water fleas") and in wild populations of great tits (small birds). In both cases the relationship is essentially exponential (and could be expressed by the same basic equation as used for population growth).

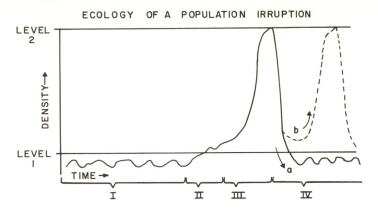

Figure 7–20. Population dynamics of the psyllid insect, *Cardiaspina albitextura*, which feeds on *Eucalyptus blakelyi* trees. The population density normally remains stabilized at a low level by the combined density-independent and density-dependent action of weather, parasites, and predators; but occasionally the population "escapes" from natural control and "irrupts" to a high density, resulting in extensive defoliation of trees. Density level 1 is that below which natural control operates effectively. Level 2 is the density at which food supply and oviposition sites become strongly limiting. (Redrawn from Clark, 1964; see also Clark, Greier, Hughs, and Morris, 1967, page 158.)

An example of genetic feedback in interspecific population regulation is shown in Figure 7–33 (see page 221).

The whole subject of population regulation can be nicely summarized by the graphic model of Figure 7–20. This graph is based on the dynamics of a particular species population (an Australian psyllid insect living on eucalyptus trees) but illustrates many of the principles that we have outlined. Normally, the population is stabilized at a low level (level 1, time I in Figure 7–20), which is well below the level that is reached when all the food and space resources are exploited. Regulation at this lower-than-saturation level is accomplished by prevailing weather, density-independent predation, and parasitism on the nymphal stages, and density-dependent predation by birds on adults. Occasionally, the stabilizing process fails, usually as a result of unusually low temperatures that reduce percentage parasitism, and the density rises above (i.e., "escapes from") the controlled level (time period II in Figure 7–20). A rapid, irruptive, or J-shaped growth-type increase then occurs (time period III) because (1) parasitism on nymphs becomes ineffective because of a rapid increase in hyperparasites (parasites that parasitize the psyllid parasites) and (2) bird predators cannot increase as rapidly as the insects (recall the vast difference in the natality of insects and birds shown in Table 7–2). The temporarily unlimited growth is halted at level 2 as the nymphs run out of food and the adults run out of oviposition

sites. The population then "crashes" as the trees are defoliated and as the predation by birds, ants, and parasites begins to catch up (time period IV, Figure 7–20). If psyllid numbers are reduced below level 1, the population again comes under control and is likely to remain low for some years, as shown by (a) in Figure 7–20. However, as shown in (b), another irruption may be generated if density does not fall below the control level.

This model also illustrates some of the difficulties in the practical control of insect pest irruptions, the frequency of which, as we have already indicated, is increased by man-made stresses, many of which act like the "unusual" weather by hampering the natural control mechanism. Often the outbreak develops so quickly that it is not detected until the exponential growth is well underway and it is too late for treatment to be effective. Obviously, control measures might help prevent a second outbreak if the treatment was specific for the target insect. Application of a broad-spectrum insecticide in time period IV, however, could do more harm than good since parasites and predators would also be killed, thus increasing, rather than decreasing, the possibility of another outbreak. In many cases no treatment at all would be preferable to "shotgun type" spraying of insecticide, carried out without knowledge of the phase of the population cycle or the condition of other populations involved in the natural control mechanism.

11. POPULATION DISPERSAL

Statement

Population dispersal* is the movement of individuals or their disseminules or propagules (seeds, spores, larvae, etc.) into or out of the population or population area. It takes three forms: *emigration*—one-way outward movement; *immigration*—one-way inward movement; and *migration*—periodic departure and return.

Dispersal supplements natality and mortality in shaping population growth form and density. In most cases some individuals or their reproductive products are constantly entering or leaving the population. Often this gradual type of dispersal has little effect on the total population (especially if the unit of population size is large), either because emigrations balance immigrations or because gains and losses are compensated for by changes in natality and mortality. In other cases, however, mass dispersal occurs involving rapid changes with corresponding effect on the population. Dispersal is greatly influenced by barriers and by the inherent power of movement of individuals (or disseminules), which is indicated by the term *vagility*. Dispersal is the means by which new or depopulated areas are colonized and equilibrium diversity is established. It is also an important component in gene flow and the process of speciation. Dispersal of small organisms and passive propagules generally takes an exponential form in that density decreases by a constant amount for equal multiples of the distance from the source. Dispersal of large, active animals deviates from this pattern and may take the form of "set distance" dispersal, normally distributed dispersal, or other forms.

Explanation and Examples

In much of our discussion so far in this chapter we have considered the population as if it were a unit isolated from other populations. Although this is not usually true (in nature, at least), it is true that exchanges between populations are often less important than some of the "internal" processes that we have been discussing. Since there are many popular misconceptions about "stocking" (i.e., artificial immigration) in natural populations, it is important to determine the part dispersal really plays in specific cases.

The effect that dispersal will have on a population depends first on the status of the growth form of the population (whether it is at or near the carrying-capacity level, or is actively growing or declining) and second on the rate of dispersal. Let us consider these in order.

If a population is "well stocked" and in balance with the limiting factors of the environment (i.e., at asymptotic level), moderate immigration or emigration will have little general or permanent effect; gains or losses by dispersal simply result in compensating changes in natality and mortality. If a population is well above or below carrying capacity, dispersal may have more pronounced effects. Immigration, for example, may speed up population growth or, in case of extreme reduction, prevent extinction. We have already seen how emigration of lemmings or snowy owls from overcrowded regions is a factor in bringing about the sharp decline or "crash" characteristic of the population fluctuations of those species. Gause (1934) produced predator-prey oscillations in protozoa by regular introduction of "immigrants" in cultures that exhibited no oscillations in the absence of immigrations. Mass dispersal can change the structure of a balanced population in other ways. Thus, the introduction of a large number of bluegill fingerlings into a pond in which the bluegill population has reached or is approaching carrying capacity may result in decreased growth throughout the population and a smaller average size of fish. Even though the biomass density remains unchanged, individual size of fish may be so reduced that fishing is poor, probably to the disappointment of the person who thought that additional "stocking" would improve fishing!

Dispersal is much influenced by the presence or absence of barriers, and the *vagility*, which is defined as the inherent power of movement. Vagility is often greater than is commonly realized. Although birds and insects are noted for their ability to "get around," many plants and lesser forms of

---

* Dispersion should not be confused with dispersal. Dispersion, especially as used in statistics, refers to the internal pattern of a population, e.g., the distribution of items around the mean.

animals actually have greater dispersal powers. Studies of the life found floating in the air, or "aerial plankton," have revealed a surprising number of organisms, not only spores, seeds, and microorganisms, but animals such as spiders which float for miles attached to threadlike parachutes of their own making.

Wolfenbarger (1946) has summarized a large volume of literature on dispersal. He found that most small organisms or disseminules tended to disperse as the logarithm of the distance from the source. When distance units were placed logarithmically and incidence units (y-axis) uniformly (semilogarithmic plot), a straight line was often obtained. The slope of this regression line then becomes a quantitative measure of dispersal useful for comparing different populations (comparing, for example, the rate of decrease in bacteria and insects, or spores and mosquitoes). Exponential dispersal can be dealt with in the same manner as exponential mortality, using the same exponential equations as previously discussed in this chapter. Knowledge of relative vagility and the ability to estimate the probability of organisms moving a given distance is often of great practical importance, as, for example, in determining how far from human habitations control measures for mosquitoes would need to be undertaken.

Organisms with well-developed nervous systems and directed movement, of course, may not follow the pattern of logarithmic decrease from the source. Bees and birds, for example, may pass by food near the hive or nest in favor of food at greater distances. Likewise, migrations and other movements directed along specific pathways or to specific points will result in population dispersal patterns characteristic for each species.

MacArthur and Wilson (1967) discuss three dispersal patterns in connection with movement of propagules from mainland to islands or from island to island: (1) the exponential pattern in which the fraction of individuals continuing with increase in distance falls off at the rate of e^{-x} (as just described above), (2) the normally distributed pattern in which the fraction falls off with distance at the rate of e^{-x^2}, and (3) a uniform pattern, which might be called a "set distance" dispersal in that propagules move a certain distance and stop as in a bird flying from one island to another or homing to a summer and a winter area. Examples of the first two patterns are shown in Figure 7–21.

Migration, as defined above, is a special and very remarkable type of population dispersion often involving the mass movement of entire populations. This can occur, of course, only in motile organisms, and is best developed in arthropods and vertebrates. Seasonal and diurnal migrations not only make possible occupation of regions that would be unfavorable in the absence of migration but also enable animals to maintain a higher average density and activity rate. Nonmigratory populations often must undergo considerable reduction in density or assume some form of dormancy during unfavorable periods. The orientation and navigation of long-distance migrants (birds, fish, etc.) constitute a popular field for experiments and theories, but are little understood as yet. Some theories and patterns are briefly reviewed elsewhere in this volume (see page 249 and Section 6, Chapter 14).

Figure 7–22 is a good illustration of emigration and migration as it affects a single species population. The barn owl is migra-

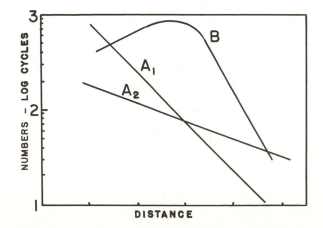

Figure 7–21. Two types of dispersal patterns. *A.* Exponential pattern. *B.* Normally distributed pattern. The exponential patterns are modeled after mosquitoes. *Aedes sollicitans* (A_1) and *A. vexans* (A_2), and the normally distributed dispersal after horseflies (Tabanidae). (*B*, Data of MacCreary and Stearns in Wolfenbarger, 1946.)

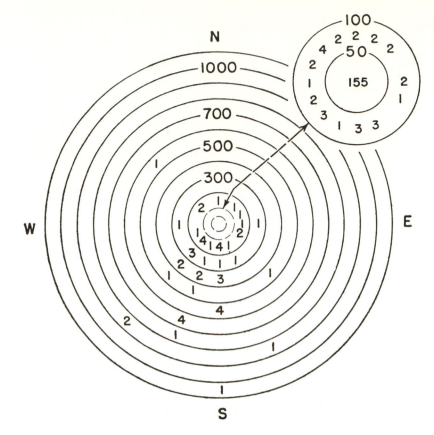

Figure 7-22. Dispersal of banded barn owls in relation to their hatching places. Distance of movement in terms of miles is indicated by concentric circles. The number of birds recovered in any given direction is indicated by the numbers between the circles. Within 100 miles dispersal is nondirectional; beyond this distance dispersal is definitely directional. (After Stewart, 1952.)

tory in northern United States but nonmigratory in the southern part of the country. In both regions young birds tend to disperse from the nesting site. As shown in Figure 7-22, this dispersal of young is random in all directions, but is restricted to about 100 miles (this figure would be a measure of the emigration vagility). Migratory movement, however, is definitely directional (southward) and involves longer distances.

Quite apart from its effect on population size and composition, dispersal brings about gene exchange between populations and hence is important from the standpoint of population genetics and speciation, as we will have occasion to note in the next two chapters. Finally, we are not concerned here with effects of immigration of a new species into a community or the effect of dispersal of one species as it may affect another. These important relations are best considered under the heading of competition.

12. POPULATION ENERGY FLOW OR BIOENERGETICS

Statement

Energy flow (= rate of assimilation) in a population provides the most reliable basis for (1) evaluating observed fluctuations in density and (2) determining the role of a population within its community.

Explanation and Example

The concepts of energy flow and productivity outlined in Chapter 3 apply to the population level as well as to the ecosystem level. As was brought out in that chapter (see especially Section 6 and Table 3-14), numbers (numerical density) overemphasize and weights (biomass density) underemphasize the importance of small organ-

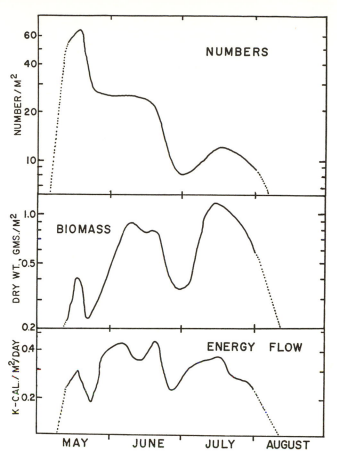

Figure 7–23. Numbers, biomass (dry weight), and energy flow (population assimilation rate) per square meter in a population of salt marsh grasshoppers (*Orchelimum fidicinium*) living in a *Spartina alterniflora* marsh near the University of Georgia's Marine Institute, Sapelo Island, Georgia. (After E. P. Odum and Alfred E. Smalley, 1959.)

isms, while the reverse is true with large organisms. If data on numbers and biomass can be used to calculate rates of change and energy flow, then a more reliable estimate of the importance of a population in its community can be obtained. Figure 7–23 illustrates a study in which this was done for a population of grasshoppers living in a salt marsh community. In this relatively simple case, the animal is a strict herbivore, is sedentary, has but one generation per year and lives in a rather uniform stand composed of one species of plant which provides its sole source of food and shelter. The numbers and biomass per square meter were determined at 3 to 4 day intervals. From these data, population growth, or production, was determined by adding the increase in weight of the living population to the growth of individuals that died during the census interval. Production was then converted to kilocalories per m² per day. Oxygen consumption (respiration) of adults and different sized nymphs in relation to temperature was then determined in the laboratory. From

these data population respiration of the average standing crop during each interval was calculated and adjusted to the actual temperature of the environment. Oxygen consumption was converted to calories by means of an oxycaloric coefficient (see Ivlev, 1934). The total population assimilation rate was then obtained by adding production to respiration.

As may be seen from Figure 7–23, numbers were at a peak in mid-May, when numerous very small nymphs hatched out from overwintering eggs, and then declined rapidly during the season. In the particular year of the study there were two periods of heavy mortality, each of which was followed by accelerated growth of survivors, and also by a small amount of recruitment (new eggs hatching in early summer, immigration during late summer). The peak of energy flow did not correspond with either the peak in numbers or the peak in biomass, but occurred during the period when the population was composed of a medium number of medium sized nymphs, all growing very rapidly. In other words, the greatest impact of the population on the marsh (in terms of consumption of the grass) did not coincide with maximum numbers or biomass, although more nearly with the latter than the former. Also, note that while numbers and biomass varied and fluctuated five- to six-fold, energy flow varied only two-fold. Since metabolism-per-gram of small nymphs was several times greater than that of adults, the high number but small biomass population of the spring was about equal to the low number but large biomass population of late summer. From the bottom graph in Figure 7–23 we see that approximately one-third of a kcal per day was assimilated by the population living on a square meter for about 2½ months. The total energy flow per m² for the season was approximately 28 kcal or about 7 grams dry weight of grass (4 kcal per gram). Since these grasshoppers assimilate only about a third of what they eat (assimilation efficiency about 33 per cent), then about 21 grams per m² were probably removed from the growing plants. It should be emphasized that the 14 grams not assimilated did not change trophic levels but were available as food for other primary consumers, such as bacteria, fungi or detritus-feeding animals.

From this example we see that even within the same species density or "standing crop" data can be misleading if used as a measure of importance; only by integrating numbers and biomass and adding respiration (which accounts for most of the energy flow in many situations) can a true picture be obtained, especially if we wish to compare populations in which individuals differ greatly in size (as ants and deer, for example). Finally, it should be emphasized that although an animal was used in the above example, the energy flow approach is equally applicable to all populations (plant, animal, or microbe) in all environments.

Another study of population bioenergetics is illustrated by the energy-flow diagram in Figure 7–24. Marsh wrens live in the same salt marsh as the grasshoppers described above, but during the breeding season they are restricted to the strips of tall grass along the margins of tidal creeks. Here their ecological density is about 100 birds per hectare (standing crop of about 0.03 gm dry weight or 0.18 kcal per square meter)

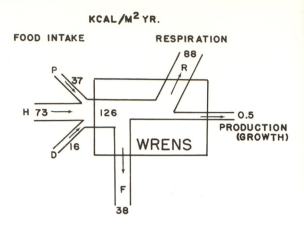

Figure 7–24. Annual energy budget of a population of marsh wrens living in a Georgia salt marsh, which has a net primary production of about 10,000 kcal/m². P = predatory insects and spiders; H = herbivorous insects; D = detritus feeders (small snails and crabs); F = energy lost in feces; R = energy lost in respiration. (After Kale, 1965.)

which equals the total density of an entire bird population of a productive forest. Since the wren is the only passerine bird in this habitat, the single species population occupies the niche (see page 234 for definition of this term) that would be filled by a multispecies population in most habitats. As shown in the diagram, the annual energy requirement of the population is supported by a diet of approximately 60 per cent herbivorous insects, 30 per cent predatory insects and spiders, and 10 per cent detritus feeders (small snails and crabs). During the nesting season the wrens eat one-fourth of the standing crop of insects and spiders in the nesting area each day, this high rate of predation being possible because of high production of the prey and the immigration of prey from surrounding areas of marsh not occupied by the wrens. While the wren population energy requirements are only about 2 per cent of the net primary production of the marsh, their impact on the primary and secondary consumer trophic levels in the grazing food chain is quite large. This study by Kale (1965) is one of the most complete studies yet made that includes not only the detailed energetics of the population but also that of the ecosystem of which it is a functional part. Such studies reveal the true "importance" of the population to a degree that could not be determined by census information alone (i.e., by knowing only "N").

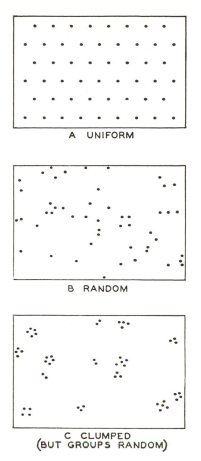

Figure 7-25. Three basic patterns of the distribution of individuals, pairs or other units, in a population.

13. POPULATION STRUCTURE: INTERNAL DISTRIBUTION PATTERNS (DISPERSION)

Statement

Individuals in a population may be distributed according to three broad patterns (Fig. 7-25): (1) random, (2) uniform (more regular than random), and (3) clumped (irregular, nonrandom). Random distribution is relatively rare in nature, occurring where the environment is very uniform and there is no tendency to aggregate. Uniform distribution may occur where competition between individuals is severe or where there is positive antagonism which promotes even spacing. Clumping of varying degree represents by far the commonest pattern, almost the rule, when individuals are considered.

However, if individuals of a population tend to form groups of a certain size—for example, pairs in animals, or vegetative clones in plants—the distribution of the *groups* may more nearly approach random. Determination of the type of distribution, of the degree of clumping (if any), and of the size and permanence of groups are necessary if a real understanding of the nature of a population is to be obtained, and especially if density is to be correctly measured. Thus, sample methods and statistical analyses which would be quite sound for random or uniform distribution might be entirely inadequate or misleading when applied to strongly clumped distributions.

Explanation

The three patterns of distribution or "intrapopulation dispersion" are shown in a simplified manner in Figure 7-25. Each rectangle contains approximately the same number of individuals. In the case of clumped distribution (C), the groups could be the same or of varying size (as shown), and they could be randomly distributed (as shown), uniformly distributed, or themselves aggregated or clumped, with large unoccupied spaces. In other words, we might consider that there are five types of distribution: (1) uniform, (2) random, (3) random clumped, (4) uniform clumped, and (5) aggregated clumped. All these types are undoubtedly found in nature. It is obvious from examining Figure 7-25 that a small sample drawn from the three populations could yield very different results. A small sample from a population with a clumped distribution would tend to give either too high or too low a density when the number in the sample is multiplied to obtain the total population. Thus "clumped" populations require larger and more carefully planned sample techniques than nonclumped ones.

It should be recalled that random distribution is one that follows the so-called "normal" or bell-shaped curve on which standard statistical methods are based. This type of distribution is to be expected in nature when many small factors are acting together on the population. When a few major factors are dominating, as is the usual case (recall principle of limiting factors), and when there is such a strong tendency for plants and animals to aggregate for (or as a result of) reproductive and other purposes, there is little reason to expect a completely random distribution in nature. However, as Cole (1946) suggests, nonrandom or "contagious" distributions of organisms can often be considered to be made up of intermingled random distributions of groups containing various numbers of individuals (as in Figure 7–25C), or the groups could turn out to be uniformly distributed (or at least more regular than random). In other words, to take an extreme case, it would be much better to determine the number of ant colonies (i.e., using the colony as the population unit) by a sample method and then determine the average number of individuals per colony, than it would be to try to measure the number of individuals directly by random samples.

A number of methods have been suggested that may be used to determine the type of spacing and degree of clumping between individuals in a population (where it is not self-evident), but there is much that must still be done in solving this important problem. Two methods may be mentioned as examples. One method is to compare the actual frequency of occurrence of different sized groups obtained in a series of samples with a "Poisson" series that gives the frequency with which groups of 0, 1, 2, 3, 4, etc., individuals will be encountered together if the distribution is random. Thus, if the occurrence of small sized groups (including blanks) and large sized groups is more frequent and the occurrence of middle sized groups less frequent than expected, then the distribution is clumped. The opposite is found in uniform distribution. Statistical tests can be used to determine whether the observed deviation from the Poisson curve is significant, but this general method has the disadvantage that sample size may influence the results. An example of the use of the Poisson method to test for random distribution in spiders is shown in Table 7–7. In all but 3 of 11 quadrats spiders were randomly distributed. The nonrandom distributions occurred in quadrats in which the vegetation was least uniform.

A general property of random distributions is that the variance (V) equals the mean (m); variance greater than the mean indicates a clumped distribution, and less than mean a uniform (regular) pattern. Thus,

Table 7–7. *Random and Nonrandom Spatial Distribution of Spiders and Clams*

A. THE NUMBER AND DISTRIBUTION OF WOLF SPIDERS (LYCOSIDAE) ON 0.1 HECTARE QUADRATS IN AN OLD-FIELD HABITAT[*]

Species	Quadrat	Number per Quadrat	Chi Square from Poisson Distribution
Lycosa timuqua	1	31	8.90[†]
	2	19	9.58[†]
	3	15	5.51
	4	16	0.09
	5	45	0.78
	6	134	1.14
L. carolinensis	2	16	0.09
	5	23	4.04
	6	15	0.05
L. rabida	3	70	17.30[†]
	4	16	0.09

B. MEAN, VARIANCE, AND SPATIAL DISTRIBUTION OF TWO SPECIES OF SMALL CLAMS ON AN INTERTIDAL MUDFLAT IN CONNECTICUT[‡]

Species and Age	Mean	Variance	Variance-mean Ratio[§]
Mulinia lateralis			
All ages	0.27	0.26	Random
Gemma gemma			
All ages	5.75	11.83	Clumped
1st year	4.43	7.72	Clumped
2nd year	1.41	1.66	Random

[*] After Kuenzler, 1958.
[†] Significant at 5 per cent level, i.e., nonrandom; in all other quadrats the distribution was random.
[‡] After Jackson, 1968.
[§] Where not significantly different (5 per cent confidence level) from one, random distribution indicated; significantly greater than one, clumped (aggregated) distribution indicated.

in a random distribution:

$$V/m = 1; \text{ standard error} = \sqrt{2/n - 1}$$

If, on the application of standard significance tests, the variance/mean ratio is found to be significantly greater than 1, the distribution is clumped; if it is significantly less, the distribution is regular; if not different from 1, the distribution is random. This approach is also illustrated in Table 7–7.

Another method, suggested by Dice (1952), involves actually measuring the distance between individuals in some standardized way. When the square root of the distance is plotted against frequency, the shape of the resulting frequency polygon indicates the distribution pattern. A symmetrical polygon (a normal bell-shaped curve, in other words) indicates random distribution, a polygon skewed to the right a uniform distribution, and one skewed to the left a clumped distribution (individuals coming closer together than expected). A numerical measure of the degree of "skewness" may be computed. This method, of course, would be most applicable to plants or stationary animals, but it could be used to determine spacing between animal colonies or domiciles (fox dens, rodent burrows, bird nests, and so forth).

For more of the statistics of spatial distribution of units within a population, see Greig-Smith (1957), Skellum (1952), and Pielou (1960).

Examples

Park (1934) found that flour beetle larvae were usually distributed throughout their very uniform environment in a random manner, since observed distribution corresponded with the Poisson distribution. "Lone wolf" type of parasites or predators, such as one of the species of spiders in Table 7–7, sometimes show a random distribution (and they often engage in "random searching" behavior for their hosts or prey). Jackson (1968) has reported that individuals of the clam, *Mulinia lateralis,* are randomly distributed on an intertidal mudflat, as were second year individuals of *Gemma gemma,* but not first year individuals, nor the total population of *Gemma;* they were clumped because of the ovoviviparous reproduction (i.e., young retained in the body of female during larval development). Results of this study are shown in Table 7–7*B*. Note that variance is close to the mean in the randomly distributed population, but significantly larger in the clumped population. The mudflat environment is very homogeneous and interspecific competition not severe—two aspects favoring random dispersion.

Forest trees which have reached sufficient height to form a part of the forest crown may show a uniform distribution according to Dice (1952), because competition for sunlight is so great that the trees tend to be spaced at very regular intervals, "more regular than random." A cornfield, orchard, or pine plantation, of course, would be a better example. Desert shrubs often are very regularly spaced almost as if planted in rows, apparently because of the intense competition (which may include antibiotics) in the low moisture environment (see Figure 14–15). As will be discussed in Section 15 of this chapter, "territoriality" tends to produce a more regular than random distribution. As for clumped distributions, the reader will not have to look far to find ample examples! Of several forest floor invertebrates studied by Cole (1946 and 1946*a*), only spiders showed a random distribution, while he reports on another study in which only 4 out of 44 plants showed a random distribution. All the rest showed varying degrees of clumping. This gives us something of an idea of what to expect in natural situations. Further attention will be given to aggregations in the next section.

14. POPULATION STRUCTURE: AGGREGATION AND ALLEE'S PRINCIPLE

Statement

As indicated in the previous section, varying degrees of "clumping" are characteristic of the internal structure of most populations at one time or another. Such clumping is the result of the aggregation of individuals (1) in response to local habitat differences, (2) in response to daily and seasonal weather changes, (3) as the result of reproductive processes, or (4) as the result of social attractions (in higher animals). The degree of aggregation to be found in a given species population, therefore, depends on the specific nature of the habitat (whether uniform or discontinuous), the weather or other physical factors, the type of reproductive pattern characteristic of the species, and the degree of sociality. Aggregation may increase competition between individuals for nutrients, food, or space, but this is often more than counterbalanced by increased survival of the group. Individuals in groups often experience a lower mortality rate during unfavorable periods or during attacks by other organisms than do isolated individuals, because the surface area exposed to the environment is less in proportion to the mass and because the group may be able to favorably modify microclimate or microhabitat. The degree of aggregation, as well as the overall density, which results in optimum population growth and survival, varies with species and conditions; therefore, undercrowding (or lack of aggregation), as well as overcrowding, may be limiting. This is Allee's principle.

Explanation and Examples

In plants, aggregation may occur in response to the first three factors listed above, whereas in higher animals spectacular aggregations may be the result of all four factors as illustrated, for example, by the great herds of reindeer or caribou that occur in the very uneven arctic habitat. In a study of the distribution of three species of plants in a Michigan old-field, Cain and Evans (1952) found that part of the study area was too wet for any of the three species, illustrating how physical habitat contributes to nonrandom distribution. Within the habitable area the three species varied in the degree of aggregation, the goldenrod (*Solidago*) showing almost a random distribution, whereas the *Lespedeza* plants were highly clumped. The latter commonly reproduce vegetatively, new individuals coming up from the root stocks; thus the pattern of reproduction, in part at least, accounts for the aggregation.

In plants in general, and probably in some of the lower animal groups, it is a rather well-defined ecological principle that aggregation is inversely related to the mobility of disseminules (seeds, spores, etc.) (Weaver and Clements, 1929; recall, also, the random and nonrandom dispersal patterns discussed in Section 11). For example, driving about the country all of us have frequently observed that in old-fields cedars, persimmons, and other plants with nonmobile seeds are nearly always clumped near a parent or along fences and other places where birds or other animals have deposited the seeds in groups. On the other hand, ragweeds and grasses, and even pine trees which have light seeds widely distributed by the wind, are, by comparison, much more evenly distributed over the old-fields.

Group survival value is an important characteristic that may result from aggregation. A group of plants may be able to withstand

the action of wind better than isolated individuals or be able to reduce water loss more effectively. With green plants, however, the deleterious effects of competition for light and nutrients generally soon overbalance the advantages of the group. The most marked group survival values are to be found in animals. Allee (1931, 1938, and 1951) has conducted many experiments in this field and summarized the extensive writings on the subject. He found, for example, that groups of fish could withstand a given dose of poison introduced into the water much better than isolated individuals. Isolated individuals were more resistant to poison when placed in water formerly occupied by a group of fish than when placed in water not so "biologically conditioned"; in the former case, mucus and other secretions aided in counteracting the poisons, thus revealing something of the mechanism of group action in this case. Bees provide another example of group survival value; a hive or cluster of bees can generate and retain enough heat in the mass for survival of all the individuals at temperatures low enough to kill all the bees if each were isolated. Colonial birds often fail to reproduce successfully when the colony size becomes small (Darling, 1938). Allee points out that these types of primitive cooperation (protocooperation) found even in very primitive phyla are the beginning of social organization, which shows varying degrees of development in the animal kingdom, culminating in the intelligent group behavior of human beings (which, we hope, has survival value!). Allee's principle is diagrammatically illustrated in Figure 7–26.

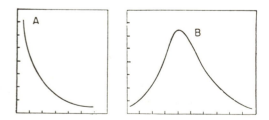

Figure 7–26. Illustration of Allee's principle. In some populations growth and survival is greatest when population size is small (A), whereas in others intraspecific protocooperation results in an intermediate sized population being the most favorable (B). In the latter instance "undercrowding" is as detrimental as "overcrowding." (From Allee *et al.,* 1949.)

Actual social aggregations as seen in the social insects and vertebrates (as contrasted with passive aggregation in response to some common environmental factor) have a definite organization involving social hierarchies and individual specializations. A social hierarchy may take the form of a "peck-order" (so-called because the phenomena were first described in chickens) with clear-cut dominance-subordinance between individuals often in linear order (like a military order of command from general to private!), or it may take the form of a more complicated pattern of leadership, dominance, and cooperation as occurs in well-knit groups of birds and insects that behave almost as a single unit. In such units there may be a definite leader, but as often as not no one individual actually leads; individuals in the group follow whichever individual acts forthrightly as if it "knew what it was about" (Scott, 1956). As pointed out in Section 10, these kinds of social organizations not only benefit the population as a whole, but they may be regulatory in preventing overgrowth.

The remarkable organizations of social insects are in a class by themselves and have been much studied (especially by W. M. Wheeler and A. E. Emerson; see Allee *et al.*, 1949, Chapter 24). The most highly developed insect societies are found among the termites (order Isoptera) and ants and bees (order Hymenoptera). A division of labor is accomplished in the most specialized species by three castes — reproductive (queens and kings), workers, and soldiers, each morphologically specialized to perform the functions of reproduction, food getting, and protection. As will be discussed in the next chapter, this kind of adaptation leads to group selection, not only within a species but in groups of closely linked species.

Allee's principle is very relevant to man. It is obvious that "urban aggregation" is beneficial to man, but only up to a point. As shown in Figure 7–26, there comes a point where increasing density becomes detrimental even in populations that benefit from a great deal of intraspecific cooperation. A real effort must be made to determine objectively what is the optimum size of cities. Cities, as well as bee or termite colonies, can get too big for their own good!

15. POPULATION STRUCTURE: ISOLATION AND TERRITORIALITY

Statement

Forces which bring about isolation or the spacing of individuals, pairs, or small groups in a population are perhaps not as widespread as those favoring aggregation but they are nevertheless very important, especially in terms of population regulation (intraspecific, interspecific, or both). Isolation usually is the result of (1) interindividual competition for resources in short supply or (2) direct antagonism. In both cases a random or a uniform distribution may result, as outlined in Section 13 because close neighbors are eliminated or driven away. Individuals, pairs, or family groups of vertebrates and the higher invertebrates commonly restrict their activities to a definite area, called the *home range*. If this area is actively defended, it is called a *territory*. Territoriality seems to be most pronounced in vertebrates and certain arthropods that have complicated reproductive behavior patterns involving nest building, egg laying, and care and protection of young. However, at the risk of offending semantic purists we are including under the heading of *territoriality* any active mechanism that spaces individuals or groups apart from one another, which means that we can talk about territoriality in plants and microorganisms as well as in animals. In higher animals the isolating mechanism is likely to be behavioral (neural), while in lower animals and plants it is chemical (i.e., accomplished by antibiotics or "allelopathic" substances, which will be considered in a later section). In summary, isolation of this sort reduces competition, conserves energy during critical periods, and prevents overcrowding and exhaustion of food supply in the case of animals, or nutrients, water, or light in the case of plants. In other words, territoriality tends to regulate populations at a level below the saturation level. In this sense, then, territoriality is a general ecological phenomenon not restricted to any one taxonomic group such as birds (in which the phenomenon happens to be well studied).

Explanation and Examples

Just as aggregation may increase competition, but has other advantages, so spacing of individuals in a population may reduce the competition for the necessities of life, but perhaps at the expense of the advantages of cooperative group action. Presumably, which pattern survives in a long-term evolutionary sense in a particular case depends on which gives the greatest long-time advantage. In any event, we find both patterns to be frequent in nature; in fact, some species populations alternate from one to the other (robins, for example, isolate into territories during the breeding season and aggregate into flocks in the winter), and thus obtain advantages from both arrangements! Again, different ages and sexes may show opposite patterns at the same time (adults isolate, young aggregate, for example).

The role of intraspecific competition in bringing about spacing, as in forest trees and desert shrubs, has already been discussed (Sections 10 and 13). We need only to call attention to the spacing which results from active antagonism as exhibited in higher animals. Many animals isolate themselves and restrict their major activities to definite areas or home ranges, which may vary from a few square inches to many square miles in the case of a puma. Since home ranges often overlap, only partial spacing is achieved (Fig. 7–27, left). Territoriality achieves the ultimate in spacing (Fig. 7–27, right). Territories of birds have been classified by Nice (1941) into several basic

types: (A) entire mating, feeding, and breeding area defended, (B) mating and nesting but not feeding area defended, (C) mating area only defended, (D) nest only defended, (E) nonbreeding areas defended. In type "A" the defended area may be quite large, larger than needed in terms of food supply for the pair and its young. For example, the tiny gnatcatcher (which weighs about 3 grams) establishes a territory averaging 4.6 acres but obtains all the food it needs in a much smaller area around the nest (Root, 1969). In most territorial behavior actual fighting over boundaries is held to a mimimum. Owners advertise their land or location in space by song or displays and potential intruders generally avoid entering an established domain. The fact that the area defended by birds is often larger at the beginning of the nesting cycle than later when the demand for food is greatest, and the fact that many territorial species of birds, fish, and reptiles do not defend the feeding area at all give support to the idea that reproductive isolation and control has greater survival value than the isolation of a food supply as such. Territoriality not only tends to prevent overpopulation, but it also provides for a quick recovery should unusual mortality of breeders occur, because a number of nonbreeding individuals are usually present that are unable to find space; these quickly fill in whenever a territory-holder is lost.

A good place to observe territorial behavior in the spring is in the shallow water of a pond containing bass or sunfish. Males may be observed defending a "nest" or

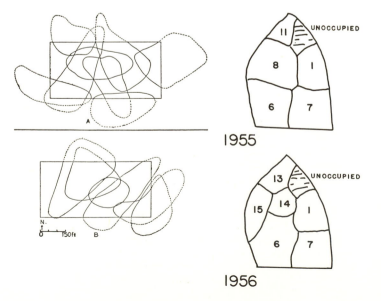

1955

1956

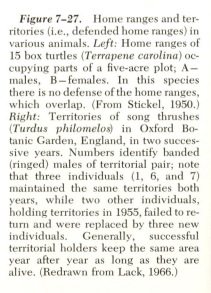

Figure 7–27. Home ranges and territories (i.e., defended home ranges) in various animals. *Left:* Home ranges of 15 box turtles (*Terrapene carolina*) occupying parts of a five-acre plot; A—males, B—females. In this species there is no defense of the home ranges, which overlap. (From Stickel, 1950.) *Right:* Territories of song thrushes (*Turdus philomelos*) in Oxford Botanic Garden, England, in two successive years. Numbers identify banded (ringed) males of territorial pair; note that three individuals (1, 6, and 7) maintained the same territories both years, while two other individuals, holding territories in 1955, failed to return and were replaced by three new individuals. Generally, successful territorial holders keep the same area year after year as long as they are alive. (Redrawn from Lack, 1966.)

"bed" (a scooped-out place on the bottom) and its vicinity. When another individual approaches the male displays his bright fin colors and charges toward the intruder. Since the male defends the area before as well as after the female lays the eggs, the defense is clearly in terms of space, and not just defense of offspring.

The extent to which man is territorial by virtue of inherent behavior, and the extent to which man can learn to use land-use control and planning as safeguards against overpopulation are both being actively debated at present. We shall come back to this in later chapters. Robert Ardrey in his best-selling book *The Territorial Imperative* (1967) has some intriguing ideas on this subject.

16. TYPES OF INTERACTION BETWEEN TWO SPECIES

Statement

Theoretically, populations of two species may interact in basic ways that correspond to combinations of 0, +, and −, as follows:

00, −−, ++, +0, −0, and +−. Three of these combinations (++, −−, and +−) are commonly subdivided, resulting in nine important interactions which have been demonstrated (see Burkholder's adaptation (1952) of Haskel's (1949) classification scheme). These (see Table 7–8) are as follows: (1) *neutralism*, in which neither population is affected by association with the other; (2) *mutual inhibition competition type* in which both populations actively inhibit each other; (3) *competition resource use type* in which each population adversely affects the other in the struggle for resources in short supply; (4) *amensalism*, in which one population is inhibited and the other not affected; (5) *parasitism* and (6) *predation*, in which one population adversely affects the other by direct attack but is nevertheless dependent on the other; (7) *commensalism*, in which one population is benefited but the other is not affected; (8) *protocooperation*, in which both populations benefit by the association but relations are not obligatory; and (9) *mutualism*, in which growth and survival of both populations is benefited and neither can survive under natural conditions without

Table 7–8. *Analysis of Two-species Population Interactions*

0 INDICATES NO SIGNIFICANT INTERACTION
+ INDICATES GROWTH, SURVIVAL, OR OTHER POPULATION ATTRIBUTE BENEFITED (POSITIVE TERM ADDED TO GROWTH EQUATION)
− INDICATES POPULATION GROWTH OR OTHER ATTRIBUTE INHIBITED (NEGATIVE TERM ADDED TO GROWTH EQUATION)

Type of Interaction*	Species 1	Species 2	General Nature of Interaction
1. Neutralism	0	0	Neither population affects the other
2. Competition: Direct Interference type	−	−	Direct inhibition of each species by the other
3. Competition: Resource use type	−	−	Indirect inhibition when common resource is in short supply
4. Amensalism	−	0	Population 1 inhibited, 2 not affected
5. Parasitism	+	−	Population 1, the parasite, generally smaller than 2, the host
6. Predation	+	−	Population 1, the predator, generally larger than 2, the prey
7. Commensalism	+	0	Population 1, the commensal, benefits while 2, the host, is not affected
8. Protocooperation	+	+	Interaction favorable to both but not obligatory
9. Mutualism	+	+	Interaction favorable to both and obligatory

* Types 2 through 4 can be classed as "negative interactions," types 7 through 9 as "positive interaction," and 5 and 6 as both.

the other. In terms of population growth and survival, these interactions involve adding positive, negative, or zero terms to basic population growth equations as written in Section 8.

All these population interactions are likely to occur in the average community and may be readily recognized and studied, at least qualitatively, even in complex communities. For a given species pair the type of interaction may change under different conditions or during successive stages in their life histories. Thus, two species might exhibit parasitism at one time, commensalism at another, and be completely neutral at still another time. Simplified communities and laboratory experiments provide means for singling out and studying quantitatively the various interactions. Also, deductive mathematical models derived from such studies enable us to analyze factors not ordinarily separable from the others.

In terms of the overall picture of the ecosystem the nine types of interactions can be reduced to two broad types, namely, the *negative interactions* and the *positive interactions*. Two principles regarding these categories are especially worthy of emphasis: (1) *In the evolution and development of ecosystems negative interactions tend to be minimized in favor of positive symbiosis that enhances the survival of the interacting species* (see Chapter 9). (2) *Recent or new associations are more likely to develop severe negative coactions than are older associations* (see Section 17).

Explanation

A familiar situation is the action of one population affecting the growth or death rate of another population. Thus, the members of one population may eat members of the other population, compete for foods, excrete harmful wastes, or otherwise interfere with the other population. Likewise, populations may help one another, the interaction being either one-way or reciprocal. Interactions of these sorts fall into several definite categories as listed in the above Statement and shown in Table 7–8. Before discussing actual cases, it will be helpful to diagram hypothetical and somewhat simplified cases in order to see how these interactions can operate to influence the growth and survival of populations. As previously pointed out, growth equation "mod-

els" make definitions more precise, clarify thinking, and enable us to determine how factors may operate in complex natural situations.

If the growth of one population can be diagrammed and described by an equation (see Section 8), the influence of another population may be expressed by a term that modifies the growth of the first population. Various terms can be substituted according to the type of interaction. For example, in competition the growth rate of each population is equal to the unlimited rate minus its own self-crowding effects (which increase as its population increases) minus the detrimental effects of the other species, N_2 (which also increase as the numbers of both species, N and N_2, increase), or

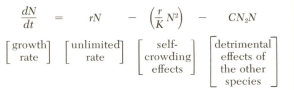

$$\frac{dN}{dt} = rN - \left(\frac{r}{K}N^2\right) - CN_2N$$

$$\begin{bmatrix}\text{growth}\\\text{rate}\end{bmatrix}\begin{bmatrix}\text{unlimited}\\\text{rate}\end{bmatrix}\begin{bmatrix}\text{self-}\\\text{crowding}\\\text{effects}\end{bmatrix}\begin{bmatrix}\text{detrimental}\\\text{effects of}\\\text{the other}\\\text{species}\end{bmatrix}$$

This equation will be recognized as the logistic equation given on page 184, except for the addition of the last term, "minus detrimental effects of the other species." There are several possible results of this kind of interaction. If "C" is small for both species so that the interspecific depressing effects are less than intraspecific (self-limiting) effects, the growth rate and perhaps the final density of both species would be depressed slightly; but both species would probably be able to live together. In other words, the depressing effects of the other species would be less important than the competition within the species. Also, if the species exhibit exponential growth (with self-limiting factor absent from the equation), then interspecific competition might provide the "leveling off" function missing from the species' own growth form. However, if "C" is large, the species exerting the largest effect will eliminate its competitor or force it into another habitat. Thus, theoretically, species which have similar requirements cannot live together because strong competition will likely develop, causing one of them to be eliminated. Our models point up some of the possibilities; we shall see in the next section how these possibilities actually work out.

When both species of interacting populations have beneficial effects on each other instead of detrimental ones, a positive term is added to the growth equations. In such cases both populations grow and prosper, reaching equilibrium levels that are mutually beneficial. If beneficial effects of the other population (the positive term in the equation) are necessary for growth and survival of both populations, the relation is known as mutualism. If, on the other hand, the beneficial effects only increase the size or growth rate of the population but are not necessary for growth or survival, the relationship comes under the heading of cooperation or protocooperation. (Since the cooperation indicated is not necessarily the result of conscious or "intelligent" reasoning, the latter term is probably to be preferred.) In both protocooperation and mutualism the outcome is similar, i.e., the growth of either population is less or zero without the presence of the other population. When an equilibrium level is reached the two populations exist together stably, usually in a definite proportion.

Consideration of population interactions as shown in Table 7–8, or in terms of growth equations, avoids confusion which often results when terms and definitions alone are considered. Thus, the term *symbiosis* is sometimes used in the same sense as mutualism; sometimes the term is used to cover commensalism and parasitism as well. Since symbiosis literally means "living together," the word is used in this book in this broad sense without regard to the nature of the relationship. The term "parasite" and the science of parasitology are generally considered to deal with any small organism that lives on or in another organism regardless of whether its effect is negative, positive, or neutral. Various nouns have been proposed for the same type of interaction, adding to the confusion. When relations are diagrammed, however, there is little doubt as to the type of interaction being considered; the word or "label" then becomes secondary to the mechanism and its result. Even if we were never able to apply equations to actual situations, these "mathematical models" are still highly useful in clarifying one's thinking, in opening the way for quantitative expression, and in helping to avoid cumbersomely worded definitions which have plagued the early history of ecology!

One final point. Note that the word "harmful" was not used in describing negative interactions. Competition and predation de-

crease the growth rate of affected populations, but this does not necessarily mean that the interaction is harmful from the long-term survival or evolutionary standpoint. In fact, negative interactions can increase the rate of natural selection, resulting in new adaptation. As we have already seen, predators and parasites are often beneficial to populations which lack self-regulation in that they may prevent overpopulation which might result in self-destruction (see Section 8).

Finally, the tendency for the severity of negative interactions to be reduced and positive ones to be increased, as outlined in the Statement, is a concept of utmost importance to man. This idea will be developed in the next several sections.

17. NEGATIVE INTERACTIONS: INTERSPECIFIC COMPETITION

Statement

Competition in the broadest sense refers to the interaction of two organisms striving for the same thing. Interspecific competition is any interaction between two or more species populations which adversely affects their growth and survival. It can take two forms, as shown in Table 7–8. The tendency for competition to bring about an ecological separation of closely related, or otherwise similar, species is known as the *competitive exclusion principle*.

Explanation

A great deal has been written about interspecific competition by ecologists, geneticists, and evolutionists. In most cases the word "competition" is used with reference to situations in which negative influences are due to a shortage of resources used by both species. This leaves the more direct reciprocal interferences, such as mutual predation or the secretion of harmful substances, to be placed in another category (as we have done in Table 7–8) even though there is no generally accepted term for these types of interactions. Interspecific chemical messengers (or allelochemic substances; see Whittaker, 1970), which provide a competitive advantage for one species against another, have been termed *allelopathic* substances (see pages 23 and 220) or antagonistic *allomones* (Brown *et al.*, 1970). The author is of the firm opinion that words such as competition, community, population, which are not only widely used in science but in the general language as well, should be precisely defined (as we have done), but should remain broad in scope. It is much less confusing to everyone to leave the basic term broad, and then "subclassify" by means of adjectives or modifying phrases where a more restrictive meaning is in order. For example, we can speak of competition for resources, antibiotic competition, or competition for light. Or we could subdivide quantitatively according to the severity of the depressing effect as Philip (1955) has done with the use of models. Thus, in general, we shall strive to avoid specialized terms except where a very specialized meaning is needed.

The competitive interaction often involves space, food or nutrients, light, waste materials, susceptibility to carnivores, disease, and so forth, and many other types of mutual interactions. The results of competition are of the greatest interest and have been much studied as one of the mechanisms of natural selection. Interspecific competition can result in equilibrium adjustments by two species, or it can result in one species population replacing another or forcing it to occupy another space or to use another food, whatever is the basis of competitive action. It is often observed that closely related organisms having similar habits or life forms often do not occur in the same places. If they do occur in the same places, they use different food, are active at different times, or are otherwise occupying somewhat different niches. The concept of the ecological niche will be more fully defined in the next chapter; but we need to point out here that the niche involves not only the physical space occupied by an organism, but also its place in the community, including its energy source and period of activity. No two species can have exactly the same niche, of course, and still be different, but species, especially if they are closely related (and hence have similar morphological and physiological characteristics), are often so similar that they have virtually the same niche requirements. Also, severe competition may occur where niches overlap to a partial extent. Experimental and observational research has shown that in a high proportion of cases there is one species to a niche. The explanation for the widely observed ecological separation of closely related (or otherwise similar) species has come to be known as the *competitive exclusion principle* (Hardin, 1960), or *Gause's principle* (after the Russian biologist who first confirmed the principle experimentally; see Figure 7–28).

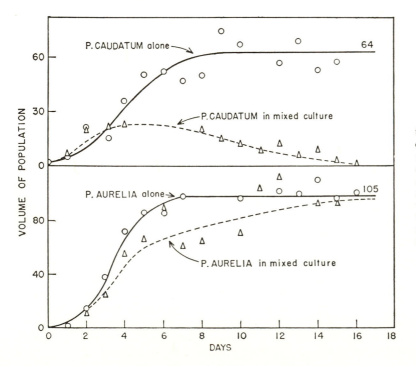

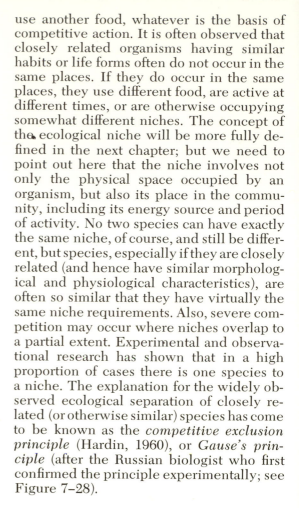

Figure 7–28. Competition between two closely related species of protozoa that have similar niches. When separate, *Paramecium caudatum* and *Paramecium aurelia* exhibit normal sigmoid growth in controlled cultures with constant food supply; when together, *P. caudatum* is eliminated. (From Allee *et al.*, 1949, after Gause.)

Some of the most widely debated theoretical aspects of competition theory revolve around what have become known as the Lotka-Volterra equations (so-called because the equations were proposed as models by Lotka and Volterra in separate publications in 1925–1926). These equations consist of a pair of differential equations similar to the one outlined in the preceding section. Such equations are useful for modelling predator-prey, parasite-host, competition, or other two-species interactions. In terms of competition within a limited space where each population has a definite K, or equilibrium level, the simultaneous growth equations can be written in the following forms:

$$\frac{dN_1}{dt} = r_1 N_1 \frac{K_1 - N_1 - \alpha N_2}{K_1}$$

$$\frac{dN_2}{dt} = r_2 N_2 \frac{K_2 - N_2 - \beta N_1}{K_2}$$

where N_1 and N_2 are number of species 1 and 2 respectively, α is the competition coefficient indicating the inhibitory effect of species 2 on 1, and β is the corresponding competition coefficient signifying the inhibition of 2 by 1. Note that the logistic equation has again been used as a basis for the model. The result of this interaction is that the species with the greatest inhibitory effect on the other will eliminate it from the space, although as Slobodkin (1962) has shown the two species might theoretically coexist if the competition coefficients were very small in relation to the ratios of saturation densities (K_1/K_2 and K_2/K_1).

To understand competition we need not only consider conditions and population attributes that lead to competitive exclusion, but also situations under which similar species coexist, since large numbers of species do share common resources in the open systems of nature (especially in mature, stable ecosystems). In Table 7–9 and Figure 7–29 we present what might be called the *"Tribolium-Trifolium"* model, which includes an experimental demonstration of exclusion in paired species of beetles (*Tribolium*) and one of coexistence in two species of clover (*Trifolium*).

One of the most thorough-going and long-term experimental studies of interspecific competition is that carried out in the laboratory of Dr. Thomas Park of the University of Chicago. Park, his students, and associates work with flour beetles, especially those belonging to the genus *Tribolium* (for a brief review of this work, see Park, 1962). These small beetles can complete their entire life history in a very simple and homogeneous habitat, namely, a jar of flour or wheat bran. The medium in this case is both food and habitat for larvae and adults. If fresh medium is added at regular intervals, a population of beetles can be maintained for a long time. In energy flow terminology (Chapter 3) this experimental setup may be described as a stabilized heterotrophic ecosystem in which imports of food energy

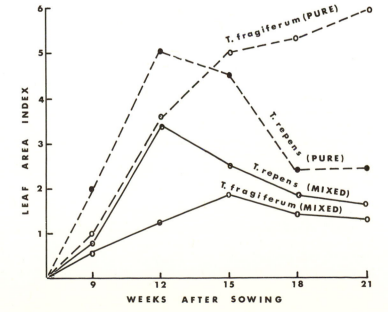

Figure 7–29. The case for coexistence in populations of clover (*Trifolium*). The graph shows population growth of two species of clover in pure (i.e., growing alone) and in mixed stands. Note that the two species in pure stands have a different growth form, reaching maturity at different times. Because of this and other differences, the two species are able to coexist in mixed stands but at reduced density even though they interfere with one another. Leaf area index, which is used as an index of biomass density, is the ratio of leaf surface area to soil surface area (cm² per cm²); see Chapter 3, page 48, and Chapter 14, page 376. (Redrawn from Harper and Chatsworth, 1963.)

*Table 7–9. The Case For Competitive Exclusion in Populations of Flour Beetles (Tribolium)**

CLIMATE	TEMPER-ATURE (°C.)	RELATIVE HUMID-ITY (%)	RESULTS OF INTERSPECIFIC COMPETITION (%)†	
			Tribolium castaneum Wins	*Tribolium confusum* Wins
Hot-wet	34	70	100	0
Hot-dry	34	30	10	90
Warm-wet	29	70	86	14
Warm-dry	29	30	13	87
Cool-wet	24	70	31	69
Cool-dry	24	30	0	100

* Data from Park, 1954.

† Twenty to 30 replicate experiments for each of the six conditions. Each species can survive at any of the climates when alone in the culture, but only one species survives when both are present in the culture. The percentages indicate the proportion of replicates in which each species remained after elimination of the other species.

balance respiratory losses. It is a highly artificial ecosystem, but one of practical interest in regard to insect damage to stored grain that is to be used for human food.

The investigators found that when two different species of *Tribolium* are placed in this homogeneous little universe, invariably one species is eliminated sooner or later while the other continues to thrive. One species always "wins," or to put it another way, two species of *Tribolium* cannot survive in this particular ecosystem which, by definition, contains only one niche for flour beetles. The relative number of individuals of each species originally placed in the culture (i.e., the stocking rate) does not affect the eventual outcome, but the "climate" imposed on the ecosystem does have a great effect on which species of the pair wins out. As shown in Table 7–9, one species (*T. castaneum*) always wins under conditions of high temperature and humidity, while the other (*T. confusum*) always wins under "cool-dry" conditions, even though either species can live indefinitely at any of the six climates provided it is alone in the culture. Under intermediate conditions each species has a certain probability of winning (for example, the probability is 0.86 that *T. castaneum* would win under "warm-wet" conditions). Population attributes, as measured

in one-species cultures, help explain some of the outcome of the competitive action. For example, the species with highest rate of increase (r) under the conditions of existence in question was usually found to win if the species difference in r was rather large. If growth rates differed only moderately, the one with the highest rate did not always win. The presence of a virus in one population could easily tip the balance. Also, genetic strains within the population differ greatly in "competitive ability."

From the *Tribolium* model it is easy to construct conditions that could result in coexistence instead of exclusion. If the cultures were alternately placed in hot-wet and cool-dry conditions (to simulate seasonal weather changes), the advantage one species would have over the other might not continue long enough for extinction of either. If the culture system was "open" and individuals of the dominant species were to immigrate (or be removed as by a predator) at a considerable rate, the competitive interaction might be so reduced that both species could coexist. One can think of many other circumstances that would favor coexistence.

Some of the most interesting experiments in plant competition are those reported by J. L. Harper and associates at the University College of North Wales (see Harper, 1961; Harper *et al.*, 1963; Chatsworth and Harper, 1962). The results of one of these studies, as shown in Figure 7–29, illustrate how a difference in growth form allows two species of clover to coexist in the same environment (i.e., same light, temperature, soil, and so forth). Of the two species, *Trifolium repens* grows faster and reaches a peak in leaf density sooner. However, *T. fragiferum* has longer petioles and higher leaves and is able to overtop the faster growing species, especially after *T. repens* has passed its peak, and thus avoids being shaded out. In mixed stands, therefore, each species inhibits the other, but they both are able to complete their life cycle and produce seed, even though each coexists at a reduced density (however, the combined density in mixed stands of two species was approximately equal to the density in pure stands). In this case the two species, although competing strongly for light, were able to coexist because of differences in morphology and the timing of growth maxima. Harper (1961) concludes that two species of plants can persist together if the populations are independently controlled by one or more of the following mechanisms: (1) different nutritional requirements (legume and nonlegume, for example), (2) different causes of mortality (differential sensitivity to grazing, for example), (3) sensitivity to different toxins, and (4) sensitivity to the same controlling factor (light, water, and so forth) at different times (as in case of the clover example just described).

Examples

The results of one of Gause's original experiments are illustrated in Figure 7–28. This is, we might say, a "classic" example of competitive exclusion. Two closely related ciliate protozoans, *Paramecium caudatum* and *Paramecium aurelia,* when in separate culture exhibited typical sigmoid population growth and maintained a constant population level in culture medium that was maintained constant with a fixed density of food items (bacteria which did not themselves multiply in the media and thus could be added at frequent intervals to keep food density constant). When both protozoans were placed in the same culture, however, *P. aurelia* alone survived after 16 days. In this case neither organism attacked the other or secreted harmful substances; *P. aurelia* populations simply had more rapid growth rate (higher intrinsic rate of increase) and thus "out-competed" *P. caudatum* for the limited amount of food under the existing conditions. On the other hand, *Paramecium caudatum* and *Paramecium bursaria* were both able to survive and reach a stable equilibrium in the same culture medium because although they were competing for the same food, *P. bursaria* occupied a different part of the culture where it could feed on bacteria without competing with *P. caudatum.* Thus, the habitat feature of the niche of these two species proved to be sufficiently different, even though their food was identical. Other examples could be cited in which two closely related species exist in the same habitat but take different food (see Figure 7–32), thus resulting in equally effective separation of niches.

An extensive series of experiments have been completed by Frank (1952 and 1957), who used species of cladocera (water fleas). As may be seen in Figure 7–30, one species eliminated the other when very closely related forms belonging to the same genus

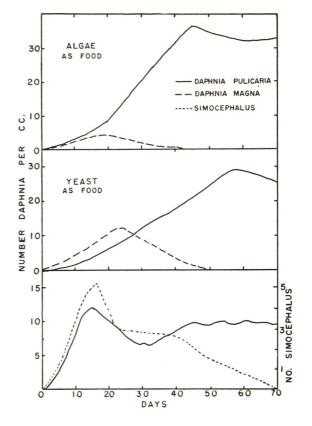

Figure 7–30. Competition between paired species of cladocera in cultures. *Daphnia pulicaria* eliminates the closely related *D. magna,* but with yeast as food for both, *D. magna* persists for a longer period since this food is less favorable for the dominant species. Competition is less severe between *Daphnia* and *Simocephalus,* which have overlapping niches. In mixed culture both species populations undergo normal growth form and persist together for 40 days, after which *Simocephalus* is gradually eliminated. (Upper two graphs redrawn from Frank, 1957; lower graph redrawn from Frank, 1952.)

were matched; however, the pattern of interaction varied with food. When yeast was used (which would support either species when grown separately), the losing species had a temporary advantage and was able to increase for about 20 days. When less similar species were compared, such as those belonging to different genera, both populations underwent more or less normal sigmoid growth in mixed culture and maintained an upper asymptote level for about 30 days (Fig. 7–30, lower graph). Gradually, thereafter, the *Simocephalus* population declined and disappeared from the culture. This case appears to be a good example of overlapping niches. The *Cladocera* experiments thus gave results intermediate between the *Tribolium-Trifolium* model (Table 7–9 and Figure 7–29). In nature, where competition would not likely be as severe, nor as long-continued, as in the small laboratory cultures, the two species would have less difficulty coexisting in the same habitat.

Another example of how habitat diversification can reduce competition so as to allow coexistence instead of exclusion is described by Crombie (1947). He found that *Tribolium* exterminates *Oryzaephilus* (another genus of flour beetle), when both live together in flour, because *Tribolium* is more active in destroying immature stages of other species. However, if glass tubes are placed in the flour into which immature stages of the smaller *Oryzaephilus* may escape, then both populations survive. Thus, when a simple "one-niche" environment is changed to a "two-niche" environment, competition is reduced sufficiently for the support of two species. This is also an example of the "direct interference" type of competition (Table 7–8).

So much for laboratory examples. It is readily conceded that crowding may be greater in laboratory experiments and, hence, competition exaggerated. Interspecific competition in plants in the field has been much studied and is generally believed to be an important factor in bringing about a succession of species (as will be described in Chapter 9). Keever (1955) describes an interesting situation in which a species of tall weed that occupies first year fallow fields in almost pure stands was gradually replaced in these fields by another species previously unknown in the region. The two species, although belonging to different genera, have very similar life histories (time of flowering, seeding) and life forms and were thus brought into intense competition.

Griggs (1956) has made an interesting study of plant competition on a rocky mountain fellfield where most of the species grew in isolated tussocks. Tussocks of one species were often invaded by other species. Griggs was able to prepare a list of species in the order of their ability to invade other tussocks. This "competitive ladder" did not prove to be the same as the order of succession because ability to invade and ability to completely replace were not entirely correlated.

We have already noted that competition between individuals of the same species is one of the most important density-dependent factors in nature, and the same can be said of interspecific competition. Competition appears to be extremely important in determining the distribution of closely related species, and Gause's rule—no two species in the same niche—seems to hold as well for field as for laboratory, although much of the evidence is circumstantial. In nature, closely related species, or species that have very similar requirements, usually occupy different geographical areas or different habitats in the same area or otherwise avoid competition by differences in daily or seasonal activity or in food. Competitive interaction can cause morphological changes (through natural selection) that enhance ecological separation (see the concept of "character displacement," Section 3, Chapter 8). For example, in middle Europe six species of titmice (small birds of the genus *Parus*) coexist, segregated partly by habitat

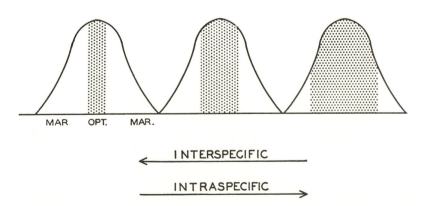

Figure 7–31. The effect of competition on the habitat distribution of birds. When intraspecific competition dominates, the species spreads out and occupies less favorable (marginal) areas; where interspecific competition is intense the species tends to be restricted to a narrower range comprising the optimum conditions. (Modified from Svardson, 1949.)

and partly by feeding areas and size of prey, which is reflected in small differences in length and width of the bill. In North America it is rare for more than two species to be found in the same locality, even though seven species are present on the continent as a whole. Lack (1969) suggests that "the American species of tits are at an earlier stage in their evolution than the European, and their differences in beak, body size, and feeding behavior are adaptations to their respective habitats, and are not yet adaptations for permitting coexistence in the same habitat." The idea that evolutionary development favors coexistence and diversity will be considered again in Chapter 9.

The general theory regarding the role that competition plays in habitat selection is summarized in Figure 7–31. The curves represent the range of habitat which can be tolerated by the species, with optimum and marginal conditions indicated. Where there is competition with other closely related or ecologically similar species the range of habitat conditions which the species occupies generally becomes restricted to the optimum (i.e., to the most favorable conditions under which the species has an advantage in some manner over its competitors). Where interspecific competition is less severe, then intraspecific competition generally brings about a wider habitat choice. Islands are good places to observe the tendency for wider habitat selection to occur when potential competitors fail to colonize. For example, meadow voles (*Microtus*) often occupy forest habitats on islands where their forest competitor, the red-backed vole (*Clethrionomys*), is absent (see Cameron, 1964). Crowell (1962) found that the cardinal was more abundant and occupied more marginal habitat in Bermuda where many of its mainland competitors are absent.

There are many cases which seem at first to be exceptions to Gause's rule but which, on careful study, prove otherwise. A good example of this is the case of two similar fish-eating birds of Britain, the cormorant (*Phalacrocorax carbo*) and the shag (*P. aristotelis*) studied by Lack (1945). These two species commonly feed in the same waters and nest on the same cliffs, yet close study showed that actual nest sites were different and, as shown in Figure 7–32, the food was basically different. Thus the shag feeds in the upper waters on free-swimming fish and eels, whereas the cormorant is more of a bottom feeder, taking flatfish (flounders) and bottom invertebrates (shrimp, etc.).

Just because closely related species are sharply separated in nature does not, of course, mean that competition is actually operating continuously to keep them separated; the two species may have evolved different requirements or preferences which effectively keep them out of competition. A single example each from the plant and animal kingdom will suffice to illustrate. In Europe, one species of *Rhododendron*, namely, *R. hirsutum*, is found on calcareous soils while another species, *R. ferrugineum*, is found on acid soils. The requirements of the two species are such that neither can live at all in the opposite type of soil so that there is never any actual competition between them (see Braun-Blanquet, 1932). Teal (1958) has made an experimental study of habitat selection of species of fiddler crabs (*Uca*) which are usually separated in their occurrence in salt marshes. One species, *U. pugilator*, is found on open sandy flats, while another, *U. pugnax*, is found on muddy substrates covered with marsh grass. Teal found that one species would tend not to invade the habitat of the other even in the absence of the other, because each species

Figure 7–32. Food habits of two closely related species of aquatic birds, the cormorant (*Phalacrocorax carbo*) and the shag (*P. aristotelis*), which are found together during the breeding season. Food habits indicate that although the habitat is similar the food is different; therefore, the niche of the two species is different and they are not actually in direct competition. (Data from Lack, 1945.)

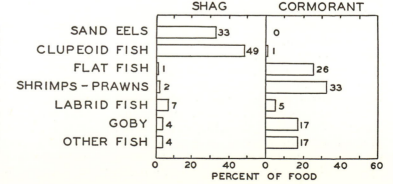

would dig burrows only in its preferred substrate. The absence of active competition, of course, does not mean that competition in the past is to be ruled out as a factor in bringing about the isolating behavior.

We may do well to close this discussion of examples by adopting the three tentative models proposed by Philip (1955) as a basis for future observation, analysis, and experimentation. These are: (1) imperfect competition, where interspecific effects are less than intraspecific effects; interspecific competition is a limiting factor but not to the extent of complete elimination of one species; (2) perfect competition as in the unmodified Gause or Lotka-Volterra model in which one species is invariably eliminated from the niche by a gradual process as crowding occurs; and (3) hyperperfect competition in which the depressing effects are great and immediately effective, as in the production of antibiotics. A striking example of direct interference or "exaggerated" competition in plants will be given in the next section.

18. NEGATIVE INTERACTIONS: PREDATION, PARASITISM, AND ANTIBIOSIS

Statement

As already indicated, predation and parasitism are examples of interactions between two populations which result in negative effects on the growth and survival of one of the populations (negative term in growth equation of one of the populations, see Table 7–8, types 5 and 6). A similar result occurs when one population produces a substance harmful to a competing population. The term *antibiosis* is commonly used for such interaction, and the term *allelopathy* (= harmful to the other) has been proposed for chemical inhibition by plants (Muller, 1966).

A cardinal principle is that the negative effects tend to be quantitatively small where the interacting populations have had a common evolutionary history in a relatively stable ecosystem. In other words, natural selection tends to lead to reduction in detrimental effects or to the elimination of the interaction altogether, since continued severe depression of a prey or host population by the predator or parasite population can

only lead to the extinction of one or both populations. Consequently, severe interaction is most frequently observed when the interaction is of recent origin (when two populations first become associated) or when there have been large-scale or sudden changes (perhaps temporary) in the ecosystem (as might be produced by man). This leads to what we might call "the principle of the instant pathogen," which explains why man's frequent unplanned or ill-planned introductions or manipulations so often lead to epidemics.

Explanation

It is difficult to approach the subject of parasitism and predation objectively. We all have a natural aversion to parasitic organisms, whether bacteria or tapeworms. Likewise, although man himself is the greatest predator the world has known (and also the greatest perpetrator of epidemics in nature), he tends to condemn all other predators without bothering to find out if they are really detrimental to his interests or not. The idea that "the only good hawk is a dead hawk" is widely held, but, as we shall see, it is by no means a true generalization.

The best way to be objective is to consider predation and parasitism from the population rather than from the individual standpoint. Predators and parasites certainly kill and injure individuals, and they depress in some measure at least the growth rate of populations or reduce the total population size. But does this always mean that populations would be better off without predators or parasites? From the long-term view, are predators and parasites the sole beneficiaries of the association? As we have pointed out in the discussion of population regulation (Section 10, see especially Figure 7–20), predators and parasites play a role in keeping herbivorous insects at a low density, but they may be ineffective when the host population erupts or "escapes" from density-dependent control. Deer populations are often cited as examples of populations that tend to erupt when predator pressure is reduced. A widely cited example of this is the Kaibab deer herd, which, as originally described by Leopold (1943) based on estimates by Rasmussen (1941), allegedly increased from 4000 (on 700,000 acres on the north side of the Grand Canyon in Arizona) in 1907 to 100,000 in 1924, coin-

cident with an organized government predator removal campaign. Caughley (1970) has
reexamined this case and concludes that
while there is no doubt that deer did increase, overgraze, and then decline, there
is doubt about the extent of the overpopulation and no real evidence that it was due
solely to removal of predators; cattle and fire
may have also played a part. He believes
that eruptions of ungulate populations are
more likely to result from changes in habitat
or food that enable the population to "escape" from the usual mortality control (as
in the model in Figure 7–20). One thing is
clear: the most violent eruptions occur
when a species is introduced into a new
area where there are both unexploited resources *and* a lack of negative interactions.
The population explosion of rabbits introduced into Australia is, of course, a well-
known example among the literally thousands of cases of severe oscillations directly
caused by man.

We now come to the most important generalization of all, namely, that negative
interactions become less negative with time
if the ecosystem is sufficiently stable and
spatially diverse to allow reciprocal adaptations. As already indicated in Section 9,
simple parasite-host or predator-prey populations introduced into experimental microecosystems oscillate violently with a certain
probability of extinction. The Lotka-Volterra
equation models of predator-prey interaction produce a perpetual undamped oscillation *unless second-order terms are added
that induce self-limitations capable of
dampening the oscillation* (see Lewontin,
1969). Pimentel (1968; Pimentel and Stone,
1968) has shown experimentally that such
second-order terms can take the form of
reciprocal adaptations, or genetic feedback.
As shown in Figure 7–33, violent oscillations occur when a host, the house fly, and
a parasitic wasp are first placed together in
a limited culture system. When individuals

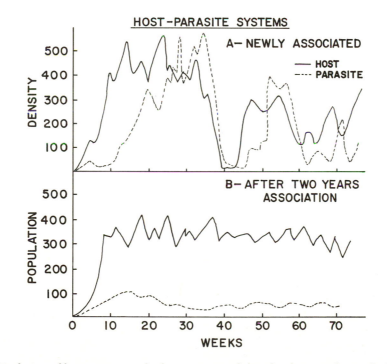

Figure 7–33. Evolution of homeostasis in the host-parasite relationship between house fly (*Musca domestica*)
and parasitic wasp (*Nasonia vitropennis*) populations in a laboratory multicell population cage consisting of 30
plastic boxes interconnected with tubes especially designed to slow down parasite dispersal. *A.* Newly associated
populations (wild stocks brought together for the first time) oscillated violently, as first the host (fly) and then the
parasite (wasp) density increased and "crashed." *B.* Populations derived from colonies in which the two species
had been associated for two years coexisted in a more stable equilibrium without "crashes." The adaptive resistance that had evolved in the host was indicated by the fact that the natality of the parasite was greatly reduced
(46 progeny per female as compared to 133 in the newly associated system), and the parasite population leveled
off at a low density. The experiment demonstrates how genetic feedback can function both as a regulatory and
stabilizing mechanism in population systems. (Redrawn from Pimentel and Stone, 1968; the lower graph (*B*) is
a composite of two experimental populations as shown in their Figures 2 and 3. Density figures are number per
cell in the 30-cell cage.)

selected from cultures that had managed to survive the violent oscillations for two years were then reestablished in new cultures, it was evident that an ecological homeostasis had evolved in which both populations had "powered down," so to speak, and were now able to coexist in a much more stable equilibrium.

In the real world of man and nature, time and circumstances may not favor such reciprocal adaptation by new associations, so that there is always the danger that the negative reaction may be irreversible in that it leads to the extinction of the host. The story of the chestnut blight in America is a case in which the question of adaptation or extinction hangs in the balance, and there is little man can do but observe.

Originally, the American chestnut was an important member of the forests of the Appalachian region of eastern North America. It had its share of parasites, diseases, and predators. Likewise, the oriental chestnut trees in China—a different but related species—had their share of parasites, and so on, including the fungus *Endothia parasitica*, which attacks the bark of the stems. In 1904 the fungus was accidentally introduced into the United States. The American chestnut

proved to be unresistant to this new parasite; finally, by 1952 all the large chestnuts had been killed, their gaunt gray trunks being a characteristic feature of Appalachian forests (Fig. 7–34). The chestnut continues to sprout up from the roots, and such sprouts may produce fruits before they are killed back, but no one can say whether the ultimate outcome will be complete extinction or adaptation. For all practical purposes the chestnut has been removed, for the time being at least, as a major influence in the forest.

The above examples are not just cases hand-picked to "prove a point." If the student will do a little reading in the library, he can find hundreds of similar examples which show: (1) that where parasites and predators have long been associated with their respective hosts and prey, the effect is moderate, neutral, or even beneficial from the long term view, and (2) that newly acquired parasites or predators are the most damaging. In fact, if one makes a list of the diseases, parasites, and insect pests that cause the greatest loss to agriculture or are most pathogenic to man himself, the list will include a large number of species which have recently been introduced into a new

Figure 7–34. Results of the chestnut blight in the southern Appalachian region (Georgia), an example of the extreme effect which a parasitic organism (fungus) introduced from the Old World may have on a newly acquired host (American chestnut tree).

area, as in the case of the chestnut blight, or acquired a new host or prey.

The *principle of the instant pathogen*, which was briefly mentioned in the Statement, can now be rephrased as follows:

Pathogenicity or pestilence is often induced by (1) sudden or rapid introduction of an organism with a potentially high intrinsic rate of increase into an ecosystem in which adaptive control mechanisms for it are weak or lacking, or (2) by abrupt or stressful environmental changes which reduce the energy available for feedback control, or otherwise impair the capacity for self-control. Man is prone to produce "instant pests" because he "introduces" and "disturbs" (often inadvertently) on a large scale and at rapid rates that do not allow time for complex adjustments to take place.

The lesson for man, of course, is to beware of new negative interactions and to avoid sponsoring new ones any faster than is absolutely necessary.

Although predation and parasitism are similar from the ecological standpoint, the extremes in the series, the large predator and the small internal parasite, do exhibit important differences other than size. Parasitic or pathogenic organisms usually have a higher biotic potential than do predators. They are often more specialized in structure, metabolism, host specificity, and life history, as is necessitated by their special environment and the problem of dispersal from one host to another.

Of special interest are organisms which are intermediate between predators and parasites, for example, the so-called parasitic insects. These forms often have the ability of consuming the entire prey individual, as does the predator, and yet they have the host specificity and high biotic potential of the parasite. Man has been able to propagate some of these organisms artificially and utilize them in the control of insect pests. Now that man has learned from sad experience that chemical control of insects has serious limitations, control by use of parasites will become more important in the future. In general, attempts to make similar use of large unspecialized predators have not been successful. Can you give reasons why this has been the case?

On the more positive side man is slowly learning how to be what Slobodkin (1962) calls a "prudent predator," that is, one who does not exterminate his prey by overexploitation. As we shall see in Chapter 10, mathematical models can help determine the rate of harvest of game and fish populations that will maintain an optimum sustained yield. The problem of the optimum yield is discussed by Beverton and Holt (1957), Ricker (1958), Menshitkin (1964), Slobodkin (1968), Silliman (1969), Wagner (1969), and many others. Theoretically, if sigmoid growth is symmetrical, as in the logistic model, the highest growth rate, dN/dt, occurs when density is $K/2$ (one-half the saturation density). However, growth form is often skewed so that the peak of the hump-backed or parabolic growth rate curve may not be midway between 0 and K. According to Wagner (1970), the maximum sustained yield density is often somewhat less than half the unexploited equilibrium density.

The problem can be attacked experimentally by setting up test populations in micro-ecosystems. One such experimental model is shown in Figure 7–35 in which guppies (*Lebistes reticulatus*), small aquarium fish, were used to "mimic" a commercial fish population being exploited by man. As shown, the maximum sustained yield was obtained when one-third of the population was harvested each reproductive period, which reduced the equilibrium density to slightly less than half the unexploited one. Within the limits of the experiment these ratios tended to be independent of the carrying capacity of the system, which was varied at three levels by manipulating the food supply.

One-species models often prove to be oversimplifications because they do not take into consideration competing species that may respond to the reduced density of the harvested species by increasing their density and using up food or other resources needed to sustain the exploited species. It is very easy for a "top predator" such as man (or a major grazer such as the cow) to tip the balance in a competitive equilibrium so that the exploited species is replaced by another species that the predator or grazer may not be prepared to utilize. In the real world, examples of such shifts (as will be noted below) are being documented with increasing frequency as man becomes more "efficient" as fisherman, hunter, and harvester of plants. This leads us to issue a statement that is both challenge and warning: *One species harvest systems, as well as monocultural systems* (such as one-crop agriculture), *are inherently unstable,* because when stressed they are vulnerable to competition, disease, parasitism, predation, or other negative interactions. As was pointed out in Chapter 3, optimum yield may be less than the maximum when the cost of maintaining "order" in inherently unstable systems is considered.

The stress of predation or harvest often affects the size of individuals in the exploited populations. Thus, harvesting at the maximum sustained yield-level usually results in reducing the average size of fish, just as maximizing timber yields for volume of wood reduces the size of tree and quality of wood. As reiterated many times in this volume, one cannot have maximum quality and quantity at the same time. Brooks and Dodson (1965) describe how large species of zooplankton are replaced by smaller species when zooplankton-feeding fish are introduced into lakes that formerly lacked such direct predators. In this case in which the ecosystem is relatively small, both the size and species composition of a whole trophic level may be controlled by one or a few species of predators.

These considerations bring us back, as always, to the role of negative interactions in the ecosystem as a whole. As was brought out earlier (Section 10) the importance of biotic control is a function of position in the physical gradient in which ecosystems develop. In Figure 7–36 we present a "barnacle model" based on the experimental studies of J. H. Connell. The intertidal zone on a rocky seacoast provides a miniature gradient from physically stressed to more biologically controlled environment (see Chapter 12, page 337, for a description of this habitat). In his study in Scotland, Connell (1961) found that the larvae of two species of barnacles settled over a wide range of the intertidal zone but survived as adults in a much more restricted range. The larger species (*Balanus*) was found to be restricted to the lower part of the zone because it was unable to tolerate long periods of exposure, while the smaller species (*Chthamalus*) was excluded from the lower zone by competition with the larger species, and by predators that are more active below the high tide mark. As shown in Figure 7–36, the physical stress of desiccation was the main controlling factor in the upper part of the gradient, while interspecific competi-

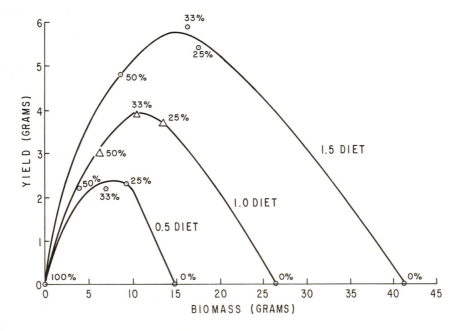

Figure 7–35. Biomass and yield in test populations of the guppy exploited at different rates (shown as per cent removal per reproductive period) at three different diet levels. The highest yields were obtained when about one-third of the population was harvested per reproductive period and mean biomass reduced to less than one-half of that of the unexploited population (yield curves skewed to the left). (After Silliman, 1969.)

tion and predation were controlling factors in the lower zones. This model can be considered to apply to more extensive gradients such as an arctic-to-tropic or a high-to-low altitude gradient provided one bears in mind that all "models" are to varying degrees oversimplifications.

No discussion of negative interactions would be complete without mentioning a brief but provocative paper by Hairston, Smith, and Slobodkin (1960), who suggest that population control at the herbivore (primary consumer) trophic level is fundamentally different from that at other trophic levels. In their words: "Populations of producers, carnivores, and decomposers are limited by their respective resources in the classical density-dependent fashion," and "interspecific competition must necessarily exist among members of these three groups." In contrast, "herbivores are seldom food-limited, appear most often to be predator-limited, and therefore are not likely to compete for common resources." This theory is based on the general observation that (1) plants in natural ecosystems are but lightly grazed (thus allowing them to build up biomass structure), and (2) many species of herbivores coexist without evident competition exclusion (see Ross, 1957; Broadhead, 1958). Undoubtedly, the generalization is far too sweeping, but it has stimulated study and argument and will un-

doubtedly continue to do so for some time to come.

Examples

The following sequence of studies illustrates how interactions between competitors and predators (or parasites) affect density and diversity, with special reference to "man the predator."

1. Larkin (1963) by computer simulation showed how different levels of exploitation will alter equilibrium levels of coexisting species.

2. Slobodkin (1964) showed that experimental removal (i.e., predation by the experimentor) of *Hydra* in two-species laboratory cultures prevented density from reaching exclusion levels, thus enabling two species to coexist where only one could do so in the absence of predation.

3. Paine (1966) found that experimental removal of predators on intertidal rocks (where space is limited) greatly reduced the diversity of the herbivores (i.e., algae grazers) because interspecific competition intensified to the point of exclusion; he predicted that in the continued absence of predators the number of species would eventually be reduced to one, as in the *Tribolium* model (a case in which reduction in grazing pressure resulted in a decrease

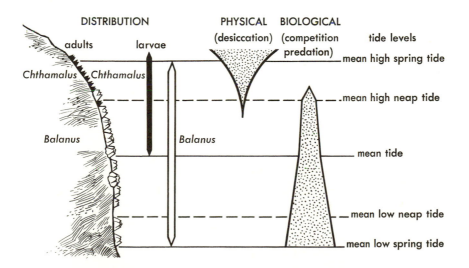

Figure 7–36. Factors that control the distribution of two species of barnacles in an intertidal gradient. The young in each species settle over a wide range but survive to adulthood in a more restricted range. Physical factors such as desiccation control upward limits of *Balanus*, while biological factors such as competition and predation control downward distribution of *Chthamalus*. (After Odum, 1963; from Connell, 1961.)

in diversity of plants was noted on page 151).

4. Murphy (1966 and 1967) diagnosed the decline of the heavily fished Pacific sardine (*Sardinops caerulea*) and the subsequent increase of the anchovy (*Engraulis mordox*), which has a very similar ecology, as a case of replacement of an overexploited species by a competing species that was not being exploited. As shown in Figure 7–37, the increase in anchovies parallels a predicted increase based on simple competition equations. This is a good example of the failure of a one-species fishery. Even though the sardine is no longer fished extensively, the shift may be permanent since the anchovy now has the upper hand in the competition (exploitation of the latter might, of course, shift the balance back again).

5. Smith (1966) describes how a succession of species-specific exploitations combined with introductions and eutrophication have resulted in successive rises and falls of commercial fish in Lake Michigan. First there was the lake trout that supported a stable fishery for half a century, but this was virtually eliminated by the combined assault of overharvest, attack by the introduced parasitic lamprey, and eutrophication. Then in rapid succession lake herring, lake whitefish, chubs, and the exotic alewife exhibited population growth and decline as each in turn was exploited and gave way under the pressure of a competitor, predator, or parasite. Recently coho salmon have been introduced and are thriving on a diet of alewifes, much to the delight of the sports fisherman. One can predict that this bonanza will soon run its course unless harvest and pollution can come under better control.

A final example of negative interaction will be taken from the work of C. H. Muller and associates who investigated allelopathic, or antibiotic, inhibitors produced by shrubs in the California chaparral (see page 396 for an account of this community type). These investigators have not only looked into the chemical nature and physiological action of the inhibitory substances, but they have also shown that these play an

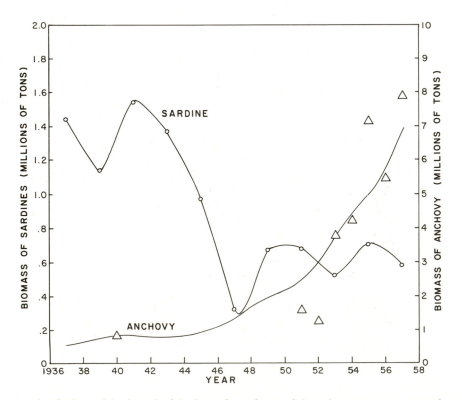

Figure 7–37. The decline of the heavily fished Pacific sardine and the subsequent increase in the unexploited anchovy (as ecological equivalent competitor). For the sardine, the points connected with the line represent actual population estimates. For the anchovy, the triangles are population estimates, and the curve was generated by analog computer from Volterra competition equations. (After Silliman, 1969; data from Murphy, 1964, and Murphy and Isaacs, 1964.)

important part in regulating the composition and dynamics of the community (see Muller, 1966 and 1969; Muller *et al.*, 1964 and 1968). Figure 7–38 shows how volatile terpenes produced by two species of aromatic shrubs inhibit the growth of herbaceous plants. The volatile toxins (notably cineole and camphor) are produced in the leaves and accumulate in the soil during the dry season to such an extent that when the rainy season comes, germination or subsequent growth of seedlings is inhibited in a wide belt around each shrub group. Other shrubs produce water soluble antibiotics of

Figure 7–38. *Upper:* Aerial view of aromatic shrubs *Salvia leucophylla* and *Artemisia californica* invading an annual grassland in the Santa Inez Valley of California and exhibiting biochemical inhibition. *Lower:* Closeup showing the zonation effect of volatile toxins produced by *Salvia* shrubs seen to the center-left of *A*. Between *A* and *B* is a zone two meters wide bare of all herbs except for a few minute, inhibited seedlings (the root systems of the shrubs which extend under part of this zone are thus free from competition with other species). Between *B* and *C* is a zone of inhibited grassland consisting of smaller plants and fewer species than in the uninhibited grassland seen to the right of *C*. (Photos courtesy of Dr. C. H. Muller, University of California, Santa Barbara.)

a different chemical nature (phenols and alkaloids, for example), which also favor shrub dominance. However, periodic fires, which are an integral part of the chaparral ecosystem, effectively remove the source of the toxins, denaturing those accumulated in the soil, and triggering the germination of fire-adapted seeds. Accordingly, fire is followed in the next rainy season by a conspicuous blooming of annuals which continue to appear each spring until shrubs grow back and the toxins again become effective. Very few herbs are left in the mature chaparral. The interaction of fire and antibiotics thus perpetuates cycle changes in composition that are the adaptive feature of this type of ecosystem. Whittaker (1970), in a review of botanical inhibitors, concludes: "Higher plants synthesize substantial quantities of substances repellent or inhibitory to other organisms. Allelopathic effects have a significant influence on the rate and species sequence of plant succession and on species composition of stable communities. Chemical interactions affect species diversity of natural communities in both directions; strong dominance and intense allelopathic effects contribute to low species diversity of some communities, whereas variety of chemical accommodations are part of the basis (as aspects of niche differentiation) of the high species diversity of others." Antibiosis, of course, is not restricted to higher plants; numerous examples among microorganisms are known, as illustrated by penicillin, the bacterial inhibitor produced by bread mold and now widely used in medicine. Referring to Table 7–8, we see that chemical antibiosis can be considered to be a form of the direct interference type of competition, or in its milder forms, as amensalism.

The role of antibiotics in ecological succession will be considered in Chapter 9.

19. POSITIVE INTERACTIONS: COMMENSALISM, COOPERATION, AND MUTUALISM

Statement

Associations between two species populations which result in positive effects are exceedingly widespread and probably as important as competition, parasitism, and so forth, in determining the nature of populations and communities. Positive interactions may be conveniently considered in an evolutionary series as follows: commensalism—one population benefited; protocooperation—both populations benefited; and mutualism—both populations benefit and have become completely dependent on each other.

Explanation

The widespread acceptance of Darwin's idea of "survival of the fittest" as an important means of bringing about natural selection has directed attention to the competitive aspects of nature. As a result, the importance of cooperation between species in nature has perhaps been underestimated. At least, positive interactions have not been subjected to as much quantitative study as have negative interactions. As in a balanced equation, it seems reasonable to assume that negative and positive relations between populations eventually tend to balance one another if the ecosystem is to achieve any kind of stability.

Commensalism represents a simple type of positive interaction and perhaps represents the first step toward the development of beneficial relations. It is especially common between sessile plants and animals on the one hand and motile organisms on the other. The ocean is an especially good place to observe commensalism. Practically every worm burrow, shellfish, or sponge contains various "uninvited guests," organisms which require the shelter of the host but do neither harm nor good in return. Perhaps you have opened oysters, or been served oysters on the half shell, and observed a small delicate crab in the mantle cavity. These are usually "commensal crabs," although sometimes they overdo their "guest" status by partaking of the host's tissues (Christensen and McDermott, 1958). Dales (1957), in his review of marine commensalism, lists 13 species that live as guests in the burrows of large sea worms (*Erechis*) and burrowing shrimp (*Callianassa* and *Upogebia*). This array of commensal fish, clams, polychaete worms, and crabs lives by snatching surplus or rejected food or waste materials from the host. Many of the commensals are not host-specific but some apparently are found associated with only one species of host.

It is but a short step to a situation in which both organisms gain by an association or interaction of some kind, in which case we have protocooperation. The late W. C. Allee (1938 and 1951) studied and wrote extensively on this subject. He believed that the beginnings of cooperation between species are to be found throughout nature. He was able to document many of them and to demonstrate mutual advantages by experiments. Returning to the sea for an example, crabs and coelenterates often associate with mutual benefit. The coelenterates grow on the backs of the crabs (or are sometimes "planted" there by the crabs), providing camouflage and protection (since coelenterates have stinging cells). In turn, the coelenterates are transported about and obtain particles of food when the crab captures and eats another animal.

In the above instance the crab is not absolutely dependent on the coelenterate, nor vice versa. A further step in the process of cooperation results when each population becomes completely dependent on the other. Such cases have been called *mutualism,* or *obligate symbiosis.* Often quite diverse kinds of organisms are associated. In fact, instances of mutualism are most likely to develop between organisms with widely different requirements. (As we have seen, organisms with similar requirements are more likely to get involved in negative interactions.) The most important examples of mutualism develop between autotrophs and heterotrophs. This is not surprising since these two components of the ecosystem must ultimately achieve some kind of balanced symbiosis. Examples that would be labeled as mutualistic go beyond such general community interdependence to the point at which one particular kind of heterotroph becomes completely dependent on a particular kind of autotroph for food, and the latter becomes dependent on the protection, mineral cycling, or other vital function provided by the heterotroph. The partnership between nitrogen-fixing bacteria and legumes is an example that has been discussed in detail in Chapter 4. Mutualism is also common between microorganisms that can digest cellulose and other resistant plant residues and animals that do not have the necessary enzyme systems for this. Obligate symbiosis between ungulates and

rumen bacteria is an example discussed in Chapter 19. As previously suggested, mutualism seems to replace parasitism as ecosystems evolve toward maturity, and it seems to be especially important when some aspect of the environment is limiting (infertile soil, for example) to an extent that mutual cooperation has a strong selective advantage. The following specific examples will illustrate some of these aspects.

Examples

Obligate symbiosis between cellulose-digesting microorganisms and animals may be illustrated by two examples. The general importance of such mutualism in the detritus food chain was discussed in Chapter 3 (page 72). The termite-intestinal flagellate partnership is a well-studied case, first worked out by Cleveland (1924 and 1926). Without the specialized flagellates (a "cluster" of species of the order Hypermastigina), many species of termites are unable to digest the wood they ingest, as shown by the fact that they starve to death when the flagellates are experimentally removed. The symbionts are so well coordinated with their host that they respond to the latter's molting hormones by encysting, thus insuring transmission and reinfection when the termite molts its gut lining and then ingests it. In this case the symbionts live inside the body of the host, but even more intimate interdependence may develop with the microorganism partners living outside the body of the animal host, and such associations may actually represent a more advanced stage in the evolution of mutualism (less chance that the relationships might revert to parasitism!). An example is the very interesting and well-documented case of tropical attine ants which cultivate fungal gardens in their nests. The ants fertilize, tend, and harvest the fungal crop in much the same manner as would an efficient human farmer. It has recently been shown (Martin, 1970) that the ant-fungal system short-circuits and speeds up the natural decomposition of leaves. As described on page 367, a succession of microorganisms is normally required to decompose leaf litter with basidiomycete fungi normally appearing during the late stages of decomposition. However, when leaves are "fertilized" by

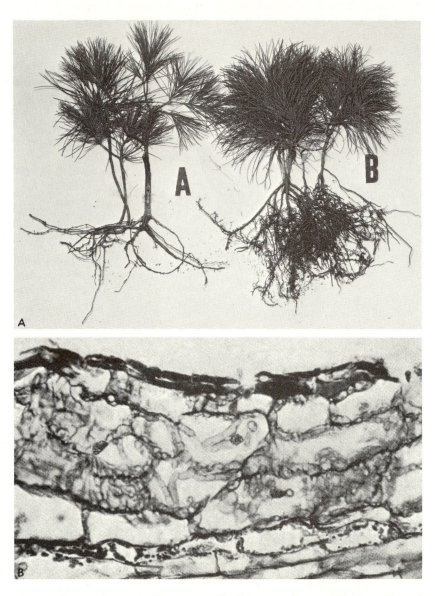

Figure 7–39. Types of mycorrhizae. *A.* Three-year-old white pine (*Pinus strobus*) seedlings devoid of mycorrhizae (left) and with a prolific development of ectotrophic mycorrhizae (right). *B.* Endotrophic mycorrhizae showing fungal mycelia inside cells of the root. [Illustration continued on opposite page.]

Figure 7–39. *C.* Peritrophic (extramatrical) mycorrhizae forming clusters or masses around the roots of a spruce (*Picea pungens*) seedling. *D.* Peritrophic mycorrhizae forming a sheath or dense fungal mantle around roots imbedded in organic litter of a tropical forest. (Photos courtesy of Professor S. A. Wilde, University of Wisconsin.)

ant excreta in the fungal gardens, these fungi are able to thrive on fresh leaves as a rapidly growing monoculture that provides food for the ants (note that a lot of "ant energy" is required to maintain this monoculture just as is required in intensive crop culture by man). Martin summarizes the situation as follows: "By cultivating as a food crop a cellulose-degrading organism the ants gain access to the vast cellulose reserves of the rain forest for indirect use as a nutrient. What termites accomplish by their endo-symbiotic association with cellulose-degrading microorganisms, the attine ants have achieved through their more complex ecto-symbiotic association with a cellulose-degrading fungus. In biochemical terms the contribution of the fungus to the ant is the enzymatic apparatus for degrading cellulose. The fecal material of the ant contains proteolytic enzymes which the fungus lacks so that the ants contribute their enzymatic apparatus to degrade protein. The symbiosis can be viewed as a metabolic alliance in which the carbon and nitrogen metabolisms of the two organisms have been integrated." Coprophagy, or the reingestion of feces, which appears to be characteristic of detritivores (see page 31), can probably be viewed as a much less elaborate but much more widespread case of mutualism that couples the carbon and nitrogen metabolism of microorganisms and animals (i.e., the concept of the "external rumen," which was suggested earlier).

Mineral cycling as well as food production is enhanced by symbiosis between microorganisms and plants. Prime examples are *mycorrhizae* (= fungus-root), comprising the mycelia of fungi which live in mutualistic association with the living roots of plants (not to be confused with parasitic fungi that kill roots). As in the case of nitrogen-fixing bacteria and legumes, the fungi interact with root tissue to form "organs" that increase the ability of the plant to extract minerals from the soil. In return, of course, the fungi are supplied some of the photosynthate by the plant. Mycorrhizae take several forms as illustrated in Figure 7–39:

1. Ectotrophic mycorrhizae, mostly basidiomycetes, which form rootlike extensions that grow out from the cortex of the root (Fig. 7–39A); while conspicuous, these apparently are not the dominant nor most effective type in mineral deficient soils.

2. Endotrophic mycorrhizae, mostly phycomycetes, which penetrate into the cells of the root (Fig. 7–39B); these are widespread in tree roots but are difficult to culture and study.

3. Peritrophic (or extramatrical) mycorrhizae, which form mantles or clusters around the roots (Figs. 7–39C and D), but mycelia do not penetrate the epidermis of the root; these are thought to be very important in creating a favorable chemical "rhizospheric environment" that transforms insoluble, or unavailable, minerals to forms that can be taken up by the roots (see Wilde, 1968).

Many trees will not grow without mycorrhizae. Forest trees transplanted to prairie soil, or introduced into a different region, often fail to grow unless inoculated with the fungal symbionts. Pine trees with healthy mycorrhizal associates grow vigorously in soil so poor by conventional agricultural standards that corn or wheat could not survive. The fungi are able to metabolize "unavailable" phosphorus and other minerals by chelation (see pages 29 and 31 for explanation of this process) or by other means that are as yet not well understood. When labeled minerals (radioactive tracer phosphorus, for example) are added to the soil, as much as 90 per cent may be quickly taken up by the mycorrhizal mass, then slowly released to the plant. The role of mycorrhizae in direct mineral recycle, its importance in the tropics, and the need for crops with such built-in recycle systems were all emphasized in Sections 7 and 8, Chapter 4 (and again in Chapter 14, page 367). For additional information on microorganism-root mutualisms, see Harley (1959), Rovira (1965), and Wilde (1968).

Lichens, as is well known, are an association of specific fungi and algae that is so intimate in terms of functional interdependence and so integrated morphologically that a sort of third kind of organism is formed which resembles neither of its components. Lichens are usually classified as single "species" even though they are composed of two unrelated species. While the components can often be cultured separately in the laboratory, the integrated unit is difficult to culture despite the fact that it is able to exist in nature under harsh conditions (lichens are often the dominant plant on bare rock or in the arctic). Lichens are also interesting because within the group

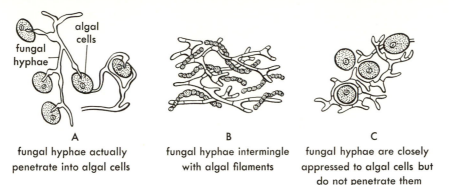

Figure 7–40.   A trend in evolution from parasitism to mutualism in the lichens. In some primitive lichens the fungi actually penetrate the algal cells, as in diagram *A*. In the more advanced species the two organisms live in greater harmony for mutual benefit, as in *B* and *C*. (After Odum, 1963.)

one sees evidence of an evolution from parasitism to mutualism. In some of the more primitive lichens, for example, the fungi actually penetrate the algal cells, as shown in Figure 7–40*A*, and are thus essentially parasites of the algae. In the more advanced species, the fungal mycelia or hyphae do not break into the algal cells, but the two live in close harmony (Fig. 7–40*B* and *C*). The "lichen model" in Figure 7–40 is perhaps a symbolic one for man. Until now man has generally acted as a parasite on his autotrophic environment, taking what he needs with little regard to the welfare of his host. Great cities are planned and grow without any regard for the fact that they are parasites on the countryside which must somehow supply food, water, air, and degrade huge quantities of wastes. Obviously it is time for man to evolve to the mutualism stage in his

relations with nature since he is a dependent heterotroph and his culture is even more dependent and increasingly demanding of resources. If man does not learn to live mutualistically with nature, then, like the "unwise" or "unadapted" parasite, he may exploit his host to the point of destroying himself.

For general reviews of symbiotic associations, see the symposia edited by Nutman and Masse (1963) and Henry (1966). The relationship between man and domesticated plants and animals, which might be considered as a special form of mutualism, is discussed in the next chapter. In Chapter 12 coral-algal associations are discussed as another important case of mutualistic relationships that enhance nutrient recycle and productivity of a whole ecosystem.

THE SPECIES AND THE INDIVIDUAL IN THE ECOSYSTEM

Chapter 8

1. CONCEPTS OF HABITAT AND ECOLOGICAL NICHE

Statement

The *habitat* of an organism is the place where it lives, or the place where one would go to find it. The *ecological niche*, on the other hand, is a more inclusive term that includes not only the physical space occupied by an organism, but also its functional role in the community (as, for example, its trophic position) and its position in environmental gradients of temperature, moisture, pH, soil, and other conditions of existence. These three aspects of the ecological niche can be conveniently designated as the *spatial or habitat niche*, the *trophic niche*, and the *multidimensional* or *hypervolume niche*. Consequently, the ecological niche of an organism depends not only on where it lives but also on what it does (how it transforms energy, behaves, responds to and modifies its physical and biotic environment), and how it is constrained by other species. By analogy, it may be said that the habitat is the organism's "address," and the niche is its "profession," biologically speaking. Since a description of the complete ecological niche for a species would include an infinite set of biological characteristics and physical parameters, the concept is most useful, and quantitatively most applicable, in terms of *differences* between species (or the same species at two or more locations) in one or a few major (i.e., operationally significant) features.

Explanation

The term habitat is widely used, not only in ecology but elsewhere. It is generally understood to mean simply the place where an organism lives. Thus, the habitat of the water "backswimmer," *Notonecta*, is the shallow, vegetation-choked areas (littoral zone) of ponds and lakes; that is where one would go to collect this particular organism. The habitat of a *Trillium* plant is a moist, shaded situation in a mature deciduous forest; that is where one would go to find *Trillium* plants. Different species in the genus *Notonecta* or *Trillium* may occur in the same general habitat but exhibit small differences in location, in which event we would say that the *microhabitat* is different. Other species in these genera exhibit large habitat, or *macrohabitat*, differences.

Habitat may also refer to the place occupied by an entire community. For example, the habitat of the "sand sage grassland community" is the series of ridges of sandy soil occurring along the north sides of rivers in the southern Great Plains region of the United States. Habitat in this case consists mostly of physical or abiotic complexes, whereas habitat as used with reference to *Notonecta* and *Trillium*, mentioned above, includes living as well as nonliving objects. Thus, the habitat of an organism or group of organisms (population) includes other organisms as well as the abiotic environment. A description of the habitat of the community would include only the latter. It is important to recognize these two possible uses of the term habitat in order to avoid confusion.

Ecological niche is a more recent concept and is not so generally understood outside the field of ecology. We had occasion to introduce the concept in the preceding chapter in connection with the discussion of competition (see page 214). Broad, albeit useful, terms, such as niche, are difficult to define and quantitate; the best approach is to consider the component concepts historically. Joseph Grinnell (1917 and 1928) used the word niche "to stand for the concept of the ultimate distributional unit, within which each species is held by its structural and instinctive limitations . . . no two species in the same general territory can occupy for long identically the same ecological niche." (Incidentally, the latter statement predates Gause's experimental demonstration of the competition exclusion principle; see Chapter 7, page 214.) Thus, Grinnell thought of the niche mostly in terms of the microhabitat, or what we would now call the *spatial niche*. Charles Elton (1927 and later publications) in England was one of the first to begin using the term "niche" in the sense of the "functional status of an organism in its community." As a result of his very great influence on ecological thinking, it has become generally accepted that niche is by no means a synonym for habitat. Since Elton placed the emphasis on energy relations, his version of the concept might be considered the *trophic niche*. In 1957, G. E. Hutchinson suggested that the niche could be visualized as a multidimensional space or hypervolume within which the environment permits an individual or species to survive indefinitely. Hutchinson's niche, which we can designate as the *multidimensional* or *hypervolume niche*, is amenable to measurement and mathematical manipulation. For example, the two-dimensional climographs shown in Figure 5–10, which picture the "climatic niche" of a species of bird and a fruit fly, could be expanded as a series of coordinates to include other environmental dimensions. Hutchinson has also made a distinction between the *fundamental niche* —the maximum "abstractly inhabited hypervolume" when the species is not constrained by competition with others—and the *realized niche*—a smaller hypervolume occupied under biotic constraints. For an additional discussion of the hypervolume niche concept, see Hutchinson, 1965.

Let us return, for the moment, to the simple analogy of the "address" and the "profession" mentioned earlier. If we wished to become acquainted with some person in our human community we would need to know, first of all, his address, that is, where he could be found. To really get to know him, however, we would have to learn more than the neighborhood where he lives or works. We would want to know something about his occupation, his interests, his associates, and the part he plays in general community life. So it is with the study of organisms; learning the habitat is just the beginning. To determine the organism's status in the natural community we would need to know something of its activities, especially its nutrition and energy source; also its rate of metabolism and growth, its effect on other organisms with which it comes into contact, and the extent to which it modifies or is capable of modifying important operations in the ecosystem.

MacArthur (1968) has pointed out that the ecological term "niche" and the genetic term "phenotype" are parallel concepts in that they both involve an infinite number of attributes, both include some or all of the same measurements, and both are most useful in determining differences between individuals and species. Thus, niches of similar species associated together in the same habitat can be compared with precision when the comparison involves only a few operationally significant measurements. MacArthur goes on to compare the niches of four species of American warblers (Parulidae) which all breed in the same macrohabitat, a spruce forest, but forage and nest in different parts of the spruce tree. He constructs a mathematical model of the situation, which consists of a set of competition equations in a matrix, a "model building" technique described in Chapter 10. From this model all kinds of predictions can be made as to what would happen to the niches if the relative abundance of any one of the species should change, or should one or more of the species not be present in the forest. Another approach to quantitating niche comparisons is suggested by Maguire (1967), who sets up "versatility indices" computed by summing ranges of tolerance for chemical factors for each species of protozoa that coexist in the same pond.

Measurements of morphological features of larger plants and animals can often be used as indices in niche comparisons. Van Valen (1965), for example, finds that the length and breadth of the bill in birds (the bill, of course, reflects the type of food eaten) is an index of "niche width" (see Figure 7–30 for a diagram of niche width in relation to intra- and interspecific competition); the coefficient of variation in bill width was found to be greater in island populations of six species of birds than in mainland populations, corresponding with the wider niche (wider variety of habitat occupied and food eaten) on islands (see also page 219). Within the same species, competition is often greatly reduced when different stages in the organism's life history occupy different niches; for example, the tadpole functions as a herbivore and the adult frog as a carnivore in the same pond. Niche segregation may even occur between the sexes. In woodpeckers of the genus *Dendrocopus,* male and female differ in bill size and foraging behavior (Ligon, 1968). In hawks, as well as in many insects, the sexes differ markedly in size, and, therefore, in the dimensions of their food niche.

Species-abundance patterns within trophic levels, taxonomic groups (birds, insects, and so on), and whole communities, as discussed in Section 4, Chapter 6, provide clues to the nature of niche relationships in groups of species that are closely associated ecologically in the same macrohabitat. As was briefly noted on page 153, there are several simple, or first-order, mathematical statements that mimic species-abundance distributions. Since the way in which the component species "divide up" the available niche space or "hypervolume" will greatly influence the type of species-abundance distribution, one can sometimes infer something about niches of the individual species from the type of species-abundance distribution exhibited by the group. For example, if the most abundant species is twice as numerous as the next most abundant one, which in turn has twice the density of the third, and so on, one would get a straight line when the number of individuals per species is plotted on a log scale on the ordinate against species sequence (from the most abundant to least abundant) on the abscissa, as shown in Figure 8–1 (Curve A). From this one might assume that the first species occupies half of the available niche space, the second, half of the remaining space (25 per cent of total), and so on. If, on the other hand, the niche space is divided up in random, contiguous, nonoverlapping segments, an entirely different curve would be obtained, as also shown in Figure 8–1 (Curve B). These two possibilities represent what seem to be extremes, with most natural distributions showing up as some kind of intermediate sigmoid curve (curve C in Figure 8–1), suggesting a more complex pattern of niche differentiation and overlap (Whittaker, 1965). MacArthur (1957 and 1960) suggests that three niche arrangements might be detected by examining species-

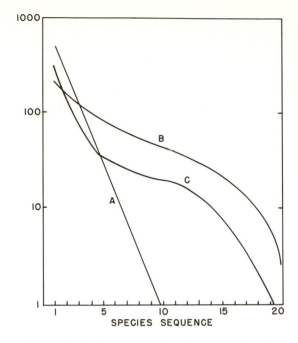

Figure 8–1. Dominance-diversity curves for a hypothetical sample of 1000 individuals in 20 species from a community. Number of individuals in the species (ordinate) are plotted against species number in sequence from the most abundant to the least abundant (abscissa). A. Geometric series. B. Nonoverlapping, random niche hypothesis. C. Intermediate sigmoid pattern, see text for explanation. (Redrawn from Whittaker, 1965.)

abundance distributions, namely, the contiguous, nonoverlapping, random pattern just noted, a discrete pattern (nonoverlapping niches that are not adjacent), and an overlapping pattern. Groups exhibiting intense interspecific competition and territorial behavior such as forest birds tend to conform to the nonoverlapping niche hypothesis, but many populations, especially those belonging to the same basic trophic level, exhibit patterns suggesting niche overlap and competition patterns other than strict competition exclusion. Whittaker (1965) finds that curves approximating the simple geometric series (A in Figure 8–1) are found in some plant communities in rigorous environments, but that plant populations in most mature communities overlap in use of space and resources, as already indicated in Section 3, Chapter 6. Land plants as well as many animals apparently coexist under conditions of partial rather than direct competition; some of the adaptations that promote niche differentiation without competition exclusion from a

Table 8–1. *Niche Segregation in Millepedes (Diplopoda) as Indicated by Percentage Occurrence in the Indicated Microhabitat on the Forest Floor of a Maple-oak Forest**

MICROHABITATS	SPECIES						
	Euryurus erythropygus	*Pseudopolydesmus serratus*	*Narceus americanus*	*Scytonotus granulatus*	*Fontaria virginiensis*	*Cleidogonia caesioannularis*	*Abacion lacterium*
Heartwood at center of logs	93.9†	0	0	0	0	0	0
Superficial wood of logs	0	66.7	4.3	6.7	0	14.3	0
Outer surface of logs beneath bark	0	20.8	71.4	0	0	0	0
Under log, but on log surface	3.0	8.3	6.9	60.0	0	0	15.8
Under log, but on ground surface	3.0	4.2	12.5	0	97.1	14.3	36.8
Within leaves of litter	0	0	0	26.7	0	42.8	0
Beneath litter on ground surface	0	0	4.7	6.7	2.9	28.6	47.4

* After O'Neill, 1967.

† Italicized figures represent microhabitats in which the species predominate.

habitat are considered in subsequent sections (see also the *"Tribolium-Trifolium* model," Table 7–9 and Figure 7–29).

Examples

Examples of the spatial niche, trophic niche, and multidimensional (hypervolume) niche are shown in Table 8–1 and Figures 8–2, 8–3, and 8–4. Microhabitat segregation of seven species of millepedes is indicated by the data in Table 8–1. All of these species live in the same general habitat, namely the forest floor of a maple-oak forest in Illinois, and all belong to the same basic trophic level, that is, they are detritus feeders. As shown, each species predominates in a different microhabitat. Presumedly each species is using a somewhat different energy source since the stage of decomposition and microflora will change in the gradient from the center of the log to the position underneath the leaf litter. Figure 8–2 pictures two similar aquatic bugs which can be collected in the same sample from a small vegetation-choked pond but which occupy very different trophic niches. The backswimmer (*Notonecta*) is an active predator that swims about grasping and eating other animals in its general size range. In contrast, the water boatman (*Corixa*), even though it looks very much like the backswimmer, plays a very different role in the community since it feeds largely on decaying vegetation. Figure 7–32 shows another example of niche separation based on food habits. The ecological literature is replete with similar examples of coexisting species that use a different energy source.

Two diagrams, one general (Fig. 8–3) and one specific (Fig. 8–4), illustrate the concept of the multifactor niche. When niches overlap to a large extent, as shown in Figure 8–3B, natural selection may favor a process of niche displacement (as indicated by the arrows), involving change in habitat, food selection, morphology, physiology, or behavior of one or both species; more about this later. The diagram (Fig. 8–4) of the climatic niche of a small bird, the cardinal, shows the limits of tolerance and the most

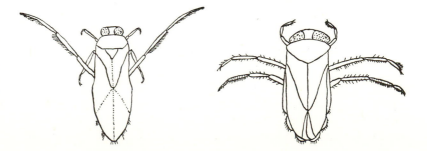

Figure 8–2. *Notonecta* (left) and *Corixa* (right), two aquatic bugs (Hemiptera) which may live in the same habitat but occupy different trophic niches because of differences in food habits.

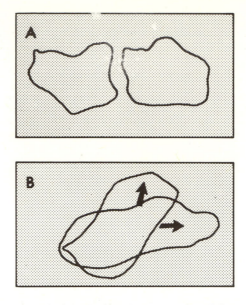

Figure 8–3. Schematic representation of the hypervolume concept of the ecological niche. The background dots represent environmental factors (temperature, food sources, minerals, other organisms) projected onto a plane. The irregular polygons enclose sets of factors that are operationally significant for a species population. In *A* two species occupy nonoverlapping niches, while in *B* niches of two species overlap to such an extent that severe competition for shared resources may result in elimination of one species or a divergence of niches as indicated by the arrows. (Redrawn from Bruce Wallace and A. M. Srb. *Adaptation.* Prentice-Hall, Englewood Cliffs, New Jersey.)

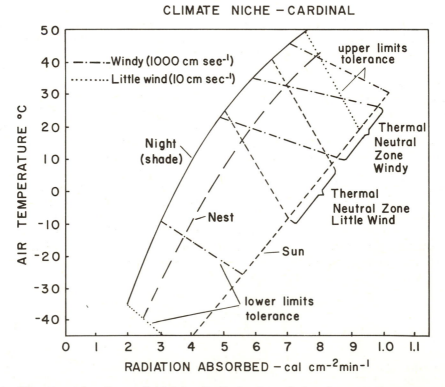

Figure 8–4. Climatic niche of a small bird (cardinal) showing relations between air temperature, radiation absorbed per cm² of animal surface and wind speed as they shape the conditions within which the organism can survive. Thermal neutral conditions are those in which the bird is not under thermal stress. Note how wind and exposure greatly alter the limits of tolerance. (Redrawn and simplified from Gates, 1969a.)

comfortable zone, in which maximum energy is available for activities other than maintenance and survival. Behavioral adaptation (for example, remaining in sheltered places during cold, windy days) enables the bird to avoid the upper and lower limits of tolerance, provided food is available to supply the necessary energy.

2. ECOLOGICAL EQUIVALENTS

Statement

Organisms that occupy the same or similar ecological niches in different geographical regions are known as *ecological equivalents*. Species that occupy equivalent niches tend to be closely related taxonomically in regions which are contiguous, but are often not closely related in regions which are widely separated or isolated from one another.

Explanation and Examples

As will be described in more detail in Part 2, the species composition of communities differs widely in different biogeographical regions, but similar ecosystems develop wherever the physical habitat is similar regardless of geographical location. The equivalent functional niches are occupied by whatever biological groups happen to make up the fauna and flora of the region. Thus, a grassland type of ecosystem develops wherever there is a grassland climate (see page 121), but the species of grass and grazers may be quite different especially where regions are widely separated. The following tabular comparison of large grazing herbivores of the grasslands of four continents will illustrate the concept of ecological equivalence:

North America	Eurasia	Africa	Australia
bison, pronghorn antelope	saga antelope, wild horses, wild asses	numerous species of antelope, zebra, etc.	the large kangaroos

We may say that the kangaroos of Australia are ecologically equivalent to the bison and pronghorn antelope of North America; or perhaps we should say "were," since man's cattle and sheep now largely replace them in both regions. The native grasses on which the herbivores feed present a very similar appearance throughout the world even though species, and often genera and whole families, may be sharply restricted to a given continent, or biogeographical region within a continent (see Figure 14–1). Ecological equivalents in three trophic niches of four coastal regions are shown in Table 8–2 (an example from the marine habitat). Note that equivalent grazing periwinkles belong to the same genus, but the equivalent benthic carnivores, which incidentally are major seafood resources in their respective regions, belong to different genera.

3. CHARACTER DISPLACEMENT: SYMPATRY AND ALLOPATRY

Statement

Species that occur in different geographical regions (or are separated by a spatial barrier) are said to be *allopatric*, while those

Table 8–2. *Ecological Equivalents in Three Major Niches of Four Coastal Zones of North and Central America*

NICHE	TROPICAL	UPPER WEST COAST	GULF COAST	UPPER EAST COAST
Grazer on intertidal rocks (periwinkles)	*Littorina ziczac*	*L. danaxis* *L. scutelata*	*L. irrorata*	*L. littorea*
Benthic carnivore	Spiny lobster (*Panulirus*)	King crab (*Paralithodes*)	Stone crab (*Menippe*)	Lobster (*Homarus*)
Plankton-feeding fish	Anchovy	Pacific herring, sardine	Menhaden, threadfin	Atlantic herring, alewife

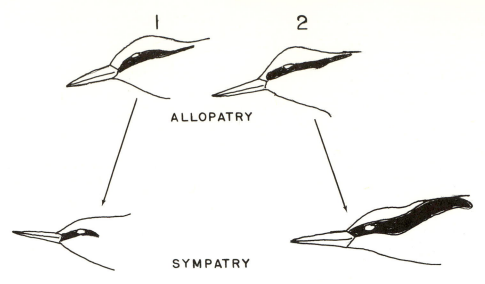

Figure 8–5. An example of character displacement. Two species of nuthatches (1 and 2) are very similar in allopatry (separated geographically), but diverge in bill size and facial pigmentation in sympatry (when occurring together). (Sketches based on photographed specimens in Vaurie, 1951.)

occurring in the same area (but not necessarily the same niche) are said to be *sympatric*. Differences in closely related species are often accentuated (i.e., diverge) in sympatry and weakened (i.e., converge) in allopatry by an evolutionary process known as *character displacement*.

Explanation and Examples

Brown and Wilson (1956) explain the phenomenon of *character displacement* as follows:

Two closely related species have overlapping ranges. In the parts of the ranges where one species occurs alone, the populations of that species are similar to the other species and may even be difficult to distinguish from it. In the area of overlap, where the two species occur together, the populations are more divergent and easily distinguished, i.e., they "displace" one another in one or more characters. The characters involved can be morphological, ecological, behavioral, or physiological; they are assumed to be genetically based.[*]

It is important to emphasize the dual nature of this pattern in that species show displacement when they are *sympatric* (= joint fatherland) and active convergence when they are *allopatric* (= different father-

lands). Character displacement has two adaptive values: (1) it enhances niche displacement (see previous section and Figure 8–3), thus reducing competition, and (2) it enhances genetic segregation by maintaining species distinctiveness (i.e., preventing hybridization) and thereby maintains a greater species diversity in the community (than would be the case if one species eliminated the other, or if they hybridized into one species).

Among the examples cited by Brown and Wilson (1956) is the case of two species of nuthatches (*Sitta*) as described by Vaurie (1951). Where the two species are allopatric, each is so similar to the other that specimens can be distinguished only by an expert in bird taxonomy. Where the two species are sympatric, there is a striking divergence in morphology so that they can be distinguished at a glance; in one species the bill and a black facial stripe become enlarged, while in the other these characters are reduced in size, as shown in Figure 8–5. Thus, the accentuated difference in bill size reduces food niche overlap, and the conspicuous difference in the facial strip enhances species recognition and prevents interbreeding—or at least cuts down on energy that might be wasted in unsuccessful or unfruitful mating with the wrong species.

Another striking example occurs among the famous "Darwin's finches" on the Galápagos islands (recall that Darwin's visit

[*] From Brown, W. L., Jr., and Wilson, E. O.: Systematic Zoology, 5(2), 1956.

to these islands off the west coast of South America greatly influenced the development of his theory of natural selection). As described by Lack (1947), two species of ground finches (*Geospiza*), occurring together on the larger islands, differ in body size and beak proportions. On some of the smaller islands, however, one or the other species may occur alone, in which case each species is almost exactly intermediate in size and bill configuration (so much so that they were originally thought to be hybrids). Thus, each allopatric species converged towards the other, partially filling, as it were, the other niche. MacArthur and Levins (1967) have suggested that character convergence, instead of displacement, might occur when a third competing species invades the territory occupied by two species already competing in overlapping niches.

4. NATURAL SELECTION: ALLOPATRIC AND SYMPATRIC SPECIATION

Statement

The species is a natural biological unit tied together by the sharing of a common gene pool (Merrell, 1962). Speciation, or the formation of new species and the development of species diversity, occurs when gene flow within the common pool is interrupted by an isolating mechanism. When isolation occurs through geographic separation of populations descended from a common ancestor, *allopatric speciation* may result. When isolation occurs through ecological or genetic means within the same area, *sympatric speciation* is a possibility.

Explanation and Examples

The theory of biological evolution, involving Darwinian natural selection and genetic mutation at the species level, is so well known to students and so well described in basic biology texts and numerous small volumes (see, for example, Savage, 1969; Wallace and Srb, 1961; Volpe, 1967), that we need not review it here. To build a background for the more difficult and more controversial aspects of evolution at the eco-

system level, to be discussed in the next chapter, we do need to focus on the contrasting concepts of allopatric and sympatric speciation.

Allopatric speciation has been generally assumed to be the primary mechanism by which species arise. According to this conventional view, two segments of a freely interbreeding population become separated spatially (as on an island or separated by a mountain range). In time sufficient genetic differences accumulate in isolation so that the segments will no longer interchange genes (interbreed) when they come together again, and thereby coexist as distinct species in different niches (or perhaps these differences are further accentuated by the additional process of character displacement). The Galápagos finches, noted in the preceding section, are a classic example of allopatric speciation. From a common ancestor a whole group of species evolved in isolation on the different islands and adaptively radiated so that a variety of potential niches was eventually exploited on reinvasion; species now present include slender-billed insect-eaters and thick-billed seed-eaters, ground- and tree-feeders, large and small bodied finches, and even a woodpecker-like finch (which, although hardly able to compete with a real woodpecker, survives in the absence of invasion by woodpecker stock).

Evidence is mounting that sympatric speciation may be more widespread and important than previously believed. It was first clearly demonstrated in higher plants in which such genetic isolating mechanisms such as polyploidy (duplication of chromosome sets that can bring about immediate genetic isolation), hybridization, self-fertilization, and asexual reproduction are more common than in animals. The great salt marsh "invasion" of Britain is a good example of almost "instant" speciation, resulting from a breakdown of geographical isolation caused by man, followed by hybridization and polyploidy. When the American salt marsh grass *Spartina alterniflora* was introduced into the British Isles, it crossed with the native species, *S. maritima*, to produce a new polyploid species, *S. townsendii*, which has now invaded formerly bare tidal mud flats not occupied by native species. As the ecological isolating mechanisms involving niche segregation (such as discussed in preceding sections) are becoming better understood, it is apparent that the potential for "on the spot" or sympatric speciation is not restricted to internal genetic mechanisms peculiar to plants. Ehrlich and Raven (1969) cite numerous cases of homing, restricted dispersal of propagules, colonization, and the like as evidence to support the contention that gene flow in nature is more restricted than commonly thought. This is to say that behavioral and reproductive patterns (see Section 5, Chapter 6 for a discussion of pattern in communities) tend to break up a species population into genetically isolated segments so that gene exchange between colonies or local communities takes place at a much lower rate than within the colonies. These authors also contend that differentiation may occur even when populations are freely exchanging genes because segments of the population are very often subjected to different ecological selection pressure. For example, a predator may exert selective pressure on one segment of a population and not on another. Thus, local interbreeding populations, as well as the species as a whole, can be the evolutionary units that bring about adaptation and diversity in communities. However, natural selection at and below the species level can be only part of the story, as we shall see in the next chapter. To understand evolution it is clear that we must consider the entire spectrum from the local population to the large functional ecosystem.

While we generally regard evolution as a slow process, something that has mostly occurred in the dim past, certain kinds of speciation, as we have seen in the case of the marsh grass, can be very rapid and are very much a "now" thing, especially since man has begun to change environments and rearrange barriers at a rapid rate in his newly found role as the "mighty geological agent" (about which we spoke in Chapter 2). Massive stress on the landscape including removal of natural vegetation not only encourages invasion by exotic species which become "weeds" (that is, thrive in places where they are not "wanted"), but such massive and rapid alterations also may set the stage for the rapid evolution of new species of weeds that are even better adapted to the new environment than any preexisting species (see Baker, 1965). One of the most widely cited examples of rapid natural

Table 8–3. *Survival of Dark and Light Pigmented Moths Released in Dark Woods (Tree Trunks Darkened by Industrial Pollution) and Light Woods (Tree Trunks Covered by Growth of Light-colored Lichens)**

	DARK WOODS		LIGHT WOODS	
	Dark Moths	*Light Moths*	*Dark Moths*	*Light Moths*
Number released	154	64	473	496
Per cent recaptured (as a measure of survival)	53	25	6	12

* Data from Kettlewell, 1956.

selection is what has come to be known as "industrial melanism," or the case of the dark pigmented tree-trunk moths that have evolved in industrial areas of England where the bark of trees has become greatly darkened by industrial pollution (which kills the lichens that give normal bark a light appearance). As shown in Table 8–3, Kettlewell (1956) demonstrated experimentally that dark moths survive better in dark (polluted) woods and light moths in natural woods, presumedly because predation by birds is selective for the individuals not protectively colored. This example leads us naturally into the subject of direct or purposeful selection by man.

5. ARTIFICIAL SELECTION: DOMESTICATION

Statement

Selection carried out by man for the purpose of adapting plants and animals to his needs is known as artificial selection. Domestication of plants and animals involves more than modifying the genetics of a species because reciprocal adaptations between the domesticated species and the domesticator (usually man) are required, which lead to a special form of mutualism (see Section 19, Chapter 7). Domestication may fail in its long-term purpose unless the mutualistic relationship is also adaptive at the ecosystem level, or can be so adapted by purposeful regulation.

Table 8–4. *Increase in Yield of Winter Wheat in the Netherlands by Artificial Selection of Varieties with Increased Grain/Straw Ratios but Without an Increase in Total Dry Matter Production* [*]

| | | YIELD-KILOGRAMS/HECTARE | | | |
VARIETY	PERIOD	Total	Grain	Straw	Grain/ Straw Ratio
Wilhelmina	1902–1932	12,600	6,426	6,174	0.51
Juliana	1934–1947	12,430	6,836	5,594	0.55
Staring	1948–1961	13,900	8,201	5,699	0.59
Felix	1958–1961	12,830	7,698	5,132	0.60
Heines VII	1953–1955	11,860	7,828	4,032	0.66

[*] After Van Dablen, 1962.

Explanation and Examples

Because man is egotistical he falls into the trap of thinking that when he domesticates another organism through artificial selection he is merely "bending" nature to suit his purpose. In actual fact, of course, domestication is a two-way street that brings about changes (ecological and social, if not genetic) in man as well as in the domesticated organism. Thus, man is just as dependent on the corn plant as the corn is dependent on man. A society that depends on corn develops a very different culture than one dependent on herding cattle. It is a real question as to who becomes the slave of whom! The same question can be posed regarding man and his machines. Perhaps the relationship between man and his tractor is not too different from that of man and his plowhorse, except that the tractor requires a greater energy input (fuel) and produces more poisonous wastes! Domestication, then, is really a special form of mutualism; we could have logically discussed it in Chapter 7, but we have somewhat arbitrarily placed it in this chapter because of its special relation to speciation.

Table 8–4 is an example of one type of artificial selection in crops that is part of the basis for the "green revolution" directed towards feeding the starving millions. As already emphasized in Chapter 3 and elsewhere, selection for yield of the edible part of the crop does not necessarily mean an increase in primary production. Above a certain point increased yield must come at a sacrifice of some other adaptive use of energy; in this case increased yield of wheat in a sequence of artificially selected varieties accompanies a decrease in "straw," which represents the plant's self-protective maintenance equipment. Highly bred strains, therefore, are not only vulnerable to disease, but require special care in which machines (as well as potent and polluting chemicals) replace men and animals on the farm, thus requiring profound changes in the social, economic, and political structure of human society. These and other potential "ecologic backlashes," and means of avoiding them, are considered in Chapter 15. The point is that artificial selection has profound effects on whole ecosystems because of the compensatory adjustments that occur by natural selection.

Dr. I. Lehr Brisbin[*] has some interesting thoughts on animal domestication, and he has prepared the following four paragraphs for use in this text:

Domestication is a state of relationship existing between two populations in which one population (the domesticator) does two things to the other (domesticated) population; first, the domesticator population acts to prevent natural selection from acting on the gene pool of the domesticated population and secondly, the domesticator population imposes some regime of artificial selection which then acts to determine the future gene-pool composition of the domesticated population in the absence of natural selection. Thus, we can define a *wild population* as

---

[*] Institute of Ecology, University of Georgia.

one whose future gene-pool composition is under the direct control of natural selection-mutation interaction in the Darwinian sense; a *domestic population* as one whose future gene-pool composition is under the direct control of some regime of artificial selection, imposed upon it externally by some other population, and a *feral population* as one whose future gene-pool composition was at one time under the control of some regime of artificial selection, but is at present under the direct control of natural selection-mutation interaction.

The domestication interaction has two types of components which may, for the sake of convenience, be designated as type "A" and type "B" components. Type "A" components concern the effects of the domesticator population upon the domesticated population. The most common of these effects involve changes in structure or morphology of the domesticated individuals and are discussed at length by Zeuner (1963). It is of interest to note that with the exception of a tendency toward hypersexuality, very little is known about functional type "A" effects. There is also a need to study type "A" effects at the population as well as at the individual level of organization.

The less familiar type "B" components of the domestication relationship concern the effects of the domesticated population upon the domesticator population. These effects, together with the type "A" effects, thus constitute a feedback-control mechanism of interpopulation regulation within the domestication relationship. A good example of type "B" effects at the population level is given by Downs and Ekvall (1965), whose study suggests that the environment may be considered as providing a template that produces through various feedback-control mechanisms a corresponding template in the structure of the domesticated populations of the area. In Downs and Ekvall's study in Tibet, the breeds of domestic animals showed a basically tripartite nature in response to the tripartite environmental template. For example, the mountain peak regions contained populations of yaks and nomad horses, while the lowland cow and donkey were the breeds of the valley regions. The transition regions were characterized by hybrid forms such as the dzo (yak × lowland cow) and mule (nomad horse × donkey). The domesticated populations impart a corresponding template to the domesticator populations since human social types also showed a basically tripartite nature. The mountain peak regions were inhabited by the Drog-ba (nomads) and the valley regions by the Rong-ba (farmers). The transition zone was inhabited by the Sama-drog, an intermediate group of people characteristic of that zone.

Domesticated populations may be considered to act as an intermediary or buffering mechanism interposed between the domesticator populations and the environment. They thus enable these domesticator populations to make finer and more thorough ecological adjustments to the regime of environmental conditions with which they must cope. In this sense, the process of domestication serves the domesticator populations as a tool for carrying out environmental adjustments, and this consideration of domestication as a tool sheds considerable ecological light on studies of the biological origins of the domestication process in early humans (Zeuner, 1963). As was the case with other tools such as fire, weapons, and clothing, domestication must have conferred considerable selective advantages on those groups of early humans which were proficient in its practice. This selective advantage could only have occurred, however, under environmental conditions which limited populations of early humans from obtaining those benefits conferred by the domesticated populations by other more culturally simple means such as hunting, fishing, and gathering. It is thus significant that no important domestications seem to have occurred among those species of animals contemporary with early man in the African savannah. Indeed, it seems to be a general rule that no important domestications were ever made by early man in tropical regions, where a stable supply of food could be obtained by hunting, fishing, and gathering. It was primarily in the region of the Middle East and central Asia where the most important domestication relationships were initiated between early men and populations of prey animals such as sheep, goats, cattle, and swine. The domestications of these species occurred in environments which undoubtedly imposed periods of low prey availability on those groups relying solely on a hunting economy, while those groups practicing domestication undoubtedly had a more stable and reliable food supply.

In summary, domestication is a special and very important type of mutualism that brings about profound changes in the ecosystem because the relationship affects a large number of other species and processes (nutrient cycling, energy flow, soil structure, and so forth) not directly involved in the interaction between the domesticator and the domesticated. As a purposeful endeavor by man, domestication can fail in its long-term objectives if the feedback constraints of natural selection that were removed by artificial selection are not compensated for by purposeful (i.e., intelligent) artificial feedback constraints. Thus, man and his cow will destroy the environment by overgrazing unless the relationship is regulated in terms of the whole ecosystem so as to be truly mutualistic (beneficial to both) rather than exploitative. Also, some of man's worst problems have been caused by domesticated plants and animals which "escape" (i.e., become feral) back into nature and become major pests. Man's myopia in this regard is what Garrett Hardin (1968) calls the "tragedy of the commons." As long as a pasture (or any other resource) is considered to be unlimited and for common use by everybody without constraints, then, as Hardin points out, overuse is inevitable since the individual gains a temporary advantage by overstocking (or overuse) and only at some later time does he as an individual begin to suffer from the collective consequences of overuse. If man does not legislate temperance, as it were, in the function of powerful man–domestic animal, man–domestic plant, and man–machine combinations, then he ultimately must face the consequences of natural selection which all too frequently result in the extermination of the "intemperate species."

6. BIOLOGICAL CLOCKS

Statement

Organisms possess a physiological mechanism for measuring time which is known as the *biological clock*. The most common and perhaps basic manifestation is the *circadian rhythm* (circa = about; dies = day), or the ability to time and repeat functions at about 24-hour intervals even in the absence of the conspicuous diurnal clues such as light. Other timed events are related to lunar periodicities (which control tides, for example) and seasonal cycles. The mechanism of the biological clock has not been resolved, but two theories are currently proposed: (1) the endogenous timer hypothesis (i.e., the "clock" is an internal device that can measure time without environmental clues) and (2) the external timing hypothesis (i.e., the internal clock is timed by external signals from the environment). Regardless of the mechanism, the ecological, or selective, advantage of the biological clock is undisputed since it couples environmental and physiological rhythms and enables organisms to anticipate daily, seasonal, and other periodicities in light, temperature, tides, and so forth.

Explanation

The coupling of natural periodicities and community action has already been discussed in Section 4, Chapter 5 and again in Chapter 6, page 156. Adaptation, it was emphasized, goes beyond mere tolerance of environmental fluctuations to active participation in the periodicity as a means of coordinating and regulating vital functions. In Chapter 5 photoperiodism was discussed in some detail as an example. In this section we focus on the mysterious (that is, unknown) ways in which the individual phases its rhythms with that of the environment.

Several features of the circadian rhythm are illustrated by the diagram in Figure 8–6. In its natural habitat a nocturnal animal such as the flying squirrel (*Glaucomys*) or the white-footed mouse (*Peromyscus*) remains in its nest during the daylight hours and is active at night. The rhythmic activity is exactly 24 hours in length and is entrained or "locked into" the daily light-dark cycle generated by the rotation of the earth on its axis. If the animal is brought into the laboratory and placed under constant darkness, the daily rhythm continues, but the onset of activity will likely shift slightly each day, indicating a persistent or "free running" period a little shorter or a little longer than 24 hours. In Figure 8–6 the period in complete darkness is shown to be about 20 minutes longer than 24 hours so that after 20 days the animal is leaving its nest for its "nightly" round six hours later than at the start of the experiment. When the light-dark cycle is reestablished, the activity is entrained again to conform to the cycle. Fluctuations in temperature have little effect on circadian rhythms. Experiments cannot prove that the clock is entirely internal since not all environmental clues are eliminated by placing the animal under constant conditions with respect to the major factors such as light and temperature. Subtle atmospheric and geophysical fluctuations pervade that might be detected by the organism.

Persistent rhythms have been detected in a wide variety of organisms from algae to man. Even a potato stored in a dark bin may show both a daily and a monthly rise and fall in oxygen consumption. Some organisms seem to have both a solar-day and a lunar-day clock. It should be pointed out that many diurnal or other rhythms are not

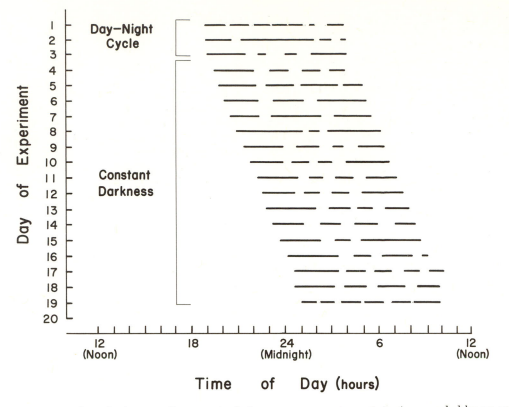

Figure 8–6. Circadian rhythm in a flying squirrel. Spontaneous running activity (as recorded by an exercise wheel) under a regular light-dark cycle repeats itself each day at the same time, but under constant darkness the rhythm proves to be "free-running" in that the activity period drifts, thus indicating that the innate period is approximately, but not exactly, 24 hours in length (slightly longer than 24 hours in this example). (Diagram based on data of DeCoursey, 1961.)

persistent in the absence of the environmental change. The vertical migration of plankton, as pictured in Figure 6–9, for example, is entirely extrinsic in that it does not occur in constant light or constant darkness.

The two theories, as outlined in the Statement, can be clarified in terms of an analogy with man-made clocks. A spring-driven clock with an internal balance wheel oscillator would be analogous to the endogenous biological clock, while an electric clock that is timed by alternating current pulses coming from a distant generator exemplifies the external timer hypothesis. Proponents of the endogenous theory (Pittendrigh, 1961; Hastings, 1969) suggest that the basic timer might be an oscillating biochemical reaction in the cell, or perhaps in the endocrine system, but so far evidence is only indirect. Pye (1969), for example, reports that glycolysis in yeast cells pulses in a rhythmic manner, but no known rhythm in yeast correlates with this metabolic oscillation. Frank Brown (1969), a proponent of the idea of an external pacemaker for the internal clock, has found that organisms respond to daily fluctuations in the earth's geomagnetic field. He suggests that plants and animals use such subtle fluctuations as a time "grid" on which they encode the information necessary for timing their physiological rhythms. The two views about biological clocks are fully outlined in a beautifully illustrated little book by Brown, Hastings, and Palmer (1970). Other reviews and symposia include: Cloudsley-Thompson (1961), Sweeney (1963), Harker (1964), Aschoff (1965), Bunning (1967).

Whatever the mechanism, the innate (i.e., inherited, not learned from experience) biological clock not only couples internal and external rhythms, it also enables the individual to anticipate changes. Thus, when the nocturnal mouse leaves its nest, it is fully alert to the vicissitudes of the environment because its physiological systems have already been programmed to "go"!

Air travel and space exploration has made man conscious of the fact that he also has circadian rhythms. Flying rapidly from one distant time zone to another is stressful to most people; several days may be required before the physiological rhythms of sleep, digestion, and so on are coordinated with the local day-night cycle. It is clear that 24-hour activity patterns must be maintained in space travel and that special entraining mechanisms will be necessary for long space journeys.

7. BASIC BEHAVIORAL PATTERNS

Statement

Behavior in the broadest sense is the overt action an organism takes to adjust to environmental circumstance so as to insure its survival. It is also an important means by which individuals become integrated into organized and regulated societies and communities. Behavior can be considered as a complex of six components that vary in importance according to the kind of organism: (1) tropisms, (2) taxes, (3) reflexes, (4) instincts, (5) learning, and (6) reasoning. The term tropism is now generally restricted to directed movements or orientations in organisms such as plants that lack nervous systems, while the other five components, listed above in more or less an evolutionary sequence (see also Figure 8–7), are associated with animals that have complex nervous and sensory systems. Originally, ethologists (ethology, from the word "ethos" meaning custom, is the science of behavior) tended to make a sharp distinction between "innate" behavior (components 1 through 4 above) and "acquired" behavior (components 5 and 6), but it is now evident that learned behavior is built on complex patterns of reflexes, instincts, and other inherited behavior patterns, including circadian and other innate body rhythms that were discussed in the preceding section.

Explanation and Examples

In Chapter 7, and in the preceding sections of this chapter, the important role of behavior was brought out in the discussions of population regulation, habitat selection, territoriality, aggregation, predator-prey interaction, competition, niche differentiation, natural selection, and all the other aspects of population ecology. For further discussions along these lines, Klopfer's little book, *Behavioral Aspects of Ecology* (1962), is recommended reading. Following the widely read works of Murchison (1935), Allee (1938), Lorenz (1935 and 1952), and Tinbergen (1951 and 1953), the study of behavior, or ethology, has become an important interdisciplinary science that more or less attempts to link together physiology, ecology, and psychology. At first, the psychologists, who study man's behavior, and the zoologists, who study animal behavior, were poles apart on the matter of "innate" versus "learned" behavior; but as in so many other aspects of biology, it has become apparent to all that man is a part of, not apart from, nature. The resulting integration is producing many large textbooks (for example, Scott, 1958; Ektin, 1964; Marler and Hamilton, 1966), many small textbooks (for example, Thorpe, 1963; Dethier and Stellar, 1964; Davis, 1966; van der Kloot, 1968), reviews (see Klopfer and Hailman, 1967), and popular books (for example, Lorenz, 1966; Morris, 1967), not to mention symposium volumes which are too numerous to cite. As in other developing sciences, a descriptive phase is followed by an experimental and theoretical phase that attempts to discover basic mechanisms and functions. As human beings we have two prime motivations (to use a behavioral term) in studying behavior: (1) to find the neurological, hormonal, and genetic roots of the complex and often too aggressive behavior of man and (2) to understand how the individual organism adapts to the environment, especially how the reciprocal interactions between organisms (cooperation, mutualism, domestication, and so on) lead to social organization. The former aspect is beyond the scope of this book, and the space allotted to behavior will allow us to outline only very briefly a few basic concepts regarding the latter in the next two sections.

As shown in Figure 8–7, the six components of behavior can be viewed in a sort of evolutionary sequence; however, note that even in the higher vertebrates and man the more primitive components (instincts, reflexes) still play a part in determining how

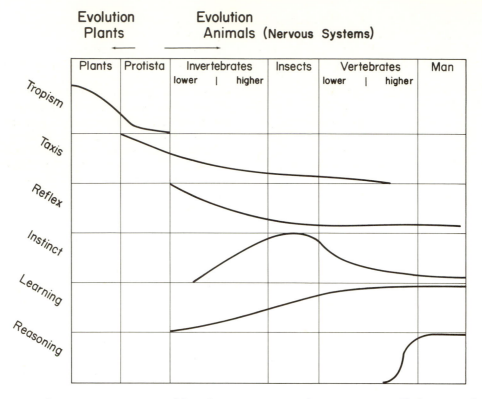

Figure 8–7. Schematic representation of the relative importance of six components of behavior in phylogeny. (Modified from Dethier and Stellar, 1964.)

individuals act, perhaps a more important part than egotistical man is willing to admit. Tropisms (tropos = turn or change) are directed movements and orientations found in plants, such as the turning of the sunflower to face the sun (phototropism), the vertical orientation of leaves of trees on a hot sunny day (heliotropism) or the downward growth of roots (geotropism). Since such adaptive behavior occurs in the absence of nervous systems, it usually involves a part of rather than the whole organism, and hormones provide the chief coordinating mechanism. We have already seen how simple growth responses (see page 216), or the secretion of antibiotics (see page 227), can be as effective in terms of adaptation and survival for the plant as is the most complex behavior for animals.

The term taxis (= arrangement) is now generally used to refer to stimulus-response movements that are so readily observed in the lower animals (Fig. 8–7). Fraenkel and Gunn (1940) in their pioneer review of these basic behaviors distinguished between: (1) undirected reactions such as a general avoidance of unfavorable environment (which they called "kineses"), (2) directed reactions (taxes, sensu strictu) with orientation directly towards or away from the stimulus (moth flying towards a light), and (3) transverse orientations, or movement at some angle to the direction of stimulus, such as the well-known light compass orientation and direction finding of bees (see von Frisch, 1955). No hard and fast line can be drawn between taxes and reflexes, but reflexes, in general, are considered to be stimulus responses of specific body organs or parts; both can be modified by experience. The well-known conditioned reflex is a beginning of "learned" behavior. Once thought to be restricted to higher animals, it now appears that lowly worms and perhaps even protozoa can be "conditioned."

The feeding behavior of hydra, a simple coelenterate that lacks a central nervous system, illustrates how plantlike tropisms and simple directed responses are coordinated by means of a hormone produced outside the hydra's body (i.e., an "environmental hormones"; see Chapter 2, page 32).

As described by Lenhoff (1968), when a prey organism is contacted and penetrated by the stinging cells of one of the tentacles, a substance called reduced glutathione diffuses from the prey, bringing about a rapid and vigorous coordination of all of the tentacles which then move the prey towards the centrally located mouth. Experiments showed that glutathione was a very specific *releaser* for the feeding behavior.

Instinctive behavior that accounts for such a large part of what insects and lower vertebrates do consists of encoded sequences of stereotyped behavior, such as the sequence of nest building, food gathering, courtship, mating, egg laying, and protection of the young that comprise the reproductive cycle of a paper wasp or a bird. Finally, learned and reasoned behavior increase in importance almost in direct proportion to the enlargement of the brain, particularly the cerebral cortex. Only in the higher primates and man does reasoning, involving problem solving and formulation of concepts, become a major component of behavior.

To illustrate how innate and acquired behavior merge we can cite the well-known case of *imprinting* as originally described by Lorenz (1935 and 1937). Newly hatched ducks or geese have an innate tendency to follow a parent, but they must "learn" to recognize the parent through association. If a person or even an inanimate moving model is substituted for the natural parent during the first few weeks of life, the image of the substitute is "imprinted" to such an extent that the young subsequently will follow the substitute in preference to a real parent. Also, song in many birds is apparently a combination of innate and learned behavior.

The spectacular homing of birds and fish is an instinctive type of behavior which, nevertheless, probably involves one or more conditioned or learned components. Salmon, for example, are able to detect the home stream by the "smell" or "taste" of the water, which was encoded into their memory during their early life in the headwaters (see Hasler, 1965). Apparently each watershed releases to the stream a slightly different organic component (not yet identified) that can be detected by the fish's sensory system even when the stream's chemistry is considerably altered by man-made pollution!

8. REGULATORY AND COMPENSATORY BEHAVIOR

Statement

Organisms regulate their internal environments and their external microenvironment by behavioral as well as by physiological means. Behavior, therefore, is an important component of factor compensation and ecotypic development; these concepts were described in Section 2, Chapter 5.

Explanation and Examples

Behavioral responses, as we have seen, are an important part of an organism's adjustment to its environment. In many cases the selection or creation of an optimum environment is accomplished as much by behavioral means as by internal physiological regulation. For example, birds and mammals are the only kinds of organisms that regulate their body temperature by internal means; hence, these animals are called *homeotherms* or *endotherms*. However, other animals (i.e., *poikilotherms* or *exotherms*) often control their body temperature almost as effectively by behavioral regulation. Bogert (1949) was among the first to show that the body temperature of reptiles often remains fairly constant because the animals move in and out of burrows in such a way that their internal temperature is maintained within fairly constant, optimum range. As the techniques of "remote sensing" become more sophisticated, miniature temperature-sensitive radio transmitters can be implanted in the tissues and the temperature monitored continuously in animals in their normal habitat. Such studies are revealing that so-called "cold-blooded" animals exhibit remarkable thermoregulatory behavior. In one study (McGinnis and Dickson, 1967), the desert lizard *Dipsosaurus* maintained its body temperature between 31° and 39°C. even though the range in the environment was more than twice as great. In another study in the laboratory, another kind of lizard (*Tiliqua*) was able to keep its body temperature between 30° and 37°C. by moving back and forth between environments having temperatures ranging from 15° to 45° (Hammel *et al.*, 1967).

9. SOCIAL BEHAVIOR

Statement

A communication network, some form of dominance hierarchy, learning, and a balance between contradictory behaviors (competition versus cooperation, aggression versus passiveness, aggregation versus isolation, and so forth) are the necessary ingredients for social organization, whether it be monospecific (involving individuals of the same species) or polyspecific (involving several species).

Explanation

In the well-knit social organizations of higher animals some form of social hierarchy is usually evident in which one or more individuals are dominant or leaders. Such organization can be quite rigid, as in the well-known "peck order" of chickens (or as in a human family unit or a military organization), or more flexible with leadership shifting from one individual to another (as in nonmilitary human society in general); see Allee *et al.* (1949) for a review of social hierarchies. Once such a hierarchy has been established and is recognized by

Table 8–5. The Basic Messages of Vertebrate Communication in Terms of "Information Carried" by Displays and Vocalizations*

Identification: species, sex, etc.
Probability: uncertainty as to action (attack or flee?).
General set: maintenance, foraging, preening, resting.
Locomotion: given at beginning and ending of locomotion.
Attack
Escape
Nonagonistic subset: expresses anxiety.
Association: close social association but not physical contact.
Bond-limited subset: between mates, parent-offspring.
Play
Copulation
Frustration: substitute actions, etc.

---

* Compiled from W. John Smith, 1969.

all individuals, the energy that might be wasted in conflicts can be devoted to the collective function of the group. Some form of audio-visual communication, of course, is necessary. Smith (1969) has analyzed the vocalization and displays of wild birds and mammals and has concluded that all the tremendous variety can be reduced to 12 "messages," as listed in Table 8–5. Note that most of the "information carried" in these messages had to do with coordinating activities of individuals and therefore with social organization. The least understood aspect of social behavior has to do with resolving the conflicts between basic drives that are contradictory. Thus, the need to isolate (i.e., need for "privacy") conflicts with the need to aggregate. We have already reviewed the selective advantages of each in the preceding chapter (Sections 14 and 15). Aggressive behavior saves energy in some situations, wastes it in others; as is becoming painfully evident in human society, what has always been regarded as "good" for the individual is not always "good" for the group. The effects of crowding on behavior are especially important and especially relevant to man at this point in his history. The remarkable changes in behavior that accompany changes in density of "cyclic" animals was discussed in Section 9, Chapter 7. In human society, crowding seems both to increase and decrease aggression. Perhaps the "militants" and the "hippies" represent the opposite reaction to urban crowding. The ecologist tends to feel that man should maintain and develop an aggressive territorialism that will prevent overuse and destruction of his environment. On the other hand, many social scientists feel that man should (or will) become increasingly passive and tolerate crowding in urban centers in the same manner that densely packed domestic animals tolerate each other in a feed lot. We cannot hope to adequately treat such a subject in this brief review. We would recommend that every student read and ponder the following two essays: (1) Tinbergen (1968), "On War and Peace in Animals and Man" and (2) Calhoun (1962), "Population Density and Social Pathology."

Chapter 9

DEVELOPMENT AND EVOLUTION OF THE ECOSYSTEM

1. THE STRATEGY OF ECOSYSTEM DEVELOPMENT[*]

Statement

Ecosystem development, or what is more often known as *ecological succession,* may be defined in terms of the following three parameters: (1) It is an orderly process of community development that involves changes in species structure and community processes with time; it is reasonably directional and, therefore, predictable. (2) It results from modification of the physical environment by the community; that is, succession is community-controlled even though the physical environment determines the pattern, the rate of change, and often sets limits as to how far development can go. (3) It culminates in a stabilized ecosystem in which maximum biomass (or high information content) and symbiotic function between organisms are maintained per unit of available energy flow. The whole sequence of communities that replaces one another in a given area is called the *sere;* the relatively transitory communities are variously called *seral stages* or *developmental stages* or *pioneer stages,* while the terminal stabilized system is known as the *climax.* Species replacement in the sere occurs because populations tend to modify the physical environment, making conditions favorable for other populations until an equilibrium between biotic and abiotic is achieved.

In a word, the "strategy" of succession as a short-term process is basically the same as the "strategy" of long-term evolutionary development of the biosphere, namely, increased control of, or homeostasis with, the physical environment in the sense of achieving maximum protection from its perturbations. The development of ecosystems has many parallels in the developmental biology of organisms, and also in the development of human society.

Explanation

The descriptive studies of succession on sand dunes, grasslands, forests, marine shores, or other sites, and more recent functional considerations, have led to the basic theory contained in the definition given above. H. T. Odum and Pinkerton (1955), building on Lotka's (1925) "law of maximum energy in biological systems," were the first to point out that succession involves a fundamental shift in energy flows as increasing energy is relegated to maintenance. Margalef (1963, 1968) has more recently documented this bioenergetic basis for succession and has extended the concept.

Changes that occur in major structural and functional characteristics of a developing ecosystem are listed in Table 9–1. Twenty-four attributes of ecological systems are grouped for convenience of discussion under six headings. Trends are emphasized by contrasting the situation in early and late development. The degree of absolute change, the rate of change, and the time required to reach a steady state may vary not only with different climatic and physiographic situations, but also with different ecosystem attributes in the same physical environment. Where good data are available, rate-of-change curves are usually convex, with changes occurring most rapidly at the beginning, but bimodal or cyclic patterns may also occur.

---

[*] This section, and also Section 3, is adapted from the author's paper "The Strategy of the Ecosystem Development," first published in *Science,* Vol. 164, pp. 262–270, April 18, 1969.

Bioenergetics of Ecosystem
Development

Attributes 1 through 5 in Table 9–1 represent the bioenergetics of the ecosystem. In the early stages of ecological succession, or in "young nature" so to speak, the rate of primary production or total (gross) photosynthesis (P) exceeds the rate of community respiration (R), so that the P/R ratio is greater than 1. In the special instance of organic pollution, the P/R ratio is typically less than 1. The term *heterotrophic succession* is often used for a developmental sequence, in which R is greater than P at the beginning— in contrast to *autotrophic succession*, in which the ratio is reversed in the early stages. In both cases, however, the theory

Table 9–1. *A Tabular Model of Ecological Succession: Trends to Be Expected in the Development of Ecosystems**

Ecosystem Attributes	Developmental Stages	Mature Stages
Community energetics		
1. Gross production/community respiration (P/R ratio)	Greater or less than 1	Approaches 1
2. Gross production/standing crop biomass (P/B ratio)	High	Low
3. Biomass supported/unit energy flow (B/E ratio)	Low	High
4. Net community production (yield)	High	Low
5. Food chains	Linear, predominantly grazing	Weblike, predominantly detritus
Community structure		
6. Total organic matter	Small	Large
7. Inorganic nutrients	Extrabiotic	Intrabiotic
8. Species diversity—variety component	Low	High
9. Species diversity—equitability component	Low	High
10. Biochemical diversity	Low	High
11. Stratification and spatial heterogeneity (pattern diversity)	Poorly organized	Well-organized
Life history		
12. Niche specialization	Broad	Narrow
13. Size of organism	Small	Large
14. Life cycles	Short, simple	Long, complex
Nutrient cycling		
15. Mineral cycles	Open	Closed
16. Nutrient exchange rate, between organisms and environment	Rapid	Slow
17. Role of detritus in nutrient regeneration	Unimportant	Important
Selection pressure		
18. Growth form	For rapid growth ("*r*-selection")	For feedback control ("*K*-selection")
19. Production	Quantity	Quality
Overall homeostasis		
20. Internal symbiosis	Undeveloped	Developed
21. Nutrient conservation	Poor	Good
22. Stability (resistance to external perturbations)	Poor	Good
23. Entropy	High	Low
24. Information	Low	High

* From E. P. Odum, in *Science, 164*:262–270, April 18, 1969. Copyright 1969 by the American Association for the Advancement of Science.

is that P/R approaches 1 as succession occurs. In other words, the energy fixed tends to be balanced by the energy cost of maintenance (that is, total community respiration) in the mature or "climax" ecosystem. The P/R ratio, therefore, should be an excellent functional index of the relative maturity of the system. The P/R relationships of a number of familiar ecosystems are graphically displayed in Figure 9–1, with the direction of autotrophic and heterotrophic succession also indicated.

As long as P exceeds R, organic matter and biomass (B) will accumulate in the system (Table 9–1, item 6), with the result that ratio P/B will tend to decrease or, conversely, the B/P, B/R, or B/E ratios (where E = P + R) will increase (Table 9–1, items 2 and 3). Recall that the reciprocals of these ratios were discussed in Chapter 3 in terms of thermodynamic order functions (page 38). Theoretically, then, the amount of standing-crop biomass supported by the available energy flow (E) increases to a maximum in the mature or climax stages (Table 9–1, item 3). As a consequence, the net community production, or yield, in an annual cycle is large in young nature and small or zero in mature nature (Table 9–1, item 4).

Comparison of Succession in a Laboratory Microcosm and a Forest

One can readily observe bioenergetic changes by initiating succession in experimental laboratory microecosystems of the type derived from natural systems as described in Chapter 2 (page 21 and Figure 2–7A). In Figure 9–2 the general pattern of a 100-day autotrophic succession in a microcosm based on data of Cooke (1967) is compared with a hypothetical model of a 100-year forest succession as presented by Kira and Shidei (1967).

During the first 40 to 60 days in a typical microcosm experiment, daytime net production (P) exceeds nighttime respiration (R), so that biomass (B) accumulates in the system. After an early "bloom" at about 30 days, both rates decline and become approximately equal at 60 to 80 days. The B/P ratio, in terms of grams of carbon supported per gram of daily carbon production, increases from less than 20 to more than 100 as the steady state is reached. Not only is autotrophic and heterotrophic metabolism balanced in the climax, but a large organic structure is supported by small daily production and respiratory rates.

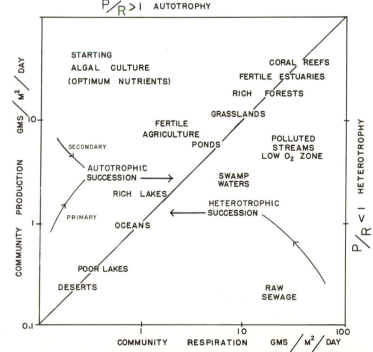

Figure 9–1. Position of various community types in a classification based on community metabolism. Gross production (P) exceeds community respiration (R) on the left side of the diagonal line (P/R greater than 1 = autotrophy), while the reverse situation holds on the right (P/R less than 1 = heterotrophy). The latter communities import organic matter or live on previous storage or accumulation. The direction of autotrophic and heterotrophic succession is shown by the arrows. Over a year's average, communities along the diagonal line tend to consume about what they make and can be considered to be metabolic climaxes. (Redrawn from H. T. Odum, 1956.)

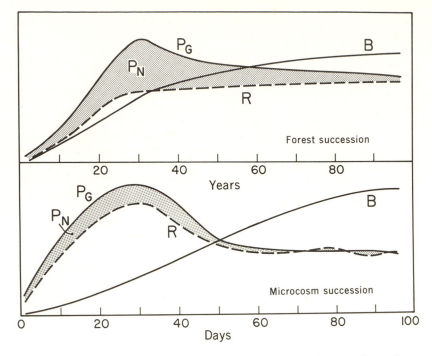

Figure 9–2. Comparison of the energetics of ecosystem development in a forest (redrawn from Kira and Shidei, 1967) and a laboratory microcosm (redrawn from Cooke, 1967). P_G, gross production; P_N, net production; R, total community respiration; B, total biomass.

While direct projection from the small laboratory microecosystem to open nature may not be entirely valid, there is evidence that the same basic trends that are seen in the laboratory are characteristic of succession on land and in large bodies of water. Seasonal successions also often follow the same pattern — an early seasonal bloom characterized by rapid growth of a few dominant species followed by the development later in the season of high B/P ratios, increased diversity, and a relatively steady, although temporary, state in terms of P and R (Margalef, 1963). Open systems may not experience a decline, at maturity, in total or gross productivity, as the space-limited microcosms do, but the general pattern of bioenergetic change in the latter seems to mimic nature quite well.

Table 9–1, as already emphasized, refers to changes that are brought about by *biological processes within the ecosystem in question*. Imported materials or energy, or geological forces acting on the system, can, of course, reverse the trends shown in Table 9–1. For example, eutrophication of a lake, whether natural or cultural, results when nutrients are imported into the lake from outside the lake — that is, from the watershed. This is equivalent to adding nutrients to the laboratory microecosystem or fertilizing a field; the system is pushed back, in successional terms, to a younger or "bloom" state. Recent studies on lake sediments (Mackereth, 1965; Cowgill and Hutchinson, 1964; Harrison, 1962), as well as theoretical considerations, have indicated that lakes can and do progress to a more oligotrophic condition when the nutrient input from the watershed slows or ceases. Thus, there is hope that the troublesome cultural eutrophication of our waters can be reversed if the inflow of nutrients from the watershed can be greatly reduced. An example is the "recovery" of Lake Washington in Seattle (see page 439). Most of all, however, this situation again emphasizes that it is the entire drainage or catchment basin, not just the lake or stream, that must be considered the ecosystem unit if we are to deal successfully with our water pollution problems (recall page 16).

To avoid confusion in these important matters it is important to distinguish between *autogenic* processes (that is, biotic processes within the system) and *allogenic* processes (such as geochemical forces acting from without). Thus, Table 9–1 describes *autogenic succession* only. *Allogenic succession* will be considered later, but it can be pointed out for now that small lakes or other systems that are transitory in a geological sense often show trends opposite from those of Table 9–1 because the *effect of allogenic processes exceeds that of autogenic ones.* Such ecosystems not only may not stabilize, but they may become extinct. Such is the ultimate fate of man-made lakes subjected to man-made erosion! (See, for example, Fig. 15–6K, page 430.)

Food Chains and Food Webs

As the ecosystem develops, subtle changes in the network pattern of food chains may be expected. The manner in which organisms are linked together through food tends to be relatively simple and linear in the very early stages of succession. Furthermore, heterotrophic utilization of net production tends to occur predominantly by way of grazing food chains —that is, plant-herbivore-carnivore sequences. In contrast, food chains become complex webs in mature stages, with the bulk of biological energy flow following detritus pathways (Table 9–1, item 5), as already described in detail in Chapter 3. The time involved in an uninterrupted succession allows for increasingly intimate associations and reciprocal adaptations between plants and animals, which lead to the development of many mechanisms that reduce grazing, such as the development of indigestible supporting tissues (cellulose, lignin, and so on), feedback control between plants and herbivores (Pimentel, 1961), and increasing predator pressure on herbivores. These and other mechanisms described in Chapters 6 and 7 enable the biological community to maintain the large and complex organic structure that mitigates perturbations of the physical environment. Severe stress or rapid changes brought about by outside forces can, of course, rob the system of these protective mechanisms and allow irruptive, cancerous growths of certain species to occur, as man too often finds to his sorrow. Examples of stress-induced pest irruptions are documented elsewhere (see pages 119, 220, and 458).

Diversity and Succession

Perhaps the most controversial of the successional trends pertains to diversity, a subject already dealt with in Chapter 6. Four components of diversity are listed in Table 9–1, items 8 through 11.

The variety of species, expressed as a species-number ratio or a species-area ratio, tends to increase during the early stages of community development. The behavior of the "evenness" component of diversity is less well known. While an increase in the variety of species together with reduced dominance by any one species or small group of species (that is, increased evenness or decreased redundancy) can be accepted as a general probability during succession, there are other community changes that may work against these trends. An increase in the size of organisms, an increase in the length and complexity of life histories, and an increase in interspecific competition that may result in competitive exclusion of species (Table 9–1, items 12 to 14) are trends that may reduce the number of species that can live in a given area. In the bloom stage of succession organisms tend to be small and to have simple life histories and rapid rates of reproduction. Changes in size appear to be a consequence of, or an adaptation to, a shift in nutrients from inorganic to organic (Table 9–1, item 7). In a mineral and nutrient-rich environment, small size is of selective advantage, especially to autotrophs, because of the greater surface-to-volume ratio. As the ecosystem develops, however, inorganic nutrients tend to become more and more tied up in the biomass (that is, to become intrabiotic), so that the selective advantage shifts to larger organisms (either larger individuals of the same species or larger species, or both), which have greater storage capacities and more complex life histories. They are thus adapted to exploiting seasonal or periodic releases of nutrients or other resources. The question of whether the seemingly direct relationship between organism size and stability is the result of feedback or is merely fortuitous remains unanswered (see Bonner, 1963; Frank, 1968).

Thus, whether or not species diversity continues to increase during succession will depend on whether the increase in potential niches resulting from increased biomass, stratification (Table 9–1, item 9), and other consequences of biological organization exceeds the countereffects of increasing size and competition. No one has yet been able to catalogue all the species in any sizable area, much less follow *total* species diversity in a successional series. Data are so far available only for segments of the community (trees, birds, and so on). The results of one of the most comprehensive studies yet made on changes in diversity of vegetation in a forest sere are shown in Figure 9–3. Each size class or stratum is represented by a different curve which indicates that each component reaches a peak in diversity at the time of maximum development of that component. Margalef (1963) postulates that diversity will tend to peak during the early or middle stages of succession and then decline in the climax, and a suggestion of such a trend is shown in Figure 9–3.

Species variety, equitability, and stratification are only three aspects of diversity which change during succession. Perhaps an even more important trend is an increase in the diversity of organic compounds, not only of those within the biomass but also of those excreted and secreted into the media (air, soil, water) as by-products of the increasing community metabolism. An increase in such "biochemical diversity" (Table 9–1, item 10) is illustrated by the increase in the variety of plant pigments along a successional gradient in aquatic situations, as described by Margalef (1967). Biochemical diversity within populations, or within systems as a whole, has not yet been systematically studied to the degree that the subject of species diversity has been.

Consequently, few generalizations can be made, except that it seems safe to say that as succession progresses, organic extrametabolites probably serve increasingly important functions as regulators which stabilize the growth and composition of the ecosystem. Such metabolites may, in fact, be extremely important in preventing populations from overshooting the equilibrial density, thereby reducing oscillations and contributions to stability.

The cause-and-effect relationship between diversity and stability is not clear and needs to be investigated from many angles. If it can be shown that biotic diversity does indeed enhance physical stability in the ecosystem, or is the result of it, then we would have an important guide for conservation practice. Preservation of hedgerows, woodlots, noneconomic species, noneutrophicated waters, and other biotic variety in man's landscape could then be justified on scientific as well as esthetic grounds, even though such preservation often must result in some reduction in the production of food or other immediate consumer needs. In other words, is variety only the spice of life or is it a necessity for the long life of the total ecosystem comprising man and nature?

Nutrient Cycling

An important trend in successional development is the closing or "tightening" of the biogeochemical cycling of major nutrients, such as nitrogen, phosphorus, and calcium (Table 9–1, items 15 to 17). Mature systems, as compared to developing ones, have a greater capacity to entrap and hold

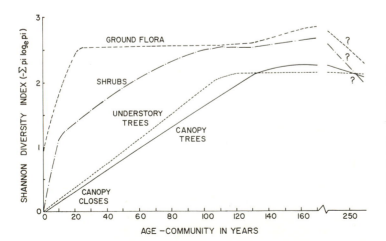

Figure 9–3. Species/numbers diversity in four size classes (strata) in a forest sere on the piedmont region of Georgia, starting with an abandoned crop field (age zero). See Chapter 6, page 149, for an explanation of the Shannon index. Too few forest stands more than 200 years old were available to establish reliable means for mature climaxes, but a downward trend is indicated by available data. (Curves drawn from unpublished data of Carl Monk, Institute of Ecology, University of Georgia.)

nutrients for cycling within the system. The watershed calcium cycle in Figure 4–6 is a good example; only very small amounts of nutrients are lost from mature systems as compared to immature or disturbed ones.

Selection Pressure: Quantity versus Quality

MacArthur and Wilson (1967) have reviewed stages of colonization of islands which provide direct parallels with stages in ecological succession on continents. Species with high rates of reproduction and growth, they find, are more likely to survive in the early, uncrowded stages of island colonization. In contrast, selection pressure favors species with lower growth potential but better capabilities for competitive survival under the equilibrium density of late stages. Using the terminology of growth equations, where r is the intrinsic rate of increase and K is the upper asymptote or equilibrium population size (see Chapter 7), we may say that "r selection" predominates in early colonization, with "K selection" prevailing as more and more species and individuals attempt to colonize (Table 9–1, item 18). The same situation is even seen within the species in certain "cyclic" northern insects in which "active" genetic strains found at low densities are replaced at high densities by "sluggish" strains that are adapted to crowding, as has already been described (see page 195).

Genetic changes involving the whole biota may be presumed to accompany the successional gradient, since, as described above, quantity production characterizes the young ecosystem while quality production and feedback control are the trademarks of the mature system (Table 9–1, item 19). Selection at the ecosystem level may be primarily interspecific, since species replacement is a characteristic of successional series, or seres, as will be illustrated below. However, in most well-studied seres there seem to be a few early successional species that are able to persist through to late stages. Whether genetic changes contribute to adaptation in such species has not been determined, so far as we know, but studies on population genetics of *Drosophila* suggest that changes in genetic composition could be important in population regulation (Ayala, 1968). Certainly, the human population, if it survives beyond its present rapid growth stage, is destined to be more and more affected by such selection pressures as adaptation to crowding becomes essential.

Overall Homeostasis

This brief review of ecosystem development emphasizes the complex nature of processes that interact. While one may well question whether all the trends described are characteristic of all types of ecosystems, there can be little doubt that the net result of community actions is an increase in symbiosis, nutrient conservation, stability, and information content (Table 9–1, items 20 to 24). The overall strategy, as was stated at the beginning of this section, is directed toward achieving as large and diverse an organic structure as is possible within the limits set by the available energy input and the prevailing physical conditions of existence (soil, water, climate, and so on). As studies of biotic communities become more functional and sophisticated, one is impressed with the importance of mutualism, parasitism, predation, commensalism, and other forms of symbiosis discussed in Chapter 7. Partnership between unrelated species is especially noteworthy as, for example, that between corals (coelenterates) and algae or between mycorrhizae and trees. In many cases, at least, biotic controls of grazing, population density, and nutrient cycling provide the chief negative feedback mechanisms that contribute to stability in the mature system by preventing overshoots and destructive oscillations. The intriguing question is: do mature ecosystems age as organisms do? In other words, after a long period of relative stability or "adulthood," do ecosystems again develop unbalanced metabolism and become more vulnerable to diseases and other perturbations? Some examples of what appear to be "catastrophic" or "cyclic" climaxes, as cited in the next section, perhaps support this analogy between development of the individual and the community.

While the basic assumption that species replace one another in a successional gradient "because populations tend to modify the physical environment making conditions favorable for other populations until an equilibrium between biotic and abiotic is achieved" (see "Statement" at beginning of this section) is certainly valid, it may be an oversimplification since very little is actually known about the chemical nature of developmental processes. A number of cases have now been documented showing that some species not only make conditions favorable for others, but actually create unfavorable conditions for themselves, thus hastening the replacement process. For example, the early pioneer annual "weeds" of grassland and old-field succession frequently produce antibiotics which accumulate in the soil and inhibit seedling growth in subsequent years. Whittaker (1970) has reviewed these studies and what is known about the chemical nature of the regulators.

Examples

If development begins on an area that has not been previously occupied by a community (such as a newly exposed rock or sand surface, or a lava flow), the process is known as *primary succession*. If community development is proceeding in an area from which a community was removed (such as an abandoned crop field or a cutover forest), the process is appropriately called *secondary succession*. Secondary succession is usually more rapid because some organisms or their disseminules are already present, and previously occupied territory is more receptive to community development than sterile areas. Primary and secondary development are shown in Tables 9–2 and 9–3, respectively. As shown in Figure 9–1, primary succession tends to begin at a lower level of productivity than does secondary succession.

The first example, Table 9–2, illustrates primary ecological succession of plant and certain invertebrate components of the community on the Lake Michigan dunes. Lake Michigan was once much larger than it is at present. In retreating to its present boundaries, it left successively younger and younger sand dunes. Because of the sand substrate, succession is slow and a series of communities of various ages are available — pioneer stages at the lake shore and increasingly older seral stages as one proceeds away from the shore. It was in this "natural laboratory of succession" that H. C. Cowles (1899) made his pioneer studies of plants and Shelford (1913) his studies of animal

succession. Olson (1958) has restudied ecosystem development on these dunes and has given us updated information on rates and processes. Because of encroachment of heavy industry, conservationists are hard pressed in their efforts to preserve some part of the "Indiana Dunes"; the public should support these efforts because these areas not only have a priceless natural beauty that can be enjoyed by urban dwellers, but they also constitute a natural "teaching laboratory" in which the "visual display" of ecological succession is dramatic.

The pioneer communities on the dunes consist of grasses (*Ammophila, Agropyron, Calamovilfa*), willow, cherry, and cottonwood trees, and animals such as tiger beetles, burrowing spiders, and grasshoppers. The pioneer community is followed by forest communities as shown, each of which has changing animal populations. Although it began on a very dry and sterile sort of habitat, development of the community eventually results in a beech–maple forest, moist and cool in contrast with the bare dune. The deep humus-rich soil, with earthworms and snails, contrasts with the dry sand on which it developed. Thus, the original relatively unhospitable pile of sand is eventually transformed completely by the action of a succession of communities.

Succession in the early stages on dunes is often arrested when the wind piles up the sand over the plants and the dune begins to move, entirely covering the vegetation in its path. Here we have an example of the arresting or reversing effect of "allogenic" perturbations as discussed in the general explanation section above. Eventually, however, as the dune moves inland it becomes stabilized and pioneer grasses and trees again become established. Using modern methods of carbon dating, Olson (1958) estimates that about 1000 years is required to reach a forest climax on the Lake Michigan dunes. (Contrast this with the 200 year sere of secondary development shown in Figures 9–3 and 9–4.) He also points out that a beech–maple forest may only be achieved on moist sites, and that an oak forest may be the terminal stage on the more elevated or exposed slopes. (Topographic climaxes will be discussed further in Section 2.)

The second example illustrates how changes in breeding birds (Table 9–3) parallel that of dominant plants in secondary succession following the abandonment

Table 9–2. *Primary Ecological Succession of Plants and Invertebrates on the Lake Michigan Dunes**

| Invertebrates of Ground Strata | Cotton-wood—Beach Grass | SERAL STAGES | | | Climax Beech–Maple Forest |
		Jack Pine Forest	Black Oak Dry Forest	Oak and Oak–Hickory Moist Forest	
White tiger beetle (*Cicindela lepida*)	✿ ✿				
Sand spider (*Trochosa cinerea*)	✿ ✿				
White grasshopper (*Trimerotropis maritima*)	✿ ✿				
Long-horn grasshopper (*Psinidia fenestralis*)	✿ ✿	✿ ✿			
Burrowing spider (*Geolycosa pikei*)	✿ ✿	✿ ✿			
Digger wasps (*Bembex* and *Microbembex*)	✿ ✿	✿ ✿			
Bronze tiger beetle (*C. scutellaris*)		✿ ✿			
Ant (*Lasius niger*)		✿ ✿			
Migratory locust (*Melanoplus*)		✿ ✿			
Sand locusts (*Ageneotettix* and *Spharagemon*)		✿ ✿			
Digger wasp (*Sphex*)		✿ ✿	✿ ✿		
Ant-lion (*Cryptoleon*)			✿ ✿		
Flatbug (*Neuroctenus*)			✿ ✿		
Grasshoppers (six species not listed above)			✿ ✿		
Wireworms (*Elateridae*)			✿ ✿	✿ ✿	✿ ✿
Snail (*Mesodon thyroides*)			✿ ✿	✿ ✿	✿ ✿
Green tiger beetle (*C. sexguttata*)				✿ ✿	✿ ✿
Millipedes (*Fontaria* and *Spirobolus*)				✿ ✿	✿ ✿
Centipedes (*Lithobius, Geophilus, Lysiopetalum*)				✿ ✿	✿ ✿
Camel cricket (*Ceuthophilus*)				✿ ✿	✿ ✿
Ants (*Camponotus, Lasius umbratus*)				✿ ✿	✿ ✿
Betsy beetle (*Passalus*)				✿ ✿	✿ ✿
Sowbugs (*Porcellio*)				✿ ✿	✿ ✿
Earthworms (*Lumbricidae*)				✿ ✿	✿ ✿
Woodroaches (*Blattidae*)				✿ ✿	✿ ✿
Grouse locust (*Tettigidae*)					✿ ✿
Cranefly larvae (*Tipulidae*)					✿ ✿
Wood snails (7 species not found in previous stages)					✿ ✿

* From Shelford, 1913. A few species of invertebrates are listed to illustrate the general pattern of change; for more complete listing see his tables L to LV.

of upland agricultural fields in southeastern United States (Fig. 9–4). Note that the most pronounced change in bird populations occurs as the life form of the plant dominants changes (herbs, shrubs, the pine, and hardwood). No species of plant or bird is able to thrive from one end of the sere to the other; species have their maxima at different points in the time gradient. Although plants are the most important organisms which bring about changes, birds are by no means entirely passive agents in the community, since the major plant dominants of the shrub and hardwood stages depend on birds or

Table 9–3.   *Distribution of Breeding Passerine Birds in a Secondary Upland Sere, Piedmont Region, Georgia**

Plant Dominants Age in Years of Study Area Bird Species (having a density of 5 or more in a given stage) †	Forbs 1–2	Grass 2–3	Grass- shrub 15	20	25	Pine Forest 35	60	100	Oak– Hickory Climax, 150–200
Grasshopper sparrow	10	30	25						
Meadowlark	5	10	15	2					
Field sparrow			35	48	25	8	3		
Yellowthroat			15	18					
Yellow-breasted chat			5	16					
Cardinal			5	4	9	10	14	20	23
Towhee			5	8	13	10	15	15	
Bachman's sparrow				8	6	4			
Prairie warbler				6	6				
White-eyed vireo				8		4	5		
Pine warbler					16	34	43	55	
Summer tanager					6	13	13	15	10
Carolina wren						4	5	20	10
Carolina chickadee						2	5	5	5
Blue-gray gnatcatcher						2	13		13
Brown-headed nuthatch							2	5	
Wood pewee							10	1	3
Hummingbird							9	10	10
Tufted titmouse							6	10	15
Yellow-throated vireo							3	5	7
Hooded warbler							3	30	11
Red-eyed vireo							3	10	43
Hairy woodpecker							1	3	5
Downy woodpecker							1	2	5
Crested flycatcher							1	10	6
Wood thrush							1	5	23
Yellow-billed cuckoo								1	9
Black and white warbler									8
Kentucky warbler									5
Acadian flycatcher									5
Totals: (including rare species not listed above)	15	40	110	136	87	93	158	239	228

* After Johnston and E. P. Odum (1956). Figures are occupied territories or estimated pairs per 100 acres.

   † By density, the "dominant" species for each stage are as follows:

    1. Forb and grass stage: Grasshopper sparrow and meadowlark.
    2. Grass-shrub stage: Field sparrow, yellow throat, and meadowlark.
    3. Young pine forest (25–60 years): Pine warbler, towhee, and summer tanager.
    4. Old pine forest (with well-developed deciduous understory): Pine warbler, Carolina wren, hooded warbler, and cardinal.
    5. Oak-hickory climax: Red-eyed vireo, wood thrush, and cardinal.

other animals to disperse seeds into new areas. The final result, or climax, is an oak–hickory forest instead of a beech–maple forest, as in the previous example, because of differences in regional climate. Keever (1950) has shown that organic matter and detritus produced by the first year dominant horseweeds greatly inhibit growth of this species in year 2 (see Figure 9–4, bar diagram), thus speeding up the replacement by asters. The probable widespread occurrence of such "chemically induced" succession was mentioned in the preceding section. In summary, while climate and other physical factors control the composition of the communities and determine the climax, the communities themselves play the major role in bringing about succession.

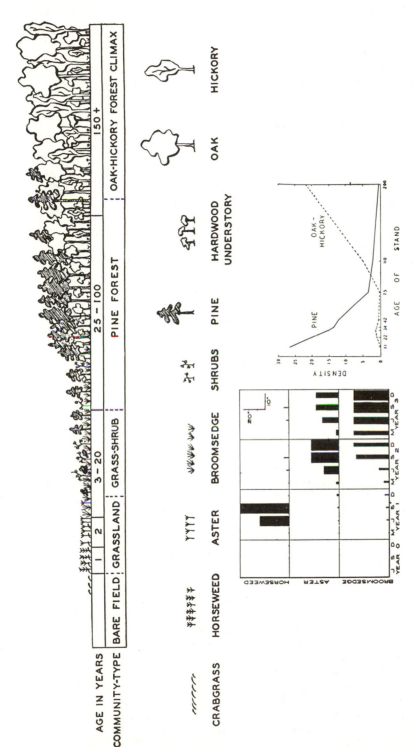

Figure 9–4. Secondary succession on the piedmont region of southeastern United States. The principal plant dominants of the upland sere which follows abandonment of crop land (cotton, corn, etc.) are shown in pictorial fashion in the upper diagram (after E. P. Odum). The lower charts contain quantitative data. On the left, the relative size of the three pioneer plants which reach dominance in successive years, namely, horseweed (*Leptilon = Erigeron*), aster (*Aster*), and broomsedge (*Andropogon*), is depicted (after Keever, 1950), while the gradual change from pine to hardwood dominance is indicated on the right (after Oosting, 1942). In the diagram at left, depicting relative size of the three plants, height of the columns represents average height of plants in inches, and width of columns represents relative diameter of stems. In the diagram at the right, the density figures are number per 100 m².

Secondary plant succession is equally as striking in grassland regions as in forests, even though only herbaceous plants are involved. In 1917 Shantz described succession on the abandoned wagon roads used by pioneers crossing the grasslands of central and western United States, and virtually the same sequence has been described many times since. While the species vary geographically the same pattern everywhere holds. This pattern involves four successive stages: (1) annual weed stage (2 to 5 years), (2) short-lived grass stage (3 to 10 years), (3) early perennial grass stage (10 to 20 years) and (4) climax grass stage (reached in 20 to 40 years). Thus, starting from bare or plowed ground, 20 to 40 years are required for nature to "build" a climax grassland, the time depending on the limiting effect of moisture, grazing, etc. A series of dry years or overgrazing causes the succession to go backwards towards the annual weed stage; how far back depends on the severity of the effect.

A very large and specialized terminology has been built up during several decades of describing ecological succession on land. Many of the terms were originally proposed by Frederic E. Clements in his pioneer monograph on plant succession (1916) and his subsequent voluminous papers and books. As is so often the case in the early descriptive phase of science, many of these terms do not contribute very much to our understanding of the processes involved. Consequently, we can leave most of these terms for the advanced student to discuss and decide on which should be retained.

Succession is also equally apparent in aquatic as well as terrestrial habitats. However, as already emphasized, the community development process in shallow water ecosystems (ponds, small lakes, coastal waters) is usually complicated by strong inputs of materials and energy which may speed up, arrest, or reverse the normal trend of community development. Thus, the small ponds formed between the dunes created by retreating Lake Michigan (see Table 9–2) soon fill up with organic matter and sediments and become substrates on which a land succession will continue. The complex nature of such changes can be best seen in artificial ponds and impounded lakes. When a reservoir is created by flooding rich soil, or an area with a large amount of organic matter (as is the case when a forested area

is flooded), the first stage in development is a highly productive "bloom" stage characterized by rapid decomposition, high microbial activity, abundant nutrients, low oxygen on the bottom, but often rapid and vigorous growth of fish. Fishermen are very happy with this stage! However, when the stored nutrients are dispersed and the accumulated food used up, the reservoir stabilizes at a lower rate of productivity, greater benthic oxygen, and lower fish yields. Fishermen become unhappy with this stage! If the watershed is well protected by mature vegetation, or if the soils of the watershed are infertile, then the stabilized stage may last for some time—a "climax" of sorts. However, erosion and various man-accelerated nutrient inputs usually produce a continuing series of "transient states" until the basin fills up. Impoundments on impoverished watersheds or primary sterile sites will, of course, exhibit a reverse pattern of low productivity at the start. Failure to recognize the basic nature of ecological succession and the relationships between the watershed and the impoundment has resulted in many failures and disappointments in man's attempts to maintain such artificial ecosystems. As will be described in Section 3 of this chapter, shallow water systems can be "pulse stabilized" at high productivity levels by high energy water level fluctuations.

Because the oceans are, generally speaking, in a steady state, and because they have been chemically and biologically stabilized for centuries, oceanographers have not been concerned with ecological succession. However, with pollution threatening to disturb equilibria, the interaction of autogenic and allogenic processes will undoubtedly receive greater attention by marine scientists. Studies in coastal waters have already contributed to theory, as is illustrated by the important work of Margalef already cited. In another paper Margalef (1967a) summarizes his observation of the changes that take place in a successional gradient in the coastal water column as follows:

1. Average size of cell and relative abundance of mobile forms among the phytoplankton increase.

2. Productivity, or rate of multiplication, slows down.

3. Chemical composition of the phytoplankton, as exemplified by the plant pigments, changes.

4. The composition of the zooplankton shifts from passive filter feeders to more active and selective hunters in response to a shift from numerous small suspended food particles to scarcer food concentrated in bigger units and dispersed in a more organized (stratified) environment.

5. In the later stages of succession total energy transfer may be lower but its efficiency seems to be improved (this kind of efficiency was discussed in Chapter 3, page 77).

The succession of organisms on substrates has received a great deal of attention because of the practical importance of "fouling" of ship bottoms and piers by barnacles and other sessile marine organisms. Studies in marine waters have also supported the *"ectocrine" theory of succession* (see page 32 for explanation of this term), which holds that organic excretions from one set of populations stimulate replacement by another set of populations (see Smayda, 1963).

A final example, that of heterotrophic succession in a hay infusion culture, is taken from the classic experiments of Woodruff. When a culture medium made by boiling hay is allowed to stand, a thriving culture of bacteria develop. If some pond water (containing seed stock of various protozoa) is then added, Woodruff found that a definite succession of protozoan populations with successive dominants occurred, as shown in Figure 9–5. In this situation energy is maximal at the beginning and then declines. Unless new medium is added, or an autotrophic regime takes over, the system eventually runs down and all the organisms die or go into resting stages (spores, cysts, etc.). This is quite different from the autotrophic successions described in the previous examples in which energy flow is maintained indefinitely. The hay infusion microcosm is a good model for the kind of succession that takes place in decaying logs, animal carcasses, fecal pellets, and in the secondary stages of sewage treatment. It might also be considered a model for the "downhill" succession that must be associated with man's exploitation of fossil fuels! In all these examples we are dealing with a series of transient stages in a declining energy gradient with no possibility of a steady state being achieved.

Heterotrophic and autotrophic successions can be combined in a laboratory microecosystem model if samples from a derived system are added to media enriched with organic matter. Succession in such a system has been described, and energy flow measured, by Gorden and colleagues (1969). The system first becomes "cloudy" as het-

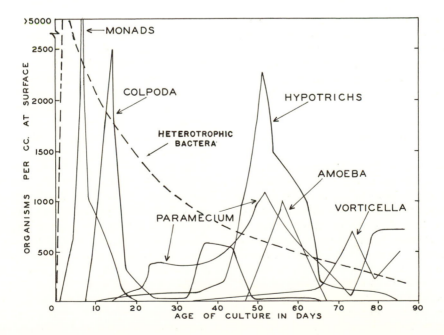

Figure 9–5. Succession in a hay infusion culture with dominance by successive species. This is a laboratory example of hetertrophic succession. (After Woodruff, 1912, with hypothetical curve for heterotrophic bacteria added.)

erotrophic bacteria "bloom," and then it turns bright green as nutrients and growth substances (especially the vitamin thiamine) required by algae are released by the activities of the bacteria. This, of course, is a good model of cultural eutrophication.

2. CONCEPT OF THE CLIMAX

Statement

The final or stable community in a developmental series (sere) is the climax community; it is self-perpetuating and in equilibrium with the physical habitat. Presumedly, in a climax community, in contrast to a developmental or other unstable community, there is no net annual accumulation of organic matter. That is, the annual production and import is balanced by the annual community consumption and export (see Figures 9–1 and 9–2). For a given region it is convenient, although somewhat arbitrary, to recognize (1) a single *climatic climax*, which is in equilibrium with the general climate, and (2) a varying number of *edaphic climaxes*, which are modified by local conditions of the substrate. The former is the theoretical community toward which all successional development in a given region is tending; it is realized where physical conditions of the substrate are not so extreme as to modify the effects of the prevailing regional climate. Succession ends in an edaphic climax where topography, soil, water, fire, or other disturbance are such that the climatic climax cannot develop.

Explanation and Examples

The assumption that autogenic development eventually produces a stable community is generally accepted as based on sound observation and theory. However, there have been two schools of interpretation. According to the "mono-climax" idea, any region has only one climax toward which all communities are developing, however slowly. According to the "poly-climax" idea, it is unrealistic to assume that all communities in a given climatic region will end up the same when conditions of physical habitat are by no means uniform. Nor are all habitats capable of being molded to a common level by the community within a

reasonable length of time as measured in terms of the life-span of man (or of a few multiples thereof!). A good compromise between these viewpoints is to recognize a single theoretical climatic climax and a variable number of edaphic climaxes, depending on the variation in the substrate. As was indicated in Chapter 7, analysis of complex situations in the light of "constants" and "variables" is sound procedure. Thus, the climatic climax is the theoretical constant against which observed conditions may be compared. The degree of deviation, if any, from the theoretical climax can be measured and the factors responsible for the deviation can be more readily determined when there is a basic "yardstick" available for comparison. One of the important justifications for preservation of wilderness areas is that we ought to have natural climaxes in every major geographic region available for comparison with various man-altered landscapes.

These concepts can best be illustrated by a specific example. Topographic situations in southern Ontario and the stable biotic communities associated with the various physical situations are shown in Figure 9–6. On level or moderately rolling areas where the soil is well drained but moist, a maple–beech community (sugar maple and beech being the dominant plants) is found to be the terminal stage in succession. Since this type of community is found again and again in the region wherever land configuration and drainage are moderate, the maple–beech community is judged to be the normal, unmodified climax of the region. Where the soil remains wetter or drier than normal (despite the action of communities), a somewhat different end-community occurs, as indicated. Still greater deviations from the climatic climax occur on steep south-facing slopes where the microclimate is warmer, or on north slopes and in deep ravines where the microclimate is colder (see Figure 5–19, page 138, which illustrates the magnitude of such differences). These latter climaxes often resemble climatic climaxes found farther south and north, respectively. Thus, as shown in Figure 9–6, we have the climatic climax where local climate and soils are normal, and various edaphic climaxes associated with different combinations of modified climates and drainages.

Theoretically, an oak–hickory community on dry soil, for example, would, if given indefinite time, gradually increase the or-

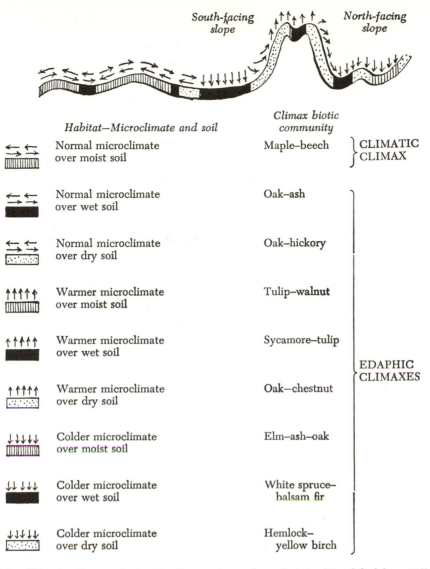

Habitat—Microclimate and soil	Climax biotic community	
Normal microclimate over moist soil	Maple–beech	CLIMATIC CLIMAX
Normal microclimate over wet soil	Oak–ash	
Normal microclimate over dry soil	Oak–hickory	
Warmer microclimate over moist soil	Tulip–walnut	
Warmer microclimate over wet soil	Sycamore–tulip	
Warmer microclimate over dry soil	Oak–chestnut	EDAPHIC CLIMAXES
Colder microclimate over moist soil	Elm–ash–oak	
Colder microclimate over wet soil	White spruce–balsam fir	
Colder microclimate over dry soil	Hemlock–yellow birch	

Figure 9–6. Climatic climax and edaphic climaxes in southern Ontario. (Simplified from Hills, 1952.)

ganic content of the soil and raise its moisture-holding properties, and thus eventually give way to the maple–beech community. Actually, we do not know whether this would occur or not, since we can see little evidence of such change and since records of undisturbed areas have not been kept for the many human generations that probably would be required. In contrast, a maple–beech community may develop in favorable situations in 200 years or less, even beginning from a plowed field! Starting with severe physical conditions, such as a steep slope or deep ravine, it seems likely that biotic communities would never be able to overcome the "handicaps" and the climatic climax would not be realized. In any event,

it is more practical to consider communities in such situations as edaphic climaxes which will remain quite stable in terms of man's life span, and probably until there is a change in the regional climate or a geological change in the substrate. In other words, the question of climatic and edaphic climaxes comes back to the point emphasized in the previous section. The orderly process of changes, which we define as ecological succession, results from the changes in the environment brought about by the organisms themselves. The more extreme the physical substrate, the more difficult modification of environment becomes and the more likely that community development will stop (or at least slow down to an imperceptible "crawl") without achieving an equilibrium condition with the regional climate.

A dramatic example of a climax controlled by special soil conditions is shown in Figure 9–7. In a certain area on the coast of northern California giant redwood forests occur side by side with pigmy forests of tiny, stunted trees. As shown in the figure, the same sandstone substrate underlies both forests, but the pigmy forest occurs where an impervious hardpan close to the surface greatly restricts root development as well as water and nutrient movement. The vegetation that reaches an equilibrium, or climax, condition with this special situation is almost totally different in species composition and structure from that of adjacent areas which lack the hardpan.

An interesting climax concept is the idea of the *catastrophic climax* or *cyclic climax*. The California chaparral vegetation (see picture on page 397) might be considered as an example. The biotic development results in a shrub climax that is especially vulnerable to "catastrophe," in this case, wildfire which removes not only the mature vegetation but also the antibiotics it produces. A rapid development of herbaceous vegetation follows until the shrub dominance is reestablished. This sort of ecosystem, then, has a natural cyclic climax controlled by the interaction of fire and antibiotics (see Chapter 7, page 228). Another example is the spruce forest in which periodic outbreaks of budworms kill the large mature trees, bringing on a vigorous growth

of young ones which largely escape damage. McDonald (1965) believes that spruce and budworm form a natural self-perpetuating system, and that attempts to control the outbreaks with insecticides is wasted effort that does more harm than good (see page 199).

Regions vary considerably in the percentage of area that is capable of supporting climatic climax communities. On the deep soils of the central plains early settlers found a large proportion of the land covered with a climax grassland. In contrast, on the sandy, "geologically young" lower coastal plain of the southeast the climatic climax (a broad-leaved evergreen forest) was originally as rare as it is today. Most of the coastal plain is occupied by edaphic climaxes or their seral stages. (See also pages 386 to 388.) In contrast, the oceans, which occupy geological ancient basins, can be considered to be in a climax state in so far as community development is concerned, as already indicated.

Usually, species composition has been used as a criterion to determine if a given community is climax or not. However, this alone is often not a good criterion, because species composition can undergo appreciable changes in response to seasons and short-term weather fluctuations even though the ecosystem as a whole remains stable. As already indicated, the P/R ratio, or other functional criterion, may provide a better index. Various statistical ratios, for example, pigment ratios (see page 62), turnover times (longer in climax than in developmental stages), coefficients of variation,

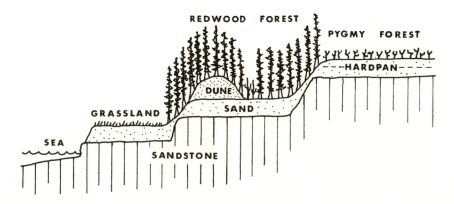

Figure 9–7. Edaphic climaxes on the coast of northern California (Mendocino region). Although the parent material (C-horizon) is the same (beach deposits and sandstone), forests of tall redwood and dwarf conifers grow side by side on adjacent marine terraces. The stunted nature of the pygmy forest is due to an iron-cemented B-horizon hardpan located about 18 inches below the surface. The soil above the impervious hardpan is extremely acid (pH 2.8 to 3.9) and low in Ca, Mg, K, P, and other nutrients. At least one of the dominant pygmy pines is an ecotype (see page 109) adapted especially to this extreme soil condition. (After Jenny, Arkley, and Schultz, 1969.)

indices of similarity, and other indicators of stability should also be useful.

Man, of course, has much to say about the progress of succession and the achievement of climaxes. When a stable community, which is not the climatic or edaphic climax for the given site, is maintained by man or his domestic animals, it may conveniently be designated as a *disclimax* (= *disturbance climax*) or *anthropogenic subclimax* (= man-generated). For example, overgrazing by stock may produce a desert community of creosote bushes, mesquite, and cactus where the local climate actually would allow a grassland to maintain itself. The desert community would be the disclimax, the grassland the climatic climax. In this case the desert community is evidence of poor management by man, whereas the same desert community in a region with a true desert climate would be a natural condition. An interesting combination of edaphic and disturbance climaxes occupies extensive areas in the California grassland region where introduced annual species have almost entirely replaced native prairie grasses. In an intensive study of these grasslands, McNaughton (1968) found that biomass, productivity, dominance-diversity relationships, and floristic composition of stabilized communities on sandstone soils differed markedly from those on serpentine soils which have a very low calcium and very high magnesium content (see page 112).

Agricultural ecosystems that have been stabilized for long periods of time can certainly be regarded as climaxes (or disclimaxes, if you prefer). Unfortunately, many crop systems, especially as currently managed in the tropics and on irrigated deserts, are by no means stable since they are subject to erosion, leaching, salt accumulation, and pest irruptions. As already indicated, maintaining high productivity in such systems requires an increasing "subsidy" by man. More about this in the next section.

3. RELEVANCE OF ECOSYSTEM DEVELOPMENT THEORY TO HUMAN ECOLOGY

Statement

The principles of ecosystem development bear importantly on the relationships between man and nature because the strategy of "maximum protection" (that is, trying to achieve maximum support of complex biomass structure), which characterizes ecological development, often conflicts with man's goal of "maximum production" (trying to obtain the highest possible yield). Recognition of the ecological basis for this conflict between man and nature is a first step in establishing rational land-use policies.

Explanation

Figure 9–2 depicts a basic conflict between the strategies of man and of nature. The "bloom-type" relationships, as exhibited by the 30-day microcosm or the 30-year forest, illustrate man's present idea of how nature should be directed. For example, the goal of agriculture or intensive forestry, as now generally practiced, is to achieve high rates of production of readily harvestable products with little standing crop left to accumulate on the landscape—in other words, a high P/B efficiency. Nature's strategy, on the other hand, as seen in the outcome of the successional process, is directed toward the reverse efficiency—a high B/P ratio, as is depicted by the relationship at the right in Figure 9–2. Man has generally been preoccupied with obtaining as much "production" from the landscape as possible, by developing and maintaining early successional types of ecosystems, usually monocultures. But, of course, man does not live by food and fiber alone; he also needs a balanced CO_2–O_2 atmosphere, the climatic buffer provided by oceans and masses of vegetation, and clean (that is, unproductive) water for cultural and industrial uses. Many essential life-cycle resources, not to mention recreational and esthetic needs, are best provided for man by the less "productive" landscapes. In other words, the landscape is not just a supply depot but is also the *oikos*—the home—in which we must live. Until recently mankind has more or less taken for granted the gas-exchange, water-purification, nutrient-cycling, and other protective functions of self-maintaining ecosystems, that is, until his numbers and his environmental manipulations became great enough to affect regional and global balances. The most pleasant and certainly the safest landscape to live in is one containing a variety of crops, forests, lakes, streams, roadsides, marshes, seashores, and "waste places"—in other words, a mixture of communities of different ecological ages. As individuals we more or less instinctively surround our houses with protective, non-edible cover (trees, shrubs, grass) at the same time that we strive to coax extra bushels from our cornfield. We all consider the cornfield a "good thing," of course, but most of us would not want to live there, and it would certainly be suicidal to cover the whole land area of the biosphere with corn-fields, since the boom and bust oscillation in such a situation would be severe.

Since it is impossible to maximize for conflicting uses in the same system, two possible solutions to the dilemma suggest themselves. We could continually compromise between quantity of yield and quality of living space, or we can deliberately plan to compartmentalize the landscape so as to simultaneously maintain highly productive and predominantly protective types as separate units subject to different management strategies (strategies ranging, for example, from intensive cropping on the one hand to wilderness management on the other). If ecosystem development theory is valid and applicable to planning, then the so-called multiple-use strategy, about which we hear so much, will work only through one or both of these approaches, because in most cases the projected multiple uses conflict with one another. It is appropriate, then, to examine some examples of the compromise and the compartmental strategies.

Pulse Stability

A more or less regular but acute physical perturbation imposed from without can maintain an ecosystem at some intermediate point in the developmental sequence, resulting in, so to speak, a compromise between youth and maturity. What one might term "fluctuating water level ecosystems" are good examples. Estuaries, and inter-tidal zones in general, are maintained in an early, relatively fertile stage by the tides, which provide the energy for rapid nutrient cycling. Likewise, freshwater marshes, such as the Florida Everglades, are held at an early successional stage by the seasonal fluctuations in water levels. The dry-season drawdown speeds up aerobic decomposition of accumulated organic matter, releas-

ing nutrients that, on reflooding, support a wet-season bloom in productivity. The life histories of many organisms are intimately coupled to this periodicity. The wood stork, for example, breeds when the water levels are falling, and the small fish on which it feeds become concentrated and easy to catch in the drying pools. If the water level remains high during the usual dry season or fails to rise in the wet season, the stork will not nest (Kahl, 1964). Stabilizing water levels in the Everglades by means of dikes, locks, and impoundments, as is now advocated by some, would, in the author's opinion, destroy rather than preserve the Everglades as we now know them just as surely as complete drainage would. Without periodic drawdowns and fires, the shallow basins would fill up with organic matter and succession would proceed from the present pond-and-prairie condition toward a scrub or swamp forest.

It is strange that man does not readily recognize the importance of recurrent changes in water level in a natural situation such as the Everglades when similar pulses are the basis for some of his most enduring food culture systems. Alternate filling and draining of ponds has been a standard procedure in fish culture for centuries in Europe and the Orient. The flooding, draining, and soil-aeration procedure in rice culture is another example. The rice paddy is thus the cultivated analogue of the natural marsh or the intertidal ecosystem.

Fire is another physical factor whose periodicity has been of vital importance to man and nature over the centuries. As described in Chapter 4, whole biotas, such as those of the African grasslands and the California chaparral, have become adapted to periodic fires, producing what ecologists often call "fire climaxes." For centuries man has used fire deliberately to maintain such climaxes or to set back succession to some desired point. The fire-controlled forest (see Figure 5–16) yields less wood than a tree farm does (that is, young trees, all of about the same age, planted in rows and harvested on a short rotation; see Figure 15–3A), but it provides a greater protective cover for the landscape, wood of higher quality, and a home for game birds (quail, wild turkey, and so on) that could not survive in a tree farm. The fire climax, then, is an example of a compromise between production and simplicity on the one hand and protection and diversity on the other.

It should be emphasized that pulse stability works only if there is a complete community (including not only plants but animals and microorganisms) adapted to the particular intensity and frequency of the perturbation. Adaptation (operation of the selection process) requires times measurable on the evolutionary scale. Most physical stresses introduced by man are too sudden, too violent, or too arrhythmic for adaptation to occur at the ecosystem level, so severe oscillation rather than stability results. In many cases, at least, modification of naturally adapted ecosystems for cultural purposes would seem preferable to complete redesign.

Prospects for a Detritus Agriculture

As indicated above, heterotrophic utilization of primary production in mature ecosystems involves largely a delayed consumption of detritus. There is no reason why man cannot make greater use of detritus and thus obtain food or other products from the more protective type of ecosystem. Again, this would represent a compromise, since the short-term yield could not be as great as the yield obtained by direct exploitation of the grazing food chain. A detritus agriculture, however, would have some compensating advantages. Present agricultural strategy is based on selection for rapid growth and edibility in food plants, which, of course, make them vulnerable to attack by insects and disease. Consequently, the more we select for succulence and growth, the more effort we must invest in the chemical control of pests; this effort, in turn, increases the likelihood of our poisoning useful organisms, not to mention ourselves. Why not also practice the reverse strategy—that is, select plants that are essentially unpalatable, or that produce their own systemic insecticides while they are growing, and then convert the net production into edible products by microbial and chemical enrichment in food factories? We could then devote our biochemical genius to the enrichment process instead of fouling up our living space with chemical poisons! The production of silage by fermentation of low-grade fodder is an example of such a procedure already in widespread use. The cultivation of detritus-eating fishes in the Orient is another example.

By tapping the detritus food chain man can also obtain an appreciable harvest from many natural systems without greatly modifying them or destroying their protective and esthetic value.

The Compartment Model

Successful though they often are, compromise systems are neither suitable nor desirable for the whole landscape. More emphasis needs to be placed on compartmentalization, so that growth-type, steady-state, and intermediate-type ecosystems can be linked with urban and industrial areas for mutual benefit. Knowing the transfer coefficients that define the flow of energy and the movement of materials and organisms (including man) between compartments, it should be possible to determine, through analog computer manipulation, rational limits for the size and capacity of each compartment. We might start, for example, with a simplified model, shown in Figure 9–8, consisting of four compartments of equal area, partitioned according to the basic biotic-function criterion—that is, according to whether the area is (1) productive, (2) protective, (3) a compromise between (1) and (2), or (4) urban-industrial. By continually refining the transfer coefficients on the basis of real-world situations, and by increasing and decreasing the size and capacity of each compartment through computer simulations, it would be possible to determine objectively the limits that must eventually be imposed on each compartment in order to maintain regional and global

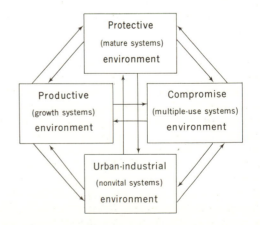

Figure 9–8. Compartment model of the basic kinds of environment required by man, partitioned according to ecosystem development and life-cycle resource criteria. See text for explanation. (After E. P. Odum, 1969.)

balances in the exchange of vital energy and of materials. A systems-analysis procedure provides at least one approach to the solution of the basic dilemma posed by the question "How do we determine when we are getting too much of a good thing?" Also it provides a means of evaluating the energy drains imposed on ecosystems by pollution, radiation, harvest, and other stresses (H. T. Odum, 1967).

Implementing any kind of compartmentalization plan, of course, would require procedures for zoning the landscape and restricting the use of some land and water areas. While the principle of zoning in cities is universally accepted, the procedures now followed do not work very well because zoning restrictions are too easily overturned by short-term economic and population pressures. Zoning the landscape would require a whole new order of thinking, including environmental and planning commissions invested with regulatory powers. Greater use of legal measures providing for tax relief, restrictions on use, scenic easements, and public ownership will be required if appreciable land and water areas are to be held in the "protective" categories. Several states (for example, New Jersey and California), in which pollution and population pressure are beginning to hurt, have made a start in this direction by enacting "open space" legislation designed to get as much unoccupied land as possible into a "protective" status so that the quality of the total environment can be preserved. Some but not all regions of the United States are fortunate in that large areas are in national forests, parks, and wildlife refuges. The fact that such areas, as well as the bordering oceans, are not quickly exploitable gives us time for the accelerated ecological study and programming needed to determine what proportions of different types of landscape provide a safe balance between man and nature. The open oceans, for example, should forever be allowed to remain protective rather than productive territory because of the biosphere's governor, as described in Chapter 2. The sea controls climates and slows down and controls the rate of decomposition and nutrient regeneration, thereby creating and maintaining the highly aerobic terrestrial environment to which the higher forms of life, such as man, are adapted. Eutrophication of the ocean in a last-ditch effort to feed the populations of

the land could well have an adverse effect on the gas and heat balances of the atmosphere.

Until we can determine more precisely how far we may safely go in expanding intensive agriculture and urban sprawl at the expense of the protective landscape, it will be good insurance to hold inviolate as much of the latter as possible. Thus, the preservation of natural areas is not a peripheral luxury for society but a capital investment from which we expect to draw interest. Also it may well be that restrictions on the use of land and water are our only practical means of avoiding overpopulation or too great an exploitation of resources, or both. Interestingly enough, restriction of land use is the analogue of a natural behavioral control mechanism known as "territoriality" by which many species of animals avoid crowding and social stress, as was discussed in detail in Chapter 7, Section 15. A blueprint for achieving such goals in human affairs is discussed in Chapter 21.

In summary, the tabular model for ecosystem development has many parallels in the development of human society itself. In the pioneer society, as in the pioneer ecosystem, high birth rates, rapid growth, high economic profits, and exploitation of accessible and unused resources are advantageous, but as the saturation level is approached, these drives must be shifted to considerations of symbiosis (that is, "civil rights," "law and order," "education," and "culture"), birth control, and the recycling of resources. A balance between youth and maturity in the socioenvironmental system is, therefore, the really basic goal that must be achieved if man as a species is to successfully pass through the present rapid-growth stage, to which he is clearly well adapted, to some kind of controlled growth-and-equilibrium stage, of which he as yet shows little understanding and to which he now shows little tendency to adapt.

4. EVOLUTION OF THE ECOSYSTEM

Statement

As in the case with short-term development as described in the preceding sections of this chapter, the long-term evolution of ecosystems is shaped by (1) allogenic (outside) forces such as geological and climatic changes and (2) autogenic (inside) processes resulting from activities of the living components of the ecosystem. The first ecosystems three billion years ago were populated by tiny anaerobic heterotrophs that lived on organic matter synthesized by abiotic processes. Following the origin and population explosion of algal autotrophs, which converted a reducing atmosphere into an oxygenic one, organisms have evolved through the long geological ages into increasingly complex and diverse systems that (1) have achieved control of the atmosphere and (2) are populated by larger and more highly organized multicellular species. Within this community component evolutionary change is believed to occur principally through *natural selection at or below the species level,* but natural selection above this level may also be important, especially (1) *coevolution,* that is, the reciprocal selection between interdependent autotrophs and heterotrophs, and (2) *group or community selection,* which leads to the maintenance of traits favorable to the group even when disadvantageous to the genetic carriers within the group.

Explanation

The broad pattern of the evolution of organisms and the oxygenic atmosphere, which make the biosphere absolutely unique among the planets of our solar system, is pictured in Figure 9–9. It is now generally believed that when life began on earth more than three billion years ago, the atmosphere contained nitrogen, ammonia, hydrogen, carbon monoxide, methane, and water vapor, but no free oxygen (see Berkner and Marshall, 1964, 1965; Drake (ed.) 1968; Tappen, 1968; and Calvin, 1969). It also contained chlorine, hydrogen sulfide, and other gases that would be poisonous to much of present-day life. The composition of the atmosphere in those early days was largely determined by the gases from volcanos, which were much more active then than now. Because of the lack of oxygen there was no ozone layer (O_2 acted on by short-wave radiation produces O_3, which in turn absorbs ultraviolet radiation) to shield out the sun's deadly ultraviolet radiation, which, therefore, penetrated to the surface of the land and water. Such radiation would kill any exposed life, but, strange to say, it was this radiation that is thought to have created a chemical evolution leading to complex organic molecules such as amino acids, which became the building blocks for primitive life. The very small amount of nonbiological oxygen produced by ultraviolet dissociations of water vapor may have provided enough ozone to form a slight shield against ultraviolet radiation. Yet as long as the atmospheric oxygen and ozone remained scarce, life could only develop under the protective cover of water. The first living organisms, then, were aquatic yeastlike anaerobes that obtained the energy necessary for respiration by the process of fermentation. Because fermentation is so much less efficient than oxidative respiration (see page 27), the primitive life could not evolve beyond the single-cell stage.

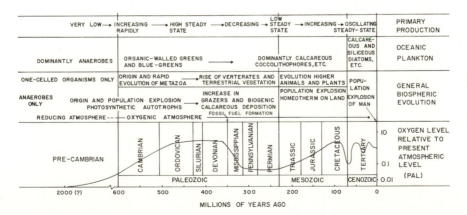

Figure 9–9. The evolution of the biosphere and its oxygenic atmosphere. See text for explanation. (Redrawn and extended on consultation with Helen Tappan, 1968.)

Primitive organisms also had a very limited food supply since they would be dependent on the slow sinking of organic materials synthesized by radiation in the upper water layers where they, the hungry microbes, could not venture! Thus, for millions of years life must have existed in a very limited and precarious condition. Berkner and Marshall (1966) picture the situation as follows: "This model of primitive ecology calls for pool depths sufficient to absorb the deadly ultraviolet, but not so deep as to cut off too much of the visible light. Life could have originated on the bottom of pools or shallow, protected seas fed, perhaps, by hot springs rich in nutrient chemicals."

Presumedly, the scarcity of organic food exerted selection pressure that led to the evolution of photosynthesis. The gradual buildup of oxygen, biologically produced in water, and its diffusion into the atmosphere brought about tremendous changes in the earth's geochemistry and made possible the rapid expansion of life and the development of larger and more complex living systems. Many minerals such as iron were precipitated from water and formed characteristic geological formations. As the oxygen in the atmosphere increased, the layer of ozone formed in the upper atmosphere thickened, the earth's surface became shielded, and life could then move up to the surface of the sea. At the same time, aerobic respiration made possible the development of complex multicellular organisms. It is thought that the first multicellular animals appeared when oxygen reached about 3 per cent of its present level (or about 0.6 per cent of the atmosphere as compared to present 20 per cent), a time now dated at about 600 million years ago, the beginning of the Cambrian period (Fig. 9–9). The term "pre-Cambrian" is used to cover that vast period of time during which only the small and primitive "single-cell" life existed. During the Cambrian there was an evolutionary explosion of new life, such as sponges, corals, worms, shellfish, seaweed, and the ancestors of seed plants and the vertebrates. Thus, the fact that the tiny green plants of the sea were able to produce an excess of oxygen over the respiration needs of all organisms made it possible to populate the whole earth in a comparatively short time. In the following periods of the Paleozoic era life not only filled all the seas, but invaded the land. Development of the green mantle of vegeta-tion provided more oxygen and food for the subsequent evolution of large creatures such as dinosaurs, mammals, and then men. At the same time calcareous and then siliceous forms were added to the organic-walled phytoplankton of the oceans (Fig. 9–9).

When oxygen use finally caught up with oxygen production sometime in the mid-Paleozoic, the concentration in the atmosphere, it is now believed, was at about its present level of 20 per cent. From the ecological viewpoint, then, biosphere evolution seems to be very much like a heterotrophic succession followed by an autotrophic regime, such as one might set up in a laboratory microcosm starting with culture medium enriched with organic matter. Since the Devonian, the geological evidence indicates some ups and downs (see Figure 9–9). During the late Paleozoic there was a marked decline (to maybe 50 per cent of present level) of O_2 and an increase in CO_2 accompanied by climatic changes. The increase in CO_2 may have triggered the vast "autotrophic bloom" that created the fossil fuels on which man's industrial civilization now depends. Following a gradual return to a high O_2–low CO_2 atmosphere, the O_2/CO_2 balance remained in what might be called an "oscillating steady state." As discussed in Chapter 2 and again in Chapter 4, man-generated CO_2 and dust pollution may be making this precarious balance still more "unsteady."

Incidentally, the story of the atmosphere, as we have briefly described it, should be told to every citizen and school child because it dramatizes man's absolute dependence on other organisms in his environment.

Figure 9–10 is a simplified model showing how the population components of the ecosystem evolve with time. Changes in the species composition of communities occur when new species evolve, old species become extinct, and surviving species change in abundance or genetic composition. During the hypothetical history of the 10 species populations shown in Figure 9–10, we start with 7 species and end up with 8 species, 2 of which are new and 5 of which have changed in abundance or genetic composition, or both! During the time period one species (no. 10) evolved and became extinct, while one species (no. 6) managed to persist without change. Many biologists believe that all evolutionary

Figure 9–10. Diagram showing several ways in which changes in species populations result in evolutionary changes in the community; for example, genetic (gene pool) change with or without a change in population size (or distribution); a change in population size without a change in the gene pool; evolution of new species and extinction of old ones. (After Valentine, 1968.)

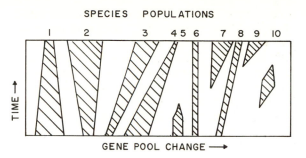

changes can be explained within the framework of the conventional natural selection–mutation theory (sometimes called neo-Darwinism), but geneticists, and especially ecologists, are intrigued by the possibilities of natural selection at the higher levels of organization. Accordingly we shall briefly consider such possibilities in the next two sections. Simpson (1969), in reviewing the "first three billion years of community evolution," concludes that the taxonomic composition of ecosystems had not yet been stabilized.

5. COEVOLUTION

Statement

Coevolution is a type of community evolution (i.e., evolutionary interactions among organisms in which exchange of genetic information among the kinds is minimal or absent) involving reciprocal selective interaction between two major groups of organisms with a close ecological relationship, such as plants and herbivores, large organisms and their microorganism symbionts, or parasites and their hosts.

Explanation

Using their studies of butterflies and plants as a basis, Ehrlich and Raven (1965) have outlined the theory of coevolution. Their hypothesis may be outlined as follows: Plants, through occasional mutations or recombinations, produce chemical compounds not directly related to basic metabolic pathways (or perhaps as waste byproducts generated in these pathways) that are not inimical to normal growth and development. Some of these compounds, by chance, serve to reduce the palatability of the plants to herbivores. Such a plant, protected from phytophagous insects, would in a sense have entered a new adaptive zone. Evolutionary radiation of the plants might follow, and eventually what began as a chance mutation or recombination might characterize an entire family or group of related families. Phytophagous insects, however, can evolve in response to physiological obstacles, as shown by man's recent experience with insecticides. Indeed, response to secondary plant substances and the evolution of resistance to insecticides seems to be intimately connected (here citing Gordon, 1961). If a mutant or recombinant appeared in a population of insects that enabled individuals to feed on the previously protected plant, selection would carry the line into a new adaptive zone, allowing it to diversify in the absence of competition with other herbivores. Thus, the diversity of plants not only may tend to augment the diversity of phytophagous animals, but the converse may also be true. In other words, the plant and the herbivore evolve together in the sense that the evolution of each is dependent on the evolution of the other. Ehrlich and Raven even go on to suggest that the plant-herbivore reciprocal selective responses may account for the high diversity of plants in the tropics (see page 149) where warm climates are especially hospitable to insects. Pimentel (1968) has used the expression "genetic feedback" for this kind of evolution which leads to population and community homeostasis within the ecosystem (see pages 197 and 221).

Examples

Coevolution has apparently resulted in a remarkable association between the bull's horn acacia plant (*Acacia cornigera*) and an ant (*Pseudomyrmex ferrugunea*) in Mexico and Central America as described by Janzen (1966, 1967). The ants live in colonies within swollen thorns of the acacia plant. If the ants are removed, severe defoliation by herbivorous insects (that would normally be preyed upon by the ants) occurs and subsequent shading by competing vegetation results in the death of the acacia in 2 to 15 months. The plant is thus dependent on the insect for protection against grazing. It would seem likely that reciprocal selection and genetic feedback are involved in all cases of mutualism, as described in Chapter 7, Section 19.

Coevolution can involve more than one step in the food chain. Brower and colleagues (1968), for example, have studied the monarch butterfly (*Danaus plexippus*), which is well known for its general unpalatability to vertebrate predators. They found that this insect is able to sequester the highly toxic glycosides present in milkweed plants on which they feed, thereby providing a highly effective defense against bird predators (not only for the caterpillar but for the adult butterfly as well). Thus, this insect has not only evolved the ability to feed on a plant that is unpalatable to other insects, but it "uses" the plant poison for its own protection against predators!

Finally, coevolution is not restricted to phagotrophic interactions. Plant populations comprising several species can be linked together in the community by their dependence on, and common evolution with, a single species of insect or hummingbird pollinator (Baker, 1963). It is easy to see how stepwise reciprocal selections can account for evolutionary trends toward diversity, interdependence, and homeostasis at the community level.

6. GROUP SELECTION*

Statement

Group selection is natural selection between groups of organisms not necessarily closely linked by mutualistic associations. Group selection theoretically leads to the maintenance of traits favorable to populations and communities, but selectively dis-

---

* This section was prepared by Dr. R. H. Crozier, Department of Zoology, University of Georgia.

advantageous to genetic carriers within populations. Conversely, it may eliminate, or keep at low frequencies, traits unfavorable to the survival of the species but selectively favorable within populations. Group selection involves extinction of populations in a process analogous to the selection of genotypes within populations by the death or reduced reproductive capacity of the appropriate types of individuals. The efficacy of group selection may, however, be enhanced by genetic drift. Group selection is a highly controversial subject among geneticists and will probably remain so for some time, as witnessed by the book-length arguments *against* by Williams (1966) and *for* by Darlington (1958) and Waddington (1962).

Explanation

Although few doubt that group selection can occur, there is considerable uncertainty over the amount of influence it can have in evolution. Williams (1966) believes that almost all group adaptations, which have been attributed to the action of group selection by population ecologists such as Wynne-Edwards (1962, 1964, 1965), can be attributed to the traditional action of individual selection. Certainly his view that it should be invoked as a mechanism only when the conclusion is inescapable has merit. Thus Orians (1969) has shown that the territorial and polygynous breeding systems of many birds, interpreted by some as group adaptations to facilitate population control, may be explained as being evolved and maintained by traditional selection at the individual level. Williams maintains that group selection is *a priori* unlikely to be an important agency countering the operation of selection within populations because of the assumed large number of generations (during which individual selection can act) before population extinction usually occurs. There is recent evidence, however, in the work of Wilson and his co-workers (see Wilson, 1969) that population extinction may take place at a high rate, allowing considerable potential for group selection to occur. Before it can be decided whether or not group selection is likely to be an important factor in the evolution of any species, the rate of extinction of local populations must be determined; this is likely to be a crucial interest of evolutionary ecology for some years to come.

A model of group selection not involving population extinction has been suggested by Wright (1945). In his model, one genotype is selectively superior to the other during the course of intrapopulation selection, but populations fixed for it or still segregating have a smaller size than those fixed for the other allele. Wright suggests that, at an appropriate population size, while some populations will become fixed for the low-population-fitness allele through selection, others will become fixed for the high-population-fitness allele through genetic drift. These latter populations will then increase in size and contribute to the flow of immigrants between populations. This disproportionate gene flow will tend, in populations still segregating, to counteract the effects of individual selection and "pull" some of these populations over to fixation for the favorable allele. The frequency of the favorable, high-population-fitness allele will thus increase in the species even though selected against within any one population. This model, however, seems to require a more delicate adjustment of factors than that required in the population-extinction model and so is not as plausible, although, as we shall see in the case of t alleles in the mouse, drift can be important in the latter case as well.

In small populations random genetic drift can interact with selection to yield results unexpected when selective pressures alone are considered. Wright and Kerr (1954) kept a number of small populations of *Drosophila melanogaster* that each initially contained equal numbers of bar-eye and wild-type alleles until they became monomorphic for one allele or the other. Although bar-eye is strongly selected against and would always be lost in large populations, a small proportion of the experimental populations did become fixed for it. This constitutes experimental proof that the course of selection can be negated some of the time by genetic drift.

Example

A clear example of group selection has been pointed out by Lewontin (1962), who carried out computer simulations of the course of three sterile t alleles in populations of the house mouse. These particular alleles lead to male sterility when homozygous, but between 85 and 99 per cent of the sperm of males heterozygous for any allele carry the allele, instead of 50 per cent as expected. The exact frequency of transmission depends on the particular allele. In large populations, the frequency of the allele should stabilize at the point at which the rate of loss in homozygotes is balanced through the replacement of alleles in the aberrant segregation ratios of heterozygous males. This point is $q = 2m - 1$, where m is the transmission ratio. For the allele t^{w2}, with $m = 0.86$ and theoretically $q = 0.70$, the observed value in one population was $q = 0.37$. Lewontin found that a breeding structure in which populations exist as small family groups of two males and six females could generate, in computer simulations, the approximate observed gene frequency. In 70 per cent of the computed families, the t allele spread to fixation either completely or at least in the males and caused extinction of the family. However, 30 per cent of families lost the t allele through genetic drift, all populations being fixed after 24 generations. The t allele would thus be lost were it not for the fact that migration in nature constantly reinfects some populations free of the alleles so that the actual frequency of any allele in a population depends on the likelihood of extinction of groups carrying it and the rate of reinfection of fresh family groups by migration.

SYSTEMS ECOLOGY: THE SYSTEMS APPROACH AND MATHEMATICAL MODELS IN ECOLOGY

Chapter 10

By Carl J. Walters

Institute of Animal Resource Ecology, The University of British Columbia, Vancouver, Canada.

> Every theory of the course of events in nature is necessarily based on some process of simplification and is to some extent, therefore, a fairy tale.
>
> Sir Napier Shaw

INTRODUCTION

We now come full circle. We started with the whole, that is, the concept of the ecosystem (Chapters 1 and 2), then proceeded to detail functional and biotic components of the ecosystem (Chapters 3 through 8) and their integrated development and evolution (Chapter 9). We now return again to the whole in terms of "formal" or mathematical treatment of the "informal" graphic and verbal models that have been used throughout this text to illustrate principles. The general concept of modeling was discussed in Section 3, Chapter 1, and it might be worthwhile to refer back to this section, and also to the section on cybernetics (page 34), as an introduction to this chapter. Diagrams of compartment and circuit models have been presented in previous chapters (see, for example, Figures 3–11 and 3–17) and frequent mention has been made of computer generation of curves that mimic or predict real-world phenomena (see, for example, Figure 7–36). The introduction to differential equations in Chapter 7, Sections 7, 8, 17, and 18, provides a necessary background for this chapter.

The application of systems analysis procedures to ecology has come to be known as *systems ecology*. As the formalized approach to holism, systems ecology is becoming a major science in its own right for two reasons: (1) extremely powerful new formal tools are now available in terms of mathematical theory, cybernetics, electronic data processing, and so forth, and (2) formal simplification of complex ecosystems provides the best hope for solutions to man's environmental problems that can no longer be trusted to trial-and-error, or one-problem—one-solution, procedures that have been chiefly relied on in the past. Since systems ecology is the wave of the future, it is appropriate that we turn to a young man trained in this new field for an introduction prepared especially for the beginning student who is not yet highly trained in mathematics and information processing. Carl Walters is a student of George Van Dyne, who, together with Jerry Olson and Bernard Patten, organized one of the first training programs for systems ecology when they were together at the Oak Ridge National Laboratory. Van Dyne and Patten have since moved to universities in which new generations of ecologists are being trained. These pioneers together with the active centers of training and research created by Ken Watt and C. S. Holling, who began by modeling population dynamics, and Howard T. Odum, who started by modeling energy flow, are revolutionizing the field of ecology, and providing a vital link with engineering where systems analyses procedures have been in use for some time. Carl Walters is thus a member of the first generation trained to teach as well as to conduct research in an area that must be widely understood and effectively applied in the future. In this chapter, Dr. Walters writes with such remarkable clarity and understanding that a potentially difficult subject has been reduced to a dialogue understandable by anyone with a background in college mathematics who has given a little time and thought to the basic principles of ecology as outlined in the preceding chapters.

1. THE NATURE OF MATHEMATICAL MODELS

Statement

Mathematical symbols provide a useful shorthand for describing complex ecological systems, and equations permit formal statements of how ecosystem components are likely to interact. The process of translating physical or biological concepts about any system into a set of mathematical relationships, and the manipulation of the mathematical system thus derived, is called systems analysis. The mathematical system is called a model and is an imperfect and abstract representation of the real world.

Explanation

Though we often think of "models" in terms of equations and computers, they can be defined more generally as any physical or abstract representations of the structure and function of real systems. Throughout this text, extensive use is made of "picture" and "verbal" models as aids to the understanding of complicated ecological processes. If described in full detail without bene-fit of some framework or outline, biological systems would appear hopelessly complex. Systems analysis is concerned with the explicit recognition, and handling of, complexity in the development of abstract models; systems analysis is simply a tool for understanding. An excellent introduction to the use of systems analysis in ecology has been published by Dale (1970).

Ability to describe and predict behavior of ecological systems by the use of models depends largely on a principle of all systems: that of *hierarchical organization* (or principle of integrative levels). This principle, discussed in Chapter 1, states simply that it is not necessary to understand precisely how a component of a system is structured from simpler subcomponents in order to predict how it will behave. Thus, it is not necessary to have full comprehension of biochemistry in order to describe the physiology of cells, nor is it necessary to understand physiology completely in order to describe the dynamics of animal populations. The concept of hierarchical organization is illustrated in Figure 10–1 in terms of "black boxes." *Understanding* in the study of systems is thought of as the ability to see how a system component is organized from simpler parts. The degree of hierarchical breakdown employed in the development of a

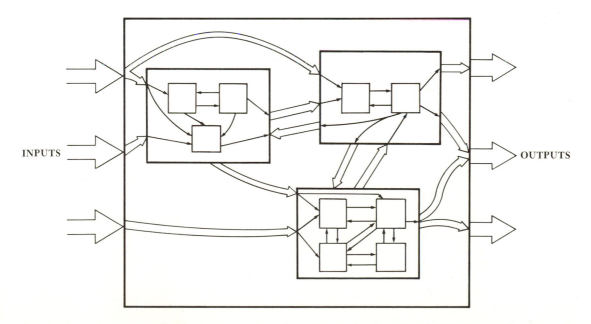

Figure 10–1. Processes and structures in ecological systems can be thought of as "black boxes" consisting of simpler black boxes, in a hierarchy of complexity. This diagram shows three levels of organization. By observing the input-output relationship for any box, one may be able to predict its behavior without understanding how it is put together from simpler components.

particular mathematical model depends on the purpose for which the model is being developed, rather than on ability to recognize natural subdivisions of the system. Although models are imperfect abstractions of real systems, they represent extremely powerful tools for the ecologist, because tentative answers and predictions regarding important matters are more important in the long run than precise treatment of unimportant details.

2. THE GOALS OF MODEL BUILDING

Statement

Models may be constructed for a variety of reasons. By providing an abstract and simplified *description* of some system, they may be used simply to guide research efforts or outline a problem for more detailed study. More often, mathematical models are developed for *prediction* of dynamic change over time. The failure of a model to predict change is in itself useful, because it may point out flaws in the conceptual framework from which the model was developed. Models can be evaluated in terms of three basic properties or goals: *realism, precision,* and *generality.* Realism refers to the degree to which the mathematical statements of the model, when translated into words, correspond to the biological concepts that they are intended to represent. Precision is the ability of the model to predict numerical change and to mimic the data on which it is based. Generality refers to the breadth of applicability of the model (the number of different situations in which it can be applied).

Explanation

Until recently, mathematical models were developed primarily in the physical sciences, in physiology, and in applied fields such as military logistics and fisheries management. In these cases, the systems being studied can be clearly defined, and models are constructed to answer specific questions. In contrast, ecological systems are often difficult to define in space and time and can be characterized in models by a host of "per-

formance measures" (energy, nutrients, population sizes, and so on); the questions asked of ecological models are often complex and based on such diffuse problems as "stability" and "trophic efficiency." Because ecosystems have highly random inputs, such as weather, it has seemed unreasonable to construct models of high predictive power when basic inputs frequently cannot be measured or predicted. Thus, ecological models are often judged in terms of generality and ability to guide research effort rather than on numerical predictive power (precision). Considering the immense complexity of interactions between plants and animals, and the difficulty of identifying and measuring these interactions, some mathematical ecologists have concluded that models cannot be both realistic and general (Levins, 1966). Like beauty, however, realism is in the eye of the beholder; some of the "box-and-arrow" models in this book may appear exceedingly complicated and realistic to the student, but absurdly simple to the practiced ecologist. The hierarchical organization of complexity in ecological systems assures that no matter how detailed the model may be, it can always be considered unrealistic if viewed in terms of lower (less abstract) levels of biological organization.

In applied ecological problems in which prediction is the goal, realism and generality are often sacrificed for precision. For example, in fisheries models it is usually necessary to predict average growth rate of individuals. Growth can be modeled accurately with equations having little basis in reality; these are sufficient for the fisheries biologist because his interest is in *yield* for a particular population, over a restricted range of population density. (For such a fishery model, see Figure 7–35.)

Two concepts related to realism and generality are *resolution* and *wholeness.* Bledsoe and Jamieson (1969) define resolution as being ". . . related to the number of attributes of a system which the model attempts to reflect." As discussed below, ecological models are constructed by representing each part of the system with a number or series of numbers. Thus, a population of animals might be represented by the number of animals in each age class, the average size of animals, and the sex ratio. Resolution in this case could be considered high; a simpler model might ignore the age distribution or sex ratio. Wholeness, as defined by Holling (1966a), refers to the number of biological processes and interactions reflected in the model. Wholeness and resolution can be thought of as subjective measures of the degree of subdivision in hierarchies of biological function and structure.

3. THE ANATOMY OF MATHEMATICAL MODELS

Statement

It is convenient to think of the mathematical model as having four basic elements. *System variables* are sets of numbers which are used to represent the *state*, or condition, of the system at any time. Ecological systems are usually thought of as consisting at any time of a series of components or compartments; in models, one or more system variables are used to characterize the state of each component. Flows or interactions between components are represented by equations called *transfer functions* or *functional relationships*. Inputs to the system, or factors affecting but not affected by the components of the system, are represented by equations called *forcing functions*. Finally, constants of the mathematical equations are called *parameters*.*

Explanation

Though system variables can be of many kinds, they usually represent biological quantities or constituents, or rates of change of these quantities. Examples are: (1) the amounts of energy at producer, consumer, and decomposer trophic levels, or the rate of energy flux between producers and consumers; (2) the number of animals in a population; (3) the amount of time needed by a predator to digest each prey. For a concise mathematical representation of the state of the system, we usually order the system variables into a list, called the *system state*

---

* Models may be either *stochastic* or *deterministic*. Stochastic models attempt to include the effects of random variability in forcing functions and parameters. Deterministic models ignore this chance variation. Stochastic models are mathematically difficult to deal with, so we will use mostly deterministic models as examples.

vector, and represent the list by a single symbol:

$$V = \begin{pmatrix} v_1 \\ v_2 \\ \cdot \\ \cdot \\ \cdot \\ v_n \end{pmatrix}$$

Here v_1 represents the value at any time of system variable 1, v_2 the value of variable 2, and so forth, and n is the number of variables included in the model. As an example, suppose we wish to make a very simple model of energy transfer in Silver Springs, Florida. Using the symbols and numbers from Figure 6–1A, we might choose biomass as a measure of energy available for transfer, and represent the state of the system as:

$$\begin{pmatrix} P \\ H \\ C \\ TC \end{pmatrix} = \begin{pmatrix} 809 \\ 37 \\ 11 \\ 1.5 \end{pmatrix}$$

Our mathematical model would then consist of a set of equations to describe the movement of energy from one component to another, and a forcing function to describe energy input (see Figure 6–1B). Parameters of this simple model would represent consumption efficiencies, respiration rates, and the like. Notice that v_1, v_2, etc., can themselves be considered sets of numbers (vectors), and used to symbolize the set of attributes which characterize each part of the system.

Transfer functions or functional relationships can take a variety of forms. Most generally, they are represented in terms of the way that each variable is *changing*, as

$$\frac{\Delta v_i}{\Delta t} = f(v_1, v_2, \ldots, v_n, F_1, F_2, \ldots, F_k)$$

Here, $\Delta v_i / \Delta t$ † is some measure of the rate of change with respect to time of system component i, and $f(v, F)$ means "a function of v_1, of $v_2, \ldots$, and of v_n, and of $F_1, \ldots, F_k$," where the v_i are system components as above and the F_i are the values of the forcing functions. *Solving* the systems model usually involves finding the values of the

---

† This type of notation was first introduced in Section 3, Chapter 7.

v_i over time, given starting values for each (*initial conditions*) and the rate equations. Another approach, employed by Holling (1966), is to choose a fixed Δv_i and solve for Δt (the amount of time required for a given change to take place). This approach is particularly useful in predator-prey models, where v_i may be the number of prey, and $\Delta v_i = 1$ represents the consumption of a single prey. For the Silver Springs example above, F_1 might be the rate of solar energy input and F_2 might be water temperature or the rate of detritus input. A simple transfer function for herbivore biomass in Silver Springs would be

$$\frac{\Delta v_2}{\Delta t} = \frac{\Delta H}{\Delta t} = f(P, H, C, TC, F_1, F_2)$$
$$= [k_{12}P - k_{22}H - k_{23}HC]F_2$$

This equation says that (1) the overall rate of change is proportional to temperature, F_2; (2) plant intake rate is proportional to plant biomass available ($k_{12}P$); (3) there is a loss rate (respiration, excretion, death) proportional to herbivore biomass ($k_{22}H$); (4) biomass is lost to carnivores at a rate proportional to both herbivore and carnivore biomass ($k_{23}HC$). Here, k_{12}, k_{22}, and k_{23} are proportionality constants and are the parameters of the transfer function. This "submodel" for herbivore biomass change is not realistic; it is presented to illustrate only the basic idea of an equation used to relate several variables. In Section 6 of this chapter, we will take a closer look at approaches to development of system equations and estimation of parameters.

As a model is being developed, it is often seen that some variables are nearly constant over the time scale being examined by the model, and that some parameters should be considered variable over time. Thus, the distinction between these parts of a model is artificial and refers to a particular set of equations that represents one stage in the analysis of a system. Likewise, forcing functions can be considered the outputs (effects) of components that are not included in the model for reasons of economy or lack of interest. Almost always, we deal with "open" systems that receive input from and deliver output to some larger "system of systems." The recognition that a particular system under study is embedded in some larger system (a lake embedded in a forest embedded in the biosphere, for example) need not be

explicit; an example is the "logistic" growth model (Chapter 7, Section 8, page 184), in which populations are assumed to have an "unlimited specific growth rate" (r) that is an implicit (unstated) function of habitat and interaction with other organisms.

4. BASIC MATHEMATICAL TOOLS IN MODEL BUILDING

Statement

The equations, or functional relationships, that define a mathematical model can take a variety of forms. In this section, we will examine the three major kinds of mathematical tools that are most frequently used in the development of models. The first, *set theory and transformations,* can be used to represent any kind of model. Set theory is employed in the development of *state-change of state* models. Here we simply list the qualitative conditions or "states" that a system might assume, and the model consists of a *transformation rule* to specify what state the system will assume next, given that it is in any state now. The second tool, *matrix algebra,* is concerned with the description and manipulation of lists and tables of numbers. Matrices provide a general symbolic way of laying out system relationships; techniques of matrix manipulation are the basis of many models. The third tool, *difference and differential equations,* is used to develop models that describe quantitatively the way systems change over time. The Silver Springs example in the last section was of this type.

Explanation

A set can be thought of as a list of elements; it is represented by brackets that enclose symbols denoting the elements of the set. An example is the alphabet, a set of 26 symbols denoting different basic sounds:

[a, b, c, d, . . . , z]

This set is *finite;* others, such as "all integer numbers," contain an infinite number of elements. The group of state variables in a systems model forms a set, as do the equations that represent these variables. A population is a set of animals or plants in which each element (individual organism) can be identified in a complicated way by examination of morphological, physiological, and behavioral attributes. The attributes that are used to define the elements of a set form another set in themselves, and so forth.

The use of sets to describe changes in the state of a system is a fundamental part of *cybernetics*. An excellent introduction to this field is given by Ashby (1963). Suppose we develop a set whose elements symbolize each of several possible conditions that an ecosystem might assume over time in "primary" successional development (see page 258) on a rock substrate: "state A" might for example represent rock with lichens and decomposer organisms; "state B," thin soil with grasses and small herbivores, and so forth. This set of ecosystem states can be represented just as sounds are represented by the alphabet. Next, we can pick an appropriate time interval and write down for each symbolic state the state that we believe the system would change to during that time interval. Symbolizing the direction of change by an arrow, we might get

$$A \rightarrow B$$
$$B \rightarrow C$$
$$C \rightarrow F$$
$$F \rightarrow D$$
$$D \rightarrow C$$

Notice that a change is specified for *each possible* starting condition, rather than for some single initial state. The set of transitions from one state to another is called the *state-change of state*, or *state transition* model, and is often written

$$T : \downarrow \begin{matrix} A & B & C & F & D \\ B & C & F & D & C \end{matrix}$$

In this representation, starting states are placed on the top line, and resulting states (after one time interval) are put below. The symbol T denotes the overall set of possible transitions. Once T has been specified from knowledge of single transitions, long-term behavior of the system can be examined by following successive changes, or *transients*:

$$A \rightarrow B \rightarrow C \rightarrow F \rightarrow D$$

One may recognize the existence of loops, or *sinks*, that had gone unnoticed in the piecemeal building of the model; such a sink is illustrated in the $C \rightarrow F \rightarrow D$ transient. To visualize this in terms of a field

situation let A represent a pioneer lichen community on a granite outcrop, B an intermediate community, C a mature community with spring and summer flora, F the mature community with fall flora, and D the mature community in winter phase that looks bare but contains dormant roots, seeds, and soil that will revert to C the following spring. The model would thus picture both the unidirectional annual changes in succession and the cyclic seasonal transitions. The primary value of state-change of state models is in the organization of understanding of a system; by constructing them, the model builder can find out nothing essentially *new* about the system.

Matrix algebra is a broad class of mathematical techniques used for dealing with information that can be organized into two-way tables. A *matrix* is simply a set of numbers or symbols organized into rows and columns. A 3×3 (3 rows by 3 columns) matrix X is shown below to illustrate the usual notation employed.

$$X = \begin{bmatrix} x_{11} & x_{12} & x_{13} \\ x_{21} & x_{22} & x_{23} \\ x_{31} & x_{32} & x_{33} \end{bmatrix}$$

Each element in the matrix is referred to as x_{ij}, where the *subscript i* denotes the row position and the *subscript j* denotes column position. A matrix with only one row or column is called a *vector*. We will not look in detail at the operations that can be performed on matrices; the student is referred to Searle (1966). One very useful operation is matrix multiplication. If one wants to multiply a matrix X by a matrix Y (and if X has the same number of columns as Y has rows), the product XY will also be a matrix. Each element in the product matrix is *defined* as the *sum of products* of the elements in the *i*th row of X times corresponding elements in the *j*th column of Y. Here is a simple example to illustrate the procedure:

$$\begin{bmatrix} 1 & 3 & 2 \\ 0 & 4 & 0 \\ 3 & 2 & 2 \end{bmatrix} \begin{bmatrix} 1 \\ 2 \\ 1 \end{bmatrix} = \begin{bmatrix} 1 \times 1 + 3 \times 2 + 2 \times 1 \\ 0 \times 1 + 4 \times 2 + 0 \times 1 \\ 3 \times 1 + 2 \times 2 + 2 \times 1 \end{bmatrix} = \begin{bmatrix} 9 \\ 8 \\ 9 \end{bmatrix}$$

$$\quad X \qquad\quad Y \qquad\qquad XY \qquad\qquad\quad XY$$

Having a symbolic operation that takes products and sums them is particularly useful in ecological modeling. For example, the sum of input and output rates defines energy transfer through an ecological component; in some cases it may be reasonable to represent each rate as proportional to (equal to a constant times) the amount of energy present in "donor" or "receiver" components. In these cases the elements k_{ij} of a matrix K can be used to represent the proportionality constants of transfer between components i and components j, and the elements v_i of a vector V used to represent the amounts of energy in each system component. The overall transfer rate for each component i is then represented by the ith element in KV.

In organizing information on a particular system, matrices can be used simply to indicate which system components are directly interrelated. If, for example, n system components have been identified, an $n x n$ matrix, I, can be used to represent all possible interactions between components i and components j. I is called an *interaction table*. An "X" can be placed in the i, j element of I to indicate that component i affects component j directly, or an "O" to indicate that i and j are not directly related. Here is an over-simplified example to illustrate the technique:

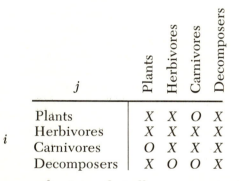

		Plants	Herbivores	Carnivores	Decomposers
	j				
	Plants	X	X	O	X
i	Herbivores	X	X	X	X
	Carnivores	O	X	X	X
	Decomposers	X	O	O	X

The example states that all components affect themselves (I_{ii} = "X"), but carnivores do not directly affect plants ($I_{31} = 0$) and vice versa ($I_{13} = 0$). Herbivores and carnivores affect decomposers (by providing energy), but decomposers do not directly affect herbivores or carnivores ($I_{42}, I_{43} = 0$). Such interaction tables can be made much larger and are useful in the same way as state transition models.

Difference and differential equations describe rates of system change; a simple introduction to the concept of rates is given in Chapter 7, Section 3. The basic idea of difference-differential equations is that they represent the way in which a variable changes as some function of its size *as size is changing*. Difference equations describe changes that occur over discrete intervals of time. Letting V_t be the value of a particular variable at time t, the general difference equation can be written as:

$$V_{t+1} = f(V_t, t)$$

which states that the value of the variable after one unit of time will be some function of its original value, and of time. One can, of course, let V_t be a vector and the $f(V_t, t)$ be a set of equations. Usually, the form of $f(V_t, t)$ is complicated and V_t must be found by computer, which uses repetitive calculations beginning with an initial value V_0. V_{t+1} can also be made a function of V_{t-1}, V_{t-2}, and so on; thus, difference equations are valuable in the representation of time lags.

Differential equations describe changes that occur continuously over time. The general form of these equations is:

$$\frac{dV}{dt} = f(V, t)$$

Again, the dV/dt and $f(V, t)$ can be vectors. When $f(V, t)$ contains only terms of the form kV, and k is a constant, the equation is said to be "linear." The simple growth equation, $dN/dt = rN$, discussed in Sections 7 and 8, Chapter 7, is of this linear form where r is the constant, k, and N is the systems variable. When variables occur in products (for example, kV_1V_2) or other functions of one another, the equation is called "nonlinear." Some of the earliest mathematical models of energy flow in ecosystems were developed by making use of matrix algebra and simple linear differential equations (Garfinkel, 1962; Patten, 1965). Energy in the ecosystem was considered as flowing through gross "compartments" (plants, herbivores, carnivores, decomposers). The amount of energy, v_i, in each compartment was assumed to change according to the simple differential equation:

$$\frac{dv_i}{dt} = \sum_j k_{ij} v_j$$

where the k_{ij} are instantaneous rates of flow from compartment j to compartment i; the k_{ii} term is negative and represents the total rate of loss from compartment i to compartments receiving energy from it. Notice that there is a sum of input-output terms for each compartment in the system; writing out the whole model would be quite tedius. We can use matrix representation to make the symbolism more concise by noting that matrix multiplication is defined in terms of sums of products such as $k_{ij}v_j$. The model can thus be written in matrix form as:

$$
\begin{bmatrix}
\dfrac{dv_1}{dt} \\[2ex]
\dfrac{dv_2}{dt} \\[2ex]
\cdot \\
\cdot \\
\cdot \\
\dfrac{dv_n}{dt}
\end{bmatrix}
=
\begin{bmatrix}
k_{11} & k_{12} & \ldots & k_{1n} \\[1ex]
k_{21} & k_{22} & \ldots & k_{2n} \\
\cdot & \cdot & & \cdot \\
\cdot & \cdot & \cdot & \cdot \\
\cdot & \cdot & \cdot & \cdot \\
k_{n1} & k_{n2} & \ldots & k_{nn}
\end{bmatrix}
\begin{bmatrix}
v_1 \\[1ex]
v_2 \\
\cdot \\
\cdot \\
\cdot \\
v_n
\end{bmatrix}
$$

or simply

$$\frac{dV}{dt} = KV$$

The simplicity of this model allows many of its properties to be determined by mathematical analysis, without the aid of computers. However, it approaches the limit of complexity for "analytical" techniques; most ecological models are more complex and are examined primarily by computer simulation techniques.

5. ANALYSIS OF MODEL PROPERTIES

Statement

The properties of mathematical models are explored with a variety of techniques. By examining the mathematical system, insight may often be obtained about corresponding properties of the real system. Of primary interest are questions of feedback and control, stability, and sensitivity of one part of the system to changes in another part.

Explanation

Many of the techniques for the development and examination of systems models have come from the field of "control systems theory" (Milsum, 1966; Milhorn, 1966; see also Chapter 2, Section 4). The concept of the feedback loops, illustrated in Figure 10–2, is central to this theory. As first defined in Chapter 2 (see page 34), feedback refers to the response of a system component to a change in its own size. Positive feedback mechanisms are those that encourage increase in size as size increases, as in exponential growth of a population (see page 182). Negative feedback mechanisms discourage further increase as size increases, i.e., they control, as in the logistic population model (see page 184). Models help to determine the relative effectiveness of different feedback mechanisms in promoting (or decreasing) stability of a system, because it is possible to vary the parameters or equations that represent these mechanisms. A model may be constructed explicitly in terms of feedback mechanisms, but more often the feedback mechanisms within a system are identified after the model is constructed as "emergent properties" of simpler interactions between system components.

Stability properties are often explored with the aid of *phase diagrams*, such as that shown for a host-parasite system in Figure 10–3. In the upper diagrams the equation parameters and variables are such that oscillations in density increase in successive generations; in the lower diagrams a different set of values results in oscillations that decrease with time in a manner similar to the "genetic feedback" model that was described in Chapter 7 (Fig. 7–32). Phase diagrams are generated by plotting the values of system variables against one another, with relative change over time represented by a line connecting the coordinate points of successive time values for the variables as shown in the two right hand diagrams, Figure 10–3. Damped oscillation of variables over a period of time is indicated by a line that spirals inwards to an "equilibrium point" (lower right graph). Instability is indicated by a line that spirals outward (upper right graph). By starting with different initial values of the variables, those system states that will lead to stability can be found. The sets of starting states for which the system will not "break down" define *stability regions* on the phase diagram. Size and shape of the stability region can be examined as parameters and equations in the model are changed.

A variety of observations on the behavior of models comes under the general heading

A. BLACK BOX DIAGRAM OF A BASIC FEEDBACK SYSTEM

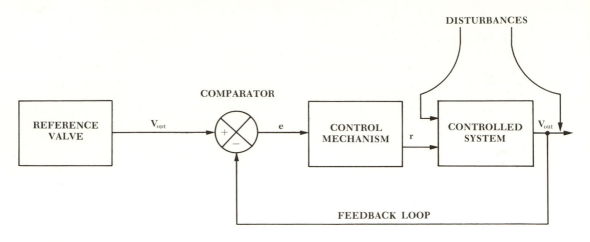

B. BEHAVIOR OF FEEDBACK SYSTEM IN RESPONSE TO DISTURBANCE

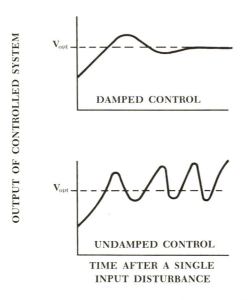

Figure 10–2. A basic feedback system as viewed in control theory (adapted from Waterman, 1968). A reference value for the state or output rate of the controlled system, V_{opt}, is compared by some mechanism (comparator) to present system value, V_{out}. The difference is an error, e, that is used by the control mechanism to provide a modified input rate, r, to the controlled system. The reference value might itself be some function of V_{out}.

of *sensitivity analysis.* By varying the forcing functions of a model, *input-output sensitivity* can be examined. In models of energy flow we can look at the effects of changing primary production on potential yield of carnivores. For example, referring to the energy flow model of a fish pond in Figure 3–11, how would the yield of game fish (bass) be affected by a 25 per cent increase in primary production (the forcing function) brought about by spending x number of dollars on additional fertilizer? The model might show that very little of the extra primary energy would reach the top carnivore level because of the prominent "side chain" involving the carnivorous insect larvae that

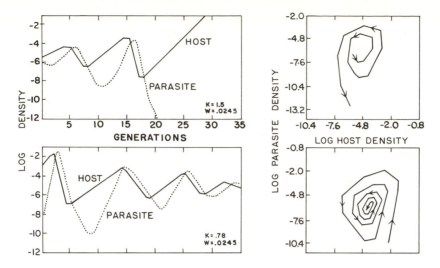

Figure 10–3. An experimental components model of host-parasite interaction. (From Holling and Ewing, 1969.) The left-hand graphs show population sizes over time for an unstable situation (top graph) and a stable situation (bottom graph). The right-hand graphs are "phase diagrams," showing population sizes as they change with respect to one another over time. Arrows indicate the direction of change over time.

are a part of this particular ecosystem. Or, it might show that the extra yield is not worth the money expended, especially in view of the additional instability created by increasing the forcing function. In another approach we might look for the amount of energy input necessary to maintain an additional trophic level. For example, how large must a microecosystem, or an island, be in order to support a carnivore population on a long-term basis without wiping out the prey part of the system (recall that in the microecosystem example cited in Chapter 2 a "grazing enclosure" was necessary to support even two trophic levels; see Figure 2–6II).

Another type of sensitivity is that of model components to one another, measured either in terms of (1) change in equilibrium or average value of one with another, or (2) stability of one as a function of stability in another. For example, in a model of nutrient cycling (Fig. 4–7) one might force some change in the value of the plant uptake component and watch what this does to the detritus component. Finally, we can look at the sensitivity of overall performance (stability, equilibrium values, and so forth) to changes in parameters and system equations. This kind of sensitivity analysis is particularly valuable in suggesting areas for more careful field measurement and experimental work. Key components or interactions often emerge from the analysis of

complex systems. These key factors can be thought of as occurring at points of convergence in a network of lines displaying causal interrelationships between system components. In summary, sensitivity analysis is a good approach to take when one does not know what the strategy is, or what strategy to take to bring about a desired result. By changing the weights of different components in some preconceived manner we can see which components in the model are *sensitive* to each change.

6. APPROACHES TO THE DEVELOPMENT OF MODELS

Statement

There are no fixed rules or criteria to guide the activities of mathematical model building. In principle, any mathematical model can be considered an extension, generalization, or special case of any other model. In practice, at least two fairly distinct strategies have been taken in ecological modeling. The *compartmental system approach* emphasizes the quantities of energy and materials in ecosystem "compartments." The examples so far in this chapter are based on this approach, as were the "box-and-flow-channel" diagrams widely used in Chapters 3 and 4. The "electrical analog" models, as illustrated in Figure 3–17, are of the compartment type, but emphasize analogies between ecological, electrical, and hydrological systems much as one might emphasize the electrical and plumbing circuits in a systems model of a building or a complex piece of man-made machinery (see H. T. Odum, 1960*a*, 1962). Compartment models are usually devel-

oped as systems of fairly simple differential equations. The *experimental components approach* emphasizes detailed analysis of ecological processes (predation, competition, and the like). In this approach, modeling effort is focused on interactions and systems equations, rather than on the identification of quantitative measures to describe system state. Models developed with

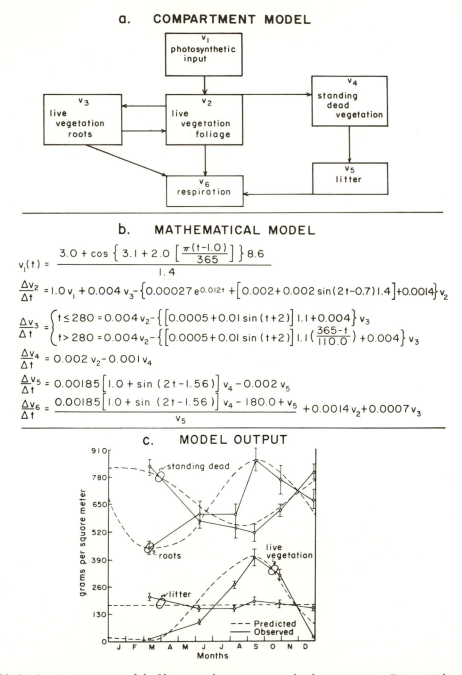

Figure 10–4. A compartment model of biomass change in a grassland ecosystem. *a.* Diagram identifying compartments of the system and showing direction of energy flow. *b.* Forcing function [$v_1(t)$] and system equations. *c.* Output from model, showing its ability to fit actual data. (From VanDyne, 1969.)

the experimental components approach have tended to be realistic and precise; compartment models have tended to be general, but not realistic. The parameters of compartment models are usually estimated from observational data on compartment sizes over time, taken without disturbing the system under study. As the name suggests, parameters in experimental components models usually come from experimental work on processes isolated from the ecosystem; field data are used only as a general check to insure that experimentally treated processes are properly combined in the model. The emphasis in compartment modeling has been on description and summary of data. Emphasis in experimental components modeling has been on prediction of the reaction of systems to disturbance and manipulations. Features from both approaches are combined in large-scale, integrated studies of whole ecosystems.

Explanation

The compartmental approach to ecological modeling is described by Patten (1971) and Van Dyne (1969). Workers using the approach have usually been interested in the gross dynamics of whole ecosystems as energy processing or nutrient cycling units. Ecosystems are seen as consisting of compartments or pools of energy or nutrients. Complicated processes associated with the populations making up each pool are assumed to counterbalance one another, resulting in simple behavior of the pool as a whole. Data for compartment models may come from experimental work, but they are usually obtained by simple measurement

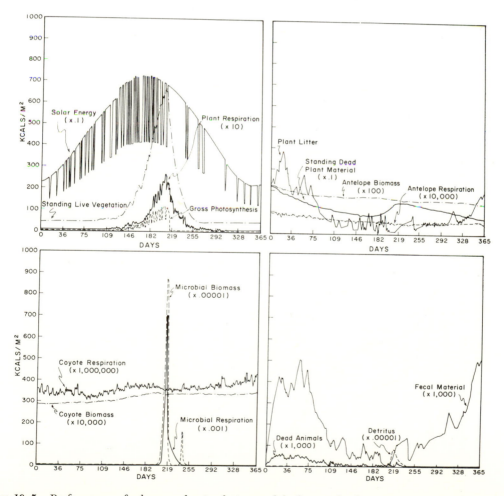

Figure 10–5. Performance of a large-scale simulation model of a grassland ecosystem. Data input was in the form of rate parameters and initial values for day 0; the curves shown are not real data. A much less realistic and complex version of this model is shown in Figure 10–4. (From VanDyne, 1969.)

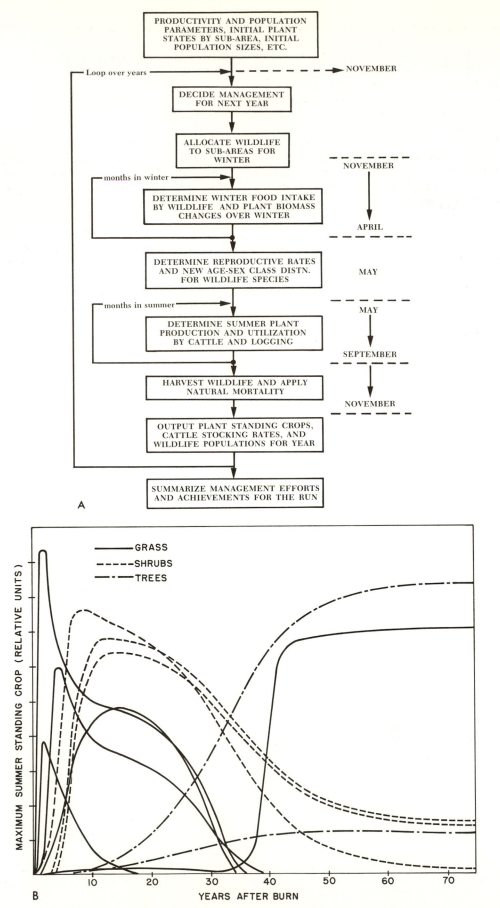

Figure 10-6 (see opposite page for legend).

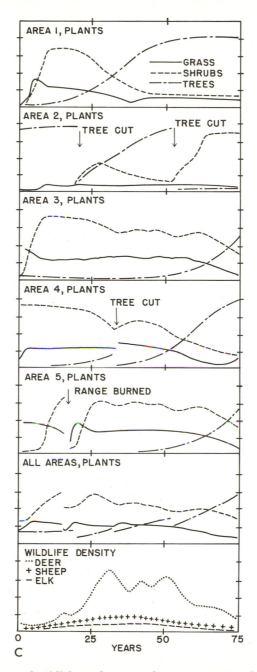

Figure 10–6. A model of plant and wildlife production and management in a forest and range area of western North America. *A.* Flow chart showing the sequence of computer calculations; each box represents one or more system equations. *B.* Simulated plant succession following a severe burn on a management area having no large grazers. Each curve represents the relative biomass of one plant species. Five species of grasses, two of shrubs, and two of trees are graphed. *C.* Summary output from a 75-year simulation on five management areas (upper 5 panels) and on the total area (lower 2 panels). Land management activities, such as tree cutting and range burning, were performed on several areas (indicated by the arrows), and the projected effects of these on plant cover and wildlife populations are indicated.

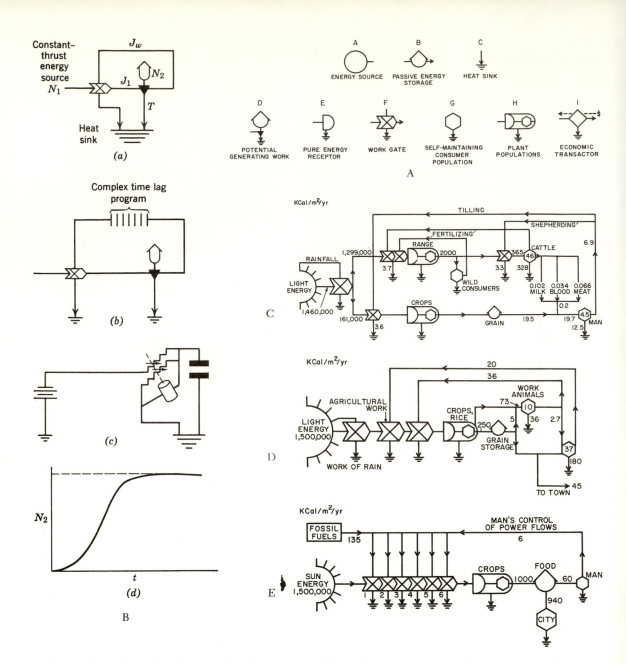

Figure 10–7. Examples of electrical analog circuit models. *A.* Energy network symbols (see text for further explanation). *B.* Circuit diagrams of the logistic population growth model and the equations for each response. (*a*) Biological circuit for a population with multiplicative loopback. (*b*) Biofilter: resonant or dampened phase shift. (*c*) Passive analog. (*d*) Graph of logistic growth.

Logistic derived from circuit:

$$J = L(N_1 - N_2) \quad \text{Inflow}$$
$$J_w = k_w N_2 \quad \text{Loopback work flow}$$
$$L = kJ_w \quad \text{Work controlling conductivity, } L = 1/R$$
$$T = dN_2/dt \quad 50\% \text{ Entropy tax, when positive}$$
$$dN_2/dt = J - T - J_w \quad \text{Balance of flows}$$

Combining equations:

$$dN_2/N_2 dt = (k_w/2)\,[k(N_1 - N_2) - 1]$$

Energy-dependent carrying capacity $= (kN_1 - 1)/k$.

C. Energy circuit for a tribal cattle system in Uganda. *D.* Energy circuits for man in unsubsidized monsoon agriculture in India largely based on solar energy, but with man-controlled energy flow acting through crops and cattle. *E.* Energy circuit for fuel subsidized industrial agriculture where high yields are based on large inflows of fossil fuels which (1) replace the work formerly done by man and animals and (2) do away with the network of animals and plants that are "nursed" in the two preceding systems, Figure 10–7C and D. (*A, C, D, E* after H. T. Odum, 1967*a*; *B* after H. T. Odum, 1967.)

over time of compartment sizes. Parameter estimates are usually found by repeatedly solving the equations, while varying parameter estimates to obtain the best fit to size-time data. A particular model would be considered inappropriate if it could not be made to "fit" data. Opponents of the compartment approach argue that it is not "scientific" because in ordinary experimental science one tries to find data that will *disprove* rather than demonstrate a hypothesized relationship. A simple six-compartment model is given in Figure 10–4 to illustrate the approach. In this model the primary producer component of the grassland ecosystem is divided into four units (above ground foliage, roots, standing dead, and litter) coupled with two energy components (photosynthetic input, P, and respiration, R). The equations for each compartment as it interacts seasonally with other compartments are shown in Figure 10–4b. Finally, the computer output of the mathematical model is shown in Figure 10–4c, showing how close the "predicted" fits the "observed" changes in the components as measured in the field. The energy model described at the end of Section 4 is another example.

The experimental components approach was first proposed formally by Holling (1966). Stated simply, the approach involves breaking down of ecological processes into very simple subprocesses, or "experimental components." Each component is then examined experimentally, and is represented by a simple equation or set of equations. Integration of the basic component models into a model of the whole process or set of processes is treated as a computer "bookkeeping" problem. Extensive use is made of complex difference equations, and computers are an absolute necessity for examining model properties. In this way it is relatively easy to represent time lags, discontinuities, and threshold responses. Watt (1968) discusses several applications of the experimental components approach in analyzing resource management problems. The approach has been used largely to analyze processes that occur at the level of individual organisms, and inferences have been to changes in species populations. The problems of applying such an approach to whole ecosystems would indeed be formidable. An example of the application of experimental components analysis to the study of predator (parasite)-prey (host) re-

lationships is given in Figure 10–3. In this complex model, Holling used equations representing many components of the overall predation process, such as hunger level and searching rate of the predator, prey density, and interference between predators.

An interesting variant on the predator-prey model is presented in a recent paper by Holling (1969) who considers land developers and speculators as the "predator" and the land as the "prey"! Two models were generated, one based on small-scale speculation (numerous small land "predators") and the other on large-scale speculation (50 per cent land under control of large-scale development). A "diagnostic perturbation" of a sudden increase in population and land demand was used to measure each model's stability. The large-scale speculation resulted in less oscillation in terms of population density, buyers' satisfaction, and amount of developed land. This model, of course, does not consider the long-term effects of large-scale development in terms of pollution, utilities, and human satisfaction in the next generation. However, it does make a case for large-scale *planned* development that includes environmental quality constraints. Certainly, unplanned, piecemeal development that eats away at the landscape in a haphazard manner, is just as self-defeating as is an "unprogrammed" predator who exploits prey without any feedback control.

Models illustrating combinations of experimental components and compartmental approaches are shown in Figures 10–5 and 10–6. The grassland ecosystem model (Fig. 10–5) was developed in conjunction with the International Biological Program, and was designed for investigation of major aspects of energy flow. Equations for solar energy, temperature, and precipitation are used as basic forcing functions. Each compartment whose simulated change over time is shown in Figure 10–5 is represented in the model by at least one difference or differential equation. The plant production-wildlife population model (Fig. 10–6) was developed to examine land use in forest areas of North America. Plant productivity and succession are simulated with a kind of compartment model emphasizing nonlinear differential equations. Basic plant productivity parameters are taken from experimental studies, while the overall successional trend is generated by "playing" with the values of parameters representing competition between plants. Difference equations and a bookkeeping procedure are employed to represent age structure, birth rates, and death rates in the wildlife populations. Wildlife feeding and the effect of food intake on reproduction and death are simulated from experimental data, using both difference and differential equations. These last two examples are presented not so much to show precisely how models are developed, but to hint at the degree of biological complexity that can be represented in mathematical terms.

Four examples of the electrical analog version of the compartment model are shown in Figure 10–7 (see also Figure 3–17). Compartments are represented by special symbols as pictured in Figure 10–7A; these include symbols for energy sources, primary producers, consumers, storage, heat sinks (where part of the potential energy is diverted into heat as a necessary function of the transfer of energy from one form to another), and work gates (a work flow that facilitates a second work flow, mathematically a product function, as indicated by the box containing an "X"). As explained by H. T. Odum (1967a, p. 59):

Each symbol has a mathematical definition and for each symbol there is a graph of functional response of output and input. Since each symbol represents something definite mathematically, the network diagram is also a statement of the computer program that has to be written for simulation in digital computers. When the response of the group of connected parts is to be studied, one may also model the system with electrical units on the passive principle where the flow of electrical current simulates the flow of carbon, and heat energy losses in the real system are simulated by heat losses in the electrical system. The voltage simulates potential energy.

If the system is constructed of electrical hardware, one may vary the input energy pulse and determine the pattern of arrival of energy at any point in the system, as, for example, to man at the end of the food chain. If something important is omitted, the response will differ from that of the real system. One can then add features or change patterns, thus gradually developing the model until it does simulate the real system.

Figure 10–7B shows how the logistic population model, which was considered in some detail in Section 8, Chapter 7, can be depicted as an electrical analog circuit with either a direct linear or a time-lag feedback loop that levels off population growth at some carrying capacity, k. Figure 10–7C through E shows three agricultural systems that form an evolutionary series from a tribal system in Africa to the fossil fuel subsidized industrial agriculture of the United States. The loopbacks in these diagrams show the quantitative importance of the work of man, animals, and fossil fuel in maintaining the flow of food to man. The concept of energy subsidy has already been emphasized in Chapter 3. As shown in Figure 10–7D, the so-called sacred cattle (work animal) of India play an important role in cultivating and fertilizing crops in "unsubsidized" agriculture; removing them would require substitution of energy from fossil fuels and expensive machinery. If we consider as round figure estimates that one square mile equals 2.5 million square meters and 1 million kilocalories as a food energy requirement to support one man for one year, then the "outputs" (in round figures) from the three agricultural models can be listed as follows:

| | | Persons supported per square mile | |
	Kcal/ M²/yr	on farm	in city
Tribal agriculture	20	50	0
Unsubsidized agriculture	245	600	100
Subsidized industrial agriculture	1000	150	2350

Serious ecological, social, and economic "backlashes" are created by a rapid shift from unsubsidized to subsidized agriculture, which increases environmental pollution and forces large numbers of small farmers into the city where there may be food but inadequate jobs, housing, and means of maintaining human dignity for the individual. Models such as these help to estimate the cost of such changes which must be paid somehow if society as a whole is to benefit. What is needed are models that consider the country and the city as one integrated and mutualistic life-support system. The common practice of "suboptimizing" a model for only part of the system, as, for example, considering only food production without taking into consideration the environmental stress, economic, and social consequences of increased production, can be extremely misleading and dangerous if such models are taken to be the "whole truth."

Part 2

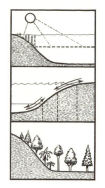

THE
HABITAT APPROACH

INTRODUCTION

In Part 1 the subject of ecology is organized from the viewpoint of principles and concepts as they apply to different levels of organization. The individual, the population, the community, and the ecosystem are the convenient levels that were used (see Chapter 1, Section 2). This method of presentation brings together the central themes which unify the subject of ecology and establishes a sound basis for applications to be made in the interest of mankind. In addition to this sort of broad basic approach, attention must also be focused on a "first hand" examination of definitely delimited areas of the earth's surface if we are to achieve a detailed understanding of the world in which we live. Experience in teaching (and we are here including "self-teaching") has shown that a "lecture–laboratory" or "theory-practice" procedure is sound. Thus, Part 1 is the "theory" and Part 2 is the "laboratory" in which we meet our subject on intimate terms, learn some of the necessary "jargon," and test the theories in the field.

In Part 1 we have strongly emphasized the functional aspects of ecology, that is, how systems of nature "work." Although no sharp distinction can or should be made between function and structure, Part 2 will emphasize structure, that is, how nature "looks," with appropriate cross references to Part 1. In other words we shall point out what the student may see on his field trips as he begins to think in terms of critical studies of nature.

By studying a particular habitat we become acquainted with organisms and physical factors actually associated in a particular ecosystem. This helps mitigate the pitfalls that may follow excessive generalization. Also, we obtain some insight into the methods, instrumentation, and technical difficulties applicable to specific situations!

If Part 1 has not been read, reference should be made to the sections on the concept of the ecosystem (Chapter 2, Section 1), habitat and niche (Chapter 8, Section 1), biogeochemical cycles (Chapter 4, Section 1), principles of limiting factors (Chapter 5, Section 3), the biotic community (Chapter 6, Section 1), and ecosystem development (Chapter 9, Section 1) as background for the discussions in Part 2.

There are four major habitats in the biosphere, namely, marine, estuarine, fresh water, and terrestrial. Since most biologists postulate that life began in the ocean, it would be logical to start study with the marine habitat. In actual practice, however, it is best to start with a freshwater habitat for several reasons. In the first place, examples of freshwater habitats are available wherever man lives. Many freshwater habitats are small, and, therefore, are more readily accessible throughout with the use of relatively simple equipment. Finally, there are fewer kinds of organisms in small bodies of fresh water than in the ocean, making it easier for the beginner to comprehend something of the nature of the natural community without an overburden of effort in learning to identify organisms in a large number of classes and phyla. For these reasons, this section begins with the freshwater environment.

1. THE FRESHWATER ENVIRONMENT: TYPES AND LIMITING FACTORS

Since water is both an essential and the most abundant substance in protoplasm, it might be said that all life is "aquatic." However, in practice we speak of an aquatic habitat as one in which water is the principal *external* as well as internal medium. Freshwater habitats may be conveniently considered in two series, as follows:

Standing-water, or lentic (*lenis*, calm), habitats: lake, pond, swamp, or bog
Running-water, or lotic (*lotus*, washed), habitats: spring, stream (brook-creek), or river

Examples of some of these habitats are shown in Figures 11–1 and 11–2. There are no sharp boundaries between the two series or between categories within a series. Geological change tends to produce a gradient in the direction indicated, while biological processes often work to stabilize or slow down the processes of lake filling and stream erosion (Chapter 9, Section 1, and Chapter 4, Section 3). Man tends to speed up the geological processes at the expense of biological ones, too often to his own detriment (see concept of "Man, the mighty geological agent," page 35). Thus, lakes tend to fill up while streams tend to cut down to base level and thus change as a result of the action of the water. When base level is reached, current is reduced, sedimentation occurs, and a base-level meandering river results which represents a more or less "climax" state. However, as deltas are built up by deposition of silt, uplifts may eventually occur elsewhere, thus starting the erosion cycle all over again. The complex interaction of "autogenic" (successional)

and "allogenic" processes in the freshwater habitat was discussed in considerable detail in Chapter 9.

Freshwater habitats occupy a relatively small portion of the earth's surface as compared to marine and terrestrial habitats, but their importance to man is far greater than their area for the following reasons: (1) They are the most convenient and cheapest source of water for domestic and industrial needs (we can and probably will get more water from the sea, but at great cost in terms of energy required and salt pollution created). (2) The freshwater components are the "bottle-neck" in the hydrological cycle (see Figure 4–8B). (3) Freshwater ecosystems provide the more convenient and cheapest waste disposal systems. Because man is abusing this natural resource so, it is clear that major effort to reduce this stress must come quickly; otherwise, water will become *the* limiting factor for man, the species!

Limiting factors that are likely to be especially important in fresh water, and hence those which we would wish to measure in any thoroughgoing study of an aquatic ecosystem, are as follows:

Temperature

Water has several unique thermal properties that combine to minimize temperature changes; thus the range of variation is smaller and changes occur more slowly in water than in air. The most important of these thermal properties are: (1) High specific heat, that is, a relatively large amount of heat is involved in changing the temperature of water. One gram-calorie (gcal) of heat is required to raise one milliliter (or one gram) of water one degree centigrade (between 15° and 16°). Only ammonia and a few other substances have values higher than 1. (2) High latent heat of fusion. Eighty calories are required to change 1 gram of ice into water with no change in temperature (and vice versa). (3) Highest known

---

* The study of fresh waters in all their aspects — physical, chemical, geological, and biological — is termed *limnology*. Freshwater ecology emphasizes the organisms-in-environment relationships in the freshwater habitat in the context of the ecosystem principle.

Figure 11–1.   Three types of standing water (lentic) habitats. *A.* A pond with rooted aquatic plants in the littoral zone. *B.* A managed farm pond without rooted plants in the littoral zone ("clean" margin pond). (*A* and *C* are U. S. Forest Service Photos; *B* is a USDA Soil Conservation Service Photo.)

latent heat of evaporation. Five hundred and thirty-six calories per gram are absorbed during evaporation which occurs more or less continually from vegetation, water, and ice surfaces. As was pointed out in Chapter 3, Section 1, a major portion of the incoming solar radiation is dissipated in the evaporation of water from the ecosystems of the world, and it is this energy flow that moderates climates and makes possible the development of life in all its fantastic diversity. (4) Water has its greatest density at 4°C.; it expands and hence becomes lighter both above and below this temperature. This unique property prevents lakes from freezing solid.

Although temperature is thus less variable in water than in air, it is nevertheless a major limiting factor because aquatic organisms often have narrow tolerances (stenothermal, see Chapter 5, Section 2). Thus, even moderate thermal pollution by man can have widespread effects (see Chapter 17). Also, temperature changes produce characteristic patterns of circulation and stratification (to be described later), which greatly influence aquatic life. Large bodies of water greatly modify the climate of adjacent areas of land (see page 123).

The measurement of temperature in water is most conveniently and efficiently done with electronic sensors such as thermistors.

Figure 11–1. C. A chain of high altitude, oligotrophic lakes of glacial origin in Alaska.

Direct reading and recording thermisters now make it easy for the beginning student to take a "temperature profile" of aquatic habitats.

Transparency

Penetration of light is often limited by suspended materials, restricting the photosynthetic zone wherever aquatic habitats have appreciable depth. Turbidity, especially when caused by clay and silt particles, is often important as a limiting factor. Conversely, when turbidity is the result of living organisms, measurements of transparency become indices of productivity. Transparency can be measured with a very simple instrument called a *Secchi disk* (after A. Secchi, an Italian, who introduced it in 1865), which consists of a white disk about 20 cm in diameter that is lowered from the surface until it just disappears from view. The depth of visual disappearance becomes the *Secchi disk transparency,* which will range from a few centimeters in very turbid bodies of water to 40 meters in a very clear, unproductive high altitude lake such as Crater Lake in Crater Lake National Park, Oregon. For the well-studied Wisconsin lakes Secchi disk transparency represents the zone of light penetration down to about

5 per cent of the solar radiation reaching the surface. While photosynthesis does occur at lower intensities, the 5 per cent level marks the lower limit of the major photosynthetic zone. Although it is obvious that modern photosensitive instruments will provide more accurate data on light penetration, the Secchi disk is still considered a useful tool by limnologists (Hutchinson, 1957, page 399). Fish pond managers often use this technique to adjust the level of fertilization to produce a good but not too great a growth of phytoplankton. The Secchi disk and the thermister are two simple, inexpensive instruments that the beginning student can use to get a rough picture of the all-important temperature and light relationships in ponds and lakes.

The exponential absorption of light by water was discussed in Chapter 3 (page 40). For a picture of how transparency is correlated with progressive eutrophication and its subsequent reversal, see Figure 16–5.

Current

Since water is "dense," the direct action of current is a very important limiting factor, especially in streams. Also, currents often largely determine the distribution of vital gases, salts, and small organisms.

Figure 11-2. Two streams illustrating the close interdependence of streams and the terrestrial watershed. *A.* Small woodland stream, whose biota is almost entirely dependent on the import of leaf and other organic detritus from the forest. *B.* A salmon stream in Alaska at a time of a salmon "run" when biomass and mineral nutrients are being "exported" by air-breathing predators (bears and gulls) that are feeding on the fish. (*A,* Soil Conservation Service Photo; *B,* U. S. Dept. Interior, Fish and Wildlife Service Photo.)

Concentration of Respiratory Gases

In rather stark contrast to the marine environment oxygen and carbon dioxide concentration are often limiting in the freshwater environment (see Chapter 5, Section 5). In this "age of pollution" dissolved oxygen concentration (D.O.) and the biological oxygen demand (B.O.D.) are becoming the most frequently measured and most in-

tensively studied physical factors. The measurement of dissolved oxygen was discussed in some detail in Chapter 2 (page 14) and its use as an index of productivity in Chapter 3 (pages 57–58). For a picture of pollution induced "oxygen sag" and its consequences in terms of the biota, see Figure 16–6. Since O_2 and CO_2 usually behave reciprocally, pollution ecologists are more and more concerned with the enrichment rather than the limiting effect of carbon dioxide in fresh water (see pages 106 and 126).

Concentration of Biogenic Salts

Nitrates and phosphates seem to be limiting to some extent in nearly all freshwater ecosystems (see Chapter 5, Section 5, subdivision 6; see also Chapter 5, Section 3, example 3). In soft-water lakes and streams, calcium and other salts also are likely to be limiting. Except for certain mineral springs, even the hardest fresh waters have a salt content or salinity of less than 0.5 part per thousand, compared with 30 to 37 parts per thousand for sea water (see Figure 12–3, page 328).

Two other characteristics of freshwater habitats may influence the number and distribution of species present (or qualitative richness of biota). Since freshwater habitats are often isolated from each other by land and sea, organisms with little means of dispersal over these barriers may have failed to become established in places otherwise favorable. Fish are especially subject to this limitation; streams, for example, only a few miles apart by land but isolated by water, may have their niches occupied by different species. On the other hand, most small organisms—algae, crustacea, protozoa, and bacteria, for example—have amazing powers of dispersal (see Chapter 7, Section 11). Thus, one may find the same kind of water flea (*Daphnia*, for example) in a pond in the United States as in England. A manual of freshwater invertebrates written for the British Isles, for example, serves almost as well for the United States. At least down to the family and generic level, the lower plants and invertebrates of fresh water show a great degree of cosmopolitanism.

Freshwater organisms have a definite problem "to solve" in regard to osmoregulation. Since the concentration of salts is greater in the internal fluids of the body or cells than in the freshwater environment (i.e., fluids are hypertonic), either water tends to enter the body by osmosis if membranes are readily permeable to water (Fig. 11–3A), or salts must be concentrated if membranes are relatively impermeable. Freshwater animals, such as protozoa with their thin cell membranes and fish with their gills (Fig. 11–3A), must have efficient means of excreting water (accomplished by contractile vacuoles in protozoa and by kidneys in fish) or the body would literally swell up and burst! Difficulties of osmoregulation may explain, partially at least, why a great number of marine animals—entire phyla, in fact—have never been able to invade the freshwater environment. In contrast, bony fish (also marine birds and mammals) whose body fluids have a salt content lower than that of sea water (i.e., hypotonic) have been able to reinvade the sea by evolving metabolic osmoregulation that involves excretion of salt and retention of water, as shown in Figure 11–3B.

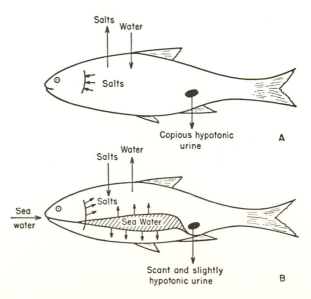

Figure 11–3. Osmoregulation in fresh water (A), compared with marine (B) bony fish. (From Florkin and Morgulis, 1949, after Baldwin.)

2. ECOLOGICAL CLASSIFICATION OF FRESHWATER ORGANISMS

Since organisms in fresh water (or in any other natural habitat) are not arranged in taxonomic order (such as is followed in taxonomic texts or systematic museums), some sort of classification on an ecological basis is useful. First, organisms may be classified as to major niches based on their position in the energy or food chain (see Chapter 2, Section 1, Figure 2–2, and Chapter 3, Sections 2–4) as:

Autotrophs (Producers):

green plants and chemosynthetic microorganisms.

Phagotrophs (Macroconsumers):

primary, secondary, etc.; herbivores, predators, parasites, etc.

Saprotrophs (Microconsumers or Decomposers):

subclassified according to nature of the organic substrate decomposed.

Within these trophic levels it is generally instructive to recognize those species which act as major dominants (see Chapter 6, Section 2).

Secondly, organisms in water may be classified as to their *life form* or *life habit*, based on their mode of life, as follows:

Benthos:

organisms attached or resting on the bottom or living in the bottom sediments. The animal benthos may be conveniently subdivided according to mode of feeding into *filter-feeders* and *deposit-feeders* (a clam and a snail, respectively, would be examples).

Periphyton or *Aufwuchs* *:

organisms (both plant and animal) attached or clinging to stems

and leaves of rooted plants or other surfaces projecting above the bottom.

Plankton:

floating organisms whose movements are more or less dependent on currents. While some of the zooplankton exhibit active swimming movements that aid in maintaining vertical position, plankton as a whole is unable to move against appreciable currents. In practice, *net plankton* is that which is caught in a fine-meshed net which is towed slowly through the water; *nannoplankton* is too small to be caught in a net and must be extracted from water collected in a bottle or by means of a pump.

Nekton:

swimming organisms able to navigate at will (and hence capable of avoiding plankton nets, water bottles, etc.). Fish, amphibians, large swimming insects, and so forth.

Neuston:

organisms resting or swimming on the surface.

Finally, organisms may be classified as to region or subhabitat. In the ponds and lakes three zones are generally evident as shown diagrammatically in Figure 11–4:

Littoral zone:

the shallow-water region with light penetration to the bottom; typically occupied by rooted plants in natural ponds and lakes, but not necessarily in

---

* The German word, *Aufwuchs*, proposed by Ruttner (1953), is perhaps more appropriate than the English "periphyton."

"managed" ponds (see Figure 11–1).

Limnetic zone: the open-water zone to the depth of effective light penetration, called the *compensation level* which is the depth at which photosynthesis just balances respiration. In general, this level will be at the depth at which light intensity is about 1 per cent of full sunlight intensity (compare with "Secchi disk transparency" depth discussed in the previous section). The community in this zone is composed only of plankton, nekton, and sometimes neuston. This zone is absent in small, shallow ponds. The term *euphotic zone* refers to the total illuminated stratum including littoral and limnetic.

Profundal zone: the bottom and deepwater area which is beyond the depth of effective light penetration. This zone is often absent in ponds.

In streams two major zones are generally evident:

Rapids zone: shallow water where velocity of current is great enough to keep the bottom clear of silt and other loose materials, thus providing a firm substrate. This zone is occupied largely by specialized benthic or periphytic organisms which become firmly attached or cling to a firm substrate, and by strong swimmers such as darters (fish).

Pool zone: deeper water where velocity of current is reduced and silt and other loose materials tend to settle to the bottom, thus providing a soft bottom, unfavorable for surface benthos but favorable for burrowing forms, nekton, and, in some cases, plankton.

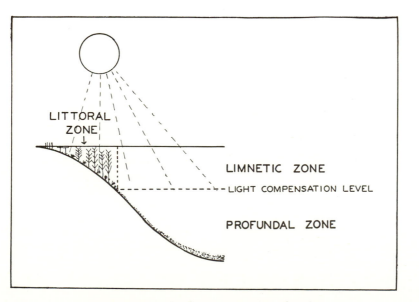

Figure 11–4. The three major zones of a lake.

Two points should be emphasized. To clarify the ecological role or niche of a given organism, its position in all three classifications should be determined. Thus a diatom that lives suspended in the open-water zone would be classified as a plankton producer of the limnetic zone. If, furthermore, it proved to be very abundant during the spring and scarce at other times, we would say that it was a major dominant among the producers of the limnetic community during the spring or vernal season. A second important point to note is that ecological classification of a particular species may be different at different stages in its life history. Thus, an animal might be a primary consumer while in the larval stage and a secondary consumer as an adult (tadpole–frog, for example); or an animal might be a member of the profundal community during the larval stage (for example, a chironomid larva) and then leave the water entirely as an adult. In this respect ecological classification is quite different from taxonomic classification, which, of course, does not change with changes in life history stages. Other factors (competition, tolerance, and so forth) which affect an organism's ecological niche were discussed in Chapter 8.

3. THE FRESHWATER BIOTA (FLORA AND FAUNA)

The major divisions of plants and many of the major animal phyla are represented by one or more genera living in freshwater communities. Considering the freshwater environment as a whole, the algae are the most important producers, with the aquatic spermatophytes ranking second. Except for the pond weeds and duck weeds, most aquatic higher plants are members of diverse families in which the majority of species are terrestrial.

Among the animal consumers, four groups will likely comprise the bulk of the biomass in most freshwater ecosystems, namely, mollusks, aquatic insects, crustacea, and fish. The annelids, rotifers, protozoa, and helminths would generally rank lower in importance, although in specific instances any of these groups may loom large in the "economy" of the system.

Among the saprotrophs, the water bacteria and the aquatic fungi seem to be of equal importance in performing the vital

role of reducing organic matter to inorganic form which may then be used again by the producers. As emphasized in Chapter 4, Section 7, bacteria and fungi are most important in the zones in which there is a large amount of organic detritus (and in waters polluted with organic materials); they are less numerous in unpolluted limnetic waters. The distribution and activities of microorganisms in aquatic habitats is discussed in Chapter 19.

In summary, the beginning student should first become familiar with algae, bacteria and fungi, the aquatic spermatophytes, the crustacea, the aquatic insects, mollusks and fish. These are the key "actors" in freshwater ecosystems.

4. LENTIC COMMUNITIES

The general zonation characteristic of ponds and lakes has been diagrammed in Figure 11–4. Characteristic organisms of the various zones are illustrated in Figures 11–5, 11–6, 11–7. A brief account, emphasizing community organization in these zones, follows. For aid in identification and for accounts of life histories, reference should be made to standard treatises on freshwater biology, monographs, and manuals as are available for local use.

Nature of Communities in the Littoral Zone

1. PRODUCERS. Within the littoral zone producers are of two main types: rooted or benthic plants, belonging mostly to the division Spermatophyta (seed plants), and phytoplankton, or floating green plants, which are mostly algae (see Figure 11–5). Sometimes the duckweeds, which are neuston spermatophytes, not attached to the bottom, are important; in fact, in some ponds they may form almost a continuous sheet on the surface at certain seasons and virtually "shade out" other green plants. When a pond or lake is polluted with excess nutrients, the filamentous-type of algae (Fig. 11–5, items 8 and 9) often develop huge "blooms" that rise to the surface, buoyed up by entrapped oxygen. Then the oxygen produced by photosynthesis largely escapes into the air, and when the bloom dies, oxygen in the water is used up, often stress-

ing or killing the fish. It is important that laymen understand this process, because on first consideration one would think that a rapid growth of algae would increase the dissolved oxygen content of the water; in the sequence just described the net result is the opposite.

Typically, rooted aquatics form concentric zones within the littoral zone, one group replacing another as the depth of the water changes (either in space or in time). A representative arrangement proceeding from shallow to deeper water may be briefly described as follows (in any given body of

water it should not be assumed that all three zones will be present or that they will be arranged in the order listed):

(a) Zone of emergent vegetation: rooted plants with principal photosynthetic surfaces projecting above the water. Thus, carbon dioxide for food manufacture is obtained from air but other raw materials are obtained from beneath the water surface. Rooted aquatics often "recover" nutrients from deep in the anaerobic sediments and thus provide a useful "nutrient pump" for the ecosystem (see page 96). Cattails, belonging to several species of the genus *Typha*,

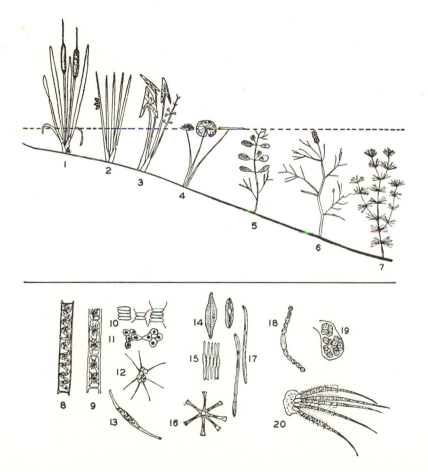

Figure 11–5. Some producers of lentic communities, including emergent, floating, and submergent rooted littoral plants (1–7), filamentous algae (8–9), and phytoplankton (10–20). The phytoplankton include representative green algae (10–13), diatoms (14–17), and blue-green algae (18–20). Note that phytoplankton exhibit "flotation" adaptations which enable them to remain suspended or at least decrease markedly the rate of sinking (these organisms, of course, have no power of movement of their own)—for example, reduction in integumentary material, flotation process, and colonial life-habit, which increases surface area, and gas vacuoles indicated by black spots in the blue-green algae cells (18–20). Organisms diagrammed are: 1, cattail (*Typha*); 2, bulrush (*Scirpus*); 3, arrowhead (*Sagittaria*); 4, water lily (*Nymphaea*); 5 and 6, two species of pond weeds (*Potamogeton diversifolia*, *P. pectinatus*); 7, muskgrass (*Chara*); 8, *Spirogyra*; 9, *Zygnema*; 10, *Scenedesmus*; 11, *Coelastrum*; 12, *Richteriella*; 13, *Closterium* (a desmid); 14, *Navicula*; 15, *Fragilaria*; 16, *Asterionella* (which floats in the water like a parachute); 17, *Nitzschia*; 18, *Anabaena*; 19, *Microcystis*; 20, *Gloeotrichia* (19 and 20 represent parts of colonies enclosed in a gelatinous matrix). (8 to 17 redrawn from Needham and Needham, 1941; 18 to 20 redrawn from Ruttner, 1963.)

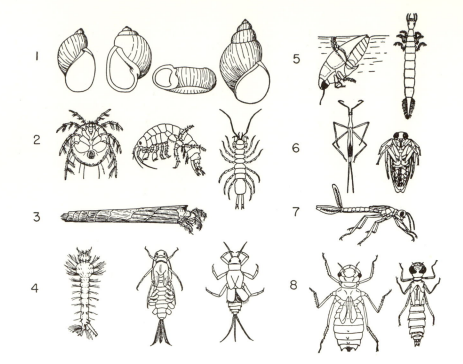

Figure 11-6. Some representative animals of the littoral zone of ponds and lakes. Series 1 to 4 are primarily herbivorous forms (primary consumers); series 5 to 8 are predators (secondary consumers). 1. Pond snails (left to right): *Lymnaea* (*pseudosuccinea*) *columella; Physa gyrina; Helisoma trivolvis; Campeloma decisum.* 2. Small arthropods living on or near the bottom or associated with plants or detritus (left to right): a water mite, or Hydracarina (*Mideopsis*); an amphipod (*Gammarus*); an isopod (*Asellus*). 3. A pond caddis fly larva (*Triaenodes*), with its thin, light portable case. 4. (left to right) A mosquito larva (*Culex pipiens*); a clinging or periphytic mayfly nymph (*Cloeon*); a benthic mayfly nymph (*Caenis*)—note gill covers which protect gills from silt. 5. A predatory diving beetle, *Dytiscus*, adult and (right) larva. 6. Two predaceous Hemipterans, a water scorpion, *Ranatra* (Nepidae), and (right) a backswimmer, *Notonecta*. 7. A damsel fly nymph, *Lestes* (Odonata-Zygoptera); note three caudal gills. 8. Two dragonfly nymphs (Odonata–Anisoptera), *Helocordulia*, a long-legged sprawling type (benthos), and (right) *Aeschna*, a slender climbing type (periphyton). (Redrawn from Robert W. Pennak, "Fresh-water Invertebrates of the United States," 1953, The Ronald Press Company.)

are a widespread dominant producer and may be considered a "type" for this niche. They occur in a wide latitudinal range because of their proclivity to form ecotypes and races (see page 109). Other plants in this category include bulrushes (*Scirpus*), arrowheads (*Sagittaria*), bur reeds (*Sparganium*), spike rushes (*Eleocharis*), and pickerelweeds (*Pontederia*). The emergent plants, together with those on the moist shore, form an important link between water and land environments. They are used for food and shelter by amphibious animals— muskrats, for example—and provide a convenient means of entry and exit into the water for aquatic insects which spend part of their lives in the water and part on land.

(b) Zone of rooted plants with floating leaves. The water lilies (*Nymphaea*, about 4 species) are the "type" in this zone in the eastern half of the United States, but other

plants (water shield—*Brasenia*—for example) have a similar life form. This zone is similar, ecologically, to the previous one, but the horizontal photosynthetic surfaces may more effectively reduce light penetration into water. The undersurfaces of the lily pads provide convenient resting places and places for egg deposition by animals.

(c) Zone of submergent vegetation: rooted or fixed plants completely or largely submerged. Leaves tend to be thin and finely divided and adapted for exchange of nutrients with the water. The pond weeds or Potamogetonaceae are usually prominent in this zone. The genus *Potamogeton* is, in fact, one of the largest genera of rooted aquatic plants, having about 65 species which occur in all temperate parts of the world. Other genera in the pond-weed family (*Ruppia, Zannichellia*) are likewise widespread and may be more important

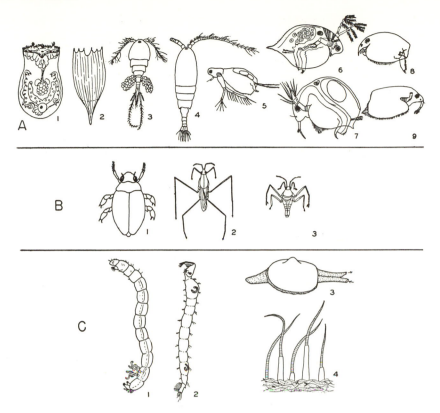

Figure 11–7. A. Representative zooplankton. Rotifers: 1, *Asplanchna;* 2, *Notholca* (lorica only). Copepods: 3, a cyclopoid copepod, *Macrocyclops;* 4, a calanoid copepod, *Senecella.* Cladocera (representative of each of five families): 5, *Diaphanosoma* (Sididae); 6, *Daphnia* (Daphnidae); 7, *Bosmina* (Bosminidae); 8, *Pleuroxus* (Chydoridae); 9, *Acantholeberis* (Macrothricidae).

B. Zooneuston. 1, a whirligig beetle, *Dineutes* (Gyrinidae); 2, a water strider, *Gerris* (Gerridae); 3, a broad-shouldered water strider, *Rhagovelia* (Veliidae).

C. Some characteristic profundal zone types. 1, a chironomid larva, or blood worm, *Tendipes* (note prolegs and abdominal gills); 2, a "phantom larva," *Chaoborus* (note two air sacs which apparently aid the animal in performing vertical migrations); 3, a "pea-shell" clam, *Musculium* (Sphaeriidae), with "foot" and two branchial siphons extended; 4, *Tubifex*, a red annelid which builds tubes on the bottom and vigorously waves its posterior end around in the water. (Redrawn from Robert W. Pennak, "Fresh-water Invertebrates of the United States," 1953, The Ronald Press Company.)

locally than species of *Potamogeton.* Other important submerged aquatics in the United States include coontail (*Ceratophyllum*), water milfoils (*Myriophyllum*), waterweed (*Elodea* or *Anacharis*), naiads (*Najas*), and wild celery (*Vallisneria*). "Muskgrass," *Chara,* and related genera *Nitella* and *Tolypella* are generally classed as algae, yet are attached to the bottom and have a life form resembling that of higher plants. *Chara* may thus be ecologically classed along with the above submerged producers. It often marks the inner boundary of the littoral zone, since it is able to grow in deep water.

In all kinds of shallow water (ponds, sluggish rivers, and marshes) root aquatics tend to become more important in warm climates. Primary production in emergent aquatics has been reported to be quite high (Penfound, 1956; Westlake, 1963; see also the section on salt marsh grass, Chapter 13). In southeastern United States and in the tropics they grow so well that they are often considered pests because they choke up waterways and hamper boating and fishing. Man's first instinct in such cases is to use poison ("weed killers" etc.), but it might be more prudent in many cases to try to harvest or graze them, especially when it has been shown that aquatic vegetation often has a high nutritive value (Boyd, 1968).

The nonrooted producers of the littoral zone comprise numerous species of algae. Many species are found floating throughout both littoral and limnetic zones (plankton), but some, especially those which are attached to or associated with rooted plants, are especially characteristic of the littoral zone. Likewise, many species have special adaptations for increasing buoyancy and hence are characteristic of the limnetic zone. As shown in Figure 11–5, the principal types of algae are:

(1) Diatoms (*Bacillariaceae*), with box-like silica shells and yellow or brown pigment in the chromatophores masking the green chlorophyll. Diatoms are good indicators of water quality (see Figures 6–6*B* and 16–5).

(2) Green algae (*Chlorophyta*), which includes single-celled forms such as desmids, filamentous forms either floating or attached, and various floating colonial forms. In these forms the chlorophyll is not masked by other pigment; consequently, populations have a bright green appearance.

(3) The blue-green algae (*Cyanophyta*), rather simple single-celled or colonial algae with diffuse chlorophyll (not concentrated into chromatoplasts) masked by blue-green pigment. This group is often of great ecological importance because of the enormous biomass that may develop in polluted ponds and lakes. As was indicated in Chapter 4, many blue-green algae are able to fix gaseous nitrogen into nitrates and thus perform the same role in water as is performed by bacteria in the soil. Many species of blue-greens are resistant to grazing (which is one reason for large biomass build-ups). Metabolites excreted from the cells and breakdown products released during decay are often toxic and impart bad tastes and odors to drinking water; hence this group is not popular with waterworks engineers! In addition to the genera illustrated in Figure 11–5, we might mention *Oscillatoria* and *Rivularia*, which may cover the bottom or be found attached to stems and leaves of submerged spermatophytes.

Filamentous green algae or "pond scums," such as *Spirogyra, Zygnema, Oedogonium,* are frequently studied in elementary botany courses. Other filamentous forms assume a periphyton life form, each kind of rooted plant often having characteristic algal encrustments, the two organisms apparently being associated in a commensal or mutualistic relationship (see Chapter 7, Section 16). *Chara* seems to form an especially favorable substrate and is nearly always covered with a crust of other algae. A sample of the free-floating phyto-plankton of the littoral zone would likely reveal numerous diatoms, desmids, and other green algae and chlorophyll-containing (hence holophytic, or "producing") protozoa, such as the familiar laboratory form *Euglena* and its many relatives.

2. Consumers. The littoral zone is the home of a greater variety of animals than are the other zones. All five "life habits" are well represented, and all phyla which have freshwater representatives are likely to be present. Some animals, especially the periphyton, exhibit a zonation paralleling that of the rooted plants, but many species occur more or less throughout the littoral zone. Vertical rather than horizontal zonation is more striking in animals. Some of the characteristic animals of littoral regions are shown in Figure 11–6. Among the periphyton forms, for example, pond snails, damsel fly nymphs and climbing dragonfly nymphs (Odonata), rotifers, flatworms, bryozoa, hydra, and midge larvae rest on or are attached to stems and leaves of the large plants. Snails feed on plants or *Aufwuchs;* midge larvae are also primary consumers but obtain their food from the detritus. The dragonfly and damsel fly larvae are exclusively carnivorous, using their hinged labium to good advantage to capture sizable prey which chance to pass by their lookout posts. Another group containing both primary and secondary consumers may be found resting or moving on the bottom or beneath silt or plant debris—for example, sprawling Odonata nymphs (which have flattened rather than cylindrical bodies), crayfish, isopods, and certain mayfly nymphs. Descending more deeply into the bottom mud are burrowing Odonata and Ephemeroptera (which either maintain burrows or have extended parts of the body reaching to the surface of the mud for breathing), clams, true worms (Annelida), snails, and especially chironomids (midges) and other Diptera larvae which live in minute burrows.

The nekton of the littoral zone is often rich in species and numbers. Adult and larval diving beetles and various adult Hemiptera are conspicuous; some of these, especially the dytiscids and notonectids, are carnivorous, whereas the hydrophylids, haliplid beetles, and corixid "bugs" are partly, at least, herbivorous or scavengers. Various Diptera larvae and pupae remain suspended in water, often near the surface.

Many of this group of animals obtain air from the surface, often carrying a bubble on the underside of the body or under the wings for use under water.

Amphibious vertebrates, frogs, salamanders, turtles, and water snakes are almost exclusively members of the littoral zone community. Tadpoles of frogs and toads are important primary consumers, feeding on algae and other plant material, whereas adults move up a trophic level or two. Cold-blooded vertebrates increase in importance as one goes south. For example, the population density and, hence, the ecological importance of turtles, frogs, and water snakes in the ponds of Louisiana and Florida are astonishing to those who are familiar only with northern ponds.

Pond fish generally move freely between the littoral and the limnetic zone, but most species spend a large part of their time in the littoral zone; many species establish territories and breed there (see Chapter 7, Section 15). Nearly every pond has one or more species of the "sunfish" or "bream" family (Centrarchidae). In the southern United States, top minnows, especially *Gambusia*, are abundant in the vegetation zones. Some species of bass, pike (*Esox*), or gar represent the end of the food chain as far as the pond ecosystem is concerned (see Chapter 3).

The zooplankton of the littoral zone is rather characteristic and differs from that of the limnetic zone in preponderance of heavier, less buoyant crustacea which often cling to plants or rest on the bottom when not actively moving their appendages. Important groups of littoral zooplankton (Fig. 11–7A) are large, weak-swimming species of Cladocera ("water fleas"), such as some species of *Daphnia* and *Simocephalus*, some species of copepods of the family Cyclopoidea and all of the Harpacticoidea, many families of ostracods, and some rotifers.

Finally, the littoral neuston (Fig. 11–7B) consists of three surface insects that are familiar to everyone who has even casually observed a pond: (1) whirligig beetles of the family Gyrinidae, the black "lucky bugs" of fishermen—these beetles are unique in that the eye is divided into two parts, one half for "seeing" above the water and the other half for underwater vision; (2) large water striders, family Gerridae; and (3) the smaller, "broad-shouldered" water striders, family Veliidae. Not so conspicuous are numerous protozoa and other microorganisms which are associated with the surface film (both above and below it).

Another remarkable microscopic community occurs among the sand grains at the water's edge (often called the psammolittoral habitat). Examination of this seemingly barren area reveals a remarkable number of algae, protozoa, tardigrades, nematodes, and harpacticoid copepods (see Pennak, 1940; Neel, 1948).

Nature of Communities in the Limnetic Zone

Phytoplankton producers of the open-water zone consist of algae of the three groups previously listed and the algaelike green flagellates, chiefly the dinoflagellates, Euglenidae and Volvocidae. Most of the limnetic forms are microscopic and hence do not attract the attention of the casual observer, even though they often color the water green. Yet phytoplankton may exceed rooted plants in food production per unit area. Typical open-water phytoplankton types are shown in Figure 11–5. Many of these forms have processes or other adaptation to aid in floating. Turbulence, or upward-current movements of water caused by temperature differences, aids in keeping phytoplankton near the surface where photosynthesis is most effective. A characteristic feature of limnetic phytoplankton in northern temperate lakes is the marked seasonal variation in population density (Fig. 11–8). Very high densities which appear quickly and persist for a short time are called "blooms" or phytoplankton "pulses" (see pages 47, 253, and 332). In the northern United States ponds and lakes often exhibit a large early spring bloom and another, usually smaller, pulse in the autumn. The spring pulse, which limnologists sometimes call the "spring flowering," typically involves the diatoms and is apparently the result of the following combination of circumstances. During the winter low water temperatures and reduced light result in a low rate of photosynthesis so that regenerated nutrients accumulate unused. With the advent of favorable temperature and light conditions the phytoplankton organisms, which have a high biotic potential, increase rapidly since nutrients are not limiting for the moment. Soon, however, nutrients are exhausted and

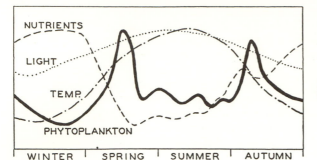

Figure 11–8. The probable mechanism for phytoplankton pulses in temperate-zone ponds and lakes. See text for explanation.

the bloom disappears. When nutrients again begin to accumulate, nitrogen-fixing blue-green algae, such as *Anabaena,* often are responsible for autumn blooms, these organisms being able to continue to increase rapidly despite a reduction of dissolved nitrogen—that is, until phosphorus, low temperature, or some other factor becomes limiting and halts the population growth.

The limnetic zooplankton is made up of few species but the numbers of individuals may be large. Copepods, cladocerans, and rotifers are generally of first importance, and the species are largely different from those found in the littoral zone. The long-antennaed or calanoid copepods (*Diaptomus,* a common genus) are especially characteristic, although the middle-length antennaed forms (*Cyclops*) may be more abundant in smaller bodies of water. Limnetic cladocera consist of highly transparent floating forms, such as *Diaphanosoma, Sida,* and *Bosmina.* Two characteristic genera of plankton rotifers are shown in Figure 11–7A. Many of the zooplanktonic crustacea are "strainers," filtering bacteria, detrital particles, and phytoplankton by means of combs of setae on their thoracic appendages. These organisms thus "graze" the plants in somewhat the same manner as cattle graze vegetation on land. Other zooplankton are predators. As might be expected, the zooplankton may exhibit pulses at the same time or immediately following those of the phytoplankton, since they largely depend on the latter. Some of the zooplankton organisms can perhaps use dissolved organic matter, but particulate food is believed to be their main energy source.

Vertical diel migration is a very characteristic feature of lake limnetic zooplankton, as has been described in Chapter 6, Section 7; Figure 6–9.

Copepods and cladocera demonstrate an interesting contrast in life history and method of reproduction. Both have achieved equal success, as it were, in a niche in which rapid reproduction is necessary for survival. Cladocera reproduce parthenogenetically, eggs developing in a "brood chamber," which is a space between the body and the carapace shell that envelops the body of the female. Development is direct, with no larval stage. Males appear only at rare intervals, usually at the onset of unfavorable conditions. Fertilized eggs develop into ephippial, or "winter," eggs, which have a resistant shell and are capable of surviving in a dry pond. Copepods, on the other hand, do not reproduce parthenogenetically, but the female is able to store enough sperm at one copulation to last for many batches of eggs. Thus, copepods are able to compete in the matter of rapid increase with other plankton which exhibit parthenogenesis or asexual reproduction. Copepods have a larval stage, called a nauplius, which is entirely free-living. Thus, copepods and cladocerans, the "codominants" of the limnetic primary-consumer groups, illustrate parallel adaptations that accomplish the same ultimate objective.

The limnetic nekton in fresh water consists almost entirely of fish. In ponds the fish of the limnetic zone are the same as those of the littoral zone, but in larger bodies of water a few species may be restricted to the limnetic zone. Most freshwater fish as adults feed on fairly sizable animals, that is, not on microscopic plankton. A few species, the gizzard shads (*Dorosoma* and *Signalosa*), for example, have "strainers" and are plankton feeders. In the large storage lakes of the TVA system gizzard shads form an important link between producers and game fish; their presence in these lakes enables game fish, such as bass and pike, to exist on a shorter food chain and to be more independent of the littoral zone, which may be withdrawn from use by fish during sea-

sonal "drawdowns." The larger species of
shad are less desirable because they grow
too big for predatory fish to eat and thereby
divert food energy away from the latter by
continuing to feed at a lower trophic level.

Nature of Communities in the Profundal Zone

Since there is no light, the inhabitants
of the profundal zone depend on the limnetic and littoral zone for basic food materials. In return, the profundal zone provides "rejuvenated" nutrients, which are
carried by currents and swimming animals
to other zones (see Chapter 2, Section 3).
The variety of life in the profundal zone, as
might be expected, is not great, but what is
there is important. The major community
constituents are bacteria and fungi, which
are especially abundant in the water–mud
interphase where organic matter accumulates, and three groups of animal consumers
(Fig. 11–7C): (1) blood worms, or hemo-
globin-containing chironomid larvae and
annelids, (2) small clams of the family
Sphaeriidae, and (3) "phantom larvae," or
Chaoborus (*Corethra*). The first two groups
are benthic forms; the last are plankton
forms that regularly move up into the limnetic zone at night and down to the bottom
during the day (see Figure 11–9). The red
annelid worms often increase in waters polluted with domestic sewage; in highly sep-
tic areas beds of these so-called "sludge
worms" may be the only macroscopic organism present. *Chaoborus* larvae are remarkable in having four air sacs, two at each
end of the body, which apparently act as
floats and also provide a reserve supply of
oxygen. These larvae are only "part-time"
members of the plankton, the adult being a
land-dwelling midge (Diptera). Most plankton organisms in fresh water remain throughout their entire life history in this life form
(i.e., holoplankton), in sharp contrast to
organisms of coastal marine plankton, many
of whom are only part-time members (i.e.,
meroplankton). All animals of the profundal
zone are adapted to withstand periods of
low oxygen concentration, whereas many
bacteria are able to carry on without oxygen
(anaerobic). The importance of the boundary zone between oxidized and reduced
sediments was alluded to in Chapter 2 (page
28) and will be considered again in the
Chapter on marine ecology.

5. LAKES

Hutchinson in his monographic *Treatise
on Limnology* (1957) remarks, "Lakes seem,
on the scale of years or of human life spans,
permanent features of landscapes, but they
are geologically transitory, usually born of
catastrophes, to mature and die quietly and
imperceptibly. The catastrophic origin of
lakes, in ice ages or periods of intense tectonic or volcanic activity, implies a localized distribution of lake basins over the land
masses of the earth, for the events, however
grandiose, that have produced the basins
have never acted on all the lands simultaneously or equally. Lakes therefore tend to
be grouped in *lake districts*." To this we
could add that man is feverously building
lakes (usually called impoundments) all
over the world, including areas where there
are no natural lakes. While not exactly born
of catastrophe, man-made lakes are probably also transitory in the geological sense
(see also Chapter 9, Section 4).

As was indicated in Section 1 of this chapter, no sharp distinction can be made between lakes and ponds. There are important ecological differences, however, other
than overall size. In lakes the limnetic and
profundal zones are relatively large, compared with the littoral zone. The reverse is
true of the bodies of water which are generally designated as ponds. Thus, the limnetic zone is the chief "producing" region
(region where light energy is fixed into food)
for the lake as a whole. The phytoplankton
and the nature of the bottom and its biota
are of primary interest in the study of lakes.
On the other hand, the littoral zone is the
chief "producing" region for ponds, and the
communities in this zone are of first interest.
Circulation of water in ponds is generally
such that limited stratification of temperature or oxygen occurs; lakes in the temperate
zone, unless very shallow, tend to become
stratified at certain seasons. Let us examine
this feature in some detail.

Stratification in Lakes—The Classic Temperate Zone Pattern

The typical seasonal cycle, illustrated in
Figure 11–9, may be described as follows.
During the summer the top waters become
warmer than the bottom waters; as a result,
only the warm top layer circulates, and it
does not mix with the more viscous colder

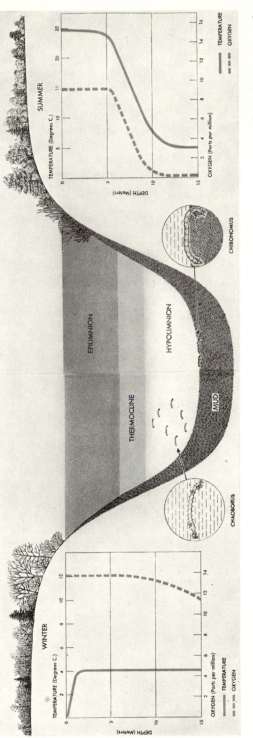

Figure 11-9. Thermal stratification in a north temperate lake (Linsley Pond, Conn.). Summer conditions are shown on the right, winter conditions on the left. Note that in summer a warm oxygen-rich circulating layer of water, the epilimnion, is separated from the cold oxygen-poor hypolimnion waters by a broad zone, called the thermocline, which is characterized by a rapid change in temperature and oxygen with increasing depth. Two typical hypolimnion organisms are shown (see also Figure 11-7). (After Deevey, 1951.)

water, creating a zone with a steep temperature gradient in between called the *thermocline*. The upper, warm circulating water is the *epilimnion* ("surface lake"), and the colder, noncirculating water is the *hypolimnion* ("*under* lake"). In Figure 11–9 note the strong drop in temperature at the thermocline in the hot months of summer. If the thermocline is below the range of effective light penetration (i.e., compensation level), as is often the case, the oxygen supply becomes depleted in the hypolimnion since both the green plant and the surface source is cut off. Note also in Figure 11–9 how the oxygen supply in the hypolimnion of Linsley Pond disappears in the summer. This is often referred to as the period of summer stagnation in the hypolimnion.

With the onset of cooler weather, the temperature of epilimnion drops until it is the same as that of the hypolimnion. Then the water of the entire lake begins circulating and oxygen is again returned to the depths during the "fall overturn." As the surface water cools below 4°C., it expands, becomes lighter, remains on the surface and freezes, if the regional climate is a cold one, bringing on winter stratification. In winter the oxygen supply is usually not greatly reduced because bacterial decomposition and respiration of organisms are not so great at low temperatures, and water holds more oxygen at low temperatures. Winter stagnation, therefore, is not generally so severe (see Figure 11–9, chart at left). An exception to this generalization may occur when snow covers the ice and prevents photosynthesis, resulting in oxygen depletion for the entire lake and "winter kill" of the fish.

In the spring, as ice melts and water becomes warmer, it becomes heavier and sinks to the bottom. Thus, when the surface temperature rises to 4°C., the lake takes another "deep breath," as it were—the *spring overturn*.

This classic picture of two seasonal overturns is typical for many lakes in America and Eurasia, but is by no means universal even in the temperate zone. Details of thermal stratification were first worked out in Swiss lakes between 1850 and 1900 by Simony and Forel, the latter often called the "father of limnology." In 1904, Birge, who later joined with Juday in the famous team of Birge and Juday, first demonstrated thermal stratification in Wisconsin lakes. In general, the deeper the lake, the slower the stratification and the thicker the hypolimnion.

The extent of depletion of oxygen in the hypolimnion during summer stratification depends on the amount of decaying matter and on the depth of the thermocline. Productively "rich" lakes generally are subject to greater oxygen depletion during the summer than "poor" lakes, because the "rain" of organic matter from the limnetic and littoral zones into the profundal zone is greater in productively rich lakes. Cultural eutrophication hastens oxygen depletion in the profundal zone as described in Chapter 3 (page 58); see also Figure 16–5, page 442. Thus, fishes which are stenothermal, low-temperature tolerant can survive only in "poor" lakes in which cold bottom waters do not become depleted of oxygen. Such species were the first to disappear on eutrophication of the Great Lakes in the United States. As already indicated, lower organisms (in contrast to fish) of the profundal zone are adapted to withstand oxygen deficiency for appreciable periods.

If waters of a lake are very transparent and will permit growth of phytoplankton below the thermocline (in the upper part of the hypolimnion), oxygen may be present here even in greater abundance than on the surface because, as indicated above, cold water holds more oxygen. We see, therefore, that the euphotic zone does not necessarily coincide with the epilimnion. The former is based on light penetration (is the "producer" zone), the latter on temperature. Often, however, they coincide roughly during the summer stagnation period.

Thermal Stratification in the Tropics

Subtropical lakes having surface temperatures that never fall below 4°C. generally show a distinct thermal gradient from top to bottom, but experience only one general circulation period per year, which comes in winter. Tropical lakes with high surface temperatures (20 to 30°C.) exhibit weak gradients and little seasonal change in temperature at any depth. Water density differences resulting from even the slight thermal gradient, however, may produce a stable stratification on a more or less year-around basis. General circulation, therefore, is irregular, occurring mostly in cooler seasons. Very deep tropical lakes tend to remain only partly mixed. As will be described later (see page 409), the construction of large, deep impoundments in the tropics is causing major "ecological disasters" because of a lack of understanding of differences in tropical and temperate land and water ecosystems.

In terms of these important water circulation patterns most of the lakes of the world can be conveniently assigned to one of the following categories (Hutchinson, 1957):

1. *Dimictic* (mictic = mixed). Two seasonal periods of free circulation, or overturns, as described in the preceding section.

2. *Cold monomictic.* Water never above 4°C. (polar regions); seasonal overturn in summer.

3. *Warm monomictic.* Water never below 4°C. (warm temperate or subtropical); one period of circulation in winter.

4. *Polymictic.* More or less continually circulating with only short, if any, stagnation periods. (High altitude, equatorial.)

5. *Oligomictic.* Rarely (or very slowly) mixed (thermally stable) as in many tropical lakes.

6. *Meromictic.* Permanently stratified, most commonly as a result of chemical difference in hypo- and epilimnial waters as described on page 313.

Geographical Distribution of Lakes

Natural lakes are most numerous in regions which have been subject to geological change in comparatively recent times, say within the past 20,000 years. Thus lakes abound in glaciated regions of northern Europe, Canada, and northern United States. Such lakes were formed as the last glaciers retreated about 10,000 to 12,000 years ago. These are the lakes which have been most studied by European and American limnologists. Natural lakes are also numerous in regions of recent uplift from the sea, as in Florida, and in regions subject to recent volcanic activity, as in the western Cascades. Volcanic lakes, formed either in extinct craters or in valleys dammed by volcanic action, are among the most beautiful in the world. On the other hand, natural lakes are rare in the geologically ancient and well-dissected Appalachian and Piedmont regions of eastern United States and in large sections of the unglaciated great plains. Thus, we see that geologi-

cal history of a region determines whether lakes will be naturally present and will also greatly influence the type of lake by determining the basic minerals which are available for incorporation into the lake ecosystem. Rainfall, of course, is important, but sizable lakes may occur even in deserts.

Classification of Lakes

Investigation of lakes over the world has shown that they possess a great variety of combinations of properties, making it difficult to select a basis for a natural classification. Hutchinson, in his 1957 monograph, lists no fewer than 75 lake types based on geomorphology and origin. However, a good introduction to the fascinating subject of world lake ecology can be obtained by considering three categories: (1) the oligotrophic–eutrophic series of ordinary clearwater lakes based on productivity, (2) special lake types, and (3) impoundments.

1. OLIGOTROPHIC — EUTROPHIC SERIES. Lakes in all regions may be classified according to primary productivity (see Chapter 3, Section 3), as outlined by the pioneer German limnologist, Thienemann. The productivity or "fertility" of a lake depends on nutrients received from regional drainage, on geological age, and on the depth. A simplified classification is as follows, with direction of geological development indicated by arrows (the interaction of geological and community development processes was thoroughly discussed in Chapter 9, Section 1):

	Conc. Nutrients Low	Conc. Nutrients High
Shallow	Morphometric Eutrophy →	Eutrophic
	↑	↑
Deep	Oligotrophic	Morphometric Oligotrophy

Typical oligotrophic ("few foods") lakes are deep, with the hypolimnion larger than the epilimnion, and have low primary productivity. Littoral plants are scarce and plankton density is low, although the number of species may be large; plankton blooms are rare since nutrients rarely accumulate sufficiently to produce a population eruption of phytoplankton. Because of low productivity of the waters above, the hypolimnion is not subject to severe oxygen depletion; hence stenothermal, cold-water bottom fishes such as lake trout and cisco are characteristic of and often restricted to the hypolimnion of oligotrophic lakes. There are also a few plankton forms characteristic of such lakes (*Mysis*, for example). In short, oligotrophic lakes are still "geologically young" and have changed but little since the time of formation. In contrast, *eutrophic* ("good foods") lakes are shallower and have a greater primary productivity. Littoral vegetation is more abundant, plankton populations are denser, and "blooms" are characteristic. Because of the heavy organic content, summer stagnation may be severe enough to exclude cold-water fishes. Lake Mendota and Linsley Pond (see Figure 11–9) are examples of eutrophic lakes which have been much studied; the Great Lakes and the Finger Lakes of New York are typical oligotrophic lakes.

The general trend of increasing productivity with decreasing depth is illustrated by the data in Table 11–1. These data also illustrate the short turnover time of lake phytoplankton (see Chapter 2, page 17) and the fact that standing crop biomass may be more influenced by the size of individuals than by the productivity; in other words, another illustration of the principle that biomass is not necessarily correlated with rate of production; see Chapter 3, page 44.

2. SPECIAL LAKE TYPES. Seven special lake types may be mentioned here.

(a) *Dystrophic lakes: brown water, humic, and bog lakes.* Generally have high concentrations of humic acid in water; bog lakes have peat-filled margins (where pH is usually low) and develop into peat bogs.

(b) *Deep, ancient lakes with endemic fauna.* Lake Baikal in Russia is the most famous of ancient lakes. It is the deepest lake in the world and was formed by earth movement during the Mesozoic era (age of reptiles). Ninety-eight per cent of 384 species of arthropods are endemic (found nowhere else), including 291 species of amphipods. Eighty-one per cent of 36 species of fish are endemic. This lake is often called the "Australia of freshwater," because of its endemic fauna (see Brooks, 1950). Recently it has been reported that this lake is threatened by industrial pollution.

Table 11–1. *Ratio of Standing Crop Biomass and Daily Primary Production Rate in Seven Swiss Lakes*[*]

LAKE	DEPTH (m)	PRODUCTIVITY (P) (Kcal/m²/day)	BIOMASS (B) (Kcal/m²)	B/P (turnover in days)
1	215	5	20†	4.0
2	151	4	3‡	0.75
3	134	4	7	1.75
4	84	6	24†	4.0
5	46	8	5‡	0.62
6	36	9	12	1.33
7	16	11	10	0.91

[*] From Findenegg, 1966.
† Population dominated by large species of algae (net plankton).
‡ Population dominated by small species of algae (nannoplankton).

(c) *Desert salt lakes:* Occur in sedimentary drainages in arid climates where evaporation exceeds precipitation (thus resulting in salt concentration). Example: Great Salt Lake, Utah. Contain community composed of a few species (but sometimes abundant) which are able to tolerate high salinity. Brine shrimp (*Artemia*) are characteristic.

(d) *Desert alkali lakes:* Occur in igneous drainages in arid climates; high pH and concentration of carbonates. Example: Pyramid Lake, Nevada.

(e) *Volcanic lakes:* Acid or alkaline lakes associated with active volcanic regions (and receiving waters from the magma); extreme chemical conditions, restricted biota. Examples: some Japanese and Philippine lakes.

(f) *Chemically stratified; meromictic (= partly mixed) lakes:* In contrast to most lakes in which bottom and surface waters mix periodically (i.e., *holomictic* or "wholly mixed" lakes), some lakes become permanently stratified by the intrusion of saline water or salts liberated from sediments, creating a permanent density difference between surface and bottom waters. In this case the boundary between circulating and noncirculating layers is a "chemocline," instead of a thermocline. Free oxygen and aerobic organisms will, of course, be absent in bottom waters of such lakes. Big Soda Lake in Nevada and the Hemmelsdorfersee in Germany are examples.

(g) *Polar lakes:* Surface temperatures remain below 4°C., or rise above it for only a brief period during the ice-free summer when circulation can take place. Plankton populations grow rapidly during this period, often storing fat for the long winter.

3. IMPOUNDMENTS. Artificial lakes, of course, vary according to the region and to the nature of the drainage. Generally, they are characterized by fluctuating water levels and high turbidity. Production of benthos is often less in impoundments than in natural lakes. Successional changes in primary productivity and fish yields that occur when large streams are impounded have been outlined in Chapter 9, page 262.

The heat budget of impoundments may differ greatly from that of natural lakes, depending on the design of the dam. If water is released from the bottom, as would be the case with dams designed for hydroelectric power generation, cold, nutrient-rich but oxygen-poor water is exported downstream while warm water is retained in the lake. The impoundment then becomes a *heat trap and nutrient exporter,* in contrast to natural lakes which discharge from the surface and, therefore, function as *nutrient traps and heat exporters.* Accordingly, the type of discharge greatly affects downstream conditions. Figure 11–10 compares water temperatures in two lakes in the same drainage basin in Montana, one with surface discharge and the other with deep-water discharge. During the hottest part of the

summer, temperatures in the thermocline and hypolimnion regions (below about 10 meters) were 5 to 8°C. lower in the lake with surface discharge. The release of cold bottom water makes it possible to develop a downstream trout fishery south of the normal range of trout, but further north such water may be too cold for good fish growth. In addition to thermal considerations, Wright (1967) lists the following effects of dams with deep-water penstocks:

1. Water is released with a higher salinity than would be obtained from surface water withdrawal.

2. Essential nutrients are lost from the reservoir, thus tending to deplete the productive capacity of the reservoir and at the same time causing eutrophication downstream.

3. Evaporative loss is increased as a result of storing warm incoming water and releasing cold hypolimnial water.

4. Low dissolved oxygen in the discharged water reduces the capacity of the stream to receive organic pollutants.

5. The discharge of hydrogen sulfide and other reduced substances lowers downstream water quality and, in extreme cases, results in fish-kills.

Some other problems in designing and managing impoundments in the face of conflicting multiple-use demands are discussed elsewhere (see pages 409 and 416–417).

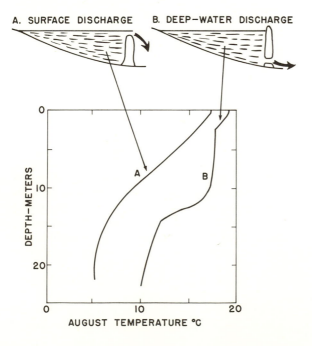

Figure 11–10. August temperature depth curves for two lakes in Montana, one (A) with surface discharge and one (B) with deep-water discharge. The dam on Lake A was formed by an earthquake-induced landslide. Lake B is an impoundment with a man-made dam. Both bodies of water are in the same drainage basin with A downstream from B. (Redrawn from Wright, 1967.)

6. PONDS

As indicated in the previous section, ponds are small bodies of water in which the littoral zone is relatively large and the limnetic and profundal regions are small or absent. Stratification is of minor importance. Ponds may be found in most regions of adequate rainfall. They are continually being formed, for example, as a stream shifts position, leaving the former bed isolated as a body of standing water or "oxbow." Because of the accumulation of organic materials and periodic flooding, flood-plain ox-bow ponds may be quite productive, as evidenced by the number of fishermen they attract. Natural ponds are also numerous in limestone regions when depressions or "sinks" develop because of solution of the underlying strata.

Temporary ponds, that is, ponds that are dry for part of the year, are especially interesting and support a unique community. Organisms in such ponds must be able to survive in a dormant stage during dry periods or be able to move in and out of ponds, as can amphibians and adult aquatic insects. Some temporary pond animals are shown in Figure 11–11. The fairy shrimps (Eubranchiopoda) are especially remarkable crustaceans which are well adapted and largely restricted to temporary ponds. The eggs survive in the dry soil for many months, whereas development and reproduction occur in a short time in late winter and spring while water is present. As is true of other marginal habitats, the temporary pond is a favorable place for those organisms adapted to it, because interspecific competition and predation are reduced. Even though a temporary pond contains water for only a few weeks, a definite seasonal succession of organisms may occur, thus enabling a surprisingly large variety of organisms to utilize a very limited amount of physical habitat (see Figure 11–11).

Ponds created by damming of a stream or basin by man, or by animals such as the beaver, are among the most numerous. Prior to about 1920, most man-made ponds in the United States were "mill ponds," formed by damming sizable streams for the purpose of providing power for small mills. At the present time a very large number of "farm ponds" are being constructed which differ from "mill ponds" in that stream water is largely detoured around the pond, or the pond is constructed in a basin without a permanent stream, in order to prevent loss of nutrients and the silting up of the basin. Such ponds have relatively little water flowing through them and are often artificially fertilized, stocked with game fishes, and managed so as to discourage rooted vegetation. Thus, producers are all phytoplankton and, in a sense, food-chain relations resemble those of lakes. However, such ponds are rarely deep enough for stratification. The contrast between a managed "farm pond" and a pond with a vegetated littoral zone is shown in Figure 11–1. A "farm pond" can be much more efficient than an unmanaged pond or mill pond from the standpoint of production of a large bio-

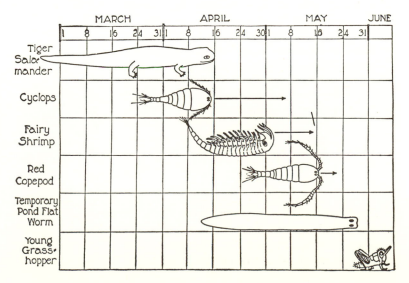

Figure 11–11. Succession of animals in a temporary pond in Illinois. The length of the animal's body plus the length of the arrow indicates the dates between which the adults of each of the five species were found. The drying up of the pond and appearance of land organisms is indicated by the young grasshopper. (From Welch, modified after Shelford, 1919.)

mass of fish, but it is less interesting to the naturalist because the variety of organisms is reduced to favor large numbers of the desired species. For a picture of the "energetics" of a managed fish pond, see Figure 3–11. Fish production in ponds is discussed on pages 71 and 416.

Beaver ponds were a characteristic feature of much of the North American continent when European man arrived. A beaver pond generally has a short ecological life history, as such ponds are abandoned when the supply of food trees in the vicinity becomes reduced. Thus, under primeval conditions, the beaver was a very important factor in opening up the forests and in maintaining both terrestrial and aquatic seral stages. Along large rivers and on coastal plains beaver often do not construct ponds but live in holes in the bank of streams and thus become essentially stream animals.

As man "conquered" the continent, the beaver was all but extirpated over large areas, but it is making a comeback, especially as marginal farmland is abandoned in favor of forest or urban development. Beaver ponds are a very useful component of natural areas since they act as water reservoirs, fire breaks, and provide habitat for fur-bearers and fish. However, flooding of forest or farmland often relegates the poor beaver to the "pest" category and brings on what I like to call the "beaver dam syndrome," meaning that man rushes into precipitous action, blows up the dam, and thereby creates a different and greater stress on the landscape. According to Wilde *et al.* (1950), sudden drainage, which lowers the water table and dries out the soil, is especially damaging to surrounding vegetation whose root systems have become adapted to moist soil. This is another example of a "shock" type stress being more limiting than a gradual change. It seems that damming and undamming streams is a privilege to be enjoyed by man but not by beaver.

7. LOTIC (RUNNING-WATER) COMMUNITIES

General Comparison of Lotic and Lentic Habitats

One does not have to be an expert, nor does one need to sample all of the varied manifestations of life, to appreciate differences between standing- and running-water habitats. Comparison of a stream and a pond or lake makes an excellent ecological study, one that will bring out important principles. An ecology class, for example, can very profitably spend a couple of hours in a field investigation of an example of each type of habitat. Results will probably be most satisfactory if the class is divided into "teams," each assigned to sample the physicochemical and biological aspects of a significant portion or zone. Thus, everyone contributes to the study and no one is left standing on the bank! If this field sampling is followed by a little laboratory work, many of the significant features of both environments will become apparent via the comparison and contrast method. When the results are tabulated, it will be found that the biotic community in streams is quite different from that in ponds, even if the identification of lower plants and invertebrates is carried only to genus (or only to family). Discovery of differences in organisms leads naturally into a consideration of the chief differences in the physical and chemical limiting factors. This, in turn, should suggest basic differences in food-chain arrangements and ways in which productivity in the two types of ecosystems might be studied. All of this, it is hoped, should whet the appetite for a return to the pond and stream for further observations, or even promote explorations into the literature!

While many ponds and lakes have been well studied as whole ecosystems, there have been very few such treatments of streams, perhaps because most are by and large incomplete systems, as will be noted later. There have been a number of excellent studies on food-chain energetics of streams with emphasis on fish, as, for example, Allen (1951), Horton (1961), and Gerking (1962). The River Thames in England has been rather well studied by a group of investigators (see Mann, 1964, 1965, 1969). Cummins *et al.* (1966) have worked out the trophic relationships in a small woodland stream, and this paper provides a good reference for class study. Experimental streams can be set up in the laboratory or out-of-doors, and such "microecosystems" are valuable tools for study (see H. T. Odum and Hoskins, 1957; McIntire and Phinney, 1965; Davis and Warren, 1965). Since most streams in the vicinity of urban areas are at least to some extent polluted, the little book *Biology of Polluted Waters* by Hynes (1960) is a good reference for the beginning student.

In general, differences between streams and ponds revolve around a triad of conditions: (1) current is much more of a major controlling and limiting factor in streams; (2) land–water interchange is relatively more extensive in streams, resulting in a more "open" ecosystem and a "heterotrophic" type of community metabolism (see Figure 3–10, page 69); and (3) oxygen tension is generally more uniform in streams, and there is little or no thermal or chemical stratification. Let us briefly discuss each of these in order.

1. *Current.* Although possession of a definite and continuous current is of course a prime characteristic of lotic habitats, streams and lakes are not sharply divided in this regard. Thus, velocity of the current varies greatly in different parts of the same stream (both longitudinally and transversely to the axis of flow) and from one time to another. In large streams or rivers the current may be so reduced that virtually standing-water conditions result. Contrariwise, wave action along rocky or sandy shores of lakes (especially in the absence of rooted plants which might buffer the wave action) may virtually duplicate stream conditions. Consequently, what would generally be considered pond organisms are often found in quiet pools of streams, and stream animals may be found in the wave-tossed portion of lakes. Nevertheless, current is the most important primary factor which (1) makes for a big difference between stream and pond life and (2) governs differences in various parts of a given stream. Hence it is certainly a factor worth primary consideration and measurement.

The velocity of current is determined by the steepness of the surface gradient, the roughness of the stream bed, and the depth and width of the stream bed. Several types of current meters have been devised, but it is difficult to measure velocity under stones and in crevices where organisms live. At any given point there is a "microstratification" of current. Where fish are concerned or where the general nature of the stream community over an appreciable stretch is being considered, the surface gradient alone gives a good index to average current conditions. The gradient in feet per mile

(or smaller units), for example, can easily be determined from topographic maps or can be measured with simple surveying instruments in the field. In Ohio, for example, Trautman (1942) found that distribution of bass and other fish was well correlated with gradient of flow. Small-mouth bass, for example, were found largely in stream sections having a gradient of 7 to 20 feet per mile; they were almost never collected where the gradient was below 3 or above 25 feet per mile.

2. *Land–Water Interchange.* Since the depth of water and the cross-section area of streams is much less than that of lakes, the land–water surface junction is relatively great in proportion to the size of the stream habitat. This means that streams are more intimately associated with the surrounding land (Fig. 11–2) than are most standing bodies of water. In fact, most streams depend on land areas and on connected ponds, backwaters, and lakes for a large portion of their basic energy supply. Streams, to be sure, have producers of their own, such as fixed filamentous green algae, encrusted diatoms, and aquatic mosses, but these are usually inadequate to support the large array of consumers found in streams. Many of the primary consumers in streams are detritus feeders that depend in large part on organic materials which are swept or fall in from terrestrial vegetation (see Hynes, 1963; Minshall, 1967). Sometimes plankton and detritus coming into the stream from quieter waters are important. Also, streams "export" energy in the form of emerging insects and stream life removed by air-breathing predators (see Figure 11–2). Thus, streams form an open ecosystem that is interdigitated with terrestrial and lentic systems. For this reason measurement of productivity must include adjacent land and standing water systems. The importance of this "watershed" ecosystem concept was emphasized in Chapter 2. Mann (1969) estimates that the Thames River, which is very productive and densely populated with fish and mollusks, depends for perhaps half its energy flow on "allochthonous" material, i.e., organic matter from outside the stream such as leaves and sewage solids, and, therefore, on the detritus rather than the grazing food chain (see Chapter 3). In a Georgia stream receiving only a small amount of domestic sewage and no industrial wastes Nelson and Scott (1962) found that the "primary

consumer organisms derived 66 per cent of their energy from allochthonous organic matter, largely leaf material.

3. *Oxygen.* Although stream organisms face more extreme conditions in regard to current and temperature than do pond organisms, oxygen is not as likely to be as variable under natural conditions in streams. Because of the small depth, large surface exposed to the air, and constant motion, streams generally contain an abundant supply of oxygen, even when there are no green plants. For this reason stream animals generally have a narrow tolerance and are especially sensitive to reduced oxygen. Therefore, stream communities are especially susceptible to and quickly modified by any type of organic pollution which reduces the oxygen supply (see Figure 16–6). As the public is now well aware, streams are the first victims of urbanization; restoring their quality will require a commitment of money and human effort that is not yet evident. More about this in Part 3.

Nature of Lotic Communities

As was outlined in the discussion of aquatic zonation in Section 2 of this chapter, streams generally exhibit two major habitats, namely, rapids and pools. Consequently, broadly speaking we may first think in terms of two stream community types, rapids communities and pool communities. Some of the characteristic organisms of these two community types are shown in Figure 11–12. Within these broad categories the type of bottom, whether sand, pebbles, clay, bedrock, or rubble rock, is very important in determining the nature of the communities and the population density of community dominants. As streams work down to base-level conditions, the distinction between rapids and pools becomes less and less until a channel habitat is developed in large rivers. The biota of a river channel resembles that of the rapids, except that population distribution is highly "clumped," owing to the frequent absence of firm substrates.

Current is the major limiting factor in rapids, but the hard bottom, especially if composed of stones, may offer favorable surfaces for organisms (both plant and animal) to attach or cling. The soft, continually shifting bottom of pool areas generally lim-

its smaller benthic organisms to burrowing forms, but the deeper, more slowly moving water is more favorable for nekton, neuston, and plankton. The species composition of rapids communities is likely to be 100 per cent different from that of the quiet water zones of ponds and lakes. Consequently, we generally think of organisms in rapids communities as being "typical" stream organisms. Pool communities, on the other hand, may be expected to contain some organisms which also occur in ponds. For example, the gyrinid beetles "gyrate" as well on the surface of a quiet pool as on the surface of the littoral zones of ponds, while bluegills, typical pond fish, also reside in deeper pools of streams.

Other conditions being equal, sand or soft silt is generally the least favorable type of bottom and supports the smallest number of species and individuals of benthic plants and animals. Clay bottom is generally more favorable than sand; flat or rubble rocks generally produce the largest variety and highest density of bottom organisms. Generally, benthic invertebrates have a higher density in rapids communities, whereas stream nekton and burrowing forms, such as clams, burrowing Odonata and Ephemeroptera, are more abundant in pools. Stream fish commonly find refuges in pools and feed in or at the base of rapids, thus linking pool and rapids communities.

It might be assumed that plankton would be absent from streams, since such organisms are at the mercy of the current. Although it is true that plankton is much less important in stream economy, compared with its dominant position in lake ecosystems, streams do have plankton. In small streams plankton, if present, originates in lakes, ponds, or backwaters connected with streams and is soon destroyed as it passes through rapids. Only in slow-moving parts of streams and in large rivers is plankton able to multiply and thus become an integral part of the community. Despite the transitory nature of much of the stream plankton, it may supply a not unimportant source of food in some rivers. In the Illinois River plankton is continually replenished in large quantities from a series of productive flood-plain lakes which are joined to the river. In the Mississippi River and its large tributaries, at least one species of fish, the primitive and unique *Polyodon*, feeds largely on zooplankton.

Organisms in rapids communities, and to a lesser extent those inhabiting pool communities, show adaptations for maintaining position in swift water. Some of the most important of these are:

1. *Permanent attachment to a firm substrate*, such as a stone, log, or leaf mass. In this category would be included the chief producer plants of streams, which are (1) attached green algae, such as *Cladophora*, with its long trailing filaments; (2) encrusting diatoms which cover various surfaces; and (3) aquatic mosses of the genus *Fontinalis* and others which cover stones even in the swiftest current. Also, a few animals

become fixed, such as freshwater sponges and caddis fly larvae, which cement their cases to stones.

2. *Hooks and suckers.* A great many rapids animals have either hooks or suckers that enable them to grip even seemingly smooth surfaces. Note how many of the animals in Figure 11–12 are so equipped. The two Diptera larvae, *Simulium* and *Blepharocera*, and the caddis, *Hydropsyche* (see Figure 11–12), are especially remarkable in this connection and are often the only animals able to withstand the pounding of swift rapids and waterfalls. *Simulium* not only has a sucker at the posterior end of

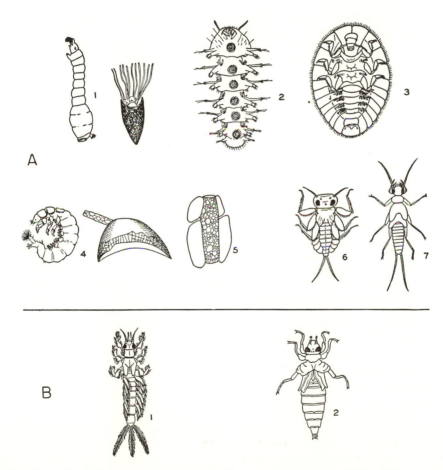

Figure 11–12. Some characteristic stream animals. A. Representative genera of the rapids community, illustrating various adaptations for living in swift current. This group is often aptly called the "torrential fauna." 1. A black-fly larva, *Simulium* (Simuliidae), and its pupa (right) in a cocoon attached to a rock; note sucker at posterior end of the larva and the "head net" used for straining food from the water. 2. A blepharocerid larva, *Bibiocephala* (note the row of ventral suckers). 3. A "water penny," *Psephenus*, the larva of the riffle beetle (Psephenidae). 4. A net-spinning caddis, *Hydropsyche*, with its net (open end faces upstream) and case. 5. The case of a caddis, *Goera*, with "ballast" rocks attached to the side. 6. A mayfly nymph, *Iron*, and 7, a stonefly nymph, *Isogenus*, both with flattened, streamlined bodies adapted to clinging to the undersides of stones. (*Note:* most stonefly nymphs have two "tails" and most mayfly nymphs have three "tails," as shown in B-1. *Iron* is atypical in this respect.)
B. Two burrowing types living in stream banks or bottoms of pools. 1. A burrowing mayfly nymph, *Hexagenia*. 2. A burrowing dragonfly nymph, *Progomphus*. (Redrawn from Robert W. Pennak, "Fresh-water Invertebrates of the United States," 1953, The Ronald Press Company.)

the body but also attaches itself by means of a silken thread. If dislodged, the legless larvae are kept from being swept very far down stream by their "safety rope" and work their way back up the thread to a favorable attachment place. In addition to hooks, *Hydropsyche*, the net-spinning caddis, cements a net around itself that acts not only as a shelter but also as a food trap for animal and plant material suspended in the water.

3. *Sticky undersurfaces.* Many animals are able to adhere to surfaces by their sticky undersurfaces. Snails and flatworms are good examples.

4. *Streamlined bodies.* Nearly all stream animals, from insect larvae to fish, exhibit definite streamlining, which means that the body is more or less egg-shaped, broadly rounded in front and tapering posteriorly, to offer minimum resistance to water flowing over it.

5. *Flattened bodies.* In addition to streamlining, many rapids animals exhibit extremely flattened bodies which enable them to find refuge under stones and in crevices. Thus, the body of stonefly and mayfly nymphs living in swift water is much flatter than the body of nymphs of related species living in ponds.

6. *Positive rheotaxis* (*rheo*, current; *taxis*, arrangement). Stream animals almost invariably orient themselves upstream and, if capable of swimming movements, continually move against the current. This is an inherent behavior pattern. In contrast, many lake animals, when placed in a current of water, merely drift with the current and make no attempt to orient themselves or move against it. An inherent behavior pattern for positive rheotaxis is just as important an adaptation as the morphological features cited above.

7. *Positive thigmotaxis* (*thigmo*, touch, contact). Many stream animals have an inherent behavior pattern to cling close to a surface or to keep the body in close contact with the surface. Thus, when a group of stream stonefly nymphs are placed in a dish, they seek to make contact with the underside of sticks, debris, or whatever is available, even clinging to each other if no other surface is available.

Thienemann (1926), the German limnologist, in his delightful essay on "the brook," has pointed out that adaptation for stream life among animals might have arisen in two ways. First, specialized structures and physiological responses may have arisen in widely different orders and families not naturally adapted to running water through natural selection. Phylogenetic development from simple skin folds to complicated suckers in the diptera would be an example. Second, animals may already have possessed favorable form or function and could therefore occupy swift water without further change. Thienemann called this preadaptation the "taking-advantage principle" (*Ausnutzungsprinzip*). Snails as a group have a sticky "foot," regardless of where they live. Thus snails, flatworms, and certain other organisms merely need to "take advantage" of their basic attribute in order to become a part of stream communities.

8. LONGITUDINAL ZONATION IN STREAMS

In lakes and ponds the prominent zonation is horizontal, whereas in streams it is longitudinal. Thus, in lakes, successive zones from the middle to the shore represent, as it were, successively older geological stages in the lake-filling process. Likewise in streams we find increasingly older stages from source to mouth. Changes are more pronounced in the upper part of streams because the gradient, volume of flow, and chemical composition change rapidly. The change in composition of communities is likely to be more pronounced in the first mile than in the last fifty miles.

The longitudinal distribution of fish in a stream may be selected as a specific example. Shelford (1911a) and Thompson and Hunt (1930) made such studies in Illinois streams, in which the gradient of flow was not greatly different from source to mouth. Headwaters species generally exhibited wide tolerances and were found throughout the length of the stream, whereas other species occupied successive sections of the stream. Thompson and Hunt found that the number of individuals decreased downstream, but the size of the fish increased, so that biomass density remained about the same. Table 11–2 shows the distribution of fish in a stream in the Virginia mountains where a distinct longitudinal change is related to a very steep gradient of tempera-

*Table 11–2.   Longitudinal Distribution of Fishes in Little Stony Creek**

Stations	1	2	3	4	5	6	7	8	9	10	11	12	13	14
pH	5.6	5.6	5.8	5.8	5.9	6.2	6.4	6.6	7.0	7.0	7.1	7.2	7.2	7.4
Temperature (°C.)	15	15	16	16	17	18	18	18	18	19	19	20	20	21
Salvelinus f. fontinalis	X	X	X	X	X	X	X	X	X	X				
Rhinichthys atratulus obtusus						X	X	X		X		X	X	
Catonotus f. flabellaris							X			X			X	X
Salmo gairdnerii irideus								X	X	X	X	X	X	X
Cottus b. bairdii												X	X	X
Campostoma anomalum													X	X
Notropis albeolus														X
Rhinichthys cataractae														X
Catostomus c. commersonnii														X

° From Burton and Odum (1945). Stations are approximately one mile apart.

ture, current velocity (highest at upper end of stream), and pH. Low temperatures and swift current limit the fish along the upper high altitude length of Little Stony Creek to relatively few species of trout, darters, and swift-water minnows. Longitudinal replacement of species is not a matter of a uniform, continuous change; specific conditions and populations may reappear at intervals, as is indicated by the discontinuous distribution of some species.

A very instructive study of longitudinal zonation of stream invertebrates can be made by an ecology class. Stations may be selected at intervals along the upper reaches of a small stream and teams assigned to make collections with comparable techniques. Results may then be tabulated as in Table 11–2 and subjected to various kinds of gradient analyses as described in Chapter 6, Section 3.

9. SPRINGS

Springs are the aquatic ecologist's natural constant temperature laboratory. Because of the relative constancy of the chemical composition, velocity of water, and temperature, in comparison with lakes, rivers, marine environments, and terrestrial communities, springs hold a position of importance as study areas that is out of proportion to their size and number. An account of the community metabolism in a large Florida spring has been given in previous chapters (see especially pages 141–143 and

279). Such studies pioneered the development of what is often called the "trophic-dynamic" approach to the study of ecosystems.

Many springs, such as the large Florida limestone springs, seem to be in a steady state, with rapid growth of organisms but a constant standing crop (H. T. Odum, 1957). In the spring itself the organisms do not modify their environment, and thus cause succession, because as the water is altered by photosynthesis and respiration, it passes downstream and is replaced by new water of the same properties from underground. These conditions make it possible to study whole communities under constant, known conditions. The cost of a similar large "chemostat" set-up in the laboratory would be prohibitive.

Some of the types of springs about which knowledge is available are: (1) hot springs, usually with a high salinity, of volcanic areas, such as Iceland, New Zealand, western United States, and North Africa; (2) the large hard-water springs in limestone districts of Florida, Denmark, and North Germany, having the average temperature of the region in which they occur; and (3) small soft-water springs emerging through shales, sandstones, and crystalline rocks; because of their small size these springs are more affected by the surrounding environment, and the community of organisms is almost entirely dependent on organic inputs of terrestrial origin (see Teal, 1957).

Hot springs provide steady-state microcosms under extreme conditions that are of special interest to biologists in connection with theories on the origin of life and the possibility of life on the planets (see Chapter 20). Studies on the hot springs have established the upper ecological temperature tolerances for survival and propagation for types of organisms as follows:

	Degrees F.	Degrees C.
Bacteria	190	88
Blue-green algae	176	80
Protozoa	129	54
Insects	122	50
Fishes	122	50

Two points should be made regarding these figures. (1) Many microorganisms can survive temperatures higher than 90°C. for brief periods, but no known species can self-sustain indefinitely above this temperature. (2) The optimum is always less (see Figures 5–1, 5–3). The optimum for blue-greens in Yellowstone hot springs is 54°C. In the 40 to 57°C. temperature zone of hot springs in the Lower Geyser Basin of Yellowstone National Park, Wyoming, blue-green algae form thick mats that support a unique community now being subjected to intensive study by R. J. Wiegert and M. L. and T. D. Brock (see Brock, 1967a, and Brock, Wiegert, and Brock, 1969) and their students. Seven species of blue-greens (Myxophyceae), three species of brine flies (Ephydridae), three species of long-legged flies (Dolichopodidae), two species of water mites (Parasitengona), two species of midges (Chiromidae), and one species of parasitic wasp (Chalcididae) have so far been found in this algal mat ecosystem (unpublished report, R. J. Wiegert). Bacteria are represented by many unicellular forms but are not nearly so numerous as in the hotter zones. The main food chain consists of blue-greens, as the producers, a brine fly, *Paracoenia turbida,* the principal grazer (herbivore), and a dolichopodid fly (*Trachytrechus angustipennis*), the chief predator, which feeds on both larvae and adults of the brine fly. Adult water mites feed on eggs of the flies, while the young parasitize the fly larvae, as do the wasp larvae. Thus, this simple hot spring system has all the ingredients for biotic balance. Strangely enough the flies are not particularly adapted to high temperatures (optimum 30 to 35°C.); eggs and larvae develop in cool spots where the mat projects above the water, while adults escape high temperatures by walking around on the mat surfaces.

In a constant temperature environment as provided by large springs, it is possible to separate the effects of light and temperature. Thus, in Iceland hot springs, Tuxen (1944) found that the temperature remained constant and favorable for the growth of algae during the long winter, but there was little light for photosynthesis. Consequently, the algal population density decreased in winter and the faunal population was reduced to a low level for lack of food and oxygen. A similar though less pronounced rhythm related to seasonal variations in light has been demonstrated for constant temperature Florida springs (H. T. Odum, 1957).

The community composition of a spring run changes with changing conditions as one goes downstream, as is the case in any stream (see the preceding section). However, conditions in a large spring run remain relatively more constant at any given point, producing a kind of succession in steady states. Runs from hot springs provide a natural ecological gradient from high temperature and high salinity to conditions of lower temperature and lower salinity.

The constant temperature properties of springs permit organisms to exist in regions where they do not otherwise occur. Thus, arctic insects occur in German springs because of the lower summer temperatures there, and organisms of warmer climates occur in the Icelandic hot springs. There seems little question that springs have provided refuges for aquatic organisms during geological periods when climatic changes have occurred.

Quantitative relationships between surface fresh waters and the circulating ground waters that keep the springs and streams running are shown in Figure 4–8B. While the underground pool is estimated to be 10 times larger than the surface inland water pool, it is far smaller than that in the polar ice caps. The rate of recharge of the circulating ground waters is but poorly known and needs to be investigated (study opportunities provided by tracer tritium was mentioned on page 98), but it is definitely very much slower than for most surface waters. As he has done with most everything else, man is exploiting this priceless resource as if it were limitless. Because of low recharge rates massive pumping from deep wells quickly lowers the water table. Use of underground water for industrial purposes should no longer be permitted when there is abundant surface water that could be used and recycled—or unless water is returned to the underground aquifers (= water bearing strata). Also, as already mentioned, storing water underground has many advantages over building large impoundments that suffer huge water losses because of evaporation and that destroy valuable agricultural and forest land as well.

MARINE ECOLOGY* Chapter 12

1. THE MARINE ENVIRONMENT

For centuries man regarded the sea as a restless surface which first hindered, then aided, his efforts to explore the world. He also learned that the sea was a source of food which could be harvested, by dint of great effort, to supplement the products of land and fresh water. Biologists early became intrigued with the amazing variety of life to be found along the shores and among the coral reefs. Study by the sea became a traditional part of advanced training in the biological sciences. Yet the great bulk of the sea remained largely a mysterious realm which might harbor all manner of sea serpents, for all anyone knew. It was only in 1872 that HMS *Challenger,* one of the first ships to be specifically equipped for the study of the sea, put forth on her now famous voyages. Since that time the study of the sea has proceeded with increasing tempo, aided by many oceanographic vessels manned by scientists of many nations as well as by marine laboratories on shore. Progress has been especially rapid since about 1930, with the development of sophisticated electronic gear, underwater cameras and television, and nets for high speed sampling at great depths.

The oceans as physical and chemical systems are becoming increasingly well understood, and knowledge of life in the sea is expanding slowly. Ideas on the origin and geological history of the oceans have moved from the realm of speculation onto a basis of firm theory. The key role of the sea in controlling the world's climates, the atmosphere, and the functioning of major mineral cycles has been emphasized and extensively documented in Chapters 2, 3, and 4. As man crowds and exploits the continents, he naturally "turns to the sea" (as widely read author Athelstan Spilhaus expresses it) in search of more minerals, food, and

even living space. As was pointedly emphasized in Chapter 3, much of the sea is "semidesert" and will not yield much food without expensive "energy subsidies." Likewise, usable mineral deposits are largely concentrated around the continental margins because of the way in which the richest ore-bearing rocks have been deposited; they are not limitless, nor are they easy to get without grave risk of pollution damage (witness increasingly frequent oil spills, see Smith, 1968). As geologist Preston Cloud (1969) has said: "A 'mineral cornucopia' beneath the sea thus exists only in hyperbole." While international cooperation in the study and exploration of the sea is advancing, international efforts to regulate the use of the sea are nil, and major advances in international marine law are urgent. The mounting ecological evidence sounds a warning to man that he must consider the oceans as an integral part of his total life-support system, and not as an inert "supply depot" that is merely there for the taking. For this reason, the study of the seas should be a required course for the citizens and students of all nations.

Sverdrup, Johnson, and Fleming (1942) and Hedgpeth (1957) have authored works that are still useful for reference, as is the collection of technical papers edited by Hill (1962–1963). Likewise the smaller books by Coker (1947), Colman (1950), and Moore (1958) remain as useful introductions for the beginner. Many recent texts in oceanography, as, for example, Dietrich (1963), King (1967), and Weyl (1970), emphasize physical aspects; there is a need for an up-to-date book on biological oceanography. The special issue of *Scientific American* (September, 1969) provides a semipopular introduction to "the oceans." In this issue Roger Ravelle tells of the fascination of the profession:

"Oceanographers have the best of two worlds —both the sea and the land. Yet many of them, like many sailors, find it extraordinarily satisfying to be far from the nearest coast on one of the small, oily and uncomfortable ships of their trade, even in the midst of a vicious storm, let

---

* *Oceanography* deals with study of the sea in all of its aspects—physical, chemical, geological, and biological. *Marine ecology* emphasizes the totality, or pattern, of relationships between organisms and the sea environment.

alone on one of those wonderful days in the Tropics when the sea and air are smiling and calm. I think the chief reason is that on ship-board both the past and the future disappear. Little can be done to remedy the mistakes of yesterday; no planning for tomorrow can reckon with the unpredictability of ships and the sea. To live in the present is the essence of being a seaman."[*]

The features of the sea which are of major ecological interest may be listed as follows:

1. The sea is big; it covers 70 per cent of the earth's surface.

2. The sea is deep (see Figure 12–4), and life extends to all its depths. Although apparently there are no abiotic zones in the ocean, life is much denser around the margins of continents and islands.

3. The sea is continuous, not separated as are land and freshwater habitats. All the oceans are connected. Temperature, salinity, and depth are the chief barriers to free movement of marine organisms.

4. The sea is in continuous circulation; air temperature differences between poles and equator set up strong winds such as the trade winds (blowing steadily in the same direction the year around) which, together with rotation of the earth, create definite currents. In addition to the wind-driven currents on the surface, deeper currents result from variations in temperature and salinity, which create differences in density. The interaction of wind stress, Coriolis force, thermohaline currents, and the physical configuration of the basin is very complex and need not concern us here (see Von Arx (1962), or other reference in physical oceanography). So effective is the circulation that oxygen depletion or "stagnation," such as often occurs in freshwater lakes, is comparatively rare in the ocean depths.

In Figure 12–1 the major surface currents of the world are shown. Noteworthy are the equatorial currents running east and west and coastal currents running north and south. Well known are the Gulf Stream and North Atlantic Drift that bring warm water and moderate climate to high latitudes of Europe; also the California Current that moves cold water southward, creating the fog belt characteristic of this coast. In summary, the major currents act as giant pinwheels (or gyres) which run clockwise in the northern hemisphere and counterclockwise in the southern hemisphere.

An important process called *upwelling* occurs where winds consistently move surface water away from precipitous coastal slopes, bringing to the surface cold water rich in nutrients which have been accumulating in the depths. The most productive marine areas are located in regions of upwelling, which occur largely on western coasts, as evidenced by the large fisheries in such regions. Upwelling produced by the Peru Current creates one of the richest fisheries in the world (see Table 3–11 for an estimate of the yield from this fishery). In addition, this upwelling supports large populations of seabirds that deposit countless tons of nitrate- and phosphate-rich guano on coastal islands. Were it not for these currents, the upwellings, and the deep currents resulting from temperature and salinity differences in the water itself, bodies and materials would pass permanently into the depths, carrying nutrients beyond the reach of "producers" in the photic surface regions. As it is, nutrients do get "lost" in the deep sediments for long periods (see Chapter 4).

Another water movement that contributes to coastal fertility is what I have called *out-welling* (E. P. Odum, 1968a), which occurs where nutrient-rich estuarine waters move out to sea (see next chapter).

5. The sea is dominated by waves of many kinds (Fig. 12–2) and by tides produced by the pull of moon and sun. Tides are especially important in the shoreward zones where marine life is often especially varied and dense. Tides are chiefly responsible for the marked periodicities in these communities and entrain the "lunar-day" biological clocks as discussed in Chapter 8. Since tides have a periodicity of about 12½ hours, high tides occur in most localities twice daily, being about 50 minutes later on successive days. Every two weeks when sun and moon are "working together" the amplitude of tides is increased—the so-called *spring* tides when high tides are very high and low tides very low—whereas midway in the fortnightly periods the range between low and high tide is smallest—the so-called *neap tides* when sun and moon tend to cancel one another. The tidal range varies from less than 1 foot in the open sea to 50 feet in certain enclosed bays. There are many factors which modify tides so that tidal patterns vary in different places over

---

[*] From "The Ocean," by Roger Revelle. Copyright © 1969 by Scientific American, Inc. All rights reserved.

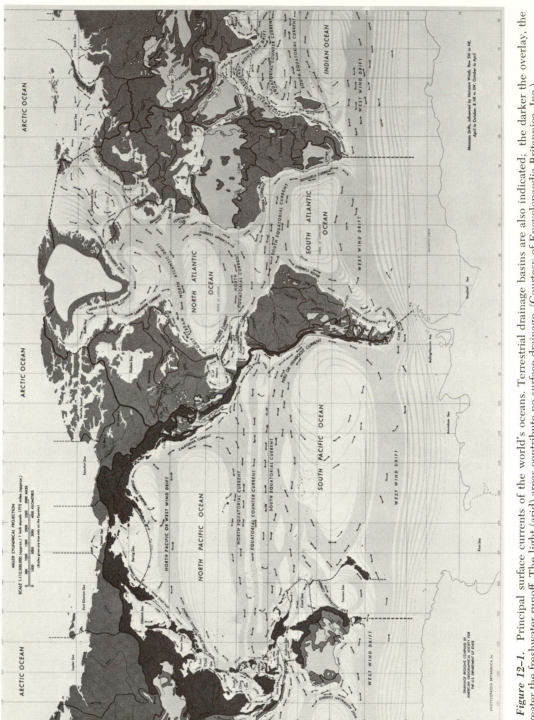

Figure 12–1. Principal surface currents of the world's oceans. Terrestrial drainage basins are also indicated; the darker the overlay, the greater the freshwater runoff. The light (arid) areas contribute no surface drainage. (Courtesy of Encyclopaedia Britannica, Inc.)

the world. The very first thing a marine ecologist does when working in the coastal zone is to consult the local tide tables!

6. The sea is salty. The average salinity or salt content is 35 parts of salts by weight per 1000 parts of water, or 3.5 per cent. This is usually written: 35‰ (= parts per thousand; recall that salinity of fresh water is less than 0.5‰). About 27‰ is sodium chloride and most of the rest consists of magnesium, calcium, and potassium salts. Since the salts dissociate into ions, the best way

Figure 12–2. The ocean is a very "physical" ecosystem. *A.* Interchanges between the sea surface and the air are especially important in regulating the climates of the world and the composition of the atmosphere. *B.* Life aboard an oceanographic vessel; lowering a sampling device. (Photos from Woods Hole Oceanographic Institute.)

to picture the chemistry of the sea is as follows (in parts per thousand = grams per kilogram):

Positive Ions (Cations)		Negative Ions (Anions)	
Sodium	10.7	Chloride	19.3
Magnesium	1.3	Sulfate	2.7
Calcium	0.4	Bicarbonate	0.1
Potassium	0.4	Carbonate	0.007
		Bromide	0.07

Since the proportion of the radicals remains virtually constant, total salinity may be computed by determining the chloride content (which is easier than determining total salinity). Thus, 19‰ chlorinity approximates 35‰ salinity. The electrical dissociation force of the cations exceeds that of the anions (by about 2.4 milliequivalents), which accounts for the alkaline nature of sea water (normally pH = 8.2). Sea water is also strongly buffered (resistant to pH change). In addition to the ions listed above, sea water, of course, contains numerous other elements (theoretically all known ones) including biogenic ions that are often in such low concentration that they are limiting to primary production (see pages 114 and 127). All these other ions constitute less than 1‰ of the ocean's salinity. As would be expected, the residence time (see page 94 for explanation of this term) of salts is much longer than for the water itself (estimated to be of the order of 10^7 and 10^4 years respectively; see Weyl, 1970).

Since temperature and salinity represent two of the important limiting factors in the sea, it is instructive to plot them together in the form of *hydroclimographs*, as shown in Figure 12–3. Each polygon represents a specific locality; each point on a polygon is the monthly average temperature plotted against salinity, 1 being January and 12 December. Compare these graphs with the temperature–moisture climographs of terrestrial habitats (Fig. 5–10, page 125). In Figure 12–3 note that salinity varies within very narrow limits in the open ocean but is seasonally quite variable in the estuarine (brackish) waters of bays and river mouths. Organisms of the open ocean are usually stenohaline (i.e., have narrow limits of tolerance to changes in salinity; see Chapter 5, Section 2), whereas organisms of inshore brackish waters are generally euryhaline. Most marine organisms have an internal salt content isotonic with sea water, and hence osmoregulation poses no problem except where salinity is subject to change. As already described (page 299), marine bony

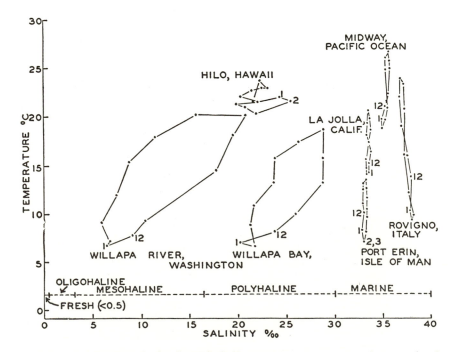

Figure 12–3. Temperature–salinity hydroclimographs for estuarine (brackish) and marine localities. Note that seasonal variation in both temperature and salinity is pronounced in estuarine habitats, whereas salinity is virtually constant in true marine habitats. (Redrawn from Hedgpeth, 1951.)

fish have a lower concentration of salts in blood and tissues (i.e., they are hypotonic) but regulate by ingestion of water and active excretion of salt through the gills.

7. The concentration of dissolved nutrients is low and constitutes an important limiting factor in determining the size of marine populations. Whereas the concentration of sodium chloride and other salts mentioned in paragraph 6 is measured in parts per thousand, nitrates, phosphates, and other nutrients are so diluted that they are measured in parts per billion.* Also, their residence time is very much shorter so that the concentration of these vital biogenic salts varies greatly from place to place and from season to season. Despite the fact that nutrients are continually being washed into the sea, their importance as limiting factors is no less in marine than in terrestrial or freshwater environments. As mentioned in Chapter 2, the basic reason for the general biological infertility of the open oceans is the very small size of the autotrophic zone relative to the heterotrophic–nutrient regenerating zone. Perhaps as an adaptation to this situation oceanic plankton have evolved a "short-circuit" nutrient recycle mechanism rather different from that in shallow water or on land, as was described in Chapter 4 (see page 105 and Figure 4–11). Therefore, the low concentration of nutrients does not necessarily indicate total scarcity, since these materials are in such "demand" by organisms, which are efficient in uptake so that they may be taken out of circulation as rapidly as they are released. The classic description of nutrient cycling in the sea, as was so well described by Harvey in 1955, thus needs to be somewhat modified in light of more recent work (see Pomeroy, 1970). As previously indicated, currents prevent permanent loss of many nutrients, although carbon and silicon may be lost by deposition of shells on the bottom of the ocean. Only in a few places of vigorous upwelling are nutrients so abundant at times that phytoplankton cannot exhaust them (i.e., nutrients are not limiting).

8. Paradoxically, the ocean and some groups of organisms that live in it are older than the ocean floor, which is constantly being altered and renewed by tectonic and sedimentary processes. Furthermore, the sea floor is apparently slowly spreading outward from midocean ridges, pushing continents apart as it progresses. While the theory of "continental drift" is an old one, only recently has it become generally accepted by geologists. According to this theory, for example, the continents of North and South America and Africa were once all joined together and have slowly drifted apart, greatly enlarging the Atlantic basin. Biologists have been intrigued by this theory since it seems to explain certain aspects of animal distribution. In addition to this marginal spreading, the level of the ocean has varied considerably as glaciers have waxed and waned. About 15,000 years ago the shore of the east coast of the United States stood 100 or more miles out to sea; as discussed in Chapter 2, the coast would move far inland of its present location if all of the ice caps melted.

2. THE MARINE BIOTA

The marine biota is varied; consequently it would be difficult to list "dominant" groups, as was done for fresh water (see Chapter 11, Section 3). Coelenterates, sponges, echinoderms, annelids, and various minor phyla that are absent or poorly represented in fresh water are very important in the ecology of the sea. The bacteria, algae, crustacea, and fish play a dominant role in both aquatic environments, with diatoms, green flagellates, and copepods equally prominent in both. The variety of algae (the brown and red algae being chiefly marine), crustacea, mollusks, and fish is greater in the sea. On the other hand, seed plants (spermatophytes) are of little importance in the sea, except for eel grass (Zostera) and a few other species in coastal waters. Insects are absent, except in brackish water, the crustacea being the "insects of the sea," ecologically speaking. The great richness of the marine biota can be illustrated by comparing a sample of marine plankton with a comparable one taken from a large lake.

Some of the more conspicuous marine organisms are shown in Figure 12–6 arranged so as to illustrate the trophic and depth relationships that link all of the biota into one vast ecosystem. This diagram does not show the microplankton or microben-

---

* In practice, nutrients are measured in microgram-atoms per liter or in milligram-atoms per cubic meter.

thos, the importance of which will be brought out in later discussion.

3. ZONATION IN THE SEA

A zonal classification that is similar to that outlined for ponds and lakes (see Chapter 11, Section 2) is also applied to the sea, except that it is customary to use a different set of terms for habitats, as shown in Figure 12–4. Something of the complex nature of the ocean floor is also depicted in the diagram, including the midocean ridges from which the continents are presumed to have drifted (Section 1, above). The same "way of life" terms as defined in the chapter on freshwater ecosystems (see page 300), namely, plankton, nekton, and benthos, are used with reference to the sea. An additional term, *pelagic*, is widely used to include the plankton plus the nekton and the neuston (the latter generally unimportant), or all of the life in the open water.

Generally, a continental shelf extends for a distance offshore, beyond which the bottom drops off steeply as the continental slope then levels off somewhat (the continental rise) before dropping down to a deeper, but more level, plain. The shallow water zone on the continental shelf is the *neritic* ("near shore") zone. Zonation within the *intertidal* zone (zone between high and low tides; also called the littoral zone) will be considered in a subsequent section. The region of the open ocean beyond the con-

tinental shelf is designated as the *oceanic* region; the region of the continental slope and rise is the *bathyal* zone, which as shown in Figure 12–4 may be "geologically active" with trenches and canyons subject to underwater erosion and avalanches. The area of the ocean "deeps," or the *abyssal* region, may lie anywhere from 2000 to 5000 meters down. Trenches may drop below 6000 meters (these very deep areas are sometimes known as the *hadal* zone). Bruun (1957a) has called the abyssal the "world's largest ecological unit." It is, of course, an incomplete ecosystem despite its extent because the primary energy source lies far above it.

The important vertical zonation is determined by light penetration with a compensation zone (see page 301) separating an upper thin *euphotic* zone (i.e., the "producing" region) from a vastly thicker *aphotic* zone. As indicated in Figure 12–4, the euphotic zone goes deeper in the clear waters of the oceanic zone, perhaps down to 100 to 200 meters, than it does in more turbid (and richer) coastal waters, where effective light penetration rarely exceeds 30 meters (see Figure 3–3). The oceanic portion of the aphotic zone is sometimes divided into vertical zones as shown in Figure 12–6.

Within these primary zones, which are based largely on physical factors, well-marked secondary zonation, both horizontal and vertical, is generally evident from the distribution of communities. Communities in each of the primary zones, except the euphotic, have two rather distinct vertical

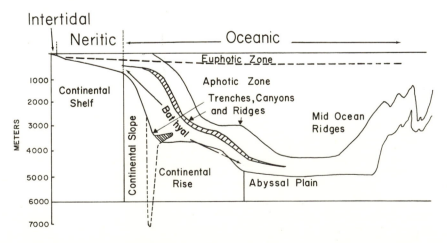

Figure 12–4. Horizontal and vertical zonation in the sea. Certain geological features of the ocean floor, such as trenches (which may go down below 6000 meters), canyons, ridges, the abyssal plain, and mid-ocean ridges that rise like great mountain peaks, are also shown in a transect across the western Atlantic. (Diagram based on Heezen, Tharp, and Ewing, 1959.)

components: the benthic or bottom dwellers (benthos) and the pelagic. As in the large lakes, the producer plants of the sea come in small packages, i.e., the microscopic phytoplankton, although large seaweeds (multicellular algae) are important in some coastal areas. Primary consumers consequently are largely zooplankton. Medium-sized animals are either plankton feeders or feed on detritus derived from plankton, while larger animals are mostly carnivorous. There are few large, strictly herbivorous animals corresponding to the deer, cattle, and horses of the land.

The sea, in contrast with both land and fresh water, contains a varied and important group of sessile (fixed) animals, many of which have a plantlike appearance (as indicated by such common names as "sea anemone," "sea pansy," and so forth). Zonation of such animals on the sea bottom is often as striking as zonation of trees on a mountain (as illustrated in subsequent sections), and provides a basis for the classification of communities just as do large plants on land. To carry the analogy further, they provide shelter for many smaller organisms, as do plants on land. Commensalism and mutualism are widespread and important interactions between many marine species (see Chapter 7). Fixed marine animals, and benthos in general in the neritic zone, usually have a pelagic stage as part of their life history. Therefore, from the community standpoint, benthic life is a part of zonal communities rather than a major community rank in itself.

4. QUANTITATIVE STUDY OF PLANKTON

Not only does plankton occupy the key role in the ocean ecosystem, but it lends itself to quantitative sampling. Much work in marine ecology has centered around the study of plankton. In 1830 J. Vaughan Thompson and in 1845 Johannes Müller used what is now called a plankton net (the name "plankton" was not proposed until 1887). Müller was studying the life history of the starfish and towed a fine-meshed net through the sea in an effort to capture larvae of the starfish. He was impressed with the great wealth of floating life which had been more or less overlooked. Müller transmitted his enthusiasm to Ernst Haeckel, who, along with other contemporary biologists, became greatly excited over this new world of life to be obtained by towing "Müller nets" through the water. Thus, Müller and Haeckel inadvertently became pioneer ecologists, and it was Haeckel who later, in 1869, coined the word ecology itself!

Plankton nets are now generally made of bolting silk or nylon, the strands of which are held firmly by binding twists. Such nets vary from 18 to 200 meshes per inch. In quantitative studies a closing net is used so that a given depth can be sampled without contamination while the net is being lowered and raised. A closing net equipped with a metering device that measures the amount of water filtered is called a Clarke-Bumpus sampler (named after two marine biologists). Even the finest silk net catches only a portion of the plankton biomass (*net plankton*); the small phytoplankton organisms as well as bacteria and protozoa pass through the smallest meshes (*nannoplankton—nanno*, dwarf; note that these terms are the same as used in freshwater ecology, see page 300). Net plankton and nannoplankton are also called respectively (and perhaps more appropriately) *macro-* and *microplankton.*

Studies relating the metabolism and composition of plankton populations in the water masses have verified what has long been suspected (see Atkins, 1945; Knight-Jones, 1951; Wood and Davis, 1956), namely: *the most important photosynthetic organisms are not the relatively large net plankton, but the nannoplankton, especially tiny green flagellates 2 to 25 microns in size.* While this seems to be especially true of the oceanic euphotic zone, the nannoplankton may also dominate the metabolism of coastal waters (Yentsch and Ryther, 1959). Of course, this is another example of the principle (as first stated in Chapter 2) that size of organisms is not a good indicator of importance in systems having short biomass turnover times and rapid nutrient cycling rates. Tiny flagellates (that is, 5μ or so in size), most of them colorless but some containing chlorophyll, have also been found in abundance in the aphotic zone at depths of 1000 meters or more. These are presumed to live heterotrophically, at least most of the time, utilizing dissolved organic matter originating in the photic zone. These flagellates may be one of the key food-chain links between primary production in the photic zone and the zooplankton and benthos of the aphotic zone (the other link may be downward streaming aggregates formed from dissolved organics, as will be discussed later). That small nannoplankton may account for a large part of total plankton respiration (as well as photosynthesis) in the oceanic zone is indicated by the studies of Pomeroy and Johannes (1966) who found that flagellated organisms too small to be concentrated in a plankton net accounted for 94 to 99 per cent of the total respiration of plankton from Gulf Stream and Sargasso Sea waters.

From the foregoing it is evident that the plankton net is a useful sampling tool for the zooplankton but not for the phytoplankton. Plankton "gently" concentrated (to avoid breaking the fragile cells) with a membrane filter (the so-called millipore filter) provides a much better sample (see Dobson and Thomas, 1964). Fluorescence microscopy is useful for distinguishing between chlorophyll-bearing and colorless forms and between living cells and dead ones in concentrated samples (see Wood, 1955). As described in Chapter 3 (pages 60–61), measurement of chlorophyll and carbon uptake are the principal methods used to estimate primary production in the plankton population.

Among the more specialized equipment may be mentioned the ingenious device called the "Hardy Continuous Plankton Recorder," which may be towed behind a ship for long distances. Plankton is embedded in a strip of gauze as the water passes through it. The gauze continually rolls up into a tank of preservative as the passing water turns a propeller at the rear of the instrument, thus providing a permanent transect sample. Such a sampler, of course, collects mostly the larger zooplankton, but results of this type of transect survey have been useful in predicting the location and yield of commercial fishes (Hardy, 1958). One thing oceanic plankton surveys have shown is that the distribution is very patchy with concentrations of phytoplankton sometimes occurring in different places from concentrations of zooplankton. The latter observation has led to the idea that secretion of antibiotics results in "mutual exclusion" of plant and animal components, but this could be partly a sampling artifact in that the smaller (and hence overlooked) zooplankton would be expected to thrive in

the midst of an algal bloom. It seems probable that zooplankton are both attracted and repelled by excreted metabolites since they are often concentrated around the edges of blooms.

The important work of Gordon Riley and his coworkers should be mentioned (first summarized in a monograph by Riley, Strommel, and Bumpus, 1949, with later work and mathematical models reported by Riley, 1963 and 1967). They found that the amount and seasonal distribution of both phytoplankton and zooplankton in any region could be predicted by means of a formula based on certain important limiting factors of the environment and physiological coefficients determined from laboratory experimentation. In very simplified and nonmathematical form the formula they devised for estimating phytoplankton production is as follows:

$$
\begin{bmatrix} \text{Rate of change} \\ \text{of phytoplank-} \\ \text{ton density} \\ \text{per unit time} \end{bmatrix} \text{equals} \begin{bmatrix} \text{rate of} \\ \text{photo-} \\ \text{syn-} \\ \text{thesis} \end{bmatrix} \text{minus}
$$

$$
\begin{bmatrix} \text{rate of} \\ \text{respi-} \\ \text{ration} \end{bmatrix} \text{minus} \begin{bmatrix} \text{rate of} \\ \text{"grazing,"} \\ \text{i.e., loss to} \\ \text{herbivores} \end{bmatrix} \text{minus}
$$

$$
\begin{bmatrix} \text{rate} \\ \text{of} \\ \text{sinking} \end{bmatrix} \text{plus} \begin{bmatrix} \text{rate of up-} \\ \text{ward movement} \\ \text{due to turbu-} \\ \text{lent eddies} \end{bmatrix}
$$

Respiration is largely determined by temperature, and photosynthesis was found to be largely limited by temperature, light, and phosphate concentration. Knowing the density of herbivores, the "grazing pressure" was determined from data obtained in laboratory cultures. Although the computation is complex, the loss, if any, as a result of the sinking of plant cells below the euphotic zone can be determined from physical oceanographic data. The theoretical operation of the six major limiting factors on phytoplankton of a specific locality together with observed and calculated density are shown in Figure 12–5. In general the observed was within 25 per cent of the calculated, remarkably close considering the complexity of the situation. Riley's model has wider significance in that it demonstrated that useful, predictive models of complex ecological situations can be developed on the basis of judicious selection of only a few of the many factors involved.

This is the start of the rapidly expanding field of "systems ecology" as reviewed in Chapter 10.

5. COMMUNITIES OF THE MARINE ENVIRONMENT

Some of the major and better known groups of organisms that make up the communities of the ocean are shown in Figure 12–6. As indicated in the previous section, this picture does not do justice to the small organisms that are so important in community function. In the following descriptions the communities of the intertidal and neritic zones will be considered first, to be followed by a brief consideration of some of the contrasting features of the oceanic zone.

Composition of Communities of the Continental Shelf Region

PRODUCERS. Phytoplanktonic *diatoms* and *dinoflagellates* are the dominants of the producer trophic level almost everywhere in the shelf region, but, as already indicated, the *microflagellates* may be equally important. These are a mixed group of uncertain taxonomy that were formerly lumped under the term "phytomastigina," i.e., "plant flagellates." Examples of all three dominant groups are shown in Figure 12–7. The diatoms will tend to dominate in northern waters, while dinoflagellates are often predominant in subtropical and tropical waters. As a group the latter are among the most versatile of organisms in that most not only function as autotrophs but some species are facultative saprotrophs or phagotrophs as well! Some species are famous for fish-killing "red tides," as described on page 358. Dinoflagellates often follow diatoms in seasonal succession in temperate waters (both lakes and the ocean).

Near shore large multicellular attached algae or "sea weeds" are also locally important, mostly on rocky or other hard bottoms in shallow water. They are attached by hold-fast organs, not roots, and often form extensive "forests" or "kelp beds" just below the low-tide mark. In addition to the *green algae* (Chlorophyta), also prominent in fresh water, these attached species belong to the more or less exclusively marine

Phaeophyta, or *brown algae*, and the Rhodophyta, or *red algae*. These three groups show a depth distribution roughly in the order named (with red algae deepest). The brown and red colors are due to pigment which masks the green chlorophyll and, as described in Chapter 5 (page 119), these pigments aid in absorption of greenish-yellow light which penetrates to the greatest depth. Some of the fixed algae are of economic importance as sources of agar and other products. On northern rocky coasts "seaweed" harvest is a regular industry, and in Japan certain species are cultured for food.

Neritic phytoplankton, at least in temperate regions, undergoes a seasonal density cycle similar to that in eutrophic lakes (compare Figure 12–5 with Figure 11–8). Primary production has been discussed and quantitated in Chapter 3.

CONSUMERS: ZOOPLANKTON. Examples of the varied animal plankton are shown in Figure 12–6. Organisms which remain for

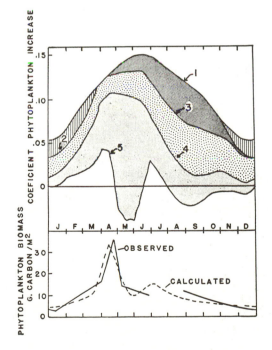

Figure 12–5. Theoretical operation of six limiting factors on phytoplankton (upper figure), and observed and calculated density during one annual cycle in the waters at Georges Bank off the coast of New England (lower figure). The limiting factors are: (1) light and temperature; (2) turbulence (which carries cells below photic zone); (3) phosphate depletion; (4) phytoplankton respiration; and (5) zooplankton grazing. Only during spring and late summer are conditions favorable for rapid growth of the population. (After Riley, 1952.)

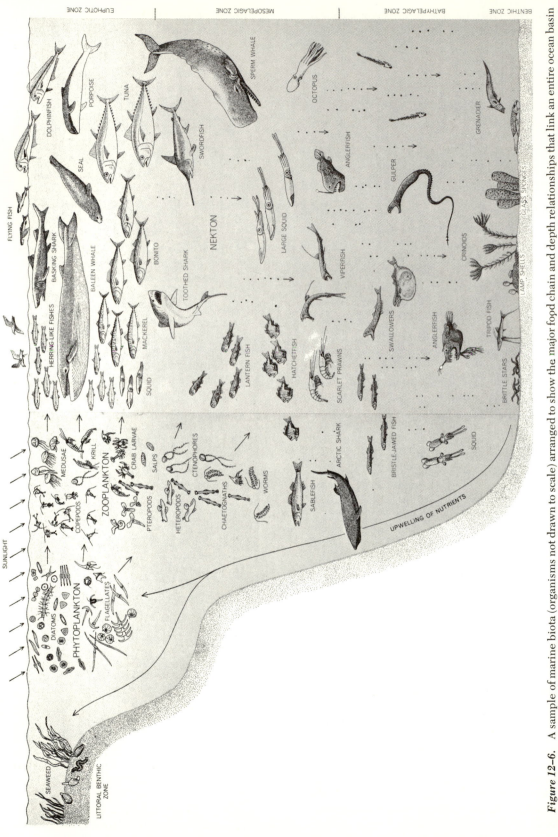

Figure 12-6. A sample of marine biota (organisms not drawn to scale) arranged to show the major food chain and depth relationships that link an entire ocean basin into a vast ecosystem. The diagram does not do justice to the microplankton and the "infauna" benthos, the importance of which is described in the text. The dots and downward arrows depict the "rain of organic detritus," which, as emphasized in the text, may not be the chief way in which food is transported from the euphotic zone to the deep zones. (From *The Nature of Oceanic Life,* by John D. Issacs, Copyright © 1969 by Scientific American, Inc. All rights reserved.)

their entire life cycle in the plankton are called *holoplankton,* or permanent plankton. Copepods (see Marshall and Orr, 1955, for a thoroughgoing study of the important genus, *Calanus*) and larger crustaceans called "krill" or euphausids are important as the link between plankton and nekton. Planktonic protozoa (not shown in Figure 12–6) include foraminiferans, radiolarians, and tintinid ciliates. Shells of the former two groups are an important part of the geological record in marine sediments. Other permanent zooplankters shown in Figure 13–6 include "wing-footed" mollusks (pteropods and heteropods), tiny jellyfish (medusae) and ctenophores, pelagic tunicates (salps), free-floating polychaete worms, and the predaceous arrow worms or chaetognaths. A considerable portion of the inshore plankton is *meroplankton,* or temporary plankton, in sharp contrast to fresh water (see Chapter 11, page 305). Most of the benthos, and much of the nekton (fish, for example) as well, in the larval stage are tiny forms which join the plankton assemblage for varying periods before settling to the bottom or becoming free-swimming organisms. Many of these temporary plankton forms have special names, as indicated in Figure 12–8, which pictures a number of examples of this unique component of inshore marine waters. As might be expected, the meroplankton varies seasonally according to the spawning habits of the parental stock, but there is sufficient overlapping to insure a quantity of meroplankton at all seasons.

A very interesting aspect of the ecology of pelagic larvae is their ability to locate the kind of bottom suitable for survival as adults. Wilson (1952, 1958) has shown by means of experiments that larvae of certain benthic polychaetes do not settle at random but respond to particular chemical conditions of the substrata to which they are adapted. When ready to metamorphose into sedentary adults, the larvae "critically examine" the bottom. If the chemical nature is "attractive," they settle; if not, they may continue their planktonic life for weeks. Metabolites released by microorganisms or by the adults are two possible sources of the "chemical messengers" that may direct the larvae to the right kind of bottom. This is a problem in "biochemical" ecology that awaits solution.

CONSUMERS: BENTHOS. As was indicated in Section 3, the marine benthos is characterized by the large number of sessile or relatively inactive animals which exhibit marked zonation in the inshore region. Bottom organisms are generally distinct for each of the three primary zones, namely, supratidal, intertidal, and subtidal which are diagrammed in Figure 12–9. At one extreme, at or above the high-tide mark, organisms must be able to withstand desiccation and air temperature changes since they are only briefly covered by water or spray. In the subtidal region, on the other hand, organisms are continually covered. The constant ebb and flow of tides between the tide marks produces an environmental gradient in regard to exposure to air and water. It is important to note that there may also be a distinct vertical zonation of the benthos. Two terms are widely used to separate the two most obvious vertical components: *epifauna* refers to organisms living on the surface, either attached or moving freely on the surface; *infauna* refers to organisms that dig into the substrate or construct tubes or burrows. Epifauna reaches maximum development in the intertidal zone, but extends throughout the ocean bottom. Infauna is more fully developed in the subtidal zone and below.

In a given region the series of benthic subcommunities which will be found to replace one another from the shore to the edge of the shelf depends largely on the type of bottom, whether sand, rock, or mud. In Figure 12–9 a transect of a typical sand beach is compared with that of a rocky substrate (man-made breakwaters and piers are similar to natural rocks) in the region of Beaufort, N.C. The most conspicuous dominants are indicated, very few of which are common to both series. Mixed in with the larger animals and seaweeds are large numbers of unicellular and filamentous algae, bacteria, and small invertebrates. A sand beach would seem at first to be a rigorous habitat but it is better populated than first appearances indicate. Most of the larger animals are specialized burrowers and the diatoms, amphipods, and other infauna which live among the sand grains would not be noticed. The interstitial fauna and flora, or "psammon" that was mentioned in the chapter on freshwater ecology (page 307), are even better developed on marine

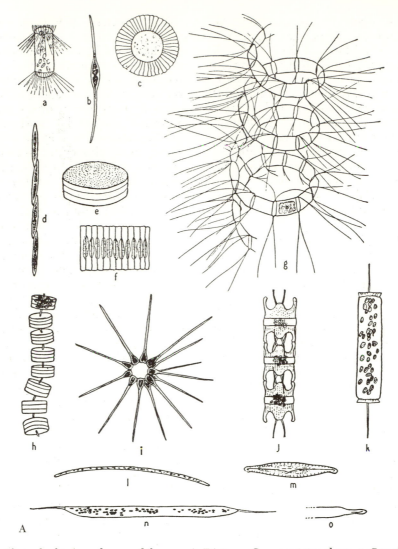

Figure 12–7. Phytoplankton producers of the sea: A. **Diatoms.** Genera pictured are: *a, Corethron; b, Nitzschia closterium; c, Planktoniella; d, Nitzschia seriata; e, Coscinodiscus; f, Fragilaria; g, Chaetoceros; h, Thalassiosira; i, Asterionella; j, Biddulphia; k, Ditylum; l, Thalassiothrix; m, Navicula; n, o, Rhizosolenia semispina,* summer and winter forms. Note that some of the genera of freshwater diatoms shown in Figure 11–5 also occur in the sea.

shores than on lake or stream shores. One of the remarkable motile sand diatoms that has an unusual light-photosynthesis response was discussed in Chapter 5 (see Figure 5–7, pages 119–120).

Some of the highly specialized sand burrowers are shown in Figure 12–10. The mole crab *Emerita,* one of the most remarkable, is capable of "sinking" backward into the sand in a few seconds. These crabs feed by extending their feathery antennae above the sand and collecting plankton from the flowing water when the tide comes in. Other animals, such as worms, feed by ingesting sand and detritus which enters their bur-

rows and extracting the food material after it is in their intestines.

Habitat selection of mollusks in a sand-mud gradient is shown in Figure 12–11. Infaunal animals often respond sharply to grain size or "texture" of the bottom. Determining the sand–silt–clay ratio has considerable predictive value as to the kinds of organisms to be expected. The method of feeding by the benthos undergoes an interesting change along a sand–mud gradient; filter-feeding predominates in and on the sandy substrate, while deposit-feeding is most common on silty or muddy substrates.

Although the diagrams in Figure 12–9

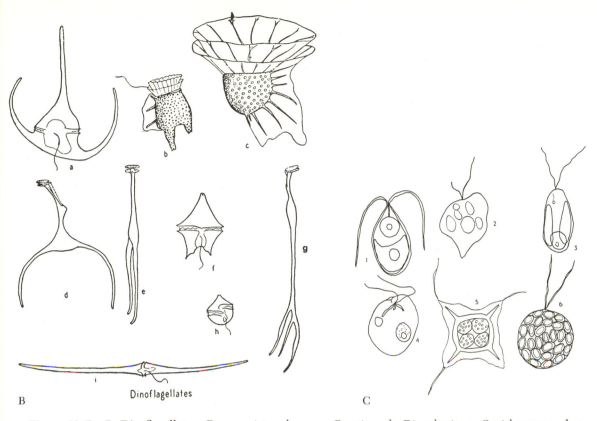

B

Dinoflagellates

C

Figure 12-7. **B. Dinoflagellates.** Genera pictured are: *a, Ceratium; b, Dinophysis; c, Ornithocercus; d, e, Triposolenia,* front and side views; *f, Peridinium; g, Amphisolenia; h, Goniaulax; i, Ceratium.* **C. Microflagellates.** Genera pictured are: 1. *Dunaliella* (phytomonad); 2. *Chloramoeba* (xanthomonad); 3. *Isochrysis* (chrysomonad); 4. *Protochrysis* (cryptomonad); 5. *Dictyocha* (silicoflagellate); 6. *Pontosphaera* (coccolithophore). (*A* and *B* from Sverdrup, Johnson, and Fleming, "The Oceans, Their Physics, Chemistry, and General Biology," 1946, published by Prentice-Hall, Inc. *C* from Wood, "Marine Microbial Ecology," 1965, Chapman and Hall, London.)

represent a specific locality, the same general arrangements can be found on any coast —only the species will be different. For example, rocky coasts generally exhibit three distinct zones: (1) a periwinkle zone (high-tide area), a barnacle or mussel zone, and a seaweed zone (low-tide area). See Stephenson and Stephenson (1949, 1952) for descriptions and photographs of these zones in different parts of the world. One should not be misled by their names into thinking that these zones are separate aggregations of animals and plants; actually there are plenty of algae (small, inconspicuous species) in the periwinkle and barnacle zones and many animals in the seaweed zone. A sort of "functional homeostasis" in the face of marked structural change in the intertidal gradient is indicated by the fact that the total chlorophyll per square meter is not greatly different in the three zones, as shown in Table 12-1. The inter-

tidal region provides a good site for a class study on relationships between community structure and community function, since organisms are easy to census (being mostly sessile), and samples of the whole community can be placed in dark and light bottles or taken into the laboratory for metabolism measurements. The rocky intertidal zone is also a good place for the experimental study of interaction of physical factors with competition and predation in shaping community structure, as was described in Chapter 7 (see Figure 7-36, page 225).

In the deeper waters of the neritic zone populations are not usually arranged in concentric zones but more likely form a "patchwork" or mosaic. One of the classic studies is that of C. G. J. Petersen (1914–1918), who thoroughly investigated the benthos of neritic waters of the important fishing grounds between Denmark and Norway. He found that there were eight primary "associations"

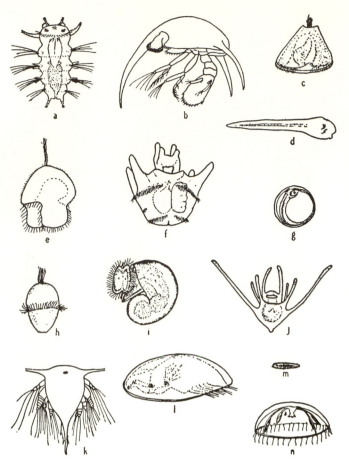

Figure 12–8. Larvae which make up the temporary plankton (meroplankton) of the sea. *a*, chaetate larva of the annelid *Platynereis*; *b*, zoea of the sand crab *Emerita*; *c*, cyphonautes larva of a bryozoan; *d*, tadpole larva of a sessile tunicate; *e*, pilidium larva of a nemertean worm; *f*, advanced pluteus larva of a sea urchin; *g*, fish egg with embryo; *h*, trochophore larva of a scaleworm; *i*, veliger larva of a snail. *j*, pluteus larva of a brittle starfish; *k*, nauplius larva of a barnacle; *l*, cypris larva of a barnacle; *m*, planula larva of a coelenterate; *n*, medusa stage of a sessile hydroid. (Sverdrup, Johnson, and Fleming, "The Oceans, Their Physics, Chemistry, and General Biology," 1946, published by Prentice-Hall, Inc.)

Table 12–1. *Community Chlorophyll Content of the Three Major Zones on the Rocky Intertidal at Woods Hole, Mass.**

		DOMINANTS		CHLOROPHYLL A (gms/m²)
ZONE	BIOMASS	*Plants*	*Animals*	
Supratidal or Periwinkle Zone	Very small	*Calothrix*†	Periwinkle (*Littorina*)	0.50
Barnacle Zone	Medium	*Gomontia*,‡ *Rivularia*† and *Fucus*§	Barnacles (*Balanus* and *Chthamalus*)	0.87
Seaweed Zone	Large	*Chrondris* and other species of brown and red seaweeds	Numerous species of mollusks and crustaceans	1.07

* Although the differences in chlorophyll content between zones is statistically significant, the only two-fold variation in the means is in striking contrast to the many-fold difference in biomass, and the completely different species composition. (Data from Gifford and E. P. Odum, 1961.)

† Blue-green algae.

‡ A green alga.

§ A brown alga.

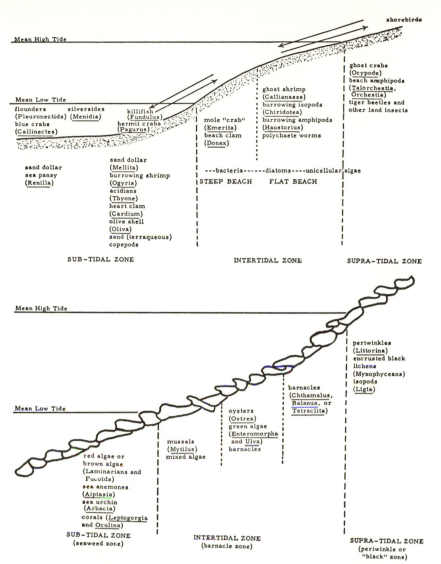

Figure 12–9. Transects of a sandy beach (upper) and of a rocky shore (lower) at Beaufort, N.C., showing zones and characteristic dominants. (Upper diagram based on data of Pearse, Humm, and Wharton, 1942; lower based on data of Stephenson and Stephenson, 1952.)

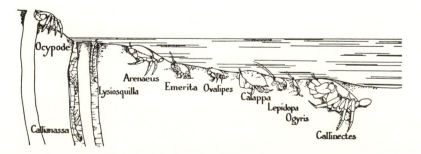

Figure 12–10. Burrowing crustaceans in resting positions in the sand of the neritic region.

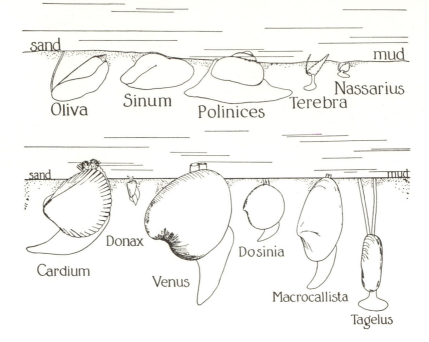

Figure 12–11. Burrowing gastropods (upper) and clams (lower) which are characteristic of sand (left) and mud flats (right) of the neritic region. (After Pearse, Humm, and Wharton, 1942.)

in this vast area, each with one or more dominant species. Studies of benthic aggregations were continued on a world-wide basis by Thorson (1955), who found what he called "parallel level bottom communities" inhabiting the same type of bottom at similar depths in widely scattered geographical areas. Furthermore, these parallel communities were often dominated by species of the same genus. For example, benthic populations in shallow water with a mixed sand–mud bottom often featured clams of the genus *Macoma* regardless of whether located in the Baltic or on the west coast of North America. Similarly, shallow-water sand bottoms might be dominated by clams of the genus *Venus* and deeper, muddy bottoms by brittle-stars of the genus *Amphiura*. This sort of ecological equivalence was discussed in general terms in Chapter 6.

High density of predators is a seemingly incongruous feature of many ocean bottom communities. However, Thorson found that many predators, such as brittle-stars, do not feed for long periods during reproduction thus allowing pelagic larvae of clams and other prey to settle and grow to a size which makes them less vulnerable to predation. Here again we have an example in which standing crop data alone would be misleading as to what "goes on" in the community.

The development of underwater cameras has extended our knowledge of marine bottom communities not only on the continental shelf but in deeper waters as well. Three photographs taken at increasing depths are shown in Figure 12–12. Such photographs have revealed animals not yet collected, and they also provide a means of estimating the density of the larger epifauna in their undisturbed habitat. Yet not even the sharpest photographs can show the infauna, which, in the soft sediments that cover most of the ocean bottoms, are most likely to be the most important of the two benthic components.

As has been the case with plankton, the smaller components of the marine benthos have become better known and their importance appreciated as a result of the investigations of the 1960s. Sanders (1960, 1968), for example, found that the use of improved dredges and finer screens for sorting revealed a diversity of small infaunal polychaetes, crustaceans, bivalves, and other invertebrates heretofore overlooked. The density of the benthos in the soft sediments decreased with increasing depth (6000 individuals per m² on continental shelf down to 25 to 100 per m² on the abyssal plain), as might be expected because of declining productivity with depth, but the species diversity proved to be greater in the abyssal than in the shelf habitat (Sanders and Hessler, 1969). This major finding supports the theory that diversity is related to stability and is not dependent on productivity. These results are graphed in Figure 6–6A.

CONSUMERS: NEKTON AND NEUSTON (see page 300 for explanation of these terms). In addition fish of the larger crustacea, turtles, mammals (whales, seals, and so forth), and marine birds comprise the active swimmers and surface dwellers. Individuals of this group commonly, but not necessarily, range over a considerable area, as is characteristic of secondary and tertiary consumers in general. Nevertheless, the nekton (and even the birds) are limited by the same "invisible barriers" of temperature, salinity, and nutrients, and by the same type of bottom, as are organisms with less power of movement. Also, strange to say, although the individual range

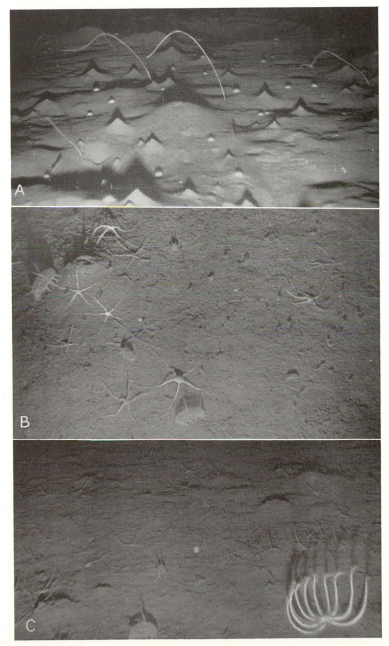

Figure 12–12. Photographs (by means of a special underwater camera called the benthograph) of the ocean bottom at three different depths off southern California (San Diego Trough). *A,* at 295 feet. Note abundant sea urchins (probably *Lytechinus*) appearing as globular, light-colored bodies, and the long curved sea whips (probably *Acanthoptilum*). Burrowing worms have built the conical piles of sediment at the mouth of their burrows. (Emery, 1952.) *B,* at 3600 feet. Vertical photograph of about 36 square feet of bottom which is composed at this point of green silty mud having a high organic content. Note the numerous brittle starfish (Ophiuroidea) and several large sea cucumbers (Holothuroidea). The latter have not been identified as to species as they have never been dredged from the sea and have only been seen in bottom photographs! (Official Navy photo, courtesy U. S. Navy Electronics Laboratory, San Diego.) *C,* at 4350 feet. Note in the right foreground the ten arms of what is probably comatulid crinoid (a relative of the starfish which is attached to the bottom by a stalk-like part). Small worm tubes and brittle starfish litter the surface, and two sea cucumbers may be seen in the left foreground. Continual activity of burrowing animals keeps the sea bottom "bumpy." Bottom edge of the picture represents a distance of about 6 feet. (Emery, 1952.)

of nekton may be great, the geographic range of a species may be less than that of many invertebrates.

As indicated in Figure 12–9, silversides, killifish, and flounders are characteristic of waters at the low-tide mark on the mid-Atlantic coast. These and other species of the same habitat move back and forth with the tides, feeding on benthos of the intertidal zone when it is covered with water. Likewise, shore birds move back and forth on the intertidal zone hunting for food when it is uncovered. It is remarkable that anything is left after these alternate attacks from land and sea! The flounders and rays are the most specialized of bottom fishes, their bodies and coloring blending with the sand and mud. Some species of the former are remarkable for their ability to change color and "match" their background.

In addition to the bottom-feeders, plankton-feeders are found in neritic communities. Members of the herring family (Clupeidae), including the herring, menhaden, sardine (pilchard), and anchovy, are especially important. Plankton is strained from the water by means of a built-in "net" formed by the gill rakers. Since even the finest filter apparatus, like the finest silk net, fails to catch most of the phytoplankton, the adult fishes subsist largely on zooplankton and are thus secondary consumers. Several studies have shown that when the fish passes through a stretch of water it does not take the same sample as does a net drawn through the same area; apparently the fish is able to selectively reject some forms and to actively seek others. Species of plankton feeders can thus avoid competition not only if "meshes" of the filters are different but also if active selection of items is different.

Although some of the bottom fish lay their eggs on the bottom and guard them as do many freshwater fish, most marine fish, including the more pelagic ones, lay large numbers of eggs which float (aided by oil droplets and other flotation adaptations) and receive no attention from the parent. Two other characteristics of pelagic fish are important: (1) their tendency to aggregate or "school," undoubtedly a valuable asset in the shelterless open water, and (2) their tendency to perform definite seasonal migrations. The latter may be due in part to the fact that the eggs and later the larvae drift helplessly in the currents until the fish become large enough to make the return trip to the breeding grounds.

The great commercial fisheries of the world are almost entirely located in or not far from the continental shelf.* Since, as already indicated, the highest primary productivity occurs in regions of cold-water upwelling, it is not surprising that the largest commercial fisheries occur in such regions. Although a large number of fish are of commercial importance, a relatively few species make up the bulk of the world commercial catch. On the Atlantic coast, for example, the northern food fishes are principally herring, cod, haddock, and halibut, whereas toward the south mullet, sea bass, weakfish, drums, and Spanish mackerel are sought. The 1967 FAO Statistics (see Holt, 1969) lists six species as making up half of the total world harvest of marine fish as follows (listed in order of total weight harvested): Peruvian anchoveta, Atlantic herring, Atlantic cod, mackerel, Alaska walleye pollock, and South African pilchard. Other important groups are the flounderlike fish (including soles, plaices, halibut), the salmons, and the tunas (including bonitos). The cod, pollock, and flounder groups are bottom types and are sought with trawls; fish of the other groups are caught in the upper waters with nets or seines of various types. Note that most of these are plankton feeders (relatively short food chain). The major world fishing areas, their yields, and the future prospects are discussed in Chapter 15 (see also Tables 3–10B and 3–11).

In addition to the predatory fish, such as sharks, marine birds are important tertiary consumers of the sea. Sea birds (also seals and sea turtles), to be sure, are a link between land and sea, since they must breed on land, but their food comes from the sea. Therefore, these air breathers are just as much a part of the food chains of the sea as are the fish and invertebrates on which they feed. The role of birds in "completing" the nitrogen and phosphorus cycles has been mentioned in Chapter 4 (Figure 4–2). As is true of marine organisms, birds are concentrated near shore and especially in the productive regions. Shore birds frequent the supratidal and intertidal zones; cormorants, sea ducks, and pelicans the subtidal areas; and petrels and shearwaters the lower neritic zone farther out to sea. Birds

---

* Fisheries for the wide-ranging, "top" carnivore such as tuna and mackerel in deep water are an exception, but here again the best areas are in regions of upwelling.

even show a vertical distribution in feeding areas in the water, as was illustrated in Figure 7–32.

BACTERIA. According to Zobell (1963), the density of bacteria in sea water ranges from less than one per liter in the open ocean to a maximum of 10^8 per milliliter inshore. As discussed elsewhere (page 105), bacteria may not be important in nutrient regeneration in the water column. In the sediments, however, bacteria probably play the same role as they do in soils. Density in marine sediments ranges from less than 10 to 10^8 per gram of surface sediment depending on the organic content. As elsewhere, detritus feeders in sediments are presumed to be obtaining most of their food energy from digesting the bacteria, protozoa, and other microorganisms that are associated with the ingested detritus (Zhukova, 1963). Fungi and yeasts are not important in the sea except where there is a lot of macrophytic detritus. Some recent advances in the study of marine microbiology are considered in Chapter 19.

A Marine Sediment Profile

Figure 12–13 is an idealized sediment profile showing the chemical and physical changes that occur in the transition between the upper oxidized layers and the reduced zone in which the supply of gaseous oxygen becomes depleted. These changing conditions greatly affect the distribution of organisms whose metabolites in turn have much to do with the chemical nature of the zones. The oxidation-reduction, or redox, potential, is measured on a millivolt scale called an Eh scale, which is analogous to pH in that it is a measure of electron activity while pH is a measure of proton ac-

tivity (see Hewitt, 1950, and Zobell, 1946, for further explanation of redox potential). In the redox discontinuity zone the Eh drops rapidly and becomes negative in the fully reduced, or sulfide, zone. The importance of the interaction between aerobic and anaerobic layers in controlling the sulfur cycle, the chemical composition of the sea, and the gaseous balance of the atmosphere has already been stressed (Chapters 2 and 4), as has the "microbial recovery" of nutrients resulting from the upward diffusion of reduced gases (H_2S, NH_3, CH_4, and H_2).

The bulk of the benthic animals occurs in the oxidized layer; these include the polychaetes and bivalves previously mentioned and also a host of very small animals such as harpacticoid copepods, turbellarian flatworms, gastrotricts, tartigrades, rotifers, ciliates, and nematodes. If light is available, algae will be present. The redox discontinuity layer is the home of chemosynthetic bacteria and, if light is present, the photosynthetic bacteria. Nematodes are likely to be the only metazoans in this region. Only the anaerobic bacteria, such as sulfate reducers and methane bacteria, the anaerobic protozoa (ciliates which feed on the bacteria) and perhaps a few nematodes can live in the completely reduced zone. It is these organisms, of course, that produce upward diffusing gases. The burrows of large macrofauna (worms, clams, and crabs) do extend deep into this zone as do roots of sea grasses and salt marsh grasses in shallow coastal waters. The importance of the latter as "nutrient pumps" was mentioned in Chapter 4 (page 96). An excellent account of the ecology of marine sediments is given by Fenchel (1969; see especially pages 44 to 78).

The oxidized zone may be quite thin in

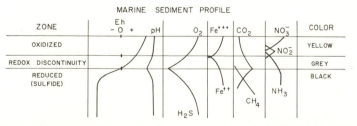

Figure 12–13. Marine sediment profile showing three zones that are often distinguishable by color. The Eh (redox potential) and pH profiles, and the vertical distribution of some compounds and ions are shown. The redox potential is minus in the reduced zone, 0 to +200 millivolts in the discontinuity transition region, and above 200 in the fully oxidized zone. Note that oxygen, carbon dioxide, and nitrate are replaced by hydrogen sulfide, methane, and ammonia in the reduced sediments. The distribution of organisms and their ecological roles in the profile are described in the text. (Redrawn from Fenchel, 1969.)

muddy or silty bottoms. Of course, if the water above the sediments becomes depleted of oxygen, then the reduced zone moves to the surface and extends upwards into the bottom waters. In recent years pollution from the affluent Scandinavian countries has turned the bottom of the Baltic Sea into one vast sulfide zone. There are a few ocean basins, such as the Black Sea, that are permanently anaerobic at the bottom.

Mangroves and Coral Reefs

Two very interesting and distinctive shallow-water marine communities of tropical and subtropical waters which deserve special mention are mangrove swamps and coral reefs. Both are important "land builders" which help form islands and extend shores. Mangroves are among the few emergent land plants that tolerate the salinities of the open sea. Different species form zones in the intertidal region, or even beyond. Next time you take a trip to south Florida or the Florida Keys, note that the red mangrove (*Rhizophora mangle*) forms the outermost zone. It has extensive prop roots which reduce tidal currents, cause extensive deposition of mud and silt, and provide surfaces for attachment of marine organisms (Fig. 12–14A). Its seeds sprout while still on the tree, the seedlings drop off and float in the water until they lodge in shallow water where well-developed roots may take hold, perhaps to start a new island! The black mangrove (*Avicennia nitida*) forms a zone nearer shore; its roots stick up above the mud like a bunch of asparagus (see second zone, Figure 12–14D). To observe the full development of the mangrove community, however, one needs to go further south where growth forms are larger and zonation more complex. Figure 12–14B shows a tropical mangrove forest in Panama, while Figure 12–14C is a close-up of the prop roots that penetrate deep into the anaerobic muds (note also the numerous crab burrows). As previously indicated, this penetration is believed to be important in mineral cycling necessary for maintaining the high primary productivity exhibited by the mangrove community (Golley, Odum, and Wilson, 1962). Zonation along a tropical mangrove shore is diagramed in Figure 12–14D. According to Davis (1940), mangroves are not only im-

portant in extending coasts and building islands, but also in protecting coasts from excessive erosion which might otherwise be produced by fierce tropical storms. Leaf detritus from mangroves has been shown to contribute a major energy input into fisheries (Heald and W. E. Odum, 1970). This example was used to illustrate the "detritus food chain" in Chapter 3 (Fig. 3–12).

Coral reefs are widely distributed in shallow waters of warm seas. As phrased by Johannes (1970), they are "among the most biologically productive, taxonomically diverse and esthetically celebrated of all communities." No student of ecology, whatever may be his special interest, should go through life without at least once donning face mask and snorkel and exploring a coral reef!

As originally described by Darwin, reefs are of three types: (1) barrier reefs along continents, (2) fringing reefs around islands, and (3) atolls, which are horseshoe-shaped ridges of reefs and islands with a lagoon in the center. Geological explorations and coring in the 1940s and 1950s have shown that Pacific reefs for the most part have developed on basaltic rock thrust up near the sea surface by submarine volcanic activity (Ladd and Tracey, 1949). Biological deposition of calcium carbonate is the means by which the reef builds up to sea level. Islands can form in the reef either as a result of a lowering of the sea level (or elevation of the volcanic underpinnings), or as a result of pounding surf and wind action that piles up broken reef chunks and coral sand above sea level so that terrestrial plants can start development.

C. M. Yonge began biological studies on the Great Barrier Reef off Australia in the 1920s and has continued to publish on the physiology and ecology of corals to this day; for reviews of this life-long work see Yonge, 1963, 1968. Japanese workers have made important contributions to the physiology of reef corals. In recent years the emphasis has shifted to studies of the community metabolism and mineral cycling of the reef community as a whole (i.e., the "holistic" approach) and to detailed studies of the remarkable coral-algal symbiosis that is clearly a major reason for the evolutionary "success" of the reef ecosystem. Let us consider briefly these two aspects.

Although corals are animals (phylum Coelenterata), a coral reef is not a heterotrophic community, but a complete ecosystem with a trophic structure that includes a large biomass of green plants (see Figure 3–15B). As described below many reefs are energetically self-sustaining but are beautifully organized to use, hoard, or recycle (as the case may be) such inputs as are received from the surrounding waters. Furthermore, although stony corals (anthozoans of the order Scleractina, but including a few species of other coelenterate groups) are major contributors to the limestone substrate, the calcareous red algae (so-called lithothamnia, especially the genus *Porolithon*) may be of equal or greater importance, especially on the seaward side of the reef since they are better able to tolerate the pounding surf. These algae contribute not only to the reef structure, but to its primary production as well (Marsh, 1968). The so-called coral reef, then, is really a coral-algal reef.

Figure 12–15 illustrates the intimate association between plant and animal components in a coral colony. One kind of algae, the so-called zooxanthellae, live in the tissues of the coral polyp (hence "endozoic"), while other kinds live in the calcareous skeleton around and below the animal bodies. Still other species of algae, both fleshy and calcareous types, are found everywhere over the limestone substrate. At night the coral polyps extend their tentacles and capture such plankton as may pass over the reef from adjacent ocean waters. Nighttime is also a period of activity for lobsters and many other invertebrates that spend the day in dark recesses of the reef. To get a full picture of life on a reef one should "skin dive" at night as well as during the day. During the day the virtually continuous sheet of algae absorbs the tropical sunlight and manufactures food at a rapid rate. Large schools of brightly colored fish graze on the algae or feed on the downstream drift of organisms and detritus. Still further downstream in deeper waters lurk the top predators, the sharks and moray eels. As would be expected, zonation is a characteristic feature of species distribution as it is on other types of "rocky" shores (see Odum and Odum, 1957).

By using the diurnal oxygen curve method (see page 58), Sargent and Austin (1949), Odum and Odum (1955), and Kohn and Helfrich (1957) found the primary productivity of coral reefs to be very high and the P/R ratio to be close to 1, indicating that the reef as a whole approaches a metabolic climax (see Chapter 9, especially Table 9–1 and Figure 9–1). *Flowing water and efficient biological recycling of nutrients are believed to be two major factors responsible for high productivity.* Based on their input-output analysis of a Pacific atoll reef, which proved to be an oasis in a desert ocean, Odum and Odum (1955) suggested that since there was not enough oceanic zooplankton (previously assumed to be the exclusive food of corals) to support the existing population, corals must be obtaining much of their food energy from symbiotic algae, which in turn must be receiving nutrients from the corals and recycling them. The controversy created by this theory has stimulated additional studies that have clarified the nature of mutualistic relationships between corals and algae. Although investigators are by no means agreed on many aspects we can at least present a "progress report" as follows:

1. The endozoic algae, which consist of small, round yellow cells in the endoderm of the polyp, have been shown to be dinoflagellates and have been placed in the new genus *Symbiodinium* by Freudenthal (1962). They have a flagellated, free-swimming stage that provides a means for distribution from one host to another, and they have now been cultured in the laboratory. Similar endozoic algae occur in other marine animals, as, for example, the giant clam *Tridacna*. McLaughlin and Zahl (1966) have reviewed culture experiments and general life history, while Halldal (1968) has studied light relations and photosynthetic capacities.

2. Direct transfer of organic materials from endozoic algae to the animal tissues has been demonstrated by the use of tracers (Muscatine and Hand, 1958; Muscatine, 1961, 1967; Goreau and Goreau, 1960) and by electron microscopy (Kawaguti, 1964). It is suspected that the coral's dependence on algal photosynthate for food energy is greater in reefs located in the open oceans than in reefs located in richer coastal waters where zooplankton are more numerous.

3. Additional field studies, as, for example, that of Johannes *et al.* (1970) on a Bermuda platform reef, have shown that the quantity of zooplankton in overlying waters is often inadequate to support coral energy requirements.

4. While Franzisket (1964) has reported that four species of corals increased in weight when held for two months in the

Figure 12–14. The mangrove ecosystem. A. The red mangrove on the west coast of Florida. Note prop roots that provide attachment surfaces for oysters and other sessile organisms. B. A mangrove forest in Panama. The biomass of prop roots alone is equal to the total woody tissue in many other kinds of forests (Golley and McGinnis, 1970).

light in seawater from which all particulate food was excluded, most experimentators find that corals grow faster and survive better when provided with some "live" food that can be captured and ingested. Johannes *et al.* (1970) suggest that the elaborate anatomical adaptations for catching zooplankton are important not so much as a "calorie catcher" but as a means of obtaining scarce nutrients such as phosphorus needed by both the coral and its algal symbionts. Once ingested by the polyp, such nutrients could

C

D   LOW   HIGH   EQUINOCTIAL STORM
     TIDE   TIDE   TIDE   TIDE

Figure 12–14.   *C.* Close-up of the prop root system in the Panama mangrove forest, showing how they penetrate deeply into the anaerobic muds; note large number of crab burrows which also extend into the reduced sediment zone. (Photos from the Institute of Ecology, University of Georgia.) *D.* Zonation in a tropical mangrove swamp. The five zones between low and storm tide are: *Rhizophora, Avicennia, Laguncularia, Hibiscus,* and *Acrostichum.* (From Dansereau, "Biogeography; an Ecological Perspective," 1957, Ronald Press, N.Y.)

be recycled repeatedly between coral and algae. Additional evidence that nutrient conservation may be enhanced by the symbiosis is provided by the studies of Pomeroy and Kuenzler (1969) who found that corals lose phosphorus very much more slowly than do similar-sized marine animals that do not have endozoic algae; the assumption is that phosphorus is recycled between plant and animal component within the colony. Such efficient reuse, of course, means that high productivity can be maintained in the system despite low phosphorus levels in the water environment (see Chapter 4, Section 5).

5. The work of T. F. and N. I. Goreau (see

Figure 12–15. *A.* Cross section of a coral colony, or "head," showing the intimate relationships between the coral animal (polyp) and several kinds of algae. See text for explanation. (Redrawn from Odum and Odum, 1955.) *B.* Underwater photograph of branching coral structures along the edge of a protected reef where currents are gentle. More massive coral "heads" predominate on the more exposed (seaward) parts of coral reefs. (Photo courtesy of Amikam Shuv, Tel Aviv University, Israel.)

summary, 1963) has shown that endozoic algae greatly increase the coral's ability to build its skeleton; calcification averaged ten times greater in light than in darkness and is greatly slowed when corals are experimentally deprived of their endozoic algae. Removal of CO_2 by algal photosynthesis apparently assists in the production of cal-

cium carbonate. Goreau believes that the symbiotic algae contribute more to skeleton formation than they do to the nutrition of the polyp.

6. The filamentous algae enmeshed in the skeleton of living corals (see Figure 12–15) are adapted to low light intensity (i.e., shade-adapted), and, as a consequence, are

rich in chlorophyll (see light-chlorophyll model, Figure 3–5). They often give the living coral mass a greenish color. In contrast to the endozoic algae, which are dinoflagellates, the skeletal algae are members of the Chlorophyta (green algae), order Siphonales. Despite recent studies by Kanwisher and Wainwright (1967), Halldal (1968) and Franzisket (1968), it is not known to what extent the skeletal algae are mutualistic with the coral host, or how much they contribute to the primary production of the reef community. Franzisket believes that the contribution from the skeletal algae is small because of their low rate of photosynthesis.

7. Corals produce large amounts of mucus which protects the delicate animals from siltation, while perhaps also providing the reef community with another means of trapping particulate nutrients. Johannes (1967) observed that large quantities of mucus slough off into the water where it forms aggregates with other organic material, thus providing a nutritious source of food for various consumers.

All in all, man has much to learn from the coral reef about "recycling" and how to prosper in a world of scarce resources. The "message," of course, is to establish better "symbiosis" with the plants and animals on which we depend.

The fact that the coral reef is a stable, species diverse, well-adapted ecosystem with a high degree of internal symbiosis does not make it immune to man's perturbations. Sewage and industrial wastes, oil spills, siltation and water stagnation brought about by dredging and filling, thermal pollution, and flooding with low salinity or silt-laden water resulting from poor land management are all beginning to take their toll. Unexpectedly, coral reefs are also threatened by a predator population explosion, something that is supposed not to happen in a well-ordered, climax system! The culprit is the "crown of thorns" starfish (*Acanthaster planci*). During the early studies on the Great Barrier Reef in the late 1920s only one specimen was collected (Yonge, 1963); today (1970) they are numerous there; they threaten the integrity of the whole reef and are spreading to other parts of the world. The cause of this epidemic is unknown, but pollution or other man-induced stress is believed to be at the root of the trouble. Will another "paradise" be lost? Coral reefs are worth saving for their recreation value alone; without them the warm water "skin diving" industry might as well close down.

Communities of the Oceanic Region

Communities in this region, of course, consist entirely of organisms with a pelagic and benthic way of life. Some oceanic species are also common in the neritic zone, but many seem to be restricted to the oceanic zone. Oceanic phytoplankton is predominantly "microplankton," as already indicated, and the zooplankton is largely "holoplankton." The large shrimplike euphausids or "krill" are important food-chain links. It is a striking characteristic of both zooplankton and fish that they are either extremely transparent or blue in color, both conditions making them all but invisible. Flotation aids such as spines, fat droplets, gelatinous capsules, and air bladders are obvious adaptations to life in the open water, as are also the "recycle" adaptations to life in nutrient-poor waters as discussed in Section 5, Chapter 4. The euphotic zone extends to greater depths in the open sea, but the total primary production under a square meter of surface is still small. However, because the open oceans are so extensive, they play an extremely important role in global O_2–CO_2 balances.

Oceanic birds are a characteristic group the easy observation of which often breaks the monotony of long ocean voyages. Petrels, albatrosses, frigate birds, and some species of terns, tropic birds, and boobies are, except during the breeding season, independent of land. One might assume that these birds would not know the meaning of "zonation," but this is far from true. As Murphy (1936) aptly expressed it: "The majority of oceanic birds are bound as peons to their own specific types of surface water." Likewise, the abundance of oceanic birds depends, indirectly, on the abundance of plankton. The same can be said for whales, which are, perhaps, the most remarkable of all marine animals, since they are as a group (order Cetacea) the chief air-breathing vertebrates that are completely independent of land. Whales belong to two types, the whalebone or baleen whales, which feed on zooplankton by means of great strainers, and the toothed whales, such as the sperm whale, which feed on nekton. Both kinds of whales are shown in Figure 12–6, as are the smaller cetaceans, which we call porpoises (also called dolphins, but not to be confused with a bony marine fish of the same name).

As already emphasized, the density of life tends to decrease with increasing depth, but "in-habitat" (i.e., within a given type of water mass or bottom) species diversity has now been shown to be high and apparently correlated with the stability of the physical environment. It now appears that at least three different mechanisms account for the transportation of food to the great depths: (1) The "rain of detritus" (see Fig-

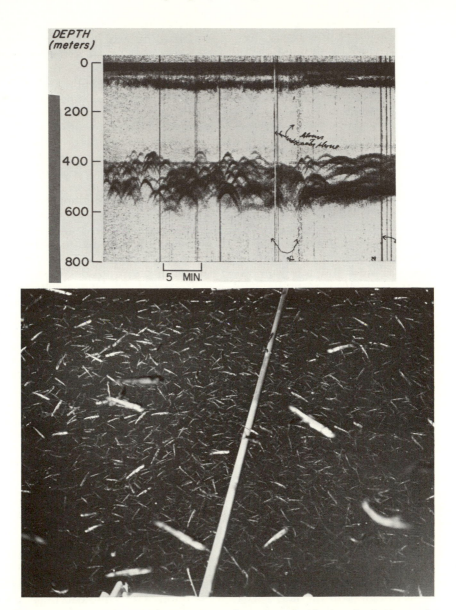

Figure 12–16. The mystery of a sound-scattering or deep scattering layer solved. Upper photo is an echo-sounder record made on a ship on the surface over continental slope waters of the western North Atlantic; the lower picture is a simultaneous photo made from an oceanographic submarine showing a school of small fish, *Ceratoscopelus maderensis,* that proved to be the sonar target which produced the sound-scattering record. The fish schools were 5 to 10 meters thick, 10 to 100 meters in diameter, and on centers 100 to 200 meters apart. Fish collected from the schools averaged 6 centimeters long. Density within the school was estimated to be 10 to 15 fish per cubic meter. The discrete hyperbolic echo-sequences (upper photo) are unusual, since most sequences are flatter, and are believed to result from the high density of individuals in the school. (Photos courtesy of R. H. Bachus, Woods Hole Oceanographic Institution. For a published description, see Bachus *et al.,* 1968.)

ure 12–6) was formerly assumed to be the chief means, but recent work has shown that the rate of sinking is so slow that most detritus originating at the surface would be completely decayed or dissolved before it reached the bottom, (2) transport by saprotrophic plankton, as, for example, the *coccolithophores* (see item 6, Figure 12–7C), which are abundant between the photic zone and the bottom, (3) formation of particulate food, or organic aggregates, from dissolved organic matter (the so-called "bubble detritus" mentioned in Chapter 2; see Baylor and Sutcliffe, 1963; Riley, 1963). Since the amount of dissolved organic matter exceeds the amount of particulate matter in sea water by about ten times, the potential for conversion into ingestible food is considerable, (4) export of organic matter from coastal zones (where P/R ratio is often greater than 1).

Large areas of the bottom of the ocean are covered with finely divided sediments, commonly called "oozes." Siliceous shells, especially those of diatoms, are conspicuous in northern waters, while calcareous shells, especially of the protozoan *Globigerina*, predominate in other regions. In very deep areas few shells reach the bottom, which is more or less bare "red clay." The deep-sea epibenthos features curious species of crustacea, echinoderms, and mollusks. Correlated with the soft "footing," many of the bottom animals have long appendages, abundant spines, stalks, or other means of support, as illustrated by the tripod fish, lamp shells, and crinoids shown in Figure 12–6.

The pelagic life, especially the nekton, of the bathyal and abyssal zones remains, perhaps, the least known of all oceanic life, for the obvious reason that it is difficult to devise nets fast enough to catch active forms at great depths. This is the region of "sea serpents," if there are any. In fact, the giant squid (see Figure 12–6), which is definitely fact and not fancy, could easily be mistaken for a sea serpent should it appear on the surface, since its tentacles are 30 to 40 feet long.

Deep-sea fish, a number of which are shown in Figure 12–6, are a curious lot. Even though there is not enough light for photosynthesis in the bathyal zone, some light does penetrate deeply, especially in clear tropical seas. Thus, we find that some deep-sea fish exhibit greatly enlarged eyes; others have very small ones, of little apparent use. Few, however, lose their eyes completely. Many animals of the depths produce their own light by means of luminescent organs (note "lantern fish" and "hatchet fish" in Figure 12–6), and some fish use a "light" (attached to a movable spine) as bait to attract their prey (note two kinds of "angler fish" in Figure 12–6). Biological light production in the sea has been reviewed by Clarke and Denton (1962). Another interesting characteristic of deep-sea fish is the enormous mouth and the ability to swallow prey larger than themselves (note "gulpers," "swallowers," and "viper-fish" in Figure 12–6). Meals are few and far between in the deeps, and fish are adapted to make the best of their opportunities!

The "echo-sounder," which has proved so useful in sounding the bottom for navigation purposes, is sensitive enough to record the location of concentrations of animals ("false bottoms," "phantom bottoms," or "deep-scattering layers"). Figure 12–16 shows an echo-sounding record and a simultaneous photograph revealing a school of small fish (each about 6 centimeters long) responsible for the sound-scattering layer. Fish with swim-bladders apparently produce most of the reflecting layers, although concentrations of larger invertebrates are capable of sound-reflecting. During the day the deep-scattering layers are found at depths down to 600 or even 1000 meters; at night the layers often move upward in the same manner as the well-known vertical migrations of zooplankton (see Figure 6–9). The widespread occurrence of these fishy sound-scattering layers attests to the fact that small fish occur in widely scattered but densely packed schools (i.e., the "crude" density of the ocean as a whole is very low, but the "ecological" density within a school may be high; see page 166).

ESTUARINE ECOLOGY

Chapter 13

1. DEFINITION AND TYPES

An estuary (*aestus*, tide), according to a definition modified from Pritchard (1967), is a semi-enclosed coastal body of water which has a free connection with the open sea; it is thus strongly affected by tidal action, and within it sea water is mixed (and usually measurably diluted) with fresh water from land drainage. River mouths, coastal bays, tidal marshes, and bodies of water behind barrier beaches are examples. Estuaries could be considered as transition zones or ecotones (see page 157) between the fresh water and marine habitats, but many of their most important physical and biological attributes are not transitional, but unique. Furthermore, uses and abuses of this zone by man are becoming so critical that it is important that the unique features of estuaries become widely understood. For these reasons, the estuarine habitat is elevated to full "chapter status" in this edition. A recommended reference source on estuaries is the volume edited by Lauff (1967).

As shown in Figure 12–3, estuarine or brackish water may be classified as oligo-, meso-, or polyhaline, according to the average salinity. This only tells part of the story, since salinity at any given location varies during the day, month, and year. Except for certain tropical estuaries, variability is a key characteristic, and organisms living in this habitat must have wide tolerances (i.e., euryhaline and eurythermal). While the physical conditions in estuaries are often stressful, and the species diversity correspondingly low, the food conditions are so favorable that the region is packed with life. In general, estuaries belong to that important class of "fluctuating water-level ecosystems." As explained in Chapter 9, Section 3 and further documented in this chapter, such systems are "pulse-stabilized" in a "youthful" stage in regard to productivity.

The literature is replete with classification schemes for estuarine types. It is important to recognize that schemes differ because classifiers select different bases for their classification. To illustrate, three different classifications will be presented based on (1) geomorphology, (2) water circulation and stratification, and (3) systems energetics.

From the geomorphological standpoint Pritchard (1967) has found it convenient to consider four subdivisions of estuaries as follows:

1. *Drowned river valleys* are most extensively developed along coastlines with relatively low and wide coastal plains. Chesapeake Bay, on the mid-Atlantic coast of the United States, is a good example.

2. *Fjord-type estuaries* are deep, U-shaped coastal indentures gouged out by glaciers and generally with a shallow sill at their mouths formed by terminal glacial deposits. The famous fjords of Norway and similar ones along the coast of British Columbia and Alaska are good examples.

3. *Bar-built estuaries* are shallow basins, often partly exposed at low tide, enclosed by a chain of offshore bars or barrier islands, broken at intervals by inlets (thus insuring "a free connection with the sea"). Sometimes the sand bars are deposited offshore, but in other instances the barriers may represent former coastal dunes that have become isolated by recent gradual rises in sea levels. In the former case the estuary develops from a former marine area, while in the latter case it develops on a flooded area of former coastal plain (see Hoyt, 1967, for a discussion of these two theories for the formation of barrier islands and their estuaries). The "sounds" behind the "outer banks" of North Carolina (Cape Hatteras National Seashore Park) and the salt marsh estuaries inshore from the Georgia "sea islands" are well-studied examples of the bar-built type of estuary (Fig. 13–1).

4. *Estuaries produced by tectonic processes* are coastal indentures formed by geological faulting or by local subsidence, often with a large inflow of fresh water. San Francisco Bay is a good example of this type of estuary.

River delta estuaries, found at the mouths of large rivers such as the Mississippi or the Nile, could be considered sufficiently distinct from Pritchard's four types to warrant

Figure 13–1. A barrier island type of estuary in Georgia. *A.* Looking in from the sea toward the barrier beach, a series of dunes—the older ones covered with mature forest—and the wide band of salt marsh estuaries that lie between the outer barriers and the mainland (too far in the distance to be seen in this photograph). *B.* Close-in view of the *Spartina* salt marsh, showing the tall grass, numerous tidal creeks and sounds, and the algal-rich mud flats (small one shown in center foreground). Note also the first stage in the formation of *Spartina* detritus (right foreground) that eventually will nourish many square miles of water (see text and Figure 13–4).

The upper photo emphasizes the interdependent sequence of sea, beach, sea island, and estuary. The salt marsh estuary is a sediment and nutrient trap that constantly adjusts to currents and sediment loads; without it the beautiful outer white sand beaches would be mudded and eroded. The barriers are constantly changing, eroding here (as on the north end of the island or right side of the picture) and building up there (as on the left foreground of the picture). Without vegetative protection on the dunes the rate of erosion can easily exceed the rate of formation of new beaches. The sea island shown in the picture (Wassaw Island) has recently been set aside as a national wilderness area (i.e., to remain "undeveloped") open to anyone who wants to visit, enjoy, or study the intricate physical and biological mechanisms that make this landscape uniquely beautiful and fertile. (Upper photo by Floyd Jillson, Atlanta Journal and Constitution Magazine; lower photo by U. S. Forest Service.)

consideration as a fifth major category. In these situations, semi-enclosed bays, channels, and brackish marshes are formed by shifting silt depositions.

Water circulation and stratification patterns provide a useful basis for estuarine classification, just at they do for lake classification (see Chapter 11, Section 5). On a hydrographic basis, estuaries can be placed in three broad categories (see Pritchard, 1952, 1955, 1967a):

1. *Highly stratified or "salt-wedge" estuary.* Where the flow of river water is strongly dominant over tidal action, as in the mouth of a large river, fresh water tends to overflow the heavier salt water, which therefore forms a "wedge" extending along the bottom for a considerable distance upstream. Because of Coriolis force the fresh water will tend to flow more strongly along the right shore as the observer faces the sea in the northern hemisphere (reversed, of course, for southern hemisphere). Such a stratified, or two-layered, estuary will exhibit a salinity profile with a "halocline," or zone of sharp change in salinity from top to bottom. The mouth of the Mississippi River is an example of the salt-wedge type.

2. *The partially mixed or moderately stratified estuary.* Where fresh water and tidal inflow are more nearly equal, the dominant mixing agent is turbulence, caused by the periodicity in the tidal action. The vertical salinity profile is less steep as more of the energy is dissipated in vertical mixing, thus creating a complex pattern of layers and water masses. Figure 13–2 is a simplified diagram of this type. Chesapeake Bay is a good example.

3. *The completely mixed or vertically homogenous estuary.* When tidal action is strongly dominant and vigorous, the water tends to be well mixed from top to bottom and the salinity relatively high (approaching that of the ocean). Major variations in salinity and temperature, when present in

this type, are horizontal rather than vertical. Bar-built and other estuaries on coastlines which lack large rivers are examples.

The *hypersaline estuary* is a special type worthy of mention here. Where the inflow of fresh water is small, the tidal amplitude low, and the evaporation very high, the salinity of enclosed bays may rise above that of the sea, at least during some seasons. The Upper Laguna Madre and other coastal lagoons of Texas are well-studied examples (see H. T. Odum, 1967); here the salinity may rise to 60‰ (recall that the salinity of the sea is around 35‰). Despite the severe physical conditions, not only are these bays inhabited by adapted organisms, but they may be biologically productive systems.

It is self-evident that these different circulation patterns and gradients will greatly influence the distribution of individual species, but so long as there are adapted populations the overall productivity need not be greatly affected by these differences. More about this later.

From an entirely different viewpoint, that of ecosystem energetics, H. T. Odum and colleagues (1969) have suggested the following classification, which would include not only the large estuarine bays and sounds, but all kinds of coastal ecological systems as well.

1. *Physically stressed systems of wide latitudinal range* are subjected to high-energy breaking waves, strong tidal currents, severe temperature or salinity shocks, low nighttime oxygen, or high rates of sedimentation. Rocky sea fronts, intertidal rocks, sand beaches, high velocity tidal channels, sedimentary deltas, and hypersaline lagoons would be included in this category. The cold, rocky coasts of western North America, pounded by heavy surf, and the warm, hypersaline bays of Texas are good examples of naturally stressed systems in two widely different climatic zones. Man-made canals connecting waters of greatly

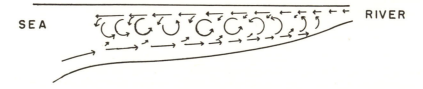

Figure 13–2. Schematic diagram of circulation in a partially mixed estuary, showing how mixing of lighter fresh water with heavier sea water tends to produce a "nutrient trap," which retains and recirculates nutrients within the estuary. In many estuaries biological factors are more important than physical factors in producing a nutrient trap.

different physical nature (such as the Cape Cod Canal) are also good examples. The proposed new sea-level Panama Canal would create a large new system of the stressed type since cold, upwelling waters of the Pacific would alternate with warm Caribbean waters surging through the canal. Such systems are generally characterized by low species diversity at any one place since few species can meet the physiological demands required to adapt to the severe oscillatory physical stresses. However, within the intertidal area sharp zonation of species and seasonal replacement of communities often result because adaptation is more efficiently accomplished by species replacement along a gradient than by adaptation within the species (recall discussions on zonation and gradient analysis in Chapters 5, 7, and 12). Thus, lists of species that occur in such systems may be quite long, even though species numbers diversity at any one time or in any one habitat (as expressed by one of the diversity indices discussed in Chapter 6) can be expected to be very low. As was discussed in Chapter 3 (see page 84), energy inputs in the form of tides, currents, or heat can be either a net stress on, or a net subsidy for, the biotic community, depending on severity and periodicity of the input. In contrast to types 3 and 4 listed below, systems organized around high-energy stresses are those in which the accessory energy is more detrimental than helpful and adaptation comes at a high metabolic cost to the community. (Again we ask: Is man irrevocably committed to covering the whole world with such stressed systems?)

2. *Natural arctic ecosystems with ice stress* are exemplified by glacial fjords, winter ice stressed intertidal zones, and under-ice communities on arctic coasts. Arctic (and Antarctic) coasts and embayments constitute a special class of physically stressed ecosystems in which light (available mostly during the very short summer season) and low temperature are strongly limiting, as is the physical "crunch" of the ice itself.

It may be noted that the impact of certain kinds of man-made pollution can be quite different on systems already adapted to physical stress as compared with systems (types 3 and 4 below) that are not so adapted. Thus, a thermal discharge from an atomic power plant that would be disastrous to a warm water estuary might actually reduce the stress, and thereby increase productivity and diversity, in an arctic estuary; but, we hasten to add, what effect will the warmer water have on currents and climates?

3. *Natural temperate coastal ecosystems with seasonal programming* include many of the best studied estuaries and shores of temperate North America, Europe, and Japan. Most of the drowned river valley, bar-built, and embayment types of estuaries (Pritchard's classification) that lie in the temperate latitudes would be included in this category. Regular seasonal pulses in primary productivity and in the reproductive and behavioral activities of animals are characteristic—often timed, or "seasonally programmed," by photoperiod or lunar periodicities, or both. The more subdued tides, waves, and currents in the semi-enclosed basins provide energy subsidies rather than stresses, while the communities in the deeper sounds and offshore waters often benefit from large imports of organic matter and nutrients from fertile shallow zones. As will be discussed later, temperate estuaries are naturally fertile but very vulnerable to damage by pollution, dredging, diking, filling, and other alterations that are all too commonplace in highly industrialized regions. Some interesting and important habitats or "subsystems" of temperate estuaries include tide pools, salt marshes, eel grass (*Zostera*) beds, seaweed bottoms, kelp beds, oyster reefs, and mud flats, which harbor dense populations of clams and sea worms.

4. *Natural tropical coastal ecosystems of high diversity.* Characteristically, temperature, salinity, and other physical factor stresses are low so that much energy of special adaptation can go into diversity and organizational behavior rather than into "antithermal maintenance." As in other tropical ecosystems, these contain many species and a great deal of chemical diversity within the species. Bright colors are often associated with complex life histories, intricate behavior patterns and a general high degree of interspecific symbiosis (i.e., "living together"). Again, the temperate zone technology associated with "monoculture" is not very well adapted to the use and management of these very different kinds of ecosystems. Unique subsystems include: mangrove swamps with special

roots adapted to salt water and anaerobic muds (see Figure 12–14); stable inshore plankton communities dominated by dino-flagellates (often phosphorescent) adapted to high light intensity and organic nutrients; and underwater tropical meadows character-ized by turtle grasses (*Thalassia*) and ben-thic algae. In shallow waters where light in-tensity is high and temperature and salinity uniform, as for example in the South Pa-cific, coral reefs often form "living" barrier islands that provide the "semi-enclosures" for the development of tropical estuaries. The ecology of coral reefs has already been discussed on pages 344–349.

5. *Emerging new systems associated with man.* While it is urgent that pollution in estuaries be reduced and that secondary and tertiary treatments of wastes (see page 435 for definitions of these terms) become nearly universal, it is likely that estuaries in urban and industrial areas will always have to bear some burden of pollution. It is important, therefore, that we recognize as a special category those estuaries which develop adaptations for man-made wastes. These need to be very carefully studied in order to determine limits of tolerance and to delimit those organisms and biological mechanisms that can be encouraged to as-sist man in waste treatment. Estuaries have varying capacities to handle "degradable" material, depending on the size of the sys-tem, flow patterns, type of estuary, and the climatic zone. Materials such as treated sewage and pulpmill wastes, seafood and other food processing wastes, petroleum wastes and dredging spoil can be decom-posed and dispersed provided that (1) the system is not also stressed by poisons (in-secticides, acids, etc.) and that (2) the rate of input is controlled at low to moderate levels and not subjected to sudden "shocks" produced by periodic massive dumping. Thus, low levels of oil or thermal pollution can be contained by an adapted system, but a massive oil spill is disastrous (and inex-cusable!) especially to large organisms such as fish and birds. Of all the man-made changes, impounding estuarine waters — that is, cutting off the "free connection with the open sea" (a key part of our definition of an estuary) — has perhaps the greatest ef-fect. It must be recognized that impounded waters are a completely different type of ecosystem, one that does not have nearly the natural capacity for waste treatment.

Even impounding waters for seafood cul-ture must be very carefully planned since man must supply by mechanical means some of the aeration, disease control, and food production formerly supplied by the free-flowing system. Mariculture, like agri-culture, has its hidden costs and is not a "free" gift of nature (recall the discussion of this important point in Chapter 3).

2. BIOTA AND PRODUCTIVITY

Typically, estuarine communities are composed of a mixture of endemic species (i.e., species restricted to the estuarine zone) and those which come in from the sea, plus a very few species with the osmoregu-latory capabilities for penetrating to or from the freshwater environment. Even the biota of hypersaline estuaries is of marine origin and not at all derived from the brine shrimp–brine fly community of inland salt lakes and high-salinity springs. Seafood populations are good examples of mixed endemic and marine species. For example, the spotted sea trout (*Cynoscion nubulosus*) is largely restricted to estuaries, while men-haden (*Brevoortia,* sp.) are found in es-tuaries mostly in the juvenile stage. Like-wise, most commercial species of oysters and crabs are primarily estuarine, while several kinds of commercially important shrimp live and spawn as adults offshore and come into the estuaries as larvae, as is shown in the life history diagram of Figure 13–3. It is quite common, in fact, for coastal nekton to use estuaries as *nursery grounds* where young growth stages can take advan-tage of the protection and abundant food. Since man often harvests such species off-shore, the vital life history and energetic connections with the nearby estuary have not always been appreciated. Anadromous fishes such as salmon and eels also depend on estuaries where they may reside for con-siderable lengths of time during their mi-grations from salt to fresh water. *The de-pendency of so many important commercial and sport fisheries on estuaries is one of the major economic reasons for preservation of these habitats.* The most productive, and hence most important, part of the nursery ground is the intertidal and adjacent shal-low-water zones, which of course are the first to suffer from ill-planned encroach-ments by man (Fig. 13–6).

Among the small organisms that constitute the base of the food chain, replacement of species in seasonal gradients provides for an efficient adaptation to seasonal changes in physical factors that are characteristic of temperate zone estuaries, as was described in Chapter 5 (page 109). In general, the holoplankton component comprises relatively few species while the meroplankton (see page 335 for definition of these terms) tends to be more diverse, reflecting the variety of benthic habitats. Consumers are often versatile in their feeding behavior. The ubiquitous mullet (*Mugil*), species of which are found in estuaries throughout the world, can feed at several trophic levels (W. E. Odum, 1970a). The marine sediment profile described in the preceding chapter (Fig. 12–13) is especially well developed in estuaries and is of key importance in systems metabolism. Because of high organic content of estuarine sediments, the sulfur biogeochemical cycle, as described in Chapter 4, plays an important role.

As shown in Table 3–7, estuaries as a class of habitats rank along with tropical rain forests and coral reefs as naturally productive ecosystems. *Characteristically, estuaries tend to be more productive than either the sea on one side or the freshwater drainage on the other.* Energy subsidies that contribute to such productivity were discussed in Chapter 3 (see especially Figure 3–17), while these and other factors have already been reconsidered in this chapter. We can now summarize the reasons for high productivity as follows (see E. P. Odum, 1961; Schelske and Odum, 1961):

1. *An estuary is a nutrient trap* that is partly physical (especially in stratified types, see Figure 13–2) and partly biological. As on coral reefs, retention and rapid recycling of nutrients by the benthos, formation of organic aggregates and detritus, and the recovery of nutrients from deep sediments by microbial activity and deeply penetrating plant roots or burrowing animals create a sort of "self-enriching" system (see Kuenzler, 1961; Pomeroy et al., 1965, 1969). As already pointed out, this natural tendency toward eutrophication makes estuaries especially vulnerable to pollution, since pollutants get "trapped" just as do useful nutrients (see the discussion of DDT, pages 74–75, Table 3–12).

2. *Estuaries benefit from a diversity of producer types "programmed" for virtually year around photosynthesis.* Estuaries often have all three types of producers that power our world, namely, macrophytes (seaweeds, sea grasses and marsh grasses), benthic microphytes, and phytoplankton. The zonation of these three production units in a Georgia estuary is shown in Figure 13–4. In this location the salt marsh grass *Spartina alterniflora* is the major producer; the microbial-enriched grass detritus "feeds" consumers in the creeks and sounds (see account of detritus food chains, Chapter 3). This role may be taken by eel grasses (*Zostera*) or seaweeds in colder waters, and by turtle grasses (*Thalassia* and related genera) in warm waters. The latter make important contributions to productivity of subtropical and tropical lagoons (see H. T. Odum, 1957a, 1967; Wood, Odum, and Zieman, 1969). Sea grasses often support large populations of

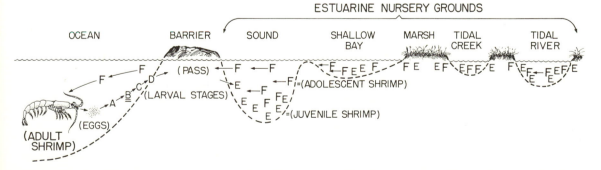

Figure 13–3. Life history of shrimp that use estuaries as nursery grounds. Adult shrimp spawn offshore and the young larval stages (*A*, nauplius; *B*, protozoea; *C*, mysis; *D*, postmysis) move shoreward into the semienclosed estuaries where the juvenile (*E*) and adolescent (*F*) stages find the food and protection they need for rapid growth in the shallow bays, creeks, or marshes. The maturing shrimp then move back into the deeper waters of the sounds and adjacent ocean where they are harvested by commercial trawlers.

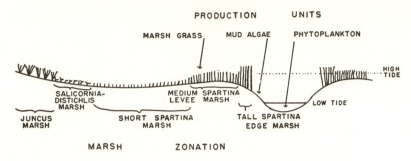

Figure 13–4. Zonation in a salt marsh in a Georgia estuary showing the three distinct types of producers which contribute to the metabolism of the system. A large portion of the marsh is covered by a single species of plant, *Spartina alterniflora*, which, however, exhibits distinct life forms or ecotypes according to physiographic conditions. The *Spartina* zones are just as different in inherent productivity and associated animal populations as are zones composed of different species. Functionally, the entire marsh is a unit since most of the marsh grass is not consumed until it has been broken down into detritus by tidal and bacterial action and transported to all parts of the estuary.

epiphytic algae (*aufwuchs*, or periphyton, see page 300) and small fauna that provide important food for grazing fish and other nekton. As shown in Figure 13–5, one may observe a "microsuccession" on sea grass blades with largest biomass and greatest diversity on the older blades.

The importance of the small benthic algae, which grow not only on the macrophytes (Fig. 13–5) and sessile animals, but on, or in, all kinds of bottoms (rock, sand, mud), is often overlooked. For example, Pomeroy (1959) estimated that the "mud algae" in the Georgia estuaries account for as much as a third of the total annual primary production. In summer, he found that the rate of photosynthesis was highest when the tide was high (the organisms were "water-cooled" as it were), while in winter it was highest when the creek banks were exposed at low tide and the sediments quickly warmed by the sun. As a result, the rate of production remained relatively constant during the year, another good example of functional homeostasis in the face of marked seasonal changes in light and temperature. Williams (1968), studying estuarine benthic diatoms, found that many form sediment tubes in which they move up and down according to light and temperature regimes (an example of adjustment by plant "behavior"). For an energy-flow diagram of the salt marsh estuary, see Teal (1962).

3. The importance of *tidal action in creating a "subsidized" fluctuating water-level ecosystem* has already been documented. In general, the higher the tidal amplitude the greater the production potential, provided that the ensuing currents are not too

abrasive. The back-and-forth movement of water does a great deal of "work," removing wastes and transporting food and nutrients so that organisms can maintain a sessile existence, which does not require the expenditure of much metabolic energy for excretion and food gathering. At what velocity currents change from subsidies to stresses is not as well known as it should be.

Estuaries are subject, as are other eutrophic systems, to blooms which sometimes "get out of hand" or become temporarily "cancerous." *Red tides*, large blooms of certain red-pigmented dinoflagellates (of the genera *Gonyaulax, Gymnodinium,* among others), are well-publicized examples. Red blooms often occur in patches in estuaries without doing any harm (Ragotzkie and Pomeroy, 1957), but in some areas blooms of monstrous proportions develop periodically and extend out into coastal waters where they can cause mass mortality of fish and other nekton as a result of toxins produced by the flagellates. According to Provasoli (1963) the neurotoxin produced by *Gonyaulax catenella* is one of the most potent poisons known. As in fresh water, large blooms of blue-green algae can also produce toxins. On the other hand many red tides are not toxic; the blooms of some species have even been reported to be readily eaten by fish (W. E. Odum, 1968*a*) and to be nutritious when fed to rats (Patton *et al.*, 1967). While the causes of red tides are not understood, most investigators believe that they develop when stable water conditions in fertile areas bring about a concentration of organic nutrients and growth substances (known to be required by dinoflagellates)

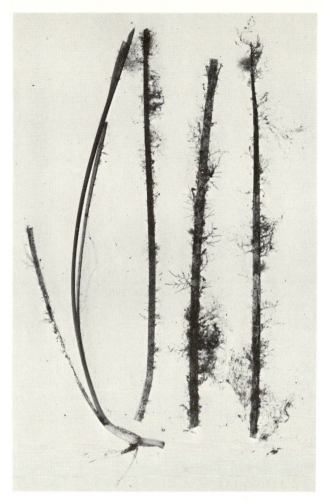

Figure 13–5. A single blade of eel grass (*Zostera*) cut into four pieces to show the development of epiphytic algae on the older portions (to the right in the picture). In addition to the multicellular algal strands visible to the eye, numerous unicellular diatoms and protozoa together with small metazoan animals form almost a complete microcommunity whose gross production equals or exceeds that of the uncolonized young or proximal part of the eel grass plant; see E. P. Odum (1966). (Photo by Dr. E. J. Kuenzler.)

that were perhaps produced by preceding blooms of other phytoplankton. While red tides are "natural" phenomena that can occur in widely scattered coastal areas (and sometimes in the open ocean), it is suspected that organic pollution can increase the frequency and severity of toxic blooms. Ryther, 1955, Hutner and McLaughlin, 1958, and Wood, 1965, present various theories.

A persistent idea about estuaries is that the inflow of river water containing fertilizers washed in from the land makes an important contribution to estuarine productivity. While this may be the case where land drainage is very rich (as in the Nile Delta before large dams blocked the sediment flow), rivers in general do not "fertilize" the estuary. In fact, river-mouth estuaries are often less productive than bays or lagoons which lack large inflows but have a well-developed benthic flora. Riley (1968) has summarized evidence that the nutrients concentrated and recycled by estuaries come originally from the sea. At the same time estuaries often generate more fertility than they can use (P exceeds R), resulting in the export or *outwelling* of nutrient and organic detritus into the ocean, as was mentioned in the previous chapter (page 325). For example, so much organic matter is produced in Georgia salt marsh estuaries and combined with so much sediment that the contained nutrients cannot be fully utilized in the estuary because of low light penetration. When these nutrients reach the clearer waters offshore they can be utilized. As shown in Table 13–1, coastal waters adjacent to such fertile estuaries may be much more productive than coastal waters at the mouths of large rivers. The export of meroplankton and detritus from estuaries can particularly enhance the secondary productivity of coastal waters. In summary, it can

Table 13–1. *Carbon–14 Primary Productivity of Coastal Waters off a Fertile Estuary as Compared to That at the Mouths of Two Large Rivers* *

AREA	PRIMARY PRODUCTION (gms carbon/ m^2/year)	SOURCE OF DATA
Coastal water off a large Georgia salt marsh estuary; no river in vicinity	547	J. P. Thomas, 1966
Coastal water off Mississippi River	288†	W. H. Thomas et al., 1960
Columbia River Mouth	88 ⎫	
Columbia River Plume	60 ⎬	Anderson, 1964
Upwelling north of the Columbia River Mouth	152 ⎭	

* Adapted from E. P. Odum, 1968a.
† Calculated from measured daily rate of 0.8 gm C/m^2.

be stated that all coastal waters capable of supporting intensive fisheries probably benefit from either (1) outwelling from shallow water "production zones" or (2) upwelling from nutrient-rich bottom waters, or (3) both (see E. P. Odum, 1968a).

3. FOOD PRODUCTION POTENTIAL

The potential high productivity of estuaries has often not been appreciated by man, who has frequently classed them as "worthless" areas suitable only for the dumping of waste materials or useful only if drained or filled and converted to terrestrial use. Figure 13–6 illustrates a particularly unfortunate modification that destroys the most productive zone and creates residential property that is vulnerable to storms. When the doubled costs (i.e., the original construction cost and subsequent cost of maintenance and repair of storm damage paid by the taxpayer) of such changes and the high potential of the unmodified estuary for seafood protein and waste treatment are considered, it is clear that utilization in the natural state is preferable. Many states are enacting legislation to preserve this "best and highest use."

Two examples will illustrate the potential for seafood production in estuaries that are left in more or less their natural state. Ac-

cording to Hopkins and Andrews (1970), the commercially valuable clam, *Rangia cuneata*, produces annually 2900 kg meat per ha and 13,900 kg shell per ha in certain Texas estuaries. Assuming 2 kcal per gm wet weight, this yield is about 580 kcal per m^2, which compares well with the yield of fish from the most intensively managed and fertilized artificial ponds (compare Table 3–11), if one bears in mind, of course, that the clam bed requires an energy input from adjacent waters. Raft culture of oysters, as practiced in Japan, can increase yields five to ten times over that obtained by harvesting wild populations. However, this type of culture, in which oysters are suspended on lines hanging from floats, requires a great deal of hand labor (i.e., human energy input). According to Bardach (1968), a 500-m^2 raft in the best oyster growing estuaries can yield 4 metric tons (wet weight) of shucked oyster meat annually. Apparently these rafts can occupy as much as one-quarter of the water area in protected bays without inducing self-pollution, in which case 2,000 kcal per m^2 per year of protein food might be obtained. Furukawa (1968), in a review of the history and techniques of Japanese oyster culture, reports that the raft method has now virtually replaced all other methods of shellfish culture in that country. The production of oyster meats in Hiroshima Bay alone rose from 20 thousand tons (wet weight) in 1950 to 240 thousand tons in 1965, the latter being a greater yield than the total harvest of natural oysters for all of Japan.

Before you get out your slide rule and start calculating from the above figures how much oyster meat might be produced by the world's estuaries, remember that such intensive local harvests depend on many acres (the exact number is undetermined) of adjacent water. Oyster "farms," therefore, have to be spaced apart and their density ultimately regulated if the natural productivity capacity, as well as other necessary uses of estuaries, is not to be destroyed by "too much of a good thing." Provided we keep yield expectation within reasonable ecological bounds, raft culture of oysters and other shellfish is a sensible way to harvest the natural productivity of estuaries. Since oysters are sensitive to pollution, having an economic investment in such culture can also be a deterrent to pollution.

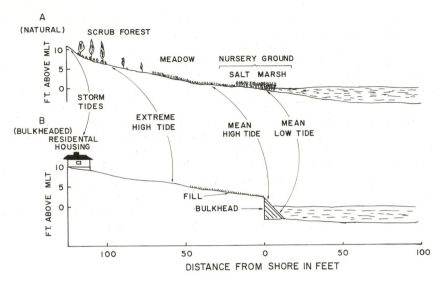

Figure 13–6. "Bulkheading" destroys the most important part of the estuarine "nursery ground" and encourages the building of housing developments that are vulnerable to hurricanes and other storms. Diagram redrawn from Mock (1966), who reports that 10 months of intensive sampling yielded 2.5 times more brown shrimp and 14 times more white shrimp (the two chief commercial species) from a natural area in the Texas estuary (A in the diagram) than could be harvested from a bulkheaded area in the same estuary (B in the diagram). This is only one of numerous kinds of modification undertaken in the name of "progress" and "development" which unwittingly destroys natural resources often at great cost to taxpayers, since most such operations are not paid for by "private" developers but by state or federal funds from the "rivers and harbors pork barrel political system" (see Allen, 1964, and W. E. Odum, 1970b).

Oyster farming in the United States has had its ups and downs. Formerly a thriving business in many areas, pollution, dredging and filling, and a depletion of "seed" stocks has all but wiped out the industry. As the human population of the world starves for protein, interest in Japanese-style raft-culture of shellfish and in pond culture of shrimp and fish (also well developed in Japan and other eastern countries; see Hickling, 1962) will undoubtedly increase in this country. In order to realize the potential, we must understand the urgency of two factors: (1) the physical destruction of estuaries must be halted and their biological fitness restored through pollution abatement, and (2) it must be widely recognized that not only is estuarine farming based on entirely different principles than is dry land farming, but consideration must always be given to other uses with which maricultural practices may conflict. Thus, unlike a cornfield, an estuary must serve uses other than the growing of food (i.e., recreation, navigation, and so on). Recently, the National Science Foundation has launched a program of "sea grants" to universities with the hope that use-oriented research will do for mariculture what the

"land grant" colleges did for agriculture during the past century. One thing is certain: marine and estuarine farming is not something the inexperienced and untrained person or group should be encouraged to undertake.

4. SUMMARY

The estuary is a good example of a coupled system that achieves a good balance between physical and biotic components, and thereby a high rate of biological productivity. It consists of several basic subsystems linked together by the ebb and flow of water that is driven by the hydrological cycle (river inflow) and the tidal cycle, both of which provide "energy subsidies" for the system as a whole. The chief subsystems are: (1) the shallow water production zones in which the rate of primary production exceeds the rate of community respirations. These include reefs, banks, seaweed or sea grass beds, algal mats, and salt marshes. This subsystem exports energy and nutrients to deeper water of the estuary and adjacent coastal shelf; (2) the sedimentary subsystem in the deeper channels, sounds, and lagoons in which respiration exceeds production and in which the particulate and dissolved organic matter from the production zone is used. Here nutrients are regenerated, recycled, and stored, and vitamins and growth regulators are manufactured; (3) the plankton and nekton, which moves freely between the two fixed subsystems producing, converting, and transporting nutrients and energy while responding to diurnal, tidal, and seasonal periodicities; this subsystem is able to react quickly to local abundance and scarcity of available resources.

From man's viewpoint estuaries must always be considered as a "multiple use" environment, which means that compromises on conflicting uses must be made in terms of the welfare of the whole (see page 268 and Figure 9–8). Since "everybody" (man and organisms) lives downstream from everybody else in an estuary, modification or pollution at one point affects distant points in both tidal directions and even in the adjacent oceans. Accordingly, *the entire estuarine ecosystem must be studied, monitored, managed, and zoned and human uses regulated in terms of the whole. Otherwise estuaries can only suffer the "tragedy of the commons"* (see page 245 and Hardin, 1968).

Chapter 14

TERRESTRIAL
ECOLOGY

1. THE TERRESTRIAL ENVIRONMENT

We now come to the land, generally conceded to be the most variable, in terms of both time and geography, of the three major environments. Although we do not wish to belabor the point, the contrast between the open-water ecosystem, such as the ocean with its small plant biomass, and the terrestrial ecosystem with its large plant biomass, has been a point of emphasis throughout Part 1 (see especially Chapter 2). Because of the fixed, conspicuous biological structure, ecological studies in the terrestrial environment have tended to emphasize the principles of population and community organization and the processes of autogenic development (i.e., ecological succession). These principles have been fully discussed in Chapters 6 through 9, and many terrestrial examples given. The productivity of terrestrial ecosystems has been considered in Chapter 3, and the general physical characteristics of the land environment have been outlined in Chapter 5, Section 5. Therefore, this chapter will be concerned mainly with the composition of and the geographical variation in terrestrial communities, with notes on some of the especially characteristic metabolic features of terrestrial ecosystems.

The following points should be kept in mind in comparing land with water as a habitat:

1. Moisture itself becomes a major limiting factor on land. Terrestrial organisms are constantly confronted with the problem of dehydration. Transpiration, or evaporation of water from plant surfaces, is an energy dissipating process unique to the terrestrial environment (see page 20).

2. Temperature variations and extremes are more pronounced in the air than in the water medium.

3. On the other hand, the rapid circulation of air throughout the globe results in a ready mixing and a remarkably constant content of oxygen and carbon dioxide (at least until man enters the picture!).

4. Although soil offers solid support, air does not. Strong skeletons have been evolved in both land plants and animals, and also special means of locomotion have been evolved in the latter.

5. Land, unlike the ocean, is not continuous; there are important geographical barriers to free movement.

6. The nature of the substrate, although important in water (as indicated in Chapters 11 and 12), is especially vital in terrestrial environments. Soil, not air, is the source of highly variable nutrients (nitrates, phosphates, and so forth); it is a highly developed ecological subsystem, as will be noted later (see Chapter 5, Section 4, subdivision 8; also Chapter 20).

In summary, we may think of climate (temperature, moisture, light, etc.) and substrate (physiography, soil, etc.) as the two groups of factors which together with population interactions determine the nature of terrestrial communities and ecosystems.

2. THE TERRESTRIAL BIOTA AND BIOGEOGRAPHIC REGIONS

Evolution on land has featured the development of the higher taxonomic categories of both the plant and the animal kingdoms. Thus, the most complex and specialized of all organisms, namely, the seed plants, the insects, and the warm-blooded vertebrates, are dominant on land today. The last named, of course, include a growing human population which, year by year, exerts a greater impact on all of the biosphere but especially on the terrestrial ecosystems. This does not mean that the lowly (from evolutionary standpoint) forms such as bacteria, fungi, protozoa, and so on are absent or unimportant; microorganisms play the same vital roles in all ecosystems.

Although man and his closest associates (domestic plants and animals, rats, fleas, and pathogenic bacteria!) show a wide distribution over the earth, each continental area tends to have its own special flora and fauna. Islands often differ greatly from the mainland. The fascinating subject of biogeography thus has special relevance in the evolution of terrestrial communities. Alfred Russell Wallace, who with Darwin coauthored one of the first statements of natural selection, early realized this and set up one of the first systems of biogeographical regions. As shown in Figure 14–1, floristic kingdoms as envisioned by the plant geographer are very similar to faunal regions as mapped by the animal geographer. The main difference is the recognition by the former of the Cape region of South Africa as a distinct major region. Although small in area the Union of South Africa has an exceptionally rich flora of over 1500 genera, 500 (30 per cent) of which are endemic, that is, found nowhere else. Many of the unique species have been widely cultivated in European gardens. When both plants and animals are considered, the Australian region, of course, is the most isolated; South America ranks next. Both these areas have a large number of endemic

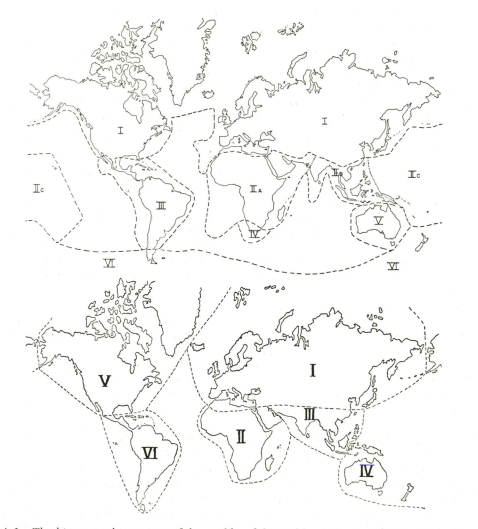

Figure 14–1. The biogeographic regions of the world as delimited by students of the flowering plants (upper map) and by animal geographers (lower map). The floristic regions or "kingdoms" as mapped by Good (1953) are as follows (upper map): I, Boreal; II, Paleotropical with subdivisions A—African, B—Indo-Malaysian, and C—Polynesian; III, Neotropical; IV, South African; V, Australian; VI, Antarctic. The zoogeographic regions (after deBeaufort, 1951) are as follows (lower map): I, Palearctic; II, Ethiopian (African); III, Oriental; IV, Australian; V, Neoarctic; VI, Neotropical. Regions I and V are often united as the Holarctic region, because considerable interchange of fauna has occurred between them.

species. Madagascar, which has long been separated from Africa, is sometimes considered a separate region.

As was discussed in Chapter 8, organisms that occupy the same ecological niche in similar communities of different biogeographical regions are known as *ecological equivalents*, although they may not even be closely related from a taxonomic standpoint. Cacti (family Cactaceae), for example, which are so prominent in New World deserts (especially neotropical region) are completely absent from the Old World, but in African deserts species of the Euphorbiaceae look exactly like cacti, having developed a similar spiny and succulent (water storing) life form. Equally striking examples are common in the animal kingdom. The point to be emphasized is that the discontinuity of land environment results in similar communities being stocked with different species. Compare Figure 14–1 with the map of the communities, Figure 14–7. For general reviews, see the books by Newbigin (1936) and Dansereau (1957) on biogeography; Cain (1944), Polunin (1960), and Good (1964) on plant geography; Hubbs (ed.) (1958), Hesse *et al.* (1951), and Udvardy (1969) on animal geography. An important contribution to island biogeography, previously cited in Chapter 9, is the small but potent book by MacArthur and Wilson (1967).

As with almost everything else ecological, man wittingly and unwittingly modifies geographical distribution of plants, animals, and microbes. He constantly experiments with introductions, even though many "backlash" and he suffers enormous economic losses from pests which, as was pointed out in Chapter 7, are so often displaced species. Remote islands and continents have experienced almost wholesale replacement of endemic species by introduced varieties. Most of the song birds, for example, that one sees in the settled parts of Hawaii are introduced. Elton (1958) has prepared an excellent review of "the ecology of invasions."

3. GENERAL STRUCTURE OF TERRESTRIAL COMMUNITIES

So varied are terrestrial organisms that a simplified classification of life forms and life habits similar to the benthos-plankton-nekton series (see Chapter 11, Section 2) generally used with reference to aquatic organisms is not practicable. Such a classification for terrestrial communities would be useful, but must await further study. We can, however, fall back on the basic trophic classification in reviewing the biotic structure of terrestrial communities. The general classification of major food niches, i.e., the autotroph-heterotroph series, is quite applicable to land.

AUTOTROPHS. The outstanding feature of terrestrial communities, of course, is the presence and usually the dominance of large rooted green plants which not only are the chief food makers but provide shelter for other organisms and play an important role in holding and modifying the earth's surface. Although there are soil algae of importance, there is nothing on land to compare with the phytoplankton of water environments. Unlike so many of the latter, which require vitamins or other organic nutrients, the basic producers of the land are strict, or obligate, autotrophs, requiring only light and mineral nutrients. Land plants, however, may depend in other ways on microorganisms for their nutrition, as we have seen in the example of symbiotic mycorrhizae (see pages 103 and 232). The *vegetation*, which is the term generally used for all the plants of an area, is such a characteristic feature that we generally classify and name land communities on the basis of it rather than on the basis of physical environment as is often convenient in aquatic situations.

A large number of life forms are represented which adapt land plants to almost every conceivable situation. Terms such as "herbaceous" and "woody" or the "tree," "shrub," "grass," and "forb" sequence (the latter including nongrassy herbs) are, of course, widely used and provide the broad basis for the recognition of major terrestrial communities, which are described later in this chapter. Still other terms refer to adaptations along environmental gradients, as, for example, "hydrophyte" (wet), "mesophyte" (moist), "zerophyte" (dry), and "halophyte" (salt). From the more detailed floristic standpoint, one of the most widely used classifications of life form is that originally proposed by Raunkaier (1934). Raunkaier's life forms are based on the position of the renewal bud or organ, and the corresponding protection provided during unfavorable cold or dry periods. The six primary categories, shown in Figure 14–2, are as follows:

Epiphytes, air plants; no roots in the soil.

Phanerophytes, aerial plants; renewal buds exposed on upright shoots. Five subgroups include trees, shrubs, stem succulents, herbaceous stems and lianas (vines).

Chamaephytes, surface plants; renewal bud at the surface of the ground.

Hemi-cryptophytes, tussock plants; bud in or just below soil surface.

Cryptophytes or *Geophytes,* earth plants; bud below surface on a bulb or rhizome.

Therophytes, annuals; complete life cycle from seed in one vegetative period; survive unfavorable seasons as seeds.

In general, the series represents one of increasing adaptation to adverse conditions of temperature and moisture. As shown in Figure 14–2, the majority of species in the tropical rain forest are phanerophytes and epiphytes, whereas more northern forests contain a higher proportion of "protected" life forms. The flora of extreme deserts and alpine areas would likely be composed mostly of annuals. However, in the study of local situations one should guard against assuming that the proportion of species in the different categories is an indicator of climate since edaphic factors and the stage in succession greatly influence life form composition (Cain, 1950). Raunkaier's life-form "spectrum" is most useful as an ecological descriptive device when the categories are weighted on a quantitative or community basis, that is, numbers of individuals as well as numbers of species considered (see Cain, 1945; Stern and Buell, 1951). In many deserts, for example, the majority of species may be annuals, but a few species of shrubs often make up the most important part of the vegetation from the standpoints of the standing crop and of the annual production of dry matter. In other words, the life-form spectrum of the vegetation (community) and the life-form spectrum of the flora are not necessarily the same. It would be well at this point to emphasize the difference between the ecological term "vegetation," which refers to the plant cover as it actually occurs on an area, and "flora," which refers to a list of taxonomic entities to be found in an area. Inter-

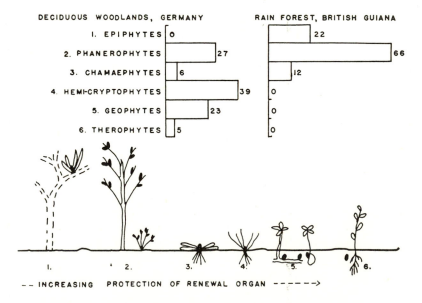

Figure 14–2. Raunkaier's life-forms of terrestrial plants. The six life-forms are diagrammed in the lower sketch with the renewal buds (or seeds in No. 6) shown as oval black bodies. The upper bar graphs compare a temperate and a tropical forest in regard to the per cent of species in the flora which belong to the six life-forms. Note that life-forms of the rain forest (where there are no unfavorable cold or dry periods) all have exposed buds, while flora of the northern forest contains a large percentage of life-forms with protected renewal organs. (Upper graph redrawn from Richards, 1952; lower diagram redrawn from Braun-Blanquet, 1932.)

estingly enough, although the term "fauna" is a parallel to flora for animals, there is no generally used parallel for the ecological distribution of animals, although the term "faunation" has been proposed (see Udvardy, 1969).

PHAGOTROPHIC (MACRO-) CONSUMERS. Correlated with the large number of niches provided by the vegetation, land communities have an extremely varied array of animal consumers. Primary consumers include not only small organisms such as insects but very large herbivores, such as the hoofed mammals. The latter are a unique feature of land with but few parallels in aquatic communities (large plant-eating turtles, for example). Thus, the "grazers" of land are a far cry in size and structure from "grazers" of water, i.e., the zooplankton. Since terrestrial autotrophs make a lot of food of nutritionally low utility (cellulose, lignin, etc.), detritivores are very prominent features of terrestrial communities (see Chapter 3; also page 19).

The variety and abundance of insects and other arthropods (which fill every conceivable niche), of course, is another important feature of terrestrial communities. The study of insect ecology has had some interesting oscillations. In the 19th and early part of the 20th centuries natural history and life history were emphasized, and the important studies on population biology were begun, as reviewed in Chapter 7. With the discovery of DDT, entomological laboratories turned primarily into chemical laboratories, and the ecological study of insects was almost forgotten except by those who worked on laboratory cultures. Now, of course, with the interest in biological control of insects revived, entomology departments are once again becoming well stocked with ecologists, and field studies are again in vogue. For an excellent review of insect ecology, see the book by Clark, Geier, Hughes, and Morris (1967).

Among other groups that have been very intensively studied in the field we should mention the birds, which are found in practically every land community and provide especially favorable material for ecological study.

THE SAPROTROPHS OR MICROCONSUMERS. Organisms which carry on the "mineralization of organic matter" and perform other valuable functions, as outlined in Chapter 2, in the terrestrial environment are chiefly the bacteria and fungi but also include protozoa and other small animals. The key roles of microbial specialists, such as the nitrogen-fixing bacteria, the mycorrhizal fungi, and the anaerobic bacteria, were discussed in Chapters 2, 4, and 7. What are often called the *decomposer microorganisms* can be considered a distinct functional or ecological group, which includes the following four taxonomic entities: (1) the fungi, including the yeasts and molds; (2) the heterotrophic bacteria, including spore-formers and nonspore-formers; (3) the actinomycetes, which are filamentous, or "thread," bacteria having certain morphological features of fungi; and (4) the soil protozoa, including amoebae, ciliates, and especially the colorless flagellates. These decomposers may be found throughout terrestrial communities, but they are especially concentrated in the uppermost soil layers (including litter). Here, the decomposition of plant residues, which accounts for such a large proportion of community respiration in so many terrestrial ecosystems, involves a sequence of microorganisms that may be shown as follows (see Kononova, 1961, page 134):

$$\text{mold fungi and non-spore-forming bacteria} \rightarrow \text{spore-forming bacteria} \rightarrow$$

$$\text{cellulose myxobacteria} \rightarrow \text{actinomycetes}$$

The first two groups utilize the more readily decomposable organic substances such as sugars, amino acids, and simple proteins. The cellulose bacteria then work on the more resistant compounds, while the actinomycetes are especially associated with *humus*, which, as described in Chapter 2, is the dark, yellow-brown, very resistant end product of the second stage of decomposition called *humification* (the first stage we designated as the formation of particulate detritus). A possible model for the structure of a humic acid molecule is shown (page 368) to illustrate key features such as the aromatic or phenolic benzene rings (labeled 1), cyclic nitrogen (2), nitrogen side chains (3), and carbohydrate residues (4), which make humic substances relatively difficult to decompose:

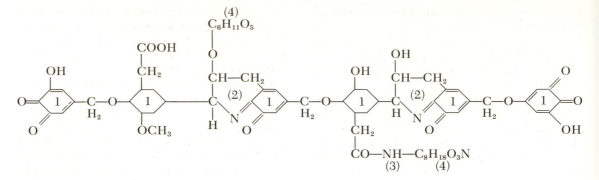

The third stage in decomposition, namely, the *mineralization of humus,* occurs very slowly in cold regions and more rapidly in warm regions, or when soil is exposed to air, as in plowing. As discussed on page 29, very little is known about the organisms and chemical processes involved in the breakdown of humus, but it is assumed that the microbes and their enzymes are different from those involved in the first two stages of decomposition (for additional discussion, see Chapter 19).

The role of soil protozoa is not well understood, but as discussed in Chapter 2 (see especially Figure 2–10*B*) they are known to play several important roles. Even those protozoa (such as ciliates) that prey on bacteria may speed up decomposition by stimulating the growth and metabolism of the prey! Where wood (lignin) is involved, a somewhat different sequence of organisms than that shown above may ensue. Fungi seem to play a more important role in the breakdown of lignin.

As stressed throughout this book, neither numbers nor biomass are good indicators of what small organisms are doing or how fast they are doing it. Microbial decomposers number in the range of 10^{12} to 10^{15} per square meter and have a biomass perhaps on the order of 1 to 10^3 grams (dry weight) per m^2 (see Table 2–2) in productive terrestrial ecosystems (much fewer, of course, in deserts or other limiting environments). Not only is measuring the standing crop very difficult, but such measurements provide very little information of ecological significance; of much greater importance are measurements of function, such as respiration, or rates of substrate decomposition, some aspects of which are discussed in the next section of this chapter and in Chapter 19.

Temperature and water are especially important in regulating the activity of decomposers; since these factors are more variable on land than in aquatic habitats, it is easy to see why decomposition in soil often proceeds in a sporadic fashion. Most bacteria and fungi require a microenvironment with a higher water content than that required by the roots of higher plants, for example. Consequently, in regions with prolonged dry periods (or prolonged cold periods), annual production in the ecosystem very often exceeds the annual decomposition, even in "climax" vegetation. As pointed out on page 135, periodic fires in these situations act as "decomposers," removing the excess accumulation of dead wood and litter. Complete prevention of fire in ecosystems such as the California chaparral may not be in the best interest of man or the ecosystem; light periodic fires assist the microbial decomposers in their work and prevent big fires that set back the succession too far and destroy human property.

The two basic strata (see page 8) that comprise all complete ecosystems, namely, the autotrophic and the heterotrophic, are well marked in the terrestrial environment. "Vegetation" and "soil" are common, everyday words for these two layers in the terrestrial ecosystem. Let us briefly consider each as a subsystem, proceeding from the ground up.

4. THE SOIL SUBSYSTEM

The physiography of the soil profile has been described and pictured in Chapter 5 (see especially Figures 5–11, 5–12), so we need only consider here the structure of the soil community and something of its metabolism. It is convenient to consider soil organisms in terms of size classes since size-metabolism relationships (see Section 5, Chapter 3) will determine sampling and other study procedures. Three size groups are commonly recognized (see Fenton, 1947):

1. The *microbiota*, which includes the soil algae (mostly green and blue-green types), the bacteria, fungi, and protozoa. This component was reviewed in the preceding section, and its contribution to total soil metabolism will be considered later. The heterotrophic microbiota are, in general, the principal link between the plant residues and the soil animals in the detritus food chain.

2. The *mesobiota*, which includes the nematodes, the small oligochaete worms (enchytraeids), the smaller insect larvae, and especially what are loosely called the microarthropods; of the latter, the soil mites (Acarina) and springtails (Collembola) are often the most abundant of forms which remain permanently in the soil. Figure 14–3 pictures some of the characteristic organisms that are readily extracted from the soil by means of a device known as a Berlese funnel or the similar Tullgren funnel. In these devices, organisms are forced down through a core of soil (removed from the field without disturbing normal texture, animal burrows, and so forth) by heat and light until they fall into a vial of preservative. Like the plankton net, the Berlese funnel is selective in that it does not sample all components. Nematodes, which are usually abundant, are but little removed by the light and heat treatment. However, if the funnel is filled with warm water and the soil enclosed in a mesh or gauze, the nematodes move out of the soil and drop by gravity to the bottom on the funnel. This type of device is called a Baermann funnel. A modification in which heat and water are applied from below works well for enchytraeids. Numbers and biomass of three important groups of the mesobiota in two contrasting soils of Denmark are shown in Table 14–1. These data were obtained by using the extraction method most efficient for each component.

From Table 14–1 we see that the microarthropods and the enchytraeids are to be numbered in the thousands per square meter, but the nematodes are counted in the millions per square meter. The biomass, however, of the three groups in the Danish soils is not too different, ranging from 1 to 13 grams per square meter. Soil arthropods extracted from a Michigan "old field" (abandoned agricultural land) numbered around 150 thousand per m^2 with an estimated living weight of about 1 gram (Hairston and Byers, 1954). While the mesobiota are primarily detritus-bacterial feeders, some, especially among the mites and insects, are predators. As shown in Figure 3–15E, the predator-prey ratio among soil arthropods is not unlike that in the above-ground community of the vegetation.

In his studies of soil nematodes Overgaard-Nielsen (1949, 1949a) found density to range from about 1 to about 20 million per m^2. A large percentage of nematodes

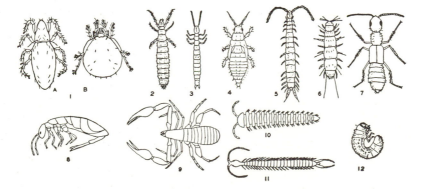

Figure 14–3. Representative arthropods of the mesobiota of litter and soil commonly obtained by the Berlese funnel sample method: 1, two oribatid mites (A—*Eulohmannia*, B—*Pelops*); 2, a proturan (*Microentomon*); 3, a japygid (*Japyx*); 4, a thrips (a thysanopteran); 5, a symphylan (*Scolopendrella*); 6, a pauropod (*Pauropus*); 7, a rove beetle (Staphylinidae); 8, a springtail or collembolan (*Entomobrya*); 9, a pseudoscorpion (or chelonecthid); 10, a millipede (or diplopod); 11, a centipede (or chilopod); 12, a scarabaeid beetle larva or grub. (1 redrawn from Baker and Wharton, 1952; 2, 3, 4, and 12 redrawn from Chu, 1949.)

Table 14–1. *Numbers and Biomass (as Live Weight) of Three*
Important Components of the Mesobiota in Two Contrasting
*Soils in Denmark**

HABITAT	ENCHYTRAEIDS		NEMATODES		MICRO-ARTHROPODS	
	Nos./m² × 1000	*Biomass gms/m²*	*Nos./m²* × 1000	*Biomass gms/m²*	*Nos./m²* × 1000	*Biomass gms/m²*
Pasture soils (mull-type)	11–45	1–3	10,000	13.5	48	2
Heath with raw humus soil (mor-type)	50	7	1,500	2	300	4–5

* Data from Overgaard-Nielsen, 1955.

apparently feed on bacteria, another large percentage (up to 40 per cent) on roots of plants and on soil algae, while not more than 2 per cent feed on other animals. Nematodes are most numerous on mineral (mull-type) soil where their biomass may be equal to that of earthworms; in this event, however, their oxygen consumption would be 10 times that of the more conspicuous earthworms (Overgaard-Nielsen, 1949a). In agricultural soils, certain species of nematodes may become serious parasites of roots of plants and are difficult to eradicate from infested soil. Crop rotation is often the best control method. In contrast to the nematodes, the microarthropods and the enchytraeids generally reach their greatest biomass in forest and organic soils, as indicated by the data in Table 14–1.

3. The *macrobiota,* which includes the roots of plants, the larger insects, earthworms (Lumbricidae), and other organisms which can easily be sorted by hand. The burrowing vertebrates such as moles, ground squirrels, and pocket gophers might also be included in this group. Very frequently the roots of plants would comprise the largest biomass component in the soil, but since their metabolism per gram would be relatively low, they may contribute less to soil respiration than do decomposers (see Table 14–3). The dry weight of roots in climax prairie is on the order of 1000 gms per m² (Weaver, 1954) and in forests 3000 gms per m² or more (see Figure 3–3B). Studies utilizing radioactive phosphorus as a tracer have indicated that a portion of the root system of grasses may be inactive at any given time (see Burton in Comar, 1957). Among the larger animals of the soil many

insects are but temporary inhabitants during hibernation or pupation. Earthworms resemble nematodes in being most abundant in mineral soils, especially calcareous clay soils where they may reach a density of 300 per m². All of the macrobiota are very important in mixing the soil and maintaining a "living sponge" consistency.

The highly mobile macroscopic invertebrates that live in the interface between litter and soil can be sampled by placing boards on the soil surface where they act as "traps" for animals seeking shelter under them. Such animals have been called the *cryptozoa* (crypto = hidden) (see Cole, 1946a). Tarpley (1967), in a two-year study in old-field habitat, found that a foot-square "cryptozoa board" sampled animals from an area of 0.5 to 1.0 square meters, depending on the density of the vegetation. He recorded 144 species that were attracted to the boards—crickets, roaches, ground beetles (Carabidae and Tenebrionidae), and spiders being the most numerous. Mean density was 16 per m² for 5 fields, and their annual estimated respiration averaged 6 kcal per m², of which 11 per cent was contributed by predatory species. Sinking cans into the soil and leaving the openings level with the surface is another method that can be used to sample this component.

The mechanical breakdown of plant litter into forms readily decomposed by microbes is one of the main "jobs" performed by soil animals. Studies in the Netherlands have shown that the diplopod *Glomeris* and numerous other soil animals assimilate only 5 to 10 per cent of dead leaf material which they ingest (that is, weight of feces = 90 to 95 per cent of ingested food) (Drift, 1958).

It was further found that feces were decomposed more rapidly by microorganisms than mechanically pulverized leaves not previously ingested by animals.

The series of pie diagrams in Figure 14–4 gives a good overall picture of the "standing crop" situation in soil, including both living and nonliving components. Thus, we see that the living components of the soil comprise but a small percentage of the total weight of soil, but because of their high rates of activity these seemingly tiny components are dominant factors in the functioning of the entire terrestrial ecosystem. What part, then, does the soil stratum play in the total community metabolism?

SOIL RESPIRATION. Three methods have been used to determine total soil metabolism (see Macfadyen, 1970):

1. The difference method, that is, subtracting the energy consumed by aboveground herbivores from net primary production. In forest and old-field communities seldom more than 5 to 10 per cent of the annual net primary production is grazed (see E. P. Odum, 1963a; Bray, 1964), leaving 90 per cent or more to be metabolized or stored in the soil-litter subsystem. In contrast, difference calculations indicate that only 40 to 60 per cent may reach the soil in heavily grazed grasslands or pastures (see Table 15–2, page 419).

2. The litter fall method. In steady-state systems, determining the amount and energy value of the detritus (litter) input into the soil subsystem is a measure of decomposition. Bray and Gorham (1964), reviewing data on litter fall in the forests of the world, found, as shown in Figure 14–5, that litter fall increases with decreasing latitude. Metabolic estimates based on litter fall, of course, do not include respiration of living roots and their associated microflora.

3. Direct measurement of CO_2 evolution from intact soils in nature should, of course, provide a means of measuring total soil respiration, including the respiration of all three groups of the biota. It is clear that early methods, which involved drawing air over the soil, gave estimates that were much

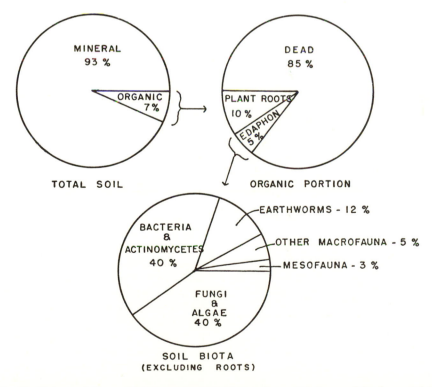

Figure 14–4. The living and the nonliving components of a meadow soil in terms of the dry weight of the "standing crop." The edaphon includes all of the soil biota exclusive of the roots of higher plants. Although living organisms make up but a small part of the weight of soil, their activity and "turnover" rates may be quite high and, therefore, their importance relatively much greater than is indicated by the size of their standing crops. (Redrawn from Tischler, 1955.)

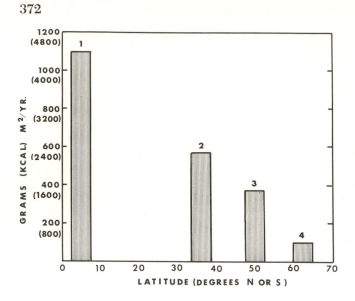

Figure 14–5. Annual litter fall in forests in relation to latitude. *1.* Equatorial forests. *2.* Warm temperate forests. *3.* Cool temperate forests. *4.* Arctic-alpine forests. (Bar graph drawn from data compiled by Bray and Gorham, 1964.)

too high, since CO_2 stored or trapped in the soil spaces was included along with the respiratory output during the period of measurement. Improved methods involve the use of inverted plastic boxes or cylinders, which are briefly capped during the actual measurement of CO_2 uptake. Recent measurements of CO_2 output, such as are shown in Table 14–2, check reasonably well with the "difference" and "litter fall" approaches when we consider that root respiration may be included in the former. Field measurements of CO_2 evolution make good ecology class exercises since all that is required is a plastic box or cylinder, some KOH to absorb the CO_2, and a simple titration apparatus.

Table 14–3 represents a preliminary attempt to determine how much of total soil respiration is due to roots and how much to the other biota. Since many microorganisms are actually an integral part of root systems, such a distinction is not particularly meaningful in practice. What is important to note is that most of the soil respiration is due to the combined root-microorganism component; the meso- and macrofauna contribute very little. Bunt (1954), for example, reports that nematodes account for only 1 per cent of soil respiration. Engelmann (1968) found that the total annual energy flow of soil arthropods was only about 2 kcal per m², less than 0.1 per cent of the soil-litter respiration shown in Table 14–3. However, this does not mean that soil animals are unimportant. When they are selectively poisoned, decomposition of litter and the recycling of minerals is measurably reduced, as shown in Figure 14–6. Consequently, insecticide poisoning of this group is of great concern (see Edwards, 1969). While a few detritus-feeding soil animals such as snails can digest cellulose, most cannot; these obtain much of their food energy by digesting microorganisms or the products of microbial transformations, obtained either through coprophagy (see page 31) or from internal symbionts. Overgaard-Nielsen (1962) sums up the situation as follows: "It is concluded that primary decomposition [of cellulose] is largely carried out by the microflora (and possibly by snails, slugs and some Diptera larvae and protozoa), while the chief importance of soil and litter invertebrates must be sought in their effects upon the chemical activity of the microflora."

Table 14–2. *Carbon Dioxide Evolution from the Forest Floor Measured by the Inverted Box Method. Means of Soil Plus Litter for Three Stands (Oak, Pine, Maple) in East Tennessee**

	LITERS CO_2/M^2	GMS CO_2/M^2	KCAL/M^2†
Daily rate – summer	3.0	6.0	16.0
Daily rate – winter	1.2	2.4	6.4
Mean annual rate	766	1,532	4,060

° After Witkamp, 1966.

† Calculated on basis of 1 liter CO_2 = 5.3 kcals (R.Q. = 0.9).

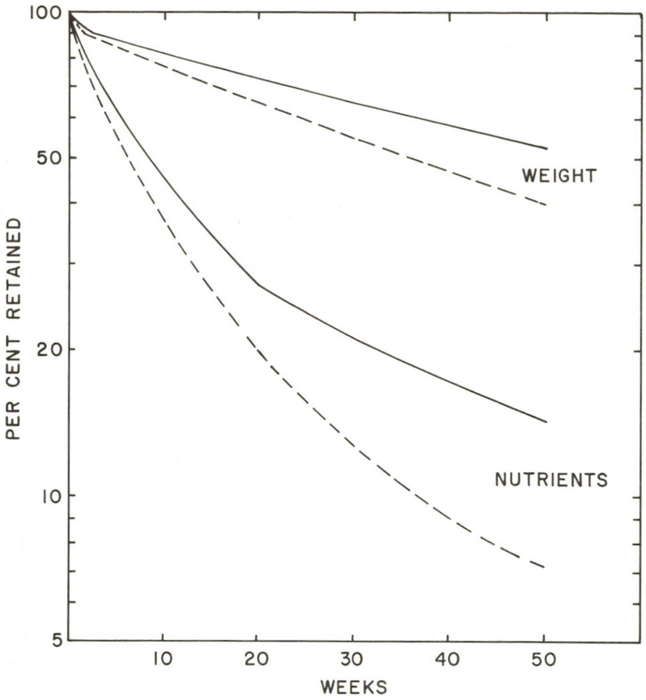

Figure 14–6. Weight loss and nutrient loss (as indicated by a cesium-134 tracer) by bagged litter placed in control plots on the forest floor (dashed lines) and in plots treated with naphthaline at a concentration that kills arthropods but does not appreciably affect bacteria and fungi (solid lines). Both the rate of decomposition and the release of mineral were considerably slowed by the treatment. (After Crossley and Witkamp, 1964.)

The use of "litter bags" has become a popular method for studying decomposition, mineral cycling, and biotic composition of the litter component of the soil subsystem (see Crossley and Hoglund, 1962). By placing samples of litter in fine-meshed nylon or fiberglass bags, decomposition rate can be determined by periodic weighing, and mineral release measured if the litter is labeled with a tracer (see Figure 14–6); also, organisms can be conveniently extracted from the bags. The litter bag, of course, is another variant of the general technique of "enclosure," as discussed in Chapter 2. As also indicated in Chapter 2, laboratory or field microcosms provide means of "getting at" complex situations via experimental simplification and amplification. Patten and Witkamp (1967) report on an interesting laboratory terrestrial microcosm composed of five components: litter, soil, microflora, millipedes, and leachate (i.e., dissolved material moving through the system). Fluxes of a radioactive tracer were measured, and transfer behavior extended through analog computer simulation. One significant finding was that the turnover time within compartments, and in the system as a whole, tended to increase as the number of compartments was increased from 2 to 5. In other words, increasing the complexity in the soil-litter subsystem increased the mineral retention.

Recommended references for soil biology include Jackson and Raw, 1966 (a good introductory reference); Gray and Parkinson (eds.), 1968 (for microbiota); Doeksen and Van der Drift (eds.), 1963 (for soil animals); and Burges and Raw (eds.), 1967 (a good advanced reference).

CAVES. Large caves provide natural constant-temperature environments (having the

Table 14–3. Estimates of the Annual Community Respiration Partitioned into Primary (Plant) and Secondary (Decomposition) Components in Two Contrasting Terrestrial Ecosystems*

	KCAL/M²/YR.	
	Beech Woods	Pasture (grazed)
Total plant respiration	4,000	470
% of above due to roots	17%	30%
Root respiration	680	140
Decomposition	1,600	2,620
Soil-litter respiration (roots + decomposition)	2,280	2,760
Root respiration as % of above	30%	5%

* After Macfadyen, 1970.

average temperature of the surface region in which they are located), as do certain large springs described in Section 9, Chapter 11. Except near the mouth (often called the "twilight zone"), the cave is populated by heterotrophs, both aquatic and terrestrial, which depend on organic matter washed in or brought in by bats or other animals that feed outside but use the cave as a roost or retreat. Because food is generally very scarce, population densities are extremely low. As on islands (see MacArthur and Wilson, 1967), species diversity appears to depend on size, age, and stability of the cave, all of which affect the balance between the rate of colonization and the rate of extinction. Caves are thus natural laboratories for the study of evolution because varying degrees of underground connection and isolation within a series of caves provide varying opportunities for speciation.

Food chains in caves resemble those of a small, shaded woodland stream or the soil subsystem, since the three primary sources of food for cave animals are: (1) particulate organic detritus, (2) dissolved organic matter absorbed on clays, and (3) bacteria. Cave bacteria appear to be representatives of noncave species. Chemosynthetic bacteria occur, but it is not known how much they may add to the energy available from imported materials. Unlike the microorganisms, cave animals are often quite different from surface relatives and have many special adaptations (including the obvious loss of eyes and pigmentation). Cave fish, for example, have a very low metabolic rate, grow very slowly, have a long life span, and lay few but very large, yolky eggs—all adaptations for surviving on meals that are few and far between.

For good reviews of the cave environment and its biota, see Barr, 1967, Poulson and White, 1969, and Culver, 1970.

5. THE VEGETATION SUBSYSTEM

Because the standing crop of producers, i.e., the vegetation, is such a conspicuous and stable feature of terrestrial environments, the composition of vegetation as such has received a great deal of attention. The quantitative study of the structure of vegetation has been called *phytosociology*, the principal aim of which is to describe vegetation, explain or predict its pattern, and classify it in a meaningful way. There have been a number of different approaches to this phase of descriptive ecology. One approach is based on the principle of succession and stems from the work of Warming, Cowles, Clements, and others. Emphasis is placed on quantitative relationships of the few species that are judged to be dominant on the theory that these largely control the community and thereby the occurrence of the large number of rarer species (see Chapter 6). Another approach which was mentioned in Chapter 6 involves the study of vegetation along a gradient and the concept of the "continuum." As stressed throughout this text, a functional approach based on productivity and mineral cycling of strata and other components is now receiving wide attention. See Ovington (1962) and Olson (1964) for comprehensive reviews of this kind of approach. Another approach to phytosociology that has not yet been mentioned in this text is that of the European school with which the name of Braun-Blanquet is associated (see the 1932 English translation by Fuller and Conard, or the revised German edition of Braun-Blanquet's *Pflanzensoziologie*, published in 1951). The following is a brief summary of underlying concepts of this school of plant ecologists based on Becking's review (1957).

The Braun-Blanquet approach is basically floristic. A complete knowledge of the flora is required since numerical relationships of all the species of higher plants, down to the smallest mosses and lichens, are considered. The aim is an objective analysis in terms of the actual floristic composition existing at the time of study; only after this composition is described are communities delimited and successional relations considered. In essence, the approach is based on two beliefs: (1) that there are distinct species combinations which repeat themselves in nature and (2) that because of the complex interaction of plants and habitat, the composition of the vegetation as a unit is more meaningful than a list of component species. Study procedures follow two steps. First comes the field analysis, which involves the selection of sample plots or quadrats and the enumeration of all plants in them. The species-area curve (see Oosting, 1956; Vestal, 1949; Goodall, 1952) is widely used to determine proper size and number of sample plots, and the sampling is usually done by strata (moss layer, herb layer, shrub layer, tree layer, and so on). Density, cover (vertical projection of aerial parts), frequency (per cent plots occupied), sociability (degree of clumping), and other indices are tabulated for each species. The second step involves the synthesis of the data to determine the degree of association of plant populations. Frequency curves are often used to determine the homogeneity or heterogeneity of a particular stand of vegetation. In formulating abstract community units or "associations," considerable emphasis is placed on indicators according to the concept of *constancy*—that is, the per cent of plots containing the species—and, especially, according to the concept of *fidelity*, which is defined as the degree of restriction of a species to a particular situation. A species with high fidelity is one with a strong preference for, or a limitation to, a particular community; to a certain extent, at least, objective statistical methods can be applied to this concept (Goodall, 1953). As we have already indicated in Chapter 5, ecological indicators of this sort are usually not the abundant and dominant species. In general, then, emphasis is placed on the numerous rare species of the community rather than on the few common species.

ROOT/SHOOT RATIO. An important and unique aspect of the structure of terrestrial ecosystems, one which underlies productivity and mineral cycling functions, is the continuous distribution of plant tissue from the top of the autotrophic stratum to the bottom of the heterotrophic stratum. The proportion of producer biomass, production or nutrients below and above ground is, therefore, of interest and can be conveniently expressed in terms of root/shoot ratios. The biomass root/shoot ratio appears to increase

Table 14–4. *The Effect of Temperature on Root/Shoot Ratios in Crops Grown Under Optimal Mineral and Water Conditions*

TEMPER-ATURE (°C.)	WHEAT	PEAS	FLAX
10	0.66	0.50	0.28
16	0.33	0.33	0.16
25	0.24	0.28	0.14

* After Van Doblen, 1962; his figures given as shoot/root ratios converted to root/shoot ratios.

Table 14–5. *Distribution of Biomass (B), Net Production (P_n), and Nitrogen (N) in Forests*

A. YOUNG (45 YR.) OAK-PINE FOREST. GRAMS DRY WEIGHT PLANT TISSUE*

	B	P_n	Turnover (B/P_n) in Yrs.
Above ground	6561	856	7.7
Below ground	3631	333	10.0
Root/shoot ratio	0.55	0.39	—
Leaves	443	382	1.2
Total woody tissue	9724	783	12.4
Leaf/wood ratio	0.046	0.49	—

B. NITROGEN CONTENT IN GMS/M² OF A TEMPERATE AND TROPICAL FOREST†

	British 55-yr. Pine Forest	Tropical Gallery Forest
Leaves	12.4	52.6
Above ground wood‡	18.5	41.2
Roots	18.4	28.2
Litter	40.9	3.9
Soil	730.8	85.3
% N above ground	3.0	44.0
% in biomass	6.0	57.8
Root/shoot§ ratio	0.60	0.30
Leaf/wood ratio	0.34	0.76
Turnover (N in biomass/annual flux)‖	4.0	?

* After Whittaker and Woodwell, 1969.
† After Ovington, 1962.
‡ Not including leaves.
§ Including leaves.
‖ Annual uptake and release 12 gms/m².

in succession. For example, Monk (1966) reports root/shoot ratios of 0.10 to 0.20 in early successional annual plants, 0.2 to 0.50 in perennial herbs and fast-growing trees such as pines, and 0.5 to 1.0 or more in slow growing climax trees such as oaks and hickories. This may be another manifestation of the increased mineral retention and stability that occurs in ecosystem development (see Chapter 9). Root/shoot ratios are also affected by temperature, as shown by the data on crop plants displayed in Table 14–4. Northern vegetation can be expected to have proportionally larger (but less active) root systems than southern vegetation. Root/shoot ratios for biomass, net primary production, and nitrogen in three forests are shown in Table 14–5. The contrast between temperate and tropical forests, as discussed in Section 6, Chapter 4 (page 102), is brought out by the comparison in Table 14–5B; as shown, a much larger proportion of nitrogen is in the biomass, and is above ground, in the tropical forest. Compare also Table 14–7, which indicates that most community respiration is below ground during the growing season in the tundra. Shown also in Table 14–5 are leaf/woody tissue ratios and turnover times. Leaves are comparable to the phytoplankton of aquatic systems in that they represent the primary photosynthetic units. In aquatic systems exchanges of nutrients and food occur by way of the water medium, while in terrestrial systems much of this exchange occurs via "direct pipelines" of living tissues (especially in tropical forests). Additional data on the relationships between biomass and production, including leaf area indices (see page 48) for a variety of Japanese forests, are summarized in Table 14–6.

TRANSPIRATION.   As was first mentioned in Chapter 2 (page 20), transpiration, or evaporation of water from plant surfaces, is a unique feature of the terrestrial ecosystem, in which it has important effects on nutrient cycling and productivity. A large part (50 to 90 per cent) of the solar energy dissipated in the evaporation of water from grassland and forest communities may be in the form of transpiration. The loss of water in transpiration can be a limiting factor, resulting in wilting especially in dry climates, in areas of intensive agriculture, or where

soils have poor water-holding capacities. Desert plants have special adaptations for reducing the rate of water loss (as discussed on page 122). On the other hand, evaporation cools the leaves and is one of the several processes that aid nutrient cycling. Other processes include transport of ions through the soil to the roots, transfer of ions across root boundaries, translocation within the plant, and foliar leaching (see Kozlowski, 1964, 1968); several of these processes require the expenditure of metabolic energy, which may exert rate-limiting control on both water and mineral transport (see Fried and Broeshart, 1967). Thus, transpiration is not a simple function of physical surface exposed; a forest does not necessarily lose more water than a grassland. Transpiration as an energy subsidy in moist forests was discussed in Chapter 3. If the air is too moist (approaching 100 per cent relative humidity), as in tropical cloud forests, trees are stunted and much of the vegetation is epiphytic, presumedly because of the lack of "transpiration pull" (see H. T. Odum and Pigeon, 1970).

Despite the many biological and physical complications, total evapotranspiration is broadly correlated with the rate of productivity. For example, Rosenzweig (1968) found that *evapotranspiration was a highly significant predictor of the annual aboveground net primary production in mature or climax terrestrial communities of all kinds (deserts, tundras, grasslands, and forests): however, the relationship was not reliable*

Table 14–6.   *Relationships Between Biomass and Net Primary Productivity (Both Expressed as kgm/m²/year) of Some Japanese forests* *

FOREST TYPE	BIOMASS (B)			NET PRO-DUCTION (P_n)	B/P_n (TURN-OVER)	LEAF AREA INDEX (CM²/CM²)
	Leaves	Total	Per Cent Leaves of Total			
Northern coniferous (*Abies* and *Picea*)						
Mature stands (4)†	0.26	26.0	1.0	2.00	13.1	—
Young regenerations (3)	0.34	8.05	4.2	1.11	7.2	—
Broad-leafed evergreen (temperate)						
Mature stands (2)	1.22	43.2	2.8	2.16	20.0	10.3
Young stands (2)	0.72	6.2	11.6	2.80	2.2	13.1
Acacia plantations (3)	0.71	9.33	7.6	2.94	3.2	—
Secondary birch forest (2)	0.32	8.77	3.6	1.08	8.1	5.2
Beech and poplar plantations (2)	0.17	2.2	7.7	0.86	2.4	3.0
Young bamboo (*Sassa*) stand (1)	0.46	1.3	3.5	1.6	0.71	5.1

* Table prepared by Dr. Kinji Hogetsu from Japanese literature to 1966. See also Kira and Shidei (1967).
† Number of stands averaged indicated in parentheses.

in unstable or developmental vegetation. He presents the following regression equation (including 5 per cent confidence interval for slope and intercept):

$$\log_{10} P_n = (1.66 \pm 0.27) \log_{10} AE - (1.66 \pm 0.07)$$

where P_n is net above-ground production in grams per square meter and AE is the annual actual evapotranspiration in millimeters. Knowing the latitude and mean monthly temperatures and precipitation (basic weather record), one can estimate AE from meteorological tables (see Thornthwaite and Mather, 1957), and then with the above equation predict what a well-adjusted, mature, natural community should be able to produce. Rosenzweig hypothesizes that the relationship between AE and P_n is due to the fact that AE measures the simultaneous availability of water and solar energy, the most important rate-limiting resources in terrestrial photosynthesis. The fact that AE and P_n are poorly correlated in developmental communities is logical, since such communities have not yet reached equilibrium conditions with their energy and water environment.

6. THE PERMEANTS OF THE TERRESTRIAL ENVIRONMENT

Corresponding to the nekton of aquatic ecosystems are the permeants, V. E. Shelford's term for the highly mobile animals such as birds, mammals, and flying insects. These move freely between strata and subsystems and between developmental and mature stages of vegetation that usually form a mosaic on most landscapes. In many cases life histories are organized so that one stage is spent in one stratum or community and another stage in an entirely different stratum or community. The permeants, therefore, often exploit the best of several worlds. This includes man, of course, who both enjoys and requires a variety of habitats, but who, as we have pointed out throughout this text, shows a distressing tendency to create a monotonous habitat for himself because of a distorted value system that places a much higher economic value on "productive" as compared with "protective" ecosystems (see Section 3, Chapter 9).

The population ecology of vertebrates and insects has been rather extensively covered in previous chapters (6 through 9). We need only consider here a couple of examples that illustrate how the permeants "link" or "couple" together subsystems and diverse communities.

As previously emphasized, large mammalian herbivores are a characteristic feature of the terrestrial environment. Many of these are ruminants, which possess a remarkable nutrient-generating "micro-ecosystem," the rumen, in which symbiotic anaerobic microorganisms break down and enrich plant material that originally has low nutritive value (see page 494; also Hungate, 1966). Under natural conditions such large herbivores are highly migratory, moving from one area to another seasonally, or in response to food availability or predator pressure. In the African grassland and savannas average ungulate densities on the order of 30 to 50 kgs per ha = 3 to 5 gms or 6 to 10 kcal per m² are indicated by recent reviews (Talbot and Talbot, 1963, Wagner, 1968) with "standing crops" up to 50 kcal per m² in especially favorable areas (Petrides, 1961). Estimated secondary production is on the order of 25 per cent biomass, or 1.5 to 2.5 kcal per m². High densities can exist without overgrazing because of migratory behavior and because the diversity of species results in a balanced use of all species and stages in the vegetation. When such animals are "fenced in" (that is, their "permeant" behavior restricted) or when they are replaced by sedentary domestic cattle, severe overgrazing often results. Through the ages man has been *exceedingly slow* to recognize the need for rotation of "pastures." The caribou-reindeer problem mentioned on page 69 and perhaps the "elk problem" in our national parks are other examples of our failure to recognize adapted behavior patterns and the necessity of replacing them with compensating human management procedures when such behavior is restricted or is "selected out" in the process of domestication (see Chapter 8, Section 5).

Migratory birds, of course, are the most spectacular permeants. North American warblers (Parulidae), for example, literally link tropical and northern coniferous forests by their seasonal migrations. Although intensively studied, many aspects of navigation and physiological adaptation are not yet understood. It has been shown, at least experimentally, that migratory birds are capable of celestial and "light compass" navigation (see review by Marler and Hamilton, 1966; also Chapter 16), but recent work indicates that biological "clocks" and less dramatic factors such as wind and other meteorological conditions may actually be more important in determining the actual time and direction of mass migratory flights, which often involve such a large number of birds that they can easily be tracked by radar. Long-range migratory birds exhibit a remarkable ability to store fat and to use it as the sole source for flight energy. Recall that a gram of fat yields 9 calories, whereas a gram of carbohydrate yields only 4 calories; thus, use of fat as "high-octane" fuel makes "adaptive" sense. Small land birds that fly nonstop across the Gulf of Mexico store in their bodies prior to the trip fat equal to three times their dry weight, and six times the energy value, of all the nonfat tissues. Furthermore, the fat is largely stored in, and used from, pre-existing adipose cells so that very little new tissue has to be laid down and then catabolized (Odum, Hicks, and Rogers, 1964). Adaptive migratory "obesity" thus differs from the unadapted human obesity (which involves tissue growth) in that the fat can be quickly used with little disruption of water, protein, and other physiological balances.

No one seems to have yet considered whether land bird migration plays any role in the cycling of nutrients, as is known to be the case with sea birds' transport of phosphorus to land (see page 91) or the upstream transport of nutrients by migratory salmon. One thing is certain; migratory birds are transporting radioactive fallout and DDT all over the world. For this reason migratory birds may be good monitors of the general level of biospheric contamination.

7. DISTRIBUTION OF MAJOR TERRESTRIAL COMMUNITIES, THE BIOMES

Regional climates interact with regional biota and substrate to produce large, easily recognizable community units, called *biomes* (Fig. 14–7). The biome is the largest land community unit which it is convenient to recognize. In a given biome the *life form* of the climatic climax vegetation (see Chap-

ter 9, Section 2) is uniform. Thus, the climax vegetation of the grassland biome is grass, although the species of dominant grasses may vary in different parts of the biome. Since the life form of the vegetation, on the one hand, reflects the major features of climate and, on the other, determines the structural nature of the habitat for animals, it provides a sound basis for a natural ecological classification (Clements and Shelford, 1939). Conversely, climatic data may be used to delimit the major vegetation formations (Holdridge, 1947).

The biome includes not only the climatic climax vegetation, which is the key to recognition, but the edaphic climaxes and the developmental stages as well, which in many cases are dominated by other life forms (see Chapter 9). Thus, grassland communities are temporary developmental stages in the deciduous forest biome where the broad-leaved deciduous tree is the climax life form. Many organisms require both the developmental and the climax stages in succession or the ecotones between them; therefore, all of the communities in a given climatic region, whether climax or not, are natural parts of the biome (see Figures 9–4 and 9–6).

The biome is identical with major "plant formation," as the term is used by plant ecologists, except that the biome is a total community unit and not a unit of vegetation alone. Animals as well as plants are considered. In general, the biome may be said to occupy a major "biotic zone" when this expression is used to mean a community zone and not a floral or faunal unit. Biome is the same thing as "major life zone" as used by European ecologists, but not the same thing as "life zone" as used in North America. In North America, "life zone" generally refers to a series of temperature zones proposed by C. Hart Merriam in 1894 and widely used by students of the birds and mammals. The original temperature criteria have been abandoned and Merriam's life zones are currently based on the distribution of organisms. These zones have, therefore, become more and more community zones, and in many cases are the same as biome divisions or subdivisions, except for terminology. For a comparison of biomes and Merriam's life zones, see E. P. Odum (1945).

The biomes of the world are shown in Figure 14–7 in a semidiagrammatic manner. Biomes and some subdivisions are shown in

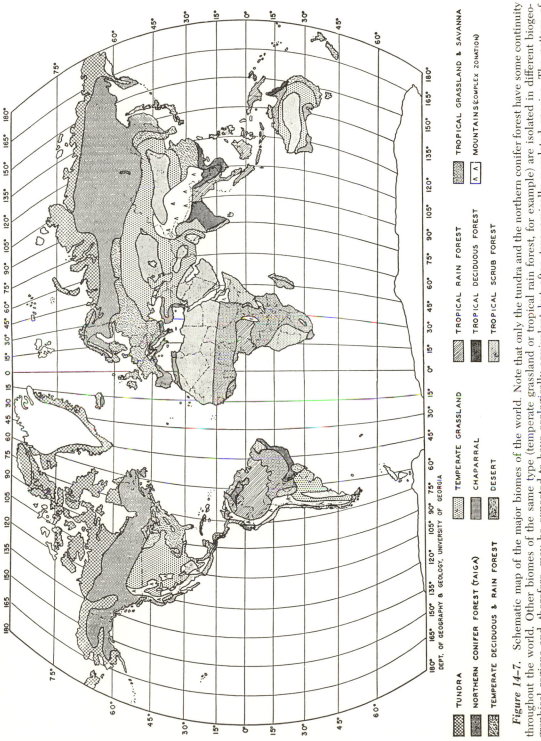

Figure 14–7. Schematic map of the major biomes of the world. Note that only the tundra and the northern conifer forest have some continuity throughout the world. Other biomes of the same type (temperate grassland or tropical rain forest, for example) are isolated in different biogeographical regions and, therefore, may be expected to have ecologically equivalent but often taxonomically unrelated species. The pattern of the major biomes is similar to but not identical with that of the primary soil groups as mapped in Figure 5–24. (Map prepared using Finch and Trewartha's [1949] map of original vegetation as a basis.)

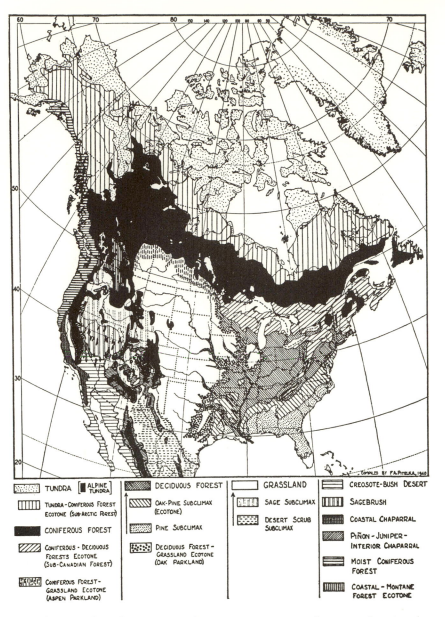

Figure 14–8. The biomes of North America, with extensive ecotones and certain subregions (zones) indicated. (After Pitelka, 1941.)

more detail for North America and Africa in Figures 14–8 and 14–9. Let us now describe briefly the principal biomes of the world. (For a brief picture atlas with short descriptions of world vegetation and soil formations, see Riley and Young, 1948.)

The Tundra

There are two tundra biomes covering large areas of the arctic, one in the Palearctic and one in the Nearctic region. Many

species occur in both since there has been a land connection between them (i.e., have a circumpolar distribution). In both continents the boundary between tundra and forest lies further north in the west where climate is moderated by warm westerly winds. Low temperatures and a short growing season (about 60 days) are the chief limiting factors. The ground remains frozen except for the upper few inches during the open season. The permanently frozen deeper soil layer is called *permafrost*. The tundra is essentially a wet arctic grassland

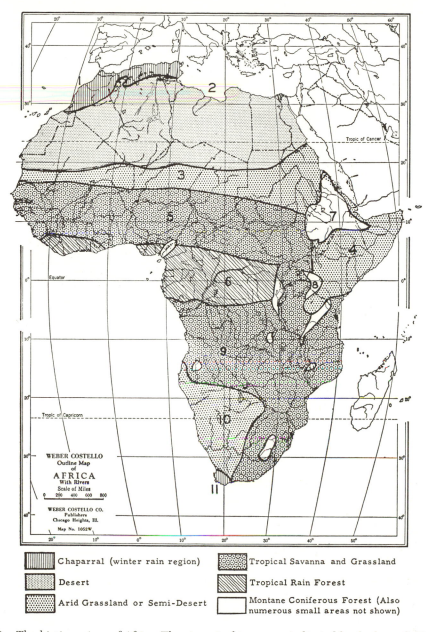

Figure 14-9. The biotic regions of Africa. The six major biomes are indicated by the legend. Note that only a relatively small part of the continent is "jungle" (i.e., tropical rain forest). The very extensive tropical savanna and grassland is the fabulous "big game" country. Some of the larger faunal areas are numbered as follows: 1, Barbary; 2, Sahara; 3, Sudanese Arid; 4, Somali Arid; 5, Northern Savanna; 6, Congo Lowland; 7, Abyssinian Highland; 8, Kenya Highland; 9, Southern Savanna; 10, Southwest Arid; 11, Cape Town "winter rain" region. All of the faunal areas are part of the Ethiopian or African biogeographic region (see Figure 14–1), except the Barbary, which is Palearctic. (Map based on sketch-map of Moreau, 1952.)

(Fig. 14-10); the vegetation consists of lichens (reindeer "moss"), grasses, sedges, and dwarf woody plants. "Low tundra" is characterized by a thick, spongy mat of living and undecayed vegetation (microbial decomposition is very slow because of low temperatures), often saturated with water and dotted with ponds when not frozen.

"High tundra," especially where there is considerable relief, may be bare except for a scanty growth of lichens and grasses. A characteristic feature of the topography is raised polygonal areas believed to be due to underlying ice wedges. Two views of the low coastal plain tundra are shown in Figure 14–10. Although the growing season is

Figure 14–10. Two views of the tundra in July on the coastal plain near the Arctic Research Laboratory, Point Barrow, Alaska. Upper photo shows a broad swale at the head of a stream system about two miles from the coast. The arctic grass *Dupontia fischeri* and the sedge *Carex aquatalis* are the dominant plants which are rooted in a peaty layer of a half-bog soil. The area is typical of what may be reasonably regarded as "climax" on low tundra sites near the coast. Shown in the picture are sample quadrats and an "enclosure" (fenced area) to keep out lemmings. During the summer when this picture was taken, lemmings reached a "high" in their cycle; at the end of August the enclosure contained 36 per cent more grass than adjacent areas, indicating the extent to which the small voles graze the vegetation when they are abundant.

Lower photo is a site about 10 miles inland showing characteristic polygonal ground. Ice wedges underlying the troughs contribute to the raised polygons. The white fruits seen in the foreground are cottongrass, *Eriophorum scheuzcheri*. (Photos by the late Royal E. Shanks, E. E. Clebsch and John Koranda.)

short, long summer photoperiods allow a respectable amount of primary production on favorable sites such as the coastal plain near Point Barrow, Alaska, as shown in Table 14–7. As in other northern ecosystems, a large portion of nutrients and primary production is below ground (see Section 5 of this chapter).

Despite the rigors of the environment (see Figure 5–18), many warm-blooded animals remain active throughout the year. These include the caribou (and its ecological equivalent, the reindeer of Eurasia), musk ox, arctic hare, arctic fox, lemming, ptarmigan, and others. Migratory birds and insects, especially (or so it seems to the visitor) the biting Diptera (mosquitoes, black flies), are conspicuous during the short summer.

The tundra food chains have been discussed (see Chapter 3, Section 4). In Chapter 7, it was pointed out that violent oscillations, or "cycles," in population density of some animals are characteristic of tundra communities. When the lemmings reach a peak of abundance the effect of their grazing on the vegetation is marked; in the region of Figure 14–10, for example, plots from which lemmings were excluded contained 36 per cent more grass in August than those heavily utilized during the spring and summer. During lemming "highs" predatory birds such as owls and jaegers are abundant and breed, whereas few predators breed at all during the years of prey scarcity (Pitelka, Tomich, and Treichel, 1955). For a brief review of the tundra as an ecosystem, see Schultz (1969).

Tundralike areas, usually called alpine tundra or alpine meadow, occur on high mountains in the temperate zone. Although there is often more snow in the alpine compared to the arctic tundra, there is no permafrost, and the photoperiod regime is different because of the more southerly latitude. Ecotypic response of plants to the different light environment was discussed in Chapter 5 (page 109).

Northern Coniferous Forest Biomes

Stretching as broad belts all the way across both North America and Eurasia (Figs. 14–7 and 14–8) are the vast northern evergreen forest regions. Extensions occur in the mountains even in the tropics. The identifying life form is the needle-leaved evergreen tree, especially the spruces, firs, and pines (Fig. 14–11). A dense shade thus exists the year around, often resulting in poor development of shrub and herb layers. However, the continuous blanket of chlorophyll present the year around results in a fairly high annual production rate despite low temperature during half of the year (see Chapter 3 and Table 14–6). The coniferous forests are among the great lumber producing regions of the world. Coniferous needles decay very slowly, and the soil develops a highly characteristic podzol profile (see Figure 5–12). The soil may contain a fair population of small organisms but comparatively few larger ones (as compared to deciduous forest or grassland soils). Many of the larger herbivorous vertebrates,

Table 14–7. *Primary Production of Coastal Plain Tundra at Point Barrow, Alaska* *

		ANNUAL PRODUCTION RATE		
		gms/m²		kcal/m²†
Gross production (P_g)			344	1550
Plant respiration (R) – Shoots	27		122	
Roots	135		608	
Total		162		730
Net production (P_n) – Shoots	82		370	
Roots	100		450	
Total		182		820
Root/shoot ratio P_g		2.1		
P_n		1.2		
R		5.0		

* After Johnson, 1969, 1970.
† Estimated on basis of 1 gram = 4.5 kcal (see Table 3–1).

Figure 14–11. Three kinds of coniferous forests. *A.* A high altitude forest of Engelmann spruce and Alpine fir in Colorado. Note the large amount of litter that accumulates because of low temperatures and long seasonal periods of snow cover. *B.* A spruce forest in Idaho with one of its chief developmental stages, aspen, a broad-leaved species whose leaves turn golden yellow in the fall (light colored stand left and center in the photo).

such as the moose, snowshoe hare, and grouse, depend, in part at least, on broad-leaved developmental communities (Fig. 14–11*B*) for their food. The seeds of the conifers provide important food for many animals such as squirrels, siskins, and cross-bills.

Like the tundra, seasonal periodicity is pronounced and populations tend to oscillate. The snowshoe hare–lynx cycles are classic examples (see Figure 7–16). Co-

niferous forests are also subject to outbreaks of bark beetles and defoliating insects (sawflies, budworms, and so forth), especially where stands have but one or two dominant species; but, as pointed out in Chapter 7 (page 192), such outbreaks are part of the continuous cycle of development to which the coniferous forest ecosystem is adapted. For an account of the conifer forest biome of North America, see Shelford and Olson (1935).

Figure 14–11. *C.* A fine example of the moist coniferous forest, often called temperate rain forest, in the Olympic National Forest, Washington. Note large size of the trees, luxuriant ground cover of ferns and other herbaceous plants, and the epiphytic mosses that festoon the limbs of trees. (U. S. Forest Service Photos.)

Moist Temperate (Mesothermal) Coniferous Forest Biome

Forests of a distinctive type occur along the west coast of North America from central California to Alaska (see Figure 14–11C), where temperatures are higher, seasonal range is relatively small and the humidity is very high. Although dominated by the conifer life form, these forests are quite different floristically and ecologically from the northern coniferous forest. Rainfall ranges from 30 to 150 inches; fog compensates for the lower rainfall in southern areas so that the humidity is everywhere high and the precipitation/evaporation ratio is extremely favorable. Oberlander (1956), for example, found that fog may account for 2 to 3 times more water than the annual precipitation and that some tall trees in position to intercept coastal fog as it moves inland may get as much as 50 inches of "rainfall" dripping down from the limbs! Because water is usually not a severe limiting factor, the forests of the west coast region are often called "temperate rain forests."

Western hemlock (*Tsuga heterophylla*), western arborvitae (*Thuja plicata*), grand fir (*Abies grandis*), and douglas fir (*Pseudotsuga*), the latter on drier sites or sub-climax on wet sites, are the big four among the dominant trees in the Puget Sound area where the forest reaches its greatest development. Southward, the magnificent redwoods (*Sequoia*) are found, and northward the sitka spruce (*Picea sitchensis*) is prominent. Unlike the drier and more northern coniferous forests, the understory vegetation is well developed wherever any light filters through, while mosses and other moisture-loving lesser plants are abundant. Epiphytic mosses are the "ecological equivalent" of the epiphytic bromeliads of moist tropical forests. The "standing crop" of producers in this biome is indeed impressive, and, as can be imagined, the production of lumber per unit area is potentially very great if the harvest-regeneration and nutrient cycles can be maintained. As with all ecosystems in which such a large percentage of nutrients may be tied up in the biomass, there is great danger that overexploitation may reduce future productivity.

For a comprehensive floristic and ecological review of western coniferous forests in general, see the monograph by Krajina (1969); this author lists 11 biogeoclimatic zones in the gradient from alpine tundra to the warm, moist coastal forests, each supporting distinct forest types.

Figure 14–12. Climatic, edaphic, and fire climaxes in the eastern United States. *A.* A virgin stand of deciduous forest in western North Carolina—a climatic climax. *B.* An edaphic climax gum (*Nyssa*) swamp forest festooned with epiphytic "Spanish moss," *Tillandsia,* on the flood plain of the Savannah River.

Temperate Deciduous Forest Biomes

Deciduous forest communities (Fig. 14–12) occupy areas with abundant, evenly distributed rainfall (30 to 60 inches) and moderate temperatures which exhibit a distinct seasonal pattern. Temperate deciduous forest originally covered eastern North America, all of Europe, and part of Japan, Australia, and the tip of South America (Fig. 14–7). Deciduous forest biomes are thus more isolated from one another than the tundras and northern coniferous forests, and species composition will, of course, reflect the degree of isolation. Since leaves are off the trees and shrubs for part of the year, the contrast between winter and summer is

Figure 14–12. *C.* A remnant of fire climax virgin longleaf pine forest on Millpond Plantation near Thomasville, Georgia. Use of frequent controlled burning has maintained the open, parklike condition and prevented the invasion of fire-tender hardwood trees. (*A*, U. S. Forest Service Photo; *B*, E. I. Du Pont de Nemours & Co.; *C,* photo by Roy Komarek, Tall Timbers Research Station.)

great. Herb and shrub layers tend to be well developed, as is also the soil biota. There are a large number of plants which produce pulpy fruits and nuts, such as acorns and beechnuts. Animals of the original forest of North America included the Virginia deer, bear, gray and fox squirrels, gray fox, bobcat, and wild turkey. The red-eyed vireo, wood thrush, tufted titmouse, ovenbird, and several woodpeckers are characteristic small birds of the climax stages. Interestingly enough, conifers, chiefly species of pines, are developmental or subclimax in many deciduous forest areas (see Figure 9–4, page 261). Characteristic grassland or "old-field" vegetation constitutes early developmental stages, which have received a great deal of ecological study.

Deciduous forest regions represent one of the most important biotic regions of the world because "white man's civilization" has achieved its greatest development in these areas. This biome is, therefore, greatly modified by man and much of it is replaced by cultivated and forest edge communities.

The deciduous forest biome of North America has a large number of important subdivisions which have different climax forest types. Some of these are:

The beech–maple forest of the north central region;

The maple–basswood forest of Wisconsin and Minnesota;

The oak–hickory forest of western and southern regions;

The oak–chestnut forest of the Appalachian mountains (now chiefly an oak forest with the chestnut wiped out by fungus diseases—see page 222);

The mixed mesophytic forest of the Appalachian plateau;

The pine edaphic forest of the southeastern coastal plain.

Each of these has distinctive features, but many organisms, especially the larger animals, range through two or more of the subdivisions (i.e., are "binding species"). As has been pointed out (see pages 131, 135, and 266), the southeastern coastal plain is peculiar in that only a very small portion of the area is or ever was actually occupied by the climatic climax vegetation, because of the immaturity of the soils and the action of fires. Although large areas are covered by edaphic climax (or disclimax) pines, most of the southeastern coastal plain is clearly a distinctive subdivision of the deciduous forest biome rather than a separate major biome. There are extensive ecotones between the deciduous forest and the northern coniferous region (the hemlock–white pine–northern hardwood region), and between the deciduous forest and the grassland (see Figure 14–8).

An excellent summary of the vegetation of deciduous forest communities has been compiled in book form by E. Lucy Braun (1950).

Broad-leaved Evergreen Subtropical Forest Biomes

Where the moisture remains high but temperature differences between winter and summer become less pronounced, the temperate deciduous forest gives way to the broad-leaved evergreen forest climax. This community is well developed in the warm-temperate marine climate of central and southern Japan (see Table 14–6) and may be seen in the "hammocks" of Florida, and in the live oak forests along the Gulf and South Atlantic coasts (Fig. 14–19). A "hammock" is an area of mature soils which has had some protection, at least from fires, thus allowing a climatic climax community to develop. Much of Florida, in common with the southeastern coastal plain in general, does not support the climatic climax but is covered with developmental stages and edaphic climaxes (pines, swamp forests, and marshes, especially) for reasons already mentioned. Thus, Davis (1943) estimated that only 140 thousand acres out of nearly 10 million acres of southern Florida is occupied by "hammock" forests.

Plant dominants of the evergreen broad-leaved forest range from the more northerly live oaks (*Quercus virginiana*), magnolias, bays, and hollies to more tropical species such as strangler-fig (*Ficus aurea*), wild tamarind (*Lysiloma*), and the gumbo limbo (*Bursera*). Palms such as the sabal or cabbage palm (*Sabal palmetto*) are also often prominent. Vines and epiphytes are characteristic. The latter group includes many ferns, the bromeliads (pineapple family), and orchids which support characteristic fauna of small animals. As is generally true of tropical and subtropical communities, dominance in the trophic groups is "shared" by more different species than is the case in northern communities For an ecological

analysis of the laurel and other broad-leaved evergreen forests of Japan, see Kimura (1960) and Kusumoto (1961), as well as the excellent summary paper of Kira and Shidei (1967).

A fine example of the climax broad-leaved forest that is being preserved is Royal Palm Hammock on Paradise Key in the Everglades National Park. It is interesting to note that although there are many plants of tropical origin in this forest, not a single species of land bird or mammal is tropical in origin (there are tropical water birds in the area), apparently because Florida has never been connected with extensive tropical communities to the south. This illustrates again the importance of geography in determining composition of communities.

Temperate Grassland Biomes

Grasslands (Fig. 14–13) cover very large areas and are extremely important from man's viewpoint. Grasslands provide natural pastures for grazing animals, and the principal agricultural food plants have been evolved by artificial selection from grasses. Conversion of moist grasslands into cultivated grain crop grasslands, as in the United States "mid-west," involves relatively little basic change in ecosystem structure and function (mainly a matter of "pushing back" succession to annual stages, as noted in Chapter 9), which may be one reason for man's success in this kind of agriculture. Man's record in using grasslands as pastures, however, is not as good. Many early civilizations developed in grassland regions in conjunction with domesticated grazing animals, but no biome type has probably been abused to a greater degree by man. Even today, the great mass of people fail to understand that natural pastures must be treated with the same care as cultivated ones, and thousands of acres continue to be converted into useless desert by overexploitation (see discussion of range management, pages 417–419, and the concept of the "tragedy of the commons," page 245).

Grasslands occur where rainfall is too low to support the forest life form but is higher than that which results in desert life forms. Generally, this means between 10 and 30 inches, depending on temperature and seasonal distribution. However, grasslands also occur in regions of forest climate where edaphic factors such as a high water table or fire favor grass in competition with woody plants. It is now generally agreed that the original prairies of Ohio, Indiana, and Illinois were "fire climaxes" (see Chapter 9). Temperate grasslands generally occur in the interior of continents (see Figure 14–7). Grassland soils are highly characteristic and contain large amounts of humus, as has been previously indicated (see Figure 5–12).

In North America the grassland biome is divided into east-west zones, i.e., tall grass, mixed grass, short grass, and bunch grass prairies; these are determined by the rainfall gradient, which is also a gradient of decreasing primary productivity. Some of the important perennial species classified according to the height of the above-ground parts are as follows:

Tall grasses (5 to 8 feet) — big bluestem (*Andropogon gerardi*), switchgrass (*Panicum virgatum*), Indian grass (*Sorghastrum nutans*), and in bottomlands the sloughgrass (*Spartina pectinata*).

Mid grasses (2 to 4 feet) — little bluestem (*A. scoparius*), needlegrass (*Stipa spartea*), dropseed (*Sporobolus heterolepis*), western wheatgrass (*Agropyron smithii*), June grass (*Koeleria cristata*), Indian rice grass (*Oryzopsis*), and many others.

Short grasses (0.5 to 1.5 feet) — buffalo grass (*Buchloe dactyloides*), blue grama (*Bouteloua gracilis*), other gramas (*Bouteloua*, sp.), the introduced bluegrass (*Poa*), and cheat grass (*Bromus*, sp.).

Roots of most species penetrate deeply (up to 6 feet) and the weight of roots of healthy climax perennials will be several times that of the tops. Weaver and Zink (1946) found that three years were required for little bluestem and grama grass to develop (starting from seedling) maximum standing crop of roots (the root/shoot ratio is higher than in most forests); after that, there was no further increase, the annual growth equaling the annual loss. The growth form of the roots is important. Some of the above species — for example, big bluestem, buffalo grass, and wheatgrass — have underground rhizomes and are thus sod-formers. Other species, such as little bluestem, June grass, and needlegrass, are bunchgrasses and grow in clumps. These two life forms may be found in all zones, but bunchgrasses predominate in the drier regions where grassland grades into desert.

The geographical origin of species is of special ecological importance in grasslands. Species of northern genera, such as *Stipa*, *Agropyron*, and *Poa*, renew growth early in the spring, reach maximum development in late spring or early summer (when seeds are produced), become semidormant in hot weather but resume growth in autumn and remain green despite frosts. Warm-season species of southern genera, on the other hand, such as *Andropogon*, *Buchloe*, and *Bouteloua*, renew growth late in the spring, but grow continuously during the summer, reaching maximum standing crop in late summer or autumn with no further growth during the latter period. From the standpoint of annual productivity of the whole ecosystem over a period of years, a mixture of cool- and warm-season grasses is favorable especially since rainfall is likely to be most abundant in spring or fall in some years and in midsummer in other years. Replacement of such adapted mixtures by "monocultures" is bound to create oscillations in productivity (another simple ecological fact not comprehended even by agriculturists!).

Forbs (composites, legumes, and so forth) generally comprise but a small part of the producer biomass in climax grasslands, but are consistently present. Certain species are of special interest as indicators (i.e., they have a high "fidelity"; see Hanson, 1950). Increased grazing or drought, or both, tends to increase the percentage of forbs which are also prominent in early seral stages. Secondary succession in the grassland biome, and the rhythmic changes in vegetation during wet and dry cycles have been described in earlier chapters (see pages 262 and 268). The automobile traveler in mid-continent United States should bear in mind that the conspicuous annual forbs of the roadsides, such as the Russian thistle tumbleweed (*Salsola*) and sunflowers (*Helianthus*), owe their luxuriance to man's continual disturbance of the soil.

The North American tall grass prairies have now been replaced by grain agriculture or cultivated pastures, or have been invaded by woody vegetation. Original or virgin tall grass prairie is hard to find, and where preserved for study (as in the University of Wisconsin Arboretum), it must be burned to preserve its prairie character.

A large proportion of mammals of the

Figure 14–13. Natural temperate grassland in central North America with two original mammalian herbivores and two of the native grass dominants. *A.* Lightly grazed grassland in the Red Rock Lakes National Wildlife Refuge, Montana, with small herd of pronghorn antelope. *B.* Short-grass grassland, Wainwright National Park, Alberta, Canada, with herd of bison. The animal in the center is wallowing; old "buffalo wallows" often may be detected in the grassland many years after the bison have been extirpated.

grassland are either running or burrowing types. Aggregation into colonies or herds is characteristic; this life habit provides some measure of protection in the open type of habitat. The important large grazing animals native to grasslands in various biogeographic regions were listed on page 239, and the importance of migratory behavior in preventing overgrazing was mentioned in the preceding section of this chapter. Burrowing rodents, such as ground squirrels, prairie dogs, and gophers, are important. Where man has reduced the population of rodent predators such as coyotes, kit foxes, and badgers, a "rodent epidemic" often results; likewise rodents increase when the grassland is overgrazed by cattle or other hoofed animals. Characteristic birds of North American grasslands include the prairie chickens, meadowlarks, longspurs, horned larks, and rodent-eating hawks; ecologically equivalent species occur in other parts of the world.

The large grazing mammals, whether native or domesticated, are as important as basic climate and soil conditions in determining the floristic composition of the community, since some species of grasses and other plants are more sensitive to grazing pressure than others. Range managers use the term "decreasers" for palatable species sensitive to grazing; their disap-

Figure 14–13. *C*. Two important climax grasses of the grassland biomes of North America. *Left:* Little blue-stem (*Andropogon scoparius*), the most important native grass of the eastern part of the grassland biome. *Right:* sideoats grama (*Bouteloua curtipendula*), a widely distributed "grama" grass in the mixed and short grass areas to the west. (*A*, U. S. Fish and Wildlife Service Photo; *B*, by E. P. Odum; *C*, U. S. Soil Conservation Service Photo.)

pearance is an indicator of grazing stress. Bradshaw (1957) has demonstrated that in the man-made grasslands of England, some of which have been overgrazed for centuries, some grass species develop ecotypes with inherent low productivity even when transplanted to ungrazed experimental gardens. Under heavy grazing pressure it is of survival value for the plant to grow slowly (even when moisture and other conditions would favor rapid growth) and thus avoid being killed completely by the grazing animal! Similar low-producing ecotypes were found in alpine regions where extreme conditions had the same effect as intense grazing. In these areas inherent productivity of species and probably of the whole ecosystem as well becomes just as much an adaptive characteristic as any structural feature. Likewise the effects of prairie fires and of primitive man must be considered in interpreting the present-day conditions (see Sauer, 1950; Malin, 1953). For general discussions of the grassland biome in North America, see Chapter 8 in Clements and Shelford (1939), Carpenter (1940), Weaver (1954), Weaver and Albertson (1956), and Malin (1956). The grassland biome has been selected for the first

major IBP (International Biological Program) study, which will involve interdisciplinary teams of researchers and a systems analysis approach (see Van Dyne, 1969; Coupland *et al.*, 1969; and Wiegert and Evans, 1967). For a mathematical model and a large-scale simulation model of a grassland ecosystem, see Figures 10–4 and 10–5, pages 287–288.

Tropical Savanna Biomes

Tropical savannas (grasslands with scattered trees or clumps of trees) are found in warm regions with 40 to 60 inches of rainfall but with a prolonged dry season when fires are an important part of the environment. The largest area of this type is in Africa (see Figure 14–9), but sizable tropical savannas or grasslands also occur in South America and Australia. Since both trees and grass must be resistant to drought and fire, the number of species in the vegetation is not large, in sharp contrast to adjacent equatorial forests. Grasses belonging to such genera as *Panicum, Pennisetum, Andropogon,* and *Imperata* provide the dominant cover, while the scattered trees are of entirely different

Figure 14–14. A view of the tropical savanna of Africa. Grass, scattered trees, which have picturesque shapes, dry season fires, and numerous species of large mammalian herbivores are unique features of this. (Photograph by Donald I. Ker, Ker & Downey Safaris Ltd., Nairobi, East Africa.)

species from those of the rain forest. In Africa, the picturesque acacias, baobab trees (*Adansonia*), arborescent euphorbias, and palms dot the landscape. Often single species of both grass and trees may be dominant over large areas.

In number and variety the population of hoofed mammals of the African savanna is unexcelled anywhere in the world (recall Section 6 of this chapter). Antelope (numerous species), wildebeest, zebra, and giraffe graze or browse and are sought by lions and other predators in areas in which the "big game" has not been replaced by man and his cattle. Insects are most abundant during the wet season when most birds nest, whereas reptiles may be more active during the dry season. Thus, seasons are regulated by rainfall rather than by temperature as in the temperate grasslands.

Land-use decisions regarding the African savannas and their diversity of ungulate grazers soon have to be made by the emerging nations of that area. Many ecologists believe it would be feasible to harvest antelope, hippopotamuses, and wildebeests on a sustained-yield basis, or perhaps to semi-domesticate them, rather than to exterminate them and substitute cattle (Petrides, 1956; Darling, 1960; Dasman and Mossman, 1962; Talbot *et al.*, 1965), not only because of the diversified use they make of natural primary production, but also because the wild animals are immune to many tropical diseases and parasites to which cattle are highly susceptible. Here again is another good place to recall the principles discussed in Chapter 3. Yields from an intensively managed cattle farm would always exceed those which could be obtained by harvesting wild populations, but the increased yield has its hidden costs and energy subsidies in the form of disease control, vegetation control and human labor and fuel. Decisions, then, should be based on total "trade-off" cost balances. Wild game ranching should certainly be given a fair trial.

A view of the African savanna country—including the three conspicuous components, namely, grass, scattered trees, and mammalian herbivores—is provided in Figure 14–14. Ecological problems of tropical grasslands are discussed in the many UNESCO reports and some of the interesting features of this biome are pictured by Aubert de la Rue *et al.* (1957).

Desert Biomes

Deserts (Fig. 14–15) generally occur in regions having less than 10 inches of rainfall, or sometimes in regions with greater rainfall that is very unevenly distributed. Scarcity of rainfall may be due to (1) high subtropical pressure, as in the Sahara and

Australian deserts, (2) geographical position in rain shadows (see Figure 5–8), as in the western North American deserts, or (3) high altitude, as in Tibetan, Bolivian, or Gobi deserts. Most deserts receive some rain during the year and have at least a sparse cover of vegetation, unless edaphic conditions of the substrate happen to be especially unfavorable. Apparently the only absolute deserts where little or no rain falls are those of the central Sahara and north Chile.

Walter (1954) has measured net production of a series of deserts and semiarid communities that lie along a rainfall gradient in Africa. As shown in Figure 14–16, the annual production of dry matter was a linear function of rainfall, at least up to 600 cms (24 inches), illustrating the sharpness with which moisture acts as an overall ("master") limiting factor. Note that annual net primary productivity of true deserts is less than 2000 kg per hectare, or a daily rate of less than 0.5 gm or 2.5 kcal per m².

When deserts are irrigated and water no longer is a limiting factor, the type of soil becomes of prime consideration. Where texture and nutrient content of the soil are favorable, irrigated deserts can be extremely productive because of the large amount of sunlight (see Table 3–4). However, the cost per pound of food produced may be high because of the high cost of development and maintenance of irrigation systems. Very large volumes of water must flow through the system; otherwise salts may accumulate in the soil (as a result of rapid evaporation rate) and become limiting. As the irrigated ecosystem "ages," increased water demands may bring on an "inflationary spiral," requiring more aqueducts to be built, higher costs of production, and greater exploitation of the underground or mountain water sources. Old World deserts are full of ruins of old irrigation systems. In many cases no one knows why they failed and why the "Garden of Eden" literally became a desert again. At least these ruins should warn us that the irrigated desert will not continue to bloom indefinitely without due attention to the basic laws of the ecosystem.

There are three life-forms of plants that are adapted to deserts: (1) the annuals, which avoid drought by growing only when there is adequate moisture (see pages 116, 366), (2) the succulents, such as the cacti, which store water, and (3) the desert shrubs, which have numerous branches ramifying from a short basal trunk bearing small, thick leaves that may be shed during the prolonged dry periods. The desert shrub presents very much the same appearance throughout the world even though species may belong to diverse taxa (another striking example of ecological equivalence). As has already been explained (see page 122), adaptation to arid conditions involves the ability to avoid wilting and remain dormant for long periods rather than a marked increase in transpiration efficiency (that is, the ratio of dry matter produced to water transpired in desert plants is somewhat but not markedly greater than for many non-desert plants).

From an ecological standpoint it is convenient to distinguish two types of deserts on the basis of temperature, namely, hot deserts and cool deserts. All desert vegetation has a highly characteristic "spaced" distribution in which individual plants are scattered thinly with large bare areas in between (see Figure 14–15). In some cases, antibiotics play a part in keeping plants spaced apart. In any event, spacing reduces competition for a scarce resource; otherwise intense competition for water might result in the death or stunting of all of the plants.[*]

In North America, the creosote bush (*Larrea*) is a widespread dominant of the southwestern hot desert (Fig. 14–15A), and sagebrush (*Artemisia*) is the chief plant over large areas of the more northern cool deserts of the Great Basin (Figure 14–15C). Bur sage (*Franseria*) is also widespread in southern areas, whereas at higher altitudes where moisture is a little greater, the giant cactus (Sahuaro) and palo verde are conspicuous components (Figure 14–15B). Eastward, a considerable amount of grass is mixed with desert shrubs to form a desert grassland type; unfortunately the grass has suffered greatly from overgrazing, fire protection, or both. In the cool deserts, especially on the alkaline soils of the internal drainage regions, saltbushes of the family

---

[*] One wonders if mankind might not learn a lesson from the desert, where the relationship between population size and resources is sharply drawn. Certainly, a human culture which allows its density to exceed resources to the extent of intellectual or physical stunting, or both, of all individuals demonstrates poor adaptation. "Birth control" in the desert is a perfectly natural thing which allows maximum development of individuals.

Figure 14–15. Three types of deserts in western North America. *A.* A low altitude "hot" desert in southern Arizona dominated by creosote bush (*Larrea*). Note the characteristic growth form of the desert shrub (numerous branches ramifying from ground level) and the rather regular spacing. *B.* An Arizona desert at a somewhat higher altitude with several kinds of cacti and a greater variety of desert shrubs and small trees. This is one of the sites selected for IBP interdisciplinary study.

Chenopodiaceae, such as shadscale (*Atriplex*), hop sage (*Grayia*), winter fat (*Eurotia*), and greasewood (*Sarcobatus*), occupy extensive zones. In fact "chenopods" are widely distributed in arid regions in other parts of the world as well. The sagebrush and shadscale communities have already been described in Chapter 6. The succulent life form, including the cacti and the arborescent yuccas and agaves, reaches its greatest development in the Mexican desert; (and in the Neotropical region), with some species of this type extending into the shrub deserts of Arizona and California; but this life form is unimportant in the cool deserts. In all deserts annual forbs and

Figure 14–15. *C.* A "cool" desert in eastern Washington dominated by sagebrush (*Artemisia*). The picture was taken during the spring at a time of peak primary production when annual grasses and forbs carpet the spaces between and around the shrubs. In progress was a radioactive tracer experiment designed to compare uptake of shrubs and annuals. (*A*, photo by R. R. Humphrey; *B*, photo by R. H. Chew; *C*, Hanford Atomic Products Operation photo.)

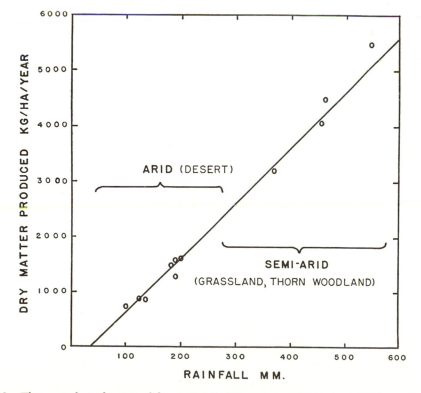

Figure 14–16. The annual production of dry matter (in kilograms per hectare per year) in a series of arid and semi-arid zones of vegetation which lie along a rainfall gradient in West Africa. (After Walter, 1954.)

grasses may make quite a show during brief wet periods. The extensive "bare ground" in deserts is not necessarily free of plants. Mosses, algae, and lichens may be present, and on sands and other finely divided soils they may form a stabilizing crust; also the bluegreen algae (often associated with lichens) are of importance as nitrogen-fixers (Shields *et al.*, 1957).

Desert animals as well as plants are adapted in various ways to lack of water. Reptiles and some insects are "pre-adapted" because of relatively impervious integuments and dry excretions (uric acid and guanine). Desert insects are "water-proofed" with substances which remain impermeable at high temperatures (Edney, 1957). While evaporation from respiratory surfaces cannot be eliminated, it is reduced to the minimum in insects by the internally invaginated spiracle system. It should be pointed out that the production of metabolic water (from breakdown of carbohydrates), often the only water available, is not in itself an adaptation; it is the conservation of this water that is adaptative, as is, in the case with tenebrionid beetles (a characteristic desert group), the ability to increase production of metabolic water at low humidities (Edney, 1957). Mammals, on the other hand, are not very well adapted as a group (because they excrete urea, which involves the loss of much water), yet certain species have developed remarkable secondary adaptations. These include rodents of the family Heteromyidae, especially the kangaroo rat (*Dipodomys*) and the pocket mouse (*Perognathus*) of the New World deserts and the jerboa (*Dipus*, family Dipodidae) of Old World deserts. These animals can live indefinitely on dry seeds and do not require drinking water. They remain in burrows during the day, and conserve water by excreting very concentrated urine and by not using water for temperature regulation. The Schmidt-Nielsens (1949) found that relative humidity in the burrow was 30 to 50 per cent, compared with 0 to 15 per cent in the desert above ground during the daytime. The relative humidity in the desert at night when the animals are above ground was about the same as that in the burrows during the day. Thus, adaptation to deserts by these rodents is as much behavioral as physiological. Other desert rodents, wood rats (*Neotoma*), for example, are unable to live solely on dry food, but survive in parts of the desert by eating succulent cacti or other plants that store water. In his review of the metabolism of desert mammals, Schmidt-Nielsen (1964) points out that these mammals are of two types, those such as the kangeroo rat or jerboa, which do not use water for temperature regulation, and those such as the camel which do. The latter must drink although they can go for long periods without water because the body tissues can tolerate elevation in body temperature and a degree of dehydration that would be fatal to most animals (camels do not store water in the hump, as is popularly supposed!). Desert birds seem to belong to this category (Bartholomew and Dawson, 1953) and probably need at least an occasional drink from dew or other sources; hence birds are most abundant where water or succulent foods are available. Man, of course, is physiologically very poorly adapted to deserts.

The vegetation of the North American deserts has been well described by Shantz (1942), Billings (1951), Shreve (1951), and many others. In going through the vast amount of literature summarized in the UNESCO reports, one is impressed with the fact that almost all of the information on the deserts of the world is purely descriptive in nature; there is less known about the actual "workings" of desert ecosystems. Since microbial decomposers will be sharply limited by dryness, one wonders if this is compensated for by a seemingly large population of rodent herbivores, which, perhaps similar to the zooplankton of the sea, play an important part in nutrient cycling. As in all ecosystems adapted to extreme conditions, a relatively large amount of net production goes into storage or reproductive organs, thus providing a food source for consumers. Among recent studies on deserts which emphasize the functional approach are the following: Chew and Chew, 1965; Niering, *et al.*, 1963.

Chaparral Biomes

In mild temperate regions with abundant winter rainfall but dry summers, the climax vegetation consists of trees or shrubs with hard, thick evergreen leaves. Under this heading we are including both the chaparral proper or "coastal chaparral" in which the shrub life-form is predominant and the "broad-sclerophyll woodland," which contains scattered trees. A "picture" of the climate of the winter rain region is shown in Figure 5–9. Chaparral communities are extensive in California and Mexico, along the shores of the Mediterranean Sea and along the southern coast of Australia (Fig. 14–17). A large number of plant species may serve as dominants, depending on the region and local conditions. Fire is an important factor that tends to perpetuate shrub dominance at the expense of trees. Thus, the chaparral may be partly, at least, a "fire disclimax," as is the long-leaf pine forest of the southeastern states.

In California, about five to six million acres of slopes and canyons are covered with chaparral. Chamiso (*Adenostoma*) and manzanita (*Arctostaphylos*) are common shrubs which often form dense thickets, and a number of evergreen oaks are characteristic, also, as either shrubs or trees. The rainy or growing season generally extends from November to May. Mule deer and many birds inhabit the chaparral during this period, then move north or to higher altitudes during the hot dry summer. Resident vertebrates are generally small and dull-colored to match the dwarf forest; the small brush

Figure 14–17. A. General view of the chaparral covered hills of California. B. Fighting a small fire in this "fire-type" vegetation. (U. S. Forest Service Photos.)

rabbits (*Sylvilagus bachmani*), wood rats, chipmunks, lizards, wren-tits, and brown towhees are characteristic. The population density of breeding birds and insects is high as the growing season comes to a close, then decreases as the vegetation dries out in late summer. This is the season when fires may sweep the slopes with incredible swiftness. Following a fire, the chaparral shrubs sprout vigorously with the first rains and may gain maximum size in 15 to 20 years. The role of fire in California chaparral has been reviewed by Sweeney (1956), who reports that fire stimulates the germination of some seeds and is essential to the persistence of certain herbaceous species in the flora. The general structure of the vegetation has been well described by Cooper (1922), while a brief discussion of both plants and animals of the chaparral biome in California is given by Cogswell (1947). (See also Figure 7–5.)

Because some of the chaparral slopes are relatively frost-free they are used for orchard culture. The greatest value to man, however, lies in watershed protection. Not only do the steep, loosely consolidated slopes erode easily, but disastrous floods are propelled into the thickly populated lowlands if the chaparral community does not remain in a healthy condition. Hellmers, Bonner, and Kelleher (1955) found that nitrogen, as well as moisture, was a limiting factor; these authors suggested that density of chaparral vegetation on slopes above cities might be increased by nitrogen fertilization.

The chaparral of the winter rain areas of the Mediterranean region is locally called "macchie," while similar vegetation in Australia is called "mellee scrub." In the Australian chaparral trees and shrubs of the genus *Eucalyptus* are dominant. It is not surprising that Australian "eucalypts" do well in California chaparral region where they have been widely introduced and largely replace the native woody vegetation in urban areas.

Piñon-juniper Biome

The piñon-juniper woodland, or the "pigmy conifers" (Fig. 14–18), appear to occupy a sufficiently large area in the interior of the Great Basin and Colorado River regions of Colorado, Utah, Arizona, New Mexico, Nevada, and west-central California and to have a sufficiently distinctive biota to be considered a biome. Alternatively, this community type could be considered a subdivision of the northern coniferous forest biome. Moisture is the critical factor, the 10 to 20 inches of unevenly distributed rainfall accounting for the parklike

Figure 14–18.   The piñon-juniper or pygmy conifer biome in Arizona. The small piñon pines and cedars form an open, parklike stand. (U. S. Forest Service Photo.)

growth of small piñon pines (*Pinus edulis* and *P. monophylla*) and cedars (several species of the genus *Juniperus*) as shown in Figure 14–18. This community occupies a wide belt between desert or grassland and the heavier coniferous forests of higher altitudes. The piñon "nuts" and cedar berries are important food for animals. The piñon jay, the gray titmouse, and the bush tit are characteristic permanent resident birds, the first two being largely restricted to this biome. For additional information on the pigmy conifers, see Woodbury (1947) and Hardy (1945).

Tropical Rain Forest Biomes

The variety of life perhaps reaches its culmination in the broad-leaved evergreen tropical rain forests which occupy low altitude zones near the equator. Rainfall exceeds 80 or 90 inches a year and is distributed over the year, usually with one or more relatively "dry" seasons (5 inches per month or less). Rain forests occur in three main areas: (1) the Amazon and Orinoco basins in South America (the largest continuous mass) and the Central American isthmus, (2) the Congo, Niger, and Zam-

bezi basins of central and western Africa, and Madagascar, and (3) the Indo–Malay–Borneo–New Guinea regions. These differ from each other in the species present (since they occupy different biogeographical regions), but the forest structure and ecology are similar in all three areas. The variation in temperature between winter and summer is less than that between night and day. Seasonal periodicities in breeding and other activities of plants and animals are largely related to variations in rainfall, or regulated by inherent rhythms. For example, some trees of the family Winteraceae apparently show continuous growth, while other species in the same family show discontinuous growth with formation of rings (Studhalter, 1955). Rain forest birds may also require periods of "rest" since reproduction often exhibits periodicity unrelated to season (Miller, 1955).

The rain forest is highly stratified. Trees generally form three layers: (1) scattered, very tall emergent trees that project above the general level of the (2) canopy layer, which forms a continuous evergreen carpet 80 to 100 feet tall, and (3) an understory stratum that becomes dense only where there is a break in the canopy. As shown in Figure 14–20, the emergent and canopy

Figure 14–19. A live oak (*Quercus virginiana*) forest with epiphytic *Tillandsia* (Spanish moss) in Louisiana, an example of a subtropical broad-leaved evergreen forest. (Standard Oil Co. photo.)

Figure 14–20. The tropical rain forest. *A.* View from above of a lowland rain forest in Panama. The tall emergent trees (with white trunks), which lose their leaves in the dry season, project above the general canopy of broad-leaved evergreen hardwoods and palms. *B.* Interior of a montane rain forest in Puerto Rico showing the abundance of epiphytes that characterize the rain forest in the mountains of the humid tropics. *C.* Close-up view of the base of a buttressed tree in this same Puerto Rican montane forest showing the above-ground mass of small roots that permeate the litter (rather than the soil as in temperate forests). (*A,* photo by George Child, Institute of Ecology, University of Georgia; *B* and *C,* photos by E. P. Odum.) (See Frontispiece for an infrared picture of rain forest.)

layers give the rain forest a "bumpy" look when viewed from the air. If rainfall during the dry season is less than 2 inches per month, the emergent trees may lose their leaves during the dry season, resulting in what may be called a *semi-evergreen rain forest*. Shrub and herb layers, often containing numbers of ferns and palms, tend to be less massive because of the dense shade, but they respond quickly to any opening in the overstory layers. The tall trees are shallow rooted and often have swollen bases or "flying buttresses." There is a profusion of climbing plants, especially woody lianas and epiphytes, which often hide the outline of the trees. The "strangler figs" and other arborescent vines are especially noteworthy. The striking difference in distribution of life-forms in the rain forest as compared with temperate forests is illustrated in Figure 14–2. The number of species of plants is very large; often there will be more species of trees in a few acres than in the entire flora of Europe.

As was discussed in Chapter 3, it seems likely that the productivity of the intact rain forest is very high so long as there is a rapid cycling of scarce nutrients. The theory of cycling and the contrast with temperate forests has been documented on pages 102–103, and 232. The tropical rain forest is perhaps the only major vegetation type in which fire is not a factor; in fact, the rain forest might be defined as the "forest that never burns" in its natural or intact condition. Man, of course, does "slash and burn" in his efforts to convert rain forest to agricultural areas, as discussed on page 103.

A much larger proportion of animals lives in the upper layers of the vegetation than in temperate forests where most of life is near the ground level. For example, 31 of 59 species of mammals in British Guiana are arboreal and 5 are amphibious, leaving only 23 which are mainly ground dwellers (Haviland, 1927). In addition to the arboreal mammals, there is an abundance of chameleons, iguanas, geckos, arboreal snakes, frogs, and birds. Ants and the orthoptera, as well as butterflies and moths, are ecologically important. Symbiosis between animals and ephiphytes is widespread. As is true of the flora, the fauna of the rain forest is incredibly rich in species. For example, in a six square mile area on Barro Colorado, a well-studied bit of rain forest in the Panama Canal Zone, there are 20,000 species of insects compared to only a few hundred in all

of France. Numerous archaic types of both animals and plants survive in the numerous niches of the unchanging environment. Many scientists feel that the rate of evolutionary change and speciation is especially high in the tropical forest regions, which, therefore, have been a source of many species that have invaded more northern communities.

Fruit and termites are staple foods for animals in the tropical forest. One reason why birds are often abundant is that so many of them, such as fruit-eating parakeets, toucans, hornbills, cotingas, trogons, and birds-of-paradise, are herbivorous. Because the "attics" of the jungle are crowded, many bird nests and insect cocoons are of a hanging type, enabling them to escape from army ants and other predators. Although there are some spectacularly bright birds and insects occupying the more open situations, the vast array of rain-forest animals are inconspicuous and many of them are nocturnal.

In the mountainous areas of the tropics is a variant of the lowland rain forest, the *montane rain forest*, which has some distinctive features. The forest becomes progressively less tall with increasing altitude, and epiphytes make up an increasingly larger proportion of the autotrophic biomass, culminating in the dwarf *cloud forests*. Possible reasons for this change in structure with altitude were discussed on page 376. A functional classification of rain forests can be based on *saturation deficit* because this determines transpiration, which in turn determines root biomass and the height of trees. Above 6000 to 10,000 feet altitude, forest vegetation in the tropics is not unlike the temperate coniferous forest.

Still another variant of the rain forest occurs along banks and flood plains of rivers and is called the *gallery forest*, or sometimes the *riverine forest*. Golley *et al.* (1969) and McGinnis *et al.* (1969) have recently completed comprehensive studies of biomass structure and the mineral and hydrologic budgets of three kinds of Panamanian rain forests: the lowland type (see Figure 14–20), the montane, and the gallery type (also mangrove forests; see Figure 12–14).

When the rain forest is removed, a secondary forest often develops that includes soft wood trees, such as *Musanga* (Africa), *Cecropia* (America), and *Macoranga* (Malaysia). The secondary forest looks lush but is quite different from the virgin forest, both ecologically and floristically. The "climax" is usually very slow to return. As we have already indicated, failure to regenerate may be related to nutrient losses. There is urgent need to preserve tracts of rain forest not only to serve as laboratories for the study of ecology and evolution, but also to provide a basis for improving man's dismal record of land use in the rain-forest regions.

For additional information on the rain-forest structure, see Richards (1952) and Bunning (1956). For a beautifully illustrated descriptive ecology of both plants and animals of the rain forest and other tropical biomes, see the book of Aubert de la Rue, Bourliere, and Harroy (1957). The prime reference on the functional ecology of the rain forest is the monograph compiled by H. T. Odum and Pigeon (1970).

Tropical Scrub and Deciduous Forest Biomes

Where moisture conditions are intermediate between desert and savanna on one hand and rain forest on the other, tropical scrub or thorn forests and tropical deciduous forests may be found. As shown in Figure 14–7, these cover very large areas. The key climatic factor is the imperfect distribution of a fairly good total rainfall. Thorn forests, which are often referred to as "the bush" in Africa or Australia and the "Caatinga" in Brazil, contain small hardwood trees often grotesquely twisted and thorny; leaves are small and are lost during dry seasons. Where the precipitation is greater or more regular a well-developed deciduous forest is found, such as the widespread monsoon forests of tropical Asia. Wet and dry seasons of approximately equal length alternate so that the seasonal appearance of "winter" and "summer" is as striking as in a temperate deciduous forest.

Zonation in Mountains

The distribution of biotic communities in mountainous regions is complicated, as would be expected in view of the diversity of physical conditions. Major communities generally appear as irregular bands, often with very narrow ecotones (or none, in the sense of a transition community with characteristics of its own; see Chapter 6). Small-scale maps such as those shown in Figures 14–7 through 14–9 are inadequate to bring out this feature. On a given mountain four or five major biomes with many zonal subdivisions may be present (Fig. 14–21). Consequently there is closer contact between biomes and more interchange of biota between different biomes than occurs in nonmountainous regions. On the other hand, similar communities are more isolated in the mountains since mountain ranges are rarely continuous. In general, many species that are characteristic of a biome in its broad, nonmountainous phase are also characteristic of the beltlike extensions in the mountains. As a result of isolation and topographic differences many other species are unique to the mountain communities. (See Bliss, 1956, for a comparison of alpine and arctic tundra.)

Importance of the Historical Approach

It should be emphasized that the biome classification as outlined and illustrated in the previous pages is based on the *potential* conditions which result from natural forces that are now operating in different parts of the world. Such a classification provides a point of reference for evaluating man's influence on the environment. Only if we know something of the natural potentialities and limiting factors can we determine best usage. Emerging new ecosystems created by man, such as grassland agriculture, irrigated deserts, water impoundments, and tropical agriculture, will persist for long periods of time only if material and energy balances are achieved between biotic and physical components. Therefore, it is important to preserve samples of natural communities in each biome for comparison with modified areas. Furthermore, as brought out in Chapter 9, Section 3, a certain proportion of the landscape must remain in "natural area" use in order to safeguard the highly modified cultivated and urban areas. Also, it is equally important that the past history of an area be carefully studied. Malin (1953) points out that "casual or random excursions into historical material to find data that appear to fit a preconceived frame of reference, or to serve a particular purpose," are to be avoided. Historical research in ecology must be conducted with an open mind and without a preconceived

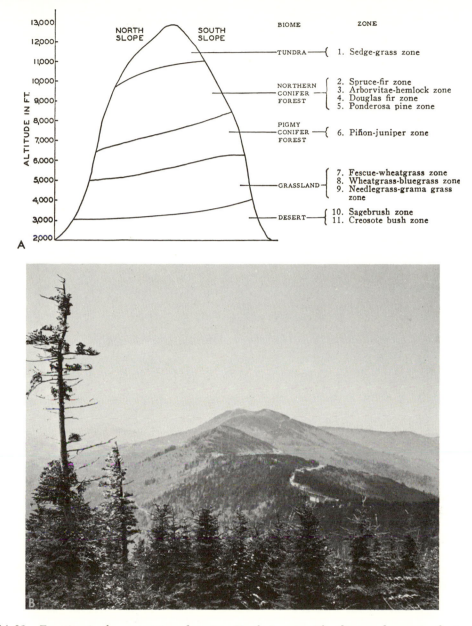

ALTITUDE IN FT.

BIOME	ZONE
TUNDRA	1. Sedge-grass zone
NORTHERN CONIFER FOREST	2. Spruce-fir zone 3. Arborvitae-hemlock zone 4. Douglas fir zone 5. Ponderosa pine zone
PIGMY CONIFER FOREST	6. Piñon-juniper zone
GRASSLAND	7. Fescue-wheatgrass zone 8. Wheatgrass-bluegrass zone 9. Needlegrass-grama grass zone
DESERT	10. Sagebrush zone 11. Creosote bush zone

Figure 14–21. Zonation in the mountains of western North America. The diagram does not refer to a specific mountain but shows general conditions which might be expected in the central Rockies (so-called "intermountain" region) of Utah. North, south, east, or west of this region zonation may be expected to vary somewhat. Thus, the creosote bush and the piñon-juniper zone are absent northward and eastward. Eastward an oak-mountain mahogany zone or westward various zones in the chaparral biome occur between the grassland and the northern conifer forest. *B.* Zonation in the eastern Blue Ridge mountains of North Carolina. The dark spruce zone along the ridge contrasts with the deciduous forest below. (*A,* diagram modified from Woodbury, 1945; zone data from Daubenmire, 1946; *B,* U. S. Forest Service Photo.)

goal if we are to fully understand the biotic communities as we find them today. What often appears on superficial examination to be undisturbed "virgin" vegetation may actually be highly modified by previous activity of man or his domestic animals. For example, what appears to be virgin "jungle" in Panama may really be secondary forest covering what once were densely populated, intensively cultivated areas. As man turns again to "reclaiming" these jungles, one wonders if anything was learned from past abuses, or how many times civilizations can rise and fall before the environment is no longer capable of recovery.

Part 3

APPLICATIONS AND TECHNOLOGY

Since the publication of the second edition of this textbook in 1959 there has been a dramatic shift in emphasis in applied ecology from the population level to the ecosystem level (or from autecology to synecology to use the older terminology; see page 6). Although the ecosystem and the holistic viewpoint were stressed even in the first edition (1953), most applications of ecological principles prior to 1960 pertained to the management or control of specific resources or species, such as water, soil, timber, game, fish, crops, insect pests, and so forth. Now, in addition to these, application centers around the ecosystem—the totality of air and water cycles, productivity, food chains, global pollution, systems analysis, and the control and management of man as well as of nature. Practice, as it were, has rapidly caught up with theory! The basic reasons for such a shift in emphasis were discussed in their ecological context in Chapter 9, Section 3. For the first time in his short history man is faced with ultimate rather than local limitations. The prediction of Thomas R. Malthus (the first edition of his famous *Essay on Population* was published in 1798) that population increases faster than the food supply takes on a new and more sinister form as we observe that "populution" (i.e., population plus pollution) increases faster than total resource requirements, not just food, can be produced and recycled. As a consequence, both the public at large and the specialist, who previously was knowledgeable about and concerned with only a very

small part of the environmental system, have become dedicated students of the basic ecosystem principles as outlined in Parts 1 and 2!

General realization that the "supply depot" and the "living space" functions of one's environment are interrelated, mutually restrictive, and not unlimited in capacity has amounted to a historic "attitude revolution" (E. P. Odum, 1969*b* and 1970*a*), which is a promising sign that man may be ready to "apply" the principles of ecological control on a large scale. As social critic Lewis Mumford has so well phrased it, "Quality in control of quantity is the great lesson of biological evolution" and "Ideological misconceptions have impelled us to promote the quantitative expansion of knowledge, power, productivity without inventing any adequate systems of control" (Mumford, 1967). From now on, then, applied ecology is primarily a matter of developing the necessary negative feedback control, as was discussed in our introduction to cybernetics in Chapter 2, pages 33–34.

Among the outpouring of rhetoric that marked the close of the 1960s there are five articles that we would wish it possible for every citizen to read. Three of these have already been mentioned: Lynn White's *The Historical Roots of Our Ecological Crisis* (see page 36), Garrett Hardin's *The Tragedy of the Commons* (see page 245), and Paul Ehrlich's *The Population Bomb* (see page 56). The other two are written by thoughtful economists: *The Economics of the Coming Spaceship Earth* by K. E. Bould-

405

ing (1966) and *Growth versus the Quality of Life* by J. Alan Wagar (1970). The consensus of these and other similar articles can be stated as follows: *Technology alone cannot solve the population and pollution dilemma; moral, economic, and legal constraints arising from full and complete public awareness that man and the landscape are a whole must also become effective.* Therefore, the applied ecology of the future will be based partly on natural science and partly on social science. In other words, C. P. Snow's (1959) two subcultures, the "scientific" community and the "humanistic" community, which have for too long remained apart (especially in academia where they are too quick to blame each other for the ills of the university as well as of society), must now integrate! (We might make an analogy here with pairs of differential equations, as discussed in Chapters 7 and 10; you cannot do anything with them, that is, apply or use them, until they are integrated.) Accordingly, success or failure in applying the principles of ecology for the benefit of man, at least for the next decade, may depend not so much on technology and environmental science as such, but on economics, law, politics, planning, and other areas in the humanities that *have up to now had very little ecological input.* For example, assimilation of environmental principles and concepts into environmental law is just now more urgent than research in oceanography insofar as the conservation of the sea is concerned. Or, urban planning along ecological lines, as outlined in McHarg's beautifully illustrated book *Design with Nature* (1969), is much more urgent than is the application of principles of population ecology to game management.

Since specific applications were frequently cited in accordance with ecological principles outlined in Part 1 and environments surveyed in Part 2, a broad overview of applied ecology will be presented in Part 3, together with some samples of the kind of advanced technology that will be needed in the near future. The positive (i.e., "resources") and the negative (i.e., "pollution") aspects will be considered first, followed by a series of chapters that review recent advances in key technology. These chapters, together with Chapter 10 ("Systems Ecology"), introduce some of the *means* by which whole ecosystems can be studied and moni-

tored so that negative feedback constraints can be established. As already emphasized, the *motivation* for the latter lies outside technology and science, and must evolve out of social behavior, education and politics. For this reason the last chapter in this book is entitled "Towards an Applied Human Ecology." In my view, at least, "human ecology" is more or less what this whole book is about. What is new in concept, and not yet developed even in theory, is the means of applying ecological principles to the management of the human population as a part of that self-contained ecosystem, the biosphere.

It is perhaps appropriate to close this introduction with a word about training, since the demand for applied ecologists will obviously be very great from now on. Increased public awareness of ecology tempts educational institutions to set up special departments and curricula (get on the bandwagon, so to speak). At the college level this could be a mistake, unless such special programs are well planned, because most universities are already "overdepartmentalized"—they already have too many departments which often teach at cross-purposes (one reason for the present student dissatisfaction with colleges). To create simply another department or degree program could do more harm than good in the long run. The unique feature and great strength of ecology lies in the fact that it is a science of synthesis which is capable of integrating the sciences and the humanities, as we have already noted. For this reason training should be interdisciplinary; but the ecologist must also be well trained in a relevant specialty. *Consequently, special courses in ecology and applied ecology are needed at all levels from grade school to graduate school, but the training of the professional student should be carried out in an open, not a closed, system.* The best advice we can give the interested student is as follows: (1) Obtain a good background education in arts and sciences with emphasis on quantitative subjects. (2) Select for a major a specialty on the basis of your interest and talents, and in terms of its relevance to man-in-environment problems; such a specialty could be biology, chemistry, or political science, or one of the more professionalized subjects such as forestry, science education, engineering, or law. (3) Enroll in an interdisciplinary degree program that will enable you to get a degree in the specialty of your choice *and* at the same time allow you the flexibility of enrolling in courses and seminars in ecology and other subjects that will help you apply your special knowledge and talents *to management at the level of the ecological system.* If your educational institution does not have a formal interdisciplinary degree mechanism, then organize one yourself by getting several professors in different departments to form an advisory committee that can help you plan, and cut departmental red tape if necessary, to meet your needs, deficiencies, and talents. (4) Select an ecologically oriented thesis topic which not only excites you as a contribution to human welfare, but which also will provide further training in synthesis. Remember that what the world needs is more and better specialists who are knowledgeable about the ecological whole!

RESORCES

ecological human law (handwritten)

Chapter 15

1. CONSERVATION OF NATURAL RESOURCES IN GENERAL

Conservation in the broadest sense has always been one of the most important applications of ecology. Unfortunately, the term "conservation" suggests "hoarding," as if the idea were simply to ration static supplies so that there would be some left for the future. In the eye of the general public the "conservationist" is too often pictured as an antisocial person who is against any kind of "development." What the real conservationist is against is *unplanned development that breaks ecological as well as human laws.* The true aim of conservation, then, is twofold: (1) to insure the preservation of a quality environment that considers esthetic and recreational as well as product needs and (2) to insure a continuous yield of useful plants, animals, and materials by establishing a balanced cycle of harvest and renewal. Thus, a "no fishing" sign on a pond may not be as good conservation as a management plan which allows for removal of several hundred pounds of fish per acre year after year. On the other hand, if the pond provides the water supply for a town, then some constraints on fishing may be the desirable conservation procedure. The short-comings of the widely advocated "multiple-use" policy and the advantages of a "compartment" plan to achieve desired balance between production and protection has already been discussed in some detail in Chapter 9, Section 3.

It is customary to divide natural resources into two categories: renewable and nonrenewable. Although it is true that coal, iron, and oil deposits are not renewable in the same sense as forests or fish, nevertheless nitrogen, iron, and energy sources are renewable just as much as living resources. Man would never lack vital materials if he would but adjust his population size and resource demands at or below the level that allows the biogeochemical cycles to operate in such a way that materials as well as organisms are "reassembled" as fast as they are "dispersed" (see Chapter 4, Section 1).

Although "hoarding" may not be the long-

time aim of good conservation, there are instances in which complete restriction of use constitutes good conservation. The setting aside of natural areas for study and esthetic enjoyment is an example. With the increase in human population, it becomes more important that adequate samples of all major natural communities be preserved undisturbed for study and enjoyment. Since man establishes his civilization and his food chains by modifying natural ecosystems (and not by creating completely new systems), it is important that we have samples of unmodified communities for study; only with such "controls" can the effects of man's modification be properly judged, and unwise practices avoided. No laboratory scientist would think of undertaking an experiment without an adequate control, yet the field ecologist is often called on to evaluate the effects of man's experiments without a control being available.

As already indicated, the shift from "special interest conservation" to "total ecosystem conservation" is helping to establish the fact in the minds of the general public that man is a part of a complex environment which must be studied, treated, and modified as a whole and not on the basis of isolated "projects." To reemphasize this point, we can do no better than quote from the writings of the late Aldo Leopold. Leopold was America's preeminent pioneer in applied ecology. The following passage, written 30 years ago (Leopold, 1941), expresses very well the general need for a sound philosophy and understanding of the ecosystem principle.

Mechanized man, having rebuilt the landscape, is now rebuilding the waters. The sober citizen who would never submit his watch or his motor to amateur tamperings freely submits his lakes to drainings, fillings, dredgings, pollutions, stabilizations, mosquito control, algae control, swimmer's itch control, and the planting of any fish able to swim. So also with rivers. We constrict them with levees and dams, and then flush them with dredgings, channelizations and floods and silt of bad farming.

The willingness of the public to accept and pay for these contradictory tamperings with the natural order arises, I think, from at least three

408

fallacies in thought. First, each of these tamperings is regarded as a separate project because it is carried out by a separate bureau or profession, and as expertly executed because its proponents are trained, each in his own narrow field. The public does not know that bureaus and professions may cancel one another, and that expertness may cancel understanding. Second, any constructed mechanism is assumed to be superior to a natural one. Steel and concrete have wrought much good, therefore anything built with them must be good. Third, we perceive organic behavior only in those organisms which we have built. We know that engines and governments are organisms; that tampering with a part may affect the whole. We do not yet know that this is true of soils and water.

Thus men too wise to tolerate hasty tinkering with our political constitution accept without a qualm the most radical amendment to our biotic constitution.

Leopold would not have been surprised by what has now become known as *ecological backlashes* or *ecological boomerangs*. We may define ecological backlash as an unforeseen detrimental consequence of an environmental modification which cancels out the projected gain or, as is too often the case, actually creates more problems than it solves. When this happens, it is a double tragedy since not only is the money spent in remaking the landscape lost to the bad investment, but additional sums must then be spent to correct all the new problems created. The reason that detrimental consequences are "unforeseen" stems both from public misconceptions, as is so well expressed in the quotation above, and from inadequate prior studies and evaluations of the impact of the technology on the environment and on the people whose lives are summarily disrupted. Farvar and Milton (1969) and Cahn (1968) describe a number of severe ecologic backlashes on the international level, and these and other cases will be documented in a forthcoming book by Milton (in press). The building of huge dams in undeveloped tropical countries is, perhaps, a good example. One such dam is on the Zambezi river in Africa. Built primarily for the purpose of producing hydroelectric power, it is also producing a whole series of "unforeseen" problems. The fish catch has not compensated for loss of grazing and agricultural lands as was "predicted" by the promoters of the project (but not by lake ecologists who know better but were not asked; see page 312). The large lake shore increased the habitat for tsetse flies with a resultant severe outbreak in cattle disease (also in human disease in the case of dams on the Nile). Displacement of people and cultures has created soil erosion and serious social upheaval as people were moved to less suitable lands or to cities that were not prepared for them. The "regulated flow" downstream from the dam turned out to be more damaging than normal flooding, which formerly enriched bottomlands each year free of charge (see concept of "fluctuating water level ecosystem," Chapter 9, page 268). As the fertility of these lands declines, expensive fertilizers, which the people cannot now afford, have to be imported. It will be many years before the full impact will be known. Other types of backlashes resulting from pesticides, and industrialized agriculture in general, will be noted in Section 3 and in Chapter 16.

While some of the more conspicuous ecological backlashes are being fostered in undeveloped countries by the technocrats of developed ones, there is also growing concern about what is happening in the rich countries. In the United States, for example, it is imperative that citizens find a way to break the vicious cycle of "pork-barrel politics," which fosters a seemingly endless series of dredgings, river channelizations, and dams far beyond any real need. Too often landscapes that are already in good shape and efficiently used by man are "remade" at great expense to the taxpayer for outdated reasons; for example, "flood control" where none is needed, or barge canals when existing rail or truck transport is going bankrupt for lack of business. It is high time that even the wealthy nations realize that they can no longer afford to spend taxpayers' money on rebuilding the countryside that is already in good shape when such money is desperately needed to rebuild cities and human societies that are in terrible shape. All of this does not mean that man should stop modifying nature; it means that careful study and planning must precede projected modification to make sure there will be a net benefit to man, not just a temporary economic benefit for a vested interest.

2. MINERAL RESOURCES

Until recently little attention was paid to conservation of mineral resources because it was assumed that there were plenty for centuries to come and that nothing could be done to save them anyway. *It is now apparent that both assumptions are dead wrong!* Cloud (1968, 1969, 1970) has inventoried supplies and reviewed prospects. He introduces two concepts (in his 1969 paper) that are useful in evaluating the situation. The first is the *demographic quotient,* which we shall designate as "Q":

$$Q = \frac{\text{total resources available}}{\substack{\text{population} \\ \text{density}} \times \substack{\text{per capita} \\ \text{consumption}}}$$

As this quotient goes down, so does the quality of modern life; it is going down at a frightening rate because available supplies can only (or eventually) go down as consumption goes up. Even if available resources could be kept constant by recycle or other means, the situation deteriorates as long as population, and especially per capita consumption, increases at a rapid rate. Thus, in the United States economic and technological growth based on exploitation of natural resources is increasing at a rate of 10 per cent per year (doubling time is about 7 years!), urban growth is increasing 6 per cent per year, while the population growth is only about 1 per cent. If the undeveloped world with its huge populations were to increase its per capita use of minerals (and the fossil fuels required to extract and use the mineral resources) to anywhere near the level of the United States, severe shortages would develop tomorrow. In developed countries the per capita demands for a moderately scarce metal such as copper is projected to triple by the year 2000! Aluminum is cited by Cloud as an example of the general situation because it is not, relatively speaking, a scarce metal. Prior to 1945, the United States produced most of the ore (bauxite) that it used, but by 1960 this country was importing three times as much ore as it was mining on its own lands. It is obvious that we can no longer afford "one-way" aluminum beer and soft-drink cans (or similar "uncycled" uses); we have to substitute or recycle, or both. The industrialized countries, in general, are no longer self-sufficient in either minerals or fossil fuels; they depend more and more on exploitation of these natural resources from the undeveloped part of the globe, where, of course, the supply is finite, cannot be increased, and, in fact, will decrease as these countries start to use their own wealth.

The other concept introduced by Cloud is the graphic model of *depletion curves,* as

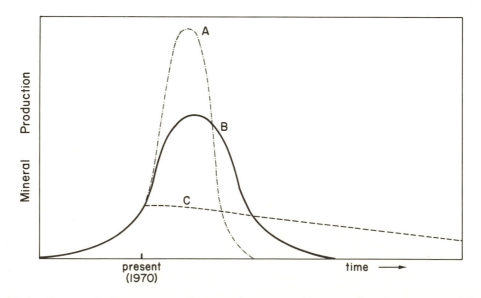

Figure 15–1. Alternate depletion patterns for mineral resources. *A.* Pattern of rapid extraction and depletion of minerals (or other nonrenewable raw materials) that will occur under the present custom of unrestricted mining, use, and throw away. Some key metals will be "mined out" before the year 2000 if this pattern persists. *B.* Depletion time can be extended by partial recycle and less wasteful use. *C.* Efficient recycle, combined with stringent conservation and substitutions (using more abundant alternate material wherever possible), can extend mineral depletion curves indefinitely. (Diagram adapted from Cloud, 1969.)

shown in Figure 15–1. With the present procedure of "mine, use, and throw away," a huge boom and bust is projected, as shown in curve A. The time scale is uncertain because of lack of data, but the "bust" could begin within this century since certain key metals such as zinc, tin, lead (needed for that electric car!), copper, and other metals could be mined out in 20 years insofar as the readily exploitable reserves are concerned. Likewise, fuels such as uranium-235 and natural gas could also be gone by then. If a program of mineral conservation involving restrictions, substitutions (using less scarce minerals wherever possible), and partial recycle were to be started now, the depletion curve could be flattened as shown in curve B. Efficient recycling combined with stringent conservation and a reduction in per capita use ("power down" by developed countries) could prolong depletion for a long time, as shown in curve C. It should be noted that even with perfect recycle depletion would still occur. Thus, if we were able to recycle 60 million tons of iron each year, about half a million tons of new iron would be needed to replace the inevitable loss from friction, rust, and so on.

Inventories and projections regarding the mineral fuels are firm and generally agreed upon. As will be noted in the next chapter, pollution rather than supply will be the limiting factor for industrial energy. As already indicated, natural gas and uranium will be soon gone, but oil and coal will last longer. In the meantime "breeder" reactors and possibly the development of fusion atomic energy should fill the energy gap. Thus, for the time being at least, biotic and mineral resources are more critical than energy; it can be hoped that these limits will actually prevent man from attempting to increase energy use to the point of literally burning up the world.

Recommended reading is the slim but potent volume entitled *Resources and Man*, published in 1969, as the summary report of a committee of the National Academy of Science (Preston Cloud, Chairman). The report urges caution in relying on the optimistic projections of some technologists in regard to: (1) the sea as an unlimited supply depot and (2) the extraction of metals from lean ores with the use of vast quantities of cheap atomic energy. As we have already noted in Chapter 12 (see page 324), most mineral wealth (and exploitable food)

is located near the shore and does not by any stretch of the imagination provide anything more than a supplement to the continental supplies. The use of vast amounts of atomic energy to extract low grade ores would convert the world into one vast strip mine and create hazardous and expensive waste disposal and pollution problems, which we hope will never have to be evaluated! The National Academy report ends with 26 recommendations that all boil down to a twofold theme: *Both human population control and better resource management that includes recycle are needed—NOW.* Other solid references on natural resources include volumes edited by Jarrett (1966) and Ciriacy-Wantrup and Parsons (1967), not to mention the earlier classic edited by Thomas (1956). The vanguard of what should be an important future output of books on the application of systems analysis (see Chapter 10) to resource management are volumes by Watt (1968), Van Dyne (ed.) (1969), Patton (ed.) (1971, Vol. 2), and H. T. Odum (1971).

3. AGRICULTURE AND FORESTRY

Principles involving limiting factors (Chapter 5) and productivity (Chapter 3) have in the past provided the chief ecological application to agriculture and forestry, but for the reasons already indicated, the *agricultural and the forestry specialist must now consider that "his" crops and forests have outputs other than food and fiber in terms of man's total ecosystem.* Crop ecology was rather thoroughly dealt with in Chapter 3, Section 3. Insofar as the temperate zone is concerned, the major limiting factors to food production have been surmounted as a result of technical advances in fertilizers, irrigation, pest control, and crop architecture through genetic selection (Chapter 8, Section 5). However, as *strongly* emphasized in Chapter 4, Section 6, and again in Chapter 14, page 402, *temperate zone agrotechnology cannot be applied to the tropics; the practice of tropical agriculture is incredibly inept, and whole new concepts based on ecological principles are urgently needed* (see also Chapter 7, page 232).

As was brought out in Chapter 3 (see especially pages 46–47), the increase in agri-

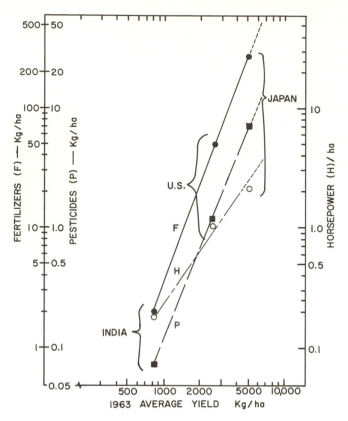

Figure 15-2. Relationships between yield of food crops (in kilograms dry weight per hectare) and requirements for fertilizer (F), pesticides (P), and horsepower (H) used in cultivation and harvest of crops. Regressions based on data from three countries (United States, India, Japan). Note that doubling the yield of food requires a 10-fold increase in use of fertilizers, pesticides, and animal or machine power. (Graph prepared from data in: The World Food Problem, A Report of the President's Science Advisory Committee, Panel on World Food Supply, the White House, 1967; volume III, pages 141, 143, and 180.)

cultural production, that is, the so-called "green revolution," has resulted chiefly from the industrialization of agriculture, which involves large fuel energy subsidies, sophisticated chemical control, and highly domesticated plant varieties. Systems models contrasting industrial and nonindustrial agriculture were discussed in Chapter 10 (see Figure 10-7C and D). Maximizing for yield without regard to other consequences is producing very serious *backlashes, both environmental and social* (see Section 1, this chapter). The former can be illustrated by the model in Figure 15-2, which shows that to double the crop yield requires a tenfold increase in fertilizer, pesticides, and horsepower. Thus, industrialized (fuel powered) agriculture (as practiced in Japan, for example) produces four times the yield per acre as does man-and-domestic-animal powered agriculture (as in India) but is 100 times as demanding of resources and energy. It is not hard to understand, then, why *agroindustry is one of the chief causes of air and water pollution.* The agricultural engineer of the future must obviously be more concerned with the quality of the rural landscape in terms of its ability to supply

clean water and air, as well as food, to the cities (not to mention recreation and other "quality" needs).

The green revolution also sows very poisonous seeds of social revolution, especially in the undeveloped countries, which are already overcrowded. Such a social backlash, as has been recently noted in the popular press,[*] is especially likely when the shift from small farms to large industrialized farms takes place too quickly and without compensatory adjustment in social and political systems. As farming becomes mechanized, small farmers and agricultural workers are driven off the land into the city where there is no work or housing for them. The gap between the rich and the poor is widened. The developed countries are not immune since their big-city ghettos also fill up with people who formerly had at least a dignified human existence working the land or engaged in small business in farm towns. The sad feature of all this is that while the consolidation of small farms into large, mechanized ones increases the food

[*] See *Time*, July 13, 1970, page 24, and *Foreign Affairs*, July, 1970, page 758.

available and increases the wealth of those few people who manage the large farms and process the food, a much larger number of people are relegated to tin shacks in the cities where they are unable to buy the food. Such are the seeds for despair, drug addiction, violence, and total social breakdown, since, as we were careful to note in Chapter 8, Section 9, man does not differ from rats or monkeys in his need for an environment that permits a stabilized social behavior. International organizations and the large philanthropic foundations that have been preoccupied with filling men's bellies without accompanying such effort with crash programs for population control and social readjustment are having second thoughts about what they are doing. In retrospect, at least, upgrading the quality of the small farm and reduction in population growth should come *before* there can be a massive shift to industrialized agriculture. It is clear that a reevaluation of basic foreign policy of the wealthy nations towards the poor nations is in order.

Unlike agriculture, a large portion of the yield in forestry has so far come from harvesting the accumulated growth of the past. As this natural reservoir is "mined," as it were, the forest industry has to adjust itself to dependence on annual growth. In the United States harvest in the western states still greatly exceeds annual growth, but in the south and east, where the old forests have all been cut, growth more nearly balances harvest.

The dilemma of forestry is illustrated by Figure 15–3, which contrasts the artificial forest (tree crop or monoculture) and the naturally developed forest. One school of foresters holds that forests should be managed as "fiber crops" in the same way as we now cultivate corn or other food crops. The philosophy here is that increased yields of forest products will be "demanded" by increased per capita use of fibers as well as increased human population density, and that the only way to meet these demands is to "farm" trees or other fiber plants. As has already been pointed out a number of times, maximizing for *quantity yield* means: (1) supplying energy subsidies (i.e., industrializing forestry), (2) monoculture with harvest of single species on short rotation (since a young forest grows faster than an older one; see Figure 9–2), (3) artificial selection for high fiber-producing varieties

that depend on high rates of fertilization and pesticide application (which, in turn, increase global pollution and the danger of disease outbreaks), and (4) a probable reduction in quality. Many foresters contend that a forest should not be treated as a "crop" because it has other important uses such as recreation, wildlife habitat, air and water sheds, and so on, that are best served by multispecies, multiaged stands harvested on a continuous but low yield basis (i.e., the "multiple-use forest," as the concept is now generally practiced in the U. S. National Forests).

The important point that needs to be widely recognized is that the tree farm, as one extreme, and the naturally developed multiple-use forest, as the other, are entirely different ecosystems in terms of the cost of maintenance and their impact on man's other necessary environmental needs. As with other resources, recycling of paper and other forest products plus stringent conservation could reduce per capita demands on the annual growth without necessarily reducing individual use. Such a reasonable strategy (and one that must sooner or later be followed) would make it unnecessary to cover the landscape with tree crops at the great risk of creating a boom and bust situation with severe backlashes of the types discussed earlier in this chapter. It would be desirable to restrict tree crops to level, fertile land and to soil types suitable for good agriculture in general; certainly the naturally adapted diverse forest is the best and safest cover for steep lands and the many thousands of other acres where soil and water conditions cannot sustain intensive culture of an artificial type for very long.

4. WILDLIFE MANAGEMENT

While the expression "wildlife" would seem to be a broad one, covering any or all noncultivated and nondomesticated life, it has in the recent past been largely used with reference to game and fur-bearing vertebrates, and to the plants and lesser animals which interact directly with the game species. Even fish, which are often "wild," have been excluded from the wildlife category and treated under other headings. The trend of the 1970s and beyond will be to

Figure 15–3. The forestry dilemma—how far to go with monoculture? The two pictures compare a pine plantation (*A*) and a naturally developed pine forest (*B*). Both stands are on "old-field" (i.e., former cropland) sites in southern United States. (U.S. Forest Service Photos.)

return to the broader concept of wildlife, since an increasingly large number of people are becoming interested in nongame species (song birds, for example) and since preservation of outdoor recreation in general depends more and more on the preservation of the totality of the wildlife ecosystem.

Wildlife management is a field of applied ecology that ranks high in the public interest. In recent years it has become a well established profession in America, attracting many young men with a love for the out-of-doors. Numbers of "wildlife technicians" are trained by the land-grant colleges, some of which have established special research and teaching units in cooperation with the State and Federal governments. As in other areas of applied ecology, such academic "segregation" from the mainstream of the sciences and the humanities has resulted in overly narrow training, sometimes producing, as Leopold has expressed it, expertness that cancels understanding (see page 409). Nevertheless, intensive study of individual game species has contributed a great deal to population

ecology as evidenced by the large number of examples involving game species that are cited in Chapter 7.

Wildlife management is concerned not only with game production land which is not suitable for anything else, but also with game "crops" which may also be produced on more productive land being used primarily for agriculture or forestry. As land becomes intensively farmed or urbanized, wildlife becomes largely a function of "edges" (see Chapter 6, Section 6 for discussion of the concept of "edge effect"). For example, Hawkins (1940) compares a Wisconsin landscape as it appeared to the first white settlers in 1838 to its appearance a century later under intensive farming. The large blocks of forest and prairie were broken up into numerous small pieces interspersed with crop fields, roads, and house sites. The original upland game birds, the ruffed grouse and the prairie chicken, have been completely replaced by the ring-necked pheasant and Hungarian partridge, introduced species that had become adapted to intensively farmed areas of Europe. Thus, as the landscape becomes

B

Figure 15–3. Continued.

more "domesticated" so does the wildlife. Under such conditions, habitat living space, disease, and predation replace food as major limiting factors. Principles pertaining to predator-prey relationships and the optimum yield concepts were extensively reviewed in Chapter 7.

This brings us to the point that efforts to foster and increase wildlife populations have been generally directed along four major lines: (1) preservation of breeding stock by means of game laws restricting the harvest and other similar measures, (2) artificial stocking, (3) habitat improvement, and (4) game farming. When game begins to get scarce, people generally think and act in the order listed above, which is sometimes unfortunate, since the third item is often more important than the first two. If suitable habitat is lacking (as indicated above), protection or stocking is useless. As the human population density increases, game and fish management as such is faced

with the same dilemma as forestry, namely, to what extent can the demand for hunting and fishing be accommodated under natural area management and to what extent will it be necessary to "farm" game and fish. The ultimate in wildlife farming involves artificial propagation of animals, stocking, and harvest within a few days or weeks of release with the hunter or fisherman paying a large fee for the privilege of shooting or fishing what amounts to a domesticated or feral population (see page 244). Like all agrobusiness, game farming has "unforeseen" problems such as predation, disease, nutritional problems, artificial selection, and so on, and requires energy subsidy in the form of supplemental food, labor, taxes, and so on.

We may well close this brief review with a comment on the introduction of exotics, that is, of species not native to the region. Since species introduced into a new environment often either fail entirely or succeed so well that they become pests (rabbits in Australia, for example), all introductions are to be considered with extreme care. In general, if there is a suitable species native to the region, it is better to concentrate on management of this species rather than to attempt to introduce a substitute. If, on the other hand, the environment has been so changed by man that native forms are unable to survive (as in the case of the Wisconsin landscape described above), introduction of a species adapted to the new environment may be in order. Certainly, introductions should be preceded by study and probably should be controlled by Federal regulations. (See Kiel, 1968.)

For more on wildlife management see Leopold's classic (1933), Dasmann's textbook (1964), and the Wildlife Society's technique manual edited by Giles (1969).

5. AQUACULTURE

The relationship between primary production and yield of food or other biotic products from aquatic ecosystems was reviewed in Chapter 3 (see especially Tables 3–7 and 3–11) and again in Chapters 11 (Sections 6 and 7), 12 (Section 5), and 13 (Section 3). Man's belated attempt to become a "prudent predator" and the difficulties of applying a theoretical optimum yield model to the harvest of wild populations of fish and shellfish were discussed in Chapter 7, Section 18. As wild populations are exploited to their limit, and reduced by overfishing, attention naturally turns to fish farming, or aquaculture, particularly since such culture can be a very efficient way to produce protein food.

Fishery science in the United States has been concerned primarily with the management of natural populations that are commercially fished and especially with species that provide sport, since sport fishing is the number one recreational sport in terms of the number of people who actively participate in it. Now, since fish that provide the best sport are carnivores, they must be produced on the end of a long food chain which greatly limits the yield per acre that can be harvested (compare figures for "stocked carnivores" versus "stocked herbivores" in Table 3–11). As shown in Figure 3–11 and Figure 3–12, at least four food chain links are required: phytoplankton, small

crustacean and insect consumers, small carnivores or detritivores (often called forage fish), and large carnivores or game fish. Management of freshwater game fishes in some habitats, such as small trout streams, involves annual stocking of hatchery-reared fish, which are quickly removed by the fisherman; but in ponds and lakes a real attempt is made to maintain a balanced population that will provide to man a sustained yield with regular inputs of inorganic fertilizers but no food or fish input (i.e., a self-reproducing system). Game fish ponds are engineered to simplify the ecosystem, that is, restrict components to those directly involved in a linear food chain leading to the desired products. This can be done by regulating the size and depth of the body of water, the rate of fertilization, and the species composition and size ratio of the fish population (see Bennett, 1962). Especially important is the ratio of forage fish (F) to the top carnivore fish (C). According to Swingle (1950), the F/C ratio in terms of biomass should be about 4 (with a range of 3 to 6). Game fish population in ponds or lakes are more likely to become "unbalanced" as a result of an increase in the F/C ratio (which results in overcrowding and great reduction in average size of fish) rather than as a result of overfishing by hook and line, which, unlike nets, take only the large individuals. The best way to maintain high yields of both game and food fishes is to periodically drain, or draw down, the body of water so as to maintain the system in a youthful, rapid growth phase as described in Chapter 9, Section 3. The practice of pond "fallowing" is an ancient art practiced for centuries in Asia and Europe (see Neess, 1946; Hickling, 1961) and is a proved application of the ecological concepts of "the fluctuating water level ecosystem" and "pulse stability" (see page 268).

As has already been emphasized, productivity of lakes is inversely proportional to depth (see Table 11–1) so that large, deep lakes cannot compare with smaller, shallow ones in the production of fish per unit of area (Rawson, 1952). Large impoundments (artificial lakes) provide good fishing for the first few years after construction (as accumulated energy in the watershed is exploited) but generally become poor as fishing lakes, as already noted (see page 262). There has been far too much emphasis on large expensive mainstream reservoirs when a series of smaller impoundments would more efficiently and cheaply serve the same purposes. Table 15–1 compares two proposed reservoir plans prepared independently by different agencies for the same region. The multiple reservoir plan would not only cost one-third as much as the one big dam for the same amount of water surface and water supply or flood control potential, but recreational fishing and most other uses, with the exception of hydroelectric power production, would be better served by small reservoirs. For power production in the future we will probably depend largely on fossil fuel and atomic energy, thereby reducing the need for new hydroelectric dams.

Aquaculture for food, as is now practiced in Japan, involves quite a different ecosystem as compared with game fish culture, the former based on a short food chain supported by heavy inputs of fertilizer, food, hatchery-reared stock, and work energy. One of the most promising opportunities for aquaculture in an affluent country is to tap the effluents from waste treatment! Heated water from power plants and certain types of diluted or partially treated domestic and industrial organic wastes flowing through a series of ponds can provide energy subsidies for adapted species of fish, mollusks, crustaceans, and other organisms that could provide human or animal food or other useful products. This is part of the "waste management park" concept as discussed in the next chapter (see also Figure 15–5). Such an organized aquaculture could help transform pollution into a resource.

The annual world harvest of marine fish has risen from 18 million metric tons (live weight) in 1938 to 55 million tons in 1967 (see FAO Fishery Statistics, 1967). About four-fifths of this comes from three areas, namely, the north Atlantic, the western and northern Pacific, and the Pacific off the western coast of South America. The recent exploitation of the Peruvian anchovies, probably the world's most naturally productive fishery (see page 71), accounts for a large part of the increased harvest in recent years. Only a relatively few species comprise the bulk of harvest (see page 342). Strange to say, only half of the harvest is used as human food; the other half becomes food for poultry and livestock. Lengthening the food chain in this manner does not make good ecological sense and is economical only so long as fish are a "free" gift of

Table 15–1. *Comparison of a Large Main-stream Reservoir Plan with an Alternate Plan for Smaller Headwaters Reservoirs Proposed for the Same Watershed**[*]

	MAIN-STREAM RESERVOIR PLAN	MULTIPLE HEADWATERS RESERVOIR PLAN
Number of reservoirs	1	34
Drainage area (sq. miles)	195	190
Flood storage (acre feet)	52,000	59,100
Surface water area for recreation (acres)	1,950	2,100
Flood pool (acres)	3,650	5,100
Bottoms inundated (acres)	1,850	1,600
Bottoms protected (acres)	3,371	8,080
Total cost	$6,000,000	$1,983,000

[*] Data from Peterson (1952); 1970 costs would be greatly inflated.

nature, harvested without the expense of fertilizers, disease and predator control, or husbandry. Thus, although the yield of food per unit area of the ocean is low as compared with animal husbandry on land, it represents an efficient use of a self-sustaining natural resource. To what extent yield of naturally produced food from the sea can be increased is controversial. Some fishery scientists feel that this yield has reached a peak, while others think it could be increased, but no more than three or four times (see Holt, 1969; Ricker, 1969). Mariculture (aquaculture in marine or estuarine waters) so far has provided an important food supplement in only a few areas such as Japan, Indonesia, and Australia (see Bardach, 1969).

Both the potential and the limitations of estuaries for the culture of shrimp, oysters, and fish were discussed in Chapter 13, Section 3. Food from the sea will be an increasingly important protein supplement to food production on land, but for the most part the oceans do not lend themselves to intensive farming for ecological reasons already documented (see pages 52 and 324).

6. RANGE MANAGEMENT

As was indicated in the discussion of the grassland biomes (Chapter 14, page 388), man seems to have an uncanny knack for abusing grassland resources. A succession of deteriorated civilizations in the grass-land regions of the world stand as mute evidence of this fact. When a man builds an artificial pasture by dint of hard labor and cold cash, he would no more think of destroying it by misuse than he would think of burning down his house. Yet as Leopold so aptly expressed it (see Section 1, this chapter), man has difficulty understanding organic cause and effect in a natural pasture that he did not build.

To determine the range's carrying capacity for grazing animals two considerations are paramount: (1) the primary productivity and (2) the per cent of the net productivity that can be removed annually and still leave the grass plant with enough reserve to enable it to maintain future productivity, and especially to sustain the stress of periodic periods of unfavorable weather (droughts, etc.). Since primary productivity is a more or less linear function of rainfall (see Figure 14–16), and knowing that less than half of the annual net production should be consumed by cattle, one should be able to calculate how many pounds of meat might be produced (using ecological transfer efficiencies; see page 76) for a given rainfall input and adjust the number of animals accordingly. However, it does not work out quite so simply because seasonal distribution of rainfall, the quality of the forage (especially protein-carbohydrate ratio), the palatability, season of growth and so on, all complicate the picture. Studies using esophageal fistulas and microdigestion techniques (see Cook, 1964; Van Dyne and Meyer, 1964) show that the cow is selective in what it eats. For these reasons, range managers have found that community or ecosystem indicators (see Chapter 5, Section 6) provide the most practical means of determining whether or not a range is being utilized properly (Humphrey, 1949). Particularly useful are plants called "decreasers" whose disappearance from the range (see page 390) provide "early warning signals" of overgrazing. When such signals are not heeded, unpalatable "weeds" (annuals) and desert shrubs (sagebrush, mesquite, etc.) invade. If overgrazing continues, essentially a man-made desert results. Once established, the desert vegetation may be hard to eradicate. Rodents and grasshoppers increase with overgrazing. Unnecessary control of predators may aggravate the rodent problem. At-

tempts to rehabilitate the range by rodent or insect control alone is a good example of what Leopold would call "isolated project management," which ignores the primary cause of the problem.

The role of fire in range management was discussed in Chapter 5 (page 136); the total interaction of fire and grazing is far from being understood. The feasibility of increasing the diversity of grazers and the possibilities of game ranching on the African grasslands were discussed in Chapter 14 (page 392).

Long-term grazing studies provide the most important experimental approach to range management. Unfortunately, most experiment station studies are not continued long enough to cover natural cycles in weather. Table 15–2 summarizes a fairly long study which compares three grazing intensities. These data show that while heavy stocking produces a short-term economic gain (more meat), the quality of both the cow and the range deteriorates. Therefore, (1) maximum stocking rate (cow population) is not the optimum for maintenance of the quality of cow and pasture and (2) short-term profit gain from overstocking comes at the expense of long-term damage to the environment. Actually, the light intensity use, in which about a third of the net production is removed by grazing, is the optimum use since the quality of range and animal is improved and the probability of severe damage during periodic droughts is greatly reduced.

The effect of trampling by grazing animals in compacting the soil can also be important. When too many animals are confined for too long, the soil may become "sod-bound" and productivity reduced, even after the animals are removed. Essentially, the consumer-regenerating portion of the ecosystem is hampered by poor aeration of the soil, and this in turn inhibits the production mechanism. Plowing or ripping with subsoil chisels has proved useful in small pastures, but such a procedure would be too expensive for large tracts of rangeland. As we have emphasized many times, the whole ecosystem, not merely the producers, must be considered. The relationship between the natural migratory behavior of wild ungulates and the management concept of pasture rotation was noted in Chapter 14, Section 6.

Because natural grasslands have evolved

Table 15–2. Grazing Intensity Model*

	INTENSITY OF GRAZING		
	Light (cow/21 acres)	*Moderate* (cow/15 acres)	*Heavy* (cow/9 acres)
Mean cow standing crop at end of growing season (lbs./acre)	321	405	648
Mean weight individual cows at end of growing season (lbs.)	1003	942	912
Mean weight at weaning of calves produced (lbs.)	382	363	354
Forage utilization; per cent aboveground vegetation grazed during growing season	37%	46%	63%
Condition of range at end of the 9-year study†	Improved	Unchanged	Deteriorated

* Data based on a 9-year study at Cottonwood, South Dakota, reported by Johnson *et al.* (1951).

† Based on relative abundance of "decreaser" plants, that is, palatable species that tend to decrease with increasing grazing pressure.

over long periods of time during which they have been subjected to varying use by animals and man, the historical aspect of the range is of great importance. Unfavorable conditions may not always be the result of recent grazing history. Fire history and other aspects need to be considered (Malin, 1953; Sauer, 1950). For further discussions of various aspects of range management, see books by Stoddart and Smith (1955), Humphrey (1962), the symposium edited by Crisp (1964), and reviews by Costello (1957), Dyksterhuis (1958), Williams (1966), and Lewis (1969).

The grazing intensity model in Table 15–2 has wider application if we substitute "man" for "cow" and "environment" for "range." The model would then illustrate: (1) that maximum stocking rate (human population) is not optimum for maintenance of the quality of man and environment and (2) short-term profit gain from overstocking comes at the expense of long-term damage to the environment. Truly the "green pastures" of biblical psalms are symbolic of the moral responsibility of man for his environment.

7. DESALINATION AND WEATHER MODIFICATION

Large-scale artificial production of water from the sea and the artificial regulation of water production from the sky represent two proposed resource modifications whose potentials for both good and evil are mind-boggling! One can project much good for humanity as a result of such resource alterations, but one can also foresee the very real possibilities of horrendous ecologic backlashes! In view of the very severe consequences of far less extensive environmental modifications, as already noted in this and other chapters, the ecologist is inclined to suggest that man is not yet ready to tinker with the global hydrological cycle (as shown in broad outline in Figure 4–8B) because (1) he does not yet have the capability of monitoring changes on the global scale and (2) he does not have the detailed knowledge of biogeochemical cycle–energy flow interaction to make accurate predictions.

There is very little hazard in small-scale desalting of ocean water for use as drinking water in densely populated coastal urban areas, although simpler and cheaper alternatives are often available, for example, recycling waste water by complete tertiary treatment (see Chapter 16). Large-scale desalination for industrial and agricultural use is something else again. For one thing another serious pollution problem is created by the large energy expenditure needed and by the accumulation of salt. All of the cost-benefit analyses that we have seen (for example, Clawson, Landsberg, and Alexander, 1969) indicate that the cost of desalted sea water even with use of so-called cheap atomic energy is a whole order of magnitude greater than the value of the water in agriculture. Desalination for agricultural use would not help the Middle East, for example, at least not until the population explosion and other social problems are first solved. The value of such desalted water for industrial use is also questionable so long as recycled water can reduce the gross input demand.

Again, all of the panel reports prepared by interdisciplinary groups of knowledgeable scientists and technologists (see, for example, the Ecological Society's report, 1966, and Cooper and Jolly, 1969), as well as numerous thoughtful writers (Waggoner, 1966; Sargent, 1967; MacDonald, 1968), urge *extreme caution* in conducting weather modification experiments *for any kind of practical gain at this time;* numerous research-oriented analyses and experiments are needed first. In many cases results opposite from those expected have been obtained. For example, while cloud seeding sometimes has increased rainfall, Lovasich and colleagues (1969) report that summer rainfall decreased rather than increased in one series of experiments. Apparently, the seeding caused an overcast or drizzle that prevented the afternoon buildup of energy necessary to trigger thunderstorms, which were normally the chief source of summer rainfall.

Since, as already noted in Chapters 2 (page 33) and 4 (page 98), man has already inadvertently modified climate both on a local and global scale, controlled weather modification experiments can perhaps be best justified as a means for obtaining the necessary knowledge to prevent harmful effects of the weather modification already in progress!

Because most people now live in cities, it is important to note that cities not only pollute the air, but they also modify climate. According to Lowry (1967), cities are "heat traps" owing to the absorption of solar radiation by vertical surfaces and the production of heat by machines. In a mid-latitude region air temperatures in cities are 1 to 3°F. higher and the humidity is 6 per cent lower than in the surrounding countryside. Because of particulate pollution cloudiness is 10 per cent greater, fog 30 to 100 per cent more frequent (higher value in winter), precipitation 10 per cent higher, sunshine 15 per cent less, and ultraviolet radiation 5–30 per cent less in the city.

8. LAND USE

When the human population of an area is small, poor land use may affect only the people who are guilty of bad judgment. As the population increases, however, everyone suffers when land is improperly used, because everyone eventually pays for rehabilitation or, as is now too often the case, everyone suffers a permanent loss of resources. For example, if grasslands in regions of low rainfall are plowed up and planted to wheat (poor land use), a "dust bowl" or temporary desert will sooner or later be the result. Rehabilitation is expensive, and all of us as taxpayers will have to pay. If the grass cover is maintained and moderately grazed (good land use), no dust bowl will be likely to develop. Likewise, if the lack of local zoning restrictions allows houses and factories to be built on flood plains (poor land use), then damage to such investment is inevitable (or can be avoided only by expensive flood control structures). If, on the other hand, flood plains are used for recreation, forestry, or agriculture (good land uses), money is added to, not subtracted from, tax dollars. Land use is thus everybody's business, and *the application of ecological principles to land-use planning is now undoubtedly the most important application of environmental science.*

So far, good land-use planning has come only after man has first severely damaged a landscape. It is as Leopold has said: Man does not seem to be able to understand a system he did not build and, therefore, he seemingly must partially destroy and rebuild before use limitations are understood. In the United States, for example, the first successful land-use planning followed an era of widespread destruction of soil resulting from ill-planned one-crop systems and exploitation of land that was unsuitable for row crops. Out of the human misery created by soil erosion and dust bowls came the soil conservation movement that is now an outstanding example of a conservation program because it involves cooperation of local people, their state university, and their state and Federal governments. Soil conservation districts have been set up along natural boundaries, such as a large watershed, and the program of land planning and education is run by local people with technical help for the land-grant universities and a Federal government bureau (the U.S. Soil Conservation Service) created by an act of Congress. The success of the soil conservation movement as a whole, and the high regard in which it is held by the public, is due in no small measure to the emphasis on the whole, that is, the whole farm and the whole watershed. In implementing the program land-use maps are prepared, as shown in a very simplified manner in Figure 15–4. Classification for use is based on natural ecological features such as soil, slope, and natural biotic communities, and the eight land types shown in Figure 15–4 are assigned carefully designated uses that can be sustained without loss of productivity. Thus, land types I and II comprise level areas with good agricultural soils that can be continuously cultivated with only simple precautions such as crop rotations and strip cropping, whereas land types III and IV (steeper slopes) require greater restrictions if cultivated, for example, periodic fallowing, perennial crops, or rotated pastures. Types V through VII are not suitable for cultivation and should be used for permanent pasture, tree crops, or retained in their natural state (for naturally developed forestry and wildlife, for example). Type VIII (steep slopes, thin soil) is productive only in its natural state, as habitat for game, furbearers, forest products, or is valued for recreational, scenic, watershed protection, or other uses; these latter uses are often more important than any "crop" the land may yield. In most cases, marshes and swamps should probably now

be classified as type VIII lands because their value as water reservoirs and wildlife habitats outweighs their value as reclaimed farmland, since increased yields can now be obtained on existing farmland. For a review of rural land-use planning, see Graham (1944).

Despite past successes, the soil conservation profession has tended to "sit on its laurels" and is failing to move with the times. For example, too much effort is now being devoted to creating more farmland by channeling streams, draining marshes and swamps, and so on, at great public expense when nothing is being done to save existing prime farmland from destruction by ill-planned urban development. The training of land management professionals in state universities is badly out-of-date and needs considerable broadening and upgrading in the terms of quantitative science, social science, pollution ecology, and human ecology. In other words, the soil conservation movement in particular, and land management science in general, needs to go beyond its present rather exclusive farm, range, or forestry orientation to the consideration of the urban-rural landscape complex where the most pressing problems now exist (see E. P. Odum, 1969a).

As we have seen, land-use planning for the urban areas is now the critical need; it is the deteriorating quality of the urban and suburban environment rather than the eroding cotton fields that threaten the entire social and economic system. Drawing up

Figure 15–4.   Land-use classification. See text for explanation. (U.S. Soil Conservation Service Photo.)

and implementing good land-use plans is infinitely more difficult in urban areas than on farms or rural watersheds because of the human social problems involved, and especially because of the tremendous difference in economic value that we now give to different usages. Farms are generally valued and bought or sold as a whole (with all land types being considered as integral and valuable parts) in contrast to urban land where commercial property is valued manyfold higher than open space land, even though in the long run both are equally important in maintaining a quality city. Therefore, successful urban land-use planning (something that has not yet really been accomplished anywhere) will require a much stronger legal, economic, and political basis than was required for bringing about the reforms of soil conservation.

Table 15–3 contrasts the consequences of planned and unplanned land use in a 45,000 acre area near a large eastern city that now has approximately 20,000 people but is projected to increase to 110,000 or perhaps 150,000 by the year 2000. As shown in column 3, judicious planning of residential and other development can preserve a third of the area as open space, including adequate space for inexpensive and efficient seminatural tertiary treatment of both industrial and domestic wastes in ponds and well-planned land fills located in a large, open space waste disposal park (the concepts of waste disposal parks will be considered in more detail in the next chapter). As shown in Figure 15–5, such a planned development could be accomplished by (1) cluster development of residential housing around village or town centers with each unit separated by broad green belts and (2) by retaining stream valleys, steep slopes, lakes, marshes, aquifer recharge areas, and waste disposal areas free from houses, buildings, and other high-density uses. Without such planning, there could

Table 15–3. *Comparison of Unplanned (Uncontrolled) and Planned (Optimum Land-use) Development of a Rapidly Growing Urban-Suburban Area**

| | YEAR 1970, POPULATION 20,000 | PROJECTED YEAR 2000 POPULATION 110,000 | |
		Unplanned (uncontrolled) Development	*Optimum Use Plan*
Developed Area	13,000 acres	38,000 acres	30,000 acres
Residential	7,500	26,000	21,300
Commercial	500	700	630
Industrial	70	300	70
Institutional	2,500	5,500	3,000
Roads	2,500	5,500	5,000
Open Space ("undeveloped area")	32,000	7,000†	15,000
Waste disposal parks	0	0	1,000‡
Recreation parks	500	2,000	5,000
Farming and forestry	11,500	0	2,000§
Natural areas	20,000	5,000	7,000‖
Total acres	45,000	45,000	45,000
Per cent open space	71%	16%	33%

* Data adapted from a plan for a Maryland urban area prepared by Wallace-McHarg Associates, 2121 Walnut Street, Philadelphia, with addition of new concepts of open space plans by E. P. Odum.

† Projected uncontrolled development for a population density 150,000 would reduce open space to zero!

‡ Land surrounding or adjacent to industrial and municipal sewage plants on which extensive waste treatment ponds and other pollution abatement facilities can be located for efficient and low cost tertiary treatment of all wastes.

§ Could include not only truck farms but demonstration farms and forest as educational laboratories for schools and colleges.

‖ Including all steep slopes, ravines, flood plains, marshes, and lakes as existing in 1963 together with samples of existing mature forest areas and marginal farmlands restored to natural-area use.

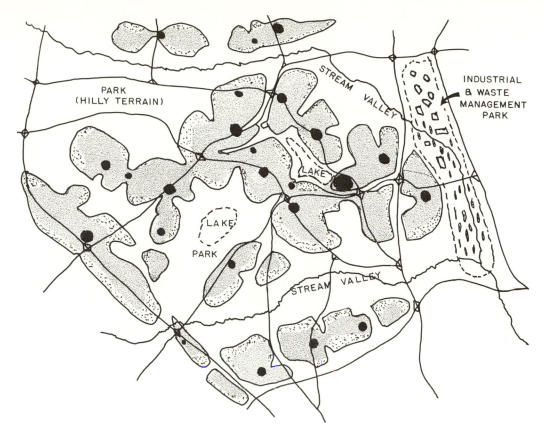

Figure 15–5. A planned urban development for 110,000 people on 45,000 acres which preserves environmental quality, natural beauty, and provides ample room for recreation and pollution abatement. A large amount of "open space" (unshaded areas in the diagram) is preserved (1) by cluster development of residential housing (two densities indicated) around town and village centers (black circles), separated by wide green belts or parks, and (2) by leaving stream valleys, lakes, and scenic areas unencumbered. Industrial plants are sited in a very large waste management park (right margin of the diagram) with room for waste treatment plants, oxidation ponds, land fills, and other means of degrading, recovering, containing, or recycling wastes and water. (Redrawn from Wallace, McHarg, Roberts and Todd Associates bulletin "Plan for the Valleys," page 59, with addition of the waste management park.)

very well be no open space by the year 2000, in which case we would have the same kind of urban blight, chronic pollution and social disorder that we now observe in the older, unplanned cities. Thus, planned cities are now as necessary as planned farms, but to project the models of Table 15–3 and Figure 15–5 to the real world of economics will require something much stronger than present day zoning concepts and a new breed of city planner who is part ecologist and part technocrat. The short-term profits that can be made by exploiting urban land are so huge that it is difficult for people to foresee the socioecologic backlashes and overshoots that accompany uncontrolled growth (see Chapter 9, Section 3). Society must quickly find a means of applying cybernetic principles to the urban machine.

McHarg's (1969) *Design with Nature* may well prove to be the most important book published in the 1960s. This beautifully illustrated volume shows how the natural landscape can provide the guidelines for quality urban development. For the first time in a single, easily readable book a documented case is made for total urban land-use planning on the basis that uncontrolled development (to paraphrase McHarg) spreading without discrimination obliterates the landscape with congestion and pollution and irrevocably destroys all that is beautiful and memorable no matter how well designed the individual home or subdivision.

Turning now to the totality of the urban and rural landscape, it is clear that (1) natural-area open space is a necessary part of man's total environment and (2) land-use

Figure 15-6. Land use—good, bad, and unfortunate. *A.* Poor agricultural land use that resulted in massive erosion, abandoned homes, and poverty. (U.S. Forest Service photo.) *B.* Good agricultural land use: strip cropping of corn and grass on type II land (see Figure 15-4). (USDA—Soil Conservation Service.)

Figure 15-6. *C.* Needless destruction and waste of wood in a logging operation. (Soil Conservation Service photo.) *D.* Understory trees, shrubs, soil, natural beauty, and water-holding capacity preserved following a well-managed logging operation. (U.S. Forest Service photo.)

Figure 15–6. *E.* Well-designed roadside seeded in a "semi-natural" community of two species of grass and a species of clover, which require only moderate fertilization and maintenance. (USDA—Soil Conservation Service.) *F.* Rehabilitation of logging roads and skid trails with a grass-clover mixture similar to that shown in *E;* unprotected logging roads often account for a major part of erosion following timber removal. (U.S. Forest Service photo.)

Figure 15–6. *G.* Reforestation of an old strip-mine area, a slow and difficult process in this instance because the land was not rehabilitated before planting of trees; many states now require miners to rehabilitate and replant immediately after mining. (U.S. Forest Service photo.) *H.* Uncontrolled acid drainage from a coal mine, a common and generally neglected problem in coal-mining regions. (U.S. Forest Service photo.)

Figure 15–6. *1.* Channelization of a stream designed to drain wetlands and move flood waters downstream more rapidly but at the risk of lowering the water table, of reducing fish and wildlife, and of downstream silting and flooding. (Georgia Game and Fish Commission photo.) *J.* Natural stream with vegetation-protected floodplain, insuring high water quality downstream at no cost to the downstream taxpayer. (Georgia Game and Fish Commission photo.)

Figure 15–6. *K.* Impoundment reservoir on an unprotected watershed that completely silted up in less than six years. (U.S. Forest Service photo.) *L.* Inexcusable (or perhaps we should say criminal) disregard for environmental design in the construction of an industrial complex. (USDA—Soil Conservation Service.)

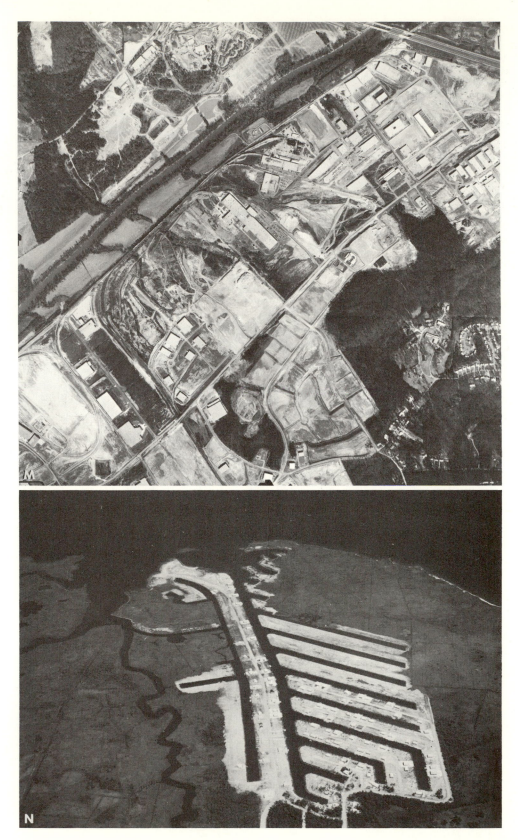

Figure 15–6. *M*. Unplanned and uncontrolled urban development, spreading without regard for protection of stream, floodplain, soil, pollution abatement, green belts, recreation, or natural beauty. (For a planned urban development that retains natural and cultural values, see Figure 15–5.) (USDA—Soil Conservation Service.) *N*. Undesirable housing development in an estuary that not only destroys seafood nursery grounds but is vulnerable to frequent storm damage (see also Figure 13–6). (U.S. Fish and Wildlife Service.)

planning can be the most effective means of preventing overpopulation of our own species by providing something similar to "territorial" control (see pages 209–211). Here is what we face in the next 50 years in the United States. As of 1970, the average population density of the U.S. (and, interestingly enough, also of the world) is *one person to ten acres of ice-free land area.* Even with the expected reduction in birth rates our population will *double in the next 30 to 50 years, leaving us with only about five acres (about two hectares) for every man, woman, and child.* Although as little as one-third of an acre can produce enough calories to sustain one person, the kind of quality diet we want—one that includes a lot of meat, fruit, and leafy vegetables—requires about 1.5 acres per person. We also need another acre per person to produce the fibers (paper, wood, cotton) and an additional half acre for roads, airports, buildings, and other entirely artificial habitats. This would leave only two acres per person for all the other diverse uses that make man something more than an "organic machine." What is often not realized is that an affluent nation requires more space and resources *per person* than an undeveloped nation so that both optimum and saturation densities come at a much lower level. For additional discussion of this "per capita" approach to determining the optimum population density for man, see E. P. Odum (1970).

We can now make a strong case for the proposition that adequate pollution-free living space, not food, should be the key to determining the optimum density for man. In other words, the size and quality of the "ecos," or "environmental house,"

should be the limiting consideration, not the number of calories we can relentlessly squeeze from the earth. This earth can feed more "warm bodies" sustained as so many domestic animals in a polluted feed lot than it can support quality human beings who have a right to a pollution-free environment, a reasonable chance for personal liberty, and a variety of options for the pursuit of happiness. A reasonable goal is to make certain that at least a third of all land remains in protective open space use. This means that such a portion of our total environment should be in national, state, or municipal parks, refuges, green belts, wilderness areas, and so on, or if in private ownership, it should be protected by scenic easements, zoning, or other definitive legal means. The dependence of the city on the countryside for all its vital resources (food, water, air, and so on) and the dependence of the country on the city for economic resources must become so widely recognized that the present political confrontation that exists between the rural and urban populations is obliterated. Somehow, ecology and economics must be merged (as was pointed out on page 39, these two words are derived from the same root). Evolving into something like what Boulding (1966) has called the "spaceship economy" must obviously take place before successful urban-rural integration can occur; in the meantime, urban and rural planners will have to do the best they can with their separate projects.

Figures 15–6A through *N* contrast good and bad land use; these photographs tell their own stories.

POLLUTION AND ENVIRONMENTAL HEALTH

Chapter 16

Pollution is an undesirable change in the physical, chemical or biological characteristics of our air, land and water that may or will harmfully affect human life or that of desirable species, our industrial processes, living conditions, and cultural assets; or that may or will waste or deteriorate our raw material resources. Pollutants are residues of the things we make, use and throw away. Pollution increases not only because as people multiply the space available to each person becomes smaller, but also because the demands per person are continually increasing, so that each throws away more year by year. As the earth becomes more crowded, there is no longer an "away." One persons's trash basket is another's living space.*

(To the "throw-away" pollutants we must add pollutants that are the inevitable by-products of transportation, industry, and agriculture; as these human activities expand, so does pollution.)

We have already made a strong case for the proposition that pollution is now the most important limiting factor for man (see especially pages 36, 406, and 431). The effort that must now be put into pollution abatement and prevention may well provide the negative feedback that will prevent man from completely raping the earth's resources, and thereby destroying himself. The problem is different only in aspect in the sharply divided world of man: in the undeveloped nations (70 per cent of the world's people) shortage of available food and resources is associated with chronic pollution and disease caused by human and animal wastes, while in the affluent or developed nations (30 per cent of the world's people) agroindustrial chemical pollution is now more serious than organic pollution. In addition, global pollution of air and water mostly emanating from the developed countries threatens everyone (see Singer, 1969).

In all the chapters of Part 1 of this text the relevance of ecological principles to both the causes of and the cures for particular pollution problems were pointed out. Since, in order to cope with pollution both on a local and global scale, the ecosystematic or holistic approach is necessary, we will attempt in this chapter to present an overview, followed by a brief summary of several of the problem areas that are attracting widespread public attention. Reforms and solutions in these especially critical areas would point the way to total solution.

By far the best textbooks on pollution are the panel reports prepared by the National Academy of Sciences or the President's Science Advisory Committee, for example, the Tukey report (1965) "Restoring the Quality of Our Environment," the Spilhaus report (1966) "Waste Management and Control," the Daddario report (1966) "Environmental Pollution," and the Miller report (1967) "Applied Science and Technological Progress." These and other reports that undoubtedly will be issued at frequent intervals in the future can be obtained at very modest cost from the Superintendent of Documents or the National Academy of Sciences, Washington, D.C. A brief "paperback" review of the different kinds of pollution and their abatement is provided by Benarde (1970). Introductions to the study of water pollution are provided by Hynes (1960), Hawkes (1963), and Warren (1971). Other references will be noted in the sections that follow.

1. THE COST OF POLLUTION

The cost of pollution is measured in three ways, all of which add up to a terrible and increasingly intolerable burden to human society: (1) The loss of resources through unnecessary wasteful exploitation, since, as the National Academy's report puts it,

---

* From the Introduction to "Waste Management and Control," a report by the Committee on Pollution, National Academy of Sciences, 1966.

"pollution is often a resource out of place." (2) The cost of pollution abatement and control, a sample projection of which is shown in Figure 16–1A; note that while the cleanup of sewage and solid wastes (refuse) is now the most expensive, the cost of abatement of the much more poisonous wastes from motor vehicles and power generation is projected to increase 100 times in the next 30 years (assuming continuation of the kind of uncontrolled urban growth that was discussed in Section 8 of the preceding chapter). (3) The cost in human health. Recognition of this aspect of pollution cost will probably do more to alert egotistical and self-centered man to the rising danger than the other kind of costs, which can be too well hidden by short-term "cost-benefit" manipulations at the local level. Figure 16–1B is a dramatic model of what is happening in the public health sector in the United States. As human mortality from infectious diseases shows a precipitous decline, mortality and sickness from environmentally related respiratory diseases and cancer has

shown an equally precipitous rise. In a recent review of the human health cost of air pollution Lave and Seskin (1970) estimate that a 50 per cent reduction in air pollution in urban areas alone could save two billion dollars annually in the aggregate cost of medical care and work hours lost in sickness, and this does not include the "cost" of human misery or death and disability caused by automobile and industrial accidents. These authors document a strong relationship between all respiratory disease and air pollution; they even go so far as to suggest that there is a great deal of evidence connecting all mortality from cancer with pollution. As environmental stresses on the human body increase, many medical scientists fear a "backlash" in infectious diseases not only because of lowered body resistance, but because viruses (which are linked with cancer in the opinion of many) and other disease organisms will increasingly slip through water treatment and food processing plants as the quality of water and food at the intake deteriorates. Both water

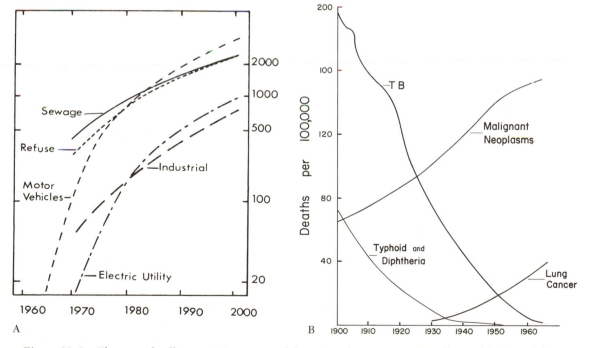

Figure 16–1. The cost of pollution. A. Projection of the cost to the consumer, in millions of dollars, of the control of primary pollutants in the Delaware River Basin. Note that sewage and refuse (solid wastes) provided the chief drain on the tax dollar in 1970, but that the cost of controlling pollution from industrial development, motor vehicles, and power generation will increase 10- to 100-fold by the year 2000 unless future expansion is planned and controlled. (Graph prepared from data in "Waste Management and Control," National Academy of Sciences (see Spilhaus), 1966.) B. The dramatic decline in mortality from infectious diseases (tuberculosis, typhoid, and diphtheria) accompanied by the equally dramatic rise in malignancies believed to be related to pollution. (Data from R. D. Grove and A. M. Hetzel: "Vital Statistics rates in the United States, 1940–1960." National Center for Health Statistics, PHS Publ. No. 1677.)

treatment and waste treatment (up to now considered as separate problems) must now be linked into a "recycle" system, as will be briefly discussed in Section 3 (see also page 87). The behavioral consequences of crowding and the breakdown in social structure that accompanies any decline in the quality of the environment has already been noted (see page 250). Maurice Visscher (1967) in another National Academy report states that mental health is probably now the major cause of human morbidity and disability.

2. THE KINDS OF POLLUTION

Classifying pollution can be as difficult and confusing as classifying lakes or other natural phenomena (see Chapter 11, page 312). Classifications according to environment (air, water, soil, etc.) and pollutant (lead, carbon dioxide, solid wastes, etc.) are, of course, widely used approaches. Large books can and will be written about each of these components. However, from the standpoint of the totality of pollution abatement (that is, from the ecosystem viewpoint) it is important that we first recognize two basic types of pollution.

First are the *nondegradable pollutants*, the materials and poisons, such as aluminum cans, mercurial salts, long-chain phenolic chemicals, and DDT, that either do not degrade or degrade only very slowly in the natural environment—in other words, substances for which there are no evolved, natural treatment processes that can keep up with rate of man-made input in the ecosystem. Such nondegradable pollutants not only accumulate but are often "biologically magnified" as they move in biogeochemical cycles and along food chains (see page 74 for explanation of the concept of "biological magnification"). Also, they frequently combine with other compounds in the environment to produce additional toxins. The only possible "abatement" for such pollution is expensive removal or extraction from the environmental life-support system. Although this is possible in a small, temporary spacecraft (see Chapter 20), removal of many such pollutants from the biosphere would be virtually impossible (how could we remove lead from the air we breathe short of requiring 200 billion people to wear gas masks?). The obvious and sensible solu-

tion (but one that is easier to define than to effect) is to outlaw the dumping of such materials into the general environment (or at least control the rate of input so as to avoid toxic buildup) or to stop production of such substances entirely (i.e., to find more degradable substitutes).

Second are the *biodegradable pollutants*, such as domestic sewage, that can be rapidly decomposed by natural processes or in engineered systems (such as a municipal sewage treatment plant) that enhance nature's great capacity to decompose and recycle. In other words, this category includes those substances for which there exist natural waste treatment mechanisms. Heat, or thermal pollution, can be considered in this category since it is dispersible by natural means, at least within the limits imposed by the total heat budget of the biosphere (see Chapter 3, Section 1).

Problems arise with the degradable type of pollution when the input into the environment exceeds the decomposition or dispersal capacity. Current problems of sewage wastes result mostly from the fact that cities have grown much faster than treatment facilities. Unlike pollution by toxic nondegradable materials, pollution by degradables is technically solvable by a combination of mechanical and biological treatment in seminatural waste disposal parks (this concept will be developed in Section 4). Again, there are limits to the total amount of organic matter that can be decomposed in a given area and an overall limit to the amount of CO_2 released into the air (see pages 33 and 98). If we are to avoid exceeding the overall limits of the biosphere, we have to preserve something like 4 to 5 acres of biologically productive land-and-freshwater space per person (plus the oceans), as was suggested in Section 8 of the preceding chapter.

The contrast in the effect of the two basic kinds of pollution on systems energetics is shown in the graphic model in Figure 16–2. Degradable pollutants that provide energy (organic matter) or nutrients (phosphates, carbonates, etc.) will increase the productivity of the ecosystem by providing a subsidy (see pages 45 and 289) when the rate of input is moderate (upper graph, Fig. 16–2). At high rates of input a critical range is reached that is frequently characterized by severe oscillations ("boom and bust" algal blooms, for example). Additional input

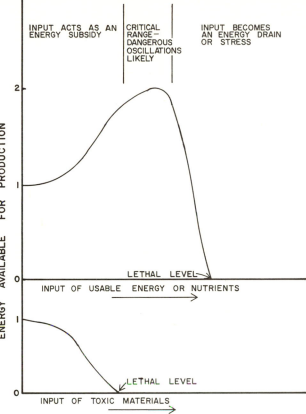

THE ENERGETICS OF POLLUTION

Figure 16–2. Schematic model of the effects of the two types of pollution – degradable organic (upper graph) and nondegradable toxic (lower graph). See text for explanation.

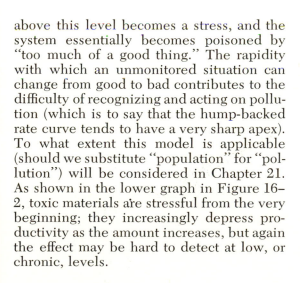

above this level becomes a stress, and the system essentially becomes poisoned by "too much of a good thing." The rapidity with which an unmonitored situation can change from good to bad contributes to the difficulty of recognizing and acting on pollution (which is to say that the hump-backed rate curve tends to have a very sharp apex). To what extent this model is applicable (should we substitute "population" for "pollution") will be considered in Chapter 21. As shown in the lower graph in Figure 16–2, toxic materials are stressful from the very beginning; they increasingly depress productivity as the amount increases, but again the effect may be hard to detect at low, or chronic, levels.

3. THE PHASES OF WASTE TREATMENT

It is customary to consider the treatment of degradable wastes in three stages: (1) *primary treatment,* a mechanical screening and sedimentation of solids (which are burned or buried); (2) *secondary treatment,* a biological reduction of organic matter; and (3) *tertiary or advanced treatment,* the chemical removal of phosphates, nitrates, organics, and other materials. The complete three-step treatment process for liquid wastes is shown in Figure 16–3. As already pointed out (see page 28), secondary treatment is carried out by an engineered biological system in which microorganisms decompose organic matter in the same manner as occurs naturally in soils and sediments. The most common design is the activated sludge system which requires electric pumps or other energy to aerate and circulate the material. Another system is the "trickling filter" system in which the primary treatment effluent moves by gravity over stones or racks of plastic surfaces that create an aerated surface resembling the rapids in a natural stream. The role of detritus-feeding invertebrates in these microbial decomposition systems was discussed on page 31. A study of a secondary

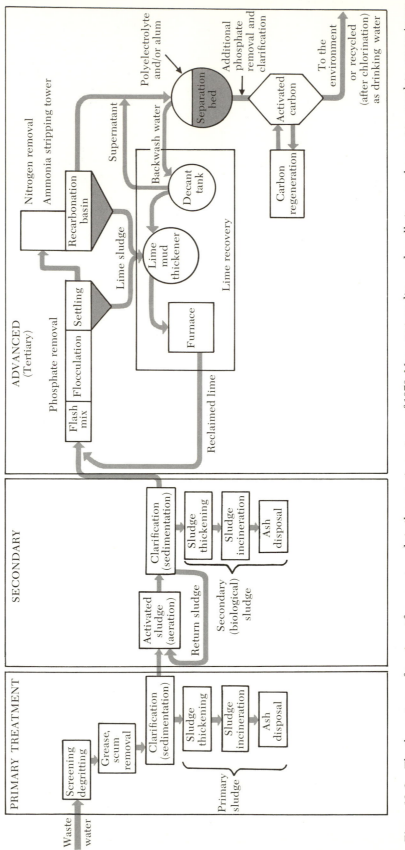

Figure 16–3. The three stages of treatment for sewage and similar organic wastes as of 1970. Many metropolitan and small city areas have yet to complete even primary treatment facilities, although most have plans to achieve secondary treatment in the next few years. Tertiary treatment must be added as rapidly as possible or cities will either choke in their own wastes, ruin the quality of the countryside, or both.

treatment plant makes an excellent (and very relevant) exercise for an ecology class. For a guide to the ecology of secondary treatment, see Hawkes (1963).

Recently primary and secondary treatment have been combined in compact "package" plants that are especially favored for suburban developments and small towns. It should be remembered that compacting the treatment into smaller space requires an increased energy input in terms of power, and also more sophisticated maintenance in terms of personnel who must operate and repair the machinery; any breakdown means instant injection of raw sewage into the environment. This, of course, illustrates again the principle that increased complexity and efficiency in the use of space requires an increase in energy expenditure for "disorder pumpout" (see page 38).

In terms of construction and maintenance costs the cheapest systems are *oxidation or waste stabilization ponds,* shallow bodies of water (4 to 5 feet deep) constructed so as to expose a maximum surface area to the air. Wastes are pumped into the bottom of the pond, and algae, which grow vigorously in the upper lighted zone, provide the aeration. Such ponds operate as aerobic-anaerobic systems in the same manner as naturally fertile lagoons (see Figure 12–13). Use of a seminatural design of this sort requires a large amount of space—about one acre for the treatment of household wastes of 100 people—and judicious maintenance. Adding devices for mechanical aeration can, of course, increase the per acre treatment capacity. Waste stabilization ponds are now widely used for treatment of sewage from suburban developments, especially in warmer climates, and they are also effective in the partial treatment of industrial wastes from paper or textile mills, food processing plants, oil refineries, and so on. Undoubtedly such ponds will be used more in the future to treat wastes from domestic animal feed lots, poultry plants, and livestock barns; such animal wastes are now largely untreated and are producing very serious pollution of waterways (recall that there are about five times more "equivalent" domestic animals than people in this world; see page 55). It should be emphasized that the waste stabilization pond is really a "conversion" system, not a complete treatment system; unsanitary organic matter is converted into sanitary algal material and nutrients that are exported to the natural environment where there must be adequate space capacity and food chains to handle it. Harvesting algae for animal food or using pond effluents for aquaculture, irrigation, and other useful purposes are obvious possibilities that need more research.

The effluent from even relatively complete secondary treatment is, of course, still highly polluting in terms of eutrophication (see page 107) and is unfit for direct human use. While the technology of secondary treatment is advanced, tertiary treatment is still mostly in the pilot plant stage. Most cities are still fighting a losing battle to provide adequate secondary treatment when they should be moving on to the complete recycle of tertiary treatment. Thousands of small towns and suburban areas have no treatment facilities at all, or only crude primary treatment. As indicated above, industrialized agriculture scarcely treats any of its burgeoning wastes (see Brady, 1967, for a review of agricultural pollution). Nature is expected to do the tertiary treatment, which she can do very effectively if allowed enough room; the trouble begins when all the "natural treatment" areas become human living space covered with additional waste-producing urban, agricultural, and industrial development. Then man must turn to artificial tertiary treatment, which is several times as expensive as conventional secondary treatment. An estimate of the relative cost of the different stages of waste treatment is given by Stephens and Weinberger (1968) as follows:

Primary treatment .... 3 to 5¢ per 1000 gallons
Secondary treatment .... 8 to 11¢ per 1000 gallons
Tertiary treatment to
    remove nutrients .... 17 to 23¢ per 1000 gallons
Tertiary treatment to pro-
    duce drinkable water...... 30 to 50¢ per 1000 gallons

The lower figures refer to the cost for large plants that process 100 million gallons per day. The capital cost of such a plant for complete tertiary treatment is of the order of 25 million dollars as compared to 20 million for secondary treatment and 10 million for primary treatment. Producing drinking water by tertiary treatment (complete recycle) is cheaper than desalination (presently estimated to cost at least one dollar per 1000 gallons), and it may soon be cheaper than piping it from distant water sources. The quality of recycled water would also be better than that which millions of urban dwellers now drink.

In less densely populated areas there is increasing interest in using terrestrial ecosystems as well as aquatic ones for tertiary treatment. This makes sense because the surface of terrestrial environments is many times greater than that of freshwater environments. Experiments in which land areas are irrigated with waste water from secondary treatment plants by means of a system of sprinklers indicate that as much as two inches of water per week can be added to the normal rainfall in eastern United States without changing the quality of the ground water (see Parizek *et al.*, 1967; Kardos, 1967; Sopper, 1968; Bouwer, 1968). That is, phosphates, nitrates, and other nutrients added at this rate were filtered out by the soil-vegetation layer. Growth of crops, pastures, and young forest plantations were also enhanced by this "spray irrigation" of waste nutrients. Obviously a land "filter" of this sort has a greater long-term capacity if there is a harvest removal of nutrients by cropping or grazing. However, past experience with all kinds of irrigation warns that gradual accumulation of nutrients or salts will occur with high-input rates. Experiments currently being conducted will probably have to be continued for many more years before the true tertiary treatment capacity of different kinds of terrestrial watersheds can be determined. In the meantime, it would be prudent to act on the principle that the optimum input will be less than the maximum input that seems to be tolerated in a 3 to 5 year experiment (see page 77 for a discussion of the principle of optimization).

4. THE STRATEGY OF WASTE MANAGEMENT AND CONTROL

Man has three basic options in dealing with waste materials: (1) dump them untreated into the nearest convenient environment such as the air, a river, lake, soil, well, or ocean; (2) contain and treat them within a delimited environmental waste management park where engineered seminatural ecosystems such as oxidation ponds, spray-irrigated forests, and land fills do most of the work of decomposition

and recycle; (3) treat them in artificial chemomechanical regeneration systems.

The first option is based on the philosophy that "the solution to pollution is dilution"; it has been and still is the principal waste disposal practice employed almost everywhere. Thus, industries and cities have tended to concentrate along waterways, which provide free sewers. Obviously, this option is no longer tenable and must be phased out as fast as possible, no matter what the cost.

The second option provides the most economical method of avoiding general environmental pollution by the fairly dilute but large volume wastes that now so badly reduce the quality of man's living space and endanger his health. Setting aside large areas for seminatural waste treatment would also preserve valuable open space that not only protects environmental quality in general, but provides other uses as well (food and fiber production, atmospheric gas exchanger, recreation, and so forth). A waste management park was included in the planned urban area illustrated by the models in Table 15–2 and Figure 15–5. Two examples of the waste park design are shown in Figure 16–4; the upper diagram shows how wastes from oil refineries are treated by means of a series of ponds, and the lower diagram shows a hypothetical plan for treatment of thermal and radioactive pollution from a new atomic power plant. In both instances the water that leaves the delimited treatment area and enters the general public environment is not in any way "polluted." In many cases, waste treatment ponds will, in the words of H. T. Odum (1967 and 1970), "self-design" to cope with the input and thus require a minimum of engineering and maintenance by man. A large, well-designed and efficiently operated sanitary land-fill for the disposal of solid wastes should be another component of the waste disposal park concept.

To exercise the sensible option of letting nature do a lot of the work, large areas of land and water have to be set aside for this purpose, which, as already indicated, also provides one of the best safeguards against the kind of over-development discussed in the preceding chapter (Section 8). Thus, new industries and municipal treatment plants should no longer be placed on the banks of streams or in the middle of congested areas where people live, but should be "zoned" or "sited" in the middle of natural areas large enough for treatment of degradable wastes and the containment of the poisonous wastes (such as radioactive wastes, acids, etc.) that should never find their way into the general environment. In the past, urban planners considered 50 to 100 acres an adequate space for an "industrial park." For self-contained waste management 1000 to 10,000 acres may be needed for a large industrial complex (see legend, Figure 16–4; also see page 466). Recycled water and recovery of useful products from wastes should more than pay for the cost of the land. Separating industry and airports from living space also pays dividends in noise abatement, as described in Section 7 of this chapter. The greatest obstacles for "designing with nature" in this manner are legal, economic, and political (see Section 6). If private industry and municipalities do not or cannot (because of inadequate laws) plan ahead, then man will be forced more and more to turn to the much more expensive and technically difficult third option of artificial treatment.

Abiotic treatment and recycling, of course, is necessary for some types of wastes, especially in densely populated industrial areas. Mechanical treatment is probably the only option for some components of air pollution which have to be stopped or reduced at the source. If this is not technically possible, a substitute energy source or industrial procedure must be sought since, as already indicated, we cannot tolerate much longer the costs of air pollution. If we are backed into a corner and have to turn to expensive artificial treatment of biodegradable wastes as well as the poisons, then who is going to pay the bill! As will be described in Chapter 20, a very sophisticated system for the mechanical treatment of wastes and for regeneration of air and water has been developed for spacecrafts, but the per capita cost is staggering.

We shall close this section on a positive note. The story of Lake Washington, a large body of water surrounded by the city of Seattle and its suburbs, provides a good demonstration of how beleaguered urban areas can reverse the insidious trend of declining water quality *by attacking the problem as a whole through a federation of city,*

county, and district governments. The sequence of deteriorating water quality, public outcry, political action, bond issues, sewage diversion, and recovery of water quality is described in detail by Edmondson (1968) and is cited in the popular press as a model for other urban centers (see, for example, an article by Earl Clark in *Harper's Magazine,* June, 1967). Four indices of water quality in Lake Washington are plotted in Figure 16–5 for the period 1933 to 1969. Two of the indices represent important physical water characteristics, and two pertain to the diversity and composition of the diatom component of the phytoplankton (i.e., they are community indices). The numbered vertical lines signify important events in the lake's history as listed in the legend of the figure. In the 1950s 11 different municipalities dumped increasing amounts of secondarily treated sewage into the lake, resulting in a progressive cultural eutrophication (i.e., nutrient enrichment, see page 16). Nuisance algal blooms and oxygen depletion in the hypolimnion (see page 58) provided warnings of the deterioration and attracted the wide public attention that paved the way for political action. In 1958, the "Metro" federation was formed and a 120 million dollar revenue bond issue put up to public vote. At first the issue failed because the city voted "yes" and the suburbs "no" (illustrating a chronic American political problem stemming from the fact that suburbs do not realize they are part of the whole urban system), but on a second try the "Metro" bond issue passed. In 1963, about a third of the sewage was diverted from the lake, and by 1968, almost all of it had been diverted. As shown in Figure 16–5, recovery of the lake has been dramatic with all four indices reversing. Edmondson (personal communication) believes that the lake will at least return to its 1930 condition in a few more years. Although large quantities of phosphates and other nutrients are still in the lake, they tend to be buried in the sediments and thereby are withdrawn from the annual biogeochemical cycle. An important aspect of this pollution abatement success story was the fact that University of Washington limnologists had for many years carried on basic studies of the lake so that trends and causes could be documented with certainty; without such information political action might have been delayed, perhaps until it

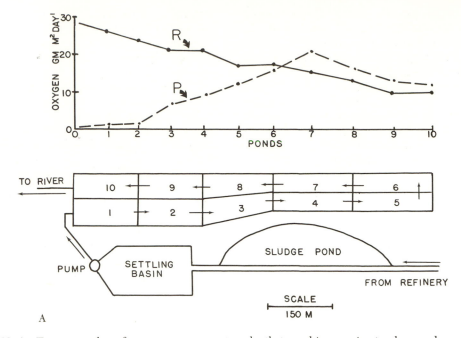

Figure 16-4. Two examples of waste management parks that combine semi-natural secondary and tertiary treatment of wastes with watershed protection, recreation, forestry, fish culture, and agriculture as by-products operations. A. Treatment of wastes from an oil refinery in Oklahoma is inexpensively accomplished by passing the refinery effluents slowly through a coupled series of basins and ponds (lower diagram). As shown in the upper graph, monitoring oxygen production (P) and utilization (R) in each pond demonstrates that the "self-designed" natural pond communities degrade the organic matter and establish a good balance between P and R by the time the water reaches the tenth pond and is released into public waterways. (Redrawn from Copeland and Dorris, 1964.)

was too late (note the precipitous increase in rate of decline between 1960 and 1963, indicating that the lake may have been "rescued" just in time).

Finally, it must be pointed out that pollution abatement in Lake Washington provided no permanent solution; the lake was saved by merely diverting the effluents to a larger body of water—Puget Sound! The next step is tertiary treatment, which will require another cycle of public education, concerted political action, and bond issues. Perhaps this time, however, it will not be necessary to wait until the ocean deteriorates! As Hasler (1969) has pointed out, case histories such as Lake Washington show that cultural eutrophication can be reversed (see also page 254).

5. MONITORING POLLUTION

Successful pollution abatement, of course, depends not only on treatment and control but also on efficient monitoring of the general environment so that we will know for sure when control measures are needed and whether those in operation are working. Monitoring takes two basic forms: (1) direct measurement of the concentration of pollutants or of key substances, such as oxygen, which are depleted by pollution and (2) the use of biological indices, which range from bioassays with microorganisms and B.O.D. measurements (see page 16) to the kind of total community indicators discussed in Chapter 5, Section 6, and Chapter 6, Section 4.

As an example of the first type of monitoring, air pollution over the large California cities is now monitored by detectors, mounted on an airplane, which daily measure and plot the concentrations of SO_2, NO_2, CO, and other pollutants, over a large regional area. Air pollution indices have become a part of weather reports in many cities. The need for world-wide monitoring of carbon dioxide has already been noted. Biological indices are widely used in monitoring water pollution. In addition to diversity indices (see Figure 6-5) and general species indicators such as shown in Figure

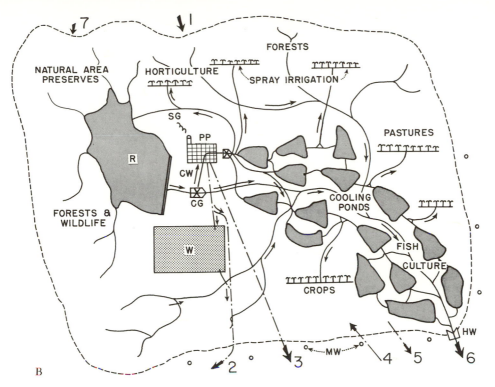

Figure 16-4. *B.* Schematic design for a waste management park for the atomic power plant (PP) of the future which is located in a natural watershed basin (outlined by the dashed lines). Waste heat (i.e., thermal pollution) in the reactor cooling water (CW) flowing from a large storage reservoir (R) is completely dissipated by evaporative cooling from the network of shallow ponds and spray irrigation systems. The warm ponds may be used for fish culture, sport fishing, or other recreation purposes. Irrigation of portions of the terrestrial watershed increases the yield of useful forest or agricultural products, while at the same time water recycles through the "living filter" of the land back into streams, ponds, and ground water. Low-level nuclear wastes and solid wastes are contained within a carefully managed land fill area (W); high-level nuclear wastes in spent fuel elements are exported to a special nuclear burial ground located outside of the management park. Stream flow, ground water, and stack gases are continuously monitored by hydrological weirs (HW), monitoring wells (MW), and stack gas control systems (SG) in order to make certain that no air or water pollution leaves the controlled area. The chief inputs and outputs for this environmental system include (see numbered marginal arrows): 1, input of sunlight and rainfall; 2, export of nuclear wastes to burial grounds; 3, electric power to cities, etc.; 4, input of nuclear and other fuels; 5, output of food, fibers, clean air, etc.; 6, downstream flow of clean water for agriculture, industry, and cities; 7, public and professional use for recreation, education, and environmental research. The size of such a complete waste management park will depend on regional climates and topography, and on the amount of electrical or other energy diverted to power cooling; but something on the order of 10,000 acres for a 2500 megawatt power plant would be the minimum needed to insure 100 per cent pollution control and allow for accidents and mechanical malfunctions as is explained in Chapter 17, page 466. However, such waste treatment capacity could also support a certain amount of light industry within the park. Heavy industry should be located within its own waste management park.

16–6, there are numerous indices of community function that can be useful—for example: the P/R ratio (see Figure 16–4A), the ratio of chlorophyll to bacterial biomass (see page 152), the mean size of organisms (pollution favors small organisms over large ones; see Oglesby, 1967, and Menhinick, 1964), the amount of hemoglobin in animal biomass as an index of low oxygen, the amount of blue-green algal pigment as an index to carbohydrate pollution, and many other indices that need to be carefully studied. Very often the community will contain

more "information" about total effects of pollution than can be deduced from the measurement of individual factors. The challenge to ecological research is to find quick ways to "read" this information!

For a review of the chemical approach, see the American Chemical Society's 1969 report, "Cleaning Our Environment; the Chemical Basis for Action," and for a review of the biological approach, see the Department of Interior's book, "The Practice of Water Pollution Biology" (edited by Machenthun, 1969).

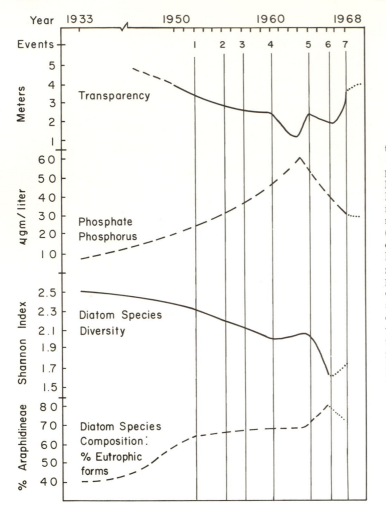

Figure 16–5. Reversal of cultural eutrophication in Lake Washington, Seattle —a story of deteriorated water quality, political action and partial recovery. Numbered events are as follows: 1. Eleven separate sewage plants discharging treated sewage into the lake. 2. First noticeable bloom of nuisance algae (*Oscillatoria*). 3. First detected oxygen depletion in bottom water (hypolimnion) during the summer. 4. Metro government sewage project legislation passed (1960). 5. First step in sewage diversion (1963). 6. Second step in sewage diversion (1965). 7. All sewage diverted (1967). The trends in four water quality indices (two physical and two biological) are shown. See text for further explanation. (Graph based on Edmondson, 1968, and Stockner and Benson, 1967).

6. ENVIRONMENTAL LAW*

The weakest link in pollution abatement strategy, as in land-use planning, is the inadequate legal protection of environmental quality and the consumer. As outlined in Chapter 9, one of the major principles relating to the development of ecological systems has to do with the distribution of energy in the system. When the ecological system is young, the major flow of energy is directed to *production*, that is, to growth and the building of a complex structure; but as the population density approaches the saturation level, the ecological system matures in the sense that a greater proportion of the available energy is shifted to *maintenance* of the complex structure that has been created. A parallel, of course, exists in the development of human society, a parallel that is justified because man and environment do constitute an ecological system. It is more than coincidental that we begin the new decade of the 1970s with mounting concern for human rights and environmental quality along with increasing unrest among young people and those who maintain our highly complex and technological society. The ecologist views all of these trends as part of a perfectly natural and predictable expression of the basic need to develop new strategies adapted to the mature system.

Up to now the greatest economic rewards and the strongest legal protection have been given to those who produce, build, pollute, and exploit nature's riches; this, we can

---

* This section is modified from the preface, written by E. P. Odum, to the book entitled *Environment Law Review – 1970*, edited by Floyd Sherrod and published by Clark Boardman and Co., New York.

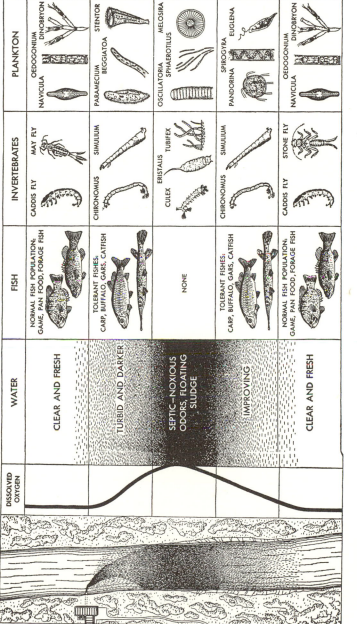

Figure 16–6. Pollution of a stream with untreated sewage and the subsequent recovery as reflected in changes in the biotic community. As the oxygen dissolved in the water decreases (curve to the left), fishes disappear and only organisms able to obtain oxygen from the surface (as in *Culex* mosquito larvae) or those which are tolerant of low oxygen concentration are found in zone of maximum organic decomposition. When bacteria have reduced all of the discharged material the stream returns to normal. (After Eliassen, Scientific American, Vol. 186, No. 3, March, 1952.)

argue, is quite proper in the pioneer stages of civilization, since man must first to some extent subdue and modify his environment in order to survive in it. Now it is obvious that at least equal rewards and protection must be given to those people, professions, and industries that maintain the quality of human existence; survival in the future depends on finding a balance between man and nature in a world of limited resources. This does not mean that man must revert to nature, but it does mean that he will need to go back to some of the good, common-sense, old-fashioned things such as return-able bottles, walking, and human concern for one's neighbors. Some things, one-way bottles for example, that we once thought represented "progress" turned out instead to be insults to both man and nature. As this transition occurs the basis for economic development shifts from exploitation to re-cycle, from throw-away to reuse, from quan-tity to quality. Legal procedures and legal education must adapt accordingly since the law, backed by strong public opinion, is the chief "negative feedback" that establishes the necessary controls. The traditional "pri-vate client-oriented" law must now be broadened to include greater emphasis on public and environmental law. University law schools, which have tended to be ultra-conservative and isolated from other aca-demic schools and departments, need to move out of the legal ivory tower and estab-lish better communication links with the environmental and social sciences and to encourage their students to seek better training in these and other relevant sub-jects. An inventory of environmental law as it is now understood and practiced (see, for example, Sherrod, 1970, and Baldwin and Page (eds.), 1970) reveals an urgent need to develop more comprehensive pro-cedures that will counter the excessive fragmentation and help resolve the con-tradictions that now make it so difficult to deal with pollution (and many other problems) on a legal basis. Not only is envi-ronmental law inadequate at the local and national levels, it does not even exist at the international level despite the obvious need to protect the atmosphere and the oceans. There is no more important area than en-vironmental law, a field that provides un-limited challenge to the motivated youth of today.

Murphy (1967), in an interesting book entitled *Governing Nature,* points out that restrictions and governmental regulations alone are inadequate to avoid pollution; there must be economic and legal incen-tives as well. He discusses effluent charges, cost internalization of product development to include waste treatment and recycle as well as production costs, tax relief for indus-tries that plan ahead on waste disposal, and other means of providing rewards for group behavior in the public interest (see also Hardin, 1968, and Crowe, 1969).

7. SOME PROBLEM AREAS

Air Pollution

The magnitude of air pollution in an industrialized country such as the United States is indicated by the data in Table 16–1. These data (1966) only show the *rela-tive importance* of pollutants and sources because the *absolute amount* increases by the year. Although the country-wide and global aspects are serious enough (see Singer, 1969), it is the local concentrations that build up over such cities as Tokyo, Los Angeles, and New York during temperature inversions (i.e., air trapped under a warm upper layer that prevents the vertical rise of pollutants) that cause the greatest imme-diate concern. As already suggested (page

Table 16–1. The Relative Magnitude of Air Pollution in the United States*

	Million Metric Tons/Yr.	
By Pollutant		
Carbon monoxide	65	(52%)
Oxides of sulfur	23	(18%)
Hydrocarbons	15	(12%)
Particulate matter	12	(10%)
Oxides of nitrogen	8	(6%)
Other gases and vapors	2	(2%)
By Source		
Transportation	74.8	(59.9%)
Industry	23.4	(18.7%)
Generation of electricity	15.7	(12.5%)
Space heating	7.8	(6.3%)
Refuse disposal	3.3	(2.6%)
Total	125.0	

* Data from the report of the National Academy of Sciences, "Waste Management and Control" (1966). Increases are projected for at least 20 years. The situa-tion will get worse before (or if?) it gets better.

432), air pollution provides the negative feedback signal that may well save industrialized society from extinction because: (1) it provides a clear danger signal that man must somehow soon "power down" in the concentrated use of industrial energy, (2) everyone contributes to it (by driving a car, using electricity, buying a product, and so on) and suffers from it, so it cannot be blamed on a convenient villain, and (3) a solution must evolve out of holistic consideration since attempts to reduce any one source, or any one pollutant, as a separate problem is not only ineffectual but might only divert that pollution to one of the other categories.

Air pollution also provides an excellent example of synergism (see page 92) in that combinations of pollutants react in the environment to produce additional pollution which greatly aggravates the total problem. For example, two components in automobile exhaust combine in the presence of sunlight to produce new and even more toxic substances, known as "photochemical smog," as follows:

$$\text{Nitrogen oxides} + \text{Hydrocarbons} \xrightarrow[\text{in sunlight}]{\text{Ultraviolet radiation}}$$

Peroxyacetyl nitrate (PAN) and Ozone (O_3)

Both secondary substances not only cause eye-watering and respiratory distress in man, but are extremely toxic to plants; ozone increases respiration of leaves, killing the plant by depleting its food, while PAN blocks the "Hill reaction" in photosynthesis, thus killing the plant by shutting down food production (see Taylor et al., 1961, and Dugger et al., 1966). The tender varieties of man's cultivated plants become early victims so that certain types of agriculture and horticulture are no longer possible in the vicinity of the big cities. Other photochemical pollutants that go under the general heading of polynuclear aromatic hydrocarbons (PAH) are known carcinogens.

Another dangerous synergism results when SO_2, which might normally be carried away and oxidized in the atmosphere, adsorbs on particulate pollution (dust, fly ash, etc.), contacts wet tissue (such as the inside of one's lungs) or moisture droplets, and turns into sulfuric acid! Such "acid" pollution is not only a health hazard, but it corrodes metal and limestone, causing millions of dollars in damage to man-made structures. There is also another kind of

synergism between cigarette smoking and air pollution. Air pollution in a city can subject the nonsmoker to the same level of blood poisoning by carbon monoxide as is experienced by a one-pack-a-day smoker (see Goldsmith and Landaw, 1968). According to Lave and Seskin (1970), the city dweller who smokes runs 10 times the risk of lung cancer as does the rural man who does not smoke.

For additional information on air pollution, see the 1965 AAAS report (Dixon, chairman) and Stern (1968).

Insecticides

To establish an objective overview of the highly controversial subject of pest control it may be helpful to think in terms of what Carroll Williams (1967) calls the "three generations of pesticides," namely, (1) the botanicals and inorganic salts (arsenicals, etc.), (2) the DDT generation (organochlorines, organophosphates, and other "broad-spectrum" poisons), and (3) the hormones ("narrow-spectrum" biochemicals) and biological controls (parasites, etc.), which aim at pinpoint control without poisoning the whole ecosystem.

The first generation pesticides were adequate to keep grandfather well fed when farms were small and diversified, farm labor plentiful, and cultural practices favorable for blocking massive buildups of pests. DDT and the other potent broad-spectrum insecticides not only ushered in an era of industrialized agriculture, but they were supposed to "solve" all pest problems forever. As is now all too evident, this optimism is partly responsible for the severe "backlash" that resulted from the almost senseless saturation of the environment with the persistent (i.e., they degrade very slowly) broad-spectrum poisons to the point that we now must phase out the use of many of them. Unheeded warnings of an entomological backlash (i.e., pest outbreaks actually induced by spraying) were voiced in the 1950s (see Soloman, 1953, and Ripper, 1956), and the poisoning of entire food chains was dramatically brought to public attention in 1962 by Rachel Carson's famous book, Silent Spring. The detailed work of Nicholson, Grezenda et al. (1964) demonstrated how whole watersheds become contaminated by the uncontrolled use of agricultural pesticides. Finally, the insidious effect of DDT and other chlorinated hydrocarbons on the nervous system and sex hormone metabolism of vertebrates (including man) is just now being documented (see page 75). In retrospect, then, it appears that the organochlorines have provided only a temporary respite, a kind of holding action, in man's continuing war with insects and other competitors, and they must now be gradually replaced by other, more ecologically sound procedures. In the meantime these substances have produced one of the world's most serious pollution problems; a paraphrase of Wurster's (1969) evaluation of the problem follows:

The chlorinated hydrocarbon insecticides, now among the world's most widely distributed synthetic chemicals, are contaminating a substantial part of the biosphere. They are dispersed throughout the environment in currents of air and water. Their movement and widespread distribution throughout the world is explained by their solubility characteristics and chemical stability, and especially their tendencies to adsorb on organic matter, to be transported in air droplets, and to become concentrated in food transfers from plants to herbivores to carnivores. Their broad toxicity indicates a potential for biological effects on many kinds of organisms. The chlorinated hydrocarbons are seriously degrading biotic communities in many parts of the world. They have been shown to destroy larval stages of valuable aquatic food organisms and to depress photosynthesis of marine phytoplankton (which could have grave effects on the gaseous balance in the atmosphere). While direct effects on the hormone balance in man have not yet been demonstrated, concentration levels in human tissue are now high enough that such effects, and also cancer and deleterious mutations, *could* occur in the future (since they have been demonstrated to occur in laboratory animals), especially if nothing is done to control and monitor the further use of these potentially hazardous chemicals.

As was noted on page 199, a fundamental difference exists between the controlled use of nonspecific poisons on crops where causes and effects are known, and the broadcast of these same poisons in forests and other natural or seminatural areas where total effects are unknown and the probability of backlash very great. Pesticide pollution has been greatly aggravated by unnecessary aerial spraying of entire landscapes. Other "unforeseen" problems arise because new insecticides are tested (often very superficially) at the organism level of organization and then used at the ecosystem level without further testing. Thus, even though a chemical kills insects in cages and does not kill a laboratory rat, this does not mean that it is safe to use in nature. Again we have a case where trouble occurs because the agricultural and commercial specialist does not know the difference between a population and an ecosystem! Examples of ecological studies in which the ecosystem is the "guinea pig" or experimental testing ground are the studies of Barrett (1968) and Malone (1968).

Brown (1961) presents an objective review of four cases of mass insect control programs. At one extreme he describes the highly successful control of fruit flies based on detailed scientific information and judicious use of chemicals. At the other extreme he cites the campaign to eradicate the imported fire ant (*Solenopsis*) as an example of (1) too little study before mass spraying was started and (2) a misdirected Federal governmental mission motivated mostly by politics and carried out, against the advice of most knowledgeable scientists, to the point of "overkill." Several millions of dollars have been spent in mass aerial spraying on the theory that "saturation bombing" could eradicate the insect once and for all. Some control has been obtained by this massive onslaught but eradication is not in sight, and aquatic and terrestrial wildlife has suffered grievously in the meantime. The tragedy of such a situation is that better control could have been obtained with far less expenditure of public funds and less general damage to the environment if the means for the individual land owner to control the ants on his own land if he so desired had been provided, or if local campaigns at the county or state levels where the problem was acute had been set up.

As the mass use of persistent broad-spectrum poisons is phased out, it is evident that the strategy of pest control will increasingly evolve into what economic entomologists call *integrated control* (see Smith and Reynolds, 1966; Smith and van den Bosch, 1967; Chant, 1966 and 1969; Kennedy, 1968; and the FAO Symposium on "Integrated Pest Control"). The concept of integrated control involves coordinated use of a mixed bag of weapons, including old-fashioned but common-sense cultural practices, judicious use of degradable or "short-lived" chemical pesticides, and greater use and simulation of nature's own control methods, i.e., biological control (see review by Kilgore and Doutts [eds.], 1967) and the third generation of pesticides as outlined at the beginning of this section. The arsenal for integrated control includes the following:

1. Predators—such as the highly effective use of lady beetles and lacewings against agricultural pests, or beetles to control weeds (see Huffaker, 1958).

2. Parasites—such as calcid wasps, which successfully control a number of major pests.

3. Pathogens—such as viruses and bacterial infections that are specific for a pest.

4. Decoy plants—cultivation of low-value crops to attract pests away from high value crops (Stern et al., 1969).

5. Rotation and diversification of crops.

6. Chemical or radiation sterilization (see Chapter 17, page 457).

7. Hormonal stimulants—such as juvenile hormones that prevent insects from completing their life cycle (see Williams, 1970).

8. Pheromones—sex lures and other biochemicals that regulate pest behavior (see page 32).

9. Degradable chemical insecticides—organic phosphates and others.

10. Artificial selection for disease and pest resistance rather than for short-term yield as such.

Truly it can be said that eternal vigilance, study, and trained professionals are part of the "disorder pumpout" in the agroecosystem. There is no "one-shot" solution, nor will there ever be one. For reviews of pesticide problems, see Moore (1966), Rudd (1964), and Mrak (1969).

Herbicides*

Like modern insecticides, herbicides were first applied on a large-scale basis shortly after World War II. Initially, they were used to clear power line right-of-ways; subsequent uses have included clearing of railroad and highway right-of-ways, weed control in agriculture and forest management, and, sadly, as crop destruction and forest defoliation agents in warfare. They have proved most valuable when used in a selective manner in agriculture and forest management situations; their usefulness becomes increasingly questionable in nonselective, blanket spraying of large areas particularly when the effects on ecosystem structure cannot be accurately predicted (note parallel in misuse of pesticides). It has been estimated that at least 50 million acres of right-of-ways in the United States have been sprayed from 1 to 30 or more times (Eggler, 1968). Although some of this spraying is necessary, much of it is of such a general, nonselective nature that it cannot be justified on either economical or ecological grounds.

Generally, herbicides fall into two groups, depending upon their mode of action. Those in the first category, which includes monuron and simazin, interfere with photosynthesis and thus cause the plant to die from lack of energy. The second group is typified by the commonly used 2,4-D(2,4-dichlorophenoxyacetic acid) and 2,4,5-T(2, 4,5-trichlorophenoxyacetic acid). The mechanisms of action in this second group are not clearly understood. Two associated, but not identical, effects are involved: defoliation and systematic herbicidal action. Oddly enough, at low concentrations these chemicals can cause increased retention of fruits and leaves and are used for this purpose in agriculture. At higher concentrations they initiate a chain of reactions that result in a weakening and eventual rupturing of the abscission layer at the base of the petiole where the leaf blade attaches to the stem. By itself such simple defoliation does not usually kill a plant, and regeneration can normally be expected to follow. In certain plants, however, there is the additional effect of drastically increased cell proliferation in tissues such as phloem, which results in blockage of nutrient transport and formation of harmful lesions. In these susceptible plants there is little chance for successful recovery. Broad-leafed herbaceous plants are particularly susceptible to 2,4-D, while 2,4,5-T and a mixture of 2,4-D and 2,4,5-T are effective on woody plants.

Effects of 2,4-D and 2,4,5-T on ecosystems are poorly understood. Obviously they are

---

* This section prepared by Dr. William E. Odum.

capable of modifying plant communities and indirectly affecting herbivores and carnivores. Knowledge of effects on aquatic systems and soil microbes is scarce. The direct toxicity to animals appears to be low. However, the production of 2,4,5-T has been characterized by the presence, often in the final product, of symmetrical 2,3,6, 7-tetrachlorodibenzo-p-dioxin, usually referred to as "dioxin." This compound has been shown to be teratogenic, or fetus-deforming, at extremely low concentrations. In addition, it has been implicated in the occurrence of severe acneform skin changes in factory workers who produce 2,4,5-T. For these reasons 2,4,5-T is considered a dangerous compound unless the final product contains no "dioxin." Furthermore, the possibility of "dioxin" formation from 2,4, 5-T or intermediate breakdown products by thermal (wood burning) or metabolic routes has not been satisfactorily investigated.

Political questions aside, the use of herbicides (2,4-D, 2,4,5-T, "picloram," and cacodylic acid) in South Viet Nam is of particular ecological interest because of the large amount of land sprayed (at least 10 per cent of the country) and the heavy doses used (commonly an order of magnitude or higher than recommended for use in the United States). Aerial spraying, usually from specially converted C-123 aircraft, was carried out between 1962 and early 1970; each plane carried a 1000 gallon tank and was capable of spraying a swath 150 meters wide by 9 kilometers long, or roughly 333 acres, in two minutes. Investigation of the sprayed areas by Fred Tschirley (1969) revealed that mangrove associations were destroyed by a single application. Semideciduous forests were damaged very little by a single treatment, but significant changes in the forest and subsequent bamboo invasion occurred in areas that had received multiple sprayings. Two of the compounds routinely used in Viet Nam are highly restricted in the United States. "Picloram" has been characterized by Galston (1970) as a herbicidal analog to DDT because of its relative persistence in soils. Cacodylic acid contains over 50 per cent arsenic and repeated use may lead to build up of arsenic in soils.

Insecticides and herbicides together are powerful "drugs" in the ecosystem since they modify the function of vital systems — the consumers and producers. It is now

being suggested that these substances be under licensed control of trained professionals, just as are drugs used to treat the human body.

Noise Pollution

Yet another serious threat to the quality of man's environment is noise pollution. If we define noise as "unwanted sound," then noise pollution is unwanted sound "dumped" into the atmosphere without regard to the adverse effects it may have. The term "noise" is also used in electronics and communication science to refer to perturbations that interfere with communication. Such noise increases with the complexity and information content of systems of all kinds. Thus, man faces a growing problem with "electronic pollution" as radio communication intensifies. In the broadest sense, then, noise pollution is another "unforeseen backlash" in the concentrated use of power.

It is now clear that high-intensity sound, such as that emitted by many industrial machines and aircraft, when continued for long periods of time is not only disturbing to man (and probably other vertebrates), but also it permanently damages hearing. Even a comparatively low level of noise such as crowd, highway, or radio noise interferes with human conversation, causes emotional and behavioral stress, and threatens the "domestic tranquility" guaranteed by our constitution. Accordingly, sound must be considered a potentially serious pollutant and a grave threat to environmental health. Therefore, measurement, abatement, regulations, and legal restrictions on noise pollution must be considered along with efforts to control the "chemical" components of air pollution.

The unit of measurement for sound is the decibel (db). This is not an absolute unit of measurement but a relative one based on the logarithm of the ratio of sound intensity (I) to a reference level (I_o), arbitrarily established as a sound pressure of 0.0002 microbars (dynes per cm^2 or an energy of about 10^{-16} watts), which originally was judged to be an intensity just audible by man. Thus,

$$bel = \log_{10} \frac{I}{I_o}$$

and

$$\text{decibel} = 10 \, \log_{10} \frac{I}{I_o}$$

Accordingly, 10, 20, and 100 decibels represents 10 times, 100 times, and 10^{10} times the threshold intensity, respectively. It is important to recognize the logarithmic nature of this scale!

The area of human hearing extends in frequency from about 20 to 20,000 cps (cycles per second) and in intensity from 0 to greater than 120 db (at which point the intensity causes physical discomfort), a 10^{12}-fold or greater range. Ordinary conversation, which is in the frequency range 250 to 10,000 cps, registers between 30 and 60 db, while noise under a jet airplane at takeoff may rise in excess of 160 db. The effect on man varies with the frequency or "pitch" of the sound. The "sound pressure level" is judged to be of greater "loudness" for the higher pitched than for the lower pitched sounds. For example, a jet plane producing sounds at a 100 db intensity is rated by most people as twice as noisy and disturbing as a prop plane producing sound at the same decibel level, because its noise output contains more high-frequency energy.

Loudness as perceived by people is expressed in units called *sones*. This is also a relative unit—1 sone equals the loudness of 40 db sound pressure at 1000 cpm. Forty db sound at 5000 cps seems twice as loud and is, therefore, assigned the value of 2 sones. On this scale 50 sones and above is too loud for comfort at any frequency within the range of hearing. In general, 85 db (10 to 50 sones, depending on frequency) can be considered the critical level for ear damage. Noise levels far below this physically damaging level may have subtle effects and be of even greater concern to the general population. People begin to complain when unwanted noise levels in residential areas reaches 35 to 40 db, and they begin to threaten community action when it reaches 50 db! A major problem in noise control lies in the difficulty of evaluating complex noise, which contains energy in a number of octave bands—that is, the kind of noise that is most often irritating. Finally, sudden noise, such as a sonic boom, produces a "startle effect" that can be more disconcerting than continuous noise. Sonic booms can also produce physical damage to property (broken windows, etc.).

The threat of noise is another compelling reason for man to preserve a *Lebensraum* larger than the minimum space necessary for his day-to-day physiological and psychological necessities. Enforced zoning and planning that separates industrial noise, highways, and so on from living space is obviously needed along with increased attention to the technology of noise abatement. As of 1970, only a few cities and states have enacted laws to control noise, and fewer still do anything about measurement or enforcement. In southern California, we are told, decibel meters are being installed along highways, and trucks and cars are being stopped not only for exceeding the speed limit, but also for exceeding the noise limit set at 82 decibels! Even more important are enforced building codes that require soundproofing in the construction of buildings and apartments. People cannot live in peace when they are crowded into cities and separated from each other by only paper-thin walls!

In metropolitan areas greenbelt vegetation, and open space in general, may have as great a value in sound amelioration as in air purification. Robinette (1969) points out that plants are efficient absorbers of noise, especially noises of high frequency. A dense evergreen hedge can reduce the noise of garbage collection by 10 db (i.e., a tenfold attenuation). Border planting along highways or streets can be effective if plantings are lower towards the noise source and higher towards the hearer, thus not only absorbing but also deflecting the noise upward. A 50-foot wide band with an inner strip of dense shrubs and an outer band of trees can be quite effective (a sort of forest edge that is also good for small wildlife).

As with most other excesses in our society, the problem of drawing lines is difficult. Sound is necessary for human existence, and much of that produced by nature (bird song) and man (music) is pleasant and purposeful. Again, as in all aspects of pollution the problem comes when there is too much of an otherwise "good thing." Then the two greatest blocks to solution are (1) the lack of public awareness of and concern for the dangers and (2) the economic pressure to delay or do nothing as long as the money is rolling in. To really solve a problem such as airport noise, it is necessary to do two things simultaneously and continuously: (1) reduce noise at its source as far as technically possible and (2) zone the area around the airport so that no one is allowed to build a house or factory within 10 miles of the airport (for their own protection under equal protection rights of government and to avoid future citizen lawsuits that disrupt economic development!) A big "green belt" of farms and forests around a jetport would be valuable not only as a noise absorber but also as an air purifier, food and fiber producer, and recreational area! Such a twofold attack on the problem is what ecologists call "ecosystem thinking." For additional discussions of noise pollution, its effects, and its abatement, see Glorig (1958), Kryter (1970), Rhodda (1967), and Burns (1969).

Noise abatement would make a good crusade for the younger generation not only because they are unwittingly contributing to the excess (amplified rock music, for example), but more importantly because an environment free of unwanted noise is likely to be a quality environment in other respects.

Other Problem Areas

Radioactive and thermal pollution are considered in the next chapter, while additional aspects of the technology of detection and control of wastes are discussed in Chapters 18, 19, and 20. Pollution as a motivating force in social, economic, and legal reforms is again emphasized in Chapter 21.

Chapter 17

RADIATION ECOLOGY

Radiation ecology is concerned with radioactive substances, radiation, and the environment. There are two rather distinct phases of radioecology which require different approaches. On the one hand we are concerned with the effects of radiation on individuals, populations, communities, and ecosystems. The other important phase of radiation ecology concerns the fate of radioactive substances released into the environment and the manner by which the ecological communities and populations control the distribution of radioactivity. The testing of atomic weapons has added on a global scale man-made radioactivity to that which is naturally present. While weapons testing has been greatly curtailed since 1962, the threat of nuclear war remains. The continued development of nuclear power for peaceful uses, which must be accelerated as the supplies of fossil fuels dwindle, means that increasing volumes of radioactive wastes must be anticipated, monitored, and controlled, as must be done for other dangerous pollutants (see Chapter 16). On the more positive side, radioactive tracers are providing very valuable tools for research. Just as the microscope in all its forms extends our ability to study structure, so tracers in all their forms extend our ability to study function. A number of ecological examples of the usefulness of tracers have been noted in Part 1 (see pages 60, 93, and 98).

The most useful sourcebooks on radioecology are the symposia volumes edited by Schultz and Klement (1963), Hungate (1966), and Nelson and Evans (1969); see also Polikarpov, 1966.

1. REVIEW OF NUCLEAR CONCEPTS AND TERMINOLOGY OF ECOLOGICAL IMPORTANCE

In order to facilitate subsequent discussion and presentation of data, some of the more important concepts and terms used in radiation ecology are very briefly discussed below. For further information, see books by Lapp and Andrews (1954), Glasstone (1958), Comer (1955), Overman and Clark (1960), and Chase and Rabinowitz (1967).

Types of Ionizing Radiations

Very high-energy radiations that are able to remove electrons from atoms and attach them to other atoms, thereby producing positive and negative *ion pairs,* are known as *ionizing radiations,* in contrast to light and most solar radiation which does not have this ionizing effect. It is believed that ionization is the chief cause of injury to protoplasm and that the damage is proportional to the number of ion pairs produced in the absorbing material. Ionizing radiations are sent out from radioactive materials on earth and are also received from space. Isotopes of elements that emit ionizing radiations are called *radionuclides* or *radioisotopes.*

Of the three ionizing radiations of primary ecological concern two are corpuscular (alpha and beta) and one is electromagnetic (gamma radiation and the related x-radiation). Corpuscular radiation consists of streams of atomic or subatomic particles which transfer their energy to whatever they strike. *Alpha particles* are parts of helium atoms and are huge on the atomic scale of things. They travel but a few centimeters in air and may be stopped by a sheet of paper or the dead layer of the skin of man, but on being stopped they produce a very large amount of ionization locally. *Beta particles* are high-speed electrons—much smaller particles that may travel a number of feet in air or up to a couple of centimeters in tissue and give up their energy over a longer path. *Ionizing electromagnetic radiations,* on the other hand, are like light, only of much shorter wavelength (see Figure 5–6). They travel great distances and readily penetrate matter, releasing their energy over long paths (the ionization is dispersed). For example, *gamma rays* penetrate biological materials easily; a given "ray" may go right through an organism without having any effect or it may produce ionization over a long path. Their effect depends on the number and energy of the rays and the distance of the organism from the source, since intensity decreases exponentially with distance. Important features of alpha, beta, and gamma radiation are diagrammed in Figure 17–1. Thus we see that the alpha, beta, gamma series is one of increasing penetration but decreasing concentration of ionization and local damage. Therefore, biologists often class radioactive substances that emit alpha or beta particles as "internal emitters" because their effect is likely to be greatest when absorbed, ingested, or otherwise deposited in or near living tissue. Conversely, radioactive substances that are primarily gamma emitters are classed as "external emitters" since they are penetrating and can produce their effect without being taken inside.

There are other types of radiation which are of at least indirect interest to the ecologist. *Neutrons* are large, uncharged particles which in themselves do not cause ionization, but, like a bull in a china shop, they wreak local havoc by bumping atoms out of their stable states. Neutrons thus induce radioactivity in nonradioactive materials or tissues through which they pass. For a given amount of absorbed energy, "fast" neutrons may do ten times, and "slow" neutrons five times, the local damage of

gamma rays. Neutrons are restricted to the vicinity of reactors or atomic explosions, but, as indicated above, they are of primary importance in the production of radioactive substances which can and do become widely distributed in nature. *X-rays* are electromagnetic radiations very similar to gamma rays, but originate from the outer electron shell rather than from the nucleus of the atom and are not sent out from radioactive substances dispersed in the environment. Since they and gamma rays have similar effects and since x-rays are easily obtained from an x-ray machine, we may conveniently use them in experimental studies on individuals, populations, or even small ecosystems. *Cosmic rays* are radiations from outer space that are mixtures of corpuscular and electromagnetic components. The intensity of cosmic rays in the biosphere is low, but, as discussed in Chapter 20, they are a major hazard in space travel. Cosmic rays and ionizing radiation from natural radioactive substances in soil and water produce what is known as *background radiation* to which the present biota is adapted. In fact, the biota may depend on this background radiation for maintaining genetic fluidity. Background varies three to four fold in various parts of the biosphere. In this chapter we are concerned primarily with the artificial radioactivity which is added to the background.

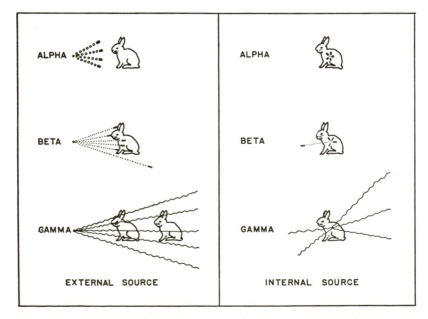

Figure 17–1. Schematic comparison of the three types of ionizing radiations of greatest ecological interest, showing relative penetration and specific ionization effect. The diagram is not intended to be quantitative.

Units of Measurements

In order to deal with radiation phenomena two types of measurement are needed: (1) a measure of the amount of a radioactive substance in terms of the number of disintegrations taking place and (2) a measure of radiation dose in terms of the energy absorbed, which is capable of causing ionization and damage.

The basic unit of the quantity of a radioactive substance is the *curie* (Ci), which is defined as the amount of material in which 3.7×10^{10} atoms disintegrate each second, or 2.2×10^{12} disintegrations per minute (dpm). The actual weight of material making up a curie is very different for a long-lived, slowly decaying isotope as compared with a rapidly decaying one. Approximately one gram of radium, for example, is a curie while very much less (about 10^{-7} grams) of newly formed radio-sodium would emit 3.7×10^{10} disintegrations a second! Since a curie represents a rather large amount of radioactivity from the biological standpoint, smaller units are widely used: *millicurie* (mCi) $= 10^{-3}$ Ci; *microcurie* (μCi) $= 10^{-6}$ Ci; *nanocurie* (nCi) (formerly called a millimicrocurie, mμc) $= 10^{-9}$ Ci; *picocurie* (pCi) (formerly called a micromicrocurie, $\mu\mu$c) $= 10^{-12}$ Ci. The possible range of activity is so tremendous that one must be careful about the position of the decimal point! The curie indicates how many alpha or beta particles or gamma rays are streaming from a radioactive source, but this information does not tell us what effect the radiation might have on organisms in the line of fire.

The other important aspect of radiation, radiation dose, has been measured with several scales. The most convenient unit for all types of radiation is the *rad*, which is defined as the absorbed dose of 100 ergs of energy per gram of tissue. The *roentgen* (R) is an older unit, which strictly speaking is to be used only for gamma and x-rays. Actually, so long as we are dealing with the effects on living organisms, the rad and the roentgen are nearly the same. A unit $1/1000$ smaller, namely the milliroentgen (mR) or the millirad (mrad), is convenient for the kind of radiation levels often encountered in the environment. It is important to emphasize that the roentgen or rad is a unit of total dose. The *dose rate* is the amount received per unit time. Thus, if an organism is receiving 10 mR per hour, the total dose

in a 24-hour period would be 240 mR or 0.240 R. As we shall see, the time over which a given dose is received is a very important consideration.

Instruments that measure ionizing radiation consist of two basic parts, (1) a detector and (2) a rate meter or electronic counter (scaler). Gaseous detectors, such as geiger tubes, are often used to measure beta radiation, while solid or liquid scintillation detectors (substances which convert the invisible radiation to visible light that is recorded by a photoelectric system) are widely used to measure gamma and other types of radiation.

Radionuclides (Radioisotopes) of Ecological Importance

There are various kinds of atoms of each elemental substance, each with a slightly different make-up, some radioactive, some not radioactive. These varieties of elements are called isotopes. Thus, there are several isotopes of the element oxygen, several isotopes of the element carbon, and so on. The isotopes that are radioactive are the unstable ones which disintegrate into other isotopes, releasing radiations at the same time. Each isotope is identified by a number, its atomic weight; each radioactive isotope, or radionuclide as they are more generally called, also has a characteristic rate of disintegration that is indicated by its half-life. Some radionuclides of ecological importance are listed in Table 17–1. One will note in Group B of Table 17–1 that calcium-45 is the radioactive isotope of calcium; it has an atomic weight of 45 and loses half its radioactivity every 160 days. Half-life is constant for a given nuclide (i.e., the rate of decay is not affected by environmental factors) and varies from a few seconds to many years, depending on the radionuclide. In general, extremely "short-lived" radionuclides are of little interest ecologically. A variable which affects the penetrating power of radiation is its energy. Most radionuclides of ecological interest have energies between 0.1 and 5 Mev (million electron volts). Relative energy of each isotope in Table 17–1 is indicated (see standard references for exact values). The greater the energy the greater the potential danger to biological material within the range of the particular type of

Table 17–1. *Radionuclides of Ecological Importance*

Group A.   Naturally occurring isotopes which contribute to background radiation.

NUCLIDE	HALF-LIFE	RADIATIONS EMITTED	
Uranium-235 (^{235}U)	7×10^8 yrs.	Alpha[3]	Gamma[0]
Uranium-238 (^{238}U)	4.5×10^9 yrs.	Alpha[3]	
Radium-226 (^{226}Ra)	1620 yrs.	Alpha[3]	Gamma[0]
Thorium-232 (^{232}Th)	1.4×10^{10} yrs.	Alpha[3]	
Potassium-40 (^{40}K)	1.3×10^9 yrs.	Beta[2]	Gamma[2]
Carbon-14 (See Group B.)			

[0] Very low energy, less than 0.2 Mev; [1] relatively low energy, 0.2–1 Mev; [2] high energy, 1–3 Mev; [3] very high energy, over 3 Mev.

Group B.   Nuclides of elements which are essential constituents of organisms, and, therefore, important as tracers in community metabolism studies as well as because of the radiation they produce.

NUCLIDE	HALF-LIFE	RADIATIONS EMITTED	
Calcium-45 (^{45}Ca)	160 days	Beta[1]	
Carbon-14 (^{14}C)	5568 yrs.	Beta[0]	
Cobalt-60 (^{60}Co)	5.27 yrs.	Beta[1]	Gamma[2]
Copper-64 (^{64}Cu)	12.8 hrs.	Beta[1]	Gamma[2]
Iodine-131 (^{131}I)	8 days	Beta[1]	Gamma[1]
Iron-59 (^{59}Fe)	45 days	Beta[1]	Gamma[2]
Hydrogen-3 (Tritium) (^{3}H)	12.4 yrs.	Beta[0]	
Manganese-54 (^{54}Mn)	300 days	Beta[2]	Gamma[2]
Phosphorus-32 (^{32}P)	14.5 days	Beta[2]	
Potassium-42 (^{42}K)	12.4 hrs.	Beta[3]	Gamma[2]
Sodium-22 (^{22}Na)	2.6 yrs.	Beta[1]	Gamma[2]
Sodium-24 (^{24}Na)	15.1 hrs.	Beta[2]	Gamma[2]
Sulfur-35 (^{35}S)	87.1 days	Beta[0]	
Zinc-65 (^{65}Zn)	250 days	Beta[1]	Gamma[2]

Also barium-140 (^{140}Ba), bromine-82 (^{82}Br), molybdenum-99 (^{99}Mo) and other trace elements.

Group C.   Nuclides important in fission products entering the environment through fallout or waste disposal.

NUCLIDE	HALF-LIFE	RADIATIONS EMITTED	
The strontium group			
Strontium-90 (^{90}Sr) and	28 yrs.	Beta[1]	
daughter yttrium-90 (^{90}Y)	2.5 days	Beta[2]	
Strontium-89 (^{89}Sr)	53 days	Beta[2]	
The cesium group			
Cesium-137 (^{137}Cs) and	33 yrs.	Beta[2]	Gamma
daughter barium-137 (^{137}Ba)	2.6 min.	Beta	Gamma[1]
Cesium-134 (^{134}Cs)	2.3 yrs.	Beta[1]	Gamma[2]
The cerium group			
Cerium-144 (^{144}Ce) and	285 days	Beta[1]	Gamma[0]
daughter praseodymium-144 (^{144}Pr)	17 min.	Beta[2]	Gamma[2]
Cerium-141 (^{141}Ce)	33 days	Beta[1]	Gamma[1]
The ruthenium group			
Ruthenium-106 (^{106}Ru) and	1 yr.	Beta[0]	
daughter rhodium-106 (^{106}Rh)	30 sec.	Beta[3]	Gamma[2]
Ruthenium-103 (^{103}Ru)	40 days	Beta[1]	Gamma[1]
Zirconium-95 (^{95}Zr) and daughter	65 days	Beta[1]	Gamma[1]
niobium-95 (^{95}Nb)	35 days	Beta[0]	Gamma[1]
Barium-140 (^{140}Ba) and daughter	12.8 days	Beta[1]	Gamma[1]
lanthanum-140 (^{140}La)	40 hrs.	Beta[2]	Gamma[2]
Neodymium-147 (^{147}Nd) and	11.3 days	Beta[1]	Gamma[1]
daughter promethium-147 (^{147}Pm)	2.6 yrs.	Beta[1]	Gamma
Yttrium-91 (^{91}Y)	61 days	Beta[2]	Gamma[1]
Plutonium-239 (^{239}Pu)	2.4×10^4 yrs.	Alpha[3]	Gamma[1]
Iodine-131 (see Group B)			
Uranium (see Group A)			

radiation. On the other hand, energetic isotopes are easier to detect in very small amounts and hence make better "tracers." For example, energetic gamma emitters such as cobalt-60, cesium-134, scandium-46, or tantalum-182 provide useful "tags" for following the movement of animals hidden from view under the bark of a tree or in the soil.

From the ecological point of view, radionuclides fall into several rather well-defined groups, as shown in Table 17–1. Naturally occurring radionuclides form one group, (A), while isotopes of metabolically important elements form another group, (B), especially important as tracers. A third important group of radionuclides, (C), are those produced by the fission of uranium and certain other elements; they involve mostly elements that are not metabolically essential (^{131}I is an exception). However, this group is the dangerous group because fission isotopes are produced in large amounts in both nuclear explosions and controlled nuclear operations, which produce power or other useful forms of energy. While most of these nuclides are not essential constituents of protoplasm they readily enter biogeochemical cycles and many of them, notably the strontium and cesium nuclides, become concentrated in the food chain, as was noted in Section 4, Chapter 4. Note that a number of isotopes in group C have a "daughter isotope" (an isotope formed during the decay of another isotope), which may be more energetic than the "parent."

It is projected that someday man will be able to harness the fusion power of the hydrogen bomb as a replacement for the fission power that is now the basis of current nuclear power developments. Such a development would eliminate the fission products but there would still remain problems involving tritium (^{3}H) and radioactivity induced by neutrons.

2. COMPARATIVE RADIOSENSITIVITY

Even before the atomic age was ushered in by the explosion of the first atomic bomb, enough work had been done with x-rays to indicate that organisms differed widely in their ability to tolerate massive doses of radiation. Comparative sensitivity of three diverse groups of organisms to single doses of x- or gamma radiation is shown in Figure 17–2. Large, single doses delivered at short time intervals (minutes or hours) are known as *acute doses,* in contrast to *chronic doses* of sublethal radiation that might be experienced continuously over a whole life cycle. The left ends of the bars indicate levels at which severe effects on reproduction (temporary or permanent sterilization, for example) may be expected in the more sensitive species of the group, and the right ends of the bars indicate levels at which a large portion (50 per cent or more) of the more resistant species would be killed outright. The arrows to the left indicate lower range of doses that would result in death or damage to sensitive life-history stages such as embryos. Thus, a dose of 200 rads will kill some insect embryos in the cleavage stage. 5,000 rads will sterilize some species of insects, but 100,000 rads may be required to kill all adult individuals of the more resistant species. In general, mammals rate as the most sensitive and microorganisms are the most resistant of organisms. Seed plants and lower vertebrates would fall somewhere between insects and mammals. Most studies have shown that rapidly dividing cells are most sensitive (which explains why sensitivity decreases

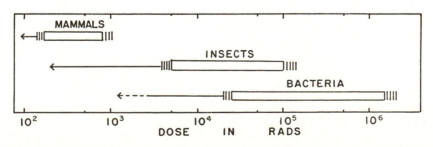

Figure 17–2. Comparative radiosensitivity of three groups of organisms to single acute doses of x- or gamma radiation. See text for explanation.

with age). Thus, any component—whether a part of an organism, a whole organism, or a population—which is undergoing rapid growth is likely to be affected by comparatively low levels of radiation regardless of taxonomic relationships.

The effects of low-level chronic doses are more difficult to measure since long-term genetic as well as somatic effects may be involved. In terms of growth response Sparrow (1962) has reported that a chronic dose of 1 R per day continued for 10 years (total dose 25,000 R) produces about the same amount of growth reduction in pine trees (which are relatively radiosensitive) as an acute dose of 60 R. Any increase in the ionizing radiation environment above background, or even a high natural background, can increase the rate of production of deleterious mutations (as can many chemicals and food additives that man now imposes on himself).

In higher plants sensitivity to ionizing radiation has been shown to be directly proportional to the size of the cell nucleus, or more specifically to chromosome volume or DNA content (Sparrow and Evans, 1961; Sparrow and Woodwell, 1962; Sparrow et al., 1963). As shown in Figure 17–3 sensitivity to radiation varies almost three orders of magnitude as does chromosome volume for a group of seed plants. Plants with large

chromosome volumes are killed by an acute dose of less than 1000 rads, while plants with small or few chromosomes may survive 50,000 rads or more. Such relationships suggest that the larger the chromosomal "target," the more likely are direct "hits" by atomic "bullets."

In higher animals no such simple and direct relationship between sensitivity and cellular structure has been found; effects on specific organ systems are more critical. Thus, mammals are very sensitive to low doses because rapidly dividing blood-making tissue in the bone marrow is especially vulnerable. A number of workers have reported that the LD_{50} (= lethal dose for 50 per cent of population) of certain wild rodents is about twice that of the laboratory white mouse or white rat (Gambino and Lindberg, 1964; Golley et al., 1965; Dunaway et al., 1969), but the reasons for this difference in closely related species has not yet been adequately explained.

Differential sensitivity is of considerable ecological interest. Should a system receive a higher level of radiation than that under which it evolved, adaptations and adjustments will occur that may include the elimination of sensitive strains or species. Examples of radiation-induced reduction in species diversity and changes in community structure are given in Section 3.

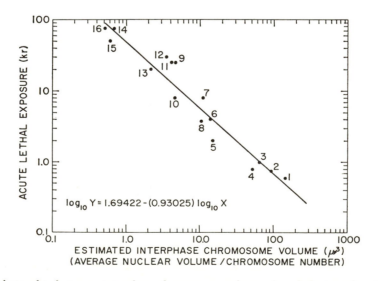

Figure 17–3. Relationship between interphase chromosome volume of seed plants and acute lethal exposure in kiloroentgens (1000 R). The species plotted are as follows: (1) *Trillium grandiflorum*, (2) *Podophyllum peltatum*, (3) *Hyacinthus* h. v. Innocence, (4) *Lilium logiflorum*, (5) *Chlorophytum elatum*, (6) *Zea mays*, (7) *Aphanostephus skirrobasis*, (8) *Crepis capillaris*, (9) *Sedum ternatum*, (10) *Lycopersicum esculentum*, (11) *Gladiolus* h. v. Friendship, (12) *Mentha spicata*, (13) *Sedum oryzifolium*, (14) *Sedum tricarpum*, (15) *Sedum alfredi* var. *nagasakianum*, (16) *Sedum rupifragum*. (After Sparrow, Schairer, and Sparrow, 1963.)

Radiation stress can alter key population interactions such as predator-prey equilibria, as shown in experiments with mites reported by Auerbach (1958), or bring on a pest irruption, an example of which is noted in the next section.

At this point we should take note of the order of magnitude of natural or background radiation doses to which organisms are accustomed, so to speak. Background radiation comes from three main sources: (1) cosmic rays, (2) potassium-40 in vivo (within living tissues), and (3) external radiation from radium and other naturally occurring radionuclides in rocks and soils. The following are estimated doses from each of these three sources in millirads per year received at five locations (see Polikarpov, 1966):

Sedimentary rock at sea level: $35 + 17 + 23 = 75$
Granitic rock at sea level: $35 + 17 + 90 = 142$
Granitic rock, 3000 m altitude: $100 + 17 + 90 = 207$
Sea surface: $35 + 28 + 1 = 67$
100 meters below sea surface: $1 + 28 + 1 = 30$

There may be no actual threshold for radiation effects. Geneticists have generally agreed that there is no threshold for genetic mutations. At present, we resort to the stopgap measure of establishing "minimum permissible levels" for both dose and amount of different radionuclides in the environment. This is a good procedure so long as it is recognized that these permissible levels do not actually represent any known thresholds. Actually, during the past decade the "permissible levels" for man have been revised downward. There is a widespread feeling that since man seems to be as radiosensitive as any other organism, all we need do is "monitor" the radiation levels and keep them low in that microenvironment where man actually lives. Loutit (1956) summed up this viewpoint as follows: "It is our belief that, if we take sufficient care radiobiologically to look after mankind, with few exceptions the rest of nature will take care of itself." This is a dangerous oversimplification. Radioactive pollution of the soil, the oceans, or other environments where man does not actually live will, nevertheless, have effects on man's vital life-support system. Most of all, as will be documented in Sections 4 and 5, any radioactive substance with a long half-life introduced into the environment anywhere in the biosphere will sooner or later find its way into man's body. To look after man radiobiologically we must take sufficient care of the ecosystem.

Differential radiation sensitivity within the species has an important practical application in insect control. As noted on page 447 in the previous chapter, radiation sterilization is one of the weapons in man's arsenal for "integrated" pest control. Male screw-worm flies, for example, can be sterilized by an acute dose of about 5000 R with little effect on the viability and behavior of the flies. Sterilized males released into the wild population will mate normally, but, of course, no offspring will be forthcoming. By flooding the natural population with large numbers of sterile males this major pest of animal husbandry has been controlled in the southern United States (Baumhover et al., 1955; Knipling, 1960). For a review of the possibilities for this kind of population control, see Bushland (1960), Knipling (1964, 1965, 1967), Cutcomp (1967), and Lawson (1967).

3. RADIATION EFFECTS AT THE ECOSYSTEM LEVEL

The effects of gamma radiation on whole communities and ecosystems have now been studied at a number of sites. Gamma sources, usually either cobalt-60 or cesium-137, of 10,000 Ci or more have been placed in fields and forests at the Brookhaven National Laboratory on Long Island (see Woodwell, 1962 and 1965), in a tropical rain forest of Puerto Rico (see H. T. Odum and Pigeon, 1970), and in a desert in Nevada (see French, 1965). The effects of unshielded reactors (which emit neutrons as well as gamma radiation) on fields and forests have been studied in Georgia (see Platt, 1965) and at the Oak Ridge National Laboratory in Tennessee (see Witherspoon, 1965 and 1969). A portable gamma source has been used to study short-term effects on a wide variety of communities at the Savannah River Ecology Laboratory in South Carolina (see McCormick and Golley, 1966; Monk, 1966b; McCormick, 1969). A lake bed community subjected to low-level chronic radiation from atomic wastes has been under study at the Oak Ridge Laboratory for many years.

Figure 17–4 summarizes the effects of the Brookhaven gamma source that was placed in an oak–pine forest (the same one whose productivity and biomass are depicted in Figure 3–3). The source was unshielded for 20 hours each day, allowing investigators to make observations and take samples during a 4-hour period each day when the source was lowered into a shielded pit. A chronic radiation gradient resulted, varying from 1000 rads at 10 meters from the source to no measurable increase above background at 140 meters, as shown by the concave curve in the upper diagram in Figure 17–4. Sedges were the most resistant plants; certain heath shrubs and grasses were slightly less resistant. Pines were considerably more sensitive than oaks (pines have larger nuclei and do not resprout

when terminal buds are killed). Growth inhibition in plants and a reduction in species diversity of animals were noted at levels as low as 2 to 5 rads per day. Although an oak forest persisted at rather high dose rates (10 to 40 rads per day), the trees were stressed and in certain zones became vulnerable to insects. In the second year of the experiment, for example, an outbreak of oak leaf aphids occurred in the zone receiving about 10 rads per day; in this zone aphids were more than 200 times as abundant as in the normal, unradiated oak forest. In summary, five zones were apparent along the radiation gradient: (1) a central zone in which no higher plants survived, (2) a sedge (*Carex*) zone, (3) a shrub zone of blueberries

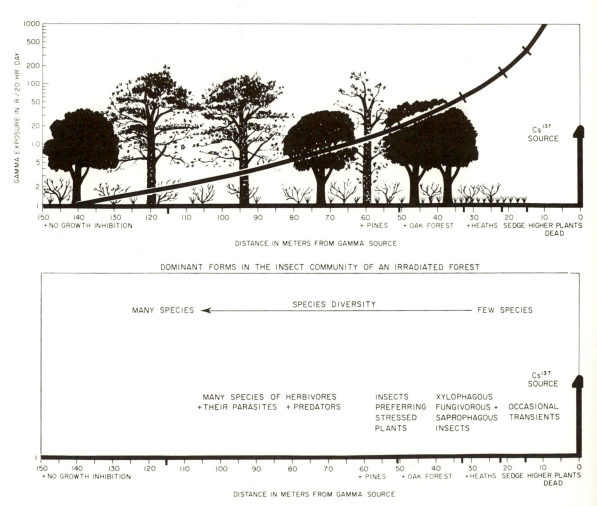

Figure 17–4. The response of an oak–pine forest to a gradient of gamma radiation from a high-level, fixed source that was unshielded for 20 hours each day for a period of two years. See text for explanation. (Used by permission of Brookhaven National Laboratory, Long Island, New York.)

and huckleberries, (4) a stressed oak forest, and (5) the intact oak–pine forest where growth inhibition was apparent but no individual plants were killed outright. Similar results have been obtained in other studies in which forest vegetation has been exposed to ionizing radiation. Where forests have been exposed to intense radiation for short periods, as at the site of the unshielded reactor in Georgia (see Platt, 1965), an old-field vegetation of annual weeds and grasses appeared after the overstory trees were apparently killed, but in the following years (with no additional radiation) many of the hardwood trees recovered by sending up dense growths of root and trunk sprouts (showing that only the above-ground parts had been killed), producing a sort of coppice vegetation that soon shaded out all of the old-field vegetation.

While, as we noted in the preceding section, one can predict the relative sensitivity of individual species of higher plants from information on chromosomal volume, there are other factors such as growth form or species interactions which may greatly modify response of species in intact communities. Herbaceous communities and early stages of succession are more resistant in general than mature forests not only because many species of the former have small nuclei, but also because there is much less "unshielded" biomass above ground, and the small herbs can recover more quickly by sprouting from seeds or from protected underground parts (see Figure 14–2). Thus, community attributes such as biomass and diversity play a role in determining vulnerability quite apart from the chromosomal volumes of individual species.

As with all kinds of stress, reduction in species diversity is associated with radiation stress. In another experiment at Brookhaven (see Woodwell, 1965) old-field vegetation was subjected to a radiation dose of 1000 rads per day. The dry matter production of the irradiated community was actually higher than that of nonradiated controls, but the species diversity was dramatically reduced. Instead of the normal mixture of many species of forbs and grasses, the irradiated old-field area developed into an almost pure stand of crabgrass (which probably would not surprise the citizen who fights crabgrass in his lawn!); recall the discussion in Section 4 of Chapter

6 about relationships between productivity, stability, and diversity.

4. THE FATE OF RADIONUCLIDES IN THE ENVIRONMENT

When radionuclides are released into the environment, they quite often become dispersed and diluted, but they may also become concentrated in living organisms and during food-chain transfers by a variety of means, which we have previously lumped under the general heading of "biological magnification" (see pages 74–75). Radioactive substances may also simply accumulate in water, soils, sediments, or air if the input exceeds the rate of natural radioactive decay. In other words we could give "nature" an apparently innocuous amount of radioactivity and have her give it back to us in a lethal package!

The ratio of a radionuclide in the organism to that in the environment is often called the *concentration factor*. A radioactive isotope behaves chemically essentially the same as the nonradioactive isotope of the same element. Therefore, the observed concentration by the organism is not the result of the radioactivity, but merely demonstrates in a measurable manner the difference between the density of the element in the environment and in the organism. Some of the earliest data on the concentrative tendencies in both aquatic and terrestrial food chains were obtained by radioecologists at the AEC Hanford plant on the Columbia river in eastern Washington state (see Foster and Rostenbach, 1954; Hanson and Kornberg, 1956; Davis and Foster, 1958). Here, trace amounts of induced radionuclides (^{32}P, etc.) and fission products (^{90}Sr, ^{137}Cs, ^{131}I, etc.) are released into the river, into waste holding ponds, and into the air. The concentration of phosphorus in the Columbia River is very low, only about 0.00003 mg per gm water (i.e., 0.003 ppm), whereas the concentration in egg yolks of ducks and geese that obtain their food from the river is about 6 mg per gm. Thus, a gram of egg yolk contains two million times more phosphorus than a gram of water in the river. We would not expect to find a concentration factor for radioactive phosphorus quite this high since, while it was moving through the food chain to the eggs, some decay would occur (this nuclide has a short half-life), thus reducing the amount. Occasionally a concentration factor as high as 1,500,000 was recorded, but the average was lower (about 200,000) (Hanson and Kornberg, 1956). Some other concentration factors reported were as follows: 250 for cesium-137 in muscle and 500 for strontium-90 in bone of waterbirds, as compared with concentration of these nuclides in the water of waste ponds in which these birds were feeding. The concentration of radioactive iodine in the thyroids of jack rabbits was 500 times that in the desert vegetation, which in turn had concentrated the nuclide released into the air in stack gases from the atomic plant. Concentration factors for strontium-90 in various parts of an aquatic food web at another site of atomic energy development are depicted in Figure 17–5.

While radioactivity does not affect the uptake of the isotope by living systems, it does, of course, have detrimental effects on active tissues once it is absorbed. The point is that allowance must be made for "ecological concentration" in establishing "maximum permissible levels" of release into the environment. Isotopes that are naturally concentrated in certain tissues (such as iodine in the thyroid or strontium in the bone) and/or those with long effective half-lives are obviously the ones to watch out for. Also, the concentration factor is likely to be greater in nutrient-poor environments than in nutrient-rich ones as will be documented in the next section. In general, concentrative tendencies can be expected to be greater in aquatic than in terrestrial ecosystems since nutrient fluxes in the "thin" media of water are more rapid than in the "thick" media of soil. For additional information on radioecological concentration processes, see Aberg and Hungate (eds.) (1967) and Polikarpov (1966).

Man's opportunity to learn more about environmental processes through the use of radioactive tracers balances to some extent the troubles he is having with environmental contamination. The uses of radioactive tracers for ecological study have been reviewed by Odum and Golley (1963), and numerous examples are to be found in the Proceedings of the two International Symposia (Schultz and Klements, 1963; Nelson and Evans, 1969). Tracers are obviously extremely useful in charting biogeochemical

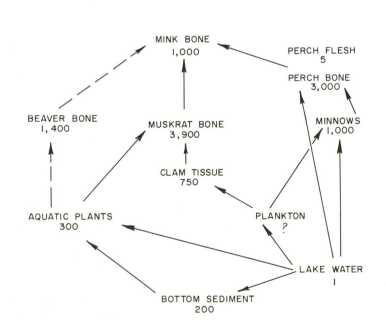

SR-90 IN PERCH LAKE FOOD WEB

Figure 17–5. Concentration of strontium-90 in various parts of the food web of a small Canadian lake receiving low-level atomic wastes. Average concentration factors are shown in terms of lake water = 1. (After Ophel, 1963; used by permission of Biology and Health Physics Division, Atomic Energy of Canada Limited, Chalk River, Ontario.)

cycles and in measuring flux rates in steady-state systems; examples of such uses were given in Chapter 4. They are also important in studies of community metabolism; carbon-14, for example, has become a basic tool for the measurement of productivity in aquatic ecosystems (see page 60). Tracers are also useful in charting the movements of organisms at the population level and in mapping food webs. Two examples will suffice to illustrate some of these possibilities.

In a study of the impact of predation on populations of cotton rats in field enclosures protected and exposed to predators Schnell (1968) "tagged" each animal with a radioactive pin inserted under the skin of the back. The tag not only enabled the investigator to locate live animals missed by the live-trapping program but also dead animals or remains left by a predator which would never be found by conventional observations. Thus, Schnell was able not only to plot accurate survivorship curves (see page 174 for explanation of this form of graphic analysis) for each population, but he could determine the exact cause of mortality for most of the animals.

Some aspects of the use of radionuclide tracers to isolate and chart food chains in intact natural communities are shown in Figure 17–6. In one study, the two dominant species of plants in a one-year abandoned field were labeled (Fig. 17–6A), and the transfer of the tracer to arthropods followed for a period of about six weeks. As shown in Figure 17–6B, "sap" feeders such as aphids became radioactive first, followed by foliage grazers and then predators. Thus, the trophic position of a particular species could be roughly determined by the shape of the uptake graph. More importantly, a map of the food-web network could be worked out, as shown in Figure 17–6C. Of the more than 100 species of insects present in the community only about 15 species were removing an appreciable amount of tracer from the dominant plants, and most of these were feeding on one of the species. The fact that one codominant was being grazed more heavily than the other was an unexpected finding, one that would not have been evident without the use of the tracer. For details of this and other similar studies, see E. P. Odum and Kuenzler, 1963; Wiegert, Odum, and Schnell, 1967; de la Cruz and Wiegert, 1967; Wiegert and

Odum, 1969; Ball, 1963; Crossley, 1963; Reichle and Crossley, 1965. Some of the limitations of tracer studies are discussed by Shure (1970).

5. THE FALLOUT PROBLEM

The radioactive dust that falls to earth after atomic explosions is called radioactive fallout. These materials mix and interact with natural particulate materials in the atmosphere (natural fallout, see Figure 4–3) and the increasing amount of man-made air pollution. The kind of radioactive fallout depends on the type of bomb. First, it may be well to distinguish between the two types of nuclear weapons, the fission bomb, in which heavy elements such as uranium and plutonium are split, with the release of energy and radioactive "fission products," and the fusion bomb, or the thermonuclear weapon in which light elements (deuterium) fuse to form a heavier element, with the release of energy and neutrons. Since an extremely high temperature (millions of degrees) is required for the latter, a fission reaction is used to "trigger" the fusion reaction. In general, the thermonuclear weapon produces less fission products and more neutrons (which induce radioactivity in the environment) than does a fission weapon per unit of energy released. According to Glasstone (1957), about 10 per cent of the energy of a nuclear weapon is in residual nuclear radiation, some of which becomes widely dispersed in the biosphere. The amount of radioactive fallout produced depends not only on the type and size of the weapon, but also on the amount of environmental material that gets mixed up in the explosion.

Fallout from weapons differs from atomic waste materials in that the radionuclides are fused with iron, silica, dust, and whatever happens to be in the vicinity to form relatively insoluble particles. These particles, which under the microscope often resemble tiny marbles of different colors, vary in size from several hundred microns to almost colloidal dimensions. The smaller particles adhere tightly to the leaves of plants where they may not only produce radiation damage to leaf tissue but may be ingested by grazing animals and dissolved by the digestive juices in the alimentary canal. Thus, this kind of fallout can

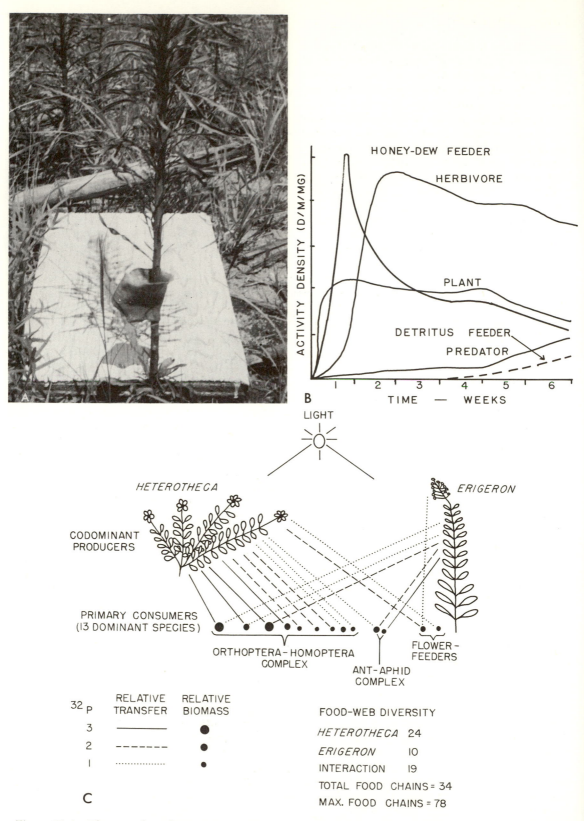

Figure 17–6. The use of a radionuclide tracer to chart food chains in intact natural communities. *A.* Labeling individual plants by using a small stem well. *B.* Pattern of uptake with time in different trophic levels. *C.* The food-web network involving two dominant species of plants and their herbivores. See text for further explanation. (Photo by Institute of Ecology, University of Georgia.)

enter the food chain directly at the herbivore, or primary consumer, trophic level.

Fallout from small atomic weapons or nuclear explosions used for peaceful purposes (excavation of harbors, canals, or surface mining) is mostly deposited in a narrow, linear pattern downwind, but some of the smallest particles may become widely dispersed and come down in rain at long distances. While the total amount of radioactivity decreases with distance from a nuclear test, it was early discovered that certain biologically significant nuclides, especially strontium-90, reached a peak in wild animal populations 50 to 100 miles from the "ground zero" of the explosion (see Nishita and Larson, 1957). This is explained by fact that ^{90}Sr has two gaseous precursors ($^{90}Kr \rightarrow {}^{90}Rb \rightarrow {}^{90}Sr$) and is therefore formed relatively late after the bomb detonation; this results in its inclusion in the smaller particles (less than 40 microns), which descend at greater distances and more readily enter food chains. Cesium-137 also has gaseous precursors and is unfortunately a significant component of the more soluble "long distance" fallout.

The large, powerful "megaton" weapons that were freely tested in the early 1960s ejected material into the stratosphere, resulting in a global contamination with a world-wide fallout that will continue for many years. The amount of fallout received by an area is roughly proportional to the rainfall. In the United States, for example, the accumulated deposition of strontium-90 by 1965 was estimated to be about 200 mCi per square mile in wet regions (eastern deciduous forest regions, for example) as compared to 80 mCi in dry regions (deserts, grasslands, etc.) (see Klement, 1965).

Studies following the weapons tests on the Pacific atolls showed that the kinds of radionuclides that enter marine food chains are rather strikingly different from those that enter terrestrial food chains (see Seymour, 1959; Palumbo, 1961). Elements of radioactive fallout that form strong complexes with organic matter such as cobalt-60, iron-59, zinc-65, and manganese-54 (which are all nuclides induced by neutron bombardment) and those that are present in particulate or colloidal form (^{144}Ce, ^{144}Pr, ^{95}Zr, and ^{106}Rh) transfer in the highest amount to marine organisms. In contrast, it is the soluble fission products such as strontium-90 and cesium-137 that are found in the highest amounts in land plants and animals. Since it was the induced isotopes, which complex with detritus, that were found in marine animals but not in marine plants or terrestrial organisms, it appears that the prominence of deposit-feeding and filter-feeding organisms in marine ecosystem food chains accounts for this difference. This is another instance in which contaminants may bypass the primary trophic level and enter directly the animal portion of food chains.

The quantity of fallout radionuclides that enters food chains and eventually becomes transferred to man depends not only on the amount received from the air (which, as was already indicated, is a direct function of precipitation) but also on the structure of the ecosystem and the nature of its biogeochemical cycles. In general, a larger proportion of fallout will enter food chains in nutrient-poor environments. In nutrient-rich environments high exchange and storage capacities of soils or sediments so dilute the fallout that relatively little uptake by plants occurs. A matlike vegetation on thin soils, such as is found on moors, heaths, granite outcrops, alpine meadows, and tundras, acts as a fallout trap (also epiphytes in tropical ecosystems), enhancing uptake by animals (see Russell, 1965), as does an active detritus food pathway. Two examples of these trends are illustrated by the data in Tables 17–2 and 17–3. Sheep on hill pastures in England accumulate 20 times as much strontium-90 in their bones as do sheep in the valleys because of the low calcium content and matlike vegetation characteristic of the hill pastures (Table 17–2). In Table 17–3 we see that the concentration of cesium-137 in deer (measured in pCi per kilogram wet weight of animal) is much higher on the sandy, low-lying coastal plain of the southeastern United States than in the adjacent Piedmont region where soils are well drained and have a high clay content. Rainfall in these two regions does not differ.

Fallout radionuclides (especially ^{90}Sr and ^{137}Cs) have been and are being passed on to man via the food chain, although concentrations in human tissues are not generally as high as those in sheep and deer. Man is somewhat protected by his position in the food chain and by food processing and cooking, which remove some of the contamination. However, in 1965, in arctic and subarctic regions (Alaska and northern Finland,

Table 17-2. *Comparison of the Amount of* ^{90}Sr *(Resulting from Fallout) in 1956 at Different Trophic Levels in Two Contrasting Ecosystems in the British Isles**

| | HILL PASTURE ACID PEAT SOIL, PH 4.3 | | | VALLEY PASTURE BROWN LOAM SOIL, PH 6.8 | | |
| | Amount ^{90}Sr | | | Amount ^{90}Sr | | |
	pCi/gm dry material	*pCi/gm calcium*	*Concentration factor*[†]	*pCi/gm dry material*	*pCi/gm calcium*	*Concentration factor*[†]
Soil (average top 4 inches)	0.112	800	1	0.038	2.6	1
Grass	2.5	2100	21	0.250	41	6.6
Sheep Bone	80	160	714	4.4	8.7	115

* Data from Bryant *et al.*, 1957.
† Ratio of the amount per gram in environment (soil) to amount per gram biological material.

for example) where reindeer or caribou meat is consumed, cesium-137 in man was 5 to 45 nCi/kgm body weight (= 5000 to 45,000 pCi/kgm) according to Hanson *et al.* (1967) and Miettinen (1969); this is every bit as high as the concentration in coastal plain deer (compare with Table 17-3). The reindeer and caribou themselves become highly contaminated by feeding on the mat-like vegetation. Thus, "exposed" human populations such as the Eskimos and Lapps are subjected to internal radiation doses appreciably higher than background, although no one knows how harmful this is. In 1965, the amount of strontium-90 in the bones of children in the United States was estimated to be 4 to 8 pCi/gm calcium in bone (see Comar, 1965); since then most surveys have indicated that this level has not risen appreciably. Again, no one can say whether this small amount is harmful, but as one of many "pollution stresses" it is certainly not doing anyone any good!

6. WASTE DISPOSAL

Although fallout problems are critical, waste disposal from peaceful uses of atomic energy is potentially a far greater problem, assuming again that we do not have an all-out atomic war. Not enough attention is now being given to the ecological aspects of waste disposal, which is the limiting factor to full exploitation of atomic energy. As Weinberg and Hammond (1970) have stated, the energy available in nuclear sources seems "essentially inexhaustible," but environmental side effects of a very large production of such energy impose the actual limit. This is another expression of the principle stated in Chapter 16: it is not energy itself that is limiting to man but rather the consequences of pollution that result from the exploitation of energy sources.

It has been customary to consider three categories of radioactive wastes:

1. *High-level wastes*—liquids or solids

Table 17-3. *Comparison of the Concentration of Cesium-137 (Resulting from Fallout) in White-tailed Deer in Coastal Plain and Piedmont Regions of Georgia and South Carolina**

| | NUMBER OF DEER | ^{137}Cs IN pCi/KGM WET WT. | |
REGION		*Mean and Standard Error*[†]	*Range*
Lower coastal plain	25	18,039 ± 2,359	2,076–54,818
Piedmont	25	3,007 ± 968	250–19,821

* Data from Jenkins and Fendley (1968).
† Difference between regions highly significant at 99 per cent level.

that must be contained since they are too dangerous to be released anywhere in the biosphere. About 100 gallons of such high-level wastes are generated by each ton of spent nuclear fuel. As of 1969, 80 million gallons were stored in 200 underground tanks at four U.S. Atomic Energy Commission sites, with 2×10^6 cubic feet of new storage space needed annually, a rate that will increase as nuclear power generation increases. Alternatives to tank storage that are being considered include (1) conversion of liquids to inert solids (ceramics) for burial in deep geological strata and (2) storage of liquids and solids in deep salt mines. The large amount of heat generated by high-level wastes compounds the problem; heat can "melt" the walls of salt mines or cause small earthquakes if injected into certain types of geological fissures.

2. *Low-level wastes*—liquids, solids, and gases that have very low radioactivity per unit volume but are far too voluminous to be contained completely; therefore, they must somehow be dispersed into the environment in such a way and at such a rate that the released equilibrium radioactivity does not appreciably raise background or become concentrated in food chains.

3. *Intermediate-level wastes*—those with radioactivity high enough to dictate local containment, but low enough so that it is possible to separate out high-level or long-lived components and handle the bulk as low-level wastes.

The uranium fuel cycle in power generation includes the following phases: (1) mining and milling, (2) refining (chemical conversion), (3) enrichment (percentage content of uranium-235 increased), (4) nuclear fuel element fabrication, (5) burnup of nuclear fuel in the reactor, (6) spent fuel reprocessing, and (7) burial or other containment of wastes. Although most of the radioactive wastes are generated in the reactor, the most difficult waste management problems occur during reprocessing (phase 6) when fission products (see Table 17–1C) are removed from the spent fuel elements. Reprocessing plants and burial grounds are located at different sites from the nuclear power plant itself, which means that there is an ever-present danger of accidents when spent fuel elements or the high-level wastes extracted from them are transported. Some low-level and intermediate-level wastes must be disposed of at the reactor site (especially when fuel elements leak or break) and during mining and fuel preparation. Thus, radioactive contamination of the environment is an ever-present threat during the entire cycle. To minimize the threat large protective areas of land must be set aside, especially for phases 5, 6, and 7. For example, nuclear burial grounds have to be quite large since one acre is required for each 50,000 cubic feet of high-level waste or 100,000 cubic feet of intermediate- or low-level materials. Such sites have to be continuously monitored to make certain that the surface water, water table, and air are not contaminated (see Figure 17–7). Land and water requirements for the power plant and its waste treatment environment are considered later.

As long as fissionable material is used as a fuel source (uranium, thorium, plutonium, etc.) large quantities of waste fission prod-

Figure 17–7. Disposal of high-level liquid wastes in the ground at the Hanford Atomic Products Operations Plant, showing the relative movement through the desert soils by significant isotopes. (Brown, Parker, and Smith, 1956.)

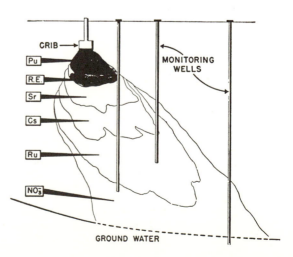

ucts (the same radionuclides involved in fallout) plus residual amounts of the fissionable materials could well be the limiting factors in exploiting the theoretically "inexhaustible" atomic energy sources. Huge "megacurie" quantities of the long-lived species (^{90}Sr, ^{137}Cs, ^{129}I, ^{99}Tc, ^{238}Pu, ^{240}Pu, ^{241}Am, ^{243}Am, and ^{244}Cm) would have to be stored. Currently used reactors are expected to be replaced in the next 15 to 20 years by "breeder" reactors, in which a catalytic burning of uranium-238, thorium-232, or perhaps lithium-6 results in a self-regeneration of fissionable material (see Weinberg and Hammond, 1970, for an evaluation of breeder reactors). Such a fuel cycle greatly reduces the fuel requirement but does not solve the waste disposal problem. Assuming that someday it will be possible to employ fusion power (see page 411), the fission products would then be eliminated but induced radionuclides would be increased — especially tritium, which could contaminate the entire global hydrological cycle. Frank Parker (1968) of Oak Ridge calculates that "release of tritium generated by a power economy, if the nuclear power were all fusion, would result in an unacceptable world-wide dosage by the year 2000." For an additional discussion of radioactive wastes, see Fox (1969).

If radioactive wastes do not prove to be the limiting factor in the exploitation of nuclear energy, then the limiting factor will be waste heat; or, as is most likely, it will be a combination of the two that will set the overall pollution constraint. What has come to be known as *thermal pollution* will become an increasingly serious problem since low utility heat is a by-product of any transfer of energy from one form to another as dictated by the second law of thermodynamics (see Section 1, Chapter 3). To some extent the shift from fossil fuel to atomic power reduces air pollution but increases water pollution, especially thermal pollution. Thus, to generate one kilowatt hour of electricity the waste heat released to the atmosphere and to cooling water is 1600 and 5300 BTU respectively for a fossil-fueled power plant, and 500 and 7600 BTU respectively for a present-day nuclear power plant (see Table 3–1 for conversion of BTU to kilocalories). Thus, an average size nuclear power plant that produces 3000 megawatts of electricity also produces waste heat at the rate of more than 20×10^9 BTU per hour.

The surface cooling capacity of water ranges from about 1.5 to 7.5 BTU per hr per ft^2 per °F. difference between air and water, depending on wind and water temperatures. Consequently, a lot of water surface is required to disperse heat, something on the order of 1.5 acres per megawatt in a temperate locality, or 4500 acres for a 3000-megawatt power station. In a 1970 task force report[*] it was recommended that each 2400-megawatt nuclear plant include 1130 acres for plant operations and containment of radioactive wastes and 7000 acres of water surface for cooling. Accordingly, if we take option 2 in the waste disposal strategy (as described on page 438), we should think in terms of a minimum area of 10,000 acres for each moderate-sized power plant in accordance with the concept of the waste disposal park (see Figure 16–4); this includes making use of by-product heat for fish culture or other useful purposes.

The use of powered cooling devices such as cooling towers can reduce the space needed, of course, but at a considerable cost, since this would essentially be taking costly option 3 in the overall waste disposal strategy. As with other wastes it is always tempting to rely on the oceans for cooling, but as another recent task force report[†] warns, the oceans can no longer be regarded as a dumping ground for all of man's wastes. While almost everyone predicts that thermal pollution will be an increasingly serious local problem, there is no agreement as to its ultimate effect on the global heat balance. For a discussion of this aspect, see Peterson (1970).

The local detrimental effects of thermal pollution on aquatic ecosystems can be listed as follows: (1) A rise in water temperature often increases the susceptibility of organisms to toxic materials (which will undoubtedly be present in waste water). (2) Critical "stenothermal" periods in life histories may be exceeded (see page 108).

[*] "Nuclear Power in the South." Report of the Southern Interstate Nuclear Board, 800 Peachtree Street, Atlanta, Georgia 30308.

[†] See "Ocean Dumping, A National Policy." A report to the President prepared by the Council on Environmental Quality, 1970.

(3) Elevated temperatures tend to foster replacement of normal algal populations by less desirable blue-greens (see page 306). (4) As water temperature rises, animals need more oxygen, yet warm water holds less (see page 126).

For additional information on biological effects of thermal pollution, see Naylor (1965), Mann (1965), Clark (1969), and Krenkel and Parker (eds.) (1969).

7. FUTURE RADIOECOLOGICAL RESEARCH

In this brief review we have tried to show that the problems of radioactivity in the environment and the thermal consequences of the use of nuclear energy will compound the already extremely critical pollution constraints on the further development of industrialized man. On the positive side we have tried to indicate some of the exciting possibilities for study made possible by iso- topes. Up to now the interdisciplinary field of radiation ecology has been primarily preoccupied with description and technology; it must now move to a position of making major contributions to the theory of ecosystems. Radiation procedures offer powerful means for solving the twofold problem of ecosystems, that of relating the one-way flow of energy to the cycling of materials and of discovering how physical and biological factors interact to control the functioning ecosystem. Only by understanding these matters in depth can man act as his own error detector and correct for perturbations, caused by his technology, that increasingly disrupt the life-support systems of the biosphere (E. P. Odum, 1965). In the not too distant future the radioecologist may well be one of those who must help decide when to contain and when to disperse the waste materials of the atomic age. If the ecologist does not know what to expect in the biological environment, who does?

REMOTE SENSING AS A TOOL FOR STUDY AND MANAGEMENT OF ECOSYSTEMS

Chapter 18

By Philip L. Johnson

School of Forest Resources, and Institute of
Ecology, University of Georgia*

Remote sensing is the acquisition of information about the biosphere by noncontact methods, usually from airplanes or satellites, in any portion of the electromagnetic spectrum. It is one of the areas of rapidly advancing technology mentioned at the beginning of this book (see page 6) that makes it possible to deal more effectively with very large ecosystems. The exciting prospect is that remote sensing will be a logical bridge between intensive ecological research on small areas and the application of principles thus revealed to planning and management of large political units such as townships, counties or states, or whole natural units such as watersheds, tropical rain forests, or ocean basins.

Photography from the ground or from aerial platforms is a familiar and conventional form of remote sensing that has been used extensively since the 1930s by foresters (Avery, 1966), geologists, and geographers (Avery and Richter, 1965) to inventory timber stands, to map geological structures, and to document land use patterns. Nearly all topographic maps are now made from stereoscopic photography. The recent introduction of a laser profilometer (Remple and Parker, 1965), which measures very small differences in elevation by the time versus distance relationship of a highly focused light pulse, promises to replace expensive ground surveying for vertical control of these maps. The same technology is being used to measure distances to the moon to the nearest foot.

In forestry, tree height, stem diameter, and, therefore, timber volume can be estimated from panchromatic aerial photographs with as little as 20 per cent error. Since the ratio of dbh (diameter breast high) of tree stems to crown diameter is consistent (for example, 1 inch: 1 foot for western conifers, and 1 inch: ¾ foot for eastern hardwoods), the total dbh or basal area of overstory stems can be estimated by measuring the crown diameters as revealed on the photographs.

Technological advances stimulated by military and space research have created a diversity of new airborne sensors (Holter and Wolfe, 1960) that capture energy from various portions of the spectrum and, therefore, have greatly increased the information gathering potential of aerial overflights. Appropriate electronic systems analyze and display the data in pictorial format or on tape for input to a computer (Peterson *et al.*, 1968). Some of these systems are electronically sophisticated and require specially trained operators. It should be emphasized that the meaning and validity of remote sensing data, especially that obtained in nonvisual portions of the spectrum, require familiarity with the ecosystem recorded. Pertinent measurements and observations obtained on the ground at the time of overflights are called "ground truths" and are essential to the success of most remote sensing research. Once the indicator value of remotely obtained data is established, ground measurements may no longer be required. Thus, a multidisciplinary approach, including the work of engineers and ecologists, is required. In attempting to inter-

---

* Present address: National Science Foundation, Washington, D.C.

468

pret remote sensing signatures formed by various spectral energies, new questions are posed about organisms and their environments. For example, we know the color of an oak tree; we do not know its spectral properties beyond the visual wavelengths. A great deal of fundamental research must be done to realize the full potential of remote sensing technology (Parker and Wolff, 1965). Two national symposia on remote sensing in ecology (BioScience vol. 17, 1967, and Johnson, ed., 1969) have focused attention on the ecological possibilities and problems of this technology.

1. PHYSICAL BASIS FOR REMOTE SENSING

To utilize the information provided by airborne sensor systems it is essential that the fundamental energy and matter relationships responsible for the images be understood. The electromagnetic spectrum has been presented in detail in Chapter 5, Section 5. The energy-matter interactions and the appropriate sensors for each spectral band are shown in Figure 5–6, page 118 (for additional information see Fritz, 1967; Krinov, 1947; Steiner and Gutermann, 1966). There are three basic types of systems available for remote sensing from airborne or satellite platforms:

1. Photography in the visible or near-visible spectrum, 380 to 1000 mμ (3800–10,000 Å),

2. optical-mechanical scanners from the ultraviolet through infrared wavelengths, 300 mμ to 40 μ, and

3. microwave for selected bands from 1 mm up to 1 meter.

All three are usually processed to present two-dimensional or pictorial displays. Radar, in contrast to energy of similar wavelength emanating from the earth, is an active system in the microwave frequencies in which the appropriate energy is generated in the aircraft and directed toward the ground. The radar return or reflected signal is captured by an antenna specific to that wavelength. The intensity of the energy returned is primarily a function of terrain aspect relative to beam direction and secondarily related to the dielectric properties of the reflecting material (Moore and Simonett,

1967). An advantage of radar imaging is its independence of weather and diurnal conditions.

All objects emit radiation. A perfect blackbody radiates energy proportional to the fourth power of its absolute temperature in degrees Kelvin. Incoming solar energy peaks at about 480 mμ (green wavelengths) and is negligible above 3 μ, while outgoing, emitted radiation peaks near 10 μ and is negligible below 3 μ. (See Chapter 3, Section 2 and Figure 5–6, page 118, for further comparison of solar and thermal radiation.) Entire populations of large animals can sometimes be accurately mapped from airplanes equipped with thermal scanners. Great advances have occurred in the last five years in developing thermally sensitive photoconductors that now permit discrimination of temperature differences as small as 0.01°C. For high sensitivity these detectors, such as lead sulfide, gold-doped germanium, or indium antimonide, must be liquid cooled with nitrogen or helium. By use of a rotating mirror a narrow field of view transverse to the flight line is focused on a detector and a signal is imaged in gray tones corresponding to different surface temperatures of the scene. Such a gray scale can be calibrated to actual rather than relative temperatures. Thus, color coded temperature contour maps can be generated corresponding to the surface heat flux of the target area at a specific time.

Frequently information from several spectral bands surpasses the sum of each band considered separately. This has led to the development of "multispectral" sensing, in which several or many spectral energies are recorded simultaneously. The data gathered may be telemetered or returned as film or magnetic tape. Such information is a function of the environment imaged, of plant and animal surfaces, or of a complex interaction of both. The biological implications of the signals recorded must be analyzed and interpreted, and there is frequently little prior experience to guide the interpreter. As is so often the case, skill outruns understanding so that an "achievement lag" occurs until such time as the ground truth and the new imagery can be coupled effectively. Only then will we know just how useful multispectral sensing really is.

In summary, potential information is a function of the matter-energy interactions peculiar to the sample. Energy absorption, emission, scattering, and reflection by any particular kind of matter are selective with regard to wavelength and are specific for that species of material, depending upon its atomic and molecular structure (Colwell *et al.*, 1963). Accordingly, each kind of biomass or physical substrate should emit a spectrum of characteristic frequency and intensity. However, the actual signal recorded is also, unfortunately, a resultant of (1) attenuation by the intervening atmosphere and (2) the fidelity of the electromechanical system employed.

Energy and Plant Relationships

Most of the earth's land surface is mantled by some kind of vegetation. Primarily it is the signature of foliar surfaces of vegetation that is perceived by remote sensors. What happens to solar energy incident upon leaves? In the visible and near-infrared wavelengths reflection or emission from the leaf's cuticle and epidermis is relatively minor and not very selective. In the red and blue ends of the visible spectrum 80 per cent or more of the energy incident on a leaf is absorbed by chlorophyll, whereas perhaps 40 per cent of the green wavelengths is reflected. Energy from the near-infrared is little affected by chloroplasts but is greatly affected by the change in refractive index between air in the intercellular voids and the hydrated cellulose of cell walls. Energy, which penetrates the leaf to the mesophyll and is reflected from it, is of greater intensity in the near-infrared than in the visible wavelengths (Colwell *et al.*, 1963). Infrared photographs such as shown in the Frontispiece are often superior to conventional color photos in revealing successional and seasonal changes that result in subtle changes in plant pigments and leaf structure. Energy absorbed in one wavelength and emitted in another (for example, fluorescence) may be important to both ultraviolet and near-infrared detectors; however, the quanta lost from plants as fluorescence is probably less than a few per cent of the available light energy. A change in plant vigor may often result in loss of turgor in foliar tissue. It appears that changes in the red-infrared reflectance of near-infrared energy may occur long before any change can be detected in the visible wavelengths.

However, the mechanism of this decrease in red or near-infrared reflectance, whether due to loss of turgor, disease mycelia, or change in cell geometry, has not been experimentally established. An excellent discussion of the physical and physiological bases of spectral properties of plants was reported by Gates *et al.* (1965) and by Knipling (1969). Detailed knowledge of these relationships accounts for the success of several recent experiments with aerial color and infrared film to detect disease in crops, such as oranges, potatoes, wheat, and conifer stands (Meyer and French, 1966), the effects of herbicides, internal water stress, and ionizing radiation effects on vegetation (Johnson, 1965). Current research suggests the possibility of predicting crop yields and losses, pigment structure of ecosystems, and perhaps even species diversity (Odum, 1969) of overstory vegetation. Where chlorophyll concentrations in the autotrophic column can be shown to be related to productivity (see page 62), it should be possible to determine such concentrations by remote sensing.

Energy and Animal Relationships

Both plants and animals are coupled to their environment by an exchange of energy at their surfaces. Since animals, particularly homeotherms, partially regulate their surface temperatures by movement and metabolic activity, the temperature of the exterior skin, fur, or feathers is an important indicator of their response to environmental factors. By measuring surface temperatures it is possible to understand the energy balance of various organisms (Gates, 1969).

For purposes of animal census a thermal scanner with a three milliradian (3-foot diameter at 1000 feet altitude) instantaneous field of view can detect about a 1°C. differential (McCullough *et al.*, 1969). This is a compromise between temperature sensitivity and resolution. To increase sensitivity a wider field of view is necessary, and consequently single animals would not likely be distinguished from background temperatures of the habitat. The best wavelength for thermal census seems to be 8 to 14 μ. With this wavelength successful population counts have been achieved for white tailed deer and some larger grazing mammals. Because sensitivity and resolution of thermal scanners are approaching the theoretical limit, there is little prospect for detecting single animals the size of a fox or dog or smaller or of species separation among larger animals. Since success is favored by a large temperature differential between animal and background, a snow cover is an ideal condition. Daytime flights were better for deer (McCullough *et al.*, 1969); and although daytime flights were preferred for interpreting thermal patterns of upland vegetation, night imagery was more successful for distinguishing types of lowland or swamp vegetation in Michigan (Weaver *et al.*, 1969).

Remote detection of individual animals, of course, can be greatly facilitated by attaching to the animal an energy source that emits an easily detectable signal. Such a procedure of attaching transmitters to animals is known as *biotelemetry*. The "transmitters" are small radios that emit long wave frequencies, but radioactive sources (high frequency, short wavelength) may also be useful. Biotelemetry permits investigators to pinpoint the location and follow the heart beat rhythm of selected individual animals which have been captured and released with a transmitter attached or implanted in the body cavity (Folk, 1967). Movement patterns of individual elk, caribou, or deer, for example, provide data on animal behavior such as size of home range, feeding times and browse preference, and response to hunting pressure by dogs or by man (Marchinton, 1969). By telemetry of physiological functions such as heart beat, the effects of environmental stress on an organism's circadian rhythms can be evaluated. In a cooperative international project, radio-instrumented reindeer or caribou and polar bears will be located by polar orbiting satellite and their positions periodically telemetered to laboratories in temperate latitudes. Ocean currents can be similarly tracked with instrumented buoys.

Radio transmitters weighing only 2.5 grams have been attached to small migratory birds, and their flights have been followed with receivers in trucks or in light aircraft. In one such study transmitters were placed on thrushes, which breed in the northern lake states and Canada, during their spring migration stopover in central Illinois. The birds were then tracked during overnight flights that took them into northern Wisconsin. The nocturnal flights began 1 to 2 hours after sunset (after the birds had gone to roost) and some continued until dawn. Flight altitudes were between 2000 and 6000 feet, and air speed was between 25 and 35 miles per hour (see Cochran, Montgomery, and Graber, 1967). The use of small transmitters to monitor body temperatures of burrowing animals was noted in Chapter 8, Section 8.

Radar at ground stations can also be used to sense the movement of animals. Some intriguing studies of insect behavior (Konrad, 1968; Glover *et al.*, 1966) and bird migration (Lack, 1959, 1962) have been made by the use of radar. In turn, migrating birds, and sometimes insects, have caused anomalous signals, whose origin was previously uncertain, on airplane monitors. The sensitivity of even routine weather station radars now in use is such that mass nocturnal bird migration can not only be monitored but the number, height, and direction of flight of individual birds can be detected (see Konrad, Hicks, and Dobson, 1968). Furthermore, recognition and tracking of mass flights of birds (and also bats leaving roosting caves) is important in avoiding dangerous collisions with aircraft. Although it is well-known that most birds migrate at night, the nature and extent of these migrations were little appreciated until it was learned how to tell "bird" signals from water droplets and other nonbiotic reflectors. Very large concentrations of migrating birds often are associated with moving weather fronts.

2. PROCESS OF INFORMATION EXTRACTION

Information on photographs or other images can be evaluated qualitatively and quantitatively. The procedure of inferring relationships by visual examination is called *photo interpretation* whereas the procedure of making photographic measurements is called *photogrammetry*.

Interpretation

Certain limitations are imposed on the image analyst. First, he is restricted to his experiences or other ground-based information (1) about analogous areas and (2) about the academic discipline to which he is applying the analysis (Lueder, 1959). Presumably, no one is potentially in a better position to extract information about vegetation from aerial photography than plant ecologists. Second, the image analyst must know the scale of the imagery for both quantitative and qualitative analyses. Size and shape are partially a function of scale, and pattern is a function of arrangement of the components; both will obviously have different meanings in satellite photography and in photomicrographs. Third, the interpreter must depend on observations of tone, texture, pattern, and resolution (Figure 18–1). Tone or density variations may be measurable in gradations of a gray body, or, in the case of nongray bodies, they may be expressed in terms of chroma and hue, tristimulus values, Munsell colors, or other color notations. Panchromatic film records nearly 225 distinguishable shades of gray, whereas color photography may potentially record up to 20,000 separate combinations of chroma and hue.

Photographic pattern of texture and spatial arrangements are frequently the most important clue to the identification, origin, or function of objects imaged (Stone, 1956). Resolution, the ratio of object size to grain size or noise level, has improved tremen-

dously with the increase in emulsion sensitivity and the improvements in optical systems and in associated electronic components. For example, it has recently been demonstrated that a single honey bee could be tracked by radar from a distance of 10 kilometers (Glover et al., 1966).

A photographic pattern may be simple or complex depending on the number of variables. If we can consider the potential information in a photograph to form an organizational hierarchy shaped like a pyramid (Johnson, 1966), then the most generalized information is at the peak and a myriad of details lie at the base. As deductive logic starts with the cap stone and proceeds to the details, the number of variables increases, confidence in our inferences from photographic indicators decreases, and the amount of ground truth or outside information required is greater. On the other hand, inductive logic commences with the details and builds generalizations. Photo reading or species identification is the first step in *inductive* reasoning. Table 18–1 is an example of *deductive* criteria applied to the photo interpretation of vegetation in which finer and finer detail is sought in four (A–D) successive steps for examining a terrain pattern. Extracting the information content of aerial images could, by analogy, be described in terms of concepts of information

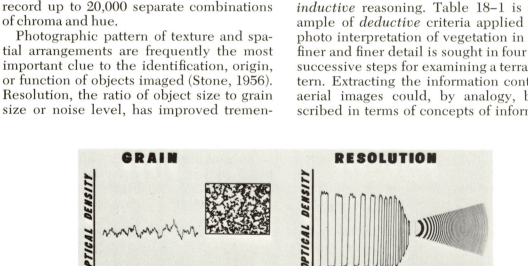

Figure 18–1. Optical density expressions of four film properties.

Table 18–1. *Deductive Approach to the Interpretation of Vegetation in Which Pattern Is Examined in Greater Detail in Four Successive Steps and at Greater Magnification*

A. General Pattern Types—satellite photography and small-scale mosaics
   Discrimination: Forest   Cultivated
                   Shrub   Barren
                   Grass   Mixed

B. Pattern Features—mosaics and small-scale prints
   Discrimination: Areal extent and distribution
                   Boundary conditions
                   Complexity

C. Pattern Elements—stereo pairs
   1. Tone and texture—Spectral class
                       (optical density)
                       Arrangement
                       Complexity and uniformity
   2. Site characteristics—Landscape, natural or
                           cultivated
                           Soil and rock type
                           Drainage type
                           Slope and exposure
   3. Structural characteristics—Canopy configuration
                                 Crown types
                                 (diversity)
                                 Density
                                 Height

D. Pattern Components—large-scale photography
   1. Local adjustments and interactions—causative factors and origin
   2. Composition—species diversity
   3. Dynamics—interactions, processes, successional trends

theory (Johnson, 1966). That is to say, as shown in Figure 18–2 photointerpretation involves an attempt to proceed from a state of disorganization of the uninterpreted photo to a state of organization in terms of useful information accumulated from the picture. Such an analogy parallels the information feedback loops discussed for biological control systems (see Figure 10–2, page 284).

The Nature of Tone and Texture

Interpretable photographic information is often expressed in terms of gradations of tone and categories of texture. The information content of these qualitative expressions is difficult to evaluate or to automate. If, however, tone and texture are equated with density, then densitometry may serve as a basis for automated pattern recognition.

Photographic information is measured by the image that is developed. The developed image may be interpreted or it may be measured by its ability to block light, i.e., its opacity, I_o/I_t. Optical density, D, is defined as the log of opacity:

$$D = \log (I_o/I_t), \qquad (1)$$

where I_o = incident intensity and I_t = transmitted intensity. Densitometers with sev-

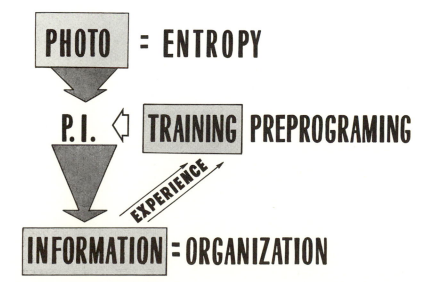

Figure 18–2. Photographic interpretation (P.I.) by analogy to an information theory model reduces the entropy of a complex pattern as information is extracted. Training is an essential input to resolve a pattern, and experience is a feedback that modulates and enhances further interpretation.

eral types of output have been developed to perform densitometric measurements. Density traces, recorded on a strip chart recorder as are other electrically generated signals, have a density (level) D, an amplitude A, and a frequency F. Therefore, a photo pattern characterized by a density trace can be represented as:

Photo Pattern $= P = D, A, F$

(two-dimensional) (2)

Sample Trace $= P_t = (\bar{d} \pm \bar{a})\bar{f}$

(one-dimensional) (3)

Frequency is to a great extent a function of scale. Since silver halide crystals on a roll film emulsion average about 0.5 to 1.0 μ^2 in area, it is possible to theoretically fix the maximum number of permutations at less than 1×10^4. Since frequency is more a function of the distance between silver grains than of their diameter, the potential number of permutations is probably lower by several orders of magnitude. The argument can be advanced that a given photo pattern in the sense of equation (2) is equivalent to an information bit in information theory, and therefore the maximum potential information of panchromatic film identifiable by densitometry might be represented as 5×10^8 bits. In this context an information bit is dimensionless, but the spatial dimensions of a pattern are important to the inferences placed on them. That is, size, shape, and arrangement are additional variables to tone and texture.

Identification of a target may also depend on the joint occurrence of several patterns, such as the sunlit pattern and shadow pattern of a single object. More complex terrain patterns may require synthesis of a number of contiguous or mutually exclusive pattern elements in which a pattern element is defined as a single discriminate following equation (2). It is also apparent that patterns are recognizable to the extent that there is a redundancy of information bits. Therefore, the photointerpreter is seldom presented with anything approaching the theoretical limits of pattern complexity.

The ecological significance associated with a photographic or imagery pattern is obviously dependent on scale and usually will require independent confirmation. The biological identity of an information bit may be a single leaf at a very large scale or perhaps an entire forest stand at scales of an earth-orbiting satellite. Since ecologists often desire information at various scales, that is, at the levels of organism, population, and ecosystem, examination of a range of scales may be a useful means of integrating information at several levels of resolution.

Photogrammetry

Automatic photointerpretation is in its infancy (Rosenfield, 1965), although great progress has been made in digitizing photographs from electrical impulses derived from telemetered images, as in the case of photographs of the surface of the moon. Discrete measurements can be obtained photogrammetrically as well as from microdensitometers. By using color filters matched to the emulsion response curve, color images or pigment saturation of film can be expressed in quantitative terms. Similarly an expression of forest crown closure — canopy cover index — has been developed by means of macrodensitometery of hemispherical photographs (Figure 18–3) taken on the ground (Johnson and Vogel, 1968). An instrument, such as shown in Figure 18–4, measures with 95 per cent confidence the per cent of vegetation silhouetted against the sky in a 90° cone that is projected on a flat plane. Light relations within a forest have been calculated from similar "fisheye" photography (Anderson, 1964). In other applications microtopography can be contoured photogrammetrically from stereo photos obtained with a bipod support (Figure 18–5). This support has been found dimensionally stable for values approaching ± 1 cm from camera heights up to 10 meters (Whittlesey, 1966). On Pleistocene sediments of the Alaskan arctic coastal plain, oriented lakes and ice wedge polygons are responsible for the existing microrelief. To evaluate minute topographic expressions of the landscape on vegetation distribution, near-surface lithology and soil processes, and ground ice volume (Brown and Johnson, 1965), topographic maps were prepared from aerial photography with 0.5 m contour interval (Brown and Johnson, 1966).

The bottleneck in extracting information from aerial photographs or scanner images has been man, the interpreter. A major advance was recently achieved with the fabrication by the University of Michigan

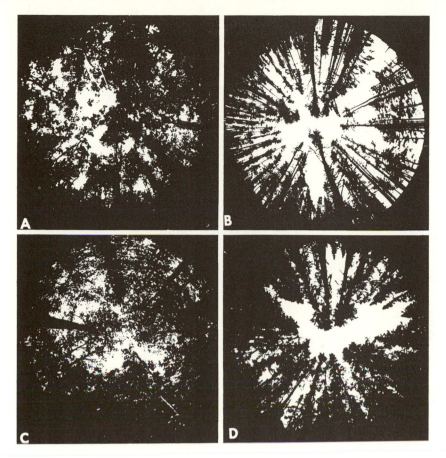

Figure 18-3. Hemispherical photography (i.e., photos taken from the ground looking straight up to the sky) of contrasting forest types with per cent canopy closure index estimated with optical density device shown in Figure 18-4. *A*, northern hardwood forest in New Hampshire, 88; *B*, boreal jack pine stand in Ontario, Canada, 54; *C*, tropical rain forest in Puerto Rico, 84; *D*, subarctic black spruce stand in interior Alaska, 72.

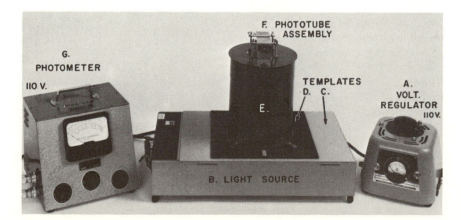

Figure 18-4. Optical density device for measuring light attenuation of transparencies. Enlargements of hemispherical or "fisheye" photographs, such as those pictured in Figure 18-3, are inserted between templates D and C. Per cent canopy closure can then be estimated from light attenuation.

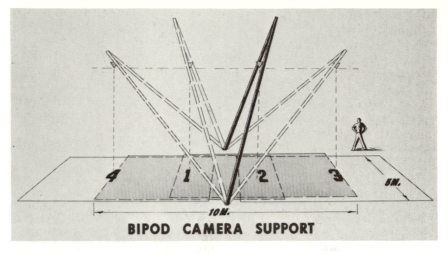

Figure 18–5. A bipod camera support used in photogrammetric stereo photography for obtaining stereoscopic photographs of the shaded area. A camera is suspended vertically at the triangle and the bipod is swung to precise positions indicated. Such a portable apparatus can be assembled and erected by two men. (Modified from Whittlesey, 1966.)

of a single-aperture, optical-mechanical scanner with 12 channels (Polcyn, 1969). This was accomplished by combining a spectrophotometer with a conventional scanner (Holter and Wolfe, 1960) in such a way that energy entering an entrance slit is split into 12 channels ranging from 400 mμ to 1000 mμ. Each spectral band is recorded separately on magnetic tape, and each channel is calibrated so that actual energy units received from the scene can be generated for each tone. Six additional channels, one in the ultraviolet and five in the thermal infrared, have been added with appropriate detectors.

Since each channel is recorded simultaneously, all images are in perfect register. The data for each spectral band or for selected bands can form input to a computer for correlation with ground truth or they can be displayed as separate pictorial images for study. For the first time the ecologist can obtain multispectral data shortly after an overflight and utilize it in quantitative form without resorting to the time-consuming task of photointerpretation.

Electronic processors and analog computers can now perform recognition and classification assignments on multispectral data. Weak signals can be masked. Figure 18–6 is one form of computer output which mapped and identified correctly crop species growing at the Purdue University Agronomy Farm in late June (Polcyn, 1969).

Automated interpretation can only accomplish what men have instructed the machine to do. Such machines may do repetitive tasks faster by utilizing a remarkable memory that is subject to immediate recall. Complete information extraction from a pattern may require logical reasoning, inference, judgment, and experience to unravel the physical, biological, and cultural aspects of a landscape and to determine how they mix to form a pattern or behave under varying climatic or managerial stresses. The trained human eye and mind are indeed a remarkable computer for which no machine can substitute. Together they offer tremendous potential for managing the biosphere.

3. ROLE OF REMOTE SENSING IN ECOLOGICAL RESEARCH

As has been abundantly documented, ecological research and planning are among the few alternatives for the survival of man's economic and cultural society that are increasingly stressed by the numbers of men and the power of his technology. Ecology as a scientific discipline has an opportunity—indeed a mandate—to bridge the gap between academia, technology, and the very real and extensive environmental problems confronting modern and future society. Toward this goal ecological inquiry can be grouped into four categories that are susceptible to remote sensing techniques:

1. Inventory and mapping of resources,
2. Quantitating the environment,
3. Describing the flow of matter and energy in the ecosystem, and
4. Evaluating change and alternative solutions for management of ecosystems.

Aerial photography and remote sensor imagery have the potential in ecology that the spectrophotometer had demonstrated in physiology. In fact, aerial photography may be thought of at the ecosystem level as analogous to the electron microscope in molecular biology; each depends upon the spectral reflectance, absorption, emission, and transmission characteristics of the respective samples.

Inventory and Mapping

Solutions of many resource problems depend on adequate assessment of physical and biological characteristics integrated over a range of areas from a few square meters to thousands of kilometers. Maps of these characteristics derived from conventional aerial photography are logical forms of communications (Figure 18–7). This approach is well illustrated by advances in geological exploration, which include magnetometer surveys, and in forest inventories (Colwell, 1961, 1968). Important biological properties of ecosystems potentially measurable by remote sensing techniques, singly or in multispectral combinations, include leaf area index; stem volume; crop acreage and yield; species and structural diversity (Olson, 1964; Miller, 1960; Wickens, 1966); weight and chlorophyll content of vegetation; certain disease and insect infestations (Norman and Fritz, 1965); kind, density, and biomass of larger animal populations; thermal and chemical pollution of aquatic systems (Schneider, 1968; Strandburg, 1966); heat, water vapor, and carbon dioxide fluxes of the earth's surfaces; evapotranspiration and water content of soils and vegetation; fire (Bjornsen, 1968); and depth and density of snow.

One of the many possible examples of vegetation mapping was done in Southeast Asia from conventional panchromatic photography of poor scale and without prior field experience with the types of vegetation that grow in Thailand (Leightly, 1965).

These maps were subsequently field checked and found to be surprisingly accurate. Consideration of the structural properties of vegetation in Thailand as a deterrent to vehicle mobility resulted in the selection of height, diameter, and spacing as primary mapping criteria (Frost *et al.*, 1965). The numerous permutations of these vegetation parameters in Thailand were grouped into 12 structural categories (Table 18–2). It was possible to map these parameters (Table 18–3) for 6000 square miles with photographic scales of 1:20,000 or smaller. It is evident that the discreteness of many of these boundaries (Figure 18–8) is a function of man's manipulation of the landscape. For example, in extensive areas of shifting agriculture and scrub forest adjacent to riceland, corn has become a major crop since 1958 with the aid of mechanical equipment (Figure 18–9). In fact, cultivation of this crop is so successful in Thailand that export of corn to Japan has replaced a substantial portion of the United States corn market to that country. The critical problem in such moist tropical areas is the degeneration of soil fertility in the absence of commercial fertilizer, which results in aban-

Table 18–2. *Predominant Vegetation Types of Thailand Based on Structural Differences Discernable on Aerial Photography*

A. Field Crops
    1. Low (< 1 m) — rice paddy, pasture land, rong agriculture
    2. High (> 1 m)
        a. row: corn, sugar cane, cassava, tobacco
        b. random: tropical grass
B. Shrub Savanna and Orchards
    3. Low
        a. row: pineapple, young fruit orchards
        b. random: shrub savanna
    4. High
        a. row: pepper fields, fruit trees
        b. random: thorny shrub
C. Tree Savanna and Plantations
    5. Open canopy: rice savanna, village platforms
    6. Closed canopy: coconut, rubber and teak plantations
D. Semievergreen Forest and Woodlands
    7. Mature dry monsoon forest, oak-pine forest
    8. Secondary shrub forest, shifting agriculture
    9. Bamboo forest
E. Evergreen Forests
    10. Evergreen Rain Forest
    11. Mangrove swamp forest
    12. Nipa palm swamp

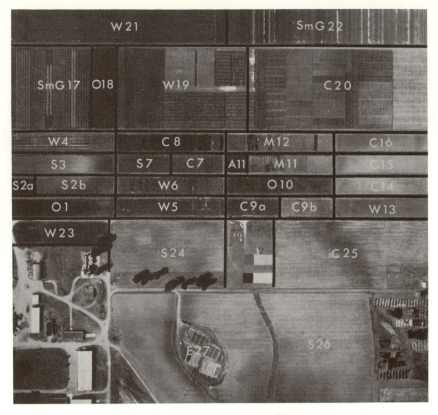

GROUND-TRUTH MAP
Portion of Purdue Agronomy Farm, Late June, 1966.

Key: W = Wheat   A = Alfalfa
        O = Oats   SmG = Small Grains
        C = Corn   M = Misc.: Legume, Soybeans,
        S = Soybeans   Sorghum, Oil Crops, or
A   Sudan

Figure 18–6.   A. Ground-truth map of a portion of Purdue Agronomy Farm, late June, 1966.

doned land and irreversible succession to thorny bamboo and coarse tropical grass.

Quantitating the Environment

Most resource problems require quantitative data in addition to definitions of what the problem is and where it is. In some cases even estimates of magnitude applied to large areas are sufficient for meaningful planning (Haefner, 1967). In more sophisticated treatments, remote sensing, particularly with the narrow bandpass instruments, can generate accurate numbers. The recent development of a laser system as an airborne profilometer for studies of microtopography is capable of portraying changes in ground elevation on the order of a few

centimeters from airplane altitudes (Rempel and Parker, 1965). Radio ice sounding techniques permit measurements of the ice-bedrock interface through ice thicknesses of 2500 meters (Rinker *et al.*, 1966). The recent development of sensitive heat and magnetic sensing tools is a product of military necessity that is rapidly finding applications in the civilian economy. One of the problems in this context has been the rapid increase in development of hardware systems and the resulting lag in man's knowledge on how such systems can be effectively utilized and with what degree of reliability. Consider, for example, the number of thermometers required to delineate cold air drainage in a valley versus an isothermal contour map obtained by remote sensing (Figure 18–10).

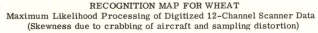

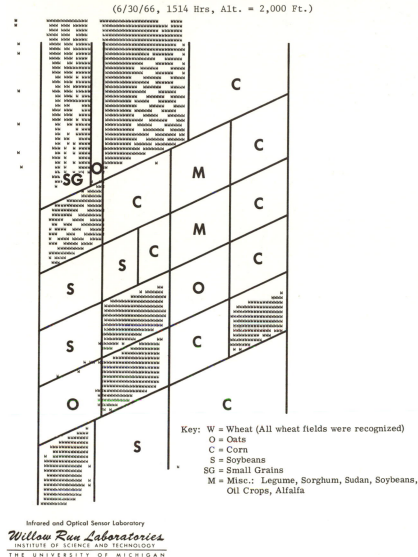

RECOGNITION MAP FOR WHEAT
Maximum Likelihood Processing of Digitized 12-Channel Scanner Data
(Skewness due to crabbing of aircraft and sampling distortion)

(6/30/66, 1514 Hrs, Alt. = 2,000 Ft.)

Key: W = Wheat (All wheat fields were recognized)
O = Oats
C = Corn
S = Soybeans
SG = Small Grains
M = Misc.: Legume, Sorghum, Sudan, Soybeans,
Oil Crops, Alfalfa

Infrared and Optical Sensor Laboratory
Willow Run Laboratories
INSTITUTE OF SCIENCE AND TECHNOLOGY
THE UNIVERSITY OF MICHIGAN
B

Figure 18–6. B. Recognition map for wheat (shaded) and other crops based on computer analysis of multi-channel imagery in the visual wavelengths. The apparent angular distortion of the fields can also be corrected by computer processing.

Flow of Matter and Energy

Failure to understand ecological processes is frequently the cause of faulty or conflicting resource planning. To appreciate function as opposed to structure in ecosystems requires the study of processes, interactions, and transfer rates between different organisms as well as between organisms and their environment. Frequently, measures of metabolic activity are desired. These are the most difficult answers to obtain from remote sensors as well as from sensors on the ground. Nevertheless, detection by airborne sensors of metabolic products or physical changes caused by active biological processes are important clues for inferences and conclusions about the direction and quantity of matter and energy flow in ecosystems (Barringer *et al.*, 1968; Lohman and Robinove, 1964) as well as for selecting sampling points for further measurements.

Aerial prediction of vegetative response to an environmental gradient was demon-

S	Salix spp.	Pm	Picea mariana
Pb	Populus balsamifera	PmL	Picea mariana/Larex laricina
Pt	Populus tremuloides	BpPg	Betula papyrifera/Picea glauca
SA	Salix spp./Alnus crispa	BnLg	Betula glandulosa/Ledum groenlandicum
Bp	Betula papyrifera	C	Carex spp.
Pg	Picea glauca	PmBn	Picea mariana/Betula glandulosa

Figure 18–7. *Top.* Mosaic illustrating a photo base map of vegetation patterns representative of Yukon River tributaries entering southeastern Yukon Flats, Alaska. *Bottom.* Mosaic of Yukon River floodplain near Circle, Alaska, illustrating a photo base map of vegetation patterns.

Table 18–3. *Parameters and Classes Used for Mapping Vegetation Structure in Thailand From Panchromatic Aerial Photography*

MAP SYMBOL

$$V = \frac{SDH}{HC}$$

Zero in the formula indicates that the applicable element could not be determined on the photography. Blank spaces in the formula indicate the applicable item does not apply or exist. Thus, overstory vegetation is characterized in the numerator and understory vegetation in the denominator.

S = STEM SPACING OF TREES

1	<4 ft.
2	4– 8 ft.
3	8–12 ft.
4	12–16 ft.
5	16–20 ft.
6	20–30 ft.
7	>30 ft.

H = HEIGHT OF PLANTS

1	<6 ft.
2	6–16 ft.
3	>16 ft.

D = DIAMETER OF WOODY PLANTS (DBH)

1	<½ in.
2	½–2 in.
3	2– 6 in.
4	6–12 in.
5	>12 in.

C = COVER (% GROUND COVER)

1	1– 10%
2	10– 25%
3	25– 50%
4	50– 75%
5	75–100%

strated in the following experiment. Photographic monitoring of the effects of chronic gamma irradiation on the Brookhaven forest, pictured in Figure 17–4 (page 458), was accomplished with four film emulsions (Johnson, 1965). The concentric zones of vegetation surrounding the ^{137}Cs source in the oak–pine stand were evaluated with qualitative and quantitative information derived from large-scale, aerial hand-held photography. Microdensitometry with an effective aperture of 40 μ provided quantitative expressions as a function of film type, leaf development, species, and the gradient of gamma irradiation. The resolution attained with this aperture and this scale photography approximated the dimensions of an average oak leaf. A good correlation (r = 0.88) was suggested between a logarithmic expression of daily gamma dosage in roentgens and optical density of panchromatic photography along this gradient (Figure 18–11). Further canopy closure photographs taken from the ground at increasing distances from the source yielded indices of overstory defoliation that correlated (r = 0.97) with the density data from panchromatic film. Thus, photographic measurements both from above and below the canopy delineated the effects of ionizing

Figure 18-8. Vegetation map and aerial photograph of vegetation structure, Chanthaburi, Thailand. See Table 18-3 for key to symbols.

Figure 18-9. *Left.* Aerial photographs of shifting cultivation taken in 1953 in Lopburi, Thailand. *Right.* The same area converted almost entirely to cornfields by 1964.

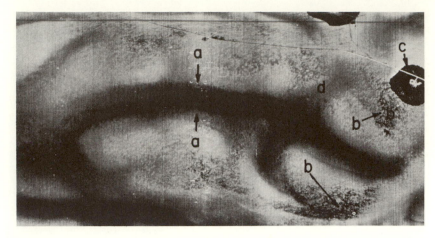

Figure 18–10. Cold air drainage in a mountain valley (a) in Pennsylvania is dramatically illustrated by thermal imagery flown at night. Ridges (b) are covered with stands of oak while northern hardwood stands occupy the valley slopes. A clearing (c) for a communication tower contains low shrubs, primarily ericaceous species.

radiation. A gray scale on infrared film had just the opposite relationship in contrast with panchromatic film; that is, the infrared film showed light tones (low density) near the source, where all vascular plants were killed, progressing to darker tones for the forest at distances beyond the influence of radiation. Color film can be similarly evaluated by measuring the pigment saturation of each of the three — red, green, and blue — emulsion layers. Thus, quantitative analysis of conventional aerial film in combination with ground data may spatially delineate a complex response within an ecosystem.

Evaluating Change and Alternative Solutions

Alteration of natural ecosystems is, of course, manifest in all resource problems. Without change, that is, without depletion,

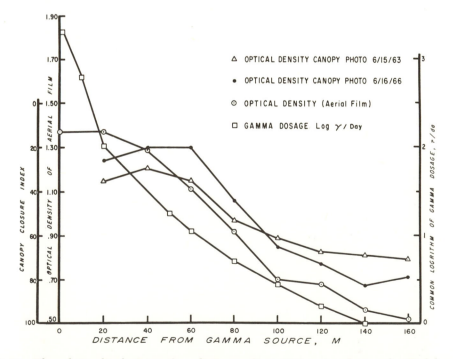

Figure 18–11. The relationship between the radiation gradient, CCI, on two dates and optical density of aerial photography with distance from a source of chromic gamma radiation in a pine–oak forest at Brookhaven National Laboratory, Long Island, N.Y. (For a diagram of this radiation gradient, see Figure 17–4.)

Table 18–4. *Anticipated Orbital Resolution of Some Selected Sensors**

SPECTRUM	SYSTEM	AREA AT 200 N MI.	GROUND RESOLUTION
Visual	15 cm FL camera	766 sq. kilometers	6–30 meters
		TEMP. SENSITIVITY	
Infrared	High resolution scanner	±0.1°K.	60 meters
Microwave	Radar, Ka band, 8 mm	±0.5°K.	0.5–1 kilometer

* Abstracted from University of Michigan publication 7219-2-F 13 entitled, *Peaceful Uses of Earth Observation Spacecraft*, vol. 1, 1966.

erosion, pollution, accrual, or epidemic, the problem is seldom recognized. This, perhaps, is the easiest type of information to procure by repetitive aerial surveillance, and it has been exploited with photography in the visual wavelengths. Thus, old conventional aerial photography may gain value with time as an index to change, but until repetitive aerial reconnaissance is widely practiced (and financed!), long-term and widespread change, natural or caused by man, will often be difficult to assess. Once trends of change or the consequences of our technology are evaluated, alternatives can often be developed from the same data. This is perhaps the most promising application for photography from spacecraft (Brock *et al.*, 1965; Lowman, 1966).

A major contribution of the technology developed to explore space will be to provide information to the natural sciences from earth-orbiting spacecraft (Badgley and Vest, 1966). Already satellites are telemetering valuable synoptic coverage of weather patterns. Once a spacecraft is in orbit, the cost of repetitive photography is minimal. Some hand-held color transparencies obtained on the Gemini missions have shown the value and quality of present space pho-

tography. For example, the swirl in the Sargasso Sea had never been observed prior to Gemini photography. As sensing systems are improved and designed specifically for space platforms, improvements in quality can be expected. Some anticipated parameters for remote sensors orbiting at 200 nautical miles are presented in Table 18–4. Perhaps the most valuable and immediate return from satellite platforms will be synoptic photographic coverage of broad areas and changing conditions amenable to assessments of our technologies.

The solution of man's confrontation with nature involves many decisions outside the province of an ecologist. To the extent that many of these decisions are rooted in biological concepts it is worth contrasting multiple-use management with ecosystem zoning. It is apparent that we can no longer afford conflicting multiple uses of some landscapes (see Chapter 9, Section 3 for a discussion of this important point). Perhaps assigning priorities and zones of use that are founded on understanding the ecosystem aided by remote sensing tools is a plausible alternate to irreversible multiple abuse.

PERSPECTIVES IN MICROBIAL ECOLOGY*

Chapter 19

By William J. Wiebe

Department of Microbiology and
Institute of Ecology, University of Georgia

Microbial contributions to nutrient and energy processes have been emphasized in the chapters of Part 1. Microorganisms, by virtue of their small size and ability to withstand adverse environmental conditions for long periods of time, are ubiquitous in the biosphere. They contain, *collectively,* an enormous number of metabolic abilities and are capable of rapid adjustment to environmental changes. Such metabolic diversity assures that most natural organic substrates introduced into any environment will be transformed, although at a slow rate in the case of humic substances or in anaerobic environments (see Chapter 2, Section 3). Perhaps because of this natural versatility man has too often assumed that any kind of waste he might find convenient to dump into the environment would somehow be transformed and "purified" by the ubiquitous microbes. Consequently, almost no one bothered to study microorganisms in nature; so until very recently microbial ecology remained a neglected subject. It is now all too evident that many man-made organics such as some pesticides, herbicides, detergents, and industrial by-products are "recalcitrant," as Alexander (1965) has expressed it, that is, slow to degrade in the natural environment. As a consequence these recalcitrant substances not only accumulate slowly to toxic levels, but they also interfere with the recycling of critical nutrients and greatly reduce the quality of air, water, and soil. Payne, Wiebe, and Christian (1970) point out that the knowledge of microbial enzyme systems is avail-

able to prejudge the prospects for decay of many chemical substances, and they suggest procedures for using this technology to establish biodegradability before chemical agents or products are cleared for general and widespread usage. However, the greatest present challenges in microbial ecology center less often on the demonstration that a particular process can or cannot occur but rather on the quantification of the *in situ* microbial response in the ecosystem.

1. A BRIEF HISTORY

In the late 19th century the essential principles and perspectives necessary for the study of microbial ecology were developed by two investigators, Winogradsky and Beijerinck. Winogradsky (1949), summarizing over 50 years of investigations, emphasized the necessity for study of microorganisms in their natural environment. Bass-Becking (1959) restated the two "rules of Beijerinck": (1) everything is everywhere and (2) the milieu selects. As Beijerinck (1921) has himself pointed out, the ecological approach to microbiology consists of two complementary phases: investigations of the conditions for the development of organisms that have for some reason, perhaps fortuitously, already come to attention, and the discovery of new organisms that appear under predetermined culture or field conditions, that is, those "selected by the milieu" either because they alone can develop or because they win out over their competitors.

There has developed in the past 30 years another point of view for the investigation of environmental microbial activity, the analysis of function and rate of activity without regard to the specific types of mi-

* The author is indebted to Dr. Holgar W. Jannasch, Woods Hole Oceanographic Institution, for permission to include in this chapter certain material from his Edgardo Baldi Memorial Lecture entitled "Current Concepts in Aquatic Microbiology" (Jannasch, 1969).

croorganisms present. This approach, which emphasizes the *quantitative effect* of microbial metabolism within an environment, is measured, for example, by total O_2 consumption, CO_2 evolution, turnover of specific organic compounds such as amino acids, or the uptake kinetics of substrates. The advances of molecular biology and biochemistry have given the microbial ecologist many tools for functional assays.

Many of the difficulties associated with studying microorganisms in their environments are, as implied, largely technical. The proper questions to ask are, to a great extent, known. It is experimental implementation for answering these questions that is difficult. In this chapter we shall emphasize the methods and techniques that have been developed to answer these questions.

2. THE QUESTION OF NUMBERS

Superficially no problem seems easier to attack than that of estimating the numbers of organisms in a sample, and yet no problem is so seductively complex to unravel experimentally when it comes to microorganisms. There are two general approaches: direct counts using the light microscope and viable counts in which macroscopic growth on a plating surface is the criterion for the presence of a bacterial cell in a sample.

Viable Counts

Robert Koch (1881) ingeniously elaborated the technique of counting bacterial cells as grown colonies on a solidified medium. His idea was based on two prerequisites: the cells to be counted must grow on the selected medium, and each colony must represent growth from a single cell in order that the counting be reproducible and accurate (i.e., the cells must not be clumped initially). Koch's technique is applicable for estimating the population of specific types of cells and for making relative comparisons between mixed populations of a specific metabolic type, but *it cannot be used to determine population density or "total bacterial counts."* The plating technique yields semiquantitative data for identification of a particular metabolic type, as for example, sulfate reducers, aerobic nonexacting heterotrophs, and nitrogen fixing bacteria. Three plate count methods are routinely used: pour plates, spread plates, and membrane filter plates. In the pour plates method molten (42°C.) agar is mixed with appropriate sample, diluted and poured into petri dishes. After appropriate incubation the colonies are counted. In the spread plate technique the sample dilutions are spread directly on the surface of cooled agar, incubated, and the colonies counted. Membrane filter counts are made by passing a measured quantity of water through a filter, generally a 0.22 to 0.45 μ diameter porosity, placing the filter on a liquid or solid medium plate, and then incubating and counting colonies. Jannasch and Jones (1959) compared several of these methods using replicated seawater samples. As shown in Table 19–1, the bacterial estimates in a milliliter of water were highest using direct microscopic counting procedures, lower with membrane microcolony counts, and lowest with pour plate counts. Each method has advantages and disadvantages, and these can change in different environments. The student may read further about these techniques in a general

Table 19–1. *Numbers of Microorganisms per Milliliter of Seawater as Indicated by Different Methods of Counting. Per Cent Error in Parentheses**

DEPTH (m)	WATER TEMPERATURE (°C.)	PLATE METHODS (5 DAYS)	PLATE METHODS (21 DAYS)	SERIAL DILUTION (5 DAYS)	SERIAL DILUTION (21 DAYS)	MACRO-COLONIES ON MF†	MICRO COLONIES ON MF†	DIRECT COUNTS ON MF†
Surface	20.1	6 (10%)	11 (10%)	7	7	8 (9%)	68 (3.1%)	244 (1.8%)
25	20.1	14 (12%)	14 (12%)	5	5	14 (7%)	31 (5.2%)	262 (2.1%)
50	19.1	9 (19%)	10 (6%)	7	7	10 (12%)	30 (8.7%)	166 (6.4%)
100	13.0	4 (14%)	6 (29%)	2	2	1 (15%)	29 (6.6%)	82 (11.3%)

* From Jannasch and Jones, 1959.
† Membrane filters.

microbiology text (Stanier, Douderoff, and Adelberg, 1970). It can be stated that generally viable counts represent only 0.1 or less of the total number of cells in autochthonous populations, that is, populations in reduced growth state as is usually the case in unpolluted nature. In such situations no direct correlation between viable count estimates and direct counts (as discussed subsequently) is seen. In other words, the media used in conventional bacteriological work is too "enriched" ("polluted," so to speak) for many natural populations that are adapted to a dilute nutrient environment.

In pollution microbiology the Most Probable Number (MPN) (see APHA, 1966) of coliforms is routinely determined. The presence of coliform bacteria in a sample is used by many health departments as an indication of the level of fecal contamination of a water supply. The term "coliform" refers to a group of bacteria that reside in the intestines of many vertebrates. Methods are available for detecting those strains of human origin. Although harmless in themselves, the presence of coliforms indicates that harder-to-detect disease microorganisms might be present. While there is great controversy concerning the applicability of this technique, the *coliform count* is presently the major method for the estimation of potable water. The method employs a statistical analysis of growth in a series of replicated dilutions to determine the approximate number of coliforms in a water sample. Although the coliform count is but a crude index, very high counts certainly indicate heavy pollution with human or animal wastes and are justification enough for placing "condemned" signs on the waterways.

Viable count estimates, then, produce relative numbers of a particular group of microorganisms but do not give "total or absolute counts." Within this restriction they can be a useful tool of analysis of the microbiology of an ecosystem. Plating methods also permit the isolation of individual strains of microorganisms for pure culture studies.

Microscopy

The other general method for estimating the number of microorganisms in a sample is direct microscopic examination. Again, a variety of techniques have been used; cells have been made fluorescent with acridine orange (Wood, 1967) and a variety of other fluorescent dyes. Various stains have been employed, and direct light microscopy has been attempted. For many protozoa and algae these techniques are adequate, although the taxonomy of the cells often cannot be established with these methods. However, three major difficulties are encountered when making direct count estimates of bacteria: (1) It is often difficult to recognize the individual bacterial cells since the "characteristic" morphology of laboratory cultivated cells often changes under *in situ* conditions. (2) Microorganisms often appear in clumps or attached to particles, and individual cells cannot be clearly distinguished. (3) About 10^6 bacteria per milliliter are needed to see one organism per field at $1000 \times$ magnification. In many environments bacterial numbers are very much less than 10^6 per ml, necessitating concentration of the microorganisms. In most instances only limited concentration can be accomplished. Also, an accurate count cannot readily be made if many of the cells are clumped or attached to detritus because of masking of individual cells and the nonrandom distribution of the population.

The recognition of bacteria in a natural sample must be considered a subjective process, and the student is cautioned to set up careful criteria for evaluation based on preliminary examination. As of this writing it is not possible to accurately distinguish living from dead cells under the microscope.

The problems associated with both viable and direct counts are well illustrated by the work of Gorden *et al.* (1969). They examined the bacterial succession in a microecosystem (see page 20 and Figure 2–6 III). In Table 19–2 viable count data show that *Bacillus sp.* increased rapidly initially, then decreased to a low but constant level. Direct microscopic counts showed, however, that after 3 days *Bacillus sp.* sporulated, thus becoming inactive in this system. In this instance the viable counts gave no indication at all of the real sequence of events and led to an *overestimation* of the numbers of active cells in the system, since the *Bacillus sp.* spores germinated and grew on the viable counts medium.

Another potentially powerful tool in visualizing microorganisms in the environment is the stereoscan electron microscope (Gray,

Table 19–2. *Succession of Microorganisms in a Derived, Closed-system Microcosm (Similar to That Shown in Figures 2–6 III and 2–7A)†*

	VIABLE BACTERIAL COUNT OF MIXED SYSTEM/ML									
	10^3	10^9	10^8	10^7	10^6	10^5		10^6	10^7	10^6
	Days									
ORGANISM°	0	1	5	10	20	30	40	50	60	70
Bacillus (vegetative)	A	A	A							
Bacillus (spores)			C	R	R	R	R	R	R	R
W		A	A			C		C	C	C
Y		C	A	C	C	C				
L		C	A	A	A	A	A	A	C	R
G		C	C	C	A	A	A	A	R	R
C		C	A	A	C		A	A	A	A
T		C					A	A	A	
B	A	A					C	A	A	A
O		R	R	R	R	R	R	C	C	C
P		R	A	R	R	R	R	C	C	C
Chlorella		C	A	A	A	A	A	C	C	C
Schizothrix		C	C	C	C	A	A	A	A	A
Scenedesmus										
Cypridopsis		R	R	R	C			C	C	C

Protocol: R rare C common A abundant

° The letters W, Y, L, etc., represent functionally distinct bacterial isolates which may or may not be "species" in the conventional taxonomic sense. Day "0" marks the time when new culture media was innoculated from an old climax culture; in 70 days the microcosm again reaches a steady state. See Figure 9–2 for a graph of the metabolism of such a microcosm succession.

† From Gorden *et al.*, 1969.

1967). Microorganisms can be examined in their physical habitat so that spatial relationships and patterns of growth can be observed. Figure 19–1 shows decomposing fungi at work on a pine needle in forest floor litter. This method works particularly well for the higher protists, but even bacteria can be successfully resolved. While presently not a strictly quantitative method, the stereoscan electron microscope has enabled us to observe organisms directly within their microhabitats.

Total Biomass Measurements

The inadequacy of direct or indirect counts of cells has led to a search for more accurate indicators of microbial biomass or standing crop. Such data are of particular importance for calculating trophic relationships. In single-species cultures the usually high population densities are measured by turbidity, dry weight, protein content, and other parameters. All of these tests are too unspecific for the conditions typically encountered in nature because of the abundant nonliving organic matter generally present. Adenosinetriphosphate (ATP) is widely distributed and biochemically essential for cell growth and maintenance, and the determination of ATP is a potentially sensitive measure of living biomass. Since it is normally *found only inside living cells* (see page 10) and is an essential component for energy transformations and biosynthetic processes, ATP is a promising index of total biomass. The principal difficulty in employing this method is the intrinsic variation in a given amount of ATP per unit weight of cells. Unfortunately, turnover time and constancy of ATP per cell (Harrison and Maitra, 1968) varies with the physiological state of cells and the type of microorganism. The published range of ATP per unit weight for a wide variety of organisms is about 50-fold. This range can probably be reduced for a particular environment since the heterogeneity of organisms should be less in a given ecosystem type

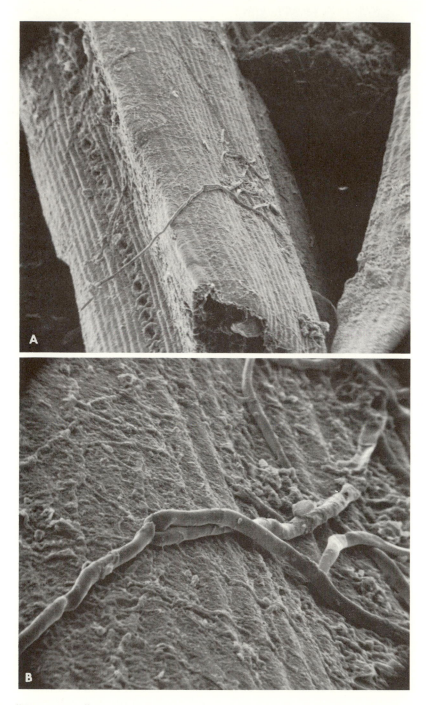

Figure 19–1. "Decomposer" in action. Scanning electronmicrograph of a fungal mycelium on a pine needle being decomposed in the forest floor litter. The upper photo shows a section of the pine needles at 100 × magnification, while the lower photo shows a close-up of the branching fungus at 500 × magnification. (Photos by Dr. Robert Todd, Institute of Ecology, University of Georgia.)

than in a miscellaneous selection of organisms from the world at large. The problem remains to determine the range of ATP levels in biomass of the major ecosystems.

At present, the completely accurate determination of cell numbers or cell biomass cannot be made in most natural systems. However, relative numbers and relative biomass data are useful for comparative purposes. Accurate standing crop data are required for estimating the microbial biomass available as food for microbial consumers in a given volume of soil or water.

3. THE QUESTION OF RECOGNITION

There are two principal reasons for investigating what kinds of organisms are present in a sample. First, it is necessary to identify the active microorganisms as a first clue to the diversity of functional types present and for comparison between samples and environments. Secondly, knowledge of types present is necessary to monitor fluctuations that take place in a system. If diversity index tests (see page 148) are to be applied or studies of succession undertaken, identification of organisms is required.

The identification and classification of most metazoans and multicellular plants is based almost exclusively on gross morphological characteristics. The higher protists are also separated on morphological bases. It is often stated that morphology of bacteria can be of little help for taxonomic purposes because of the small size of cells and because of their generally uniform shape. This statement can be applied reasonably well to the two major (i.e., most commonly studied, not necessarily most important) orders of bacteria, namely, the Pseudomonodales and Eubacteriales, and to a limited extent to several other orders. What should be noted, however, is that presently *all orders of bacteria are separated exclusively on morphological grounds.* Many families, genera, and species also are identified on morphological bases—for example, the species in the genera *Beggiatoa* and *Sorangium* (Breed *et al.,* 1957). Thus, direct morphological identification is possible for some groups of bacteria. This statement holds for many of the photosynthetic bacteria; it is less true for the chemolithotrophs, and the least valid for the heterotrophs (see pages 25–27 for definitions). Un-

fortunately the heterotrophs constitute the majority of microbes in many environments. One difficulty with these microorganisms is that their morphology at both the gross and fine structure levels can vary greatly depending on environmental conditions (Brock, 1966).

The bacterial taxonomist has been forced to supplement direct observation with tests for biochemical and physiological activities. The problems of distinguishing bacterial genera and species are too complex for present examination, but there is an excellent review in Skerman (1967). Recent revisions of microbial classification are assigning less and less importance to the poorly substantiated phylogenetic relationships and are favoring a pragmatically derived determinative system. This system has been developed with the aid of a numerical analysis of data for a computerized taxonomic scheme (Sokol and Sneath, 1963). By this method morphological, biochemical, and physiological characteristics of cultures are determined and the taxonomy established on the basis of mathematically analyzed, overall similarities. The advantage of this system is that one also acquires data on bacterial function, e.g., glucose dissimilation, amino acid requirements, and so on, in addition to a taxonomic scheme. While one cannot directly extrapolate pure culture capabilities to *in situ* capabilities either quantitatively or qualitatively, "it is justifiable, and operationally necessary, to assume that in most characteristics pertinent to the habitat the pure cultures resemble their progenitors in nature" (Hungate, 1962). For an example of a numerical analysis applied to ecology, see Lovelace, Colwell, and Tubiash (1968).

The most serious disadvantage inherent in the use of numerical taxonomic analyses in microbial ecology is the relatively large number of tests that must be run and recorded for each isolate and the large number of isolates that usually must be examined.

Two recent approaches to taxonomy deserve mention here. Both are based on the constancy of the base-pairs in deoxyribonucleic acid (DNA) in cells. In the first method total guanine and cytosine base composition is determined. The method is based on the fact that similar types of organisms contain similar DNAs; its value lies in distinguishing between superficially similar groups such as the genera *Vibrio* and *Aerobacter,* which have G + C values of

40 and 55, respectively, since many diverse groups can have the same G + C ratios. A summary of G + C values published through 1966 was compiled by Hill (1967). The second method involves direct matching of DNAs or DNA-m-RNA (messenger ribonucleic acid) from two different species and determining similarity by the overall level of homology (see Marmur, Falkow, and Mandel, 1963, and Mandel, 1969, for a more complete examination of this subject). The second method is time-consuming and difficult to perform precisely. Its routine use awaits the development of a standardized and simplified technique. However, it is useful because it directly compares the genotypes of two strains.

Finally a word should be said about the use of the immunofluorescence technique for identification of cell types in ecology. Antibodies can be produced against the surface antigens of individual bacterial species, and these antibodies can be labeled with a variety of fluorochromes. When an antibody is prepared in this manner and mixed with a cell suspension, the cell becomes fluorescent under ultraviolet light and can be seen directly with a fluorescence microscope (for details see Quinn, 1968, pages 163 through 165). This technique has been applied to the identification of specific species of cells *in situ* (Hill and Gray, 1967) and to detection of food pathogens directly in the food (Fantasia, 1969); it is a potentially useful tool for tracing a specific population within an environment.

More important than biomass determinations and the types of microorganisms, particularly in the study of energy fluxes in an ecosystem, are the measurements of activities performed by, and those capable of being performed by, a microbial population, and the direct measurements of *in situ* activities. These subjects constitute the topics in these last two sections of this review.

4. THE QUESTION OF PERFORMANCE

Pure Cultures and Enrichments

Measurements of the extent of microbial activity in an ecosystem may be made with either pure cultures or by chemically analyzing the environment. For pure culture studies we have already explained

the use of various tests for classification and pointed out that these results also delineate some of the physiological and biochemical capabilities of the bacteria. Pure cultures remain an essential and justifiable component of microbial ecological investigations, but the direct interpretation of results with reference to the environment must be made cautiously because there is no assurance that (1) activities seen in pure culture operate *in situ*, (2) cultural conditions have not repressed or induced metabolic pathways, or (3) interactions of the mixed population could not alter individual strain activities. These statements are valid even in light of Hungate's (1962) statement (see page 489).

The selective enrichment technique does reveal potential metabolic activities that may be present in the environment. For example, a recent symposium on microbial enrichments and isolations (Schlegel, 1964) has established that entirely new metabolic types of microorganisms are still to be discovered and that *by understanding the enrichment conditions for growth and isolation we gain an insight into the types of environment involved.* Newly described forms of photosynthetic bacteria illustrate the ecological significance of light absorption and pigment composition (Pfenning, 1967; Truper and Jannasch, 1968). Eimhjellen *et al.* (1966) discovered a new bacteriochlorophyll that absorbs light in the far infra red (1020 nm*), demonstrating utilization of a wavelength that was previously not considered suitable as a light energy source for photosynthesis.

The inorganic nutrient cycles, particularly the nitrogen and the sulfur cycles (see Brock, 1966), have been investigated by means of enrichment procedures. Virtually all the key organisms in these cycles were isolated by use of specific enrichments (see Chapter 4, Section 1, for an account of these key microorganisms). This method has been and will continue to be a powerful tool in metabolic studies for ecology, but additional techniques must be developed to deal more directly with natural populations.

End-product Chemical Analysis

Another approach to the problem of measuring metabolic activities of microorganisms

---

* Nanometer; 1 nm = 10^{-9} m.

is the direct chemical analysis of the environment to ascertain the "end products" of metabolism. Determination of soluble and particulate organic carbon, standing stocks of amino acids, lipids, and carbohydrates yield quantitative and qualitative measures of the standing stocks for the whole system. Standing-stock data give an instantaneous measure of the quantity and quality of compounds, and this is of importance, as we have previously indicated, for understanding what nutrients are available to living organisms at any given moment.

Such data can be deceptive, however, because they do not reflect the turnover rate of compounds, that is, the rate at which a compound is being produced and utilized. For example, Hobbie *et al.* (1968) found that the standing stocks of threonine and methionine in the St. Johns estuary in Chesapeake Bay, Virginia, were similar—1.50 and 1.31 μg per liter, respectively—but that methionine was regenerated and utilized eight times as rapidly as threonine. Over a period of time, then, methionine represents eight times the quantitative importance of threonine, although at any instant their quantities are the same. We will examine this point more fully in the following section.

In most ecosystems we can only estimate how activities are interrelated. We do know that the cycles of the various elements (see pages 88 to 92) are coupled to specific metabolic activities; in some systems—for example, shallow anaerobic ponds—the interactions of heterotrophs and autotrophs in the sulfur cycle has been successfully explained. The use of radioactive tracers (see page 93) for elucidating complex ecological pathways has been employed successfully in recent years. For example, Pomeroy *et al.* (1966) have traced the utilization and distribution of phosphorus in a salt marsh ecosystem by following the movement of introduced ^{32}P. Using this technique they established not only what organisms were responsible for phosphorus uptake but also what the food chain relationships were between organisms.

It will not be possible to measure the interactions between activities or pathway in most ecosystems until we understand quantitatively and qualitatively the nature of the various individual cycles. Hopefully mathematical modeling of these interactions will greatly facilitate the experimental approach.

5. THE QUESTION OF RATE OF FUNCTION

The measurement of the quantitative nature and flux of *in situ* activities is perhaps the most important problem in microbial ecology. The fact that "rate" is more important in small organisms than "standing state" was strongly emphasized in Chapter 3 (page 85). What can and should be measured? The initial attempts to assess natural microbial transformation rates dealt with (1) respiration or (2) degradation of certain organic energy sources or substrates for microbial growth. The total heterotrophic activity of a mixed population can be measured by determining CO_2 production or O_2 consumption—using oxygen electrodes (Kanwisher, 1959), oversized Warburg vessels, or by measuring the B.O.D. (biochemical oxygen demand; see page 16). Measurement of organic energy-source degradation has been performed by adding a known amount of substrate to a sample and calculating the loss of the substrate over time. A variety of similar techniques have been employed. In the following sections we will discuss some of the recent techniques for measuring *in situ* activities of entire mixed microbial populations.

Measurement of in situ Levels of Organic Compounds and Their Turnover Times

In some environments, such as many lakes and streams and the open ocean, total metabolic activities are very low. The necessary increase in sensitivity for measurement of many of these reactions has been achieved by applying radioactive isotope techniques (see page 460). Following the development of ^{14}C technique for measurement of photosynthetic productivity by green plants (see page 60), Kusnetsov (1958) and Sorokin (1964) attempted to determine the rates of bacterial photo- and chemosynthesis. While Kusnetsov (1967) has shown that the heterotrophic uptake of bicarbonate may be considerable, particularly in the dark, and thus causes a decrease in the amount of CO_2 released, the method promises to help establish the *in situ* bacterial contribution to CO_2 flux rates.

In 1962, Parsons and Strickland suggested adding ^{14}C-labeled organic substrates (initially glucose and acetate) to water samples and measuring the ^{14}C uptake after a short incubation period. Surprisingly, the relationship between rate of uptake and substrate concentration exhibited a saturation curve that suggested the involvement of strictly first-order kinetics, as found for permease activities of whole-cell pure culture suspensions (Kepes and Cohen, 1962). In other words, *the mixed natural population of the water sample behaved like a pure culture* in that the uptake curve was a straight line, indicative of a single-enzyme reaction. The major criticism of their work was that they added radioactive substrates at many times the levels found in the environment.

The pioneering study of Parsons and Strickland encouraged Wright and Hobbie (1966) to refine this technique. Using the Lineweaver-Burke modification of the Michaelis-Menten equation* they attempted to estimate the amount of glucose and acetate originally present in the waters of Lake Erken, Sweden, by adding a range of concentrations of radioactive substrate. They found two separate mechanisms of uptake, one for algae and one for bacteria. Specific active transport systems at low substrate levels were traced to bacteria, while a diffusion gradient at higher substrate levels was seen for the algae. This tracer method has been refined so that turnover or replacement time of a substrate can be calculated. Hobbie *et al.* (1968) reported turnover times of hours to days for several amino acids in St. Johns estuary, Chesapeake Bay. Initial attempts to apply this method to the open ocean have not met with success (Vaccaro and Jannasch, 1966), possibly because of the very low natural activity. It is worth noting that all growing bacteria can be regarded as aquatic whether they are in the ocean, in the water surrounding a soil particle, or within another living organism.

Cell and Energy Relations

Theoretical yields of energy and cells from a system can also be made. Such figures permit a maximum limit for the energy exchange within an ecosystem to be calculated. While this concept is presently in the formative stages, its potential for ecological application is very great and for this reason is being dealt with in this section in some detail.

†Some years ago, Monod (1942) observed that the growth of bacteria is directly proportional to the quantity of energy-yielding substrate provided in the culture medium. Since that time a considerable effort has been exerted to obtain growth constants for use in mathematical treatments of growth. We now know that *yield in dry weight of cells (biomass) of anaerobic bacteria is fairly constantly related to the quantity of energy made available by fermentation* as moles of ATP. The average yield (Y_{ATP}) is 10.5 gm of cells per mole of ATP generated per mole of substrate catabolized. Only in cases where the complete fermentative pathway of energy-yielding catabolism is not known do we find results seemingly at odds with the predicted Y_{ATP}.

However, Y_{ATP} cannot be used for aerobic pathways; for the number of moles of ATP generated by aerobic catabolism cannot be specified and varies considerably among the bacteria and other microorganisms. Relating growth yields to (1) moles of substrate utilized ($Y_{substrate}$), (2) equivalents of substrate carbon utilized (Y_{carbon}), or (3) moles of oxygen consumed during growth (Y_{O_2}) has not provided constant values.

Examining growth of soil pseudomonads on a large group of single organic compounds, Mayberry, Prochazka, and Payne (1967) found it useful to consider substrates according to the number of available electrons (av e$^-$) that comprised them rather than the number of carbons. For example, glucose comprises 24 av e$^-$, acetic acid 8, glycerol 14, and benzoic acid 30. (The number of av e$^-$ for any organic compound can be determined by calculating the O_2 required to completely combust one mole, and then multiplying moles of oxygen required by four—the number of electrons required to reduce one mole of oxygen.) When these workers divided the $Y_{substrate}$ values obtained in their own studies and in those of many other investigators by the av e$^-$ per mole of the specific substrate

---

* This equation defines the relationship between uptake rate and substrate concentration. For an explanation, see Fruton and Simmons (1959) or any general biochemistry text.

† This section on Y_{kcal} and $Y_{av\,e^-}$ was prepared by Dr. W. J. Payne, Department of Microbiology, University of Georgia.

employed, they obtained values ranging between 2.00 and 3.92 with a mean $Y_{\text{av } e^-}$ of 3.14 gm dry weight of cells per av e^-. In this range, 78 per cent of the values lay between 2.75 and 3.50.

Even for the growth of microorganisms in mixed cultures, $Y_{\text{av } e^-}$ is very nearly 3.14. Data from 33 experiments reported by four different investigators, who fed pure biochemicals to sewage sludge cultures and measured the $Y_{\text{substrate}}$, indicated a mean $Y_{\text{av } e^-}$ of 2.95.

These findings strongly infer that for the heterotrophic microorganisms, which grow aerobically on and totally consume a variety of substrates and produce no other end products than cells and CO_2, the yield in minimal media will be approximately 3.14 gm dry weight per av e^-, irrespective of the species and of the organic substrate utilized.

Mayberry, Prochazka, and Payne (1967) further predicted that the yield of cells from any type of bacterial growth would be constantly related to the amount of energy taken from the culture medium by both assimilation and dissimilation. They examined the heats of combustion of a large number of organic compounds—as determined by Kharash (1929)—and accepted an average figure of 26.5 kcal per av e^- as representative of organic substrates bacteria can utilize for growth. The simple relationship

$$Y_{\text{kcal}} = \frac{Y_{\text{av } e}}{E_c} = \frac{3.14 \text{ gm/av } e^-}{26.5 \text{ kcal/av } e^-} =$$

$$0.118 \text{ gm/kcal}$$

was formulated. For experimental purposes E_c, or total energy taken from the cultures, was considered to be represented by

$$E_a + E_d = E_c$$

where E_a is the energy assimilated into the structure of the cells and E_d the energy dissimilated by oxidative or fermentative metabolism. E_a is always calculable from experimental data as $Y_{\text{substrate}}$ times the heat of combustion of bacterial cells. These workers found by bomb calorimetry that the *average heat of combustion of bacterial cells is 5.3 kcal per gm.* Therefore,

$$E_a = Y_{\text{substrate}} \times 5.3 \text{ kcal/gm}$$

E_d for fermentative growth is also calculable, requiring only data from experimental fermentation balances. In the lactic acid fermentation, for example, 1 mole of glucose

with a heat of combustion of 673 kcal per mole is fermented to 2 moles of lactate with a combined heat of combustion of 2×326 or 652 kcal per mole. The energy dissimilated is then $673 - 652$, or 21 kcal per mole. In the work of Bauchop and Elsden (1956) on glucose fermentation by *Streptococcus fecalis*, where Y_{glucose} is 22 gm per mole,

$$Y_{\text{kcal}} = \frac{22 \text{ gm/mole}}{117 \text{ kcal/mole} + 21 \text{ kcal/mole}} =$$

$$0.159 \text{ gm/kcal}$$

Examination of the results of six studies employing nine different anaerobic and facultative species grown anaerobically on seven different substrates provided a mean Y_{kcal} of 0.130 gm per kcal.

Determination of Y_{kcal} for aerobic growth requires a different method for determining E_d. Values for $Y_{\text{substrate}}$ and E_a are obtained in the same manner as they were for anaerobic growth; but for aerobic growth E_d is taken as the product of moles of oxygen consumed per mole of substrate utilized during growth (O_2 per mole) multiplied by four times the average energy per av e^- for organic molecules (4 av e^-/mole $\times$ 26.5 kcal/av e^- or 106 kcal/mole). From the experiments of Mayberry, Prochazka, and Payne with a soil pseudomonad growing on benzoic acid, where O_2 per mole = 3.46,

$$Y_{\text{kcal}} = \frac{86.8 \text{ gm/mole}}{460 \text{ kcal/mole} + 366 \text{ kcal/mole}} =$$

$$0.105 \text{ gm/kcal}$$

Inspection of the results of five studies with seven different aerobically grown species of microorganisms on 26 different substrates provided a mean Y_{kcal} of 0.116 gm per kcal. Taken altogether, the mean Y_{kcal} for aerobic and anaerobic growth was found to be 0.121, which is in close agreement with the predicted 0.118 gm per kcal.

Further close agreement with the predicted value was obtained from results of the previously mentioned experiments with sewage sludge cultures that were fed pure biochemicals. The mean Y_{kcal} was calculated to be 0.110 gm per kcal for these studies.

It is thus apparent that an average of 0.118 grams of cells is very likely to be generated aerobically or anaerobically from any sort of microbial culture for every kilo-

calorie of energy removed from the culture medium by the growing cells (see Payne, 1970 for a review of this subject).

Measurement of Rates of Activities[*]

There are two broad categories of environments with regard to nutrient levels: (1) those in which substrate levels are high (gm per l or gm per kg) and (2) those which are low (mg per l or mg per kg). With regard to the flow or supply of nutrients there are also two dimensions: (1) continuous and (2) discontinuous. All intergradations, of course, occur. In general, discontinuous, high substrate-level ecosystems are the easiest to simulate and examine in the laboratory—for example, the decomposition and sequence of events of buried leaf litter. Successional studies have been performed on various food processes such as sauerkraut and pickle fermentations. They are more readily amenable to study because the high metabolic activity and high substrate-level systems are technically less difficult to measure. Such environments represent excellent models for the study of succession.

Of all of the environments that have been studied, the microbial transformation of cellulose in the vertebrate rumen is one of the best understood (Hungate, 1963). This system represents a high-nutrient, continuous environment. Activities can be described in terms of rates, with fair confidence that activity is constant. Using this principle Hungate and associates elucidated the organisms involved in cellulose transformation, the products formed, and the energy balance for the entire system. The anaerobic nature of this system is inefficient for bacterial growth (i.e., only 10 per cent of the energy is assimilated by bacteria), but the very nature of this inefficiency constitutes the reason that ruminants can subsist on such a substrate as cellulose at all. The major portion of the residual energy of microbial action consists of fatty acids that are converted from cellulose but are not further degraded. These end products are directly available for assimilation by the ruminant. Thus, "efficiency" can be a deceptive term. In this example anaerobic

metabolism is inefficient for the bacteria but highly efficient for the ruminant.

Another extensively studied high-nutrient continuous system is the activated sludge bed. Javornichy and Prokesova (1963) used this system to demonstrate the vital role that protozoa play, by cropping the bacteria, in maintaining bacterial metabolism at an actual level. Such systems are excellent for the study of metabolic rates.

Low nutrient, discontinuous environments by their very nature might not be expected to be represented in major habitats. However, many streams and lakes may be categorized as this system type. During many portions of the year organic levels are quite low, but periodically significant amounts of material enter the waters from the surrounding land after storms. The quantitative and qualitative nature of the water is altered, but even with this increase the system is a low-nutrient type as defined here. In many of these ecosystems attached bacteria form coatings on the rocks and surface structures; thus, there is a continuous component to the system, the water mass of the stream or lake, and a discontinuous component, the benthos. In studies of the microbiology of such systems both components should be considered jointly.

The most difficult type of environment to examine is the low-nutrient, continuous flow system, such as some streams and springs and the open ocean. Yet their kinetic rates are of paramount importance in view of the massive areas of the earth which they cover. Our knowledge of microbial metabolism in general bacteriology is based on studies done in the presence of relatively high concentrations of substrates (as was noted in Section 2). Recent work on metabolic regulation (Maaløe and Kjeldgaard, 1966) has shown that *direct extrapolations of data obtained under "optimal" substrate concentrations to conditions of growth at extremely low substrate levels cannot be made.* Biosynthetic abilities, as well as minimum requirements of the bacterial cell, are strongly affected by changes of the growth rate or by the concentration of the growth-limiting nutrient. Furthermore, the nature of the limiting substrate—its function as a source of energy or as an essential nutrient—determines the ratio between respiratory and biosynthetic metabolism (Herbert *et al.*, 1956). *These circumstances call for activity measurements directly within*

---

[*] Portions of this section are adapted from Jannasch (1969).

the natural habitat and experimental growth tests at substrate concentrations approximating those of the natural environment.

Brock (1967) gave an example for the first approach by determining the incorporation of labeled thymidine into the DNA of individual cells of *Leucothrix* sp. by autoradiography. The advantage of this procedure lies in its specificity, as compared, for instance, to the relatively unspecific dark uptake of labeled bicarbonate. The disadvantages of this technique are the limited applicability to species that can be cultivated in pure culture and recognized morphologically *in situ* and the errors incurred by the necessity of incubation in a test tube, which will be described later.

The second approach, the measurement of growth or metabolic activities at extremely low concentrations of the limiting nutrient, meets with other technical difficulties. In his fundamental work on bacterial growth, Monod (1942) had found that the relationship between growth rate and concentration of the limiting substrate could be described by a "data-fitting" curve, which corresponded to the Michaelis-Menten relationship between the rate of enzymatic reaction and substrate concentration. It is impossible, however, to obtain growth data for the lower section of the curve near the origin—that part of the function which is of most interest to the ecologist. The inoculated cells partially autolyze in the presence of extremely low nutrient levels, thereby releasing additional substrates and giving rise to so-called "cryptic growth," which obscures the true growth rate per substrate relationship (Postgate and Hunter, 1963).

In 1950, Monod, and independently Novick and Szilard, developed the chemostat, a basically simple device used to grow bacteria in the presence of constant environmental conditions, including the concentration of the limiting substrate. Instead of using a closed or batch system, as, for instance, a stoppered culture flask in which the concentration of the substrate changes with growth and time, an open system was used simply by continuously adding fresh medium and withdrawing an equal amount from the culture containing the growth products including cells (see Figure 2–6 I for a diagram of a simple chemostat).

Continuous culture techniques had existed before 1950, especially for the maintenance of microbial cultures or, in applied microbiology, for continuous production of cells or metabolic products. Not until it was discovered, however, that a homogeneously mixed continuous culture of an exponentially growing microorganism represents a self-adjusting system (chemostat) did this technique become an invaluable tool in quantitative studies. Passing a "transient state," in which the growth rate adjusts to the chosen concentration of the limiting substrate, the system will reach a "steady state" indicated by a constant population density. As long as the external conditions, including the composition and the flow of medium, are maintained unchanged, the growth rate will be constant and equal to the dilution rate of the system. During this steady state the important parameters for the determination of growth and metabolic activity can be expressed by simple mathematical relations (Herbert *et al.*, 1956).

The step from studying microbial activities in closed culture systems to the application of open, continuous-flow systems marked a new and significant advance for study of the microbial ecology of low-nutrient, continuous environments. A year before the chemostat was described, van Niel (1949) made the following statement:

Growth is the expression *par excellence* of the dynamic nature of living organisms. Among the general methods available for scientific investigation of dynamic phenomena, the most useful ones were those which deal with kinetic aspects. . . . Kinetic investigations on cultures of microorganisms are eminently suited for establishing relations between growth and environmental factors, especially the nature and the amount of nutrients.

The development of the chemostat made it possible to apply this reasoning in the examination of microbial ecosystem kinetics.

The principle of competition for the limiting substrate in the natural habitat holds for the chemostat as well. Since, however, the continuous removal of cells from the chemostat by dilution proceeds in an indiscriminate fashion and at the same rate for all species present, the system is strongly selective: species attaining the fastest growth rate under the conditions given will compete successfully and, eventually, will displace all other species. Thus, *the chemostat does not allow the reproduction of natural conditions. Its actual value lies in the fact that in the chemostat a steady state can be established, which is the basis for most of the kinetic studies.*

In the ecological literature the term "steady state" is used for time-dependent cyclic states of mixed populations. When its original definition, obtained in studies on chemical kinetics, is lost, the term also loses its value as a tool in the study of ecological population dynamics. In natural populations individual species may *not* be in steady state, but rather in transient states, i.e., increasing or decreasing along with season or other periodic environmental changes. Thus, *the microbiologically defined steady state of a population is a useful experimental tool and is a situation found under natural conditions only in the most stable of environments*, as, for example, in the deep sea or in soil under a mature forest. The limiting factor concept in relation to steady-state and transient-state conditions was discussed in Chapter 5, Section 1.

The natural habitat of microorganisms must be visualized as composed of as many individual open systems as there are species present, each population being controlled by its individual limiting factor and all of them being more or less interrelated by the availability of energy. The principal relationships between population density, growth rate, and concentration of the limiting nutrient in open systems are distinctly different from those of closed culture systems. For instance, in open systems the population density is at its maximum when the growth rates and nutrient concentrations are minimal. Nutrient concentrations increase with decreasing population density, and the maximum growth rate, of course, is approached when nutrients are high and density low (see "optimum yield" concept, page 223). From this point of view it becomes understandable that the process of mere sampling, the change from an open system to a closed one, confers drastic environmental changes.

Continuous culture studies have shown (Jannasch, 1963 and 1967*b*) that many bacterial isolates are unable to maintain growth in certain natural waters below a threshold concentration of the limiting nutrient. The quantity of these "leftover" concentrations, available to but not utilized by the microorganisms, is dependent on environmental factors — for example, the pH or Eh of the system.

While the classic enrichment techniques are limited to microbial species of pronounced substrate specificity (e.g., sulfide oxidizing, nitrate reducing bacteria, etc.), the selective properties of the chemostat have been used to produce enrichments and isolations of those inconspicuous metabolic types which might in fact be responsible for much of the degradation of the large variety of organic substrates in water (Jannasch, 1967*a*). These studies reaffirm that bacteriological media of the strength usually used in agar plates select for "weed" types of organisms that are incapable of metabolizing substrates under natural conditions, and hence which would be rare in unpolluted nature. Species actively growing at natural substrate concentrations tend to be outgrown by such weed species on the common test media. "Climax" microorganisms of low-nutrient environments thus might be overlooked because of the relatively high nutrient level generally found in bacteriological media.

This concept of contrasting an active and an inactive part of the microflora of a natural habitat can be found in Winogradsky's (1949) definition of "autochthonous" and "zymogenous" microorganisms, the former exhibiting the more or less constant turnover processes at low-nutrient concentrations, and the latter exhibiting requirements for high-substrate concentrations and showing vegetative growth or "bloom" only at an occasional or seasonal rise of the nutrient level. The consequence implied by this distinction is clear: if we want to study the microbial activities in low-nutrient, constant-flow environments, we have to look for the right organisms, namely, those active under natural low-nutrient conditions. These may not be the "laboratory bugs" that have received the most intensive study.

Brock and Brock (1968) studied *in situ* growth rates of benthic algae in an attempt to use a small stream as a continuous culture. Cutting off the energy source by covering a limited section of the stream with black foil and measuring the outwash rate of the algal cells from the darkened area, they calculated the original growth rates. Similar studies, but in a laboratory system, were done by feeding a chemostat with filter-sterilized or unsterilized water inoculated with the bacterial isolate to be tested. The temperature and other factors of the specific environment were reproduced experimentally. From the difference between the dilution rate of the system and the outwash rate of the organisms, their growth rate in the absence or the presence of the competing microflora in the unsupplemented natural water can be calculated with a high degree of accuracy (see also Jannasch, 1969).

6. SUMMARY

Microbial ecology as a field of study should not be separated from "general" ecology; while some answers to specific questions require specialized analytical techniques, the fundamentals of ecology must hold for microorganismic interactions. This chapter has emphasized techniques because, as stated in the introduction, the broad questions in microbial ecology are known; it is the methodology for formulating answers that is most often lacking. The need to develop *in situ* methods for the study of microbial activity in ecosystems was stressed, since the enriched pure culture techniques of conventional laboratory microbiology are not adequate for ecological assay. The student should remember that microbial ecology is not a side issue of ecology but very much a central issue, particularly with respect to understanding the cycling of elements, the bioenergetics of ecosystems, and the control of man-made pollution.

ECOLOGY OF SPACE TRAVEL

Chapter 20

*By G. Dennis Cooke**

Kent State University, Kent, Ohio

Introduction

One of the most exciting new areas in science involves the development of partially or completely regenerating ecosystems for the life support of man during long space flights or during extended exploration of extraterrestrial environments. Although adequate, temporary support systems for 1 to 3 astronauts for brief 1 to 30 day orbital or exploratory flights (such as lunar landings) have been perfected, there has yet been little effort or progress toward the development of a completely closed, stable regenerating ecosystem capable of sustaining populations of men in space for long periods. The construction of this kind of life-support system has particular relevance to ecology and to ecosystem theory since it is generally agreed that such systems must be, partly at least, bioregenerative. A second area of ecological interest pertains to "exobiology," or the question of the existence of life on other planets. Exobiology includes the likely probability that we will find eventually a primordial ecosystem, or at least a prebiotic stage in chemical evolution, such as is presumed to have existed on earth eons ago (see Chapter 9, pages 271–272).

In most broad aspects the problems of man's survival in an artificial spacecraft are the same as the problems involved in his continued survival on his earth spaceship, which is rapidly reaching critical levels of crowding. For example, detection and control of air and water pollution, adequate quantity and nutritional quality of food, what to do with accumulated toxic wastes and garbage, and the social problems created by reduced living space are common concerns of cities and spacecraft. In addition, the space traveler must cope with two new environmental problems, namely weightlessness (zero, or greatly reduced, gravity) and a radiation field that includes potentially lethal rays, which are shielded out by the earth's atmosphere. The fact that we are not now able to engineer a completely closed ecosystem that would be reliable for a long existence in space (nor can anyone predict when we will be able to do so because we have not yet given it serious attention) is striking evidence of our ignorance of, contempt for, and lack of interest in the study of vital balances that keep our own biosphere operational. Therefore, future efforts to construct a life-support system by miniaturizing the biosphere and determining the minimum ecosystem for man is a goal that is as important for the quality of human life on earth as it is for the successful exploration of the planets.

1. TYPES OF LIFE-SUPPORT SYSTEMS

Compartment models for three current types of life-support systems are diagramed in Figure 20–1. Three stages in degree of regeneration are compared with respect to the flow of energy, cycling of matter, and the type of regulation. The flow of matter in the storage or nonregenerative type is unidirectional, and the life of the system is based upon the amount of material that can be stored. Regulation is achieved entirely through external means. With partial regeneration some matter is cycled between the astronaut and the support organism or mechanical devices, but control remains almost entirely external and mechanical,

* This chapter was prepared while the author held a postdoctoral fellowship at the Institute of Ecology, University of Georgia, with support from NIH Training Grant ES 00074 and NASA contract NsG 706/11-003.

Figure 20-1. Simplified compartment models for three stages in the development of regenerative systems.

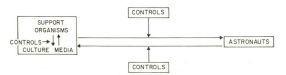

and the life of the system depends entirely upon the reliability and longevity of the hardware. By closing the support system, except for energy, there is a cycling of matter and a flow of energy which may be regulated or stabilized either through external hardware or through internal biological mechanisms, or both. Here, longevity is limited either by the reliability of the hardware or by continued homeostatic interactions among biotic components.

For all space missions so far (up to 1970), the storage or nonregenerating system has been used with only a small amount of regeneration. Here, all the materials necessary for life, including water, food, and

oxygen, are packaged aboard the vehicle, along with the equipment for the detoxification and storage (or dumping into space) of metabolic wastes. However, as the number of astronauts per vehicle and the duration of space flight increase, the weight of the storage system increases rapidly to the point where propulsion limitations become critical. The relationship between weight and duration or extent of mission is illustrated in Figure 20-2. The ordinate represents the fixed cost in weight of the support equipment, including the energy source but not the expendable fuel. The nonregenerating storage system has the least size and cost for short periods of time and re-

Figure 20-2. Relationship between weight, duration of mission, and degree of regeneration in the spacecraft. As time in space increases, regeneration becomes increasingly important. The circled 1, 2, and 3 indicate points where it pays to regenerate water, respiratory gases, and food, respectively. (Redrawn from Meyers, 1963.)

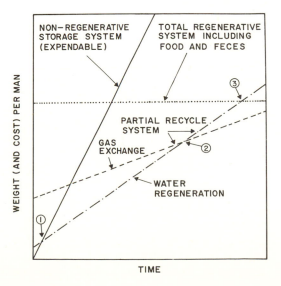

quires relatively simple controls, but the weight cost per day of the space mission increases very rapidly.

In theory, the weight penalty for a storage system can be partially or completely avoided by regenerating some or all of the astronaut's physiological requirements (Figs. 20–1, 20–2). By providing a power source, either sunlight or electric, or both, and the raw waste materials (CO_2, urine, waste water), it is possible to chemically regenerate oxygen and water. In addition, by coupling the matter excreted by the astronaut and an energy source with certain organisms (algae, bacteria, and others), respiratory gases and potable water can be regenerated and food can be grown and harvested. As regenerative steps are added, the fixed weight cost increases (Fig. 20–2), but the rate of increase per day of travel, in comparison to nonregenerating equipment, decreases. A completely regenerating support system will have to be large and expensive but it would have the same cost regardless of mission duration and is, therefore, highly desirable.

The circled numbers *1*, *2*, and *3* in Figure 20–2 indicate the points in time when it would pay, so to speak, to regenerate water, oxygen, and food, respectively. At this writing (1969) it is not possible to be very specific about the time scale on the x-axis. Point *1* is presumed to be reached in a matter of a few weeks, point *2* in a few months and point *3* perhaps in a year. As pointed out by Hock (1960), the life of a storage or partial regenerative system could be prolonged if we could induce hibernation in man!

Several types of partially or completely closed regenerative systems have been proposed for space flight. These are: (1) mechanical chemoregeneration, (2) algal photosynthetic, (3) bacterial chemosynthetic, and (4) a multispecies, multitrophic level microecosystem similar to the self-regulatory or "climax" systems of nature. The most important criteria for selecting one of these as the basis for a life-support system are power input and weight, stability, and longevity. From the standpoint of the engineer who must get the payload into orbit, the smallest total system (power, weight, area, and volume) capable of supporting life for long periods, and the one that is reliable within the probabilities of the equipment, is optimal. From the astronaut's standpoint,

once the capsule leaves the earth's atmosphere, stability and longevity of the regenerative equipment are more important than efficiency of regeneration. In terms of ecosystem development theory, as discussed in Chapter 9 (page 252), the engineer's goal is to achieve as high a P/B efficiency (where P = rate of gaseous regeneration or food production and B = weight of the regenerative machinery) as possible, like the early succession systems in nature. Unfortunately, such efficiency may be achieved at the expense of internal stability and size of living space for the astronauts who, therefore, might prefer the lower P/B efficiency (or higher B/P ratio) of the mature or climax systems of nature!

Actual research on the development of life-support systems has gone in two directions: mechanical and biological. The complex mechanical chemoregenerative system, capable of regenerating gases and water but not food and designed to dispose of wastes, is nearly operational. This system is fairly reliable and capable of supporting life for rather long periods. For very long missions, a high weight cost for mechanical chemoregeneration is imposed since the hardware is bulky and heavy, power requirements high, and food and certain gases must be stored and resupplied. Problems are also posed by the high temperature necessary to remove CO_2 and by the gradual buildup of toxins (such as carbon monoxide) that need not be worried about on short missions. For space flights involving very long periods of time in which resupply and hence storage and chemoregeneration are not possible, the other alternative, a partial or completely bioregenerating ecosystem, will have to be employed.

Research on biologically based systems is directed, at present, at the feasibility of using chemosynthetic bacteria or small photosynthetic organisms such as *Chlorella* or duckweed as "producers" for the ecosystem, since, as indicated above, engineering considerations apparently rule out consideration of larger organisms. In other words, the problem of weight versus efficiency looms again in the choice of a biological "gas exchanger." As we have seen in Chapter 3, the "size-metabolism principle" is such that the smaller the organism the higher its regenerative efficiency; but such efficiency comes at the expense of longevity of individuals (which is another way of expressing the P/B and B/P contrast mentioned earlier). The shorter the life of the individual, the more difficult it is to prevent or dampen oscillations in population size and the gene pool. A pound of chemosynthetic bacteria can remove more CO_2 from the cabin atmosphere than a pound of *Chlorella* algae, but the bacteria are harder to control. Likewise *Chlorella* are more efficient as gas exchangers on a weight basis than duckweed or other higher plants, but in this instance the *Chlorella* are harder to control.

To date (1969), no closed biologically based regenerative life support system, based on either bacteria or algae, is anywhere near an operational state, primarily because of the large size and unreliability of the external mechanical control machinery (see Figure 20–1) thus far devised and because of the tendency for a low diversity ecosystem to undergo reorganization (succession). An ecologically sound, but not yet seriously considered, alternative to the single-species microbial producer system is a heavier multispecies, mature-type ecosystem, which possesses internal self-regulatory mechanisms that could, in part at least, substitute for these external controls.

The following sections of this chapter are a more detailed examination of life-support systems from the viewpoint of the ecologist. Arguments are presented to support the use of the mechanical chemoregenerative system for short missions and to reject the use of closed ecosystems based on a single-species producer microorganism for long missions. It is suggested that only the multi-species mature ecosystem has the necessary stability for really long-term space exploration. Research on such systems also has the greatest feedback value for understanding the limits man must put on his own life here on earth.

Mechanical Chemoregeneration

The only type of regenerative life-support equipment that will be operational in the near future will be based on the mechanical chemoregeneration of gases and water and the storage of food and supplemental materials. This type of partially closed system represents a considerable weight savings over the nonregenerating type and could conceivably support manned missions for hundreds of days. According to Foster (1966), water and oxygen represent 94 per cent by weight of all the materials required for life support. By adding periodic resupply, such as would be possible for lunar exploration or from space supply stations, this system could be used for very long periods for certain types of missions. Figure 20–3 is an "Integrated Life-Support System" (ILSS), one of several possible designs of a spacecraft constructed to operate with a mechanical chemoregenerative system. This prototype physical-chemical system has been designed and tested by General Dynamics for the Langley Research Center (Armstrong, 1966; NASA CR-614). When completed, it will be capable of supporting four men on stored dry food and regenerated oxygen and water for one year, but resupply of food and certain gases is necessary every 90 days. A schematic diagram of the mechanical and chemical regeneration process is shown in Figure 20–4.

The recovery of water is the most important regenerative procedure in terms of weight savings. Water is used in the space capsule for drinking, for reconstituting food, for bathing, and in the electrolysis unit of the oxygen regenerator. About half of a man's daily intake of water reappears as urine, the rest in feces and in the vapor from perspiration and exhalation. The water content of feces is too low to justify recovery. Because of odor and the danger of bacterial contamination, fecal water will probably be vented to space.

The method of reclaiming potable water from these sources is the waste heat air evaporation method. Urine is treated with disinfectants as it is passed into an air evaporation unit containing fibrous wicks which become saturated with the urine. The hot air stream evaporates the water, leaving the suspended and dissolved impurities in the wick. The humidified hot air passes into a cooling condenser and water is removed and stored. Wash water and moist cabin air are treated in a similar manner.

The recovered water is checked for bacteria, conductivity, and chemical contamination after being cycled into holding tanks, and if water quality standards are not met, the water is reprocessed. The most serious problems of water regeneration are the collection of the condensed water in the

Figure 20–3. A simulated "Integrated Life Support System" that regenerates water and gases by mechanical and chemical means. *A*. Exterior view of the life support chamber. The small side chamber, to the right, is the air lock by which astronauts exit and enter. *B*. Atmospheric regeneration equipment. *C*. Photograph of Bosch CO_2 reduction-catalyzer plates, showing accumulated carbon which must be ejected from capsule. *D*. Water management controls with storage tanks above and below. Pictures *B* and *D* illustrate dramatically how much complicated "hardware" is necessary to substitute for the most elementary functions of nature! (Photos courtesy Langley Research Center, Langley, Virginia.)

absence of gravity and the avoidance of chemical or bacteriological contamination.

Most of the oxygen consumed by a man reappears as carbon dioxide; the rest reappears as water. The system employed to regenerate oxygen involves the reduction of carbon dioxide to water and carbon, and the water, along with some that is metabolically produced, is subjected to electrolysis, which separates the H_2 from the O, thus producing gaseous oxygen (see electrolysis unit, Figure 20–4). The cabin air is circulated over a regenerable silica gel (dehumidifier, Figure 20–4) to absorb water and the dehumidified air is passed through a molecular sieve which removes the CO_2 (CO_2 concentrator, Figure 20–4). The adsorbent bed is heated to remove the CO_2

from the concentrator unit to the CO_2 reduction reactor, and the CO_2 is reduced by a Bosch reaction according to the following equation:

$$CO_2 + 2H_2 \xrightarrow[1100°F.]{\text{Fe cat.}} 2H_2O + C$$

Water is produced as steam, cooled, and diverted to water storage, and electrolysis of some of the water yields oxygen to the cabin atmosphere and hydrogen to the reduction unit as follows:

$$2H_2O \rightarrow 2H_2 + O_2$$

Methane, carbon monoxide, and elemental carbon are by-products that have to be jettisoned from the space craft. The

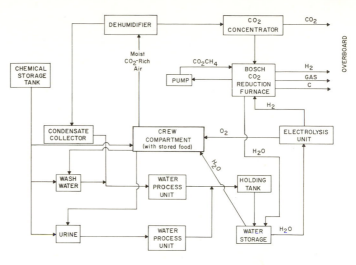

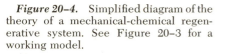

Figure 20–4. Simplified diagram of the theory of a mechanical-chemical regenerative system. See Figure 20–3 for a working model.

loss of carbon is made up from the resupply of food, but nitrogen, oxygen, and hydrogen must be stored to make up deficits from system leakage and from loss by dumping.

The reliability of the mechanical system, in so far as tested on the ground, seems to be good for missions of up to 100 days. As with any mechanical device, failure or reduced performance, necessitating on-board repair or a mission abort, can be expected with increasing time of operation due to wear. Finally, it is difficult to build a system that must handle liquids and gases at very high temperatures and pressures without leaks, which could be serious sources of contamination or fires. The procedure of dumping waste materials into space and thus contributing to space pollution or perhaps to the contamination of extraterrestrial life systems, if these exist, is not to be ignored. Although the regenerative system seems relatively simple in theory (as shown in Figure 20–4), one is impressed with the large and complex equipment (Fig. 20–3) necessary to replace the most elementary function of nature!

Bioregenerative Systems

As already indicated, research on two types of bioregenerative life-support systems, photosynthetic and chemosynthetic, are currently receiving support from NASA. Both are based on the coupling of a single regenerative or "producer" species population to the astronaut. In order to achieve the high P/B efficiency demanded by engineering considerations, strains of algae and bac-

teria having very high growth potentials have been favored, even though (as also already pointed out) such strains are largely "laboratory weeds" that are transitory in the systems of the real world (i.e., the biosphere). As shown in Figure 20–1, producer organisms could be used only as gas exchangers in the partially regenerative craft, or they could also provide food in the closed system of complete regeneration, as shown in the schematic diagram of Figure 20–5.

A gas exchanger for space capsules based on photosynthesis was independently suggested in the early 1950s by several investigators. This idea was later expanded, primarily on the basis of work by Dr. Jack Myers at the University of Texas, into a concept for a completely regenerative system (Figure 20–5) in which plants are provided with CO_2 and nutrients from heterotrophic (astronaut) metabolism and in the presence of light convert these to food and oxygen for use by the astronaut. The experiments of Eley and Myers (1964), who closed an algae-mouse system for gas exchange for 82 days, represents one of the longest couplings of a mammal and a support organism. On the basis of ecological theory, one can predict that such oversimplified systems, much like early "bloom" stages in nature, will be difficult to stabilize.

The choice of a photosynthetic organism for a life support system has centered on the alga *Chlorella*, which is small, relatively easy to culture at high density on enriched media, has very little cellulose, and has a high rate of photosynthesis per unit biomass. Duckweeds (*Spirodela, Lemna*) have received some attention primarily because

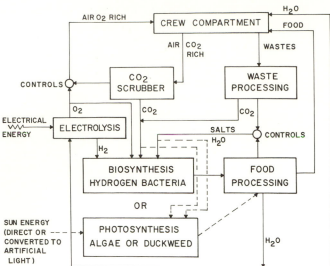

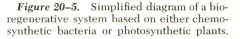

Figure 20–5. Simplified diagram of a bioregenerative system based on either chemosynthetic bacteria or photosynthetic plants.

as floating plants they exchange O_2 and CO_2 directly with air, thus reducing many of the problems of gas exchange (Jenkins, 1966). Larger plants, despite the fact that they constitute a major base for man's food chain on earth, have not been seriously considered in the United States' space program. According to available but incomplete reports, they are being considered in the Russian program.

A number of difficulties have been encountered during experiments with algae-mammal couplings; among them have been invasion of competitors or predators (bacteria or zooplankton), accumulation of algal toxins, and the presence of algal mutants with lowered rates of metabolism. These problems are to be expected of an over-simplified ecosystem without adequate internal controls. The most serious problems concern algal and mammal nutrition and gas exchange ratios, and these will be discussed here in greater detail.

The nutrition of the plant population and the astronaut is very complex and represents one of the most serious drawbacks to the use of a single algal photosynthetic system. The basic problem is to prevent succession in the very young-type ecosystem. Since algal cells must be grown in continuous culture (rather than in batches) with a constant removal of cells and a replacement of culture medium, a continuously adequate supply of plant nutrients must be provided. If the nutrient materials removed during harvesting are not replaced from feces and urine or from a nutrient storage source, the initial

high rate of photosynthesis declines. In addition, some nutrients may be lost to detritus or to accumulatory processes of the astronaut, thus limiting the life of a closed system. Any such progressive tie-up of nutrients in organic matter is exactly what would be expected during succession in natural ecosystems. If algae are to be used as astronaut food, maintenance of a constant quantity and proportion of plant nutrients in the medium will be critical since a nitrogen deficiency, for example, will result in the production of large quantities of lipid in the algal cells (Fogg, 1965). Thus, if nutrient levels are not maintained within narrow limits by complicated external machinery, the desired photosynthetic rate and the quality of the algae as food will drop.

Human feces and urine, along with a small supplemental supply of micronutrients, will probably provide an adequate source of nutrients for continued algal growth (Golueke and Oswald, 1964). However, if algae are the only regenerators of nutrients from this source, the system will not work, since algae are not known to have the ability to rapidly degrade organic matter to an elemental level. Either a mechanical device to prepare the medium or a complex bacteria-protozoa-algal nutrient regenerating system like those used to treat human sewage (Oswald and Golueke, 1964) would be needed.

In order to close the photosynthetic system for gas and food, the astronaut must derive his energy from the algae. However, humans apparently cannot use algae as 100 per cent of their diet. Algae, as well as bacteria, are too rich in protein and too low in carbohydrates. In addition to having a bad odor and taste (Powell, Nevels, and McDowell, 1964) and producing gastric disturbances (McDowell and Leveille, 1964), algae are low in certain essential amino acids (Krauss, 1962) and are poorly utilized in the gut without mechanical disruption of the cells (Dam *et al.*, 1965). Algae could probably be used as a diet supplement, but not as the sole source of energy if the astronaut and his gut flora are to be the only heterotrophs in the system.

A small change in the respiratory quotient (RQ) of the astronaut (a ratio between CO_2 produced and O_2 consumed) or a change in the plant assimilatory quotient (AQ) (a ratio between O_2 produced and CO_2 consumed) will lead to a loss or accumulation of oxygen or carbon dioxide in the system. The AQ/RQ balance must be maintained within very narrow limits for the entire duration of the mission, even though the rate of metabolism of the astronaut can be expected to vary considerably, depending on the work load, diet, diurnal rhythms, and other factors that affect metabolism. Similarly, plant AQ will change with variations in the nutrient content of the medium. Thus, Eley and Myers (1964) reported that mouse metabolism was very erratic during the 82 days of coupling with the algal population, which led to problems in AQ/RQ balance. Bowman and Thomae (1961) also reported a similar difficulty. No investigation to date has shown that the two-species ecosystem can maintain an AQ/RQ balance for long periods, even with the use of great numbers of external controls.

The instability of a unialgal population is notorious even when growing in a continuous culture system in which all cellular requirements (nutrients, light, temperature, etc.) are presumably held constant. Despite the use of rather elegant equipment, large oscillations in the population density tend to occur. The basic objection to the use of a two-species (man and support organism) life-support system is thus one of reliability. There are no alternative sources of energy, water, or oxygen (no food web diversity) for the astronaut in the event that the metabolism of the algal population is reduced or fails—a grim prospect for an astronaut several hundred thousand miles from Earth.

All of the difficulties described here for the two-species life-support system are exactly those which bring about a succession from young to more mature but slower communities in nature. The probability of long-term stable operation with only one "bloom" species of producer in the system would seem to be low, unless the expenditure of vast quantities of energy for control purposes can be accomplished. For a system, such as is diagrammed in Figure 20–5, to be practical, the controls, shown by small circles in the diagram, would have to be enormously large, complex, and energetically expensive.

The chemosynthetic bacteria-astronaut life-support system in which hydrogen bacteria would replace algae (see Figure 20–5) was proposed in the early 1960s, and, to date, very little is known about the properties of such a system. Culture methods, growth characteristics, biochemistry, and genetics of the bacteria have been the primary emphasis of research (Bongers, 1964a, 1964b; Repaske, 1962, 1966). It will be recalled from Chapter 1 that the chemosynthetic bacteria oxidize a wide range of inorganic compounds as energy sources for carbon dioxide assimilation into cellular constituents (hence they are chemoautotrophs). A number of kinds of bacteria can utilize molecular hydrogen in this manner, among them certain Pseudomonads which are placed in the genus *Hydrogenomonas*. It is important to note that these bacteria, which are being cultured and studied as possibilities for life support systems, are not obligate hydrogen fixers but are rather facultative heterotrophs that normally live in soil on organic matter. It is also important to note that when functioning as autotrophs they do not release oxygen, which, therefore, must be provided to the astronaut by other means.

The chemosynthetic ecosystem would combine features of the mechanical-regenerative and the photosynthetic system. Oxygen and hydrogen are derived by the electrolysis of water as in the mechanical regenerative system (compare Figure 20–4 and Figure 20–5), but the hydrogen, instead of being vented, is used by the hydrogen bacterium to reduce CO_2 and to form water and bacterial protoplasm. The astronaut would use the oxygen from electrolysis for respiration and, in theory, bacteria for food; he provides water, carbon dioxide and nutrients for further bacterial metabolism. A theoretically balanced, closed system could be achieved according to the following equations:

Electrolysis:

$$6H_2O \xrightarrow[\text{energy}]{\text{Electrical}} 6H_2 + 3O_2$$

Biosynthesis by hydrogen bacteria:

$$6H_2 + 2O_2 + CO_2 \rightarrow CH_2O + 5H_2O$$

Human respiration:

$$CH_2O + O_2 \rightarrow CO_2 + H_2O$$

This system would have the advantages of low weight and power requirements. According to the review by Jenkins (1966), the electrolysis-*Hydrogenomonas* system would require half the weight and several orders of magnitude less power per man than an algal bioregenerative system which uses controlled artificial illumination. However, the system would have all of the culturing and nutritional disadvantages described for the algal system, and more. There is no evidence that hydrogen bacteria will provide a suitable diet for man, and the bacteria might easily "switch" to heterotrophy rather than "work" at hydrogen fixation. Furthermore, high production strains selected in the laboratory could easily mutate back to low production strains under the stress of radiation in space, and the successful invasion of a phage could destroy the entire system. Again, we have the familiar situation of high production efficiency but low reliability. At the present time, at least, it would seem that *Hydrogenomonas* is more feasible as a supplemental CO_2 "scrubber" in a partially regenerative system than as the crucial link in a closed system.

To summarize, of the many criteria that a life-support system must meet, only two are of paramount importance. These are (1) to be amenable to spacecraft and (2) to be stable and reliable. The two simple microbial ecosystems (photosynthetic and chemosynthetic) may be small and light enough, and therefore meet the first requirement. However, these simple systems are inherently unstable and depend upon a very large investment in power for control. The probability of failure of the two-species (man and microorganism) linkage during a long space voyage due to successional processes or to stress is high. It would

seem obvious that these simple ecosystems pose a serious risk to the astronaut, and further work on them, as the basis for a life-support system, should be abandoned. Above all, the narrow engineering approach cannot be applied to bioregeneration. Organisms cannot be "designed" and "tested" like transistors or batteries to perform "one function" or to solve "one problem"; they have evolved with other organisms as a unit and they carry out a variety of functions which must dovetail with other activities of the ecosystem. The "minimum ecosystem" for man must clearly be a multispecies one.

The Case for the Multispecies Life-Support System

Although, as we have seen, man probably will not be able to live alone with only one or two species of "bloom-type" autotrophic microorganisms to provide the vital necessities, the question still remains as to what other or additional biotic and physical components will be required. In other words, what is the minimum diversity for a given time-stability requirement? We cannot now provide an answer to the question, nor will we be able to do so until the designers of bioregenerative systems become more concerned with basic ecological principles such as those pertaining to stability-diversity relationships, biogeochemical cycles, food chains, and ecosystem development (succession). The concept of maximum output for minimum size must be abandoned in favor of the ecological rule that states that "the optimum efficiency is always less than the maximum." In other words, we must "design with nature," not abort her. This requires that emphasis be placed on ecosystems in a mature or "climax" state that have a high degree of self-regulation and that are, therefore, resistant to perturbations and less likely to undergo successional changes.

The advantages of a mature multispecies life-support system have been repeatedly discussed in the American literature (Golueke, Oswald, and McGauhey, 1959; Taub, 1963; Patten, 1963a; H. T. Odum, 1963; E. P. Odum, 1966; Cooke et al., 1968; and others) and in Russian review publications (Shepelev, 1965; Ivlev, 1966). The basic drawback to the use of the mature ecosystem is, as already stated, its high biomass/productivity ratio and its consequent very large size. H. T. Odum (1963) estimated that all but about 2 per cent of the photosynthetic production of mature ecosystems goes to the respiratory maintenance of components of the system other than single large consumers. Based on these computations, he estimated *an area of two acres per man as the minimum ecosystem.* Consideration of the psychosocial needs of man as long-term members of an isolated community may require enlargement of the minimum space to, perhaps, *ten acres.* If this does prove to be the minimum ecosystem size, then we must make certain that every man, woman, and child on earth has at least ten acres as part of his or her life-support system. We need not be limited in our planning to the current limits of "payload" that can be launched into space, because it should be possible to build a large space station by coupling together individually launched space component capsules. Launching of very large payloads will be greatly facilitated by the use of space stations in orbit or on the moon where gravitational pull is very low.

Regardless of weight and area considerations, the argument for the use of a mature multispecies system is ecologically sound, provided that all basic biogeochemical cycles and energy flow pathways have been allowed to evolve as a unit. Studies on laboratory microecosystems (as discussed in Chapters 2 and 10), while not directly applicable because they are too small to include man, have demonstrated the greater stability that comes with diversity. Thus, Beyers (1962) has shown that metabolism of mixed cultures is less influenced by temperature changes than are pure cultures and that the climax state is more resistant to ionizing radiation than is the transient or early successional state (see Cooke *et al.*, 1968). H. T. Odum (1963) has summarized the position of the ecologists as follows:

In appraising the potential costs of closed system designs one has the alternative of paying for a complex ecosystem with self maintenance, respiration, and controls in the form of multiple species as ecological engineering, or in restricting the production to some reduced system like an artificial algal turbidistat and supplying the structure, maintenance, controls, and the rest of the functions as metallic-hardware engineering. Where the natural combinations of circuits and "biohardware" have already been selected for power and miniaturization for millions of years probably at thermodynamic limits, it is exceedingly questionable that better utilization of energy can be arranged for maintenance and control purposes with bulky, nonreproducing, nonself maintaining engineering.

2. EXOBIOLOGY

No completely convincing evidence for the existence of life beyond the earth, or an "exobiology," as Lederberg (1960) has expressed it, has yet been found, but the possibility is not ruled out by what we have learned about the environment of Mars and other planets that have an atmosphere. Although temperatures and other physical conditions of existence are extreme, they are not beyond the limits of tolerance of some of our toughest earth inhabitants (bacteria, viruses, lichens, etc.), especially when one considers the likelihood of the existence of milder microclimates below the surface or in protected areas. It is certain, however, that there can be no large "oxygen gulpers" such as men or dinosaurs on the other planets of our solar system because there is very little or no oxygen in their atmospheres. It is now certain that the green areas and the so-called "canals" of Mars do not represent vegetation or the work of intelligent beings. However, infrared spectroscopic observations of the Martian dark areas are interpretable in terms of organic matter, and recent unmanned spacecraft (Mariners VI and VII) orbiting the planet have detected ammonia, which could be of biological origin.

As was discussed in Chapter 9, an oxygenic atmosphere is not necessary for the existence of life. Indeed, according to the generally held theory of Oparin (1938), the origin, or "spontaneous generation," of life is most likely to occur after a period of abiotic, radiation-induced synthesis of complex organic macromolecules in a reducing atmosphere such as that believed to have been present in early Precambrian times on earth and as is now present on Mars and Jupiter. The successful synthesis of essential biochemical constituents of life under simulated primitive earth conditions by Calvin (1951), Miller (1953; 1957), Oro (1967), Fox (1965), and others lends credibility to this theory. Therefore, unmanned as well as possible future manned space-craft that travel to the planets will be looking for primitive anaerobes, or at least for their prebiotic primordia. Such expeditions may indeed be like stepping into a time machine that will take us back billions of years to the origin of life. There is, of course, the other possibility that life once existed on our neighboring planets but has become extinct, perhaps because ecosystems never became sufficiently organized to ameliorate the deadly radiation fluxes, or perhaps because some organism became overambitious and destroyed its own life-support system. In that case, we may well be looking into the future of the earth!

If and when we find extraterrestrial life, the "principle of the instant pathogen" (Chapter 7, page 223) cautions us to be very careful about cross-contamination. A planetary "bug" just might thrive on the enriched media of the earth and run amuck, or, conversely, introduced microorganisms from earth might destroy the primitive life systems of other planets. For these reasons research on the sterilization of spacecraft is being pushed. All of us, as citizens of the earth, should insist that rigid decontamination standards be applied to all spacecraft, even those returning from the moon or other lifeless bodies because even there, spores may yet be found. We have enough trouble with misplaced organisms within the biosphere, so we can hardly risk the introduction of unknowns!

For a condensed and readable prospectus on space ecology and biology, see the National Aeronautics and Space Administration's Special Publication No. 92 edited by Jenkins (1966). For background material on Mars, see Pittendrich, Vishniac, and Pearman (1966).

The Extrabiospheric Environment

In addition to the internal maintenance of nutrient cycles, regulated energy flow, and temperature control, the spacecraft ecosystem must also cope with an external physical environment that is different, at least in degree, from that found within the earth's protective atmospheric and gravitational shields. Three aspects are of immediate concern: (1) fluctuations in gravity ranging from complete weightlessness to the gravity acceleration experienced during launch; (2) three types of radiation, namely galactic cosmic radiation, Van Allen Belts, and solar flares of intense proton flux, all of which create potentially much higher energy levels than encountered at the earth's surface; and (3) the absence of regular environmental pulses to entrain circadian and other biological "clocks" that coordinate organ systems within organisms as well as organisms within ecosystems. Even if a truly independent internal cellular clock or timer does exist (see Chapter 8, page 245), biological rhythms will get out of phase without periodic entrainment by the external environment.

Although much needs to be learned about all of these new dimensions, there does not seem to be any reason why they should prove unsurmountable. To a certain extent radiation can be shielded, zero gravity can be endured (at least long enough to reach the gravity fields of other planets), gravity substitutes could be devised, and timers within the spacecraft ecosystem can be provided. As in so many situations involving new stresses, it will be the interaction or summation of factors (weightlessness plus protons, for example) that will prove to be most critical elements of space travel. A great deal of simulation, both mechanical and theoretical, will need to be done on earth before we can fully understand the impact of these factors.

SUMMARY

Present space exploration, and that planned for the next 20 years or so, involves the use of temporary storage or partially regenerative life-support systems. A completely regenerative life-support system capable of maintaining man for long periods and requiring only the input of light energy represents a closed microecosystem, which functions in the same basic way as the biosphere. Successful construction of such a "minimum ecosystem for man in space" is not yet in sight. Research and development of closed systems should be accelerated, however, because they are also of basic significance to more pressing questions concerning man's minimum ecosystem on Earth. It is becoming increasingly apparent that man on Earth needs a "lebensraum" that includes more than just the physiological requirements necessary for life. Man requires a complex and diverse environment that includes not only the physiological necessities but also an array of psychological and sociological interactions. The minimum ecosystem for man in the biosphere is more than just those things necessary to keep him alive; the limits to human population growth cannot be defined simply in terms of the amount of food, water, and oxygen available per man. An increased knowledge of the limitation of the minimum ecosystem is, therefore, as important to our survival on the Earth spaceship as it is in any satellites that we may project into space.

The contribution of ecological theory, as outlined in Chapters 1 through 10, to the building of closed systems is obvious even though such theory has not been considered in the early engineering attempts to construct regenerative systems. The principles pertaining to diversity and stability, trophic level, biogeochemical cycles, as well as those relating to ecological succession are especially relevant. The overriding principle of "the optimum is less than the maximum" is especially important since selection of biotic components for the system cannot be based on their metabolic efficiency alone. Ecological consideration clearly indicates that a closed system must contain more biological components than just man and one or two species of friendly microbes. Microscopic algae or hydrogen bacteria provide no advantage over mechanical-chemical devices for regeneration of respiratory gases and water. It is clear that attention must be focused on large, diverse, self-regulating ecosystems containing organisms that are more easily linked with man.

Certain physical environmental factors encountered in space, notably zero gravity and high radiation flux, will be new or quantitatively different from the energy environment on earth. At the present time it appears that these factors will not prove to be serious limiting factors as long as their effects on biological systems are understood and compensated for.

Although it is certain that no highly developed oxygenic ecosystem, such as now controls the earth's environment, exists on any of the planets of our solar system, the probability of primitive life, or at least the organic primordia of life, is not too remote since the reducing atmosphere and other conditions of existence on some of the planets resemble those believed to exist on earth at the time life originated.

TOWARD AN APPLIED HUMAN ECOLOGY

The optimum for quality is always less than the maximum quantity that can be sustained (see pages 23, 197, 224, 419, 506).

The earth can support more "warm bodies" sustained as so many domestic animals in a polluted feed lot than it can support quality human beings who exercise the right of a pollution-free environment, a reasonable chance for personal liberty and a variety of options for the pursuit of happiness (see pages 56, 424).

It is not energy itself that is limiting, but the pollution consequences of exploiting energy (Chapter 17, page 464). Pollution is now the most important limiting factor for man (Chapter 16, page 432).

It is man the geological agent, not so much man the animal, that is too much under the influence of positive feedback and, therefore, needs to be subjected to negative feedback (Chapter 2, page 36).

To maintain order in an ecosystem, energy must be expended to pump out disorder; both pollution and harvest are stresses that increase the cost of maintenance. The more we demand from nature, the less energy nature has for maintenance, and, therefore, the more it costs man to prevent disorder (Chapter 3, pages 37, 38, 77).

Up to now man has generally acted as a parasite on his environment, taking what he wants with little regard for the welfare of his host (i.e., his life-support system) (Chapter 7, page 233).

To resolve conflicts of use, and to maintain optimum unpolluted living space, the landscape needs to be compartmentalized (i.e., "zoned") to provide a safe balance between productive and protective ecosystems. Restrictions on the use of land and water are our only practical means of avoiding overpopulation, or overexploitation of resources, or both (see pages 270, 424).

Diversity is a necessity, not just the spice, of life (Chapter 9, page 256).

The concept of recycle must become a major goal for society (Chapter 4, pages 87, 104; also page 411).

General realization that the supply depot and the living space functions of one's environment are interrelated, mutually restricted, and not unlimited in capacity has amounted to a historic "attitude revolution" that is a promising sign that man may be ready to "apply" the principles of ecological control on a large scale (Introduction, Part 3, page 405).

Technology alone cannot solve the population and pollution dilemma; moral, economic, and legal constraints, arising from full and complete public awareness that man and the landscape are a whole, must also become effective (Introduction, Part 3, page 406).

These excerpts, selected from various chapters of this book, add up to one conclusion: *The time has come for man to manage his own population as well as the resources on which he depends,* because for the first time in his brief history he is faced with ultimate, rather than merely local, limitations. *Ecosystem management and applied human ecology thus become new ventures that require the merging of a host of disciplines and missions which up to now have been promoted independently of one another.* Ecological principles, as we have outlined them in this book, provide a basis for a theoretical model, so to speak; but just how applied human ecology is to be developed and structured so that worthwhile goals can be achieved in the real world of society can be but dimly perceived at this point in history.

1. HISTORICAL REVIEW

Sociologists, anthropologists, geographers, and animal ecologists first developed an interest in the ecological approach to the study of human society. Now, as we have seen, nearly all the disciplines and professions in both the sciences and the humanities are eager to find a common meeting ground in the area of human ecology. A brief review of the early approaches is in order as a background for outlining the probable future development of applied human ecology.

The impact of man on the landscape has always been a major emphasis in geography and anthropology. Marsh wrote a classic treatise on this theme in 1864, entitled *Man and Nature: or Physical Geography as Modified by Human Action* (see page 11). A more recent comprehensive treatment of this theme is the book *Man's Role in Changing the Face of the Earth*, edited by W. L. Thomas, Jr. (1956).

Two contrasting viewpoints regarding the relationship of human culture to environment are frequently debated: (1) The physical environment exerts a dominant influence on culture and civilization ("environmental determinism"; see Meggers, 1954), as evidenced, for example, by the differences in human customs in arid and humid regions. (2) The physical environment places only a minor limitation on the development of advanced human culture, as evidenced by the rather similar urbanized civilizations that have been achieved at various times in the past in a diversity of natural environments. Today the question might be rephrased as follows: To what extent does man's continuing trouble with deteriorating environments stem from the fact that his culture has indeed tended to be too independent of the natural environment? Several geographers have been concerned with the reconstruction of past environments and man's influence on them. Butzer (1964) has written an excellent book on Pleistocene geography and ecology with emphasis on man-land relationships in Old World prehistory. For the New World, Sauer (1966) has reconstructed the aboriginal ecology of the "early Spanish Main" (the Caribbean, Central America, and northern South America). Another example is Bennett's (1968) work on human influences on the zoogeography of Panama from the time of the early hunters and gatherers, through the rise and fall of a densely populated agricultural empire to the present time when large areas have once again reverted to second-growth jungle. Such research has clearly documented the significant impact that aboriginal peoples had on the environment and has shown that major ecological change, often to the detriment of man, is not confined to industrial societies nor to the twentieth century. The use of fire (see Section 5 (9), Chapter 5) and the domestication of plants and animals (see Section 5, Chapter 8) changed the face of the earth long before the industrial revolution. As already emphasized, domestication freed man from direct dependence on wild nature for food, but failure to control his symbionts (especially domestic and feral-grazing animals and row-crop agriculture) has resulted in widespread destruction of productive soil and vegetation (see page 245). Many of these themes are treated in *Land and Life*, a collection of articles by Carl O. Sauer, edited by Leighly (1967), while anthropological viewpoints are treated in *Culture and the Evolution of Man*, a collection of articles edited by Ashley Montagu (1960).

The books by Hawley (1950) and Quinn (1950) summarize the development of human ecology from the sociological side, which began with Galpin's work (1915) on rural sociology and with the studies on the ecology of cities made by Park, Burgess and McKenzie (1925) and their students. Urbanization continues to be the major focus of sociological research since cities are growing at a rate many times faster than that of the general population. Only recently, however, have sociologists (see Duncan, 1964) by and large embraced the ecological theme of this chapter, namely, that the decline in quality of the living space, not in the supply of energy or resources, is the critical problem; or expressed in another way: it is *how* materials and energy are used, and *how* growth and use of space is planned and controlled that determines *whether* human values are preserved or lost (see Section 8, Chapter 15). The architect Eliel Saarinen in his book *The City* (1943) traces the decline of cities to (1) replacement of creative architecture by noncreative innovation that lacks "organic order and correlation" and (2) a decline of public interest in town planning because of overemphasis on economic values. Planner Ian McHarg expresses a similar view (see page 423). In this connection the concept of *social indicators*, as a parallel to *pollution indicators* (see page 440), is important in "monitoring" the quality of urban life. Some examples of indices of the quality of domestic life as listed by Bauer (1966) include per cent of population married, per cent divorced, per cent of fatherless families, per cent on welfare, per cent of teenage unemployment, crime rate, and so on. Likewise, the per cent of population voting and number of school years completed by members of the population can serve as indices of the quality of political life. Smock (1969) suggests that if such social indicators had been well accepted and regularly measured (as, for example, are economic indicators) in the past, the urban problems might not have been allowed to reach their present crisis stage. The same, of course, can be said about pollution (see Chapter 16, Section 5).

The sociological dilemma can perhaps be summarized by considering two views of the city: (1) It is the ultimate creation of human civilization where want and strife are unknown and life, leisure, and culture can be enjoyed in comfort by men shielded from the harsh elements of the physical environment. (2) The city is a gross alteration of nature that provides a thousand ways to destroy and cheapen the basic conditions on which human life and dignity depend. As the ecologist views it, situation (1) will come only when the city functions as an integral part of the total biospheric ecosystem, and situation (2) is inevitable so long as cities are allowed to grow without negative feedback control, or are "managed" as something apart from their life-support systems.

However one wishes to view the interaction of man's "natural" and "cultural" attributes, human ecology must go beyond the principles of general ecology because man's flexibility in behavior, his ability to control his immediate surroundings, and his tendency to develop culture independently of environment are greater than those of other organisms. Howard W. Odum (1953) has defined culture as "the way people live in identified areas, times, and settings"; it is the sum total of cumulative processes and products of societal achievement. He points out that there is a basic component, or "folk culture," that is relatively constant and a "technological culture" that may change rapidly in time. William F. Ogburn (1922) introduced the concept of "cultural lag" to indicate that man's attitudes and social customs (i.e., folk culture) often do not keep pace with technological development. Unless society makes a conscious effort through education and regulation to reduce this "lag," very serious imbalances and disorders in the

societal system occur in times of rapid technological change (see pages 292 and 412).

An important sociological contribution to the integration of man and nature is to be found in the concept of *regionalism*, especially as developed by Howard W. Odum and his students (see H. W. Odum, 1936; Odum and Moore, 1938; H. W. Odum, 1951). Regionalism, as an approach to the study of society, is based on the recognition of distinct differences in both cultural and natural attributes of different areas which, nevertheless, are interdependent. Therefore, thorough inventories of man and resources at the local, state, and regional levels provide the basis for national and international coordination. As with soil conservation planning (see pages 420–421), regional social study was first motivated by the desire to upgrade "backward" regions (such as the "South" of the 1930s) so they could contribute to, rather than detract from, the total economy of a nation. However, "the primary purposes of regionalism are found in the end product of integration of regions more than in the mere study and development of regions themselves" (H. W. Odum, 1951). The concept that uniquely different cultural units function together as wholes is, of course, parallel to the ecologist's concept of the "ecosystem." One practical problem in "integrating" cultural and environmental data is that sociological statistics on which cultural indices are based refer to political units (counties, states) which frequently do not coincide with natural units (climatic zones, soil types, biomes, physiographic regions). As suggested in Chapter 2 (page 16), the watershed is a practical ecosystem unit for management that combines natural and cultural attributes.

Thomas Huxley's essay *Man's Place in Nature*, written over a hundred years ago (reprinted in 1963), typifies the nineteenth century biologist's attitude towards man and his environment. Reflecting the influence of Darwinism, these early biologists approached man's role in nature from the standpoint of his *natural lineage or kinship* with other organisms. After more than 50 years the idea of evolutionary kinship with nature began to be accepted by a majority of people, and attention then turned to the idea of man's *natural linkage* or *ecological interdependence* with the biota, a concept that is far from being universally accepted today. Charles C. Adams (1935), J. W. Brews (1935), Frazier Darling (1951), and Marsten Bates (1952) were among the first animal ecologists to write about "human ecology" as such. W. C. Allee (1938 and 1951) approached human ecology from the standpoint of the parallels between animal and human social organization. The "Allee principle" (see page 207), dealing with under- and over-crowding, is especially relevant since man must first aggregate, develop a social structure, and modify the environment in order to prosper; but as with any other species he also can suffer grievously from overpopulation.

The merging of animal behavior and psychology to form a new science of "behavior," as described in Sections 7 through 9 of Chapter 8, is an important step in the conceptual integration of man and nature. Paul Sears's *Deserts on the March* (1935) ushered in an era of conservationist books, notably Vogt's *The Road to Survival* (1948) and Osborn's *The Plundered Planet* (1948), which were very widely read and were influential in transforming Leopold's "land ethic" (see page 11) into political and legal action. Sears's later approach (see his essay, *The Ecology of Man*, 1957) was to compare cultural patterns that are well adapted to the environment with those that are not. Among the well-adapted patterns he cited farming systems in the small, prosperous European countries and the 1000 year old rice culture in the Philippines (both of which are now being replaced by agro-business systems; pages 46, 412). Among those patterns that are not well adapted he cited many examples of the failure of technology to keep in adjustment with the natural environment—for example, the building of factories and homes on floodplains of rivers only to have them washed away by floods that are bound to come sooner or later. Sears contended that a great many "natural disasters" resulting from floods, dust bowls, and so on, which one reads about in the newspapers (and pays for in money for rehabilitation or expensive "control" devices), are not really caused by nature but by man's stupidity in his treatment of nature, a point that is certainly valid today! With the publication of Rachel Carson's *Silent Spring* in 1962 conservationists and biologists turned their attention from man's destruction of the natural environment to the even more dangerous threat of pollution and the "ecological backlash," as discussed in Chapters 15 and 16. Illustrative of these trends are the popular books by Udall (1963), emphasizing land preservation; Rienow and Rienow (1967), emphasizing pollution; Whyte (1968), emphasizing planning; Dasman (1968), stressing need for diversity; and Marine (1969), stressing the shortcomings of the engineering approach and "pork-barrel" politics.

Thus, it was not until the 1960s that the viewpoints of the geographers, the social scientists, the biologists, and the resource scientists began to merge into a consensus as to just what human ecology is or ought to be. As evidence of the almost frantic search for common ground one can cite the numerous symposia, many of them international in scope, that were published in the 1960s. These brought together specialists from all the sciences and humanities, and provided a meeting ground for the applied and the basic worker who previously were not in communication. Also important are the governmental reports prepared by "interdisciplinary report-writing committees" (see E. P. Odum, 1967), as cited in chapters of Part 3 of this text. For the student interested in diverse (and divergent) viewpoints on human ecology as expressed by numerous writers, the volumes edited by Darling and Milton (1966), Bresler (1968), and Shepard and McKinley (1969) are recommended. But do not expect to find much of a synthesis in such books; for the most part this is yet to come.

2. THE POPULATION ECOLOGY OF MAN

If we conceive of population ecology in the broad sense as discussed in Chapter 7, then human ecology can be considered as the population ecology of a very special species—man! Human ecology is broader than demography, the field of human population analysis (see Thomlinson, 1965, for a good review of this field), since it deals with the relations of the population to external factors and larger units, as well as with internal dynamics. As we have strongly emphasized, populations of men, like other populations, are a part of biotic communities and ecosystems.

One of the main features, as indicated above, that distinguish human populations from other populations is the degree of dominance of which man as a group is capable (see Chapter 6, Section 2, for discussion of ecological dominance). Although dominance is an obvious concept and easy to talk about in general terms, it is difficult to measure quantitatively. Man often assumes that he is 100 per cent dominant over his surroundings when actually this may be far from the case. He may air-condition his home and office and thus think he is independent of the climate, but unless he also air-conditions all his food plants and animals, he is still going to be very much affected by hot and cold weather, droughts, and other climatic phenomena. Likewise, a farmer may think he has his cornfield under full control, yet phosphorus may be escaping down the hill to the ocean at such a rapid rate that from the standpoint of his future welfare, he is not controlling the situation at all but only disrupting it! At this point it is pertinent to refer to the philosophical discussion of the "biosphere" versus the "noosphere" (page 35) and to the quotations from Hutchinson (pages 36 and 103) and Leopold (pages 408 and 409) regarding man's place in ecosystems and the part he plays in biogeochemical cycles. Complete domination of nature is probably not possible and would be very precarious, or unstable, since man is a very "dependent" heterotroph who lives "high" on the food chain. It would be safer and much more pleasant if man accepted the idea that there is a desirable degree of ecological dependence, which means sharing the world with many other organisms instead of looking at each square inch as a possible source of food and wealth or as a site to make over into something artificial. If human behavior is indeed ultimately based on "reasoning" (see page 248), it is clear that man must (1) study and understand his own population growth form (see Chapter 7, Section 8, for an account of population growth form), (2) determine quantitatively what is the optimum size and configuration of a human population in relation to the carrying capacity of a given area, and then (3) be prepared to accept "cultural regulation" where "natural regulation" is inoperative (or insufficient or too late).

The growth form of the human population has been one of the most controversial of subjects, which explains why society, by and large, has been hesitant to take any action on it. In discussing man's "population problem" people often go back to Thomas R. Malthus, whose famous *Essay on the Principle of Population* went through six editions between 1798 and 1826. Malthus brought out the fact, which has been amply demonstrated to apply to organisms generally, that populations have an inherent "positive feedback" ability for exponential growth (see Section 7, Chapter 7). This does not mean that populations *necessarily* exceed their food supply. Malthus was not aware that there is a difference between density-dependent and density-independent factors, or that the former can operate in a negative feedback manner to prevent overpopulation (see Section 10, Chapter 7). Nor could Malthus have anticipated that total energy utilization (not just food energy) and its pollution by-products may now be the factors that limit the number of people that can inhabit the earth. Therefore, it is not fair to brand Malthus a "false prophet"; he should be credited with clarifying one important principle of population growth. What is important to note is that *density-dependent control is not now evident in the human population growth pattern;* recall from page 188 a recent study showing that the human population growth *world-wide* shows a positive correlation with density (that is, the population is growing faster as density increases, in sharp contrast to populations of most organisms in which growth rate decreases with increasing density). To be sure the birth rate, and to a lesser extent the growth rate, is beginning to decline in the industrialized countries, but unfortunately this increases the disproportionate ratio between the rich and the poor (see page 412).

One thing is certain: the human population growth form does not conform to the sigmoid or simple logistic pattern, for reasons already explained on page 185; growth will not just "automatically" level off at some steady-state level as does the growth of yeast cells in a confined vessel where individuals are immediately affected by their own waste products (see Figure 7–11, page 186). Since for man there will always be a long time lag in "self-crowding effects," and also in the effects of overuse of a resource, population density will tend to "overshoot" unless there are factors that sharply reduce growth rate *before* the deleterious effects of crowding begin to be felt. Thus, man seems to have two basic "options," as shown in Figure 21–1. The population growth can be allowed to continue unrestricted until the density exceeds some vital capacity (food, resources, space, pollution, and so on), as shown in Model 1 (Fig. 21–1); then large numbers of people will die, be killed, or suffer grievous hardships until the density is reduced (or perhaps until the limitation is raised, if this is possible). If control is not established at that point, additional overshoots could occur (see Figure 7–20). It is evident that some areas of the earth are already so overpopulated that a single unfavorable perturbation, such as a flood, typhoon, or crop failure for one season, will result in the deaths of thousands, even millions, of people. As noted in Section 1 of this chapter, "natural causes" are too often blamed for such deaths, thus absolving man of any responsibility (see Hardin, 1971). The alternative is for man to face up to the fact that such deaths are really caused by overpopulation. Once man accepts responsibility, it will be possible to anticipate limits, establish population controls (birth control, land- and water-use restrictions, resource conservation and recycle, reducing economic "growth stimulants," and so forth) so that density remains well below the critical limits, as shown in Model

2. To accomplish this, human values must control the interaction of science (which we can equate with "understanding") and technology (which we equate with "skill"), as shown in Model 3.

In 1959, when the second edition of this book was being prepared, books and papers dealing with the human population were about equally divided on the question of the danger of overpopulation. At that time thoughtful and learned men in both the natural and social sciences warned that the danger was very great, and there were equally responsible writers (again representing many disciplines) who took the opposite view. In 1970, writers are almost unanimous in expressing the need for some kind of control of human population growth. Even the most enthusiastic agriculturists admit that the "green revolution" (and other technical advances) only postpone the time when growth must be controlled, especially in view of the frightening increase in the ratio of poor people to rich people (see page 412). Controversy now takes a different form, as already noted in the introduction to Part 3 (see page 406), which perhaps can be exemplified by the following two statements:

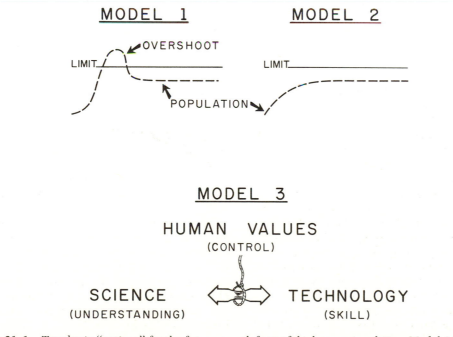

Figure 21–1. Two basic "options" for the future growth form of the human population. Model 1 is probable, at least on a partial or local scale, if growth is allowed to continue without imposed restrictions until the density "overshoots" some vital capacity or limit. Model 2 is possible if forces develop that sharply reduce the growth rate *before* the deleterious effects of crowding, pollution, or overuse of resources begin to be felt. To insure that the "option" shown by Model 2 prevails, it will be necessary for human values, rather than economic values, to control the interaction between science and technology, as shown in Model 3. See text for further discussion.

1. Technology can solve the problems of population and pollution.

2. There is no technical solution to the population and pollution dilemma; ethical, legal, political, and economic constraints are necessary.

The ecologist takes the position that both statements are correct, but each by itself is inadequate as a basis for policies, goals, or action. The second statement must be accepted if the first statement is to be realized.

3. COMPONENTS FOR AN APPLIED HUMAN ECOLOGY

As we have stressed in this chapter, the means of applying ecological principles to the solution of the human population and pollution problems do not now exist in any society, but the need to implement total ecosystem management is being very rapidly recognized throughout the world. If man is to manage man, as well as the resources on which he depends, some or all of the following reforms and procedures will have to be realized (page numbers refer to sections of this text in which the ecological context of the item is discussed):

1. Removal of all restrictions on family planning, birth control, and abortion so that the number of offspring can be restricted to those who can be loved, educated, and supported as quality individuals within the limits of local resources and space (thus making growth rate more responsive to carrying capacity, as is the case for many well-regulated animal populations; page 197).

2. Regional land-use planning (total zoning) as a means of controlling the size and distribution of the population and to insure at least one-third open space in new and expanded metropolitan areas; this requires the establishment of environmental commissions with authority to implement state, regional, and national plans that have been thoroughly studied, reviewed, and approved through democratic processes (analogous to "territorial control" in natural populations; pages 209–211).

3. Reorientation of taxation procedures in order to reduce sharply "growth stimulants" as the population density and pressure on resources increase (analogous to growth inhibition in natural populations; pages 34–35).

4. Greater emphasis in law and medicine on environmental and consumer protection (pages 442, 444).

5. A consensus on what constitutes the optimum population in order to provide a "set point" for the application of the negative feedback controls as listed in items 1 through 4 above (pages 56, 431; see also E. P. Odum, 1970; Taylor, 1970).

6. Cost internalization for whole product cycles to avoid the backlash that comes when production, pollution, and recycle costs are considered separately (page 444). Consideration of whole product cycles must also be applied to the redesign of agricultural systems so as to reduce resource waste and pollution. This means emphasizing quality, diversity, disease resistance, and so forth, rather than yield as such (see pages 46–49, 411).

7. Development of a "spaceship economy," in which the emphasis is placed on the quality of the capital stock and human resources rather than on rates of production and consumption as such, and a shift from quantity to quality boosterism in civic promotion to avoid "overselling" (pages 56, 405).

8. Recycle and stringent conservation of water and all mineral and biological resources (pages 87, 410–411).

9. By-product approach to waste disposal, involving waste management that combines tertiary treatment and recovery of wastes with recreation, watershed, and airshed protection (pages 438–441).

10. General realization that the city depends on the green countryside for all of its vital resources (air, water, food) and that the country depends on the city for most of its economic resources, so that the present suicidal political confrontation between urban and rural populations can be replaced by a total political concern for the urban-rural complex as one system (pages 23, 424, 512).

11. A shift in emphasis in systems science from the short-term "one problem–one solution" or the "technological quick-fix" to the modeling of long-term solutions to large problems (i.e., a shift in the "engineering mind" from preoccupation with the part to consideration of the whole; see Preface and page 276).

12. Greater emphasis in education (from grade school to the training of technologists) on the principle of the totality of man and environment, that is, ecosystem ecology (pages 406–407).

BIBLIOGRAPHY

NOTE: In addition to the references which are actually cited in the text, a number of other recent (1965–1971) books, summary papers, and symposium volumes that contain useful bibliographies are included in this list.

Aberg, B., and Hungate, F. P. (eds.). 1967. *Radioeco-logical Concentration Processes.* Pergamon Press, Oxford. 1040 pp.

Adams, Charles C. 1935. The relation of general ecology to human ecology. Ecology, 16:316–335.

Ahlgren, I. F., and Ahlgren, C. E. 1960. Ecological effects of forest fires. Bot. Rev., 26:483–533.

Albertson, F. W.; Tomanek, G. W.; and Riegel, Andrew. 1957. Ecology of drought cycles and grazing intensity on grasslands of Central Great Plains. Ecol. Monogr., 27:27–44.

Alexander, Martin. 1964. Biochemical ecology of soil micro-organisms. Amer. Rev. Microbiol., 18:217–252.

———. 1965. Biodegradation problems of molecular recalcitrance and micromicrobial fallibility. In: *Advances in Applied Microbiology* (W. W. Umbreit, ed.). Academic Press, New York. pp. 35–80.

———. 1971. *Microbial Ecology.* John Wiley & Sons, Inc., New York. 511 pp.

Allee, W. C. 1931. *Animal Aggregations: A Study in General Sociology.* University of Chicago Press, Chicago.

———. 1951. *Cooperation among Animals with Human Implications.* Schuman, New York (revised edition of *Social Life of Animals,* W. W. Norton & Co., Inc., New York, 1938).

Allee, W. C.; Emerson, A. E.; Park, Orlando; Park, Thomas; and Schmidt, Karl P. 1949. *Principles of Animal Ecology.* W. B. Saunders Company, Philadelphia.

Allee, W. C., and Schmidt, K. P. 1951. *Ecological Animal Geography* (2nd ed.). John Wiley & Sons, Inc., New York.

Allen, G. M. 1964. Estuarine destruction—a monument to progress. Trans. 29th N. A. Wildl. and Res. Conf. pp. 324–333.

Allen, K. R. 1951. The Horokiwi Stream: a study of a trout population. New Zealand Mar. Dept. Fish. Bull., No. 10. 238 pp.

Allison, F. E. 1935. Carbohydrate supply as a primary factor in legume symbiosis. Soil Sci., 39:123–145.

American Association for the Advancement of Science. 1965. Air Conservation (J. P. Dixon, ed.). AAAS Publ. No. 80, Washington, D.C. 335 pp.

American Public Health Association. 1965. Standard methods for examination of water and wastewater including bottom sediments and sludges (12th ed.).

Amer. Public Health Assoc., Inc., New York. 769 pp.

Anderson, G. C. 1964. The seasonal and geographic distribution of primary productivity off Washington and Oregon coasts. Limnol. Oceanogr., 9:284–302.

Anderson, G. R. 1955. Nitrogen fixation by pseudo-monas-like soil bacteria. J. Bact., 70:129–133.

Anderson, M. C. 1964. Studies of the woodland light climate. I. The photographic computation of light conditions. J. Ecol., 52:27–41.

Andrewartha, H. G. 1961. *Introduction to the Study of Animal Populations.* University of Chicago Press, Chicago. 281 pp.

Andrewartha, H. G., and Birch, L. C. 1953. The Lotka-Volterra theory of interspecific competition. Aust. J. Zool., 1:174–177.

———. 1954. *The Distribution and Abundance of Animals.* University of Chicago Press, Chicago.

Anita, N. J.; McAllister, C. D.; Parsons, T. R.; Stephens, K.; and Strickland, J. D. H. 1963. Further measurements of primary production using a large volume plastic sphere. Limnol. Oceanogr., 8:166–183.

Ardrey, Robert. 1966. *The Territorial Imperative.* Atheneum, New York. 390 pp.

Armstrong, R. C. 1966. Life support system for space flights of extended time periods. Prepared by General Dynamics, San Diego, Calif., for Langley Research Center. NASA CR-614.

Army, T. J., and Greer, F. A. 1967. Photosynthesis and crop production systems. In: *Harvesting the Sun* (San Pietro, Greer and Army, eds.). Academic Press, New York. pp. 321–332.

Aruga, Y., and Monsi, M. 1963. Chlorophyll amount as an indicator of matter production in bio-communities. Plant and Cell Physiol., 4:29–39.

Aschoff, J. 1965. *Circadian Clocks.* North-Holland Publishing Company, Amsterdam.

———. 1965a. Circadian rhythms in man. Science, 148:1427–1432.

Ashby, W. R. 1963. *An Introduction to Cybernetics.* John Wiley & Sons, Inc., New York. 295 pp.

Atkins, W. R. G. 1945. Autotrophic flagellates as the major constituent of oceanic plankton. Nature, 156:446–447.

Aubert de la Rue, Edgar; Bourliere, Francois; and Harroy, Jean-Paul. 1957. *The Tropics.* Alfred A. Knopf, Inc., New York.

Auerbach, Stanley I. 1958. The soil ecosystem and

radioactive waste disposal to the ground. Ecology, 39:522–529.

———. 1965. Radionuclide cycling: current status and future needs. Health Physics, 11:1355–1361.

Auerbach, S. I., and Crossley, D. A. 1958. Strontium-90 and cesium-137 uptake under natural conditions. Proc. Int. Conf. Peaceful Uses Atomic Energy, Geneva. Paper No. 401.

Auerbach, S. I.; Crossley, D. A.; and Engelman, M. D. 1957. Effects of gamma radiation on collembola population growth. Science, 126:614.

Auerbach, S. I.; Olson, J. S.; and Waller, H. D. 1964. Landscape investigations using cesium-137. Nature, 201:761–764.

Avery, T. E. 1966. Forester's guide to aerial photo-interpretation. U. S. Department of Agriculture, Department Handbook, 308, p. 40.

Avery, T. E., and Richter, G. 1965. An airphoto index to physical and cultural features in eastern United States. Photogr. Engr., 31:896–914.

Ayala, Francisco J. 1968. Genotype, environment and population numbers. Science, 162:1453–1459.

Azzi, Girolamo. 1956. *Agricultural Ecology.* Constable, London.

Baas-Becking, L. G. M. 1959. Geology and Microbiology. In: *Contributions to Marine Microbiology* (T. M. Skerman, ed.). New Zealand Dept. Scient. Indust. Res. Information series No. 22, New Zealand Oceanogr. Inst. Mem. No. 3, pp. 48–64.

Backus, R. H.; Craddock, J. E.; Haedrich, R. L.; Shores, D. L.; Teal, J. M.; Wing, A. S.; Mead, G. W.; and Clarke, W. D. 1968. *Ceratoscopelus maderensis:* peculiar sound-scattering layer identified with this mytophid fish. Science, 160:991–993.

Badgley, P. C., and Vest, W. L. 1966. Orbital remote sensing and natural resources. Photogr. Engr., 32: 780–790.

Baker, Edward W., and Wharton, G. W. 1952. *An Introduction to Acarology.* Macmillan Co., New York.

Baker, H. G. 1961. The adaptation of flowering plants to nocturnal and crepuscular pollinators. Quart. Rev. Biol., 36:64–73.

———. 1963. Evolutionary mechanisms in pollination biology. Science, 139:877–883.

———. 1965. Characteristics and modes of origin of weeds. In: *The Genetics of Colonizing Species* (Baker and Stebbins, eds.). Academic Press, New York.

Bakuzis, E. V. 1969. Forestry viewed in an ecosystem perspective. In: *The Ecosystem Concept in Natural Resource Management* (G. Van Dyne, ed.). Academic Press, New York. pp. 189–257.

Baldwin, M. F., and Page, J. K. (eds.). 1970. Law and the Environment. A conference, Warrenton, Va., Sept. 1968. A Conservation Foundation Publication, Walker, New York. 432 pp.

Ball, R. C., and Hooper, F. F. 1963. Translocation of phosphorus in a trout stream ecosystem. In: *Radioecology* (V. Schultz and A. W. Klement, eds.). Reinhold Publishing Company, New York. pp. 217–228.

Baltensweiler, W. 1964. *Zeiraphera griseana* Hubner (Lepidoptera: Tortricidae) in the European Alps. A contribution to the problem of cycles. Canad. Entomol., 96:792–800.

Bardach, John E. 1968. *Harvest of the Sea.* Harper and Row Publishers, Inc., New York.

———. 1969. Aquaculture. Science, 161:1098–1106.

Barick, F. B. 1950. The edge effect of the lesser vegetation of certain Adirondack forest types with particular reference to deer and grouse. Roosevelt Wildl. Bull., 9:1–146.

Barnes, Harold. 1959. *Oceanography and Marine Biology; A Book of Techniques.* Macmillan Company, New York. 218 pp.

Barr, Thomas C. 1967. Observations on the ecology of caves. Amer. Nat., 101:475–491.

Barrett, Gary W. 1969. The effects of an acute insecticide stress on a semienclosed grassland ecosystem. Ecology, 49:1019–1035.

Barringer, A. R.; Newbury, B. C.; and Moffatt, A. J. 1968. Surveillance of pollution from airborne and space platforms. Proc. 5th Sym. Remote Sensing of Environment. University of Michigan Press, Ann Arbor.

Bartholomew, George A., and Dawson, W. R. 1953. Respiratory water loss in some birds of southwestern United States. Physiol. Zool., 26:162–166.

Bartsch, A. F., and Allum, M. O. 1957. Biological factors in the treatment of raw sewage in artificial ponds. Limnol. Oceanogr., 2:77–84.

Bates, Marston. 1945. Observations on climate and seasonal distribution of mosquitoes in eastern Colombia. J. Anim. Ecol., 14:17–25.

———. 1949. *The Natural History of Mosquitoes.* Macmillan Company, New York.

———. 1952. *Where Winter Never Comes. A Study of Man and Nature in the Tropics.* Charles Scribner's Sons, New York.

———. 1957. Application of ecology to public health. Ecology, 38:60–63.

———. 1960. *The Forest and the Sea.* Random House, New York.

———. 1964. *Man in Nature* (2nd ed.). Prentice-Hall, Inc., Englewood Cliffs, N.J.

Bauchop, T., and Elsden, S. R. 1960. The growth of microorganisms in relation to their energy supply. J. Gen. Microbiol., 23:457–469.

Bauer, Raymond A. (ed.). 1966. *Social Indicators.* M.I.T. Press, Cambridge, Mass.

Baumhover, A. H.; Graham, A. J.; Hopkins, D. E.; Dudley, F. H.; New, W. D.; and Bushland, R. C. 1955. Screw-worm control through release of sterilized flies. J. Econ. Entomol., 48:462–466.

Baylor, E. R., and Sutcliffe, W. H., Jr. 1963. Dissolved organic matter in seawater as a source of particulate food. Limnol. Oceanogr., 8:369–371.

Beals, E. W. 1969. Vegetational change along altitudinal gradients. Science, 165:981–985.

Beck, Stanley D. 1960. Insects and the length of the day. Scient. Amer. 202(2):108–118.

Becking, Rudy W. 1957. The Zurich-Montpellier School of Phytosociology. Bot. Rev., 23:411–488.

Beecher, William J. 1942. Nesting birds and the vegetation substrate. Chicago Ornithological Society, Chicago.

Beeton, A. M. 1961. Environmental changes in Lake Erie. Trans. Amer. Fish. Soc., 90:153–159.

Beijerinck, M. W. 1921–1940. *Verzamelde Geschriften,* 1–6. Nijhoff, Den Haaug.

Benarde, M. A. 1970. *Our Precarious Habitat. An Integrated Approach to Understanding Man's Effect on His Environment.* W. W. Norton & Company, Inc., New York. 362 pp.

Bennett, Charles F. 1968. *Human Influences on the Zoogeography of Panama.* Ibero-Americana, Vol. 51, 115 pp.

Bennett, George W. 1962. *Management of Artificial*

Lakes and Ponds. Reinhold Publishing Company, New York.

Bennett, I. L., and Robinson, H. L. (eds.). 1967. *The World Food Problem.* A report of the President's Science Advisory Committee, Panel on the World Food Supply. Superintendent of Documents, Washington, D.C. 3 Vols.

Benton, G. S.; Blackburn, R. T.; and Snead, V. O. 1950. The role of the atmosphere in the hydrological cycle. Trans. Amer. Geophys. Union, 31:61–73.

Berkner, L. V., and Marshall, L. C. 1964. The history of growth of oxygen in the earth's atmosphere. In: *The Origin and Evolution of Atmospheres and Oceans* (D. J. Brancazio and A. G. W. Cameron, eds.). John Wiley & Sons, Inc., New York. pp. 102–126.

———. 1965. History of major atmospheric components. Proc. Natl. Acad. Sci. (Wash.), 53:1215–1226.

———. 1966. The role of oxygen. Sat. Rev., May 7, 1966. pp. 30–33.

Bertalanffy, Ludwig von. 1957. Quantitative laws in metabolism and growth. Quart. Rev. Biol., 32:217–231.

——— (ed.). 1969. *General Systems Theory; Foundations, Development, Applications.* George Braziller, Inc., New York. 290 pp.

Best, R. 1962. Production factors in the tropics. In: *Fundamentals of Dry-Matter Production and Distribution.* Neth. J. Agr. Sci., 10 (No. 5, special issue): 347–353.

Beverton, R. J. H., and Holt, S. J. 1957. On the dynamics of exploited fish populations. Great Brit. Min. Agr. Fish, Food, Fish. Invest. Ser., 2:19:1–533.

Beyers, Robert J. 1962. Relationship between temperature and the metabolism of experimental ecosystems. Science, 136:980–982.

———. 1963. The metabolism of twelve aquatic laboratory microecosystems. Ecol. Monogr., 33:281–306.

———. 1964. The microcosm approach to ecosystem biology. Amer. Biol. Teacher, 26:491–498.

———. 1965. The pattern of photosynthesis and respiration in laboratory microecosystems. In: *Primary Productivity in Aquatic Environments* (C. R. Goldman, ed.). Mem. Inst. Ital. Idrobiol., 18 suppl. University of California Press, Berkeley.

Beyers, R. J.; Larimer, J.; Odum, H. T.; Parker, R. B.; and Armstrong, N. E. 1963. Directions for determination of changes in carbon dioxide concentration from changes in pH. Publ. Inst. Mar. Sci. Univ. Texas, 9:454–489.

Billings, W. D. 1952. The environment complex in relation to plant growth and distribution. Quart. Rev. Biol., 27:251–265.

———. 1957. Physiological ecology. Ann. Rev. Plant Physiol., 8:375–392.

———. 1968. Plants, Man and the Ecosystem. Wadsworth, Belmont, Calif. 154 pp.

Birch, L. C. 1948. The intrinsic rate of natural increase of an insect population. J. Anim. Ecol., 17:15–26.

———. 1957. The meaning of competition. Amer. Nat., 41:5–18.

Birge, E. A. 1904. The thermocline and its biological significance. Trans. Amer. Micr. Soc., 25:5–33.

———. 1915. The heat budgets of American and European lakes. Trans. Wisconsin Acad. Sci. Arts Lets., 18:166–213.

Björkman, J. 1966. The effect of oxygen concentration on photosynthesis in higher plants. Physiol. Plantarum, 19:618–633.

Bjornsen, R. L. 1968. Infrared mapping of large fires. Proc. 5th Sym. Remote Sensing of Environment, University of Michigan Press, Ann Arbor.

Black, C. A. 1968. *Soil-Plant Relationships* (2nd ed.). John Wiley & Sons, Inc., New York. 792 pp.

Blackman, F. F. 1905. Optima and limiting factors. Ann. Botany, 19:281–298.

Bledsoe, L. J., and Jameson, D. A. 1969. Model structure of a grassland ecosystem. In: *The Grassland Ecosystem: A Preliminary Synthesis* (R. L. Dix and R. G. Beidleman, eds.). Range Sci. Dept. Sci. Ser. #2, Colorado State University, Fort Collins, Col. pp. 410–437.

Bliss, L. C. 1956. A comparison of plant development in microenvironments of arctic and alpine tundras. Ecol. Monogr., 26:303–337.

———. 1966. Plant productivity in Alpine microenvironments. Ecol. Monogr., 36:125–155.

Blum, Murray S. 1969. Alarm pheromones. Ann. Rev. Entomol., 14:57–80.

Bodenheimer, F. S. 1937. Population problems of social insects. Biol. Rev., 12:393–430.

———. 1938. *Problems of Animal Ecology.* Oxford University Press, London.

Bogert, C. M. 1949. Thermoregulation in reptiles, a factor in evolution. Evolution, 3:195–211.

Bogue, D. T. 1969. *Principles of Demography.* John Wiley & Sons, Inc., New York.

Bond, R. M. 1933. A contribution to the study of the natural food-cycle in aquatic environments. Bull. Bingham Oceanogr. Coll., 4:1–89.

Bongers, L. 1964. Sustaining life in space—a new approach. Aero. Med., 35:139–144.

Bonner, James. 1950. The role of toxic substances in the interaction of higher plants. Bot. Rev., 16:51–64.

Bonner, John T. 1965. *Size and Cycle: An Essay on the Structure of Biology.* Princeton University Press, Princeton, N.J. 228 pp.

Borgstrom, George. 1965. *The Hungry Planet.* Macmillan Company, New York. 487 pp.

Borkovec, A. B. 1966. Insect chemosterilants. In: *Advances in Pest Control* (R. L. Metcalf, ed.), Vol. VII. John Wiley & Sons, Inc., New York.

Bormann, F. H. 1956. Ecological implications of changes in photosynthetic response of *Pinus taeda* seedlings during ontogeny. Ecology, 37:70–75.

Bormann, F. H., and Likens, G. E. 1967. Nutrient cycling. Science, 155:424–429.

Botkin, D. B.; Woodwell, G. M.; and Tempel, Neal. 1970. Forest productivity estimated from carbon dioxide uptake. Ecology, 51:1057–1060.

Boulding, Kenneth E. 1962. *A Reconstruction of Economics.* Science Editions, New York.

———. 1964. *The Meaning of the 20th Century: The Great Transition.* Harper & Row Publishers, Inc., New York.

———. 1966. The economics of the coming spaceship earth. In: *Environmental Quality in a Growing Economy. Resources for the Future.* Johns Hopkins Press, Baltimore. pp. 3–14.

———. 1966a. Economics and ecology. In: *Future Environments of North America* (Darling and Milton, eds.). The Natural History Press, Garden City, N.Y. pp. 225–234.

Bouwer, Herman, 1968. Returning wastes to the land, a new role for agriculture. J. Soil and Water Cons., 23:164–168.

Bowman, K. O.; Hutcheson, K.; Odum, E. P.; and Shenton, L. R. 1970. Comments on the distribution of indices of diversity. In: *International Symposium on Statistical Ecology*, Vol. 3. Pennsylvania State University Press.

Bowman, R. O., and Thomas, F. W. 1961. Long-term non-toxic support of animal life with algae. Science, 134:55–56.

Boyd, C. E. 1968. Fresh-water plants: a potential source of protein. Econ. Bot., 22:359–368.

Boysen-Jensen, P. 1949. The production of matter in agricultural plants and its limitation. Det. Danske Videnskal Selskab. Biol. and Med., 21:1–28.

Bradshaw, A. D. 1957. Genecology of productivity of grasses (abstract). J. Anim. Ecol., 26:242.

Brady, Nyle C. (ed.). 1967. Agriculture and the quality of our environment. Amer. Assoc. Adv. Sci. (Wash.), Publ. 85, 460 pp.

Braun, E. Lucy. 1950. *Deciduous Forests of Eastern North America*. Blakiston Co., Philadelphia.

Braun-Blanquet, J. 1932. *Plant Sociology: the Study of Plant Communities* (Translated and edited by G. D. Fuller and H. C. Conard). McGraw-Hill Book Co., Inc., New York.

———. 1951. Pflanzensoziologie. Springer-Verlag, Vienna.

Bray, J. R. 1961. Measurement of leaf utilization as an index of minimum level of primary consumption. Oikos, 12:70–74.

———. 1962. Estimates of energy budget for a *Typha* (cattail) marsh. Science, 136:1119–1120.

———. 1963. Root production and the estimation of net productivity. Canad. Bot., 41:65–72.

———. 1964. Primary consumption in three forest canopies. Ecology, 45:165–167.

Bray, J. R., and Gorham, E. 1964. Litter production in forests of the world. In: Adv. Ecol. Res. (J. B. Cragg, ed.), 2:101–157.

Breed, R. S.; Murray, E. G. D.; and Smith, N. R. 1957. *Bergey's Manual of Determinative Bacteriology*. The Williams and Wilkins Co., Baltimore.

Bresler, Joel B. (ed.). 1968. *Environments of Man*. Addison-Wesley Publishing Co., Reading, Mass. 289 pp.

Brews, J. W. 1935. *Human Ecology*. Oxford University Press, London.

Briggs, L. J., and Shantz, H. L. 1914. Relative water requirement of plants. J. Agr. Res., 3:1–63.

Broadhead, E. 1958. The psocid fauna of larch trees in northern England. J. Anim. Ecol., 27:217–263.

Brock, G. C.; Harvey, D. I.; Kohler, R. J.; and Myskowski, M. P. 1965. Photographic considerations for aerospace. Itek Corp., Lexington, Mass. 118 pp.

Brock, M. L.; Wiegert, R. G.; and Brock, T. D. 1969. Feeding by *Paracoenia* and *Ephydra* (Diptera: Ephydridae) on microorganisms of hot springs. Ecology, 50:192–200.

Brock, T. D. 1966. *Principles of Microbial Ecology*. Prentice-Hall, Inc., Englewood Cliffs, N.J.

———. 1967. Bacterial growth rate in the sea: direct analysis by thymidine autoradiography. Science, 155:81–83.

———. 1967a. Relationship between primary productivity and standing crop along a hot spring thermal gradient. Ecology, 48:566–571.

Brock, T. D., and Brock, M. L. 1966. Temperature options for algal development in Yellowstone and Iceland hot springs. Nature, 209:733–734.

———. 1968. Measurement of steady state growth rates of a thermophilic algae directly in nature. J. Bact., 95:811–815.

Brody, Samuel. 1945. *Bioenergetics and Growth*. Reinhold Publishing Company, New York.

Brooks, John L. 1950. Speciation in ancient lakes. Quart. Rev. Biol., 25:30–60; 131–176.

Brooks, John L., and Dodson, S. I. 1965. Predation, body size and composition of plankton. Science, 150:28–35.

Brower, L. P.; Ryerson, W. N.; Coppinger, L. L.; and Glazier, S. C. 1968. Ecological chemistry and the palatability spectrum. Science, 161:1349–1350. (See also Zoologica, 49:137, 1964.)

Brown, A. H. 1953. The effects of light on respiration using isotopically enriched oxygen. Amer. J. Bot., 40:719–729.

Brown, Frank, 1969. A hypothesis for timing of circadian rhythms. Canad. J. Bot., 47:287–298.

Brown, Frank A.; Hastings, J. W.; and Palmer, J. D. 1970. *The Biological Clock — Two Views*. Academic Press, New York. 94 pp.

Brown, J., and Johnson, P. L. 1965. Pedo-ecological investigations, Barrow, Alaska. U. S. Army Cold Reg. Res. Engr. Lab. Tech. Rept.

———. 1966. U. S. Army CRREL Topographic Map, Barrow, Alaska (1:25,000). U. S. Army Cold Reg. Res. Engr. Lab., Spec. Rept. 101.

Brown, L. L., and Wilson, E. O. 1956. Character displacement. Systematic Zool., 5:49–64.

Brown, William L. 1961. Mass insect control programs: four case histories. Psyche, 68:75–109.

Brown, William L.; Eisner, L. T.; and Whittaker, R. H. 1970. Allomones and kairomones: transspecific chemical messengers. BioScience, 20:21–22.

Bruun, Anton F. 1957. Animals of the abyss. Scient. Amer., 197:50–57.

Bryant, F. J.; Chamberlain, A. C.; Morgan, A.; Spicer, C. S. 1957. Radiostrontium in soil, grass, milk and bone in U.K.; 1956 results. J. Nuc. En., 6:22–40.

Buchner, P. 1965. *Algal Symbiosis*. Chapter I. Endosymbiosis of Animals with Plant Microorganisms. John Wiley & Sons, Inc. (Interscience), New York. 3–22 pp.

Budyko, M. T. 1955. "Atlas of the Heat Balance." Leningrad. See: The heat balance of the earth's surface. Translated by N. A. Stepannova. U. S. Dept. Commerce, Washington, D.C., 1958.

Buell, M. F. 1956. Spruce-fir and maple-basswood competition in Itasca Park, Minn. Ecology, 37:606.

Bullock, T. H. 1955. Compensation for temperature in the metabolism and activity of poikotherms. Biol. Rev., 30:311–342.

Bunning, E. 1967. *The Physiological Clock* (revised 2nd ed.). Springer, Berlin.

Burges, A., and Raw, F. (eds.). 1967. *Soil Biology*. Academic Press, New York. 532 pp.

Burkholder, Paul R. 1952. Cooperation and conflict among primitive organisms. Amer. Sci., 40:601–631.

———. 1956. Studies on the nutritive value of *Spartina* grass growing in the salt marsh areas of coastal Georgia. Bull. Torrey Bot. Club, 83:327–334.

Burkholder, Paul R., and Bornside, George H. 1957. Decomposition of marsh grass by aerobic marine bacteria. Bull. Torrey Bot. Club, 84:366–383.

Burns, W. 1969. *Noise and Man*. J. B. Lippincott, Philadelphia.

Burr, G. O.; Hartt, H. E.; Brodie, H. W.; Tanimoto, T.; Kortschak, H. P.; Takahashi, D.; Ashton, F. M.; and Coleman, R. E. 1957. The sugar cane plant. Ann. Rev. Plant. Physiol., 8:275–308.

Burris, R. H. 1969. Progress in the biochemistry of nitrogen fixation. Proc. Royal Soc. Biol., 172:339–354.

Burton, G. W., and Odum, E. P. 1945. The distribution of stream fish in the vicinity of Mountain Lake, Virginia. Ecology, 26:182–193.

Bushland, R. C. 1960. Male sterilization for the control of insects. In: *Advances in Pest Control Research* (R. L. Metcalf, ed.), Vol. III. John Wiley & Sons, Inc., New York.

Butler, C. G. 1967. Insect pheromones. Biol. Rev., 42:42–87.

Butler, P. A. 1966. The problem of pesticides in estuaries. Amer. Fish. Soc. Spec. Publ., 3:110–115.

———. 1969. The significance of DDT residues in estuarine fauna. In: *Chemical Fallout* (M. W. Miller and G. G. Berg, eds.). Charles C Thomas, Springfield, Ill. pp. 205–220.

Butzer, Karl W. 1964. *Environment and Archeology, An Introduction to Pleistocene Geography.* Aldine Press, Chicago. 524 pp.

Buzzati-Traverso, Adriano A. (ed.). 1958. *Perspectives in Marine Biology.* University of California Press, Berkeley.

Cahn, Robert. 1969. Ecology and international assistance. Reprinted from the *Christian Science Monitor* by the Conservation Foundation, Washington, D.C. 16 pp.

Cain, Stanley A. 1944. *Foundations of Plant Geography.* Harper & Bros., New York. 556 pp.

———. 1945. A biological spectrum of the flora of the Great Smoky Mountain National Park. Butler Univ. Bot. Studies, 7:1–14.

———. 1950. Life-forms and phytoclimate. Bot. Rev., 16:1–32.

Cain, Stanley A., and Evans, Francis C. 1952. The distribution patterns of three plant species in an old-field community in southeastern Michigan. Contr. Lab. Vert. Biol., Univ. Mich., 52:1–11.

Caldecott, R. S., and Snyder, L. A. (eds.). 1960. Radioisotopes in the biosphere. University of Minnesota Press, Minneapolis.

Calhoun, J. B. 1962. A "behavioral sink." In: *Roots of Behavior* (E. L. Bliss, ed.). Harper and Row Publishers, Inc., New York.

———. 1962a. Population density and social pathology. Sci. Amer., 206(2):1399–1408.

Calloway, Doris H. (ed.). 1966. *Human Ecology in Space Flight.* N.Y. Acad. Sci., New York. 285 pp.

Calvin, Melvin. 1951. Reduction of carbon dioxide in aqueous solutions by ionizing radiation. Science, 114:416.

———. 1969. *Chemical Evolution; Molecular Evolution Towards the Origin of Living Systems on the Earth and Elsewhere.* Oxford University Press, New York. 278 pp.

Cameron, Austin W. 1964. Competitive exclusion between the rodent genera *Microtus* and *Clethrionomys.* Evolution, 18:630–634.

Cameron, W. M., and Pritchard, D. W. 1963. Estuaries. In: *The Sea* (M. N. Hill, ed.), Vol. II. John Wiley & Sons, Inc., New York. pp. 306–323.

Cannon, Helen L. 1952. The effect of uranium-vanadium deposits on the vegetation of the Colorado Plateau. Am. J. Sc., 250:735–770.

———. 1953. Geobotanical reconnaissance near Grants, New Mexico. Geol. Surv. Circ. 264:1–8.

———. 1954. Botanical methods of prospecting for uranium. Mining Engineering, Feb.: 217–220.

Cannon, Walter B. 1939. *The Wisdom of the Body.* W. W. Norton & Co., New York.

Carpenter, E. J. 1969. A simple, inexpensive algal chemostat. Limnol. Oceanogr., 14:720–721.

Carpenter, J. R. 1939. The biome. Amer. Midl. Nat., 21:75–91.

———. 1940. The grassland biome. Ecol. Monogr., 10:617–684.

———. 1940a. Insect outbreaks in Europe. J. Anim. Ecol., 9:108–147.

Carson, H. L. 1958. Response to selection under different conditions of recombination in *Drosophila.* Cold Spring Harbor Sym. Quant. Biol., 23:291–306.

Carson, Rachel. 1962. *Silent Spring.* Houghton Mifflin Co., Boston.

Caughley, Graeme. 1970. Eruption of ungulate populations with emphasis on Himalayan thor in New Zealand. Ecology, 51:53–72.

Chambers, K. L. (ed.). 1970. *Biochemical Coevolution.* Twenty-ninth Biology Colloquium, Oregon State University Press, Eugene.

Chang, Jen-hu. 1968. The agricultural potential of the humid tropics. Geogr. Rev., 58:334–361.

Chant, D. A. 1966. Integrated control systems. In: *Scientific Aspects of Pest Control.* National Academy of Science Publ. 1402, Washington, D.C. pp. 193–218.

Chapman, R. N. 1928. The quantitative analysis of environmental factors. Ecology, 9:111–122.

———. 1931. *Animal Ecology, with Special Reference to Insects.* McGraw-Hill Book Co., Inc., New York.

Chase, Grafton D., and Rabinowitz, Joseph L. 1967. *Principles of Radioisotope Methodology* (3rd ed.). Burgess Publishing Company, New York.

Chatworthy, J. N., and Harper, J. L. 1962. The comparative biology of closely related species living in the same area. J. Exp. Bot., 13:307–324.

Chew, R. M., and Chew, A. E. 1965. The primary productivity of a desert shrub (*Larrea tridentata*) community. Ecol. Monogr., 35:355–375.

———. 1970. Energy relationships of the mammals of a desert shrub *Larrea tridentata* community. Ecol. Monogr., 40:1–21.

Chitty, Dennis. 1960. Population processes in the vole and their relevance to general theory. Canad. J. Zool., 38:99–113.

———. 1967. The natural selection of self-regulatory behavior in animal populations. Proc. Ecol. Soc. Australia, 2:51–78.

Chovnick, A. (ed.). 1960. Biological clocks. Cold Spring Harbor Sym. Quant. Biol., Vol. 25.

Christensen, A. M., and McDermott. 1958. Life history and biology of the oyster crab, *Pinnotheres ostreum* Say. Biol. Bull., 144:146–179.

Christian, John J. 1950. The adreno-pituitary system and population cycles in mammals. J. Mamm., 31:247–259.

———. 1956. Adrenal and reproductive responses to population size in mice from freely growing populations. Ecology, 37:248–273.

———. 1959. The role of endocrine and behavioral factors in the growth of mammalian populations. In: *Symposium on Comparative Endocrinology* (Grobman, ed.). John Wiley & Sons, Inc., New York. pp. 7–97.

———. 1963. Endocrine adaptive mechanisms and the physiologic regulation of population growth. In: *Physiological Mammalogy* (W. V. Mayer and R. G. van Gelder, eds.). Academic Press, New York. pp. 189–353.

―――. 1970. Social subordination, population density, and mammalian evolution. Science, 168:84–90.

Christian, J. J., and Davis, D. E. 1964. Endocrines, behavior and populations. Science, 146:1550–1560.

Chu, H. F. 1949. *How to Know the Immature Insects.* Wm. C. Brown, Dubuque, Iowa.

Ciriancy-Wantrup, S., and Parsons, J. J. (eds.). 1967. *Natural Resources: Quality and Quantity.* University of California Press, Berkeley. 217 pp.

Clark, F. E. 1967. Bacteria in Soil. In: *Soil Biology* (N. A. Burges and F. Raw, eds.). Academic Press, New York.

Clark, John R. 1969. Thermal pollution and aquatic life. Sci. Amer., 220(3):3–11.

Clark, L. R. 1964. The population dynamics of *Cardiaspina albitextura* (Phyllidae). Aust. J. Zool., 12: 362–380.

Clark, L. R.; Geier, P. W.; Hughes, R. D.; and Morris, R. F. 1967. *The Ecology of Insect Populations in Theory and Practice.* Methuen, London. 232 pp.

Clark, P. J.; Eckstrom, P. T.; and Linden, L. C. 1964. On the number of individuals per occupation in a human society. Ecology, 45:367–372.

Clarke, F. W. 1924. The data of geochemistry. U. S. Geol. Surv. Bull. No. 228.

Clarke, George L. 1933. Diurnal migration of plankton and its correlation with changes in submarine irradiation. Biol. Bull., 65:402–436. (See also Biol. Bull. 67:432–455.)

―――. 1946. Dynamics of production in a marine area. Ecol. Monogr., 16:321–335.

―――. 1948. The nutritional value of marine zooplankton with a consideration of its use as emergency food. Ecology, 29:54–71.

―――. 1954. *Elements of Ecology.* John Wiley & Sons, Inc., New York. (Revised printing, 1965.)

Clarke, George L., and Backus, R. H. 1956. Measurement of light penetration in relation to vertical migration and records of luminescence of deep-sea animals. Deep-Sea Res., 4:1–14.

Clarke, George L., and Denton, E. J. 1962. Light and animal life. In: *The Sea* (M. N. Hill, ed.), Vol. 1. John Wiley & Sons, Inc., New York. pp. 456–468.

Clawson, M.; Landsberg, H. H.; and Alexander, L. T. 1969. Desalted seawater for agriculture: is it economic? Science, 164:1141–1148.

Clements, F. E. 1905. Research methods in ecology. Univ. Publ. Co., Lincoln, Nebraska. 199 pp.

―――. 1916. Plant succession: analysis of the development of vegetation. Publ. Carnegie Inst., Wash., 242: 1–512. (See also reprinted edition, 1928, entitled *Plant Succession and Indicators.* Wilson, New York.)

Clements, Frederic E., and Shelford, V. E. 1939. Bioecology. John Wiley & Sons, Inc., New York.

Cleveland, L. R. 1924. The physiological and symbiotic relationships between the intestinal protozoa of termites and their host, with special reference to *Reticulitermes fluipes* Kollar. Biol. Bull., 46:177–225.

―――. 1926. Symbiosis among animals with special reference to termites and their intestinal flagellates. Quart. Rev. Biol., 1:51–60.

Clifford, Paul A. 1957. Pesticide residues in fluid market milk. Publ. Health Repts. 72:729–734.

Cloud, Preston E., Jr. 1968. Realities in mineral distribution. The Texas Quarterly, II:103–126. University of Texas Press, Austin.

―――. 1969. Resources, population and the quality of life. Paper presented at AAAS Meeting, Boston, Mass. Mimeographed, 29 pages (to be published as a chapter in book in press by McGraw-Hill Book Co., Inc., N.Y.).

―――― (ed.). 1969a. *Resources and Man.* W. H. Freeman and Company, San Francisco. 259 pp.

―――. 1970. Russian roulette? (editorial). Science, 167:1323.

Cloudsley-Thompson, J. L. 1961. *Rhythmic Activity in Animal Physiology and Behavior.* Academic Press, New York.

Cochran, W. W.; Montgomery, G. G.; and Graber, R. R. 1967. Migratory flights of *Hylocichla* thrushes in spring: a radiotelemetry study. The Living Bird, 6: 213–225.

Coffin, C. C.; Hayes, F. R.; Jodrey, L. H.; and Whiteway, S. G. 1949. Exchanges of materials in a lake as studied by the addition of radioactive phosphorus. Canad. J. Res., Section C, 27:207–222.

Cogswell, Howard L. 1947. Chaparral country. Aud. Mag., 49:75–81.

Coker, Robert E. 1947. *This Great and Wide Sea.* University of North Carolina Press, Chapel Hill.

―――. 1954. *Streams, Lakes, Ponds.* University of North Carolina Press, Chapel Hill.

Cole, LaMont C. 1946. A theory for analyzing contagiously distributed populations. Ecology, 27:329–341.

―――. 1946a. A study of the cryptozoa of an Illinois woodland. Ecol. Monogr., 16:49–86.

―――. 1949. The measurement of interspecific association. Ecology, 30:411–424.

―――. 1951. Population cycles and random oscillations. J. Wildl. Mgt., 15:233–251.

―――. 1954. Some features of random cycles. J. Wildl. Managt., 18:107–109.

―――. 1954a. The population consequences of life history phenomena. Quart. Rev. Biol., 29:103–137.

―――. 1957. Sketches of general and comparative demography. Cold Spring Harbor Sym. Quant. Biol., 22:1–15.

―――. 1958. The ecosphere. Sci. Amer., 198(4):83–96.

―――. 1966. Protect the friendly microbes. In: *The Fragile Breath of Life.* 10th Anniversary Issue, Science and Humanity Supplement. Sat. Rev., May 7, 1966. pp. 46–47.

Coleman, D. C. 1970. Food webs of small arthropods of a broomsedge field studied with radioisotope-labelled fungi. Proc. IBP Symposium Methods Study in Soil Ecology, pp. 203–207. Paris.

Coleman, D. C., and McGinnis, J. T. 1970. Quantification of fungus-small arthropod food chains in the soil. Oikos, 21:134–137.

Colman, John S. 1950. *The Sea and Its Mysteries.* W. W. Norton & Company, Inc., New York.

Colwell, R. N. 1961. Some practical applications of multiband spectral reconnaissance. Amer. Sci., 29: 3–36.

―――. 1967. Remote sensing as a means of determining ecological conditions. BioScience, 17:444–449.

―――. 1968. Remote sensing of natural resources. Sci. Amer., 218(1):54–69.

Colwell, R. N.; Brewer, W.; Landis, G.; Langley, P.; Morgan, J.; Rinker, J.; Robinson, J. M.; and Sorem, A. L. 1963. Basic matter and energy relationships involved in remote reconnaissance. Photogr. Engr., 29:761–799.

Comar, C. L. 1955. *Radioisotopes in Biology and Agriculture. Principles and Practice.* McGraw-Hill Book Co., Inc., New York.

―――. 1965. The movement of fallout radionuclides through the biosphere and man. Ann. Rev. Nucl. Sci., 15:175–206.

Connell, Joseph H. 1961. The influence of interspecific competition and other factors on the distribution of the barnacle, *Chthamalus stellatus*. Ecology, 42:133–146.

Connell, Joseph H., and Orias, E. 1964. The ecological regulation of species diversity. Amer. Nat., 98:399–414.

Cook, C. W. 1964. Symposium on nutrition of forages and pastures: Collecting forage samples representative of ingested material of grazing animals for nutritional studies. J. Anim. Sci., 23:265–270.

Cooke, G. Dennis. 1967. The pattern of autotrophic succession in laboratory microecosystems. BioScience, 17:717–721.

Cooke, G. Dennis; Beyers, R. J.; and Odum, E. P. 1968. The case for the multi-species ecological system, with special reference to succession and stability. In: *Bioregenerative Systems*. NASA Spec. Publ., 165:129–139.

Coon, C. S. 1954. *The Story of Man*. Alfred A. Knopf, Inc., New York.

Cooper, Charles F. 1961. The ecology of fire. Sci. Amer., 204(4):150–160.

Cooper, Charles F., and Jolly, William C. 1969. Ecological effects of weather modification: A problem analysis. School of Natural Resources, University of Michigan, Ann Arbor. 159 pp.

Cooper, William S. 1922. The broad-sclerophyll vegetation of California; an ecological study of the chaparral and its related communities. Carnegie Inst. Wash. No. 319:1–124.

Copeland, B. J., and Dorris, T. C. 1962. Photosynthetic productivity in oil refinery effluent holding ponds. J. Water Poll. Cont. Fed., 34:1104–1111.

———. 1964. Community metabolism in ecosystems receiving oil refinery effluents. Limnol. Oceanogr., 9:431–447.

Costello, D. F. 1944. Important species of the major forage types in Colorado and Wyoming. Ecol. Monogr., 14:107–134.

———. 1957. Application of ecology to range management. Ecology, 38:49–53.

Coupland, R. T.; Zacharuk, R. Y.; and Paul, E. A. 1969. Procedures for study of grassland ecosystems. In: *The Ecosystem Concept in Natural Resource Management*. (G. Van Dyne, ed.). Academic Press, New York. pp. 25–47.

Cowgill, U. M., and Hutchinson, G. E. 1964. Cultural eutrophication in Lago Monterosi during Roman antiquity. Proc. Int. Assoc. Theor. Appl. Limnol., 15(2):644–645.

Cowles, H. C. 1899. The ecological relations of the vegetation of the sand dunes of Lake Michigan. Bot. Gaz., 27:95–117; 167–202; 281–308; 361–391.

Cox, George W. (ed.). 1969. *Readings in Conservation Ecology*. Appleton-Century-Crofts, New York. 596 pp. (paperback).

Craig, Harmon. 1954. Carbon 13 in plants and the relation between carbon 13 and carbon 14 variations in nature. J. Geol., 62:115–149.

Craighead, J. J., and Craighead, F. C., Jr. 1956. *Hawks, Owls and Wildlife*. The Stackpole Co., Harrisburg, Penna.

Crisp, D. (ed.). 1964. Grazing in terrestrial and marine environments. Blackwell Sci. Publ., Oxford. 322 pp.

Crocker, Robert L. 1952. Soil genesis and the pedogenic factors. Quart. Rev. Biol., 27:139–168.

Crombie, A. C. 1947. Interspecific competition. J. Anim. Ecol., 16:44–73.

Crossley, D. A. 1963. Consumption of vegetation by insects. In: *Radioecology* (V. Schultz and A. W. Klement, eds.). Reinhold Publishing Company, New York. pp. 427–430.

———. 1964. Biological elimination of radionuclides. Nuclear Safety, 5:265–268.

———. 1966. Radioisotope measurement of food consumption by a leaf beetle species, *Chrysomela knabi* Broun. Ecology, 47:1–8.

Crossley, D. A., and Bohnsack, K. K. 1960. The oribated mite fauna in pine litter. Ecology, 41:785–790.

Crossley, D. A., and Hoglund, Mary P. 1962. A litterbag method for the study of microarthropods inhabiting leaf litter. Ecology, 43:571–573.

Crossley, D. A., and Howden, H. F. 1961. Insect-vegetation relationships in a radioactive waste area. Ecology, 42:302–317.

Crossley, D. A., and Witkamp, Martin. 1964. Forest soil mites and mineral cycling. Acarologia, 137–445.

Crowell, K. L. 1962. Reduced interspecific competition among the birds of Bermuda. Ecology, 43:75–88.

de la Cruz, A. A., and Wiegert, R. G. 1967. 32-Phosphorus tracer studies of a horse weed aphid-ant food chain. Amer. Midl. Nat., 77:501–509.

Culver, D. C. 1970. Analysis of simple cave communities. I. Caves as islands. Evolution, 24:463–474.

Cummings, K. W. 1967. Calorific equivalents for studies in ecological energetics. Mimeographed, 52 pages. Pymatuning Lab. Ecol., University of Pittsburgh, Pittsburgh.

Cummings, K. W.; Coffman, W. P.; and Roff, P. A. 1966. Trophic relations in a small woodland stream. Verh. int. Ver. Limnol., 16:627–638.

Currie, R. I. 1958. Some observations on organic production in the northeast Atlantic. Rapp. Proc. Verb. Cons. Int. Explor. Mer., 144:96–102.

Curtis, J. T. 1955. A prairie continuum in Wisconsin. Ecology, 36:558–566.

Curtis, J. T., and McIntosh, R. P. 1951. An upland forest continuum in the prairie-forest border region of Wisconsin. Ecology, 32:476–496.

Cushing, D. H. 1964. The work of grazing in the sea. In: *Grazing in Terrestrial Environments*. Sym. Brit. Ecol. Soc., No. 4. Blackwell, Oxford. pp. 209–225.

Cutcomp, L. K. 1967. Progress in insect control by irradiation induced sterility. Pans, 13:61–70.

Cutler, D. W., and Bal, D. B. 1926. Influence of Protozoa on the process of nitrogen fixation by *Azotobacter chroococcum*. Ann. Appl. Biol., 13:516–534.

Cutting, C. L. 1952. Economic aspects of utilization of fish. Biochem. Soc. Sym. No. 6. Biochemical Society, Cambridge, England.

Daddario, E. Q. 1966. Environmental pollution. A challenge to science and technology. Rpt. of Subcommittee on Science, Research and Development, 89th Congress, U. S. Govt. Publ. 70–1770.

Daisley, K. W. 1957. Vitamin B_{12} in marine ecology. Nature, 180:142–143.

Dale, M. B. 1970. Systems analysis and ecology. Ecology, 51:2–16.

Dales, J. H. 1968. *Pollution, Property and Prices*. University of Toronto Press, Toronto. 111 pp.

Dales, R. Phillip. 1957. Commensalism. In: *Treatise on Marine Ecology and Paleoecology* (J. W. Hedgpeth, ed.), Vol. 1. Geological Society of America, Boulder, Col. pp. 391–412.

Dam, R.; Lee, S.; Fry, P.; and Fox, H. 1965. Utilization

of algae as a protein source for humans. J. Nutr., 86: 376–382.

Dansereau, Pierre. 1957. *Biogeography: an Ecological Perspective*. Ronald Press, New York.

Darling, F. Fraser, 1938. *Bird Flocks and the Breeding Cycle*. Cambridge University Press, Cambridge, England.

———. 1951. The ecological approach to the social sciences. Amer. Sc., 39:244–254.

———. 1960. Wildlife husbandry in Africa. Sci. Amer., 203(5):123–128.

Darling, F. Fraser, and Milton, J. P. (eds.). 1966. *Future Environments of North America*. Natural History Press, Garden City, N.Y. 785 pp.

Darlington, C. D. 1958. *The Evolution of Genetic Systems*. Basic Books, New York. 265 pp.

Darnell, R. M. 1958. Food habitats of fishes and larger invertebrates of Lake Pontchartrain, Louisiana, an estuarine community. Publ. Inst. Mar. Sci., Univ. Texas, 5:353–416.

———. 1967. Organic detritus in relation to the estuarine ecosystem. In: *Estuaries* (G. H. Lauff, ed.). Amer. Assoc. Adv. Sci. (Wash.). pp. 376–382.

Dasmann, R. F. 1964. *Wildlife Biology*. John Wiley & Sons, Inc., New York. 231 pp.

———. 1964a. *African Game Ranching*. Macmillan Company, New York. 75 pp.

———. 1968. *A Different Kind of Country*. Macmillan Company, New York.

———. 1968a. *Environmental Conservation* (2nd ed.). John Wiley & Sons, Inc., New York. 375 pp.

Dasmann, R. F., and Mossman, A. S. 1962. Abundance and population structure of wild ungulates in some areas of southern Rhodesia. J. Wildl. Manag., 26: 262–268.

Daubenmire, R. F. 1947. *Plants and Environment*. John Wiley & Sons, Inc., New York.

———. 1956. Climate as a determinant of vegetation distribution in eastern Washington and northern Idaho. Ecol. Monogr., 26:131–154.

———. 1959. *Plants and Environment* (2nd ed.). John Wiley, New York. 422 pp.

———. 1966. Vegetation: identification of typal communities. Science, 151:291–298.

———. 1968. Ecology of fire in grasslands. In: *Advances in Ecological Research* (J. B. Cragg, ed.), Vol. V. Academic Press, New York. pp. 209–266.

———. 1968a. *Plant Communities*. Harper & Row, New York. 300 pp.

Davidson, James. 1938. On the growth of the sheep population in Tasmania. Tr. Roy. Soc. S. Australia, 62:342–346.

Davidson, James, and Andrewartha, H. G. 1948. Annual trends in a natural population of *Thrips imaginis* (Thysanoptera). J. Anim. Ecol., 17:193–199; 200–222.

Davis, C. C. 1964. Evidence for eutrophication of Lake Erie from phytoplankton records. Limnol. Oceanogr., 9:275–283.

Davis, David E. 1951. The characteristics of global rat populations. Am. J. Pub. Health, 41:158–163.

———. 1953. The characteristics of rat populations. Quart. Rev. Biol., 28:373–401.

———. 1966. *Integral Animal Behavior*. Macmillan Company, New York. 118 pp.

Davis, G. E., and Warren, C. E. 1965. Trophic relations of a sculpin in laboratory stream communities. J. Wildl. Mgt., 29:846–871.

Davis, John H., Jr. 1940. The ecology and geologic role

of mangroves in Florida. Publ. Carnegie Inst., Wash., 517:303–412.

———. 1943. The natural features of southern Florida, especially the vegetation of the Everglades. Fla. Geol. Surv. Bull. No. 25.

Davis, J. J., and Foster, R. F. 1958. Bioaccumulation of radioisotopes through aquatic food chains. Ecology, 39:530–535.

Davis, Margaret B. 1969. Palynology and environmental history during the quaternary period. Amer. Sci., 57:317–332.

Dearborn, Walter F., and Rothney, J. W. M. 1941. *Predicting the Child's Development*. Harvard University Press, Cambridge.

DeBach, P. 1964. *Biological Control of Insect Pests and Weeds*. Reinhold Publishing Company, New York.

———. 1966. The competitive displacement and coexistence principles. Ann. Rev. Entomol., 11:183–212.

deBeaufort, L. F. 1951. *Zoogeography of the Land and Inland Waters*. Sidgwick and Jackson, London.

DeCoursey, P. J. 1961. Effect of light on the circadian activity rhythm of the flying squirrel *Glaucomys volans*. Z. Vergleich Physiol., 44:331–354.

Deevey, Edward S., Jr. 1947. Life tables for natural populations of animals. Quart. Rev. Biol., 22:283–314.

———. 1950. The probability of death. Scient. Amer., 182:58–60.

———. 1951. Life in the depths of a pond. Scient. Amer., 185:68–72.

———. 1952. Radiocarbon dating. Scient. Amer., 186: 24–28.

———. 1958. The equilibrium population. In: *The Population Ahead* (R. G. Francis, ed.). University of Minnesota Press. Minneapolis. pp. 64–86.

Delwiche, C. C. 1965. The cycling of carbon and nitrogen in the biosphere. In: *Microbiology and Soil Fertility*. Proc. 1964 Biol. Coll., Oregon State Univ. pp. 29–58.

———. 1970. The nitrogen cycle. Sci. Amer., 223(5): 137–146.

Dempster, J. P. 1960. A quantitative study of the predators on eggs and larvae of the broom beetle, *Phytodecta olivacea* using the precipitin test. J. Anim. Ecol., 29:149–167.

Dendy, J. S. 1945. Predicting depth distribution in three TVA storage type reservoirs. Trans. Amer. Fish. Soc., 75:65–71.

Dethier, V. G., and Stellar, Eliot. 1964. *Animal Behavior* (2nd ed.). Prentice-Hall, Inc., Englewood Cliffs, N.J. 118 pp.

Dice, Lee R. 1952. Measure of spacing between individuals within a population. Contr. Lab. Vert. Biol., Univ. Mich., 55:1–23.

———. 1952a. *Natural Communities*. University of Michigan Press, Ann Arbor.

Dietrich, Gunter. 1963. *General Oceanography*. John Wiley & Sons, Inc., New York. 588 pp.

Dixon, J. P. (ed.). 1965. Air Conservation. AAAS Publ. 80, Washington, D.C. 335 pp.

Dobson, Anne N., and Thomas, W. H. 1964. Concentrating plankton in a gentle fashion. Limnol. Oceanogr., 9:455–456.

Dobzhansky, Theodosius. 1967. *The Biology of Ultimate Concern*. New American Library, New York. 172 pp.

———. 1968. Adaptedness and fitness. In: *Population*

Biology and Evolution (R. C. Lewontin, ed.). Syracuse University Press. pp. 109–121.

Doeksen, J., and van der Drift, J. 1963. *Soil Organisms.* North-Holland Publishing Company, Amsterdam. 453 pp.

Dokuchaev, V. V. 1889. The zones of nature (in Russian). Akad. Nauk Moscow, Vol. 6.

Dougherty, E. C. 1959. Introduction to axenic culture of invertebrate metazoa: a goal. Ann. N.Y. Acad. Sci., 77:27–54.

Dowdy, W. W. 1947. An ecological study of the arthropoda of an oak-hickory forest with reference to stratification. Ecology, 28:418–439.

Downs, J. F., and Ekvall, R. B. 1965. Animal and social types in the exploitation of the Tibetan Plateau. In: *Man, Culture and Animals* (A. Leeds and A. P. Vayda, eds.). Amer. Assoc. Adv. Sci. (Wash.) Publ. No. 78. pp. 169–184.

Drake, Ellen T. (ed.). 1968. *Evolution and Environment.* Yale University Press, New Haven, Conn. 478 pp.

van der Drift, J. 1958. The role of the soil fauna in the decomposition of forest litter (abstract). XVth Internat. Cong. Zool., Sect. IV, Paper 3.

Droop, M. R. 1957. Vitamin B_{12} in marine ecology. Nature, 180:1041–1042.

——. 1963. Algae and invertebrates in symbiosis. In: *Symbiotic Associations* (P. S. Nutman and B. Mosse, eds.). Cambridge University Press, Cambridge. pp. 171–199.

Droop, M. R., and Wood, E. J. Ferguson (eds.). 1968. *Advances in Microbiology of the Sea*, Vol. 1. Academic Press, New York. 239 pp.

Dublin, L. I., and Lotka, A. J. 1925. On the true rate of natural increase as exemplified by the population of the United States, 1920. J. Amer. Statist. Assoc., 20: 305–339.

Duddington, C. L. 1962. Predacious fungi and the control of eelworms. In: *Viewpoints in Biology* (Carthy and Duddington, eds.). Butterworth, London. pp. 151–200.

Dugdale, R. C. 1967. Nutrient limitation in the sea: dynamics, identification, and significance. Limnol. Oceanogr., 12:685–695.

Dugdale, V. A. 1966. Aspects of the nitrogen nutrition of some naturally occurring populations of blue-green algae. In: *Environmental Requirements of Blue-Green Algae.* Pacific North-west Water Lab., Corvallis, Oregon. pp. 35–53.

Dugger, W. M., Jr.; Koukol, Jane; and Palmer, R. L. 1966. Physiological and biochemical effects of atmospheric oxidants on plants. J. Air. Poll. Cont. Assoc., 16:467–471.

Dugger, W. M., Jr., and Taylor, O. C. 1961. Interaction of light and smog components in plants. Plant Physiol., Suppl. 36, xliv.

Duncan, Otis Dudley. 1964. Social Organizations and the Ecosystem. In: *Handbook of Modern Sociology* (R. E. L. Faris, ed.). Rand McNally & Company, Chicago.

Dyksterhuis, E. J. 1946. The vegetation of the Fort Worth prairie. Ecol. Monogr., 16:1–29.

——. 1958. Ecological principles in range evaluation. Bot. Rev., 24:253–272.

Eckardt, F. E. (ed.). 1968. Functioning of terrestrial ecosystems at the primary production level. UNESCO, Paris. 516 pp.

Eckenfelder, W. W., Jr., and O'Connor, D. J. 1961. *Biological Waste Treatment.* Pergamon Publishing Company, New York. 299 pp.

Ecological Society of America (*Ad Hoc* Weather Working Group). 1966. Biological aspects of weather modification. Ecol. Soc. Amer. Bull., 47:39–78.

Edmondson, W. T. 1955. Factors affecting productivity in fertilized sea water. Papers in Marine Biology and Oceanography. Suppl. to: Deep-Sea Res., 3:451–464.

——. 1956. The relation of photosynthesis by phytoplankton to light in lakes. Ecol., 37:161–174.

——. 1961. Changes in Lake Washington following an increase in the nutrient income. Proc. Internat. Assoc. Theor. Appl. Limnol., 14:167–176.

——. 1968. Water-quality management and lake eutrophication: the Lake Washington case. In: *Water Resources Management and Public Policy* (T. H. Campbell and R. O. Sylvester, eds.). Univ. Wash. Press, Seattle, pp. 139–178.

——. 1968a. A graphical model for evaluating the use of the egg ratio for measuring birth and death rates. Oecologia, 1:1–37.

——. 1970. Phosphorus, nitrogen and algae in Lake Washington after diversion of sewage. Science, 169: 690–691.

Edmondson, W. T.; Anderson, G. C.; and Peterson, D. R. 1956. Artificial eutrophication of Lake Washington. Limnol. Oceanogr., 1:47–53.

Edney, E. B. 1957. The water relations of terrestrial arthropods. Cambridge Manage. in Exp. Biol. No. 5, Cambridge University Press, New York.

Edwards, C. A. 1966. Effect of pesticide residues on soil invertebrates and plants. In: *Ecology and the Industrial Society.* John Wiley & Sons, Inc., New York. pp. 239–261. (See also Residue Rev., 13:83–132, 1966.)

——. 1969. Soil pollutants and soil animals. Scient. Amer., 220(4):88–92; 97–99.

Edwards, C. A.; Reichle, D. E.; and Crossley, D. A. 1970. The role of soil invertebrates in turnover of organic matter and nutrients. In: *Analysis of Temperate Forest Ecosystems* (D. E. Reichle, ed.). Springer-Verlag, Berlin. pp. 147–172.

Efford, Ian E. 1969. Energy transfer in Marion Lake; with particular reference to fish feeding. Verh. Internat. Verein. Limnol., 17:104–108.

Egerton, F. N. 1968. Leeuwenhoek as a founder of animal demography. J. Hist. Biol., 1:1–22.

Egler, Frank E. 1968. Herbicides. In: *A Practical Guide to the Study of the Productivity of Large Herbivores.* IBP Handbook No. 7 (F. B. Golley and H. K. Buechner, eds.). Blackwell, Oxford. pp. 252–255

Ehrenfield, D. W. 1970. *Biological Conservation.* Holt, Rinehart & Winston, Inc., New York. 224 pp.

Ehrlich, Paul R. 1968. *The Population Bomb.* Ballantine Books, Inc., New York.

Ehrlich, Paul R., and Ehrlich, Anne H. 1970. *Population, Resources, Environment: Issues in Human Ecology.* W. H. Freeman and Co., San Francisco. 400 pp.

Ehrlich, Paul R., and Raven, Peter H. 1964. Butterflies and plants: a study in coevolution. Evolution, 18: 586–608.

——. 1969. Differentiation of populations. Science, 165:1228–1232.

Eimhjellen, K. E.; Aasmunchud, A.; and Jensen, A. 1966. A new bacterial chlorophyll. Biochem. Biophys. Res. Com., 10:232–236.

Einarsen, A. S. 1945. Some factors affecting ring-necked pheasant population density. Murrelet., 26:39–44.

Ektin, W. (ed.). 1964. Social behavior and organization among vertebrates. University of Chicago Press, Chicago.

Eldridge, E. F. 1963. Irrigation as a source of water pollution. Water Poll. Cont. Fed., 35:614–625.

Eley, J. H., Jr., and Myers, J. 1964. Study of a photosynthetic gas exchanger: A quantitative repetition of the Priestly experiment. Texas J. Sci., 16:296–333.

Eliassen, Rolf. 1952. Stream pollution. Sc. Am., 186:17–21.

Ellenberg, Heinz. 1950. Landwistschaftliche pflanzensoziolgie. Band 1. Unkrautgemeinschaftenals Zeiger für Klima und Boden. Eugen Ulmer, Stuttgart.

Elton, Charles. 1927. *Animal Ecology.* MacMillan Company, New York. (2nd ed. 1935; 3rd ed. 1947.)

———. 1933. *The Ecology of Animals.* Methuen, London.

———. 1942. *Voles, Mice and Lemmings: Problems in Population Dynamics.* Oxford University Press, London.

———. 1949. Population interspersion: an essay on animal community patterns. J. Ecol., 37:1–23.

———. 1958. *The Ecology of Invasions by Animals and Plants.* Methuen, London. 181 pp.

———. 1966. *The Pattern of Animal Communities.* Methuen, London. 432 pp.

Elton, Charles, and Miller, R. S. 1954. The ecological survey of animal communities: with a practical system of classifying habitats by structural characters. J. Ecol., 42:460–496.

Ely, Ralph L. 1957. Radioactive tracer study of sewage field in Santa Monica Bay. I.R.E. Trans. Nuclear Sc., NS-4:49–50.

Emerson, Alfred E. 1954. Dynamic homeostasis: A unifying principle in organic, social and ethical evolution. Scient. Monthly, 78(2):67–85.

Emerson, R. 1929. Relation between maximum rate of photosynthesis and chlorophyll concentration. J. Gen. Physiol., 12:609–622.

Emery, K. O. 1952. Submarine photography with the benthograph. Scient. Monthly, 75:3–11.

Emery, K. O., and Iselin, C. O. D. 1967. Human food from ocean and land. Science, 157:1279–1281.

Emery, K. O., and Rittenberg, S. C. 1952. Early diagenesis of California basin sediments in relation to origin of oil. Bull. Amer. Assoc. Petr. Geologists, 36:735–806.

Emery, K. O.; Tracey, J. I.; and Ladd, H. S. 1949. Submarine geology and topography in northern Marshalls. Amer. Geophys. Union Trans., 30:50–58.

Emlen, John T. 1952. Social behavior in nesting cliff swallows. Condor, 54:177–199.

Engelmann, M. D. 1968. The role of soil arthropods in community energetics. Amer. Zool., 8:61–69.

Epstein, S. S., and H. Shafner. 1968. Chemical mutagens in the human environment. Nature, 219:385–387

Erdean, R.; Kerry, R.; and Stephenson, W. 1956. The ecology and distribution of intertidal organisms on rocky shores of the Queensland mainland. Australian J. Mar. and Freshwater Res., 7:88–146.

Errington, Paul L. 1945. Some contributions of a 15-year local study of the northern bobwhite to a knowledge of population phenomena. Ecol. Monogr., 15:1–34.

———. 1946. Predation and vertebrate populations. Quart. Rev. Biol., 21:144–177; 221–245.

Evans, Francis C. 1956. Ecosystem as the basic unit in ecology. Science, 123:1227–1228.

Evans, Francis C., and Cain, Stanley A. 1952. Preliminary studies on the vegetation of an old field community in southeastern Michigan. Contr. Lab. Vert. Biol., Univ. Mich., 51:1–17.

Evans, Francis C., and Smith, F. E. 1952. The intrinsic rate of natural increase for the human louse, *Pediculus humanus* L. Amer. Nat., 86:299–310.

Evans, L. T. (ed.). 1963. *Environmental Control of Plant Growth.* Academic Press, New York. 449 pp.

Evenari, M. 1949. Germination inhibitors. Bot. Rev., 15:153–194.

Eyre, S. R. 1963. *Vegetation and Soils.* Aldine Publ. Co., Chicago.

Eyster, Clyde. 1964. Micronutrient requirements for green plants, especially algae. In: *Algae and Man* (D. F. Jackson, ed.). Plenum Press, New York. pp. 86–119.

Fairbrother, Nan. 1970. *New Lives, New Landscapes. Planning for the 21st Century.* Alfred A. Knopf, Inc., New York. 386 pp.

Fantasia, L. D. 1969. Accelerated immunofluorescence procedure for the detection of *Salmonella* in foods and animal by-products. Appl. Microbiol., 18:708–713.

Farner, D. S. 1961. Comparative physiology: photoperiodicity. Ann. Rev. Physiol., 23:71–96.

———. 1964. The photoperiodic control of reproductive cycles in birds. Amer. Sci., 52:137–156.

———. 1964a. Time measurement in vertebrate photoperiodism. Amer. Nat., 98:375–386.

Farvar, M. T., and Milton, John (eds.). 1969. The unforeseen international ecologic boomerang. Nat. Hist., 78:42–72.

Fautin, Reed W. 1946. Biotic communities of the northern desert shrub biome in western Utah. Ecol. Monogr., 16:251–310.

Fenchel, T. 1969. The ecology of marine microbenthos. Part IV. Ophelia, 6:1–182.

———. 1970. Studies on the decomposition of organic detritus derived from the turtle grass, *Thalassia testudinum.* Limnol. Oceanogr., 15:14–20.

Fenton, G. R. 1947. The soil fauna; with special reference to the ecosystem of forest soil. J. Anim. Ecol., 16:76–93.

Fiebleman, J. K. 1954. Theory of integrative levels. Brit. J. Phil. Sci., 5:59–66.

Filzer, Paul. 1956. Pflanzengemeinschaft und Umwelt, Ergebnisse und Probleme der Botanischen Standortforschung. Enke, Stuttgart.

Finch, Vernon C., and Trewartha, Glenn T. 1949. *Physical Elements of Geography.* McGraw-Hill Book Co., Inc., New York.

Fischer, A. G. 1960. Latitudinal variations in organic diversity. Evolution, 14:64–81.

Fishbein, L.; Flamm, W. G.; and Falk, H. L. 1970. *Chemical Mutagens.* Academic Press, New York. 364 pp.

Fisher, James, and Vevers, H. G. 1944. The breeding distribution, history and population of the north Atlantic gannet (*Sula bassana*): Changes in the world numbers of the gannet in a century. J. Anim. Ecol., 13:49–62.

Fisher, R. A.; Corbet, A. S.; and Williams, C. B. 1943. The relation between the number of species and the number of individuals in a random sample of an animal population. J. Anim. Ecol., 12:42–58.

Fleischer, W. E. 1935. The relation between chlorophyll content and rate of photosynthesis. J. Gen. Physiol., 18:573–597.

Fleming, R. H., and Laevastu, T. 1956. The influence of hydrographic conditions on the behavior of fish. FAO Fisheries Bull., 9:181–196.

Florkin, Marcel (Morgulis, Sergius, ed.). 1949. *Biochemical Evolution.* Academic Press, New York.

Fogg, G. E. 1955. *Nitrogen Fixation.* New Biology, No. 18, pp. 52–71. Penguin Books, London.

———. 1962. Extracellular products. In: *Physiology and Biochemistry of Algae* (R. A. Lewin, ed.). Academic Press, New York. pp. 475–489.

———. 1965. *Algal cultures and phytoplankton ecology.* University of Wisconsin Press, Madison. 126 pp.

Fogg, G. E., and Steward, W. D. 1965. Nitrogen fixation in blue-green algae. Sci. Progr., 53:191–201.

Forbes, S. A. 1887. The lake as a microcosm. Bull. Sc. A. Peoria. Reprinted in Ill. Nat. Hist. Surv. Bull., 15:537–550, 1925.

Fosberg, F. R. (ed.). 1963. *Man's Place in the Island Ecosystem.* Bishop Museum Press, Honolulu, Hawaii.

Foster, J. F. 1966. Life support systems and outer space. Batelle Technical Review, January, 1966.

Foster, R. F. 1958. Radioactive tracing of the movement of an essential element through an aquatic community with specific reference to radiophosphorus. Publ. della Stazione Zool. di Napoli.

Foster, R. F., and Davis, J. J. 1956. The accumulation of radioactive substances in aquatic forms. Proc. Int. Conf. Peaceful Uses Atomic Energy, Geneva, 13: 364–367.

Foster, R. F., and Rostenbach, R. E. 1954. Distribution of radioisotopes in the Columbia River. J. Amer. Water Works Assoc., 46:663–640.

Fox, Charles H. 1969. Radioactive wastes (revised ed.). U. S. Atomic Energy Commission Publ., AEC, Washington, D.C.

Fox, S. W. 1965. Simulated natural experiments in spontaneous organization of morphological units for proteinoid. In: *The Origin of Prebiological Systems and of Their Molecular Matrices.* Academic Press, New York.

Fraenkel, G. S., and Gunn, D. L. 1940. *The Orientation of Animals.* Oxford University Press, London.

Frank, Peter W. 1952. A laboratory study of intraspecies and interspecies competition in *Daphnia pulicaria* and *Simocephalus vetulus.* Physiol. Zool., 25:178–204.

———. 1957. Coactions in laboratory populations of two species of *Daphnia.* Ecology, 38:510–518.

———. 1965. The biodemography of an intertidal snail population. Ecology, 46:831–844.

———. 1968. Life histories and community stability. Ecology, 49:355–357.

Frankenberg, D., and Smith, K. L., Jr. 1967. Coprophagy in marine animals. Limnol. Oceanogr., 12:443–450.

Franzisket, L. 1964. Die Stoffwechselintensitat der Riffcorallen und ihre okologische, phylogenetische und soziologische Bedeutung. Z. vergleich Physiol., 49:91–113.

———. 1968. Zur okologie der Fadenalgen im skelett lebender riffkorallen. Zool. JB. Physiol. Bd., 74:246–253.

Freeman, T. W. 1957. *Pre-famine Ireland.* Manchester University Press.

French, N. R. 1965. Radiation and animal population: problems, progress and projections. Health Physics, 11:1557–1568.

Frere, M. H.; Menzel, R. G.; Larson, K. H.; Overstreet, Roy; and Reitemeier, R. J. 1963. The behavior of radioactive fallout in soils and plants. Publ. 1092, National Research Council, Washington, D.C.

Freudenthal, H. D. 1962. *Symbiodinium,* gen. nov., and *Symbiodinium microadriaticum,* sp. nov., a zooxanthella: taxonomy, life cycle and morphology. J. Protozool., 9:45–52.

Freudenthal, H. D., and Goreau, N. I. 1959. The physiology of skeleton formation in corals. II. Calcium deposition by hemotypic corals under various conditions in the reef. Biol. Bull., 117:239.

Frey, David G. 1963. *Limnology in North America.* University of Wisconsin Press, Madison. 734 pp.

Fried, M., and Broeshart, H. 1967. *The Soil-Plant System.* Academic Press, New York.

Friederichs, K. 1930. Die Grundfragen und Gesetzmässigkeiten der landund forstwirtschaftlichen Zoologie. Paul Parey, Berlin. 2 Vols.

Frisch, K. Von. 1950. *Bees, Their Vision, Chemical Senses, and Language.* Cornell Univ. Press, Ithaca.

———. 1955. *The Dancing Bees* (translated by D. Ilse). Harcourt, Brace & World, Inc., New York.

Fritz, N. L. 1967. Optimum methods for using infrared sensitive color films. Photogr. Engr., 33:1128–1138.

Frost, R. E.; Johnson, P. L.; Leighty, R. D.; Anderson, V. H.; Poulin, A. O.; and Rinker, J. N. 1965. Selected airphoto patterns for mobility studies in Thailand. U. S. Army Cold Regions Spec. Rept. Vol. I:1–86.

Fruton, J. S., and Simmonds, S. 1958. *General Biochemistry.* John Wiley & Sons, Inc., New York. 1077 pp.

Fry, F. E. J. 1947. Effects of the environment on animal activity. Univ. Toronto Studies, Biol. Ser. No. 55, University of Toronto Press, Toronto. pp. 1–62.

———. 1958. Temperature compensation. Ann. Rev. Physiol., 20:207–227.

Fulton, J. F. (ed.). 1955. *Howell's Textbook of Physiology* (17th ed.). W. B. Saunders Company, Philadelphia.

Furukawa, Atsushi. 1968. The raft method of oyster culture in Japan. In: *Proceedings of the Oyster Culture Workshop* (T. L. Linton, ed.). Mar. Fish. Div. Ga. Game and Fish Comm., Brunswick, Georgia. pp. 49–54.

Gaarder, T., and Gran, H. H. 1927. Investigations of the production of plankton in the Oslo Fjord. Rapp. et Proc.-Verb., Cons. Int. Explor. Mer., 42:1–48.

Galpin, C. J. 1915. The Social Anatomy of an Agricultural Community, Res. Bull., No. 34. Agr. Exp. Sta., University of Wisconsin, Madison.

Galston, A. W. 1970. Plants, people and politics. BioScience, 20:405–410.

Galtsoff, Paul S. 1956. Ecological changes affecting the productivity of oyster grounds. Trans. N. A. Wildl. Conf., 21:408–419.

Gambino, J. J., and Lindberg, R. G. 1964. Response of the pocket mouse to ionizing radiation. Rad. Res., 22:586–597.

Garfinkel, D. 1962. Digital computer simulation of ecological systems. Nature, 194:856–857.

Garfinkel, D.; MacArthur, R. H.; and Sack, R. 1964. Computer simulation and analysis of simple ecological systems. Ann. N.Y. Acad. Sci., 115:943–951.

Garren, Kenneth H. 1943. Effects of fire on the vegetation of the southeastern United States. Bot. Rev., 9:617–654.

Gates, David M. 1962. *Energy Exchange in the Bio-*

sphere. Harper and Row Publishers, Inc., New York. 151 pp.

——. 1963. The energy environment in which we live. Amer. Scient., 51:327–348.

——. 1965. Radiant energy, its receipt and disposal. Metero. Monogr., 6:1–26.

——. 1965a. Energy, plants and ecology. Ecology, 46:1–13.

——. 1969. Infrared measurement of plant and temperature and their interpretation. In: *Remote Sensing in Ecology* (P. L. Johnson, ed.). University of Georgia Press, Athens. pp. 95–107.

——. 1969a. Climate and stability. In: *Diversity and Stability in Ecological Systems* (Woodwell and Smith, eds.). Brookhaven Nat. Lab. Sym. Bio., 22: 115–127.

Gates, D. M.; Keegen, H. J.; Schlefer, J. C.; and Weidner, V. R. 1965. Spectral properties of plants. Appl. Optics, 4:11–20.

Gause, G. F. 1932. Ecology of populations. Quart. Rev. Biol., 7:27–46. (See also Quart. Rev. Biol., 11:320–336, 1936.)

——. 1934. *The Struggle for Existence.* Williams & Wilkins, Baltimore. 163 pp.

——. 1934a. Experimental analysis of Vito Volterra's mathematical theory of the struggle for existence. Science, 79:16–17.

Geiger, R. 1957. *The Climate Near the Ground* (2nd ed., revised by M. N. Stewart). Harvard University Press, Cambridge, Mass.

Gerking, S. C. 1962. Production and food utilization in a population of bluegill sunfish. Ecol. Monogr., 32: 31–78.

—— (ed.). 1967. *The Biological Basis of Freshwater Fish Production.* Blackwell, Oxford. 495 pp.

Gessner, F. 1949. Der chlorophyllgehalt in see und seine photosynthetische Valenz als geophysikalishes problem. Schweizer Zeit. f. Hydrologie, 11:378–410.

Gibson, David T. 1968. Microbial degradation of aromatic compounds. Science, 161:1093–1097.

Gifford, C. E., and Odum, E. P. 1961. Chlorophyll a content of intertidal zones on a rocky seashore. Limnol. Oceanogr., 6:83–85.

Giles, G. W. 1967. Agricultural power and equipment. In: *The World Food Problem*, Vol. III. A Report of the President's Science Advisory Committee, Panel of World Food Supply, White House, Washington, D.C. pp. 175–208.

Giles, R. H. (ed.). 1969. *Wildlife Management Technics* (3rd ed., revised). The Wildlife Society, Washington, D.C. 623 pp.

Glasstone, Samuel (ed.). 1957. *The Effects of Nuclear Weapons.* U. S. Atomic Energy Commission, Washington, D.C.

——. 1958. *Atomic Energy* (2nd ed.). D. Van Nostrand Co., New York.

Gleason, H. A. 1922. On the relation between species and area. Ecology, 3:156–162.

——. 1926. The individualistic concept of the plant association. Bull. Torrey Bot. Club, 53:7–26.

Glorig, A. 1958. *Noise and Your Ear.* Grune and Stratton, New York.

Glover, K. M.; Hardy, K. R.; Konrad, T. G.; Sullivan, W. N.; and Michaels, A. S. 1966. Radar observations of insects in free flight. Science, 154:967–972.

Goldman, C. R. 1960. Molybdenum as a factor limiting primary productivity in Castle Lake, California. Science, 132:1016–1017.

——. 1961. The contribution of alder trees (*Alnus tenuifolia*) to the primary productivity in Castle Lake, California. Ecology, 42:282–288.

——. 1962. A method of studying nutrient limiting factors *in situ* in water columns isolated by polyethylene film. Limnol. Oceanogr., 7:99–101.

——. 1964. Primary productivity and micronutrient limiting factors in some North American and New Zealand lakes. Verh. int. Ver. Limnol., 15:365–374.

—— (ed.). 1965. Primary productivity in aquatic environments. University of California Press, Berkeley. 464 pp.

Goldschmidt, V. M. 1954. *Geochemistry.* Clarendon Press, Oxford. 730 pp.

Goldsmith, John B., and Landaw, S. A. 1968. Carbon monoxide and human health. Science, 162:1352–1359.

Golley, Frank B. 1960. Energy dynamics of a food chain of an old-field community. Ecol. Monogr., 30:187–206.

——. 1960a. An index to the rate of cellulose decomposition in the soil. Ecology, 41:551–552.

——. 1961. Energy values of ecological materials. Ecology, 43:581–584.

——. 1965. Structure and function of an old-field broomsedge community. Ecol. Monogr., 35:113–131.

——. 1968. Secondary productivity in terrestrial communities. Amer. Zool., 8:53–59.

Golley, F. B., and Gentry, John B. 1964. Bioenergetics of the southern harvester ant, *Pogonomyrmex badius.* Ecology, 43:217–225.

——. 1969. Response of rodents to acute gamma radiation under field conditions. In: *Second National Symposium of Radioecology* (Nelson and Evans, eds.). Clearinghouse, Fed. Tech. Info., Springfield, Va. pp. 166–172.

Golley, F. B.; Gentry, J. B.; Menhinick, E.; and Carmon, J. L. 1965. Response of wild rodents to acute gamma radiation. Rad. Res., 24:350–356.

Golley, F. B.; McGinnis, J. T.; Clements, R. G.; Child, G. I.; and Duever, M. J. 1969. The structure of tropical forests in Panama and Columbia. BioScience, 19: 693–696.

Golley, F. B.; Odum, H. T.; and Wilson, R. F. 1962. The structure and metabolism of a Puerto Rican red mangrove forest in May. Ecology, 43:9–19.

Golueke, Clarence G. 1960. The ecology of a biotic community consisting of algae and bacteria. Ecology, 41:65–73.

Golueke, C. G., and Oswald, W. J. 1963. Closing an ecological system consisting of a mammal, algae, and non-photosynthetic microorganisms. Amer. Biol. Teach., 25:522–528.

——. 1964. Role of plants in closed systems. Ann. Rev. Plant Physiol., 15:387–408.

Golueke, C. G.; Oswald, W. J.; and McGauhey, P. H. 1959. The biological control of enclosed environments. Sewage and Indust. Wastes, 31:1125–1142.

Good, R. D. 1964. *The Geography of Flowering Plants* (3rd ed.). Longmans, London. 518 pp.

Goodall, D. W. 1952. Quantitative aspects of plant distribution. Biol. Rev., 27:194–245.

——. 1953. Objective methods for classification of vegetation, II. Fidelity and indicator value. Australian J. Bot., 1:434–456.

——. 1963. The continuum and the individualistic association. Vegetatio, 11:297–316.

Goodman, G. T.; Edwards, R. W.; and Lambert, J. M.

(eds.). 1965. *Ecology and Industrial Society.* John Wiley & Sons, Inc., New York.

Gorden, Robert W. 1969. A proposed energy budget of a soybean field. Bull. Ga. Acad. Sci., 27:41–52.

Gorden, R. W.; Beyers, R. J.; Odum, E. P.; and Eagon, E. G. 1969. Studies of a simple laboratory microecosystem: bacterial activities in a heterotrophic succession. Ecology, 50:86–100.

Gordon, H. T. 1961. Nutritional factors in insect resistance to chemicals. Ann. Rev. Entomol., 6:27–54.

Goreau, T. F. 1961. On the relation of calcification to primary production in reef-building organisms. In: *The Biology of Hydra and Some Other Coelenterates* (H. M. Lenhoff and W. F. Loomis, eds.). University of Miami Press, Miami. pp. 269–285.

———. 1963. Calcium carbonate deposition by coralline algae and corals in relation to their roles as reef-builders. Ann. N.Y. Acad. Sci., 109:127–167.

Goreau, T. F., and Goreau, N. I. 1960. Distribution of labelled carbon in reef-building corals with and without zooxanthellae. Science, 131:668–669.

Graham, Edward H. 1944. *Natural Principles of Land Use.* Oxford University Press, New York.

Graham, S. A. 1951. Developing forests resistant to insect injury. Scient. Monthly, 73:235–244.

———. 1952. *Principles of Forest Entomology* (revised ed.). McGraw-Hill Book Co., Inc., New York.

Grandfield, C. O. 1945. Alfalfa seed production as affected by organic reserves, air temperature, humidity and soil moisture. J. Agr. Res., 70:123–132.

Gray, T. R. G. 1967. Stereoscan electron microscopy of soil microorganisms. Science, 155:1668–1670.

Gray, T. R. G., and Parkinson, D. (eds.). 1968. *The Ecology of Soil Bacteria.* University of Toronto Press, Toronto.

Green, J. 1968. *The Biology of Estuarine Animals.* University of Washington Press, Seattle. 401 pp.

Greig-Smith, P. 1964. *Quantitative Plant Ecology* (2nd ed.). Butterworth, London.

Griggs, Robert F. 1956. Competition and succession on a rocky mountain fellfield. Ecology, 37:8–28.

Grinnell, Joseph. 1917. Field test of theories concerning distributional control. Amer. Nat., 51:115–128.

———. 1928. Presence and absence of animals. Univ. Calif. Chron., 30:429–450.

Grosch, Daniel S. 1965. *Biological Effects of Radiations.* Blaisdell Publishing Company, New York.

Gross, A. O. 1947. Cyclic invasion of the snowy owl and the migration of 1945–46. Auk, 64:584–601.

Grove, Robert D., and Hetzel, Alice M. 1968. *Vital Statistics Rates in the United States, 1940–1960.* U. S. Dept. of Health, Education and Welfare, Public Health Service (National Center for Health Statistics), Washington, D.C.

Grzenda, Alfred; Caver, G. J.; and Nicholson, H. P. 1964. Water pollution by insecticides in an agricultural river basin. II. The zooplankton, bottom fauna, and fish. Limnol. Oceanogr., 9:318–323.

Gunter, Gordon. 1956. Some relations of faunal distributions to salinity in estuarine waters. Ecology, 37:616–619.

Haagen-Smit, A. J.; Darley, E. F.; Zaitlin, E. F.; Hull, M.; and Noble, W. 1952. Investigation of injury to plants from air pollution in the Los Angeles area. Plant Physiol., 27:18–34.

Haefner, H. 1967. Airphoto interpretation of rural land use in Western Europe. Photogrammetria, 22:143–152.

Hairston, N. G. 1959. Species abundance and community organization. Ecology, 40:404–416.

Hairston, N. G., and Byers, G. W. 1954. The soil arthropods of a field in Southern Michigan. A study in community ecology. Contrib. Lab. Vert. Biol. Univ. Michigan, 64:1–37.

Hairston, N. G.; Smith, F. E.; and Slobodkin, L. B. 1960. Community structure, population control and competition. Amer. Nat., 94:421–425.

Halldal, Per. 1968. Photosynthetic capacities and photosynthetic action spectra of endozoic algae of the massive coral *Favia.* Biol. Bull., 134:411–424.

Hamilton, William J. 1937. The biology of microtine cycles. J. Agr. Res., 54:784–789.

Hammel, H. T.; Caldwell, F. T.; and Abrams, R. M. 1967. Regulation of body temperature in the blue-tongued lizard. Science, 156:1260–1262.

Hanson, H. C. 1950. Ecology of the grassland. II. Bot. Rev., 16:283–360.

Hanson, W. C., and Kornberg, H. A. 1956. Radioactivity in terrestrial animals near an atomic energy site. Proc. Int. Conf. Peaceful Uses Atomic Energy, Geneva, 13:385–388.

Hanson, W. C.; Watson, D. G.; and Perkins, R. W. 1967. Concentration and retention of fallout radionuclides in Alaskan arctic ecosystems. In: *Radioecological Concentration Processes* (Aberg and Hungate, eds.). Pergamon Press, Oxford. pp. 233–245.

Hardin, Garrett. 1960. The competitive exclusion principle. Science, 131:1292–1297.

———. 1963. The cybernetics of competition: a biologist's view of society. Persp. Biol. Med., 7:58–84.

———. 1968. The tragedy of the commons. Science, 162:1243–1248.

———. 1969. The economics of wilderness. Nat. Hist. Mag., 78:20, 22–26.

——— (ed.). 1969a. *Population, Evolution and Birth Control.* W. H. Freeman & Company, San Francisco. 386 pp.

———. 1971. Nobody ever dies of overpopulation (editorial). Science, 171:527.

Hardy, A. C. 1957. *The Open Sea: The World of Plankton.* Houghton Mifflin Co., New York.

———. 1958. Towards prediction in the sea. In: *Perspectives in Marine Biology* (A. Buzzati-Traverso, ed.). Univ. Calif. Press, Berkeley. pp. 159–186.

———. 1969. *The Open Sea: Fish and Fisheries.* Houghton Mifflin Co., New York.

Hardy, Ross. 1945. Breeding birds of the pigmy conifers in the Book Cliff region of Eastern Utah. Auk, 62:523–542.

Hardy, R. W. F.; Holsten, R. D.; Jackson, E. K.; and Burns, R. C. 1968. The acetylene-ethylene assay for N_2 fixation: laboratory and field evaluation. Plant Physiol., 43:1185–1207.

Harker, J. E. 1964. *The Physiology of Diurnal Rhythms.* Cambridge University Press, Cambridge.

Harley, J. L. 1952. Associations between microorganisms and higher plants (mycorrhiza). Ann. Rev. Microbiol., 6:367–386.

———. 1959. *The Biology of Mycorrhiza.* Plant Science Monographs, Leonard Hill, London, and Interscience Publ., New York. 233 pp.

Harper, John L. 1961. Approaches to the study of plant competition. In: *Mechanisms in Biological Competition.* Sym. Soc. Exp. Biol., Number XV, pp. 1–268.

———. 1968. The regulation of numbers and mass in plant populations. In: *Population Biology and Evolu-*

tion (R. C. Lewontin, ed.). Syracuse University Press. pp. 139–158.

———. 1969. The role of predation in vegetational diversity. In: *Diversity and Stability in Ecological Systems* (Woodwell and Smith, eds.). Brookhaven Sym. Biol., Number 22, pp. 48–62.

Harper, J. L., and Chatworthy, J. N. 1963. The comparative biology of closely related species. VI analysis of the growth of *Trifolium repens* and *T. fragifesum* in pure and mixed populations. J. Exp. Bot., 14:172–190.

Harper, J. L.; Williams, J. T.; and Sagar, G. R. 1965. The behavior of seeds in soil. J. Ecol., 51:273–286.

Harris, E. 1959. The nitrogen cycle in Long Island Sound. Bull. Bingham Oceanogr. Coll., 17:31–65.

Harrison, A. D. 1962. Hydrobiological studies of all saline and acid still waters in Western Cape Province. Trans. Roy. Soc. S. Africa, 36:213.

Harrison, D. E. F., and Maitra, P. 1968. The role of ATP in the control of energy metabolism in growing bacteria. J. Gen. Microbiol., 53:vii.

Hart, J. S. 1952. Lethal temperatures of fish from different latitudes. Publ. Ontario Fish. Res. Lab., 72:1–79.

———. 1957. Climatic and temperature induced changes in the energetics of Homeotherms. Rev. Canad. Biol., 16:133–174.

Hart, M. L. (ed.). 1962. Fundamentals of dry-matter production and distribution. Neth. J. Agr. Sci., 10:309–444 (special issue).

Harvey, H. W. 1950. On the production of living matter in the sea off Plymouth. J. Mar. Biol. Assoc. U.K. n.s., 29:97–137.

———. 1955. *The Chemistry and Fertility of Sea Water.* Cambridge University Press, Cambridge.

Haskell, E. F. 1949. A clarification of social science. Main Currents in Modern Thought, 7:45–51.

Hasler, A. D. 1947. Eutrophication of lakes by domestic drainage. Ecology, 28:383–395.

———. 1965. *Underwater Guideposts. Homing of Salmon.* University of Wisconsin Press, Madison. 155 pp.

———. 1969. Cultural eutrophication is reversible. BioScience, 19:425–431.

Hasting, J. W. 1959. Unicellular clocks. Ann. Rev. Microbiol., 13:297–312.

Haurwitz, Bernhard, and Austin, J. M. 1944. *Climatology.* McGraw-Hill Book Co., Inc., New York.

Haviland, Maud D. 1926. *Forest, Steppe and Tundra.* Cambridge University Press, Cambridge.

Hawkes, H. A. 1963. *The Ecology of Waste Water Treatment.* Pergamon Press, New York. 203 pp.

Hawkins, Arthur S. 1940. A wildlife history of Faville Grove, Wisconsin. Trans. Wisc. Acad. Sci. Arts Lets., 32:29–65.

Hawley, Amos H. 1950. *Human Ecology: A Theory of Community Structure.* Ronald Press, New York.

Hayes, F. R., and Coffin, C. C. 1951. Radioactive phosphorus and exchange of lake nutrients. Endeavour, 10:78–81.

Hazard, T. P., and Eddy, R. E. 1950. Modification of the sexual cycle in the brook trout (*Salvelinus fontinalis*) by control of light. Trans. Amer. Fish. Soc., 80:158–162.

Heald, Eric J. 1969. The production of organic detritus in a south Florida estuary. Dissertation, University of Miami. 110 pp.

Heald, E. J., and Odum, W. E. 1970. The contribution of mangrove swamps to Florida fisheries. Proc. Gulf and Carib. Fish. Inst., 22:130–135.

Hedgpeth, Joel W. 1951. The classification of estuarine and brackish waters and the hydrographic climate, in Report No. 11 of National Research Council Committee on a Treatise on Marine Ecology and Paleoecology, pp. 49–56.

——— (ed.). 1957. *Treatise on Marine Ecology and Paleoecology.* Vol. 1. Ecology. The Geological Society of America, New York. 1296 pp.

———. 1966. Aspects of the estuarine ecosystem. In: *A Symposium on Estuarine Fisheries.* Amer. Fish. Soc. Spec. Publ., No. 3, Washington, D.C. (Suppl. to Vol. 95 (4), Trans. Amer. Fish. Soc.). pp. 3–11.

Heese, R.; Allee, W. C.; and Schmidt, K. P. 1951. *Ecological Animal Geography* (2nd ed.). John Wiley & Sons, New York. 715 pp.

Heezen, Bruce C.; Tarp, M.; and Ewing, M. 1959. The floors of the ocean. I. North Atlantic. Geol. Soc. Amer. Spec. Paper 65. 122 pp.

Hegner, Robert. 1938. *Big Fleas Have Little Fleas, or Who's Who among the Protozoa.* Williams & Wilkins Co., Baltimore.

Heinicke, A. J., and Childers, N. F. 1937. The daily rate of photosynthesis of a young apple tree of bearing age. Mem. Cornell Univ. Agr. Exp. Sta., 201:3–52.

Hellmers, Henry; Bonner, J. F.; and Kelleher, J. M. 1955. Soil fertility: a watershed management problem in the San Gabriel Mountains of southern California. Soil Sci., 80:180–197.

Henning, Daniel H. 1968. The politics of natural resources administration. Ann. Reg. Sci. (Western Washington State College), 11:239–248.

Henry, S. M. (ed.). 1966. *Symbiosis.* Academic Press, New York. Vol. 1, 478 pp.; Vol. 2, 400 pp.

Herbert, D.; Elsworth, R.; and Telling, R. C. 1956. The continuous culture of bacteria: a theoretical and experimental study. J. Gen. Microbiol., 14:601–622.

Hesketh, J. D., and Baker, D. 1967. Light and carbon assimilation by plant communities. Crop Sci., 7:285–293.

Hewitt, E. J. 1957. Some aspects of micro-nutrient element metabolism in plants. Nature, 180:1020–1022.

Hewitt, L. F. 1950. Oxidation-reduction potentials in bacteriology and biochemistry. E. and S. Livingstone, Edinburgh. 215 pp.

Heyward, Frank. 1939. The relation of fire to stand composition of longleaf pine forests. Ecology, 20:287–304.

Heyward, Frank, and Barnette, R. M. 1934. Effect of frequent fires on the chemical composition of forest soils in the longleaf pine region. Univ. Fla. Agr. Exp. Sta. Bull. No. 265.

Hibbert, A. R. 1967. Forest treatment effects on water yield. In: *International Symposium on Forest Hydrology* (W. E. Sopper and H. W. Lull, eds.). Pergamon Press, New York. pp. 527–543.

Hickey, Joseph J. 1943. A Guide to Bird Watching. Oxford University Press, London and New York.

——— (ed.). 1969. *Peregrine Falcon Populations: Their Biology and Decline.* University of Wisconsin Press, Madison. 596 pp.

Hickey, J. J., and Anderson, D. W. 1968. Chlorinated hydrocarbons and egg shell changes in raptorial and fish-eating birds. Science, 162:271–272.

Hickling, C. F. 1948. Fish farming in the Middle and Far East. Nature, 161:748–751.

———. 1961. *Tropical Inland Fisheries.* John Wiley & Sons, New York. 287 pp.

———. 1962. *Fish Culture.* Faber and Faber, London.

Hill, L. R. 1966. An index to deoxyribonucleic acid base compositions of bacteria species. J. Gen. Microbiol., 44:419–437.

Hill, I. R., and Gray, T. R. G. 1967. Application of the fluorescent-antibody technique to an ecological study of bacteria in soil. J. Bacteriol., 93:1888–1896.

Hill, M. N. (ed.). 1962–63. *The Sea.* John Wiley & Sons, Inc. (Interscience), New York. 3 volumes.

Hills, G. A. 1952. The classification and evaluation of site for forestry. Res. Rep. No. 24, Ontario Dept. Lands and Forests.

Hinde, R. A. 1956. Ethological models and concept of drive. Brit. J. Phil. Sci., 6:321–331.

Hjort, John. 1926. Fluctuations in the year classes of important food fishes. J. du Conseil Permanent Internationale pour L'Exploration de la Mer, 1:1–38.

Hobbie, J. E.; Crawford, C. C.; and Webb, K. L. 1968. Amino acid flux in an estuary. Science, 159:1963–1964.

Hock, R. J. 1960. The potential application of hibernation to space travel. Aero. Med., 31:485–489.

Hocker, Harold W., Jr. 1956. Certain aspects of climate as related to distribution of loblolly pine. Ecology, 37:824–834.

Hogetsu, K., and Ichimura, S. 1954. Studies on the biological production of Lake Suwa. VI. The ecological studies on the production of phytoplankton. Japanese J. Bot., 14:280–303.

Holeman, J. N. 1968. The sediment yield of major rivers of the world. Water Res., 4:737–747.

Holdridge, L. R. 1947. Determination of world plant formations from simple climatic data. Science, 105:267–368.

——. 1967. *Life Zone Ecology* (2nd ed.). Trop. Res. Center, San Jose, Costa Rica. 206 pp.

Hollaender, A. (ed.). 1954. *Radiation Biology.* Vol. I. High Energy Radiation. Vol. II. Ultraviolet and Related Radiation. Vol. III. Visible and Near-visible Light. McGraw-Hill Book Co., Inc., New York.

Holling, C. S. 1961. Principles of insect predation. Ann. Rev. Entomol., 6:163–182.

——. 1964. The analysis of complex population processes. Can. Entomol., 96:335–347.

——. 1965. The functional response of predators to prey density and its role in mimicry and population regulation. Mem. Entomol. Soc. Canad., No. 45, 60 pp.

——. 1966. The functional response of invertebrate predators to prey density. Mem. Entomol. Soc. Canad., No. 48, 86 pp.

——. 1966a. The strategy of building models of complex ecological systems. In: *Systems Analysis in Ecology* (K. E. F. Watt, ed.). Academic Press, New York. pp. 195–214.

——. 1969. Stability in ecological and social systems. In: *Diversity and Stability in Ecological Systems* (Woodwell and Smith, eds.). Brookhaven Sym. Biol., No. 22, pp. 128–141.

Holling, C. S., and Ewing, Stephen. 1969. Blind-man's bluff: Exploring the response surface generated by realistic ecological simulation model. Proc. Internat. Sym. Stat. Ecol., New Haven, Conn. 70 pp.

Hollingshead, A. B. 1940. Human ecology and human society. Ecol. Monogr., 10:354–366.

Holm-Hansen, V., and Booth, C. R. 1966. The measurement of adenosine triphosphate in the ocean and its ecological significance. Limnol. Oceanogr., 11:510–519.

Holm-Hansen, V.; Sutcliffe, W. H.; and Sharp, J. 1968. Measurement of deoxyribonucleic acid in the ocean and its ecological significance. Limnol. Oceanogr., 13:507–514.

Holt, S. J. 1969. The food resources of the ocean. Scient. Amer., 221(3):178–194.

Holter, M., and Wolfe, W. 1960. Optical-mechanical scanning techniques. Univ. Michigan, Willow Run Labs., Infrared Lab. 2900–154–R.

Hooker, H. D. 1917. Liebig's law of minimum in relation to general biological problems. Science, 46:197–204.

Hood, Donald W. (ed.). 1970. Organic matter in natural waters. Inst. Mar. Sci., University of Alaska, Occ. Publ, No. 1, 625 pp.

Hopkins, S. H., and Andrews, J. D. 1970. *Rangia cuneata* on the east coast. Science, 167:868–869.

Horton, P. A. 1961. The bionomics of brown trout in a Dartmoor stream. J. Anim. Ecol., 30:331–338.

Hoyt, J. H. 1967. Barrier island formation. Bull. Geol. Soc. Amer., 78:1125–1136.

Hozumi, K.; Yoda, K.; and Kira, T. 1969. Production ecology of tropical rain forests in southwestern Cambodia. II. Photosynthetic production in an evergreen seasonal forest. Nat. and Life in S.E. Asia (Japan Soc. Promo. Sci.), 6:57–58.

Hubbs, C. L. (ed.). 1958. *Zoogeography.* Amer. Assn. Adv. Sci. (Wash.), Publ. 51, 509 pp.

Huber, B. 1952. Der Einfluss der Vegetation auf die Schwankungen des CO^2-Gehaltes der Atmosphaxe. Arch. Met. Wien., B 4:154.

Huffaker, C. B. 1957. Fundamentals of biological control of weeds. Hilgardia, 27:101–167.

——. 1959. Biological control of weeds with insects. Ann. Rev. Entomol., 4:251–276.

Humphrey, R. R. 1949. Field comments on the range condition method for forage survey. J. Range Mgt., 2:1–10.

Hungate, F. P. (ed.). 1966. *Radiation and Terrestrial Ecosystems.* Proc. Hanford Sym., May 3–5, 1965. Pergamon Press, New York, 1675 pp. (First published in Health Physics, Vol. II, No. 12, 1965.)

Hungate, R. E. 1962. Ecology of bacteria. In: *The Bacteria*, Vol. 4 (I. C. Gunsalus and R. Y. Stanier, eds.). Academic Press, New York. pp. 95–119.

——. 1963. Symbiotic associations: the rumen bacteria. In: *Symbiotic Associations.* 13th Sym. Soc. Gen. Microbiol. Cambridge University Press. pp. 266–297.

——. 1966. *The Rumen and Its Microbes.* Academic Press, New York. 533 pp.

Hunter, W. S., and Vernberg, W. B. 1955. Studies on oxygen consumption of digenetic trematodes. II. Effects of two extremes in oxygen tension. Exper. Parsit., 4:427–434.

Huntington, E. 1945. *Mainsprings of Civilization.* John Wiley & Sons, Inc., New York.

Hutcheson, Kermit. 1970. A test for comparing diversities based on the Shannon formula. J. Theor. Biol., 29:151–154.

Hutchinson, G. E. 1944. Limnological studies in Connecticut: critical examination of the supposed relationship between phytoplankton periodicity and chemical changes in lake waters. Ecology, 25:3–26.

——. 1944a. Nitrogen and biogeochemistry of the atmosphere. Amer. Scient., 32:178–195.

——. 1948. On living in the biosphere. Scient. Monthly, 67:393–398.

——. 1950. Survey of contemporary knowledge of

biogeochemistry. III. The biogeochemistry of vertebrate excretion. Bull. Amer. Mus. Nat. Hist., 96:554.

———. 1953. The concept of pattern in ecology. Proc. Acad. Nat. Sci. (Phila.), 105:1–12.

———. 1954. The biochemistry of the terrestrial atmosphere. In: *The Earth as a Planet* (G. P. Kuiper, ed.). University of Chicago Press. pp. 371–433.

———. 1957. A treatise in limnology. *Geography, Physics and Chemistry,* Vol. I. John Wiley & Sons, New York. 1015 pp.

———. 1957a. Concluding remarks. Cold Spring Harbor Sym. Quant. Biol., 22:415–427.

———. 1959. Homage to Santa Rosalina, or why are there so many kinds of animals? Am. Nat., 93:145–159.

———. 1964. The Lacustrine Microcosm reconsidered. Amer. Sci., 52:331–341.

———. 1965. The niche: an abstractly inhabited hypervolume. In: *The Ecological Theatre and the Evolutionary Play.* Yale University Press, New Haven, Conn. pp. 26–78.

———. 1967. A Treatise on Limnology. Vol. II. *Introduction to Lake Biology and the Limnoplankton.* John Wiley and Sons, Inc., New York. 1115 pp.

———. 1967a. Ecological biology in relation to the maintenance and improvement of the human environment. In: *Applied Science and Technical Progress.* Proc. Nat. Acad. Sci. (Wash.). pp. 171–184.

———. 1970. The biosphere. Scient. Amer., 223(3): 44–53.

Hutchinson, G. E., and Bowen, V. T. 1948. A direct demonstration of phosphorus cycle in a small lake. Proc. Nat. Acad. Sc., 33:148–153.

———. 1950. Limnological studies in Connecticut: quantitative radiochemical study of the phosphorus cycle in Linsley Pond. Ecology, 31:194–203.

Hutner, S. H., and McLaughlin, J. A. 1958. Poisonous tides. Scient. Amer., 199(2):92–97.

Hutner, S. H., and Provasoli, L. 1964. Nutrition of algae. Amer. Rev. Plant Physiol., 15:37–56.

Huxley, Julian. 1935. Chemical regulation and the hormone concept. Biol. Revs., 10:427.

Huxley, T. H. 1863. Evidence as to man's place in nature. London. (Reprinted with an introduction by Ashley Montague in Ann Arbor Paperbacks, University of Michigan Press, 1959. 184 pp.)

Hynes, H. B. N. 1960. *The Biology of Polluted Waters.* Liverpool University Press, Liverpool. 202 pp.

———. 1963. Imported organic matter and secondary productivity in streams. Proc. XVI Internat. Cong. Zool. (Wash.), 4:324–329.

Ichimura, Shun-ei. 1954. Ecological studies on the plankton in paddy fields. I. Seasonal fluctuations in the standing crop and productivity of plankton. Japanese J. Bot., 14:269–279.

Inove, E.; Tani, N.; Imai, K.; and Isobe, S. 1958. The aerodynamic measurement of photosynthesis over a nursery of rice plants. J. Agr. Meteor. (Japan), 14: 45–53.

———. 1958a. The aerodynamic measurement of photosynthesis over the wheatfield. J. Agr. Meteor. (Japan), 13:121.

Isaacs, John D. 1969. The nature of oceanic life. Scient. Amer., 221(3):146–162.

Ito, Yosiaki. 1959. A comparative study on survivorship curves for natural insect populations. Japanese J. Ecol., 9:107–115.

———. 1961. Factors that affect the fluctuations of animal numbers, with special reference to insect

outbreaks. Bull. Inst. Agr. Sci. (Japan), Series C, No. 13, pp. 57–89.

Ivlev, V. S. 1934. Eine Mikromethode zur Bestimmung des Kaloriengehalts von Nahrstoffen. Biochem. Ztschr., 275:49–55.

———. 1945. The biological productivity of waters. In: *Uspekhi Soureminnoi Biologii (Advances in Modern Biology),* 19:98–120. (In Russian, English translation by W. E. Ricker.) J. Fish. Res. Bd. Canad., 23: 1727–1759. 1966.

———. 1966. Balance of energy and matter in a closed biological system. In: *Physiology of Marine Animals* (V. S. Ivlev, chief ed.), pp. 107–115. Academy of Sciences, Oceanographic Commission. (In Russian). "Science" Publishing House, Moscow.

Jackson, D. F. (ed.). 1968. *Algae, Man and the Environment.* Syracuse University Press, Syracuse. 554 pp.

Jackson, J. B. C. 1968. Bivalves; spatial and size-frequency distributions of two intertidal species. Science, 161:479–480.

Jackson, R. M., and Raw, F. 1966. *Life in the Soil.* Studies in Biology, No. 2. St. Martin's Press, New York. 60 pp.

Jacot, A. P. 1940. The fauna of the soil. Quart. Rev. Biol., 15:28–58.

Jannasch, H. W. 1963. Bakterielles Wachstum bei geringen Substrathkonzentrationen. Arch. Microbiol., 45:323–342.

———. 1967. Growth of marine bacteria at limiting concentrations of organic carbon in seawater. Limnol. Oceanogr., 12:264–271.

———. 1967a. Enrichments of aquatic bacteria in continuous culture. Arch. Microbiol., 59:165–173.

———. 1969. Estimations of bacterial growth rates in natural waters. J. Bacteriol., 99:156–160.

Jannasch, H. W., and Jones, G. E. 1959. Bacterial populations in seawater as determined by different methods of enumeration. Limnol. Oceanogr., 4:128–139.

Jansen, D. H. 1966. Coevolution of mutualism between ants and acacias in Central America. Evolution, 20:249–275.

———. 1967. Interaction of the bull's horn acacia (*Acacia cornigera* L.) with an ant inhabitant (*Pseudomyrmex ferruginea* F. Smith) in eastern Mexico. Univ. Kansas Sci. Bull., 47:315–558. (See also Ecology, 48:26–35, 1967.)

Jarrett, H. (ed.). 1966. *Environmental Quality in a Growing Economy. Resources for the Future.* Johns Hopkins Press, Baltimore, Md.

Javornichy, P., and Prokesova, V. 1963. The influence of protozoa and bacteria upon the oxidation of organic substances in water. Int. Rev. ges. Hydrobiol., 48:335–350.

Jenkins, D. W. (ed.). 1966. *Significant Achievements in Space Bioscience, 1958–1964.* NASA SP-92. Supt. Doc., U. S. Government Printing Office, Washington, D.C. 128 pp.

———. 1966a. Manned space flight. In: *Significant Achievements in Space Bioscience, 1958–1964* (D. W. Jenkins, ed.). NASA SP-92. pp. 77–109.

Jenkins, James H., and Fendley, T. T. 1968. The extent of contamination, detention and health significances of high accumulation of radio-activity in southeastern game populations. Proc. 22nd Ann. Conf. Southeast Assoc. Game and Fish Commissioners, 22:89–95.

Jenny, H.; Arkley, R. J.; and Schultz, A. M. 1969. The pygmy forest-podsol ecosystem and its dune associates of the Medocino coast. Madrono, 20:60–75.

Johannes, R. E. 1964. Phosphorus excretion in marine animals: microzooplankton and nutrient regeneration. Science, 146:923–924.

———. 1965. The influence of marine protozoa on nutrient regeneration. Limnol. Oceanogr., 10:434–442.

———. 1967. Ecology of organic aggregates in the vicinity of a coral reef. Limnol. Oceanogr., 12:189–195.

———. 1968. Nutrient regeneration in lakes and oceans. In: Advances in Microbiology of the Sea. (M. Droop and E. J. Ferguson Wood, eds.). Vol. I, pp. 203–213. Academic Press, New York.

———. 1970. Coral reefs and pollution. Report prepared for FAO Technical Conference on Marine Pollution, Rome, December, 1970.

Johannes, R. E.; Coles, S. L.; and Kuenzel, N. T. 1970. The role of zooplankton in the nutrition of scleractinian corals. Limnol. Oceanogr., 15:579–586.

Johannes, R. E.; Coward, S. J.; and Webb, K. L. 1969. Are dissolved amino acids an energy source for marine invertebrates? Comp. Bioch. Physiol., 27:283–288.

Johannes, R. E., and Webb, K. L. 1970. Release of dissolved organic compounds by marine and freshwater invertebrates. Proc. Conf. on Organic Matter in Natural Waters.

Johnson, Cecil E. (ed.). 1970. Eco-crisis. John Wiley & Sons, Inc., New York. 182 pp.

Johnson, L. E.; Albee, L. R.; Smith, R. O.; and Moxon, A. L. 1951. Cows, calves and grass. South Dakota Agr. Expt. St. Bull. No. 412.

Johnson, P. L. 1965. Radioactive contamination to vegetation. Photogr. Engr., 31:984–990.

———. 1966. A consideration of methodology in photo interpretation. Proc. 4th Sym. Remote Sensing of Environment, pp. 719–725. University of Michigan Press, Ann Arbor.

———. 1969. Arctic plants, ecosystems and strategies. Arctic, 22:341–355.

——— (ed.). 1969. Remote Sensing in Ecology. University of Georgia Press, Athens. 244 pp.

———. 1970. Dynamics of carbon dioxide and productivity in an arctic biosphere. Ecology, 51:73–80.

Johnson, P. L., and Vogel, T. C. 1968. Vegetation of the Yukon Flats Region. Alaska Res. Rept. 209. Cold Regions Res. Engin. Lab., Hanover, N.H.

Johnston, David W., and Odum, Eugene P. 1956. Breeding bird populations in relation to plant succession on the Piedmont of Georgia. Ecology, 37:50–62.

Jones, J. R. Erichsen. 1949. A further ecological study of a calcareous stream in the "Black Mountain" district of south Wales. J. Anim. Ecol., 18:142–159.

Juday, Chancey. 1940. The annual energy budget of an inland lake. Ecology, 21:438–450.

———. 1942. The summer standing crop of plants and animals in four Wisconsin lakes. Trans. Wisconsin Acad. Sci., 34:103–135.

Kahl, M. Philip. 1964. The food ecology of the wood stork. Ecol. Monogr., 34:97–117.

Kale, Herbert W. 1965. Ecology and bioenergetics of the long-billed marsh wren Telmatodytes palustris griseus (Brewster) in Georgia salt marshes. Publ. Nuttall Ornithol. Club, No. 5. Cambridge, Mass. 142 pp.

Kamen, Martin D. 1953. Discoveries in nitrogen fixation. Scient. Amer., 188:38–42.

Kamen, Martin D., and Gest, H. 1949. Evidence for a

nitrogenase system in the photosynthetic bacterium Rhodospirillum rubrum. Science, 109:560.

Kanwisher, J. W. 1959. Polarographic oxygen electrode. Limnol. Oceanogr., 4:210–217.

Kanwisher, J. W., and Wainwright, S. A. 1967. Oxygen balance in some reef corals. Biol. Bull., 133:378–390.

Kardos, L. T. 1967. Waste water renovation by the land—a living filter. In: Agriculture and Quality of Our Environment (Brady, ed.). Publ. No. 85, Amer. Assoc. Adv. Sci. (Wash.). pp. 241–250.

Kaushik, N. K., and Haynes, N. B. N. 1968. Experimental study on the role of autumn-shed leaves in aquatic environments. J. Ecol., 56:229–243.

Kawaguti, S. 1964. An electron microscope proof for a path of nutritive substances from zooxanthellae to the reef coral tissue. Proc. Japanese Acad., 40:832–835.

Kawanabe, Hirayo. 1959. Food competition among fishes. Mem. College Sci. Univ. Kyoto, Ser. B., 26:253–268.

Keeling, C. D. 1970. Is carbon dioxide from fossil fuels changing man's environment? Proc. Amer. Phil. Soc., 114(1):10–17.

Keever, Catherine. 1950. Causes of succession on old fields of the Piedmont, North Carolina. Ecol. Monogr., 20:229–250.

———. 1955. Heterotheca latifolia, a new and aggressive exotic dominant in Piedmont old-field succession. Ecology, 36:732–739.

Keith, Lloyd B. 1963. Wildlife's Ten-year Cycle. University of Wisconsin Press, Madison. 201 pp.

Kendeigh, S. Charles. 1934. The role of environment in the life of birds. Ecol. Monogr., 4:299–417.

———. 1944. Measurement of bird populations. Ecol. Monogr., 14:67–106.

———. 1949. Effect of temperature and season on energy resources of the English Sparrow. Auk, 66:113–127.

———. 1961. Animal Ecology. Prentice-Hall, Englewood Cliffs, N.J. 468 pp.

———. 1969. Energy response of birds to their thermal environment. Wilson Bull., 81:441–449.

Kennedy, J. S. 1968. The motivation of integrated control. J. Appl. Ecol., 4:492–499.

Kepes, A., and Cohen, G. N. 1962. Permeation. In: The Bacteria, Vol. 4 (E. C. Gunsalus and R. Y. Stanier, eds.). Academic Press, New York. pp. 179–221.

Kershaw, K. A. 1964. Quantitative and dynamic ecology. Elsevier Publishing Co., Amsterdam.

Kettlewell, H. B. D. 1956. Further selection experiments on industrial melanism in the Lepidoptera. Heredity, 10:287–301.

Kharasch, M. S. 1929. Heats of combustion of organic compounds. Bur. Std. J. Res., 2:359–430.

Kiel, W. H. (ed.). 1968. Introduction of exotic animals; ecological and socioeconomic considerations. Texas A and M University, College Station. 25 pp.

Kilgore, W. W. 1967. Chemosterilants. In: Pest Control: Biological, Physical and Selected Chemical Methods (Kilgore and Doutt, eds.). Academic Press, New York. pp. 197–239.

Kilgore, W. W., and Doutt, R. L. (eds.). 1967. Pest Control: Biological, Physical and Selected Chemical Methods. Academic Press, New York.

Kimball, James W. 1948. Pheasant population characteristics and trends in the Dakotas. Trans. 13th N. A. Wildl. Conf., 13:291–314.

Kimura, M. 1960. Primary production of the warm temperate level forests in the southern part of Osumi Peninsula, Kyushu, Japan. Mic. Rep. Res. Inst. Nat. Res., 52:36–47.

King, C. A. M. 1967. *An Introduction to Oceanography.* McGraw-Hill Book Co., Inc., New York. 337 pp.

King, Ralph T. 1942. Is it a wise policy to introduce exotic game birds? Aud. Mag., 44:136–145; 230–236.

Kinne, Otto. 1956. Über den Einfluss des Salzgehaltes und der Temperatur auf Wachstum, Form und Vermehrung bei dem Hydroidpolypen *Cordylophora caspia* (Pallas), Thecata, Clavidae. Zool. Jahrb., 66: 565–638.

Kinne, Otto (ed.). 1970. *Marine Ecology,* Volume I, Environmental Factors. John Wiley & Sons, Inc., New York. 624 pp.

Kira, T., and Shidei, T. 1967. Primary production and turnover of organic matter in different forest ecosystems of the western Pacific. Japanese J. Ecol., 17:70–87.

Kittredge, J., Jr. 1939. The forest floor of the Chaparral of the San Gabriel Mountains, California. J. Agric. Res., 58:521–535.

Klages, K. H. W. 1942. *Ecological Crop Geography.* Macmillan Company, New York.

Kleiber, Max. 1961. *The Fire of Life.* John Wiley & Sons, Inc., New York.

Kleiber, Max, and Dougherty, J. E. 1934. The influence of environmental temperature on the utilization of food energy in baby chicks. J. Gen. Physiol., 17:701–726.

Klement, Alfred W. 1965. Radioactive fallout phenomena and mechanisms. Health Physics, 11:1265–1274.

Klopfer, P. H. 1962. *Behavioral Aspects of Ecology.* Prentice-Hall, Inc., Englewood Cliffs, N.J. 166 pp.

———. 1965. Imprinting: a reassessment. Science, 147:302–303.

Klopfer, P. H., and Hailman, J. P. 1967. *An Introduction to Animal Behavior: Ethology's First Century.* Prentice-Hall, Inc., Englewood Cliffs, N.J. 297 pp.

Klopfer, P. H., and MacArthur, R. H. 1961. On the causes of tropical species diversity: Niche overlap. Amer. Nat., 95:223–226.

Kluijver, H. N. 1951. The population ecology of the great tit, *Parus m. major* L. Ardea, 39:1–135.

Knight-Jones, E. W. 1951. Preliminary studies of nanoplankton and ultraplankton systematics and abundance by a quantitative culture method. J. Cons. Explor. Mer., 17:140–155.

Knipling, E. B. 1969. Leaf reflectance and image formation on color infrared film. In: *Remote Sensing in Ecology* (P. L. Johnson, ed.). University of Georgia Press, Athens. pp. 17–29.

Knipling, E. F. 1960. The eradication of the screwworm fly. Scient. Amer., 203(4):54–61.

———. 1963. The sterility principle. Agr. Sci. Rev., 1(1):2.

———. 1965. The sterility method of pest population control. In: *Research in Pesticides* (G. O. Chichester, ed.). Academic Press, New York. pp. 233–249.

———. 1967. Sterile technique, principles involved, current application, limitations and future applications. In: *Genetics of Insect Vectors of Disease* (Wright and Pal, eds.). Elsevier Publishing Co., Amsterdam. pp. 587–616.

Koch, R. 1881. Zur Untersuchung von pathogenen Organsmen. Mittl. Kaiserl. Ges. Amt., 1:148.

Kohn, Alan J., and Helfrich, Philip. 1957. Primary or-

ganic productivity of a Hawaiian coral reef. Limnol. Oceanogr., 2:241–251.

Komarek, E. V. 1964. The natural history of lightning. Proc. 3rd Ann. Tall Timbers Fire Ecol. Conf., Tallahassee, Fla. pp. 139–184.

———. 1967. Fire and the ecology of man. Proc. 6th Ann. Tall Timbers Fire Ecol. Conf., Tallahassee, Fla. pp. 143–170.

——— (ed.). 1969. Ecological animal control by habitat management. Tall Timbers Res. Sta., Tallahassee, Fla. 244 pp.

Kononova, M. M. 1961. *Soil Organic Matter: Its Nature, Its Role in Soil Formation and in Soil Fertility.* (Translated from Russian by T. Z. Nowankowski and G. A. Greenwood.) Pergamon Press, N.Y. 450 pp.

Konrad, T. G. 1968. Radar as a tool in meteorology, entomology and ornithology. Proc. 5th Symp. Remote Sensing of Environment. University of Michigan Press, Ann Arbor.

Konrad, T. G.; Hicks, J. J.; and Dobson, E. B. 1968. Radar characteristics of birds in flight. Science, 159: 274–280.

Kozlowski, T. T. 1964. *Water Metabolism in Plants.* Harper & Row Publishers, Inc., New York.

——— (ed.). 1968. *Water Deficits and Plant Growth.* 2 Vols. Academic Press, New York.

Krajina, V. J. 1969. Ecology of forest trees in British Columbia. In: *Ecology of Western North America* (University of British Columbia, Vancouver), 2:1–147.

Krauss, R. W. 1962. Mass culture of algae for food and other organic compounds. Amer. J. Bot., 49:425–435.

Krenkel, R. A., and Parker, F. L. (eds.). 1969. *Biological Aspects of Thermal Pollution.* Proc. Nat. Sym. Therm. Poll. Vanderbilt University Press, Nashville, Tenn.

Krinov, E. L. 1947. Spectral reflectance properties of natural formations. Academy of Sciences, U.S.S.R. (Translated by G. Belkov, Natural Resources Council of Canada, 1953.) Tech. transl. 439, p. 268.

Kucera, C. L.; Dahlman, R. C.; and Koelling, M. L. 1967. Total net productivity and turnover on an energy basis for tall grass prairie. Ecology, 48:536–541.

Kuentzel, L. E. 1969. Bacteria, carbon dioxide and algal blooms. J. Water Poll. Cont. Fed., 41:1737–1747.

Kuenzler, E. J. 1958. Niche relations of three species of Lycosid spiders. Ecology, 39:494–500.

———. 1961. Phosphorus budget of a mussel population. Limnol. Oceanogr., 6:400–415.

———. 1961a. Structure and energy flow of a mussel population. Limnol. Oceanogr., 6:191–204.

Kühnelt, W. 1950. *Bodenbiologie mit besunderer Berucksichtigung der Tierwelt.* Wien.

Kulp, J. L.; Eckelmann, W. R.; and Schulert, A. P. 1957. Strontium-90 in man. I. Science, 125:219. II. Science, 127:266–274.

Kusnetsov, S. J. 1958. A study of the size of bacterial populations and of organic matter formation due to the photo- and chemosynthesis in water bodies of different types. Verh. int. Ver. Limnol., 13:156–169.

———. 1967. Produktion der Biomasse heterotropher Bakterien und die Geschwindigkeit ihre Vermehrung im Rubinsk-Stausee. Verh. int. Ver. Limnol., 16: 1493–1500.

Kurihara, Y.; Aedie, J. M.; Hobson, P. N.; and Mann, S. O. 1968. Relationship between bacteria and ciliate protozoa in the sheep rumen. Japanese Gen. Microbiol., 51:267–288.

Kusumoto, T. 1961. An ecological analysis of the distribution of broad-leaved evergreen trees, based on dry matter production. Japanese J. Bot., 17:307–331.

Kutkuhn, J. H. 1966. The role of estuaries in the development and perpetuation of commercial shrimp resources. In: *A Symposium on Estuarine Fisheries* (R. F. Smith, ed.). Amer. Fish. Soc. Spec. Publ. No. 3, pp. 16–36.

Lack, David L. 1945. Ecology of closely related species with special reference to cormorant (*Phalacrocorax carbo*) and shag (*P. aristotelis*). J. Anim. Ecol., 14: 12–16.

———. 1947. *Darwin's Finches*. Cambridge University Press, New York.

———. 1954. *The Natural Regulation of Animal Numbers*. Oxford University Press, New York.

———. 1959. Watching migration by radar. Brit. Birds, 52:258–267.

———. 1962. Radar evidence on migratory orientation. Brit. Birds, 55:139–158.

———. 1966. *Population Studies of Birds*. Clarendon Press, Oxford. 341 pp.

———. 1969. Tit niches in two worlds; or homage to Evelyn Hutchinson. Amer. Nat., 103:43–49.

Ladd, H. S. 1961. Reef building. Science, 134:703–715.

Ladd, H. S., and Tracey, J. I. 1949. The problem of coral reefs. Sci. Mon., 69:297–305.

Lamprey, H. P. 1964. Estimation of the large mammal densities, biomass and energy exchange in the Tarangire Game Reserve and the Masai Steppe in Tanganyika. E. Afr. Wildl. J., 2:1–47.

Lane, Charles E. (ed.). *Reservoir Fishery Resources Symposium*. Amer. Fish. Soc. Washington, D.C. 569 pp.

Lang, R., and Barnes, O. K. 1942. Range forage production in relation to time and frequency of harvesting. Wyoming Agr. Exp. Sta. Bull. No. 253.

Lange, O. L.; Kock, W.; and Schulze, E. D. 1969. CO_2-gas exchange and water relationships of plants in the Negev desert at the end of the dry period. Ber. dtsch. Bot. Ges., 82:39–61.

Lange, W. 1967. Effect of carbohydrates on the symbiotic growth of planktonic blue-green algae with bacteria. Nature, 215:2177.

Langley, L. L. 1965. *Homeostasis*. Reinhold Publishing Company, New York.

Lapp, Ralph E., and Andrews, Howard L. 1954. *Nuclear Radiation Physics* (2nd ed.). Prentice-Hall, Inc., Englewood Cliffs, N.J.

Larkin, P. A. 1963. Interspecific competition and exploitation. J. Fish. Res. Bd. Canad., 20:647–678.

Larkin, P. A., and Hourston, A. S. 1964. A model for the simulation of the population biology of Pacific salmon. J. Fish. Res. Bd. Canad., 21:1245–1265.

Laskey, Amelia R. 1939. A study of nesting eastern bluebirds. Bird-Banding, 10:23–32.

Lauff, George A. (ed.). 1967. Estuaries. Amer. Assoc. Adv. Sci. Publ. No. 83. Washington, D.C. 757 pp.

Lave, Lester B., and Seskin, E. P. 1970. Air pollution and human health. Science, 169:723–733.

Lawson, F. R. 1967. Theory of control of insect populations by sexually sterile males. Ann. Entomol. Soc. Amer., 60:713–722.

Lederberg, J. 1960. Exobiology: Approaches to life beyond the earth. Science, 132:393–400.

Leeds, A., and Vayda, A. P. 1965. Man, culture and animals. The role of animals in human ecological adjustments. Amer. Assoc. Adv. Sci. Publ. No. 78. 304 pp.

Leigh, E. G. 1965. On the relation between productivity, diversity and stability of a community. Proc. Natl. Acad. Sci. (Wash.), 53:777–783.

Leighly, John (ed.). 1967. *Land and Life—A Selection from the Writings of Carl Ortwin Sauer*. University of California Press, Berkeley. 435 pp.

Leighty, R. D. 1965. Terrain mapping from aerial photography for purpose of vehicle mobility. J. Terramechanics, 2:55–67.

Lemon, E. R. 1960. Photosynthesis under field conditions. II. An aerodynamic method for determining the turbulent carbon dioxide exchange between the atmosphere and a corn field. Agron. J., 52:697–703. (See also Agron. J., 58:265–268, 1966.)

———. 1967. Aerodynamic studies of CO_2 exchange between the atmosphere and the plant. In: *Harvesting the Sun* (San Pietro, et al., eds.). Academic Press, New York. pp. 263–290.

Lenhoff, Howard M. 1968. Behavior, hormones, and hydra. Science, 161:434–442.

Lent, C. M. (ed.). 1969. Adaptations of intertidal organisms. Amer. Zool., 9:269–426.

Leopold, Aldo. 1933. *Game Management*. Charles Scribner's Sons, New York.

———. 1933a. The conservation ethic. J. Forestry, 31:634–643.

———. 1941. Lakes in relation to terrestrial life patterns. In: *A Symposium on Hydrobiology*. University of Wisconsin Press, Madison. pp. 17–22.

———. 1943. Deer irruptions. Wisconsin Cons. Bull., Aug., 1943. Reprinted in Wisconsin Cons. Dept. Publ., 321:3–11.

———. 1949. The land ethic. In: *A Sand County Almanac*. Oxford University Press, New York.

Leopold, Aldo, and Jones, S. E. 1947. A phenological record for Sauk and Dane Counties, Wisconsin, 1935–1945. Ecol. Monogr., 17:81–122.

Leopold, Luna B., and Langbein, W. B. 1962. The concept of entropy in landscape evolution. U. S. Geol. Surv. Paper 500 A.

Leopold, L. B., and Maddock, Thomas. 1954. *The Flood Control Controversy: Big Dams, Little Dams, and Land Management*. The Ronald Press, New York.

Leslie, P. H. 1945. On the use of matrices in certain population mathematics. Biometrika, 33:183–212.

Leslie, P. H., and Park, Thomas. 1949. The intrinsic rate of natural increase of *Tribolium castaneum* Herbst. Ecology, 30:469–477.

Leslie, P. H., and Ranson, R. M. 1940. The mortality, fertility, and rate of natural increase of the Vole (*Microtus agrestis*) as observed in the laboratory. J. Anim. Ecol., 9:27–52.

Levi, Herbert W. 1952. Evaluations of wildlife importations. Sci. Mon., 74:315–322.

Levins, Richard. 1966. The strategy of model building in population biology. Amer. Scient., 54:421–431.

Lewin, J. C. 1963. Heterotrophy in marine diatoms. In: *Marine Microbiology* (C. H. Oppenheimer, ed.). Charles C Thomas, Springfield, Ill. pp. 229–235.

Lewin, J. C., and Lewin, R. A. 1960. Auxotrophy and heterotrophy in marine littoral diatoms. Canad. J. Microbiol., 6:127–134.

Lewis, J. K. 1969. Range management viewed in the ecosystem framework. In: *The Ecosystem Concept in Natural Resource Management* (G. Van Dyne, ed.). Academic Press, New York. pp. 97–187.

Lewontin, R. C. 1962. Interdeme selection controlling

apolymorphism in the house mouse. Amer. Nat., 96:65–78.

———— (ed.). 1968. *Population Biology and Evolution.* Syracuse University Press, Syracuse, N.Y. 205 pp.

Libby, W. F. 1955. *Radiocarbon Dating.* University of Chicago Press, Chicago.

————. 1956. Radioactive fallout and radioactive strontium. Science, 123:657–658.

Liebig, Justus. 1840. *Chemistry in Its Application to Agriculture and Physiology.* Taylor and Walton, London. (4th ed. 1847.)

Lieth, Helmut (ed.). 1962. Die Stoffproduktion der Pflanzendecke. Gustav Fischer Verlag, Stuttgart, Germany. 156 pp.

————. 1963. The role of vegetation in the carbon dioxide content of the atmosphere. J. Geophys. Res., 68(13):3887–3898.

————. 1964. Versuch einer Kartographischen Darstellung der Produktivität der Pflanzendecke auf der Erde. In: *Geographisches Taschenbuch* 1964/65. Franz Steiner Verlag, Weisbaden. pp. 72–80.

Ligon, J. D. 1968. Sexual differences in foraging behavior in two species of *Dendrocopus* woodpeckers. Auk, 85:203–215.

Likens, G. E.; Bormann, F. H.; and Johnson, N. M. 1969. Nitrification: Importance to nutrient losses from a cutover forested ecosystem. Science, 163:1205–1206.

Likens, G. E.; Bormann, F. H.; Johnson, N. M.; and Pierce, R. S. 1967. The calcium, magnesium, potassium and sodium budgets for a small forested ecosystem. Ecology, 48:772–785.

Lindeman, Raymond L. 1941. Seasonal food-cycle dynamics in a senescent lake. Amer. Midl. Nat., 26:636–673.

————. 1942. The trophic-dynamic aspect of ecology. Ecology, 23:399–418.

Lloyd, M., and Ghelardi, R. J. 1964. A table for calculating the equitability component of species diversity. J. Anim. Ecol., 33:421–425.

Lloyd, M.; Zar, J. H.; and Karr, J. R. 1968. On the calculation of information-theoretical measures of diversity. Amer. Midl. Nat., 79:257–272.

Lohman, S. W., and Robinove, C. J. 1964. Photographic description and appraisal of water resources, Photogrammetria, 19:21–41.

Loomis, R. S.; Williama, W. A.; and Duncan, W. G. 1967. Community architecture and the productivity of terrestrial plant communities. In: *Harvesting the Sun* (A. San Pietro, et al., eds.). Academic Press, N.Y. pp. 291–308.

Loosanoff, V. L. 1954. New advances in the study of bivalve larvae. Amer. Sci., 42:607–624.

Lord, R. D. 1961. Magnitudes of reproduction in cottontail rabbits. J. Wildl. Mgt., 25:28–33.

Lorenz, K. Z. 1935. Der Kumpan in der Umwelt des Vogels. J. Ornithol., 83:137–214. (See also Auk, 54:245–273, 1937.)

————. 1950. The comparative method of studying innate behaviour patterns. In: Soc. Exp. Biol. Symp. Physiological mechanisms in animal behavior. IV. pp. 221–268. Academic Press, New York.

————. 1952. *King Soloman's Ring.* Thomas Y. Crowell, New York. (See also 1961 reprint published by William Morrow & Company, New York.)

————. 1966. *On Aggression.* Methuen, London.

————. 1966a. *Evolution and Modification of Behaviour.* Methuen, London.

Lotka, A. J. 1925. *Elements of Physical Biology.* Williams & Wilkins, Baltimore. 460 pp. Reprinted by Dover Publ., New York, 1956.

Loutit, J. F. 1956. The experimental animal for the study of biological effects of radiation. Proc. Int. Conf. Peaceful Uses Atomic Energy, Geneva, 11:3–6.

————. 1962. *Irradiation of Mice and Men.* University of Chicago Press, Chicago.

Lovasich, J. L.; Neyman, Jerzy; Scott, E. L.; and Smith, J. A. 1969. Timing of the apparent effects of cloud seeding. Science, 165:892–893.

Lovelace, B.; Colwell, R. G.; and Tubiash, H. 1968. Quantitative and qualitative study of the commensal bacterial flora of *Crassostrea virginica* in Chesapeake Bay. Proc. Natl. Shellfish Assn., 58:82–87.

Lowe, C. H., and Heath, W. G. 1969. Behavioral and physiological responses to temperature in the desert pupfish *Cyprinodon macularius.* Physiol. Zool., 42:53–59.

Lowe, D. S., and Braithwaite, J. G. N. 1966. A spectrum matching technique for enhancing image contrast. Appl. Optics, 5:893–898.

Lowman, P. D., Jr. 1966. The earth and orbit. National Geographic Magazine, pp. 654–671.

Lowry, William P. 1967. The climate of cities. Scient. Amer., 217(2):15–23.

Lucas, C. E. 1947. The ecological effects of external metabolites. Biol. Rev. Cambr. Phil. Soc., 22:270–295.

Lueder, D. R. 1959. Aerial photographic interpretation: Principles and applications. McGraw-Hill Book Co., Inc., New York. 462 pp.

Lundegårdh, H. 1954. Klima und Boden in ihrer Wirkung auf das Pflanzenleben. Gustav Fischer Verlag, Jena.

Lynch, D. L., and Cotnoir, L. J., Jr. 1956. The influence of clay minerals on the breakdown of certain organic substances. Soil Sci. Proc., 20:367–370.

Maaloe, O. N., and Kjeldgaard, O. 1966. Control of macromolecular synthesis. Benjamin Co., Inc., New York.

MacArthur, Robert. 1957. On the relative abundance of bird species. Proc. Natl. Acad. Sci. (Wash.), 45:293–295.

————. 1960. On the relative abundance of species. Amer. Nat., 94:25–36.

————. 1965. Patterns of species diversity. Biol. Rev., 40:410–533.

————. 1968. The theory of the niche. In: *Population Biology and Evolution* (R. C. Lewontin, ed.). Syracuse University Press, Syracuse. pp. 159–176.

MacArthur, R., and Connell, Joseph. 1966. *The Biology of Populations.* John Wiley & Sons, Inc., N.Y. 200 pp.

MacArthur, R., and Levins, R. 1967. The limiting similarity, convergence, and divergence of coexisting species. Amer. Nat., 101:377–385.

MacArthur, R., and MacArthur, J. 1961. On bird species diversity. Ecology, 42:594–598.

MacArthur, R., and Wilson, E. O. 1967. *The Theory of Island Biogeography.* Princeton University Press, Princeton, N.J. 203 pp.

MacDonald, G. J. F. 1968. Science and politics in rainmaking. Bull. Atom. Scient. Oct., pp. 8–14.

MacFadyen, A. 1949. The meaning of productivity in biological systems. J. Anim. Ecol., 17:75–80.

————. 1961. Metabolism of soil invertebrates in relation to soil fertility. Ann. Appl. Biol., 49:215–218.

————. 1963. *Animal Ecology: Aims and Methods* (2nd ed.). Pitman, London. 344 pp.

———. 1964. Energy flow in ecosystems and its exploitation in grazing. In: *Symposium on Grazing* (D. J. Crisp, ed.). Blackwell, Oxford. pp. 3–20.

———. 1970. Soil metabolism in relation to ecosystem energy flow and to primary and secondary production. In: *Method of Study in Soil Ecology* (J. Phillipson, ed.). UNESCO, Paris. pp. 167–172.

Machenthum, K. M. 1969. The practice of water pollution biology. Fed. Water Poll. Cont. Adm., Washington, D.C. 281 pp. (Available from the U. S. Gov. Printing Office.)

Mackereth, F. J. H. 1965. Chemical investigations of lake sediments and their interpretation. Proc. Roy. Soc. London, Series B, 161:295–309.

MacLulich, D. A. 1937. Fluctuations in the numbers of the varying hare (*Lepus americanus*). Univ. Toronto Studies, Biol. Ser., No. 43.

Maguire, Bassett. 1967. A partial analysis of the niche. Amer. Nat., 101:515–523.

Major, Jack. 1951. A functional, factorial approach to plant ecology. Ecology, 32:392–412.

Malin, James C. 1953. Soil, animal and plant relations of the grassland, historically reconsidered. Sci. Mon., 76:207–220.

———. 1956. *The Grasslands of North America: Prolegomena to Its History*. 3rd printing, with addenda. James C. Malin, Lawrence, Kan.

Malone, C. R. 1969. The effects of diazinon contamination on an old-field ecosystem. Amer. Midl. Nat., 82:1–27.

Malthus, T. R. 1798. *An Essay on the Principle of Population*. Johnson, London. (Reprinted in Everyman's Library, 1914.)

Mandel, M. 1969. New approaches to bacterial taxonomy: perspectives and prospects. Ann. Rev. Microbiol., 33:239–274.

Mann, K. H. 1964. The pattern of energy flow in the fish and invertebrate fauna of the River Thames. Verh. int. Ver. Limnol., 15:485–495.

———. 1965. Energy transformations by a population of fish in the river Thames. J. Anim. Ecol., 34:253–275.

———. 1965. Heated effluents and their effect on the invertebrate fauna of rivers. Proc. Soc. Water Treatment Exam., 14:45–53.

———. 1969. The dynamics of aquatic ecosystems. In: *Advances in Ecological Research*, Vol. 6 (J. B. Cragg, ed.). Academic Press, New York. pp. 1–81.

Manning, W. M., and Juday, R. E. 1941. The chlorophyll content and productivity of some lakes in northeastern Wisconsin. Trans. Wisconsin Acad. Sci. Arts. Lett., 33:363–393.

Marchinton, R. L. 1969. Portable radios in determination of ecological parameters of large vertebrates with reference to deer. In: *Remote Sensing in Ecology* (P. L. Johnson, ed.). University of Georgia Press, Athens. pp. 148–163.

Margalef, Ramon. 1951. Diversidad de Especies en las Comunidades Naturales. Proc. Inst. Biol. Apl., 9:5.

———. 1958. Temporal succession and spatial heterogeneity in phytoplankton. In: *Perspectives in Marine Biology* (Buzzati-Traverso, ed.). University of California Press, Berkeley. pp. 323–347.

———. 1958a. Information theory in ecology. Gen. Syst., 3:36–71.

———. 1961. Correlations entre certains caractères synthétiques des populations de phytoplancton. Hydrobiologia, 18:155–164.

———. 1961a. Communication of structure in plankton populations. Limnol. Oceanogr., 6(2):124–128.

———. 1963. Successions of populations. Adv. Frontiers of Plant Sci. (Instit. Adv. Sci. & Culture, New Delhi, India), 2:137–188.

———. 1963a. On certain unifying principles in ecology. Amer. Nat., 97:357–374.

———. 1964. Correspondence between the classic types of lakes and the structural and dynamic properties of their populations. Verh. int. Ver. Limnol., 15:169–170.

———. 1967. Concepts relative to the organization of plankton. Oceanogr. Mar. Biol. Ann. Rev., 5:257–289.

———. 1967a. The food web in the pelagic environment. Helgolander Wiss. Meeresunters, 15:548–559.

———. 1968. *Perspectives in Ecological Theory*. University of Chicago Press, Chicago. 112 pp.

Margalef, R., and Ryther, J. H. 1960. Pigment composition and productivity as related to succession in experimental populations of phytoplankton. Biol. Bull., 119:326–327.

Marine, Gene. 1969. *America the Raped*. The engineering mentality and the devastation of a continent. Simon and Schuster, New York.

Marler, P. R., and Hamilton, Jr., W. J. 1966. *Mechanisms of Animal Behavior*. John Wiley & Sons, Inc., New York. 771 pp.

Marmur, J.; Falkow, S.; and Mandel, M. 1963. New approaches to bacterial taxonomy. Ann. Rev. Microbiol., 17:329–372.

Marsh, George Perkins. 1864. *Man and Nature; or Physical Geography as Modified by Human Action*. Reprinted by Harvard University Press, Cambridge, 1965 (D. Lowenthal, ed.). For an evaluation of Marsh's classic, see Russell, Franklin, 1968. Horizon, 10:17–23.

Marsh, J. A. 1970. Primary productivity of reef-building calcareous red algae. Ecology, 51:255–263.

Marshall, S. M., and Orr, A. P. 1955. *The Biology of a Marine Copepod* Calanus finmarchicus (Gunnerus). Oliver and Boyd, London.

Mason, H. L., and Langenheim, J. H. 1957. Language analysis and the concept environment. Ecology, 38:325–340.

Mason, W. H., and Odum, E. P. 1969. The effect of coprophagy on retention and bioelimination of radionuclides of detritus-feeding animals. Proc. 2nd Nat. Sym. on Radioecology (D. J. Nelson and F. C. Evans, eds.). Clearinghouse Fed. Sci. Tech. Inf., U. S. Dept. Commerce, Springfield, Va. pp. 721–724.

Martin, Michael M. 1970. The biochemical basis of the fungus-attine ant symbiosis. Science, 169:16–20.

Maruyama, M. 1963. The second cybernetics. Deviation-amplifying mutual causal processes. Amer. Sci., 51:164–179.

Mayberry, W. R.; Prochazka, G. J.; and Payne, W. J. 1967. Growth yields of bacteria on selected organic compounds. Appl. Microbiol., 15:1332–1338.

McConnell, William J. 1962. Productivity relations in carboy microcosms. Limnol. Oceanogr., 7:335–343.

McCormick, F. J. 1969. Effects of ionizing radiation on a pine forest. In: 2nd Nat. Sym. Radioecology (D. Nelson and F. Evans, eds.). Clearinghouse Fed. Sci. Tech. Info., U. S. Dept. Commerce, Springfield, Va. pp. 78–87.

McCormick, F. J., and Platt, R. B. 1962. Effects of ionizing radiation on a natural plant community. Rad. Bot., 2:161–204.

McCormick, F. J., and Golley, F. B. 1966. Irradiation of natural vegetation—an experimental facility, procedures and dosimetry. Health Physics, 12:1467–1474.

McCormick, Robert A., and Ludwig, J. H. 1967. Climate modification by atmospheric aerosols. Science, 156:1358–1359.

McCullough, D. R.; Olson, C. E.; and Queal, L. M. 1969. Progress in large animal census by thermal mapping. In: *Remote Sensing in Ecology* (P. L. Johnson, ed.). University of Georgia Press, Athens. pp. 138–147.

McDiffett, Wayne F. 1970. The transformation of energy by a stream detritivore, *Pteronarcys scotti* (Plecoptera). Ecology, 51:975–988.

McDonald, D. R. 1965. Biological interactions associated with spruce budworm infestations. In: *Ecological Effects of Nuclear War* (G. Woodwell, ed.). Brookhaven Nat. Lab. Publ. No. 917, pp. 61–68.

McDowell, M. E., and Leveille, G. A. 1964. Algae systems. Conference on Space Nutrition and Related Waste Problems. NASA SP-70., pp. 317–322.

McGinnis, J. T.; Golley, F. B.; Clements, R. G.; Child, G. I.; and Duever, M. J. 1969. Element and hydrologic budgets of the Panamanian tropical moist forest. BioScience, 19:697–700.

McGinnis, S. M., and Dickson, L. L. 1967. Thermoregulation in the desert iguana *Dipsosaurus dorsalis*. Science, 156:1757–1759.

McHarg, Ian L. 1969. *Design with Nature*. Natural History Press, Garden City, N.Y. 197 pp.

McIntire, C. D., and Phinney, H. K. 1965. Laboratory studies of periphyton production and community in lotic environments. Ecol. Monogr., 35:237–258.

McIntosh, Robert P. 1967. The continuum concept of vegetation. Bot. Rev., 33:130–187.

———. 1967a. An index of diversity and the relation of certain concepts of diversity. Ecology, 48:392–404.

McLaughlin, J. J. A., and Zahl, P. A. 1966. Endozoic algae. In: *Symbiosis* (S. M. Henry, ed.). Academic Press, New York. pp. 257–297.

McMillan, Calvin. 1956. Nature of the plant community, 1. Uniform garden and light period studies of five grass taxa in Nebraska. Ecology, 37:330–340.

———. 1960. Ecotypes and community function. Amer. Nat., 94:245–255.

———. 1969. Ecotypes and ecosystem function. BioScience, 19:131–134.

McNaughton, S. J. 1966. Ecotype function in the *Typha* community-type. Ecol. Monogr., 36:297–325.

———. 1968. Structure and function in California grasslands. Ecology, 49:962–972.

McPherson, J. K., and Muller, Cornelius H. 1969. Allelopathic effects of *Adonostoma fasciculatum* "chamise" in California chaparral. Ecol. Monogr., 39:177–198.

Meggers, Betty J. 1954. Environmental limitation to the development of culture. Amer. Anthrop., 56:801–824.

Mellanby, Helen. 1938. *Animal Life in Fresh Water*. Methuen, London.

Menhinick, Edward F. 1962. Comparison of invertebrate populations of soil and litter of mowed grasslands in areas treated and untreated with pesticides. Ecology, 43:556–561.

———. 1963. Estimation of insect population density

in herbaceous vegetation with emphasis on removal sweeping. Ecology, 44:617–622.

———. 1964. A comparison of some species diversity indices applied to samples of field insects. Ecology, 45:859–861.

———. 1967. Structure, stability and energy flow in plants and arthropods in a sericea *Lespedeza* stand. Ecol. Monogr., 37:255–272.

Menshutkin, V. V. 1962. The realization of elementary models for fish populations on electronic computers. (In Russian.) Vopr. Ikhtiol., 4:625–631.

Menzel, D. W.; Hulbert, E. M.; and Ryther, J. H. 1963. The effects of enriching Sargasso Sea water on the production and species composition of the phytoplankton. Deep-Sea Res., 10:209–219.

Menzel, D. W., and Ryther, J. H. 1961. Nutrients limiting the production of phytoplankton in the Sargasso Sea, with special reference to iron. Deep-Sea Res., 7:276–281.

Menzel, D. W., and Spaeth, J. P. 1962. Occurrence of vitamin B_{12} in the Sargasso Sea. Limnol. Oceanogr., 7:151–154.

Merrell, D. J. 1962. *Evolution and Genetics*. Holt, Rinehart and Winston, New York.

Merriam, C. Hart. 1894. Laws of temperature control of the geographic distribution of terrestrial animals and plants. National Geographic Magazine, 6:229–238. (See also Life zones and crop zones. U. S. D. A. Bull. No. 10, 1899.)

Meslow, E. C., and Keith, L. B. 1968. Demographic parameters of a snowshoe hare population. J. Wildl. Mgt., 32:812–834.

Meyer, M. P., and French, D. W. 1967. Detection of diseased trees. Photogr. Engr., 33:1035–1040.

Michael, Donald N. 1968. *The Unprepared Society. Planning For a Precarious Future*. Basic Books, Inc., New York. 138 pp.

Miettinen, J. K. 1969. Enrichment of radioactivity by arctic ecosystems in Finnish Lapland. In: Proc. 2nd National Sym. on Radioecology (Nelson and Evans, eds.). Clearinghouse Fed. Sci., Tech. Info., Springfield, Va.

Milhorn, H. T. 1966. *The Application of Control Theory to Physiological Systems*. W. B. Saunders Company, Philadelphia. 386 pp.

Miller, Alden H. 1955. Breeding cycles in a constant equatorial environment in Columbia, South America. Acta XI Cong. Internat. Orn., Basal. pp. 495–503.

Miller, George P. (ed.). 1967. Applied science and technological progress. Nat. Acad. Sci. Rept. (Available from Supt. Doc., Washington, D.C.)

Miller, M. W., and Berg, G. G. (eds.). 1969. *Chemical Fallout*. Charles C Thomas, Springfield, Ill.

Miller, R. G. 1960. The interpretation of tropical vegetation and crops on aerial photographs. Photogrammetria, 16:230–240.

Miller, R. S. 1967. *Pattern and Process in Competition*. Adv. Ecol. Res., 4:1–74. Academic Press, New York.

Miller, S. L. 1953. A production of amino acids under possible primitive earth conditions. Science, 117:528–529.

———. 1957. Mechanism of synthesis of amino acids by electric discharge. Biochem. Biophys. Acta, 23:480.

Milne, A. 1957. Theories of natural control of insect populations. Cold Spring Habor Sym. Quant. Biol., 22:253–271.

———. 1962. On a theory of natural control of insect population. J. Theor. Biol., 3:19–50.

Milsum, J. H. 1966. Biological control systems analysis. McGraw-Hill Book Co., Inc., New York. 466 pp.

Milthrope, F. L. 1956. *The Growth of Leaves*. Butterworth Publications, London.

Milton, John (ed.). Ecological aspects of international development (Proc. Conf. at Airlie House, Warrenton, Va. 1968). Unpublished manuscripts on Mekong river development, the Kariba lake basin, the Aswan dam, and others.

Minshall, G. W. 1967. Role of allochthonous detritus in the trophic structure of a woodland spring-brook community. Ecology, 48:139–149.

Mishan, E. J. 1967. The cost of economic growth. Staples Press, London.

Mishima, Jiro, and Odum, E. P. 1963. Excretion rate of Zn^{65} by *Littorina irrorata* in relation to temperature and body size. Limnol. Oceanogr., 8:39–44.

Misra, R.; Singh, J. S.; and Singh, K. P. 1968. A new hypothesis to account for the opposite trophic-biomass structure on land and in water. Current Sci. (India), 37:382–383.

Möbius, Karl. 1877. *Die Auster und die Austernwirtschaft*. Berlin. Translated into English and published in Rept. U. S. Fish Comm., 1880, pp. 683–751.

Mock, C. R. 1966. Natural and altered estuarine habitats of penaeid shrimp. Proc. Gulf and Caribbean Fisheries Inst., 19:86–98.

Mohr, Carl O. 1940. Comparative populations of game, fur and other mammals. Amer. Midl. Nat., 24:581–584.

Möller, F. 1963. On the influence of changes in the CO_2 concentration in air on the radiation balance of the earth's surface and on the climate. J. Geophys. Res., 68:3877–3886.

Monk, Carl D. 1966. Ecological importance of root/shoot ratios. Bull. Torrey Bot. Club, 93:402–406.

———. 1966a. Root-shoot dry weights in loblolly pine. Bot. Gaz., 127:246–248.

———. 1966b. Effects of short-term gamma irradiation on an old field. Rad. Bot., 6:329–335.

———. 1966c. An ecological significance of evergreenness. Ecology, 47:504–505.

———. 1967. Tree species diversity in the eastern deciduous forest with particular reference to north central Florida. Amer. Nat., 101:173–187.

———. 1968. Successional and environmental relationships of the forest vegetation of north central Florida. Amer. Midl. Nat., 79:441–457.

Monod, J. 1942. Recherches sur la croissance des cultures bacteriennes. Hermann, Paris. 210 pp.

———. 1950. La technique de culture continuée; theorie et applications. Amer. Inst. Pasteur, 79:390–410.

Monsi, M., and Oshima, Y. 1955. A theoretical analysis of the succession of plant community based on the production of matter. Japanese J. Bot., 15:60–82.

Montagu, M. F. Ashley (ed.). 1962. *Culture and the Evolution of Man*. Oxford University Press, New York. 376 pp.

Monteith, J. L. 1962. Measurement and interpretation of carbon dioxide fluxes in the field. Netherlands J. Agri. Sci., 10 (special issue):334–346.

———. 1963. Gas exchange in plant communities. In: *Environmental Control of Plant Growth* (L. T. Evans, ed.). Academic Press, New York.

———. 1965. Light distribution and photosynthesis in field crops. Ann. Bot. N. S., 29:17–37.

Monteith, J. L., and Szeicz, G. 1960. The carbon dioxide flux over a field of sugar beets. Quart. J. Roy. Meteoro. Soc., 86:205–214.

Moore, H. B. 1958. *Marine Ecology*. John Wiley & Sons, Inc., New York. 493 pp.

Moore, N. W. (ed.). 1966. Pesticides in the environment and their effect on wildlife (A supplement to the Journal of Applied Ecology, Vol. 3). Blackwell, Oxford. 312 pp.

Moore, R. K., and Simonett, D. S. 1967. Radar remote sensing in biology. BioScience, 17:384–390.

Moran, P. A. P. 1949. The statistical analysis of the sunspot and lynx cycles. J. Anim. Ecol., 18:115–116.

Moreau, R. E. 1952. Africa since the mesozoic. Proc. Zool. Soc. London, 121:869–913.

Mori, Syuiti. 1957. Daily rhythmic activity of the sea-pen, *Cavernularia obesa valenciennes*. XV. Controlling the activity by light. Seto. Mar. Biol. Lab., 6:89–98.

Morita, R. Y., and Zobell, C. E. 1959. Deep-sea bacteria. Galathea Report, 1:139–154. Danish Science Press, Ltd., Copenhagen.

Morowitz, H. J. 1968. *Energy Flow in Biology. Biological Organization as a Problem in Thermal Physics*. Academic Press, N.Y. 179 pp. (See review by H. T. Odum, Science, 164:683–684, 1969.)

Morris, Clarence. 1964. The rights and duties of beasts and trees: a law teacher's essay for landscape architects. J. Legal Ed., 17:185–192.

Morris, D. 1967. *The Naked Ape*. Dell Publishing Company, New York.

Morris, H. D. 1968. Effect of burning on forage production of coastal Bermuda grass at varying levels of fertilization. Agron. J., 60:518–521.

Morrison, Frank B. 1947. *Feeds and Feeding* (20th ed.). Morrison, Ithaca, New York.

Motomura, I. 1932. A statistical treatment of associations. (In Japanese.) Japanese J. Zool., 44:379–383.

Moustafa, H. H., and Collins, E. B. 1968. Molar growth yields of certain lactic acid bacteria as influenced by autolysis. J. Bacteriol., 96:117–125.

Mrak, Emil M. (ed.). 1969. Report of Secretary's Commission on Pesticides and Their Relationship to Environmental Health. Pts. I and II. U. S. Dept. Health Education and Welfare, U. S. Gov. Printing Office, Washington, D.C.

Mudd, S. (ed.). 1964. The population crisis and the use of world resources. Indiana University Press, Bloomington, Ind.

Mukerjee, R. 1945. *Social Ecology*. Longmans, Green, New York.

Muller, C. H. 1966. The role of chemical inhibition (allelopathy) in vegetational composition. Bull. Torrey Bot. Club, 93:332–351.

———. 1969. Allelopathy as a factor in ecological process. Vegetatio, 18:348–357.

Muller, C. H.; Hanawalt, R. B.; and McPherson, J. K. 1968. Allelopathic control of herb growth in the fire cycle of California Chaparral. Bull. Torrey Bot. Club, 95:225–231.

Muller, C. H.; Muller, W. H.; and Haines, B. L. 1964. Volatile growth inhibitors produced by aromatic shrubs. Science, 143:471–473.

Mumford, Lewis. 1967. Quality in the control of quantity. In: *Natural Resources, Quality and Quantity* (Ciriacy-Wantrup and Parsons, eds.). University of California Press, Berkeley. pp. 7–18.

Murchison, C. (ed.). 1935. *A Handbook of Social Psychology*. Clark University Press, Worcester, Mass.

Murie, Adolph. 1944. Dall Sheep, Chapter 3 in: Wolves

of Mount McKinley. Natl. Parks Service Fauna No. 5, Washington.

Murphy, Earl F. 1967. *Governing Nature*. Quadrangle Books, Chicago.

Murphy, G. I. 1964. Notes on the harvest of anchovies. Calif. Coop. Oceanic Fish Invest. Mar. Res. Comm. Mtg., San Pedro Doc. XLIL:3.

———. 1967. Vital statistics of the Pacific sardine (*Sardinops caerulea*) and the population consequences. Ecology, 48:731–736.

Murphy, G. I., and Isaacs, J. D. 1964. Species replacement in marine ecosystems with reference to the California Current. Calif. Coop. Oceanic. Fish Invest. Mar. Res. Comm. Mtg., San Pedro Doc. VII:1–6.

Murphy, Robert C. 1936. *Oceanic Birds of South America*. 2 Vols. Macmillan Company, New York.

Muscatine, L. 1961. Symbiosis in marine and freshwater coelenterates. In: *The Biology of Hydra and Some Other Coelenterates* (H. M. Lenhoff and W. F. Loomis, eds.). University of Miami Press, Miami. pp. 255–268.

———. 1967. Glycerol excretion by symbiotic algae from corals and *Tridacna* and its control by the host. Science, 156:516–518.

Muscatine, L., and Hand, Cadet. 1958. Direct evidence for the transfer of materials from symbiotic algae to the tissues of a coelenterate. Proc. Natl. Acad. Sci. (Wash.), 44:1259–1263.

Musgrave, R. B., and Moss, D. N. 1961. Photosynthesis under field conditions: I. A portable, closed system for determining net assimilation and respiration of corn. Crop Sci., 1:37–41.

Myers, Jack. 1949. The pattern of photosynthesis in *Chlorella*. In: *Photosynthesis in Plants* (J. Franck and W. E. Loomis, eds.). Iowa State College Press, Ames. pp. 349–364.

———. 1954. Basic remarks on the use of plants as biological gas exchangers in a closed system. J. Aviation Medicine, 25:407–411.

———. 1960. The use of photosynthesis in a closed ecological system. In: *The Physics and Medicine of the Atmosphere and Space*. John Wiley & Sons, Inc., New York. pp. 387–396.

———. 1963. Space biology; ecological aspects. Introductory remarks. Amer. Biol. Teacher, 25:409–411.

National Academy of Science. 1966. Scientific Aspects of Pest Control. Natl. Acad. Sci. Publ. 1402, Washington, D.C. 470 pp.

———. 1966a. Waste Management and Control, Publ. 1400. (See Spilhaus, ed., 1966.)

———. 1969. Report of Committee on Resources and Man. (See Cloud, ed., 1969a.)

———. 1969a. Eutrophication: Causes, Consequences and Correctives. Internat. Symp. Eutrophication, Washington, D.C. 661 pp.

Naylor, E. 1965. Effects of heated effluents upon marine and estuarine organisms. Adv. Mar. Biol., 3:63–103.

Needham, James G., and Needham, Paul R. 1941. *A Guide to the Study of Fresh-water Biology*. Comstock, Ithaca, New York.

Neel, J. K. 1948. A limnological investigation of the psammon in Douglas Lake, Michigan, with special reference to shoal and shoreline dynamics. Trans. Amer. Micro. Soc., 67:1–53.

Neess, John C. 1946. Development and status of pond fertilization in central Europe. Trans. Amer. Fish. Soc., 76:335–358.

Nelson, D. J. 1967. Microchemical constituents in contemporary and pre-Columbian clam shells. In: *Quaternary Paleoecology*. Proc. VII Cong. Int. Assoc. Quaternary Research, Vol. 7, pp. 185–204.

Nelson, D. J. 1967a. Ecological behavior of radionuclides in the Clinch and Tennessee Rivers. In: *Reservoir Fishery Resource Symposium* (C. E. Lane, ed.). Amer. Fishery Soc., Washington, D.C. pp. 169–187.

Nelson, D. J., and Evans, F. C. (eds.). 1969. *Symposium on Radioecology*. Proc. 2nd Natl. Sym., Clearinghouse Fed. Sci. Tech. Info., U. S. Dept. Commerce, Springfield, Va. 774 pp.

Nelson, D. J., and Scott, D. C. 1962. Role of detritus in the productivity of a rock-outcrop community in a Piedmont stream. Limnol. Oceanogr., 7:396–413.

Newbigin, M. I. 1936. *Plant and Animal Geography*. Methuen, London. 298 pp.

Newbould, P. J. 1963. Production ecology. Sci. Progr., 51:93–104.

Newell, Richard. 1965. The role of detritus in the nutrition of two marine deposit feeders, the prosobranch *Hydrobia ulvae* and the bivalve *Macoma balthica*. Proc. Zool. Soc. London, 144:25–45.

Newsom, L. D. 1967. Consequences of insecticide use on nontarget organisms. Ann. Rev. Entomol., 12:257–286.

Ney, L. F. 1960. Gas exchange by the duckweed family. ONR Contract No. Nonr-2887 (00), Stanford Res. Inst.

Nice, Margaret M. 1941. The role of territory in bird life. Amer. Midl. Nat., 26:441–487.

Nicholson, A. J. 1933. The balance of animal populations. J. Anim. Ecol., 2:132–178.

———. 1954. An outline of the dynamics of animal populations. Australian J. Zool., 2:9–65.

———. 1957. The self-adjustment of populations to change. Cold Spring Harbor Sym. Quant. Biol., 22:153–173.

———. 1958. Dynamics of insect populations. Ann. Rev. Entomol., 3:107–136.

Nicholson, H. Page; Grzenda, A. R.; Lauer, G. J.; Cox, W. S.; and Teasley, J. I. 1964. Water pollution by insecticides in an agricultural river basin. 1. Occurrence of insecticides in river and treated municipal water. Limnol. Oceanogr., 9:310–316.

Nielsen. (See Overgaard-Nielsen and Steeman-Nielsen.)

Niering, W. A.; Whittaker, R. H.; and Lowe, C. H. 1963. The Saguaro: a population in relation to environment. Science, 142:15–23.

Nikolsky, G. V. 1963. *The Ecology of Fishes*. (Translated from Russian by L. Birkett.) Academic Press, New York.

Nishita, H., and Larson, K. H. 1957. Summary of certain trends in soil-plant relationship studies of the biological availability of fall-out debris. Univ. Calif. Los Angeles, Atomic Energy Project Report No. 401:1–67.

Nixon, S. W. 1969. A synthetic microcosm. Limnol. Oceanogr., 14:142–145.

Norman, A. G. 1957. Soil-plant relationships and plant nutrition. Amer. J. Bot., 44:67–73.

Norman, G. G., and Fritz, N. L. 1965. Infrared photography as an indicator of disease and decline in citrus trees. Proc. Fla. State Horticult. Soc., 78:59–63.

Novic, A., and Szilard, L. 1950. Description of the chemostat. Science, 112:715–716.

Nutman, P. S. 1956. The influence of the legume in root-nodule symbiosis. Biol. Rev., 31:109–151.

Nutman, P. S., and Mosse, B. (eds.). 1963. *Symbiotic*

Associations. 13th Sym. for Gen. Microbiol. Cambridge University Press, New York. 356 pp.

Oberlander, G. T. 1956. Summer fog precipitation on the San Francisco Peninsula. Ecology, 37:851–852.

Odum, Eugene P. 1945. The concept of the biome as applied to the distribution of North American birds. Wilson Bull., 57:191–201.

———. 1950. Bird populations of the Highlands (North Carolina) Plateau in relation to plant succession and avian invasion. Ecology, 31:587–605.

———. 1956. Consideration of the total environment in power reactor waste disposal. Proc. Int. Conf. Peaceful Uses Atomic Energy, Geneva, 13:350–353.

———. 1957. The ecosystem approach in the teaching of ecology illustrated with sample class data. Ecology, 38:531–535.

———. 1960. Organic production and turnover in old-field succession. Ecology, 41:34–49.

———. 1960a. Factors which regulate primary productivity and heterotrophic utilization in the ecosystem. Trans. Seminar Algae and Metropolitan Wastes. Taft Engineering Center, U. S. Public Health Service, Cincinnati, Ohio. pp. 65–71.

———. 1961. The role of tidal marshes in estuarine production. The Conservationist (New York State Conservation Dept., Albany), 15(6):12–15.

———. 1961a. Excretion rate of radio-isotopes as indices of metabolic rates: biological half-life of zinc-65 in relation to temperature, food consumption, growth and reproduction in arthropods. Biol. Bull., 121:371–372 (abstract).

———. 1962. Relationships between structure and function in the ecosystem. Japanese J. Ecol., 12:108–118.

———. 1963. *Ecology.* Modern Biology Series. Holt, Rinehart and Winston, New York. 152 pp.

———. 1963a. Primary and secondary energy flow in relation to ecosystem structure. Proc. XVI Int. Cong. Zool., Washington, D.C. pp. 336–338.

———. 1964. The new ecology. BioScience, 14:14–16.

———. 1965. Feedback between radiation ecology and general ecology. Health Physics, 11:1257–1262.

———. 1965a. Ecological effects of nuclear war (summary) (G. M. Woodwell, ed.). Brookhaven Nat. Lab., Upton, L.I., New York. pp. 69–72.

———. 1966. Regenerative systems: Discussion (A. H. Brown, discussion leader). In: *Human Ecology in Space Flight* (D. H. Calloway, ed.). The N.Y. Acad. Sci., Interdisciplinary Communications Program, New York. pp. 82–119.

———. 1967. Man and the landscape; a review of "Waste Management and Control" (A. Spilhaus, ed.). Natl. Acad. Sci. Publ. 1400. Scientist and Citizen (now Environment), 9:91–114.

———. 1968. Energy flow in ecosystems: A historical review. Amer. Zool., 8:11–18.

———. 1968a. A research challenge: Evaluating the productivity of coastal and estuarine water. Proc. 2nd Sea Grant Conf., Grad. School Oceanography, University of Rhode Island, Newport. pp. 63–64.

———. 1969. The strategy of ecosystem development. Science, 164:262–270.

———. 1969a. Air-land-water = an ecological whole. J. Soil and Water Cons., 24:4–7.

———. 1969b. The attitude lag. BioScience, 19:403.

———. 1970. Optimum population and environment: A Georgian microcosm. Current History, 58:355–359; 365.

———. 1970a. The attitude revolution. In: *Crisis of Survival.* Scott, Foresman & Co., Glenview, Ill. pp. 9–15.

———. 1971. Ecosystem theory in relation to man. In: *Ecosystems: Structure and Function.* 31st Biol. Coll., Oregon State University Press, Corvallis, Ore.

Odum, Eugene P., and Burleigh, Thomas D. 1946. Southward invasion in Georgia. Auk, 63:388–401.

Odum, Eugene P.; Connell, Clyde E.; and Davenport, L. B. 1962. Population energy flow of three primary consumer components of old-field ecosystems. Ecology, 43:88–96.

Odum, Eugene P.; Connell, Clyde E.; and Stoddard, H. L. 1961. Flight energy and estimated flight ranges of some migratory birds. Auk, 78:515–527.

Odum, Eugene P., and de la Cruz, Armando A. 1963. Detritus as a major component of ecosystems. AIBS Bulletin (now BioScience), 13:39–40.

———. 1967. Particulate organic detritus in a Georgia salt marsh-estuarine ecosystem. In: *Estuaries* (G. Lauff, ed.). Amer. Assoc. Adv. Sci. Publ., 83:383–388.

Odum, Eugene P., and Golley, Frank B. 1963. Radioactive tracers as an aid to the measurement of energy flow at the population level in nature. In: *Radioecology* (V. Schultz and A. W. Klement, eds.). Reinhold Publishing Company, New York. pp. 403–410.

Odum, Eugene P., and Johnston, David W. 1951. The house wren breeding in Georgia: an analysis of a range extension. Auk, 68:357–366.

Odum, Eugene P., and Kuenzler, Edward J. 1963. Experimental isolation of food chains in an old-field ecosystem with use of phosphorus-32. In: *Radioecology* (V. Schultz and A. W. Klement, eds.). Reinhold Publishing Company, New York. pp. 113–120.

Odum, Eugene P.; Kuenzler, E. J.; and Blunt, Marion X. 1958. Uptake of P^{32} and primary productivity in marine benthic algae. Limnol. Oceanogr., 3:340–345.

Odum, Eugene P.; Marshall, S. G.; and Marples, T. G. 1965. The caloric content of migrating birds. Ecology, 46:901–904.

Odum, Eugene P., and Odum, H. T. 1957. Zonation of corals on Japtan Reef, Eniwetok Atoll. Atoll Res. Bull., 52:1–3.

Odum, Eugene P., and Pontin, A. J. 1961. Population density of the underground ant, *Lasius flavus*, as determined by tagging with ^{32}P. Ecology, 42:186–188.

Odum, Eugene P.; Rogers, D. T.; and Hicks, D. L. 1964. Homeostasis of nonfat components of migratory birds. Science, 143:1037–1039.

Odum, Eugene P., and Smalley, A. E. 1959. Comparison of population energy flow of a herbivorous and a deposit-feeding invertebrate in a salt marsh ecosystem. Proc. Natl. Acad. Sci., 45:617–622.

Odum, Howard T. 1956. Primary production in flowing waters. Limnol. Oceanogr., 1:102–117.

———. 1956a. Efficiencies, size of organisms, and community structure. Ecology, 37:592–597.

———. 1957. Trophic structure and productivity of Silver Springs, Florida. Ecol. Monogr., 27:55–112.

———. 1957a. Primary production measurement in eleven Florida springs and a marine turtle-grass community. Limnol. Oceanogr., 2:85–97.

———. 1960. Analysis of diurnal oxygen curves for the assay of reaeration rates and metabolism in polluted marine bays. In: *Waste Disposal in the Marine Environment* (E. A. Pearson, ed.). Pergamon Press, New York. pp. 547–555.

————. 1960a. Ecological potential and analogue circuits for the ecosystem. Amer. Sci., 48:1–8.

————. 1960b. Ten classroom sessions in ecology. Amer. Biol. Teacher, 22:71–78.

————. 1962. Ecological tools and their use – Man and the ecosystem. In: *Proceedings of the Lockwood Conference on the Suburban Forest and Ecology* (Paul E. Waggoner and J. D. Ovington, eds.). The Connecticut Agricultural Experiment Station Bulletin 652. pp. 57–75.

————. 1963. Limits of remote ecosystems containing man. Amer. Biol. Teacher, 25:429–443.

————. 1967. Biological circuits and the marine systems of Texas. In: *Pollution and Marine Ecology* (T. A. Olson and F. J. Burgess, eds.). John Wiley & Sons, Inc. (Interscience), New York. pp. 99–157.

————. 1967a. Energetics of world food production. In: *The World Food Problem, A report of the President's Science Advisory Committee*. Panel on World Food Supply (I. L. Bennett, Chairman). The White House, Washington, D.C. Vol. 3, pp. 55–94.

————. 1968. Work circuits and systems stress. In: *Symposium on Primary Productivity and Mineral Cycling in Natural Ecosystems* (H. E. Young, ed.). University of Maine Press, Orono. pp. 81–138.

————. 1971. *Environment, Power and Society*. John Wiley & Sons, Inc., New York. 331 pp.

Odum, Howard T.; Beyers, Robert J.; and Armstrong, Neal E. 1963. Consequences of small storage capacity in nannoplankton pertinent to measurement of primary production in tropical waters. J. Mar. Res., 21:191–198.

Odum, Howard T.; Cantlon, J. E.; and Kornicker, L. S. 1960. An organizational hierarchy postulate for the interpretation of species-individuals distribution, species entropy and ecosystem evolution and the meaning of a species-variety index. Ecology, 41:395–399.

Odum, Howard T.; Copeland, B. J.; and McMahan, E. A. 1969. Coastal ecological systems of the United States. A report to the Federal Water Pollution Control Administration (mimeographed).

Odum, Howard T., and Hoskin, Charles M. 1957. Metabolism of a laboratory stream microcosm. Publ. Inst. Mar. Sci., Univ. Texas, 4:115–133.

Odum, Howard T., and Hoskin, C. M. 1958. Comparative studies on the metabolism of marine waters. Publ. Inst. Mar. Sci., Univ. Texas, 5:16–46.

Odum, Howard T.; McConnell, W. M.; and Abbott, W. 1958. The chlorophyll "A" of communities. Publ. Inst. Mar. Sci., Univ. Texas, 5:65–97.

Odum, Howard T., and Pigeon, R. F. (eds.). 1970. A tropical rain forest. A study of irradiation and ecology at El Verde, Puerto Rico. Nat. Tech. Info. Service, Springfield, Va. 1678 pp.

Odum, Howard T., and Odum, E. P. 1955. Trophic structure and productivity of a windward coral reef community on Eniwetok Atoll. Ecol. Monogr., 25:291–320.

Odum, Howard T., and Pinkerton, R. C. 1955. Times speed regulator, the optimum efficiency for maximum output in physical and biological systems. Amer. Sci., 43:331–343.

Odum, Howard T.; Siler, Walter L.; Beyers, R. J.; and Armstrong, Neal. 1963. Experiments in engineering marine ecosystems. Publ. Inst. Mar. Sci., Univ. Texas, 9:373–404.

Odum, Howard W. 1936. *Southern Regions of the Uni-*

ted States. University of North Carolina Press, Chapel Hill. 664 pp.

————. 1951. The promise of regionalism. Chapter 15 in: *Regionalism in America* (Merrill Jensen, ed.). University of Wisconsin Press, Madison.

————. 1953. Folk sociology as a subject for the historical study of total human society and the empirical study of group behavior. Social Forces, 31:193–223.

Odum, Howard W., and Moore, Harry E. 1938. *American Regionalism*. Henry Holt & Co., Inc., New York.

Odum, William E. 1968. The ecological significance of fine particle selection by the striped mullet *Mugil cephalus*. Limnol. Oceanogr., 13:92–98.

————. 1968a. Mullet grazing on a dinoflagellate bloom. Chesapeake Science, 9:202–204.

————. 1968b. Pesticide pollution in estuaries. Sea Frontiers (International Oceanographic Foundation, Miami, Fla.), 14:234–245.

————. 1970. Pathways of energy flow in a south Florida estuary. Dissertation. University of Miami, 180 pp.

————. 1970a. Utilization of the direct grazing and plant detritus food chains by the striped mullet *Mugil cephalus*. In: *Marine Food Chains* (J. H. Steele, ed.). Oliver and Boyd, Edinburgh. pp. 222–240.

————. 1970b. Insidious alteration of the estuarine environment. Trans. Amer. Fish. Soc., 99:836–847.

Odum, William E.; Woodwell, C. M.; and Wurster, C. F. 1969. DDT residues absorbed from organic detritus by fiddler crabs. Science, 164:576–577.

Ogburn, W. F. 1922. Social change with respect to culture and original nature. Huebsch, New York.

Oglesby, R. T. 1967. Biological and physiological basis of indicator organisms and communities. In: *Pollution and Marine Biology* (T. A. Olson, and F. J. Burgess, eds.). John Wiley & Sons, Inc. (Interscience), New York. pp. 267–269.

Olson, D. P. 1964. The use of aerial photographs in studies of marsh vegetation. Maine Agr. Expt. Sta. Bull., 13, 62 pp.

Olson, J. S. 1958. Rates of succession and soil changes on southern Lake Michigan sand dunes. Bot. Gaz., 119:125–170.

————. 1963. Energy storage and the balance of producers and decomposers in ecological systems. Ecology, 44:322–332.

————. 1964. Gross and net production of terrestrial vegetation. J. Ecol., 62:99–118.

O'Neill, R. V. 1967. Niche segregation in seven species of diplopods. Ecology, 48:983.

————. 1968. Population energetics of the millipede, *Narceus americanus*. Ecology, 49:803–809.

Oosting, Henry J. 1942. An ecological analysis of the plant communities of Piedmont, North Carolina. Amer. Midl. Nat., 28:1–126.

————. 1956. *The Study of Plant Communities* (2nd ed.). W. H. Freeman, San Francisco.

Oparin, A. I. 1938. *The Origin of Life*. Macmillan Company, New York.

Ophel, Ivan L. 1963. The fate of radiostrontium in a freshwater community. In: *Radioecology* (V. Schultz and W. Klement, eds.). Reinhold Publishing Company, New York. pp. 213–216.

Oppenheimer, C. H. (ed.). 1963. *Symposium on Marine Microbiology*. Charles C Thomas, Springfield, Ill. 769 pp.

Orians, G. H. 1969. On the evolution of mating systems in birds and mammals. Amer. Nat., 103:589–603.

Oro, J. 1963. Studies in experimental cosmochemistry. Ann. N.Y. Acad. Sci., 108:464–481.

Osborn, Fairfield. 1948. *Our Plundered Planet*. Little, Brown and Co., Boston.

Oswald, W. J., and Golueke, C. G. 1964. Fundamental factors in waste utilization in isolated systems. Dev. Ind. Microbiol., 5:196–206.

Overgaard-Nielsen, C. 1949. Studies on soil microfauna. II. The soil inhabiting nematodes. Natura Jutlandica, 2:1–131.

———. 1949a. Freeliving nematodes and soil microbiology. Proc. 4th Internat. Cong. Microbiol., Copenhagen, 1947, pp. 283–484.

———. 1955. Studies on Enchytraeidae. 2. Field studies. Natura Jutlandica, 4:5–58.

———. 1962. Carbohydrases in soil and litter invertebrates. Oikos, 13:200–215.

Overman, Ralph T., and Clark, Herbert M. 1960. *Radioisotope Techniques*. McGraw-Hill Book Co., Inc., New York.

Ovington, J. D. 1957. Dry matter production by *Pinus sylvestris*. Ann. Bot. n.s., 21:287–314.

———. 1961. Some aspects of energy flow in plantations of *Pinus sylvestris*. Ann. Bot. n.s., 25:12–20.

———. 1962. Quantitative ecology and the woodland ecosystem concept. In: Advances in Ecological Research (Cragg, ed.). Academic Press, New York. Vol. 1, pp. 103–192.

———. 1965. Organic production, turnover and mineral cycling in woodlands, Biol. Rev., Cambridge Phil. Soc., 40:295–336.

Ovington, J. D., and Madgwick, H. A. I. 1959. Distribution of organic matter and plant nutrients in a plantation of Scots Pine. Forest Sci., 5:344–355.

Packard, Charles. 1935. The relation between age and radiosensitivity of *Drosophila* eggs. Radiology, 25:223–230.

———. 1945. Roentgen radiations in biological research. Radiology, 45:522–533.

Paddock, W. C. 1970. How green the green revolution. BioScience, 20:897–902.

Paddock, William, and Paddock, Paul. 1967. *Famine 1975! America's Decision: Who Will Survive*. Little, Brown and Co., Boston. 286 pp.

Paine, R. T. 1966. Food web diversity and species diversity. Amer. Nat., 100:65–75.

Palmgren, P. 1949. Some remarks on the short-term fluctuations in the numbers of northern birds and mammals. Oikos, 1:114–121.

Palumbo, Ralph F. 1961. The difference in uptake of radioisotopes by marine and terrestrial organisms. In: *Recent Advances in Botany*. University of Toronto Press, Toronto. pp. 1367–1372.

Paris, O. H., and Pitelka, F. A. 1962. Population characteristics of the terrestrial isopod *Armedillidium vulgare* in California grassland. Ecology, 43:229–248.

Parizek, R. R.; Kardos, L. T.; Sopper, W. E.; Myers, E. A.; Davis, D. E.; Farrell, M. A.; and Nesbitt, J. B. 1967. Waste water renovation and conservation. Penn. State Univ. Studies No. 23. Pennsylvania State University, University Park. 71 pp.

Park, K.; Hood, D. W.; and Odum, H. T. 1958. Diurnal pH variation in Texas bays, and its application to primary production estimation. Publ. Inst. Mar. Sci., Univ. Texas, 5:47–62.

Park, Robert E.; Burgess, E. W.; and McKenzie, R. D. 1925. *The City*. University of Chicago Press, Chicago.

Park, Thomas. 1934. Studies in population physiology: effect of conditioned flour upon the productivity and population decline of *Tribolium confusum*. J. Exp. Zool., 68:167–182.

———. 1939. Analytical population studies in relation to general ecology. Amer. Midl. Nat., 21:235–255. (See also Quart. Rev. Biol., 16:274–293; 440–461, 1941.)

———. 1948. Experimental studies of interspecific competition. I. Competition between populations of flour beetles *Tribolium confusum* Duval and *T. castaneum* Herbst. Physiol. Zool., 18:265–308.

———. 1954. Experimental studies of interspecific competition. II. Temperature, humidity and competition in two species of Tribolium. Physiol. Zool., 27:177–238.

———. 1957. Experimental studies of interspecific competition. III. Relation of initial species proportion to competitive outcome in populations of Tribolium. Physiol. Zool., 30:22–40.

———. 1962. Beetles, competition and populations. Science, 138:1369–1375.

Parker, D. C., and Wolff, M. F. 1965. Remote sensing. Internat. Sci. and Technol., July, 43:20–31; 73.

Parker, Frank L. 1968. Radioactive wastes from fusion reactors. Science, 159:83–84.

Parker, J. R. 1930. Some effects of temperature and moisture upon *Melanoplus mexicanus* and *Camnula pellucida* Scudder (Orthoptera). Bull. Univ. Mont. Agr. Exp. Sta., 223:1–132.

Parsons, T. R., and Strickland, J. D. H. 1962. On the production of particulate organic carbon by heterotrophic processes in sea water. Deep-Sea Res., 8:211–222.

Patrick, Ruth. 1949. A proposed biological measure of stream conditions based on a survey of the Conestoga Basin, Lancaster County, Pennsylvania. Proc. Acad. Nat. Sci. Phila., 101:277–341.

———. 1953. Aquatic organisms as an aid in solving waste disposal problems. Sewage Industr. Wastes, 25:210–214.

———. 1954. Diatoms as an indication of river change. Proc. 9th Industr. Waste Conf., Purdue Univ. Enging. Extn. Ser., 87:325–330.

———. 1957. Diatoms as indicators of changes in environmental conditions. Trans. Sem. on Biol. Prob. Water Poll. R. A. Taft San. Engr. Center, Cinn., Ohio. pp. 71–83.

———. 1967. Diatom communities in estuaries. In: *Estuaries* (G. H. Lauff, ed.). Amer. Assoc. Adv. Sci. Publ. No. 83, Washington, D.C. pp. 311–315.

Patten, B. C. 1961. Preliminary method for estimating stability in plankton. Science, 134:1010–1011.

———. 1961a. Negentropy flow in communities of plankton. Limnol. Oceanogr., 6:26–30.

———. 1962. Species diversity in net plankton of Raritan Bay. J. Mar. Res., 20:57–75.

———. 1963. Plankton: Optimum diversity structure of a summer community. Science, 140:894–898.

———. 1963a. Information processing behavior of a natural plankton community. Amer. Biol. Tech., 25:489–501.

———. 1965. Community organization and energy relationships in plankton. Oak Ridge Nat. Lab. Rep. ORNL-3634. Oak Ridge, Tenn.

———. 1966. Systems ecology; a course sequence in mathematical ecology. BioScience, 16:593–598.

———. 1968. Mathematical models of plankton production. Int. Rev. ges. Hydrobiol., 53:357–408.

——— (ed.). 1971. *Systems Analysis and Simulation in Ecology*. Academic Press, New York.

Patten, B. C., and Witkamp, M. 1967. Systems analysis of [134]Cesium kinetics in terrestrial microcosms. Ecology, 48:813–824.

Patton, S.; Chandler, P. T.; Kalan, E. B.; Loeblich; Fuller, G.; and Benson, A. A. 1967. Food value of red tide (*Gonyaulax polyedra*). Science, 158:789–790.

Payne, W. J. 1970. Energy yields and growth of heterotrophs. Ann. Rev. Microbiol., 24:17–51.

Payne, W. J.; Wiebe, W. J.; and Christian, R. R. 1970. Assays for biodegradability essential to unrestricted usage of organic compounds. BioScience, 20:862–865.

Peakall, David B. 1967. Pesticide-induced enzyme breakdown of steroids in birds. Nature, 216:505–506.

———. 1970. Pesticide and the reproduction of birds. Scient. Amer., 222(4):73–78.

Pearl, Raymond. 1927. The growth of populations. Quart. Rev. Biol., 2:532–548.

———. 1930. *The Biology of Population Growth.* Alfred A. Knopf, Inc., New York.

Pearl, Raymond, and Parker, S. L. 1921. Experimental studies on the duration of life: Introductory discussion of the duration of life in *Drosophila.* Amer. Nat., 55:481–509.

Pearl, Raymond, and Reed, L. J. 1920. On the rate of growth of the population of the United States since 1790 and its mathematical representation. Proc. Natl. Acad. Sci. (Wash.), 6:275–288.

Pearse, A. S. 1939. *Animal Ecology* (2nd ed.). McGraw-Hill Book Co., Inc., New York.

Pearse, A. S.; Humm, H. J.; and Wharton, G. W. 1942. Ecology of sand beaches at Beaufort, N.C. Ecol. Monogr., 12:136–190.

Pearson, Oliver P. 1960. The oxygen consumption and bioenergetics of harvest mice. Physiol. Zool., 33:152–160.

Pendleton, Robert C. 1956. Labeling animals with radioisotopes. Ecology, 37:686–689.

Pendleton, Robert C., and Grundmann, A. W. 1954. Use of P^{32} in tracing some insect-plant relationships of the thistle, *Cirsium undulatum.* Ecology, 35:187–191.

Penfound, William T. 1956. Primary production of vascular aquatic plants. Limnol. Oceanogr., 1:92–101.

Penman, H. L. 1956. Weather and water in the growth of grass. In: *The Growth of Leaves.* (See Milthrope, 1956.)

Pennak, Robert W. 1939. The microscopic fauna of the sandy beaches. Publ. Amer. Assoc. Adv. Sc., 10:94–106.

———. 1953. *Fresh-water Invertebrates of the United States.* Ronald Press, New York.

———. 1955. Comparative limnology of eight Colorado mountain lakes. Univ. of Colorado Studies. Ser. Biol., 2:1–74.

Petersen, C. G. J. 1914; 1915; 1918. The animal associations of the sea-bottom in the North Atlantic. Kobenhavn Ber. Biol. Sta., 22:89–98; 23:1–28; 26:1–62.

Peterson, Elmer T. 1952. Insoak is the answer. Land, 11:83–88.

Peterson, Malcolm. 1970. The space available. Environment, 12:2–9.

Peterson, R. M.; Cochrane, G. R.; and Simonett, D. S. 1969. A multi-sensor study of plant communities at Horsefly Mountain, Oregon. In: *Remote Sensing in Ecology* (P. L. Johnson, ed.). University of Georgia Press, Athens. pp. 63–85.

Petrides, George A. 1950. The determination of sex and age ratios in fur animals. Amer. Midl. Nat., 43:355–382.

———. 1956. Big game densities and range carrying capacity in East Africa. N. A. Wildl. Conf. Trans., 21:525–537.

———. 1961. Review of *Introduction à L'Ecologie des Ongules de Parc National Albert* by François Burlière and Jacques Versdruren. J. Wildl. Mgt., 25:217–218.

Petrusewicz, K. (ed.). 1967. Secondary productivity of terrestrial ecosystems. Inst. Ecol. Polish Acad. Sci. Int. Biol. Program (Warsaw). Vol. I, 367 pp.; Vol. II, 879 pp.

Pfennig, Norbert. 1967. Photosynthetic bacteria. Ann. Rev. Microbiol., 21:285–324.

Philip, J. P. 1955. Note on the mathematical theory of animal population dynamics and a recent fallacy. Australian J. Zool., 3:287–294.

Phillips, E. A. 1959. *Methods of Vegetation Study.* Holt, Rinehart & Winston, Inc., New York.

Phillipson, J. (ed.). 1970. Methods of study in soil ecology. UNESCO, Paris. 303 pp.

Pianka, E. R. 1967. Lizard species diversity. Ecology, 48:333–351.

Picozzi, N. 1968. Grouse bags in relation to the management and geology of heather moors. J. Appl. Ecol., 5:483–488.

Pielou, E. C. 1960. A single mechanism to account for regular, random and aggregated populations. J. Ecol., 48:575–584.

———. 1966. The measurement of diversity in different types of biological collections. J. Theoret. Biol., 13:131–144.

———. 1966a. Species-diversity and pattern-diversity in the study of ecological succession. J. Theoret. Biol., 10:370–383.

———. 1966b. Shannon's formula as a measure of specific diversity: its use and disuse. Amer. Nat., 100:463–465.

———. 1969. *An Introduction to Mathematical Ecology.* John Wiley & Sons, Inc. (Interscience), New York. 286 pp.

Pimentel, David. 1961. Animal population regulation by the genetic feedback mechanism. Amer. Nat., 95:65–79.

———. 1961a. An ecological approach to the insecticide problem. J. Econ. Entomol., 54:108–114.

———. 1961b. Competition and the species-per-genus structure of communities. Ann. Entomol. Soc. Amer., 54:323–333.

———. 1961c. Species diversity and insect population outbreaks. Ann. Entomol. Soc. Amer., 54:76–86.

———. 1968. Population regulation and genetic feedback. Science, 159:1432–1437.

Pimentel, David, and Stone, F. A. 1968. Evolution and population ecology of parasite-host systems. Canad. Entomol., 100:655–662.

Pirt, S. J. 1957. The oxygen requirement of growing cultures of an *Aerobacter* species determined by means of the continuous culture technique. J. Gen. Microbiol., 16:59–75.

Pitelka, Frank A. 1941. Distribution of birds in relation to major biotic communities. Amer. Midl. Nat., 25:113–137.

———. 1951. Ecologic overlap and interspecific strife in breeding populations of Anna and Allen Hummingbirds. Ecology, 32:641–661.

Pitelka, F. A.; Tomich, P. Q.; and Treichel, G. W.

1955. Ecological relations of jaegers and owls as lemming predators near Barrow, Alaska. Ecol. Monogr., 25:85–117.

Pittendrigh, Colin S. 1961. Temporal organization in living systems. Harvey Lecture Series, 56:93–125. Academic Press, New York.

Pittendrigh, Colin S.; Vishniac, Wolf; and Pearman, J. P. T. 1966. Biology and the exploration of Mars. Natl. Acad. Sci. (Wash.), publ. No. 1296. 516 pp.

Plass, Gilbert N. 1955. The carbon dioxide theory of climate change. Tellus, 8:140–154.

———. 1959. Carbon dioxide and climate. Scient. Amer., 201(1):41–47.

Platt, Robert B. 1957. Field installation for automatic recording of microenvironmental gradients. Trans. Amer. Geo. Union, 38:160–170.

———. 1965. Ionizing radiation and homeostasis of ecosystems. In: Ecological Effects of Nuclear War (Woodwell, ed.). Brookhaven National Laboratory, Publ. No. 917. pp. 39–60.

Platt, Robert B., and Griffiths, John. 1964. Environmental Measurement and Interpretation. Reinhold Publishing Company, New York. 235 pp.

Polcyn, F.; Spansail, N.; and Malida, W. 1969. How multispectral sensing can help the ecologist. In: Remote Sensing in Ecology (P. L. Johnson, ed.). University of Georgia Press, Athens. pp. 194–218.

Polikarpou, G. G. 1966. Radioecology of aquatic organisms. (Translated from Russian by S. Technica and edited by Schultz and Klement.)

Polunin, N. 1960. Introduction to Plant Geography, McGraw-Hill Book Co., Inc., New York. 640 pp.

Pomeroy, L. R. 1959. Algal productivity in Georgia salt marshes. Limnol. Oceanogr., 4:386–397.

———. 1960. Residence time of dissolved phosphate in natural waters. Science, 131:1731–1732.

———. 1970. The strategy of mineral cycling. Ann. Rev. Ecol. and Systematics, 1:171–190.

Pomeroy, L. R., and Johannes, R. E. 1966. Total plankton respiration. Deep-Sea Res., 13:971–973.

———. 1968. Respiration of ultraplankton in the upper 500 meters of the ocean. Deep-Sea Res., 15:381–391.

Pomeroy, L. R.; Johannes, R. E.; Odum, E. P.; and Roffman, B. 1969. The phosphorus and zinc cycles and productivity of a salt marsh. In: Proc. 2nd Sym. on Radioecology (D. J. Nelson and F. C. Evans, eds.). Clearinghouse Fed. Sci. Tech. Info., Springfield, Va. pp. 412–419.

Pomeroy, L. R., and Kuenzler, E. J. 1969. Phosphorus turnover by coral reef animals. In: Proc. 2nd Sym. on Radioecol. (Nelson and Evans, eds.). U. S. Atomic Energy Commission, TID 4500. pp. 474–482.

Pomeroy, L. R.; Odum, E. P.; Johannes, R. E.; and Roffman, B. 1966. Flux of ^{32}P and ^{65}Zn through a salt marsh ecosystem. Proceedings of the Symposium on the Dispersal of Radioactive Wastes into Seas. Oceans and Surface Waters, Vienna. pp. 177–188.

Pomeroy, L. R.; Mathews, H. M.; and Min, H. S. 1963. Excretion of phosphate and soluble organic phosphorus compounds by zooplankton. Limnol Oceanogr., 8:50–55.

Pomeroy, L. R.; Smith, E. E.; Grant, C. M. 1965. The exchange of phosphate between estuarine water and sediment. Limnol. Oceanogr., 10:167–172.

Postgate, J. R., and Hunter, J. R. 1963. The survival of starved bacteria. J. Appl. Bacteriol., 26:295–306.

Poulson, T. L., and White, W. B. 1969. The cave environment. Science, 165:971–980.

Pound, C. E., and Egler, F. E. 1966. Vegetation management (right of way). In: McGraw-Hill Encyclopedia of Science, Vol. 14, pp. 287–288.

President's Science Advisory Committee, The White House, Environmental Pollution Panel. 1965. Restoring Quality of Our Environment (Tukey, ed.). U. S. Government Printing Office, Washington, D.C. 317 pp.

———. 1966. Effective Use of the Sea. U. S. Government Printing Office, Washington, D.C. 144 pp.

———. 1967. The World Food Problem. 3 vols. U. S. Government Printing Office, Washington, D.C.

Preston, F. W. 1948. The commonness and rarity of species. Ecology, 29:254–283.

Pritchard, D. W. 1952. Estuarine hydrography. Adv. Geophys., 1:243–280.

———. 1955. Estuarine circulation patterns. Proc. Amer. Soc. Civil Engrs., 81:717.

———. 1967. What is an estuary: physical viewpoint. In: Estuaries (G. H. Lauff, ed.). Amer. Assoc. Adv. Sci. Publ. No. 83, Washington, D.C. pp. 3–5.

———. 1967a. Observations of circulation in coastal plain estuaries. In: Estuaries (G. H. Lauff, ed.). Amer. Assoc. Adv. Sci. Publ. No. 83, Washington, D.C. pp. 37–44.

Proctor, Vernon W. 1957. Studies of algal antibiosis using Hematococcus and Chlamydomonas. Limnol. Oceanogr., 2:125–139.

Prosser, C. L. (ed.). 1967. Molecular mechanisms of temperature adaptation. Amer. Assoc. Adv. Sci., Washington, D.C.

Provasoli, Luigi. 1958. Nutrition and ecology of protozoa and algae. Ann. Rev. Microbiol., 12:279–308.

———. 1963. Organic regulation of phytoplankton fertility. In: The Sea (M. N. Hill, ed.). John Wiley & Sons, Inc. (Interscience), New York. Vol. 2, pp. 165–219.

Pye, Kendall. 1969. Indigenous rhythms in yeast. Canad. J. Bot., 47:271.

Quinn, James A. 1950. Human Ecology. Prentice-Hall, Inc., Englewood Cliffs, N.J.

Quinn, L. Y. 1968. Immunological Concepts. Iowa State University Press, Ames. 260 pp.

Rabinowitch, E. I. 1951. Photosynthesis and Related Processes. Vol. II (1):603–1208. John Wiley & Sons, Inc. (Interscience), New York.

Rabinowitch, E., and Govindjee. 1969. Photosynthesis. John Wiley & Sons, Inc., New York.

Rafter, T. A., and Fergusson, G. J. 1957. The atom bomb effect: recent increase in the C^{14} content of the atmosphere, biosphere, and surface waters of the ocean. New Zealand J. Sci. Tech., 38:871–883.

Ragotzkie, R. A. 1960. Marine marsh. In: McGraw-Hill Encyclopedia of Science and Technology. pp. 217–218.

Ragotzkie, R. A., and Bryson, R. A. 1955. Hydrology of the Duplin River, Sapelo Island, Georgia. Bull. Mar. Sci. Gulf and Carib., 5:297–314.

Ragotzkie, Robert A., and Pomeroy, Lawrence R. 1957. Life history of a dinoflagellate bloom. Limnol. Oceanogr., 2:62–69.

Rasmussen, D. I. 1941. Biotic communities of Kaibab Plateau, Arizona. Ecol. Monogr., 11:229–275.

Raunkaier, C. 1934. The Life Form of Plants and Statistical Plant Geography. Clarendon Press, Oxford.

Ravera, O. 1969. Seasonal variation of the biomass and biocoenotic structure of plankton of the Bay of Ispra (Lago Maggiore). Verh. int. Ver. Limnol., 17:237–254.

Rawson, D. S. 1952. Mean depth and the fish production of large lakes. Ecology, 33:513–521.

———. 1956. Algal indicators of trophic lake types. Limnol. Oceanogr., 1:18–25.

Raymont, J. E. G. 1966. The production of marine plankton. In: *Advances in Ecological Research* (J. B. Cragg, ed.). Academic Press, New York. Vol. 3, pp. 117–205.

Redfield, Alfred C. 1958. The biological control of chemical factors in the environment. Amer. Sci., 46:205–221.

Reichle, David (ed.). 1970. *Analysis of Temperate Forest Ecosystems.* Springer-Verlag. Heidelberg, Berlin. 304 pp.

Reichle, David, and Crossley, D. A. 1965. Radiocesium dispersion in a cryptozoan food web. Health Physics, 11:1375–1384.

Reifsnyder, W. E., and Lull, H. W. 1965. Radiant energy in relation to forests. Tech. Bull. No. 1344. U. S. Dept. Agriculture, Forest Service. 111 pp.

Remple, R. C., and Parker, A. K. 1965. An information note on an airborne laser terrain profiler and micro-relief studies. In: Proc. 3rd Sym. Remote Sensing Environment, University of Michigan. pp. 321–337.

Repaske, R. 1962. Nutritional requirements for *Hydrogenomonas eutropha.* J. Bact., 83:418–433.

———. 1966. Characteristics of hydrogen bacteria. Biotech. and Bioeng., 8:217–235.

Revelle, Roger (ed.). 1965. Atmospheric carbon dioxide. In: *Restoring the Quality of Our Environment.* Rept. Pres. Sci. Adv. Com., The White House. pp. 111–133.

———. 1966. Population and food supplies: the edge of the knife. In: *Prospects of the World Food Supply: A Symposium.* Nat. Acad. Sci. (Wash.). pp. 24–47.

Revelle, R., and Suess, Hans E. 1956. Carbon dioxide exchange between atmosphere and ocean and the question of an increase of atmospheric CO_2 during past decades. Tellus, 9:118–127.

Rhodda, M. 1967. *Noise and Society.* Oliver and Boyd, London. 113 pp.

Rice, E. L., and Kelting, R. W. 1955. The species-area curve. Ecology, 36:7–11.

Rice, E. L., and Penfound, William T. 1954. Plant succession and yield of living plant material in a plowed prairie in central Oklahoma. Ecology, 35:176–180.

Rice, T. R. 1954. Biotic influences affecting population growth of planktonic algae. Fishery Bull. (U. S. Fish and Wildlife Service), 87:227–245.

Richards, F. A., and Thompson, T. G. 1952. The estimation and characterization by pigment analysis. II. A spectrophotometric method for the estimation of plankton pigments. J. Mar. Res., 11:156–172.

Richards, P. W. 1952. *The Tropical Rain Forest.* Cambridge University Press, New York.

Ricker, W. E. 1946. Production and utilization of fish populations. Ecol. Monogr., 16:373–391.

———. 1954. Stock and recruitment. J. Fish Res. Bd. Canada, 11:559–622.

———. 1954a. Effects of compensatory mortality upon population abundance. J. Wildl. Mgt., 18:45–51.

———. 1958. Handbook of computation for biological statistics of fish populations. Fish. Res. Bd. Canad. Bull., 119:1–300.

———. 1962. Regulation of the abundance of pink salmon populations. In: *Symposium on Pink Salmon* (N. J. Wilimovsky, ed.). University of British Columbia, Vancouver.

———. 1969. Food from the sea. In: *Resources and Man* (P. Cloud, ed.). W. H. Freeman & Co., San Francisco. pp. 87–108.

Ricker, W. E., and Forester, R. E. 1948. Computation of fish production. Bull. Bingham Oceanogr. Coll., 11:173–211.

Rienow, Robert, and Rienow, L. T. 1967. *Moment in the Sun.* A Report on the Deteriorating Quality of the American Environment. Dial Press, New York. 286 pp.

Rigler, F. H. 1961. The uptake and release of inorganic phosphorus by *Daphnia magna* Straus. Limnol. Oceanogr., 6:165–174.

Rigler, R. L. 1956. A tracer study of phosphorus cycle in lake water. Ecology, 37:550–556.

Riley, Gordon A. 1939. Plankton studies: the western North Atlantic, May–June, 1939. J. Mar. Res., 2: 145–162.

———. 1941. Plankton studies: Long Island Sound. Bull. Bingham Oceanogr. Coll., 7:1–89.

———. 1943. Physiological aspects of spring diatom flowering. Bull. Bingham Oceanogr. Coll., 8:1–53.

———. 1944. The carbon metabolism and photosynthetic efficiency of the earth. Amer. Sci., 32:132–134.

———. 1952. Phytoplankton of Block Island Sound, 1949. Bull. Bingham oceanogr. Coll., 13:40–64.

———. 1952a. Biological Oceanography. In: *Survey of Biological Progress,* 2:79–104. Academic Press, New York.

———. 1956. Oceanography of Long Island Sound, 1952–54. IX. Production and utilization of organic matter. Bull. Bingham Oceanogr. Coll., 15:324–344.

———. 1963. Organic aggregates in sea water and the dynamics of their formation and utilization. Limnol. Oceanogr., 8:372–381.

———. 1967. Mathematical model of nutrient conditions in coastal waters. Bull. Bingham Oceanogr. Coll., 19:72–80.

Riley, Gordon A.; Strommel, H.; and Bumpus, D. F. 1949. Quantitative ecology of the plankton of the western Atlantic. Bull. Bingham Oceanogr. Coll., 12:1–169.

Rinker, J. N.; Evans, S.; and Robin, G. de Q. 1966. Radio ice-sounding techniques. In: Proc. 4th Sym. Remote Sensing of Environment. University of Michigan Press, Ann Arbor. pp. 793–800.

Ripper, W. E. 1956. Effect of pesticides on the balance of arthropod populations. Ann. Rev. Entomol., 1: 403–438.

Rittenberg, S. C. 1963. Marine bacteriology and the problem of mineralization. In: *Symposium on Marine Microbiology* (C. H. Oppenheimer, ed.). Charles C Thomas, Springfield, Ill. pp. 48–60.

Robinette, Gary O. 1969. The functional spectrum of plants — sound control. Grounds Maintenance, 4:42–43.

Robertson, J. S. 1957. Theory and use of tracers in determining transfer rates in biological systems. Physiol. Rev., 37:133–154.

Rodhe, W. 1955. Can plankton production proceed during winter darkness in subarctic lakes? Proc. Internat. Assoc. Theor. Appl. Limnol., 12:117–122.

Rohrman, F. A.; Steyerwand, B. G.; and Ludwig, S. H. 1967. Industrial emissions of carbon dioxide in the U.S.: a projection. Science, 156:931–932.

Romney, E. M.; Neel, J. W.; Nishita, H.; Olafson, J. H.; and Larson, K. H. 1957. Plant uptake of Sr 90, Y91, Ru106, Cs137 and Ce144 from soils. Soil Sci., 83: 369–376.

Root, Richard B. 1967. The niche exploitation pattern of the blue-gray gnatcatcher. Ecol. Monogr., 37:317–350.

———. 1969. The behavior and reproductive success of the blue-gray gnatcatcher. Condor, 71:16–31.

Rosenfeld, Azriel. 1965. Automatic imagery interpretation. Photogr. Engr., 31:240–242.

Rosenzweig, M. L. 1968. Net primary production of terrestrial communities; prediction from climatological data. Amer. Nat., 102:67–74.

Ross, H. H. 1957. Principles of natural coexistence indicated by leafhopper populations. Evolution, 11:113–129.

Rounsefell, G. A. 1946. Fish production in lakes as a guide for estimating production in proposed reservoirs. Copeia, 1946:29–40.

———. 1953. *Fishery Science—Its Methods and Applications*. John Wiley & Sons, Inc., New York.

Rovira, A. D. 1965. Interaction between plant roots and soil microorganisms. Ann. Rev. Microbiol., 19:241–266.

Rübel, Eduard. 1935. The replaceability of ecological factors and the law of the minimum. Ecology, 16:336–341.

Rudd, R. I. 1964. *Pesticides and the Living Landscape*. University of Wisconsin Press, Madison. 320 pp.

Russell, E. John. 1957. *The World of the Soil*. The New Naturalist: A Survey of British Natural History. William Collins Sons and Co., London.

Russell, E. W. 1961. *Soil Conditions and Plant Growth* (9th ed.). Longmans, London.

Russell, R. S. 1965. Interception and retention of airborne material on plants. Health Physics, 11:1305–1315.

Ruttner, Franz. 1963. *Fundamentals of Limnology* (3rd ed.). (Translated by Frey and Fry.) University of Toronto Press, Toronto. 295 pp.

Ryther, John H. 1954. The ecology of phytoplankton blooms in Moriches Bay and Great South Bay, Long Island, New York. Biol. Bull., 106:198–209.

———. 1955. Ecology of autotrophic marine dinoflagellates with reference to red water conditions. In: *The Luminescence of Biological Systems* (F. H. Johnson, ed.). Amer. Assoc. Adv. Sci., Washington, D.C. pp. 387–414.

———. 1956. Photosynthesis in the ocean as a function of light intensity. Limnol. Oceanogr., 1:61–70.

———. 1969. Photosynthesis and fish production in the sea. Science, 166:72–76.

Ryther, J. H., and Yentsch, C. S. 1957. The estimation of phytoplankton production in the ocean from chlorophyll and light data. Limnol. Oceanogr., 2:281–286.

Saarinen, Eliel. 1943. *The City: Its Growth, Its Decay, Its Future*. M.I.T. Press, Cambridge, Mass. 380 pp.

Salisbury, F. B. 1963. *The Flowering Process*. Pergamon Press, Oxford.

Sampson, Arthur W. 1952. *Range Management, Principles and Practices*. John Wiley & Sons, Inc., New York.

Sanders, H. L. 1960. Benthic studies in Buzzards Bay. III. The structure of the soft bottom community. Limnol. Oceanogr., 5:138–153.

———. 1968. Marine benthic diversity: a comparative study. Amer. Nat., 102:243–282.

Sanders, H. L., and Hessler, Robert R. 1969. Ecology of the deep-sea benthos. Science, 163:1419–1424.

Sanderson, Ivan T. 1945. *Living Treasure*. Viking, New York.

San Pietro, A.; Greer, F. A.; and Army, T. J. (eds.). 1967. *Harvesting the Sun*. Academic Press, New York. 342 pp.

Sargent, F. 1967. A dangerous game: taming the weather. Bull. Amer. Meteor. Soc., 48:452–458.

Sargent, M. C., and Austin, T. S. 1949. Organic productivity of an atoll. Amer. Geophys. Union Trans., 30:245–249.

Sauer, C. O. 1950. Grassland, climax, fire and man. J. Range Mgt., 3:16–21.

———. 1966. *The Early Spanish Main*. University of California Press, Berkeley. 306 pp.

Saunders, George W. 1957. Interrelations of dissolved organic matter and phytoplankton. Bot. Rev., 23:389–409.

Saunders, J. F. (ed.). 1968. *Bioregenerative Systems*. NASA. SP-165. Supt. Doc., U. S. Printing Office, Washington. 153 pp.

Savage, J. M. 1969. *Evolution* (2nd ed.). Holt, Rinehart and Winston, New York. 160 pp.

Sax, Karl. 1955. *Standing Room Only*. Beacon Press, Boston. 206 pp.

Schaefer, M. B. 1965. The potential harvest of the sea. Trans. Amer. Fish Soc., 94:123–128.

Schaller, Friedrich. 1968. *Soil Animals*. University of Michigan Press, Ann Arbor. 144 pp.

Schelske, Claire L., and Odum, E. P. 1961. Mechanisms maintaining high productivity in Georgia estuaries. Proc. Gulf and Carib. Fish. Inst., 14:75–80.

Schlegel, H. G. (ed.). 1965. Anreicherungskultur und Mutantenauslese. Abl. Bakt. Abt., I Suppl., Vol. I.

Schmid, Calvin F. 1950. Generalizations concerning the ecology of the American City. Amer. Soc. Rev., 40:264–281.

Schmidt-Nielsen, Bodil, and Schmidt-Nielsen, Knut. 1949. The water economy of desert mammals. Sci. Mon., 69:180–185.

———. 1952. Water metabolism of desert mammals. Physiol. Rev., 32:135–166.

Schmidt-Nielson, Knut. 1964. *Desert Animals. Physiological Problems of Heat and Water*. Oxford University Press, London. 270 pp.

Schneider, W. J. 1968. Color photographs for water resources studies. Photogr. Engr., 34:257–262.

Schrodinger, E. 1944. *What Is Life? The Physical Aspects of the Living Cell*. Cambridge University Press, Cambridge, England. 91 pp.

Schnell, Jay H. 1968. The limiting effects of natural predation on experimental cotton rat populations.

Schultz, A. M. 1964. The nutrient-recovery hypothesis for arctic microtine cycles. II. Ecosystem variables in relation to arctic microtine cycles. In: *Grazing in Terrestrial and Marine Environments* (D. J. Crisp, ed.). Blackwell, Oxford. pp. 57–68.

———. 1967. The ecosystem as a conceptual tool in the management of natural resources. In: *Natural Resources: Quality and Quantity* (Ciriancy-Wantrup and J. J. Parsons, eds.). University of California Press, Berkeley. pp. 139–161.

———. 1969. A study of an ecosystem: the arctic tundra. In: *The Ecosystem Concept in Natural Resource Management* (G. Van Dyne, ed.). Academic Press, New York. pp. 77–93.

———. 1971. *Ecosystems and Environments*. Canfield Press (Harper and Row), San Francisco.

Schultz, Vincent, and Klement, A. W. (eds.). 1963. *Radioecology*. Proc. 1st Nat. Symposium. Reinhold Publishing Company, New York. 746 pp.

Scott, J. P. (ed.). 1956. Allee memorial number. Series of papers on social organization in animals. Ecology, 37:211–273.

———. 1958. *Animal Behavior*. University of Chicago Press, Chicago.

Searle, N. E. 1965. Physiology of flowering. Ann. Rev. Plant Physiol., 16:97–118.

Searle, S. R. 1966. Matrix algebra for the biological sciences. John Wiley & Sons, Inc., New York. 296 pp.

Sears, Mary (ed.). 1964. *Progress in Oceanography*, 2 Vols. Macmillan Company, New York.

Sears, Paul B. 1935. *Deserts on the March*. University of Oklahoma Press, Norman.

———. 1957. *The Ecology of Man*. Condon Lectures. University of Oregon Press. 61 pp.

Selye, Hans. 1955. *Stress and Disease*. Science, 122:625–631.

Seymour, A. H. 1959. The distribution of radioisotopes among marine organisms in the western Pacific. Publ. della Stazione Zool. Napoli, 31 (suppl.): 25–33.

Shannon, C. E., and Weaver, W. 1963. The mathematical theory of communication. University of Illinois Press, Urbana. 117 pp.

Shantz, H. L. 1911. Natural vegetation as an indicator of the capabilities of land for crop production in the great plains area. U. S. Dept. Agr., Bureau of Plant Industry, Bull. 201.

———. 1917. Plant succession on abandoned roads in eastern Colorado. J. Ecol., 5:19–42.

———. 1942. The desert vegetation of North America. Bot. Rev., 8:195–246.

Shelford, V. E. 1911. Physiological animal geography. J. Morphol., 22:551–618.

———. 1911a. Ecological succession: stream fishes and the method of physiographic analysis. Biol. Bull., 21:9–34.

———. 1911b. Ecological succession: pond fishes. Biol. Bull., 21:127–151.

———. 1913. *Animal Communities in Temperate America*. University of Chicago Press, Chicago.

———. 1919. Nature's mobilization. Nat. Hist., 19:205–210.

———. 1929. *Laboratory and Field Ecology*. Williams & Wilkins Co., Baltimore.

———. 1943. The abundance of the collared lemming in the Churchill area, 1929–1940. Ecology, 24:472–484.

———. 1951. Fluctuations of forest animal populations in east central Illinois. Ecol. Monogr., 21:183–214.

———. 1963. *The Ecology of North America*. University of Illinois Press, Urbana. 610 pp.

Shelford, V. E., and Olson, Sigurd. 1935. Sere, climax and influent animals with special reference to the transcontinental coniferous forest of North America. Ecology, 16:375–402.

Shepard, Paul, and McKinley, Daniel (ed.). 1969. *The Subversive Science: Essays Towards an Ecology of Man*. Houghton Mifflin Company, Boston. 453 pp.

Shepelev, Ye Ya. 1965. Some aspects of human ecology in closed systems with recirculation of substances. Problems of Space Biology (Sisakyan, N. M., ed.), 4:166–175. Izd-vo Akademii Nauk. USSR Moscow, 1965. NASA TTF-386.

Shields, Lora M. 1953. Nitrogen sources of seed plants and environmental influences affecting the nitrogen cycle. Bot. Rev., 19:321–376.

Shure, D. J. 1970. Limitations in radiotracer determination of consumer trophic positions. Ecology, 51:899–901.

Silliman, R. P. 1969. Population models and test populations as research tools. BioScience, 19:524–528.

Simonson, Roy W. 1951. The soil under natural and cultural environment. J. Soil and Water Cons., 6:63–69.

Simpson, E. H. 1949. Measurement of diversity. Nature, 163:688.

Simpson, George Gaylord. 1969. The first three billion years of community evolution. In: *Diversity and Stability in Ecological Systems* (Woodwell and Smith, eds.). Brookhaven Symposium in Biology, No. 22. Brookhaven Nat. Lab., Upton, New York. pp. 162–177.

Singer, S. F. (ed.). 1970. *Global Effects of Environmental Pollution*. Springer-Verlag, New York. 210 pp.

Skellum, J. G. 1952. Studies in statistical ecology. I-spatial pattern. Biometrica, 39:346–362.

Skerman, V. D. B. 1968. *A Guide to the Identification of the Genera of Bacteria*. The Williams and Wilkins Co., Baltimore.

Slobodkin, L. B. 1954. Population dynamics in *Daphnia obtusa* Kurz. Ecol. Monogr., 24:69–88.

———. 1960. Ecological energy relationships at the population level. Amer. Nat., 95:213–236.

———. 1962. Growth and regulation of animal populations. Holt, Rinehart and Winston, New York. 184 pp.

———. 1962a. Energy in animal ecology. In: Advances in Ecological Research (J. B. Cragg, ed.). Vol. 1, pp. 69–101.

———. 1964. Experimental populations of hydrida. In: Brit. Ecol. Soc. Jubilee Symp. Suppl. to J. Ecol., 52, and J. Anim. Ecol., 33:1–244. Blackwell, Oxford. pp. 131–148.

———. 1968. How to be a predator. Amer. Zool., 8:43–51.

Smayda, T. J. 1963. Succession of phytoplankton, and the ocean as a holoclenotic environment. In: Sym. Mar. Microbiol. (C. H. Oppenheimer, ed.). Chapter 27, pp. 260–274. Charles C Thomas, Springfield, Ill.

Smith, Frank E. 1966. *The Politics of Conservation*. Pantheon Books (Random House), New York. 350 pp.

Smith, Frederick E. 1954. Quantitative aspects of population growth. In: *Dynamics of Growth Processes* (E. J. Boell, ed.). Princeton University Press, Princeton.

———. 1963. Population dynamics in *Daphnia magna* and a new model for population growth. Ecology, 44:651–653.

———. 1969. Today the environment, tomorrow the world. BioScience, 19:317–320.

———. 1970. Ecological demand and environmental response. J. For., 68:752–755.

Smith, J. E. (ed.). 1968. *"Torrey Canyon" Pollution and Marine Life*. A report by the Plymouth Laboratory of the Marine Biological Association of the United Kingdom. Cambridge University Press, New York.

Smith, Ray F., and Reynolds, H. T. 1966. Principles, definition and scope of integrated pest control. Proc. FAO Sym. Integrated Pest Control, 1:11–17.

Smith, Ray F., and van den Basch, R. 1967. Integrated control. In: *Pest Control: Biological, Physical and Selected Chemical Methods* (Kilgore and Doutt, eds.). Academic Press, New York. pp. 295–340.

Smith, Roland F. (ed.). 1966. A symposium on estuarine fisheries. Suppl. to Trans. Amer. Fish. Soc., 95(4):1–154.

Smith, S. H. 1966. Species succession and fishery exploitation in the Great Lakes. Sym. Overexploited

Animal Populations. Amer. Assoc. Adv. Sci., Washington, D.C. Mimeographed. 28 pp.

Smith, W. John. 1969. Messages of vertebrate communication. Science, 165:145–150.

Smock, Robert B. 1969. The social necessity for the concept of ecological dependency: Some elements of an applied human ecology. Paper presented Amer. Soc. Assoc. San Francisco.

Snow, C. P. 1959. *The Two Cultures and the Scientific Revolution.* Cambridge University Press, New York.

Sokol, R. R., and Sneath, P. H. A. 1963. *Principles of Numerical Taxonomy.* W. H. Freeman and Co., San Francisco.

Solomon, M. E. 1949. The natural control of animal populations. J. Anim. Ecol., 18:1–32.

———. 1953. Insect population balance and chemical control of pests. Pest outbreaks induced by spraying. Chem. Ind., (43):1143–1147.

Somero, G. N. 1969. Enzymic mechanisms of temperature compensation. Amer. Nat., 103:517–530.

Sondheimer, Ernest, and Simeone, John B. (eds.). 1969. *Chemical Ecology.* Academic Press, New York.

Sopper, William E. 1968. Waste water renovation for reuse: key to optimum use of water resources. In: *Water Research.* Pergamon Press, New York. Vol. 2, pp. 471–480.

Sorensen, T. 1948. A method of establishing groups of equal amplitude in plant society based on similarity of species content. K. Danske Vidensk. Selsk., 5:1–34.

Sorokin, J. T. 1964. On the trophic role of chemosynthesis in water bodies. Int. Rev. ges. Hydrobiol., 49:307–324.

———. 1966. On the trophic role of chemosynthesis and bacterial biosynthesis in water bodies. In: *Primary Production in Aquatic Environments* (C. R. Goldman, ed.). University of California Press, Berkeley. pp. 189–205.

Southward, A. J. 1958. The zonation of plants and animals on rocky shores. Biol. Rev., 33:137–177.

Southwood, T. R. E. 1966. *Ecological Methods, with Particular Reference to the Study of Insect Populations.* Methuen, London. 391 pp.

Sparrow, A. H. 1962. The role of the cell nucleus in determining radiosensitivity. Brookhaven Lecture Series No. 17. Brookhaven Nat. Lab. Publ. No. 766.

Sparrow, A. H., and Evans, H. J. 1961. Nuclear factors affecting radiosensitivity. 1. The influence of nuclear size and structure, chromosome complement and DNA content. In: *Fundamental Aspects of Radiosensitivity.* Brookhaven Symposia in Biology No. 14, Brookhaven Nat. Lab. pp. 76–100.

Sparrow, A. H.; Schairer, L. A.; and Sparrow, R. C. 1963. Relationship between nuclear volumes, chromosome numbers, and relative radiosensitivities. Science, 141:163–166.

Sparrow, A. H., and Woodwell, G. M. 1962. Prediction of the sensitivity of plants to chronic gamma irradiation. Rad. Bot., 2:9–26.

Spilhaus, Athelstan (ed.). 1966. *Waste Management and Control.* Publ. No. 1400. National Academy Science, Washington, D.C. 257 pp. (See review by E. P. Odum, 1967.)

Spivey, W. A. 1963. *Linear Programming—an Introduction.* Macmillan Company, New York. 167 pp.

Springer, Leonard M. 1950. Aerial census of interstate antelope herds of California, Idaho, Nevada and Oregon. J. Wildl. Mgt., 14:295–298.

Stanier, R. Y.; Douderoff, M.; and Adelberg, E. A. 1963. *The Microbial World* (2nd ed.). Prentice-Hall, Inc., Englewood Cliffs, N.J. 753 pp.

Steemann-Nielsen, E. 1952. The use of radioactive carbon (C^{14}) for measuring organic production in the sea. J. Cons. Int. Explor. Mer., 18:117–140.

———. 1954. On organic production in the ocean. J. Cons. Int. Explor. Mer., 49:309–328.

———. 1963. Fertility of the oceans: Productivity, definition and measurement. In: *The Sea* (M. N. Hill, ed.), Vol. 2. John Wiley & Sons, Inc. pp. 129–164.

Steiner, D., and Gutermann, T. 1966. Russian data on spectral reflectance of vegetation, soil, and rock types. Final Tech. Rept. on Contract No. DA-91-EUC-3863/OI-652-0106, Department of Geography, University of Zurich, Switzerland. 232 pp.

Stephen, D. G., and Weinberger, L. W. 1968. Wastewater reuse—has it arrived? J. Water Poll. Cont. Fed., 40:529.

Stephenson, T. A., and Stephenson, Anne. 1949. The universal features of zonation between tide-marks on rocky coasts. J. Ecol., 37:289–305.

———. 1952. Life between tide-marks in North America: Northern Florida and the Carolinas. J. Ecol., 40:1–49.

Stern, A. C. (ed.). 1968. *Air Pollution* (2nd ed.). 3 vols. Academic Press, New York.

Stern, V. M. R.; Mueller, V.; Sevadarian, V.; and Way, M. 1969. Lygus bug control in cotton through alfalfa interplanting. Calif. Agr., 23(2):8–10.

Stern, W. L., and Buell, M. F. 1951. Life-form spectra of New Jersey pine barren forest and Minnesota jack pine forest. Bull. Torrey Bot. Club, 78:61–65.

Steward, W. D. 1966. *Nitrogen Fixation in Plants.* Athone Press, New York.

Steward, W. D.; Fitzgerald, G. P.; and Burris, R. H. 1967. *In situ* studies on N_2 fixation using acetylene reduction technique. Proc. Natl. Acad. Sci., 58:2071–2078.

Stewart, Paul A. 1952. Dispersal, breeding, behavior, and longevity of banded barn owls in North America. Auk, 69:277–285.

Stickel, Lucille F. 1950. Populations and home range relationships of the box turtle, *Terrapene c. carolina* (Linnaeus). Ecol. Monogr., 20:351–378.

Stockner, John G., and Benson, W. W. 1967. The succession of diatom assemblages in the recent sediments of Lake Washington. Limnol. Oceanogr., 12:513–532.

Stoddard, D. R. 1965. Geography and the ecological approach. The ecosystem as a geographical principle and method. Geography, 50:242–251.

Stoddard, Herbert L. 1932. *The Bobwhite Quail, Its Habits, Preservation and Increase.* Charles Scribner's Sons, New York.

———. 1936. Relation of burning to timber and wildlife. Proc. 1st N. A. Wildl. Conf., 1:1–4.

Stoddart, Laurence A., and Smith, Arthur D. 1955. *Range Management* (2nd ed.). McGraw-Hill Book Co., Inc., New York.

Stone, K. H. 1956. Air photo interpretation procedures. Photogr. Engr., 22:123–132.

Strandburg, C. H. 1966. Water quality analysis. Photogr. Engr., 32:234–248.

Strickland, John D. H. 1965. Phytoplankton and marine primary productivity. Ann. Rev. Microbiol., 19:127–162.

Strickland, John D. H., and Parsons, T. R. 1968. A practical handbook of seawater analysis. Fish. Res. Bd. Canad., Bull. 167. 311 pp.

Suess, Hans E. 1969. Tritium geophysics as an international project. Science, 163:1405–1410.

Sukachev, V. N. 1944. (On principles of genetic classification in biocenology.) (In Russian.) Zur. Obshchei Biol., 5:213–227. (Translated and condensed by F. Raney and R. Daubenmire. Ecology, 39:364–367.)

———. 1959. The correlation between the concepts "forest ecosystem" and "forest biogeocoenose" and their importance for the classification of forests. Proc. IX Internat. Bot. Cong., Vol. II, page 387 (abstract). (See also Silva Fennica, 105:94, 1960.)

Sukachev, V. N., and Dylis, N. V. (eds.). 1964. Fundamentals of forest biogeocoenology. Bot. Inst. and Lab. For. Sci. U.S.S.R., Moscow. 474 pp. (In Russian; see review in Science, 148:828, 1965.)

———. 1968 (Russian ed. 1964). *Fundamentals of Forest Biogeocoenology*. Oliver and Boyd, London. 672 pp.

Summerhayes, V. S., and Elton, C. S. 1923. Contributions to the ecology of Spitsbergen and Bear Island. J. Ecol., 11:214–286.

Svardson, Gunnar. 1949. Competition and habitat selection in birds. Oikos, 1:157–174.

Sverdrup, H. U.; Johnson, Martin W.; and Fleming, Richard. 1942. *The Oceans, Their Physics, Chemistry, and General Biology*. Prentice-Hall, Inc., Englewood Cliffs, N. J.

Swan, L. W. 1961. The ecology of the high Himalayas. Scient. Amer., 205(4):68–78.

Swank, W. T., and Miner, N. H. 1968. Conversion of hardwood-covered watershed to white pine reduces water yield. Water Resources Res., 4:947–954.

Sweeney, B. 1963. Biological clocks in plants. Ann. Rev. Plant Physiol., 14:411–440.

Sweeney, J. R. 1956. Responses of vegetation to fire. Univ. Calif. Publ. Bot., 28:143–250.

Swift, L. W., and van Bavel, C. H. M. 1961. Mountain topography and solar energy available for evapotranspiration. J. Geophys. Res., 66:2565.

Swingle, H. S. 1950. Relationships and dynamics of balanced and unbalanced fish populations. Agricultural Experiment Station, Alabama Polytechnic Inst. pp. 1–74.

Swingle, H. S., and Smith, E. V. 1947. Management of farm fish ponds (rev. ed.). Alabama Polytechnic Inst. Agr. Exp. Sta. Bull. No. 254.

Taber, R. D., and Dasmann, Raymond. 1957. The dynamics of three natural populations of the deer, *Odocoileus hemionus columbianus*. Ecology, 38:233–246.

Tadros, T. M. 1957. Evidence of the presence of an edapho-biotic factor in the problem of serpentine tolerance. Ecology, 38:14–23.

Takahashi, M., and Ichimura, S. 1968. Vertical distribution and organic matter production of photosynthetic sulfur bacteria in Japanese lakes. Limnol. Oceanogr., 13:644–655.

Talbot, L. M. 1963. Comparison of the efficiency of wild animals and domestic livestock in the utilization of east African rangelands. I.V.C.N. (NS) no. 1, pp. 328–335.

Talbot, L. M.; Payne, W. J. A.; Ledger, H. P.; Verdcourt, L. D.; and Talbot, M. H. 1965. The meat production potential of wild animals in Africa; a review of biological knowledge. Tech. Comm. no. 16, Commonwealth Bureau of Animal Breeding and Genetics, Edinburgh (Farnhan Royal, Buck, England). 42 pp.

Talbot, L. M., and Talbot, M. H. 1963. The high biomass of wild ungulates on East African savanna. Trans. N. A. Wildl. Conf., 28:465–476.

Tamiya, Hiroshi. 1957. Mass culture of algae. Ann. Rev. Plant Physiol., 8:309–334.

Tanner, J. T. 1966. Effects of population density on growth rates of animal populations. Ecology, 47:733–745.

Tansley, A. G. 1935. The use and abuse of vegetational concepts and terms. Ecology, 16:284–307.

Tappan, Helen. 1968. Primary production, isotopes and the atmosphere. In: *Palaeogeography, Palaeoclimatology, Palaeoecology*, vol. 4. Elsevier Publishing Co., Amsterdam. pp. 187–210.

Tarpley, Wallace A. 1967. A study of the cryptozoa in an old-field ecosystem. Ph.D. Dissertation, University of Georgia, Athens.

Tarswell, C. M. (ed.). 1965. Biological Problems of Water Pollution. Third Seminar, Public Health Serv. publ. no. 999-WP-25.

Taub, Freida B. 1963. Some ecological aspects of space biology. Amer. Biol. Teacher, 25:412–421.

———. 1969. A biological model of a freshwater community in a gnotobiotic ecosystem. Limnol. Oceanogr., 14:136–142.

———. 1969. Gnotobiotic models of freshwater communities. Verh. int. Ver. Limnol., 17:485–496.

Taylor, L. R. (ed.), 1970. *The Optimum Population for Britain*. Academic Press, New York. 182 pp.

Taylor, O. C.; Dugger, W. M., Jr.; Cardiff, E. A.; and Darley, E. F. 1961. Interaction of light and atmospheric photochemical (smog) within plants. Nature, 192:814–816.

Taylor, Walter P. 1934. Significance of extreme or intermittent conditions in distribution of species and management of natural resources, with a restatement of Liebig's law of the minimum. Ecology, 15:274–379.

Taylor, W. R. 1964. Light and photosynthesis in intertidal benthic diatoms. Helgd. Wiss. Meeresunters, 10:29–37.

Teal, John M. 1957. Community metabolism in a temperate cold spring. Ecol. Monogr., 27:283–302.

———. 1958. Distribution of fiddler crabs in Georgia salt marshes. Ecology, 39:185–193.

———. 1962. Energy flow in the salt marsh ecosystem of Georgia. Ecology, 43:614–624.

Teal, J. M., and Kanwisher, J. 1966. Gas transport in the marsh grass, *Spartina alterniflora*. J. Exp. Bot., 17:355–361.

Thienemann, August. 1926. *Limnologie*. Jedermanns Bücherei, Breslau.

———. 1926a. Der Nahrungskreislauf im Wasser. Verh. deutsch. Zool. Ges., 31:29–79.

———. 1939. Grundzüge einer allgemeinen Oekologie. Arch. Hydrobiol., 35:267–285.

Thomas, J. P. 1966. The influence of the Altamaha river on primary production beyond the mouth of the river. M.S. Thesis, University of Georgia, Athens.

Thomas, Moyer D. 1955. Effect of ecological factors on photosynthesis. Annual Rev. Plant Physiol., 6:135–156.

Thomas, Moyer D., and Hill, George R. 1949. Photosynthesis under field conditions. In *Photosynthesis in Plants*. (James Franck and Walter E. Loomis, eds.). Iowa State College Press, Ames. pp. 19–52.

Thomas, William A. 1969. Accumulation and cycling of calcium by dogwood trees. Ecol. Monogr., 39:101–120.

Thomas, W. H. 1964. An experimental evaluation of the C¹⁴ method for measuring phytoplankton production using cultures of *Dunaliella primolecta* Butcher. Fishery Bull., 63:273–292.

Thomas, W. H., and Simmons, E. G. 1960. Phytoplankton production in the Mississippi delta. In: *Recent Sediments, Northwestern Gulf of Mexico.* Amer. Assoc. Petrol. Geol. Tulsa, Okla. pp. 103–116.

Thomas, William L. (ed.). 1956. *Man's Role in Changing the Face of the Earth.* University of Chicago Press, Chicago.

Thomlinson, Ralph. 1965. *Population Dynamics: Causes and Consequences of World Demographic Change.* Random House, New York. 575 pp.

Thompson, D'Arcy W. 1942. *On Growth and Form.* Cambridge University Press, Cambridge, England.

Thompson, David H., and Bennett, G. W. 1939. Fish management in small artificial lakes. Trans. 4th N. A. Wildl. Conf., 4:311–317.

Thompson, David H., and Hunt, Francis D. 1930. The fishes of Champaign County: a study of the distribution and abundance of fishes in small streams. Ill. Nat. Hist. Surv. Bull., 19:1–101.

Thornthwaite, C. W. 1931. The climates of North America according to a new classification. Geogr. Rev., 21:633–655.

———. 1933. The climates of the earth. Geogr. Rev., 28:433–440.

———. 1948. An approach to a rational classification of climate. Geogr. Rev., 38:55–94.

———. 1955. Discussions on the relationships between meteorology and oceanography. J. Mar. Res., 14:510–515.

Thornthwaite, C. W., and Mather, J. R. 1957. Instructions and tables for computing potential evapotranspiration and water balance. Drexel Inst. Technol. Lab. Climatol. Publ. Climatol., 17:231–615.

Thorpe, W. H. 1963. *Learning and instinct in animals.* Methuen, London.

Thorson, Gunnar. 1955. Modern aspects of marine level-bottom animal communities. J. Mar. Res., 14:387–397.

———. 1956. Marine level-bottom communities of recent seas, their temperature adaptation, and their "balance" between predators and food animals. Trans. N.Y. Acad. Sci., 18:693–700.

Tilly, L. J. 1968. The structure and dynamics of Cone Spring. Ecol. Monogr., 38:169–197.

Tinbergen, N. 1951. *The Study of Instinct.* Oxford University Press, New York.

———. 1953. *Social Behaviour in Animals.* Methuen, London.

———. 1968. On war and peace in animals and man. Science, 160:1411–1418.

Tischler, Wolfgang. 1955. Synökologie der Landtiere. Gustav Fischer Verlag, Stuttgart.

Tracey, J. I., Jr.; Cloud, P. E.; and Emery, K. O. 1955. Conspicuous features of organic reefs. Atoll Res. Bull., 46:1–3.

Tramer, E. J. 1969. Bird species diversity; components of Shannon's formula. Ecology, 50:927–929.

Tranquillini, W. 1959. Die Stoffproduktion der Zirbe (*Pinus cembra*) an der Waldgrenze wahrend eines Jahres. Planta, 54:107–151.

Transeau, E. N. 1926. The accumulation of energy by plants. Ohio J. Sci., 26:1–10.

Trautman, Milton B. 1942. Fish distribution and abundance correlated with stream gradients as a consideration in stocking programs. Trans. 7th N. A. Wildl. Conf., 7:221–223.

Treshow, Michael. 1970. *Environment and Plant Response.* McGraw-Hill Book Co., Inc., New York. 422 pp.

Tribe, H. T. 1957. Ecology of microorganisms in soil as observed during their development upon buried cellulose film. In: *Microbial Ecology* (see Williams and Spicer, 1957).

———. 1961. Microbiology of cellulose decomposition in soil. Soil Sci., 92:61–77.

———. 1963. The microbial component of humus. In: *Soil Organisms* (Ed. J. Packser and J. van der Drift). North-Holland Publishing Company, Amsterdam.

Truper, H. G., and Jannasch, H. W. 1968. *Chromatium buderi* nov. spec., eine neue Art der "grossen" Thiorhodaceae. Arch. Microbiol., 61:363–372.

Tschirley, Fred H. 1969. Defoliation in Vietnam. Science, 163:779–786.

Tukey, J. W. (ed.). 1965. Restoring the quality of our environment. Rept. Environ. Poll. Panel, President's Sci. Advisory Comm., The White House (available from Supt. Doc., Washington, D.C.). 317 pp.

Turner, F. B. (ed.). 1968. Energy flow in ecosystems (Refresher Course. Amer. Soc. Zool.). Amer. Zool., 8:10–69.

Tuxen, S. L. 1944. The hot springs of Iceland. In: *The Zoology of Iceland.* Ejnar Munksgaard, Copenhagen.

Twomey, Arthur C. 1936. Climographic studies of certain introduced and migratory birds. Ecology, 17:122–132.

Udall, Stewart L. 1965. *The Quiet Crisis.* Holt, Rinehart and Winston, New York. 209 pp.

Udvardy, Miklos D. F. 1969. *Dynamic Zoogeography. With special reference to land animals.* Van Nostrand Reinhold, New York. 446 pp.

UNESCO. 1955. Arid Zone Research. VI. Plant Ecology. Reviews of research. Paris.

UNESCO. 1957. Arid Zone Research. VIII. Human and Animal Ecology, Paris.

United States Department of Agriculture. 1938. Soils and Men. Yearbook of Agriculture for 1938.

Utida, Syunro. 1957. Cyclic fluctuations of population density intrinsic to the host-parasite system. Ecology, 38:442–449.

Uvarov, B. P. 1957. The aridity factor in the ecology of locust and grasshoppers of the old world. In: *Human and Animal Ecology.* Arid Zone Res. VIII. UNESCO, Paris.

Vaccaro, R. F., and Jannasch, H. W. 1966. Studies on the heterotrophic activity in seawater based on glucose assimilation. Limnol. Oceanogr., 11:596–607.

Valentine, James W. 1968. Climatic regulation of species diversification and extinction. Bull. Geol. Soc. Amer., 79:273–276.

———. 1968a. The evolution of ecological units above the population level. J. Paleontol., 42:253–267.

Vallentyne, J. R. 1960. Geochemistry of the Biosphere. In: *McGraw-Hill Encyclopedia of Science and Technology*, Vol. 2, pp. 239–245.

———. 1962. Solubility and the decomposition of organic matter in nature. Arch. Hydrobiol., 58:423–34.

———. 1963. Environmental biophysics and microbial ubiquity. Ann. N.Y. Acad. Sci., 108:342–352.

Van den Bosch, R., and Stern, V. M. 1962. The integration of chemical and biological control of arthropod pests. Ann. Rev. Entomol., 7:367.

Van der Kloot, W. G. 1968. *Behavior.* Holt, Rinehart and Winston, New York. 166 pp.

Van Dobben, W. H. 1962. Influence of temperature and light conditions on dry-matter distribution, development rate and yield in arable crops. In: *Fundamentals of Dry-Matter Production and Distribution* (M. Hart, ed.). Neth. J. Agr. Sci., 10 (No. 5, special issue):377–389.

Van Dyne, G. M. 1966. Ecosystems, systems ecology and systems ecologists. Oak Ridge National Laboratory Report 3957. 31 pp.

———. 1969. Grassland management, research and teaching viewed in a systems context. Range Science Department, Science Series No. 3, Colorado State University. 39 pp.

——— (ed.). 1969a. *The Ecosystem Concept in Natural Resource Management*. Academic Press, New York. 383 pp.

Van Dyne, G. M., and Meyer, J. H. 1964. A method for measurement of forage intake of grazing livestock using microdigestion techniques. J. Range Mgt., 17:204–208.

van Niel, C. B. 1949. The kinetics of growth of microorganisms. In: *The Chemistry and Physiology of Growth*. Princeton University Press, Princeton.

Van Valen, Leigh. 1965. Morphological variation and width of ecological niche. Amer. Nat., 99:377–390.

Varley, G. C. 1947. The natural control of population balance in the knapweed gall-fly (*Urophora jaceana*). J. Anim. Ecol., 16:139–187.

———. 1949. Population changes in German forest pests. J. Anim. Ecol., 18:117–122.

———. 1970. The concept of energy flow applied to a woodland community. In: *Quality and Quantity of Food*. Symp. Brit. Ecol. Soc., Blackwell, Oxford. pp. 389–405.

Varley, G. C., and Edwards, R. I. 1957. The bearing of parasite behaviour on the dynamics of insect host and parasite populations. J. Anim. Ecol., 26:471–477.

Vaurie, C. 1951. Adaptive differences between two sympatric species of nuthatches (*Sitta*). Proc. Int. Ornithol. Cong., 19:163–166.

Vayda, Andrew P. (ed.). 1969. *Environment and Cultural Behavior: Ecological Studies in Cultural Anthropology*. Natural History Press, Garden City, N.Y. 482 pp.

Verhulst, P. F. 1838. Notice sur la loi que la population suit dans son accroissement. Corresp. Math. et Phys., 10:113–121.

Vernadsky, W. I. 1944. Problems in biogeochemistry. II. Trans. Conn. Acad. Arts Sci., 35:493–494.

———. 1945. The biosphere and the noosphere. Amer. Sci., 33:1–12.

Vernberg, F. J., and Vernberg, W. B. 1970. *The Animal and the Environment*. Holt, Rinehart and Winston, New York. 416 pp.

Vestal, A. G. 1949. Minimum areas for different vegetations. Illinois Biol. Monogr., 20:1–129.

Viosca, Percy, Jr. 1935. Statistics on the productivity of inland waters, the master key to better fish culture. Trans. Amer. Fish. Soc., 65:350–358.

Visscher, Maurice B. 1967. Applied science and medical progress. In: *Applied Science and Technological Progress* (Rept. Committee on Science and Astronautics, G. P. Miller, Chairman). U. S. Government Printing Office, Washington, D.C. pp. 185–206.

Vogt, William. 1948. *Road to Survival*. Sloane, New York.

Volpe, E. P. 1967. *Understanding Evolution*. William C. Brown, Dubuque, Iowa. 160 pp.

Volterra, Vito. 1926. Variations and fluctuations of the number of individuals in animal species living together. In: *Animal Ecology* (R. N. Chapman, ed.). McGraw-Hill Book Co., Inc., New York. pp. 409–448.

———. 1928. Variations and fluctuations of the number of individuals in animal species living together. J. Conseil, 3:1–51.

Von Arx, William S. 1962. *An Introduction to Physical Oceanography*. Addison-Wesley Publishing Company, Inc., Reading, Mass.

Voute, A. D. 1946. Regulation of the density of the insect populations in virgin-forests and cultivated woods. Arch. Neerlandaises de Zool., 7:435–470. (Review by Varley, J. Anim. Ecol., 17:82–83.)

Waddington, C. H. 1962. *New Patterns in Genetics and Development*. Columbia University Press, New York. 271 pp.

Wadsworth, R. M.; Chapas, L. C.; Rutter, A. J.; Solomon, M. E.; and Wilson, J. W. (eds.). 1967. *The Measurement of Environmental Factors in Terrestrial Ecology*. Blackwell, Oxford. 314 pp.

Wagar, J. Alan. 1970. Growth versus the quality of life. Science, 168:1179–1184.

Waggoner, P. E. 1966. Weather modification and the living environment. In: *Future Environments of North America* (Darling and Milton, eds.). The Natural History Press, Garden City, N.Y. pp. 87–98.

Wagner, Frederic H. 1969. Ecosystem concepts in fish and game management. In: *The Ecosystem Concept in Natural Resource Management* (G. M. Van Dyne, ed.). Academic Press, New York. pp. 259–307.

Waksman, Selman A. 1932. *Principles of Soil Microbiology* (2nd ed.). Williams & Wilkins, Co., Baltimore.

———. 1941. Aquatic bacteria in relation to the cycle of organic matter in lakes. In: *Symposium on Hydrobiology*. University of Wisconsin Press, Madison. pp. 86–105.

———. 1952. *Soil Microbiology*. John Wiley & Sons, Inc., New York.

Wallace, Bruce, and Srb, A. M. 1961. *Adaptation*. Prentice-Hall, Englewood Cliffs, N.J. 113 pp.

Waloff, Z. 1966. The upsurges and recessions of the desert locust: an historical survey. Antilocust Mem. No. 8, London. 111 pp.

Walter, H. 1954. Le facteur eau dans les régions arides et sa signification pour l'organisation de la végétation dans les contrées sub-tropicales. In: *Les Divisions Ecologiques du Monde*. Centre Nationale de la Recherche Scientifique, Paris. pp. 27–39.

Wangersky, P. J., and Cunningham, W. J. 1956. On time lags in equations of growth. Proc. Natl. Acad. Sci. (Wash.), 42:699–702.

———. 1957. Time lag in population models. Cold Spring Harbor Sym. Quant. Biol., 22:329–338.

Ward, Barbara. 1966. *Spaceship Earth*. Columbia University Press, New York.

Warington, Robert. 1851. Notice of observation on the adjustment of the relations between animal and plant kingdoms. Quart. J. Chem. Soc., London, 3:52–54. (See also *On the aquarium*. Notices Royal Inst., 2:403–408, 1857.)

Warren, Charles E. 1971. *Biology and Water Pollution Control*. W. B. Saunders Company, Philadelphia. 434 pp.

Waterman, T. H. 1968. Systems theory and biology—view of a biologist. In: *Systems Theory and Biology* (M. D. Mesarovic, ed.). Proceedings of the 3rd Systems Symposium at Case Institute of Technology. Springer-Verlag, New York Inc. 403 pp.

Waters, T. F. 1965. Interpretation of invertebrate drift in streams. Ecology, 46:327–334.

Watson, D. J. 1956. Leaf growth in relation to crop yield. In: *The Growth of Leaves.* (See Milthrope, 1956.)

Watt, K. E. F. 1963. How closely does the model mimic reality? Canad. Entomol. Mem., 31:109–111.

———. 1963a. Mathematical models for five agricultural crop pests. Mem. Entomol. Soc. Canad., 32:83–91.

———. 1963b. Dynamic programming, "look ahead programming," and the strategy of insect pest control. Canad. Entomol., 95:525–536.

———. 1964. Computers and the evaluation of resource management strategies. Amer. Sci., 52:408–418.

———. 1965. Community stability and the strategy of biological control. Canad. Entomol., 97:887–895.

———. 1966. *Systems Analysis in Ecology.* Academic Press, New York. 276 pp.

———. 1968. *Ecology and Resource Management: A Quantitative Approach.* McGraw-Hill Book Co., Inc., New York. 450 pp.

Weaver, D. K.; Butler, W. E.; and Olson, C. E. 1969. Observations on interpretation of vegetation from infrared imagery. In: *Remote Sensing in Ecology* (P. L. Johnson, ed.). University of Georgia Press, Athens. pp. 132–147.

Weaver, J. E. 1954. *North American Prairie.* Johnsen Publ. Co., Lincoln, Nebraska.

Weaver, J. E., and Albertson, F. W. 1956. *Grasslands of the Great Plains: Their Nature and Use.* Johnsen Publ. Co., Lincoln, Nebraska.

Weaver, J. E., and Clements, F. E. 1938. *Plant Ecology* (2nd ed.). McGraw-Hill Book Co., Inc., New York.

Weaver, J. E., and Zink, E. 1946. Annual increase of underground materials in three range grasses. Ecology, 27:115–127.

Weimer, R. J., and Hoyt, J. H. 1964. *Callianasa major* burrows, geologic indicators of littoral and shallow neritic environments. J. Paleontol., 38:761–767.

Weinberg, Alvin M., and Hammond, R. P. 1970. Limits to the use of energy. Amer. Sci., 58:412–418.

Welch, B. L., and Welch, A. S. (eds.). 1970. *Physiological Effects of Noise.* Plenum Press, New York.

Welch, Harold. 1967. Energy flow through the major macroscopic components of an aquatic ecosystem. Ph.D. Dissertation, University of Georgia, Athens.

———. 1968. Relationships between assimilation efficiencies and growth efficiencies for aquatic consumers. Ecology, 49:755–759.

Wellington, W. G. 1957. Individual differences as a factor in population dynamics: the development of a problem. Canad. J. Zool., 35:293–323.

———. 1960. Qualitative changes in natural populations during changes in abundance. Canad. J. Zool., 38:289–314.

Wells, B. W. 1928. Plant communities of the coastal plain of North Carolina and their successional relations. Ecology, 9:230–242.

Went, Fritz W. 1957. *The Experimental Control of Plant Growth.* Chronica Botanica Co., Waltham, Mass.

Went, F. W., and Stark, N. 1968. Mycorrhiza. Bio-Science, 18:1035–1039.

Westlake, D. F. 1963. Comparisons of plant productivity. Biol. Rev., 38:385–429.

———. 1965. Some basic data for investigations of the productivity of aquatic macrophytes. In: *Primary Production in Aquatic Environments* (Goldman, ed.). University of California Press, Berkeley.

Weyl, Peter K. 1970. *Oceanography: An Introduction to the Marine Environment.* John Wiley & Sons, Inc., New York. 535 pp.

White, Lynn. 1967. The historical roots of our ecological crisis. Science, 155:1203–1207.

Whitehead, F. H. 1957. Productivity in alpine vegetation (abstract). J. Anim. Ecol., 26:241.

———. 1957a. The importance of experimental ecology to the study of British Flora. In: *Progress in the Study of British Flora* (J. E. Lousley, ed.). Arbroath, T. Buncle & Co., London. pp. 56–60.

Whittaker, R. H. 1951. A criticism of the plant association and climatic climax concepts. Northwest Sci., 25:17–31.

——— (ed.). 1954. The ecology of serpentine soils. Ecology, 35:258–288.

———. 1954a. Plant populations and the basis of plant indication. In: Angewandte Pflanzensoziologie, Veroffentlichungen des Karntner Landesinstituts fur angewandte Pflanzensoziologie in Klagenfurt, Festschrift Aichinger, Vol. 1.

———. 1956. Vegetation of the Great Smoky Mountains. Ecol. Monogr., 26:1–80.

———. 1962. Classification of natural communities. Bot. Rev., 28:1–239.

———. 1965. Dominance and diversity in land plant communities. Science, 147:250–260.

———. 1967. Gradient analysis of vegetation. Biol. Rev., 42:207–264.

———. 1969. New concepts of kingdoms of organisms. Science, 163:150–160.

———. 1970. The biochemical ecology of higher plants. In: *Chemical Ecology* (E. Sondheimer and J. B. Simeone, eds.). Academic Press, New York. pp. 43–70.

———. 1970a. *Communities and Ecosystems.* Macmillan Company, New York. 158 pp.

Whittaker, R. H., and Woodwell, G. M. 1969. Structure, production and diversity of the oak-pine forest at Brookhaven, New York. J. Ecol., 57:155–174.

Whitten, J. L. 1966. *That We May Live.* Van Nostrand, Princeton, N. J. (A congressman refutes Rachel Carson's *Silent Spring.*)

Whittlesey, J. H. 1966. Bipod camera support. Photogr. Engr., 32:1005–1010.

Whyte, William H. 1968. *The Last Landscape.* Doubleday, Garden City, N.Y. 326 pp.

Wickens, G. E. 1966. The practical application of aerial photography for ecological surveys in the savannah regions of Africa. Photogrammetria, 21:33–41.

Wiegert, R. G. 1965. Energy dynamics of the grasshopper populations in old-field and alfalfa field ecosystems. Oikos, 16:161–176.

———. 1968. Thermodynamic considerations in animal nutrition. Amer. Zool., 8:71–81.

Wiegert, R. G.; Coleman, D. C.; and Odum, E. P. 1970. Energetics of the litter-soil subsystem. In: *Methods of Study in Soil Ecology.* Proc. Paris Sym. Internat. Biol. Prog., UNESCO, Paris. pp. 93–98.

Wiegert, R. G., and Evans, F. C. 1964. Primary production and the disappearance of dead vegetation in an old-field in southeastern Michigan. Ecology, 45:49–63.

———. 1967. Investigations of secondary productivity in grasslands. In: *Secondary Productivity of Terrestrial Ecosystems* (K. Petrusewicz, ed.). Polish Acad. Sci., Warsaw. Vol. II, pp. 499–518.

Wiegert, R. G., and Odum, E. P. 1969. Radionuclide tracer measurement of food web diversity in nature. In: Proc. 2nd Natl. Sym. on Radioecology (D. J. Nelson and F. C. Evans, eds.). Clearinghouse Fed. Sci. Tech. Info., Springfield, Va. pp. 709–710.

Wiegert, R. G.; Odum, E. P.; and Schnell, J. H. 1967. Forb-arthropod food chains in a one-year experimental field. Ecology, 48:75–83.

Wiegert, R. G., and Owen, D. F. 1971. Trophic structure, available resources and population density in terrestrial vs. aquatic ecosystems. J. Theoret. Biol., 30:69–81.

Wiener, N. 1948. *Cybernetics.* Tech. Press, Cambridge, Mass.

Wilde, S. A. 1954. Mycorrhizal fungi: their distribution and effect on tree growth. Soil. Sci., 78:23–31.

———. 1968. Mycorrhizae and tree nutrition. BioScience, 18:482–484.

Wilde, S. A.; Youngberg, C. T.; and Hovind, J. H. 1950. Changes in ground water, soil fertility and forest growth produced by construction and removal of beaver dams. J. Wildl. Mgt., 14:123–128.

Wilhm, J. L. 1967. Comparison of some diversity indices applied to populations of benthic macroinvertebrates in a stream receiving organic wastes. J. Water Poll. Cont. Fed., 39:1673–1683.

Wilhm, J. L., and Dorris, T. C. 1966. Species diversity of benthic microorganisms in a stream receiving domestic and oil refinery effluents. Amer. Midl. Nat., 76:427–449.

———. 1968. Biological parameters for water quality criteria. BioScience, 18:477–481.

———. 1968a. Biomass units versus numbers of individuals in species diversity indices. Ecology, 49:153–156.

Wilimovsky, N. J. (ed.). 1962. Symposium on pink salmon. Inst. Fish., Univ. Brit. Columbia. 226 pp.

Willard, W. K. 1960. Avian uptake of fission products from an area contaminated by low level atomic wastes. Science, 135:38–40.

Williams, C. B. 1947. The logarithmic series and its application to biological problems. J. Ecol., 34:253–272.

———. 1960. The range and pattern of insect abundance. Amer. Nat., 44:137–151.

———. 1964. *Patterns in the Balance of Nature; and Related Problems in Quantitative Ecology.* Academic Press, New York. 324 pp.

Williams, Carroll M. 1967. Third-generation pesticides. Scient. Amer., 217(1):13–17.

———. 1970. Hormonal interactions between plants and insects. In: *Chemical Ecology* (Sondheimer and Simeone, eds.). Academic Press, New York. pp. 103–132.

Williams, George C. 1966. *Adaptation and Natural Selection. A Critique of Some Current Evolutionary Thought.* Princeton University Press, Princeton. 170 pp.

Williams, R. B. 1962. The ecology of diatom populations in a Georgia salt marsh. Ph.D. Dissertation, Harvard University, Cambridge, Mass. 146 pp.

———. 1969. The potential importance of *Spartina alterniflora* in conveying zinc, manganese, and iron into estuarine food chains. In: Proc. Second Symposium on Radioecology (D. J. Nelson and F. C. Evans, eds.). U. S. Atomic Energy Comm., RID 4500. 774 pp.

Williams, R. E. O., and Spicer, C. C. (ed.). 1957. *Microbial Ecology.* 7th Symposium of the Soc. Gen. Microbiology. Cambridge University Press, New York.

Williams, W. A. 1966. Range improvement as related to

net productivity, energy flow and foliage configuration. J. Range Mgt., 19:29–34. (See also J. Range Mgt., 21:355–360.)

Williamson, M. H. 1957. An elementary theory of interspecific competition. Nature, 180:1–7.

Willstatter, R., and Stroll, A. 1918. *Untersuchungen uber die Assimilation der Koblensaure.* Julius Springer, Berlin.

Wilson, D. P. 1952. The influence of the nature of the substratum on the metamorphosis of the larvae of marine animals, especially the larvae of *Ophelia bicornis.* Ann. Inst. Oceanogr., 27:49–156.

Wilson, D. P. 1958. Some problems in larval ecology related to the localized distribution of bottom animals. In: *Perspectives in Marine Biology* (A. Buzzati-Traverso, ed.). University of California Press, Berkeley. pp. 87–103.

Wilson, E. O. 1965. Chemical communication in the social insects. Science, 149:1064–1071.

———. 1969. The species equilibrium. In: *Diversity and Stability in Ecological Systems* (Woodwell and Smith, eds.). Brookhaven Symposia in Biology, No. 22 Brookhaven Nat. Lab., Upton, N.Y. pp. 38–47.

Winogradsky, S. 1949. *Microbiologie du Sol: Problèmes et Methodes.* Masson et Cie., Paris. 861 pp.

Wirth, Louis. 1928. *The Ghetto.* University of Chicago Press, Chicago.

———. 1945. Human ecology. Amer. J. Sociol., 50:483–488.

Witherspoon, J. P. 1965. Radiation damage to forest surrounding an unshielded fast reactor. Health Physics, 11:1637–1642.

———. 1969. Radiosensitivity of forest tree species to acute fast neutron radiation. In: Proc. 2nd Natl. Sym. Radioecology (D. Nelson and F. Evans, eds.). Clearinghouse Fed. Sci. Tech. Info., U. S. Dept. Commerce, Springfield, Va. pp. 120–126.

Withrow, R. B. (ed.). 1959. Photoperiodism and related phenomena in plants and animals. Publ. no. 55, Amer. Assoc. Adv. Sci. (Wash.).

Witkamp, M. 1966. Rates of carbon dioxide evolution from the forest floor. Ecology, 47:492–494.

———. 1966a. Decomposition of leaf litter in relation to environmental conditions, microflora and microbial respiration. Ecology, 47:194–201.

———. 1963. Microbial populations of leaf litter in relation to environmental conditions and decomposition. Ecology, 44:370–377.

Witkamp, M., and Drift, J. van der. 1961. Breakdown of forest litter in relation to environmental factors. Plant and Soil, 15:295–311.

Witkamp, M., and Olson, J. S. 1963. Breakdown of confined and nonconfined oak litter. Oikos, 14:138–147.

Wittwer, S. H. 1957. Nutrient uptake with special reference to foliar absorption. In: *Atomic Energy and Agriculture* (C. L. Comar, ed.). Amer. Assoc. Adv. Sci. Publ. No. 49:139–164.

Wohlschlag, Donald E. 1960. Metabolism of an antarctic fish and the phenomenon of cold adaptation. Ecology, 41:287–292.

Wolcott, G. N. 1937. An animal census of two pastures and a meadow in northern New York. Ecol. Monogr., 7:1–90.

Wolfanger, Louis A. 1930. *The Major Soil Divisions of the United States.* John Wiley & Sons, Inc., New York.

Wolfe, John N.; Wareham, Richard T.; and Scofield, Herbert T. 1949. Microclimates and macroclimates of Neotoma, a small valley in central Ohio. Ohio State University Publication, 8:1–267 (Bull. No. 41).

Wolfenbarger, D. O. 1946. Dispersion of small organisms. Amer. Midl. Nat., 35:1–152.

Wolman, A. 1965. The metabolism of cities. Scient. Amer., 213:179–190.

Wood, E. J. Ferguson. 1955. Fluorescent microscopy in marine microbiology. J. Cons. Int. Expl. Mer., 21:6–7. (See also Z. allg. Mikrobiol., 2:164–165, 1962, and Limnol. Oceanogr., 7:32–35, 1962.)

———. 1965. *Marine Microbial Ecology.* Chapman and Hall, London. 243 pp.

———. 1967. *Microbiology of Oceans and Estuaries.* Elsevier, Amsterdam. 319 pp.

Wood, E. J. Ferguson, and Davis, P. S. 1956. Importance of smaller phytoplankton elements. Nature, 177:436.

Wood, E. J. Ferguson; Odum, W. E.; and Zieman, J. C. 1969. Influence of sea grasses on the productivity of coastal lagoons. In: *Lagunas Costeras, Un Simposio.* Mem. Simp. Intern. Lagunas costeras. UNAM-UNECCO (Nov. 28–30, 1967). Mexico, D. F. pp. 495–502.

Woods, F. W. 1960. Energy flow sylviculture—a new concept for forestry. Proc. Soc. Amer. For. pp. 25–27.

Woodbury, Angus M. 1947. Distribution of pigmy conifers in Utah and northeastern Arizona. Ecology, 28:113–126.

Woodruff, L. L. 1912. Observations on the origin and sequence of the protozoan fauna of hay infusions. J. Exp. Zool., 12:205–264.

Woodwell, G. M. 1962. Effects of ionizing radiation on terrestrial ecosystems. Science, 138:572–577.

———. 1963. The ecological effects of radiation. Scient. Amer., 208(6):1–11.

——— (ed.). 1965. *Ecological Effects of Nuclear War.* Brookhaven National Laboratory Publ. no. 917, 72 pp.

———. 1965a. Effects of ionizing radiation on ecological systems. In: *Ecological Effects of Nuclear War* (Woodwell, ed.). Brookhaven National Laboratory Publ. no. 917. pp. 20–38.

———. 1967. Toxic substances and ecological cycles. Scient. Amer., 216(3):24–31.

Woodwell, G. M., and Botkin, D. B. 1970. Metabolism of terrestrial ecosystems by gas exchange techniques: The Brookhaven approach. In: *Studies in Ecology* (D. E. Reichle, ed.). Springer-Verlag, New York. pp. 73–85.

Woodwell, G. M., and Dykeman, W. D. 1966. Respiration of a forest measured by CO_2 accumulation during temperature inversions. Science, 154:1031–1034.

Woodwell, G. M., and Smith, H. H. (eds.). 1969. *Diversity and Stability in Ecological Systems.* Brookhaven, Nat. Lab. Publ. no. 22, Upton, N.Y. 264 pp.

Woodwell, G. M., and Whittaker, R. H. 1968. Primary production in terrestrial communities. Amer. Zool., 8:19–30.

Woodwell, G. M.; Wurster, C. F.; and Isaacson, P. A. 1967. DDT residues in an east coast estuary: A case of biological concentration of a persistent insecticide. Science, 156:821–824.

Wright, John C. 1967. Effects of impoundments on productivity, water chemistry, and heat budgets of rivers. In: *Reservoir Fishery Resources Symposium.* Amer. Fish. Soc., Washington, D.C. pp. 188–199.

Wright, R. T., and Hobbie, J. E. 1966. The use of glucose and acetate by bacteria and algae in aquatic ecosystems. Ecology, 47:447–464.

Wright, Sewall. 1945. Tempo and mode in evolution: a critical review. Ecology, 26:415–419.

———. 1960. Physiological genetics, ecology of populations, and natural selection. In: *Evolution After Darwin* (S. Tax, ed.). University of Chicago Press, Chicago. Vol. 1, pp. 429–475.

Wright, Sewall, and Kerr, Warwick F. 1954. Experimental studies of the distribution of gene frequencies in very small populations of *Drosophila melanogaster.* II. Bar. Evolution, 8:225–240.

Wurster, Charles F. 1969. Chlorinated hydrocarbon insecticides and the world ecosystem. Biol. Cons., 1:123–129.

Wynne-Edwards, V. C. 1962. *Animal Dispersion in Relation to Social Behavior.* Hafner, New York.

———. 1964. Group selection and kin selection. Nature, 201:1145–1147.

———. 1965. Self-regulating systems in populations of animals. Science, 147:1543–1548.

———. 1965a. Social organization as a population regulator. Sym. Zool. Soc. London, 14:173–178.

Yeatter, R. E. 1950. Effect of different preincubation temperatures on the hatchability of pheasant eggs. Science, 112:529–530.

Yentsch, C. S., and Ryther, J. H. 1959. Relative significance of the net phytoplankton and nannoplankton in the water of Vineyard Sound. J. Conseil, Conseil Perm. Intern. Exploration Mer., 24:231–238.

Yonge, C. M. 1963. The biology of coral reefs. Adv. Mar. Biol., 1:209–260.

———. 1968. Living corals. Review lecture. Proc. Roy. Soc. B., 169:329–344.

Young, H. E.; Strand, Lars; and Altenberger, Russell. 1964. Preliminary fresh and dry weight tables for seven tree species in Maine. Tech. Bull. 12, Maine Agri. Exp. Station (University of Maine, Orono).

Yount, James L. 1956. Factors that control species numbers in Silver Springs, Florida. Limnol. Oceanogr., 1:286–295.

Zelinsky, Wilbur (ed.). 1970. *Geography and a Crowding World.* Oxford University Press, New York. 601 pp.

Zeuner, F. E. 1963. *History of Domesticated Animals.* Harper & Row Publishers, Inc., New York.

Zeuthen, E. 1953. Oxygen uptake and body size in organisms. Quart. Rev. Biol., 28:1–12.

Zhukova, A. I. 1963. On the quantitative significance of microorganisms in nutrition of aquatic invertebrates. In: *Symposium on Marine Microbiology* (C. H. Oppenheimer, ed.). Charles C Thomas, Springfield, Ill.

Zippin, Calvin. 1958. The removal method of population estimation. J. Wildl. Mgt., 22:82–90.

ZoBell, Claude E. 1946. *Marine Microbiology.* Chronica Botanica, Waltham, Mass.

———. 1946a. Studies on redox potential of marine sediments. Bull. Amer. Assoc. Petrol. Geol., 30:477–513.

———. 1963. Domain of the marine microbiologist. In: *Symposium on Marine Microbiology* (C. H. Oppenheimer, ed.). Charles C Thomas, Springfield, Ill.

ZoBell, Claude E.; Sisler, F. D.; and Oppenheimer, C. H. 1953. Evidence of biochemical heating of Lake Meade mud. J. Sed. Petrol., 23:13–17.

INDEX WITH REFERENCE GLOSSARY

Numbers in **boldface** type indicate pages on which terms and concepts are most fully defined and explained. Key groups of organisms are listed, but species names are not indexed except in connection with the illustration of ecological principles. Names of persons are not indexed; see Bibliography for names of authors.